Lecture Notes in Civil Engineering

Volume 86

Lecture Notes in Civil Engineering (LNCE) publishes the latest developments in Civil Engineering - quickly, informally and in top quality. Though original research reported in proceedings and post-proceedings represents the core of LNCE, edited volumes of exceptionally high quality and interest may also be considered for publication. Volumes published in LNCE embrace all aspects and subfields of, as well as new challenges in, Civil Engineering. Topics in the series include:

- Construction and Structural Mechanics
- Building Materials
- Concrete, Steel and Timber Structures
- Geotechnical Engineering
- Earthquake Engineering
- Coastal Engineering
- Ocean and Offshore Engineering; Ships and Floating Structures
- Hydraulics, Hydrology and Water Resources Engineering
- Environmental Engineering and Sustainability
- Structural Health and Monitoring
- Surveying and Geographical Information Systems
- Indoor Environments
- Transportation and Traffic
- Risk Analysis
- Safety and Security

To submit a proposal or request further information, please contact the appropriate Springer Editor:

- Mr. Pierpaolo Riva at pierpaolo.riva@springer.com (Europe and Americas);
- Ms. Swati Meherishi at swati.meherishi@springer.com (Asia - except China, and Australia, New Zealand);
- Dr. Mengchu Huang at mengchu.huang@springer.com (China).

All books in the series now indexed by Scopus and EI Compendex database!

More information about this series at http://www.springer.com/series/15087

Madhavi Latha Gali · P. Raghuveer Rao
Editors

Geohazards

Proceedings of IGC 2018

Springer

Editors
Madhavi Latha Gali
Department of Civil Engineering
Indian Institute of Science
Bengaluru, Karnataka, India

P. Raghuveer Rao
Department of Civil Engineering
Indian Institute of Science
Bengaluru, Karnataka, India

ISSN 2366-2557 ISSN 2366-2565 (electronic)
Lecture Notes in Civil Engineering
ISBN 978-981-15-6232-7 ISBN 978-981-15-6233-4 (eBook)
https://doi.org/10.1007/978-981-15-6233-4

This Springer imprint is published by the registered company Springer Nature Singapore Pte Ltd.
The registered company address is: 152 Beach Road, #21-01/04 Gateway East, Singapore 189721, Singapore

Madhavi Latha Gali · P. Raghuveer Rao
Editors

Geohazards

Proceedings of IGC 2018

Springer

Editors
Madhavi Latha Gali
Department of Civil Engineering
Indian Institute of Science
Bengaluru, Karnataka, India

P. Raghuveer Rao
Department of Civil Engineering
Indian Institute of Science
Bengaluru, Karnataka, India

ISSN 2366-2557 ISSN 2366-2565 (electronic)
Lecture Notes in Civil Engineering
ISBN 978-981-15-6232-7 ISBN 978-981-15-6233-4 (eBook)
https://doi.org/10.1007/978-981-15-6233-4

This Springer imprint is published by the registered company Springer Nature Singapore Pte Ltd.
The registered company address is: 152 Beach Road, #21-01/04 Gateway East, Singapore 189721, Singapore

Preface

Indian Geotechnical Conference (IGC 2018) was held at the National Science Complex of the Indian Institute of Science, Bengaluru, during 13–15 December 2018. This is the annual conference of the Indian Geotechnical Society (IGS), which was established in the year 1948 with the aim to promote cooperation among the engineers, scientists and practitioners for the advancement and dissemination of knowledge in the field of geotechnical engineering. IGC 2018 was a special event since it coincided with the 70 years celebrations of IGS.

The conference was a grand event with about 700 participants. The conference was inaugurated on 13 December in the presence of the President of IGS Prof. G. L. Sivakumar Babu and the Chief Guest Prof. E. C. Shin, Vice-president Asia, International Society of Soil Mechanics and Geotechnical Engineering (ISSMGE). The conference had 14 keynote lectures and 12 theme lectures presented by eminent academicians and practitioners from different parts of the world. Totally, 313 technical papers under 12 different themes of the conference were presented during the conference in 19 oral presentation sessions and 10 digital display sessions. All the participants of the conference had a common vision of deliberating on current geotechnical engineering research and practice and to strengthen the relationship between scientists, researchers and practising engineers within the fields of geotechnical engineering and to bring focus to problems that are relevant to society's needs and develop solutions. The conference acted as a platform to academicians and field engineers to interact, share knowledge and experiences and identify potential collaborations. The conference also provided opportunity to many young students, researchers and engineers and helped them to get connected to people involved in geotechnical engineering research and practice and national and international groups and technical committees.

All papers submitted to IGC 2018 had undergone a peer-review process and subsequently revised before being accepted. To publish conference proceedings through Springer, selected papers from the conference were grouped into four different volumes, namely Geotechnical Characterization and Modelling, Construction in Geotechnical Engineering, Geohazards and Problematic Soils and Ground Improvement. This book on *Geohazards* has 54 chapters written on various hazards in geotechnical engineering, including earthquakes, landslides and liquefaction. The subject area of these chapters spans and overlaps across different fields of engineering seismology, earthquake engineering, soil–structure interaction and earthquake-resistant design. While more than three-fourths of the chapters in this book discuss earthquake-related instabilities, failures and ground improvement for seismic hazard mitigation, other chapters talk about a variety of hazards like ground squeezing, mud-pumping and rainfall-induced slope instabilities. Interesting case studies on landslides are discussed in at least five chapters. Other chapters present studies focused on site response analysis, earthquake early warning systems, probabilistic seismic hazard analysis and seismic response analysis of foundations, dams and retaining walls. Overall, this book is a package of the whole gamut of earthquake geotechnical engineering capsuled into different chapters.

We sincerely thank the Indian Geotechnical Society, especially Prof. G. L. Sivakumar Babu, President, IGS and Prof. J. T. Shahu, Honorary Secretary, IGS, for their great support in organizing the conference. We also thank the organizing committee of IGC 2018, Prof. P. V. Sivapullaiah, Conference Chair, Prof. H. N. Ramesh, Conference Vice-chair, Dr. C. R. Parthasarathy, Prof. P. Anbazhagan and Prof. K. V. Vijayendra; organizing secretaries, Prof. K. Vijaya Bhaskar Raju, Treasurer, for all their hard work, long working hours spent and responsibility shared in planning and executing various tasks of this outstanding event. The unconditional support extended by the conference advisory committee, technical committee, sponsors of the conference, keynote speakers, theme speakers, session chairs, session coordinators, student volunteers, participants, presenters and authors of the technical papers in making the conference a grand success is sincerely appreciated. We thank the entire Springer team, in particular Swati Meherishi, Rini Christy Xavier Rajasekaran, Muskan Jaiswal and Ashok Kumar, for their hard work and support in bringing out the proceedings of IGC 2018.

Bengaluru, India

Madhavi Latha Gali
P. Raghuveer Rao
(Editors)

Contents

Seismic Response of Shallow Strip Footing with Log-Spiral Failure Mechanism Using Limit Analysis 1
Litan Debnath and Sima Ghosh

A Comparative Study of Soil Slope Stability Under Seismic Loading Condition 11
Suman Hazari, Sima Ghosh, and Richi Prasad Sharma

Landslide Susceptibility Analysis and Mapping in Sastha Valley of Periyar River Basin 23
P. Preji and Biju Longhinos

Investigations and Mitigation Measures of Landslide Affected Areas in Hill Roads of East Africa—Case Study on Projects of Ethiopia 43
Avik Kumar Mandal, S. Sailesh, and Pradyot Biswas

Stability Assessment of Cut Slopes Along Shivpuri–Kaudiyala Road (NH-58), Uttarakhand Himalayas 67
Neeraj, Koushik Pandit, and Shantanu Sarkar

Multivariate Adaptive Regression Spline Based Reliability Analysis of Stability of Durgawati Earthen Dam 81
V. Kumar, P. Samui, A. Burman, and N. Himanshu

Landslides Hazard Mapping Using High-Resolution Satellite Data 95
Saloni Jain, Rakesh Khosa, and A. K. Gosain

Analysis of Rainfall-Induced Slope Failure Using Monte Carlo Simulations: A Case Study 111
A. S. S. Raghuram and B. Munwar Basha

Probabilistic Assessment of Paglajhora Landslide Using SLOPE/W 129
Abhirup Dikshit and Neelima Satyam

Stability Analysis of Hill Slopes in The Nilgiris, Tamil Nadu 137
Praveen Kumar and Krishnamurthy Premalatha

Effect of Joint Orientation on the Seismic Stability of Rock Slope with Transmission Tower 149
Mohammad Zaid, Abdullah Talib, and Md. Rehan Sadique

Landslide Hazard and Risk Assessment Along NH-108 in Parts of Lesser Himalaya, Uttarkashi, Using Weighted Overlay Method 163
Pankaj Kumar, Anupam Mital, P. K. Champati Ray, and Shovan L. Chattoraj

Seismic Response of Gravity Retaining Wall 181
Monica Joseph and Subhadeep Banerjee

Some Studies on the Pseudostatic Analyses of Water Retention Type Tailings Dams 195
Pankaj Kumar and B. V. S. Viswanadham

Behavior of Reinforced Retaining Wall Against Railway Embankment Using Midas Under Static and Seismic Loading 213
Shilpa S. Vadavadagi and Sowmiya Chawla

Analysis of Retaining Wall in Static and Seismic Condition with Inclusion of Geofoam Using Plaxis 2D 223
Pankajkumar Yadav, D. K. Singh, P. P. Dahale, and A. H. Padade

Static and Seismic Analysis of Twin Metro Underground Tunnels 241
Manendra Singh, M. N. Viladkar, and N. K. Samadhiya

Soil-Structure Interaction Analysis of a Raft Foundation Supporting RC Chimney 259
Shilpa Dixit, Srinivasa Phanikanth Vedula, and Srinivas Kakaraparthi

Evolution in Liquefaction Strength of Ganga River Sand Due to Intrusion of Non-plastic Silt 283
Nishant Nilay and Pradipta Chakrabortty

Probabilistic Approach of Liquefaction Assessment for Guwahati Based on Shear Wave Velocity Values 299
K. S. Vipin and S. D. Anitha Kumari

Generalized Solution for the Critical Soil Wedge Angle Under Seismic Passive Earth Pressure Condition 309
Priyam Chatterjee, Bikash Chandra Chattopadhyay, and Joyanta Maity

Dynamic Slope Stability Analysis of Ash Dykes 323
P. Hari Prasad, K. Vineeth Reddy, and G. Kalyan Kumar

Analysis of Pile Under Seismic Motion Using Pseudo-static Approach . . . 335
Tanumaya Mitra, Kalyan Kumar Chattopadhyay, and Ambarish Ghosh

Ground Response Analysis and Determination of Liquefaction Potential Index . . . 345
Tanumaya Mitra, Kalyan Kumar Chattopadhyay, and Ambarish Ghosh

Probabilistic Seismic Hazard Analysis of Vadodara Region . . . 353
Payal Mehta and T. P. Thaker

Response of Monopile Supported Offshore Wind Turbine in Liquefied Soil . . . 367
Sangeet Kumar Patra and Sumanta Haldar

Reliability Analysis of Single Pile in Lateral Spreading Ground: A Three-Dimensional Investigation . . . 383
J. S. Rajeswari and Rajib Sarkar

Influence of Soil–Structure Interaction in Elevated Water Tank . . . 399
Sutanuka Nath and Atanu Kumar Dutta

Seismic Ground Response Analysis for Soil Site in Johor, Malaysia . . . 411
Anurag Sahare and Deepankar Choudhury

Design of Pile Foundation System for Wharf Structure in Liquefiable Soils . . . 429
Putti Swathi Priyadarsini and Satyam Neelima

Ground Response Analysis of a Nuclear Power Plant Site in Southern India: A Nonlinear Approach . . . 441
U. Veena, Naveen James, and T. G. Sitharam

Assessment of Soil Liquefaction Resistance by Finite Element Approach . . . 457
Anasua GuhaRay and Mohammad Zaid

Liquefaction Susceptibility Mapping of Kollam Coastal Stretch, Kerala, Considering Geotechnical Parameters . . . 471
S. K. Sithara, S. Surya, Sayana Parveen, Liz Maria Damiyan, Vinayak Mohan, A. Muhammed Siddik, and S. Adarsh

Finite Element Analysis of Foundation on Layered and Homogeneous Soil Deposit Under Dynamic Loading . . . 481
Abhay Kumar Verma and Supriya Mohanty

Site Classification of Strong Motion Stations of Uttarakhand, India, Based on Standard Spectral Ratio, and Horizontal-to-Vertical Spectral Ratio Methods . . . 495
N. H. Harinarayan and Abhishek Kumar

Seismic Site Classification and Site Period Determination of NIT Silchar Using MASW . . . 507
Arindam Saha, Kallol Saha, and Ashim Kanti Dey

Comparative Study of 1D, 2D and 3D Ground Response Analysis of Pond Ash from Odisha Under Different Earthquake Motions . . . 523
M. V. Ravi Kishore Reddy, Supriya Mohanty, and Rehana Shaik

Effect of Soil Grain Size on Liquefaction Strength of Sandy Soil . . . 539
Pradipta Chakrabortty, Angshuman Das, and Anil

Seismic Vulnerability Assessment of Pile Foundation in Liquefied Soil Incorporating Ground Motion Uncertainty . . . 555
Partha Bhowmik and Rajib Saha

1 g Shake Table Study on Seismic Behaviour of Model Structure Supported by Pile Foundation in Liquefiable Soil . . . 563
Archana Kunwar, Swagata DebRoy, and Rajib Saha

Studying and Comparing the Declustered EQ Catalogue Obtained from Different Methods for Guwahati Region NE India . . . 577
Niranjan Borah and Abhishek Kumar

3D Finite Element Analysis of Seismic Behavior of Soil-Piled Raft-Structural System . . . 591
Chaidul Haque Chaudhuri and Rajib Saha

Seismic Requalification of Pile-Supported Structure: Pseudo-Static Approach . . . 601
Pradeep Kumar Dammala, Manoj Kumar Manne, and A. Murali Krishna

Earthquake Early Warning System . . . 617
Aarti Taneja, Aniket Desai, and Ravi S. Jakka

Dynamic Characteristics of Subsoil Deposit of Tripura by Nakamura Method . . . 621
Rajat Debnath and Rajib Saha

Equivalent 1D Ground Response Analysis (GRA) of Black Cotton Soil for Three Different Sites Near Indore City . . . 633
Deepshikha Shukla and Chandresh H. Solanki

Pattern Recognition to Identify Susceptible Areas in Northwestern Himalaya . . . 653
Swati Singh Rajput, Ravi S. Jakka, and Amita Sinvhal

Modelling of Response of Tunnels Excavated in Squeezing Ground Condition . . . 667
Dipaloke Majumder, M. N. Viladkar, and Mahendra Singh

Liquefaction Mitigation with Stone Columns for a Sewage Treatment Plant: A Case Study . . . 689
K. Ganesh Deepak and Jinu Mary Jacob

Behavior of 3D-Reinforced Granular Trenches Under Cyclic Loading . . . 705
M. Hema and N. Unnikrishnan

A Review on the Role of Geosynthetics in Preventing the Excessive Settlement and Mud Pumping of Ballasted Railway Track . . . 715
Ramesh Gedela and Rajagopal Karpurapu

Strain Energy-Based Modeling of Soil Liquefaction Using Data-Driven Techniques . . . 727
S. Raj Athira and S. Adarsh

Numerical Simulation of Tiered Reinforced Soil Retaining Wall Subjected to Dynamic Excitations . . . 739
Seema Kumari and Arup Bhattacharjee

Improvement of Seismic-Bearing Capacity of Foundation on Soft Clay by Granular Material . . . 751
Paramita Bhattacharya and Puja Dutta

About the Editors

Dr. Madhavi Latha Gali is a Professor in the Department of Civil Engineering, Indian Institute of Science (IISc) Bangalore, India. She completed her Ph.D. from Indian Institute of Technology Madras, and has previously worked as a post-doctoral fellow and assistant professor at IISc and IIT Guwahati respectively. Professor Latha is a member of various professional bodies including IGS, ISSMGE and ISRM, and is the Editor-in-Chief of the Indian Geotechnical Journal, and an Editorial board member in many reputed journals. Her research work focuses on fundamental aspects of soil and ground reinforcement, and she has authored 70 journal articles, 4 book chapters and has developed a web-course on Geotechnical Earthquake Engineering on the NPTEL platform, sponsored by the Ministry of Human Resources Development, Government of India.

Dr. P. Raghuveer Rao is Principal Research Scientist at Department of Civil Engineering and involved in teaching, research and consultancy in the broad area of geotechnical engineering. He has been working with the department since 1989 and has been teaching courses related to subsurface exploration and soil testing, Earth retaining structures, behavior and testing of unsaturated soils, and fundamental of soil behavior for Masters and Doctoral students. His research interests are geotechnical instrumentation, slope stability analysis, numerical modelling, mechanics of unsaturated soils, contaminant transport through soil and reinforced earth structures. He has conducted several field and laboratory tests for design of foundations of different structures like buildings, turbo-generator and water tanks. He has analyzed stability of several embankments, tailing dams and stability of large size surge shafts for a hydropower project through numerical modelling and trial wedge method. He has 21 publications in journals and conference proceedings.

Seismic Response of Shallow Strip Footing with Log-Spiral Failure Mechanism Using Limit Analysis

Litan Debnath and Sima Ghosh

Abstract Computation of seismic bearing capacity that reduced to a single coefficient $N_{\gamma}^{'}$ along with the mathematical models of shallow strip footing resting on $c-\phi$ soil is done here. Upper-bound limit analysis method in pseudo-static approach with log-spiral failure mechanisms having the base angle of left triangular rigid block at left footing edge is assumed. Effects of both the horizontal and vertical seismic acceleration coefficients have been found to always reduce the bearing capacity ratio $N_{\gamma E}^{'} \big/ N_{\gamma S}^{'}$ significantly. From the results obtained by the present method of analyses, parametric studies have been done along with the comparative study with available solutions under seismic conditions to show the efficiency of the proposed modification. From the comparative study, it is seen that the present log-spiral failure mechanism result is found to be lesser than other researchers in the seismic case.

Keywords Bearing capacity · Single coefficient · Limit analysis

1 Introduction

Estimation of bearing capacity of the foundation is an important parameter in the design of any substructures, as the bearing capacity of soils is the most important of all the topics in soil engineering. The design of a footing involves two limit states; a serviceability limit state, which generally translates into a maximum settlement or differential settlement and an ultimate limit state. The bearing capacity of strip footings, even though, was studied a lot but not in detail in every aspect in the past. It has been extensively studied by several investigators such as Terzaghi (1943), Meyerhof (1963), and Chen (1975), but not in case of seismic areas. Under earthquake condition, the design requires special attention to reduce the devastating effect of the natural hazards. The traditional method used is 'pseudo-static method' for evaluating the effect of an earthquake load on the stability of soil–foundation system which gives an approximate solution to the real dynamic nature of the complex problem.

L. Debnath (✉) · S. Ghosh
Nit Agartala, Agartala, Tripura, India
e-mail: litandbnth4@gmail.com

M. Latha Gali and P. Raghuveer Rao (eds.), *Geohazards*, Lecture Notes in Civil Engineering 86, https://doi.org/10.1007/978-981-15-6233-4_1

The approach of pseudo-static: horizontal and vertical accelerations are applied to the center of gravity of the structure, and the problem is reduced to a static case of bearing capacity. At present, researchers are mostly interested in 'pseudo-dynamic method' as this method, which incorporates time-dependent effect of applied seismic load and effect of shear and primary waves, is applied to study the effect of variation of parameters like soil friction angle, wall friction angle, time period of earthquake ground motion, seismic shear and primary wave velocities of soil and seismic peak horizontal and vertical ground accelerations on the seismic earth pressures. Sarma and Lossifelis (1990), Richards et al. (1993) have suggested a method for computing the seismic bearing capacity of strip footings in seismic areas with the consideration of inertia forces on all parts of the soil–structure system (soil and foundation). Their method is based on an approach which they have been using for the analysis of slopes; it is a limit equilibrium method based on prior assumption concerning the inter-slice forces. It is well known that these methods give approximate solutions of the failure load and that the solution with respect to the exact solution cannot be said to be an upper- or a lower-bound one. Dormieux and Pecker (1995) and Soubra (1996) used the upper-bound method of the limit analysis theory and developed upper-bound solutions of the seismic bearing capacity factors. These solutions are rigorous upper-bound ones with respect to the exact solutions for an associated flow rule Coulomb material. In this paper, an upper-bound limit analysis method is presented in which the incremental external work done and the incremental internal energy dissipation are calculated after establishing the velocity field of the kinematically admissible failure mechanism as mentioned in Chen and Liu (1990).

1.1 Method of Analysis

In this paper, an upper-bound limit analysis method with pseudo-static approach is presented for the log-spiral failure mechanism. The assumption made here is that the soil is homogeneous of effective unit weight γ having Mohr–Coulomb shear strength characteristics c and ϕ and can be considered as a rigid plastic body; for shallow foundations, $D_f < B_0$, the overburden is idealized as a surcharge $q = \gamma D_f$. This method considered allows us to get a rigorous upper-bound solution with respect to the exact solution for an associated flow rule (Figs. 1 and 2).

The upper-bound theorem states that an upper-bound solution can be obtained if the work expended by the external loads is set equal to the power dissipated internally for a kinematically admissible velocity field. Here, the incremental external work due to an external force is the external force multiplied by the corresponding incremental displacement or velocity. The incremental external work due to self-weight in a region is the vertical component of the velocity in that region multiplied by the weight of the region; and the internal energy dissipation which forms the major part of the analysis is due to the factors such as cohesion (c), angle of friction (ϕ), length of the line along which the energy dissipation is calculated, and the velocity (V) with which the zone moves. The incremental external work due to the foundation load P,

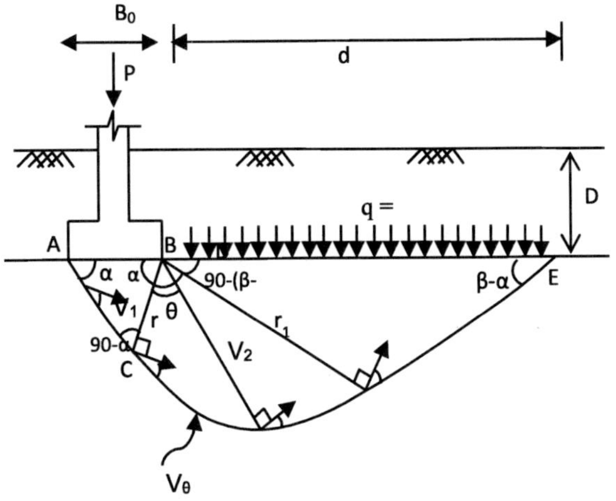

Fig. 1 Log spiral failure surface

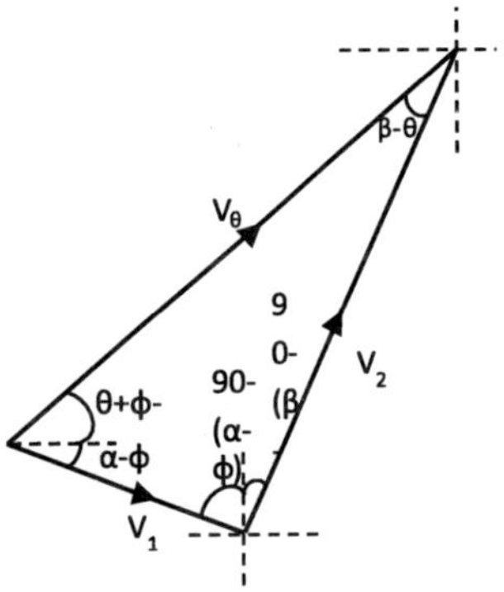

Fig. 2 Velocity hodograph

surcharge load q, weight of the soil wedges ABC and BDE, log-spiral zone BCD, and the corresponding inertial forces are

$$\Delta W_P = P\,[\sin(\alpha - \varphi)\,(1 - k_v) + \cos\,(\alpha - \varphi)\,k_h]\,V_1 \tag{1}$$

$$\Delta W_q = q\,d\left[-\sin\,(90^\circ - (\beta - \alpha + \varphi))\,(1 - k_v) + \cos\,(90^\circ - (\beta - \alpha + \varphi))\,k_h\right]V_2 \tag{2}$$

$$\Delta W_{ABC} = \gamma\frac{B_0 r_0 \sin(90^\circ - (\alpha - \varphi))}{2}.[\sin(\alpha - \varphi)\,(1 - k_v) + \cos\,(\alpha - \varphi)\,k_h]\,V_1 \tag{3}$$

$$\Delta W_{BCD} = \gamma.\int_0^\beta \frac{1}{2}.r.\mathrm{d}\theta.r.V_\theta$$

$$= \frac{\gamma}{2} \cdot \int_0^\beta r_0^2 . e^{2\theta \tan\varphi} . V_1 . e^{\theta \tan\varphi} . d\theta . [- \sin(\theta + \varphi - \alpha)(1 - k_v)$$

$$+ \cos(\theta + \varphi - \alpha)\, k] \tag{4}$$

$$\Delta W_{BDE} = \gamma . \frac{d.DE.\ \sin(\beta - \alpha)}{2} . \begin{bmatrix} - \sin(90^\circ - (\beta - \alpha + \varphi))(1 - k_v) \\ + \cos\ (90^\circ - (\beta - \alpha + \varphi))\, k_h \end{bmatrix} V_2 \tag{5}$$

The incremental internal energy dissipations along the velocity discontinuities *AC, DE, CD,* and radial line are

$$\Delta D_{AC} = c.AC. \cos\varphi . V_1 \tag{6}$$

$$\Delta D_{DE} = c.DE. \cos\varphi . V_2 \infty = c.DE. \cos\varphi . V_1 . e^{\beta \tan\varphi} \tag{7}$$

$$\Delta D_{CD} = c. \left(\int_0^\beta r_0 . e^{2\theta \tan\varphi} . V_1 . d\theta \right) . \cos\varphi = c. \left(r_0 . V_1 \int_0^\beta e^{2\theta \tan\varphi} . d\theta \right) . \cos\varphi \tag{8}$$

$$\Delta D_{\text{rad}} = \Delta D_{CD} \tag{9}$$

$$\Rightarrow \int_0^\beta r_0 . e^{2\theta \tan\varphi} . V_1 . d\theta = B_0 . \frac{\sin\alpha}{\cos\varphi} . \left[\frac{e^{2\beta \tan\varphi} - 1}{2 \tan\varphi} \right] \tag{10}$$

Equating the total incremental external work to the total incremental internal energy dissipation gives

$$\Rightarrow q_U = \left[\frac{\gamma B_0}{2} \right] \times N_\gamma' \tag{11}$$

where N_γ' is a single coefficient for simultaneous resistance of unit weight, surcharge, and cohesion.

1.2 Optimization of N_γ'

Optimization is required which is the process of finding the optimum values of the variables for a particular criterion or, in our context, the best decisions for a particular measure of performance, in this case to find the optimum value of N_γ'. MS Excel has been chosen for doing this process of finding the 'global concave optimization'. Here, the input values are the four constants k_h,k_v,$2c/\gamma B_0$, and D_f/B_0. This spreadsheet model also checks for the criteria proposed by Richards et al. (1993) to avoid the shear fluidization (i.e., the plastic flow of the material at a finite effective stress). As and when the inputs are given, the spreadsheet model calculates the optimized value of N_γ' for the ϕ value between 10° and 45° with the interval of 5° s.

1.3 Results

In this present study, the values presented are for the following considerations: angle of soil friction $\phi = 20°, 25°, 30°, 35°, 40°$, and 45°; depth to breadth ratio D_f/B_0 = 0, 0.25, 0.50, 0.75, and 1; horizontal seismic coefficient k_h = 0.1, 0.2, and 0.3; vertical seismic coefficient k_v = 0, $k_h/2$, and k_h. With the above considerations from the MS Excel spreadsheet model developed, the optimized values are found by giving four constants k_h,k_v,$2c/\gamma B_0$, and D_f/B_0 as inputs and tabulated in Table 1 for $2c/\gamma B_0 = 0.5$, respectively. This also checks for the criteria proposed by Richards et al. (1993) to avoid the shear fluidization (i.e., the plastic flow of the

Table 1 Seismic bearing capacity coefficients ($N_{\gamma E}$)

ϕ	$N'_{\gamma E}/N'_{\gamma S}$; $2c/\gamma B_0 = 0.5$					
	$\delta = 0$	$k_h = 0.1$				
	k_v	D_f/B_0				
		0	0.25	0.5	0.75	1
20	0	0.737	0.728	0.713	0.692	0.651
	$k_h/2$	0.754	0.742	0.725	0.703	–
	k_h	0.772	0.756	0.738	0.715	0.678
25	0	0.712	0.704	0.691	0.673	0.647
	$k_h/2$	0.723	0.712	0.698	0.679	0.652
	k_h	0.736	0.721	0.705	0.685	0.658
30	0	0.684	0.678	0.667	0.653	0.634
	$k_h/2$	0.691	0.682	0.669	0.654	0.634
	k_h	0.697	0.686	0.672	0.655	0.634
35	0	0.655	0.650	0.642	0.631	0.617
	$k_h/2$	0.656	0.649	0.639	0.627	0.613
	k_h	0.657	0.648	0.637	0.623	0.608
40	0	0.623	0.620	0.614	0.606	0.596
	$k_h/2$	0.619	0.614	0.607	0.599	0.588
	k_h	0.615	0.608	0.600	0.590	0.579

material at a finite effective stress). The tabulated values of the bearing capacity ratio $N'_{\gamma E} / N'_{\gamma S}$ are for the ϕ values between 20° and 40° with the interval of 5° s. For seismic condition, N'_{γ} is represented as $N'_{\gamma E}$ and static condition N'_{γ} is represented as $N'_{\gamma S}$.

1.4 Parametric Study

A detailed parametric study has been conducted using the results that are obtained from the optimization which gives the optimum seismic bearing capacity coefficient N'_{γ} which are obtained by supplying the four constants, viz., k_h, k_v, $2c / \gamma B_0$, and D_f / B_0 are as follows:

Figure 3 shows the variation of reduced reduction factor $N'_{\gamma E} / N'_{\gamma S}$ with k_h when $k_v = \frac{k_h}{2}, \delta = \frac{\varphi}{2}, \frac{D_f}{B} = 0, \frac{2c}{\gamma B_0} = 0.25$ for different value of ϕ. Due to increase in ϕ, internal resistance of the soil particles will be increased which resembles the fact that increase in reduction factor is due to increase in ϕ. Here in this plot, reduction factor increases by 37.57% due to increases in ϕ from 20° to 45° for the case of $k_h = 0.2$.

Figure 4 shows the variation of reduction factor $N'_{\gamma E} / N'_{\gamma S}$ with k_h when $\delta = \frac{\varphi}{2}, \varphi = 40°, \frac{D_f}{B_0} = 0, \frac{2c}{\gamma B_0} = 0.25$ for different value of acceleration factor k_v / k_h. From the plot, it is seen that reduction factor along the horizontal seismic acceleration is going to be decreased due to increase in k_v from 0 to k_h. Due to increase in vertical seismic acceleration, the disturbance in the soil particles increases, which allows more soil mass to participate in the vibration and hence decrease in reduction factor. Here due to increase in vertical seismic acceleration by $k_h / 2$, reduction factor decreases by 16.18% for the case of $k_h = 0.2$.

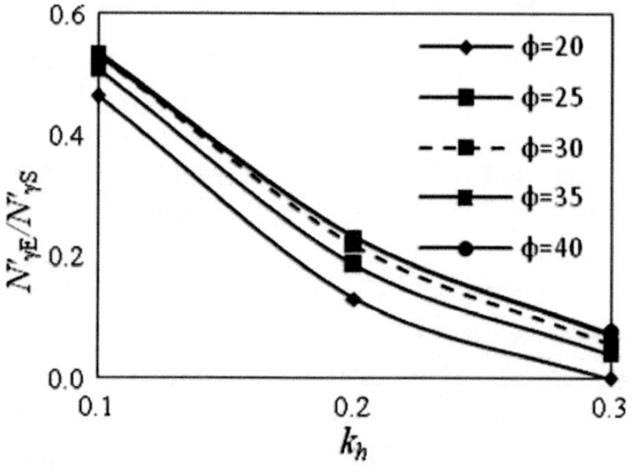

Fig. 3 Variation of $N'_{\gamma E} / N'_{\gamma S}$ with k_h for $k_v = k_h/2$, $2c/\gamma B_n = 0.25$, $D_f/B_n = 0$(LSFM)

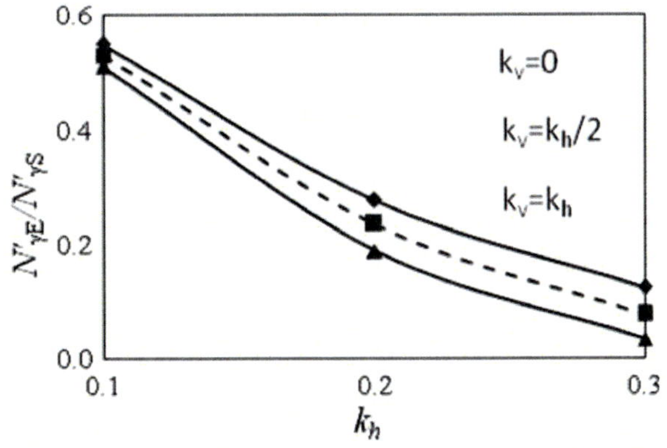

Fig. 4 Variation of $N'_{\gamma E}\Big/N'_{\gamma S}$ with k_h for $\phi = 40°,\ 2c/\gamma B_n = 0.25,\ D_f/B_n = 0$(LSFM)

Figure 5 shows the variation of reduction factor $N'_{\gamma E}\Big/N'_{\gamma S}$ with k_h when $\delta = \frac{\varphi}{2},\ \varphi = 40°,\ k_v = \frac{k_h}{2},\ \frac{2c}{\gamma B_0} = 0.25$ for different value of D_f/B_0. From the plot, it is seen that reduction factor along the horizontal seismic acceleration is going to be increased due to increase in D_f/B_0 from 0 to 1. Due to increase in D_f/B_0, the disturbance in the soil particles increases, which allows more soil mass to participate in the vibration and hence decrease in reduction factor capacity factor $N'_{\gamma E}\Big/N'_{\gamma S}$.

Figure 6 shows the variation of the reduction factor $N'_{\gamma E}\Big/N'_{\gamma S}$ with k_h when $\varphi = 40°, \delta = \frac{\varphi}{2}, k_v = \frac{k_h}{2}, \frac{D_f}{B_0} = 0$ for different value of $2c/\gamma B_0$ ratio. From the plot, it is seen that reduction factor increases by 20.89% due to increase in $2c/\gamma B_0$ by 0.25 for the case of $k_h = 0.2$. This is due to increase in cohesion, and reduction factor will be increased and increase in cohesion causes increase in intermolecular attraction among the soil particle which offers more reduction factor.

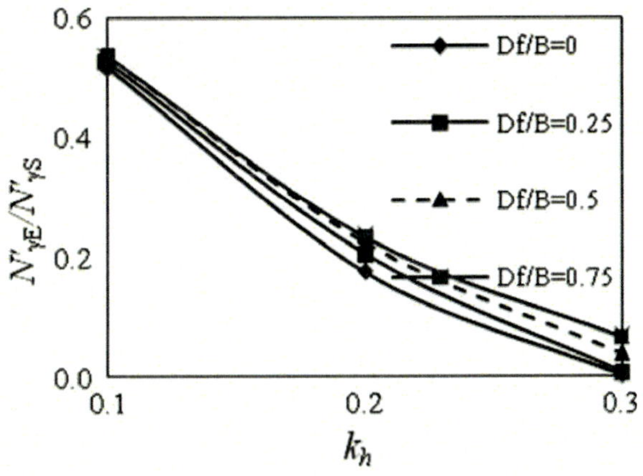

Fig. 5 Variation of $N'_{\gamma E}\Big/N'_{\gamma S}$ with k_h for $k_v = k_h/2, \phi = 40°$, $2c/\gamma B_n = 0.25$(LSFM)

Fig. 6 Variation of $N'_{\gamma E} / N'_{\gamma S}$ with k_h for $k_v = k_h/2$, $\phi = 40°$, $D_f / B_n = 0$(LSFM)

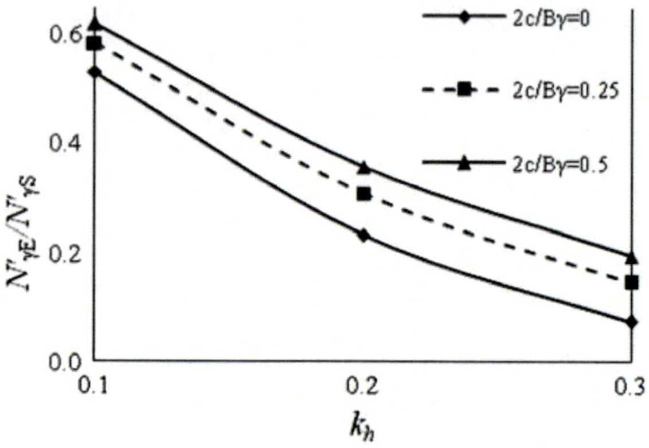

1.5 Comparison of Result

To broaden the perspectives of different authors on same or similar type of work with different approach with the analyses that have been done is compared. Figure 7 shows the comparison of results that are obtained by the process of optimizing using MS Excel spreadsheet model.

Figure 7 shows the comparative result of q_{uE} versus φ for the values of $\varphi = 15°$, $20°, 25°, 30°, 35°$, and $40°$. The present study of log-spiral mechanism and Richard's et al. (1993) results show clearly that they are in good agreement between $\varphi = 20°$, $25°$, $30°$, and $35°$. And also the curve of log-spiral mechanism is lying below the curve of IS 6403:1981 method but above the curve of Richard et al. (1993) between the above said φ Values. It can be seen that Richard et al. (1993) and IS 6403:1981 (1981) method increases in values from present log-spiral mechanism study as the following:

Fig. 7 Variation of q_{uE} with ϕ for $k_h = 0.2$, $k_v = 0.1$, $2c / \gamma B_n = 0.25$, $D_f / B_n = 0.5, \delta = 0$

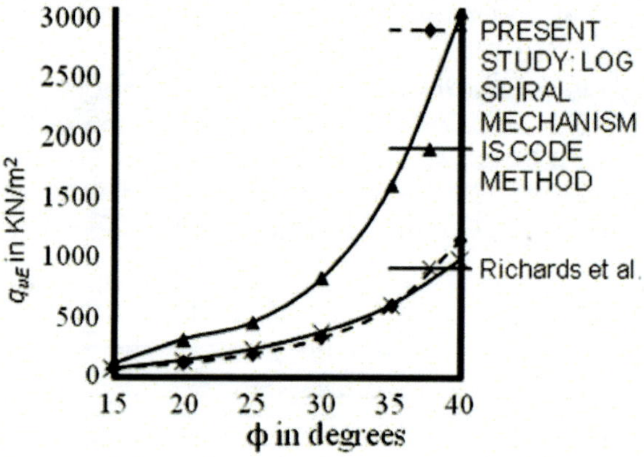

For $\varphi = 20°$, values increase by 15.59 and 162.59%.
For $\varphi = 25°$, values increase by 18.99 and 136.03%.
For $\varphi = 30°$, values increase by 13.52 and 151.44%.
For $\varphi = 35°$, values increase by 2.04 and 168.92%.

1.6 Conclusions

The limit analysis method using log-spiral failure surfaces is used to compute the seismic bearing capacity factor $N_{\gamma}^{'}$ which is reduced to a single coefficient. This factor is for horizontal shallow strip footing in soil has been obtained by pseudo-static approach for seismic forces. Extensive numerical iteration technique is carried out with the successfully developed spreadsheet model that does the process of finding the global concave optimized value to obtain the minimum $N_{\gamma}^{'}$ value. Results have been presented for seismic accelerations in both the horizontal and vertical directions, angles of soil friction, and wall friction. Both the horizontal and vertical seismic accelerations reduce the bearing capacity factors significantly. The result shows that the present analysis leads to the ultimate bearing capacity load q_{uE} values that are lesser than that of IS 6403:1981 (1981) method values and Richard et al. (1993) values. This gives very good agreement for the φ values 20°, 25°, 30°, and 35°.

References

Chen WF (1975) Limit analysis and soil plasticity. Elsevier Scientific Publishing Company, London, p 637
Chen WF, Liu XL (1990) Limit analysis in soil mechanic. Elsevier Science Publications Company, New York
Dormieux L, Pecker A (1995) Seismic bearing capacity of foundations on cohesionless soil. J Geotech Eng ASCE 121(3):300–303
IS Code (6403:1981) Indian Standard Code of Practice for Determination of Bearing Capacity of Shallow Foundations. Bureau of Indian Standards, Manak Bhavan, New Delhi
Meyerhof GG (1963) Some recent research on the bearing capacity of foundations. Can Geotech J 1(1):16–26 (Ottawa)
Richards R, Elms DG, Budhu M (1993) Seismic bearing capacity and settlements of foundations. ASCE 119:662–674
Sarma SK, Iossifelis IS (1990) Seismic bearing capacity factors of shallow strip footings. Géotechnique, The Institution of Civil Engineers, London 40(2):265–273
Soubra AH (1996) seismic bearing capacity of shallow strip footings in seismic conditions. ASCE 116(5):740–759
Terzaghi K (1943) Theoretical soil mechanics. Wiley, New York, p 510

A Comparative Study of Soil Slope Stability Under Seismic Loading Condition

Suman Hazari, Sima Ghosh, and Richi Prasad Sharma

Abstract The stability of the earth slope during seismic loading conditions is an important topic in the geotechnical engineering field. Several researchers reported stability analysis of slope under static and seismic loading conditions. This paper represents a comparative study of homogeneous $c - \phi$ soil slope stability using different dynamic approaches considering planer, circular and log-spiral rupture surfaces. Also, this paper provides a comparative study between the values of factor of safety obtained by the analytical approach and numerical solution. Results obtained from the different approaches are presented in tabular form to understand the comparison of different approaches considering different failure surfaces. It is observed that values of factor of safety obtained by the dynamic analysis considering log-spiral rupture surface are lower than the corresponding values obtained by the dynamic analyses assuming circular and planer failure surface.

Keywords $c - \phi$ soil · Seismic analysis · Logarithmic rupture surface · Factor of safety

List of Notations

$a(z, T)$	Acceleration at depth z, time t
Q_h, Q_v	Horizontal and vertical inertia forces due to seismic acceleration
b_1	Width of ith slice
b_2	Width of jth slice
c	Cohesion of soil
ϕ	Soil friction angle
N	Normal force
T	Tangential force
α	Angle of the base of the vertical slice with horizontal

S. Hazari (✉) · S. Ghosh · R. P. Sharma
Department of Civil Engineering, NIT Agartala, Jirania, Tripura, India
e-mail: sumanhazari22@gmail.com

M. Latha Gali and P. Raghuveer Rao (eds.), *Geohazards*, Lecture Notes in Civil Engineering 86, https://doi.org/10.1007/978-981-15-6233-4_2

i, j	Number of the vertical slice at different zones
r_0	Initial radius of log-spiral
r	Final radius of log-spiral
H	Height of slope
w_i, w_j	Weight of the slice
g	Acceleration due to gravity
G	Shear modulus of soil
Ω	Angular frequency of base shaking
k_h, k_v	Intensity of horizontal and vertical seismic acceleration
q	Surcharge
t	Time of vibration (seconds)
T	Period of lateral shaking (seconds)
v_s	Shear wave velocity
v_p	Primary wave velocity
β	Angle of slope with horizontal
γ	Unit weight of soil
FOS	Factor of safety
η	Wave length of the vertically propagating shear wave, Tv_S
λ	Wave length of the vertically propagating primary wave, Tv_p
ρ	Density of soil
ν	Poisson's ratio
K	Coefficient of reinforcement to maintain the stability
ζ	Damping ratio
τ	Shear resistance
η_s	Soil viscosity
$\omega_s H/v_s$	Normalized frequency of s-wave
$\omega_p H/v_p$	Normalized frequency of p-wave

1 Introduction

Determination of seismic slope stability is a crucial topic for geotechnical engineer. The traditional concept of slope stability is to evaluate the minimum factor of safety. Different methods used in the slope stability analysis are limit equilibrium method, limit analysis theorem, numerical modeling, etc. The limit equilibrium method satisfies force and/or moment equilibrium of a soil mass above the potential failure wedges. A limit analysis is used due to its physical significance and strict solving range. Numerical modeling has emerged as a tool based on the application of finite element and upper bound analysis of plasticity limit analysis. The most common limit equilibrium techniques are the Swedish circle method or method of slices in which the failure wedges are divided into the number of slices. Fellenius (1936) established a line which is called Fellenius line to locate the center of the most critical circle which gives a minimum factor of safety. Bishop (1955) upgraded the method of slices in which inter-slice forces are taken into account. Consequently, the same method with noncircular failure surfaces was upgraded by Morgenstern and Price (1965), Spencer (1967), Janbu (1954, 1973), Chowdhury (1978) and Zhu et al. (2003) based on different assumptions of inter-slice forces.

The seismic behavior of soil slope by pseudo-static approach has been reported by Terzaghi (1950). In the pseudo-static analysis, the loads coming from the earthquakes are represented by the static inertia forces which act at the center of the failure soil mass. Several investigations (Spencer 1973; Sarma 1979; Koppula 1984; Leshchinsky and San 1994 and Choudhury et al. 2007) have been carried out to analyze the stability of slope by means of limit equilibrium technique at earthquake loading conditions. The pseudo-static analysis used by several researchers does not consider the effects of time and phase lag of waves through the soil medium. Steedman and Zeng (1990) introduced a new method called pseudo-dynamic approaches for retaining walls to overcome this drawback. Several researchers (Choudhury and Nimbalkar 2005, 2006, 2007; Nimbalkar and Choudhury (2008); Ghosh 2007; Ghosh and Sharma 2010, 2012; Saha and Ghosh 2014a, b, 2015) analyzed the different earth structure using the pseudo-dynamic approach to overcome the limitation of the pseudo-static method. Chanda et al. (2015) used a pseudo-dynamic approach considering circular failure mechanism for homogeneous slope in which Fellenius line is used to locate the center of the most critical circle. The simple pseudo-dynamic method considers only waves traveling upward through soil medium which causes a violation of the free-surface boundary condition. To overcome this disadvantage, Bellezza (2014) suggested a modified pseudo-dynamic approach that satisfies the zero-stress boundary condition considering the failure surface as a planner rupture surface. Further Bellezza (2015) extended this modified pseudo-dynamic method considering both horizontal and vertical seismic accelerations. Pain et al. (2015, 2016) carried out the modified pseudo-dynamic method to analyze the seismic stability of the retaining wall as well as the uplift capacity of horizontal strip anchor. Chanda et al. (2017a) analyzed the slope made up of $c - \phi$ soil to determine the stability number as well as the reinforcement strength against factor of safety considering the limit equilibrium approach. Further, Chanda et al. (2017b) proposed a modified pseudo-dynamic approach for the calculation of the stability number as well as reinforcement strength against safety factor of the slope of circular failure surface in which Fellenius line is used to locate the center of the most critical circle.

Slope stability analysis by finite element techniques has been widely used in the last few decades. Several studied of slope stability analysis based on the finite element method are reported by Pasternack and Gao (1988), Mastoni and San (1992), Duncan (1996) Griffiths and Lane (1999).

Here, an attempt is made to a comparative study of soil slope considering planer, circular and log-spiral failure surface to evaluate the factor of safety using pseudo-static, pseudo-dynamic and modified pseudo-dynamic and numerical method. An attempt is also made to take the slope where the failure of a surface is assumed any of the planer, circular and log-spiral surface.

2 Methodology

Figures 1 and 2 show failure mechanism of slope and details of slice, respectively. Hazari et al. (2020) presented a homogeneous slope made up of $c - \phi$ soil, the angle of which is β with the horizontal and height of the slope is H. Here, the potential failure surface is circular and has been divided into a number of vertical slices. The circular portion of the failure surface (AE) is governed by the height of the soil slope (DE) and the location of the center of the circular arc (O). 'OA' and 'OE' are radius r of the circular arc passing through the center 'O'. The failure wedge ADC is divided into 'n' number of slices. The wedge is divided into two zones in which the zone ADB is divided into 'm' number of slices and zone BDE is divided into '$n - m$' number of slices. The forces acting on each slice include weight (w_i) acting vertically downward at the c.g. of the slice, cohesive force (c_i) actin along failure surface in the direction opposite to the movement of the failure mass, the normal (N_i) and tangential forces (T_i) acting on the lower boundary of the slice, reaction (R_i) inclined at an angle ϕ with the normal. Similarly, the forces acting on kth slice include weight (w_j), cohesive force (c_j), the normal (N_j) and tangential forces (T_j), reaction (R_j) inclined at an angle ϕ with the normal of the failure surface. During earthquake loading condition, the shear and primary waves transmitting through the soil mass at a velocity.

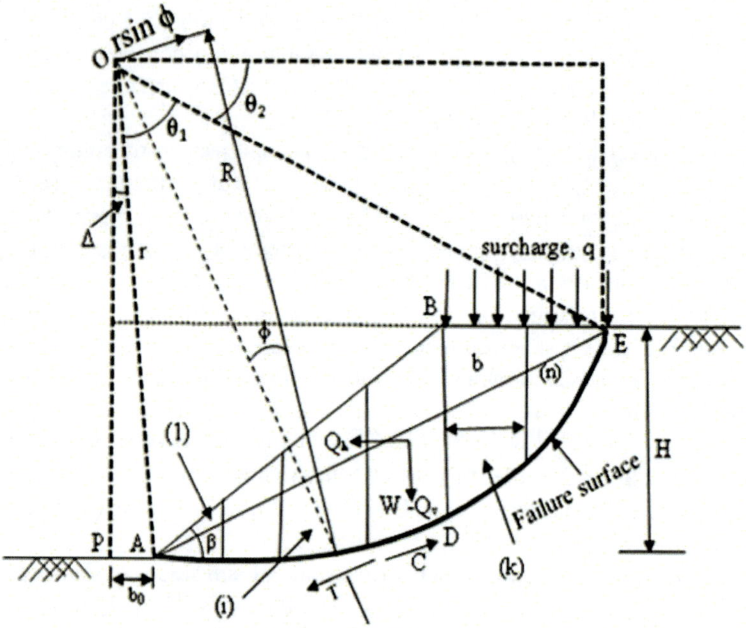

Fig. 1 Failure mechanism of the slope

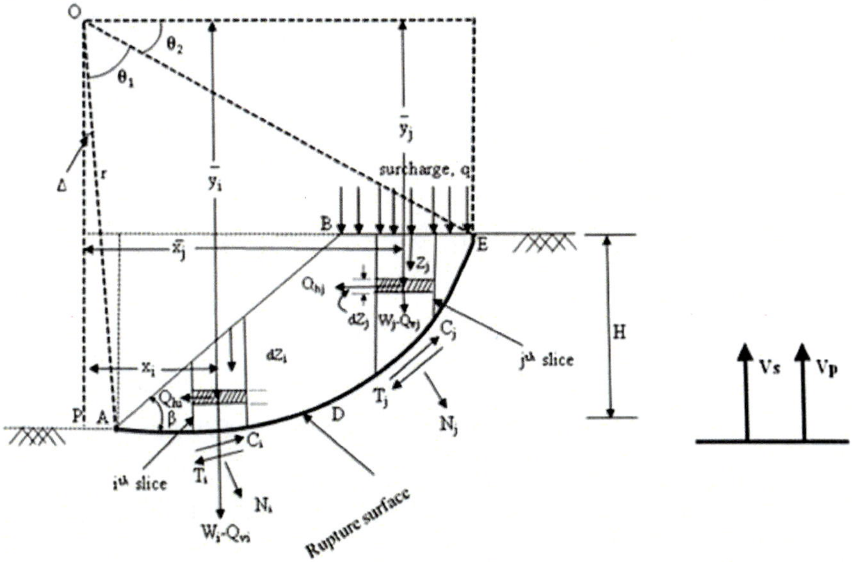

Fig. 2 Generalized slice and center of gravity

$$V_s = \sqrt{G/\rho} \tag{1}$$

$$V_p = \sqrt{2G(1-\nu)/\rho(1-2\nu)} \tag{2}$$

where G, ρ, ν and ω are shear modulus, density and Poisson's ratio of the slope material, respectively.

If the base of the slope is subjected to harmonic horizontal and vertical seismic accelerations of amplitude $k_h g$ and $k_v g$, the accelerations at any depth z below the top can be expressed as (Choudhury et al. 2005),

$$a_h(z,t) = k_h g \sin \omega\left(t - \frac{H-z}{V_s}\right) \tag{3}$$

$$a_v(z,t) = k_v g \sin \omega\left(t - \frac{H-z}{V_p}\right) \tag{4}$$

Finally, the factor of safety (FOS) is given by the following expression:

$$\text{FOS} = \frac{M_R}{M_h + M_v + M_T} \tag{5}$$

The horizontal and vertical disturbing moment M_h and M_v are expressed as,

$$M_h = (Q_{h_1} \times \overline{y}_1) + (Q_{h_2} \times \overline{y}_2) + \sum_{i=3}^{m} (Q_{h_i} \times \overline{y}_i) + \sum_{j=m+1}^{n-1} (Q_{h_i} \times qBk_h)\overline{y}_j + (Q_{h_n} \times qBk_h)\overline{y}_n \quad (6)$$

$$M_v = (Q_{v_1} \times \overline{x}_1) + (Q_{v_2} \times \overline{x}_2) + \sum_{i=3}^{m} (Q_{v_i} \times \overline{x}_i) + \sum_{j=m+1}^{n-1} (Q_{v_i} \times qBk_v)\overline{x}_j + (Q_{v_n} \times qBk_v)\overline{x}_n \quad (7)$$

The disturbing moment due to the tangential component (T) is given by

$$M_T = r\left[\sum_{i=1}^{m} (W_i \sin \alpha_i) + \sum_{j=m+1}^{n} (W_j \sin \alpha_j)\right] \quad (8)$$

The resisting moment of soil slope is given by,

$$M_R = r\left[c\left\{\sum_{i=1}^{m} (b_1 \sec \alpha_i) + \sum_{j=m+1}^{n} (b_2 \sec \alpha_j)\right\} + \sum_{i=1}^{m} \begin{matrix} (w_i \cos \alpha_i) \tan \varphi + \\ \sum_{j=m+1}^{n} (w_j \cos \alpha_j) \tan \varphi \end{matrix}\right] \quad (9)$$

Hazari et al. (2018a) analyzed the slope by taking a log-spiral rupture surface. This logarithmic spiral surface is governed by the height of slope H and the location of the center of the logarithmic spiral arc 'O'. OE and OA are the initial radius (r_o) and final radius (r), respectively, of the logarithmic spiral curve.

The generalized equation for the log-spiral curve is given by

$$r = r_0 e^{\theta_1 \tan \phi} \quad (10)$$

where r = Final radius of the log-spiral curve.

The initial radius r_0 is given by

$$r_0 = \frac{H \operatorname{cosec}(\theta_1 + \theta_2)}{e^{\theta_1 \tan \phi} - \frac{\sin \theta_2}{\sin(\theta_1 + \theta_2)}} \quad (11)$$

Similarly, factor of safety (FOS) is given by

$$\mathrm{FOS} = \frac{M_{\mathrm{res}}}{M_h + M_v + M_t} \quad (12)$$

Hazari et al. (2018b) have modified the failure mechanism that was considered by Hazari et al. (2017) by taking a log-spiral rupture surface. The simple pseudo-dynamic method considers only waves traveling upward through soil medium which causes a violation of the free-surface boundary condition. To overcome this disadvantage, Bellezza (2014) suggested a modified pseudo-dynamic approach that satisfies the zero-stress boundary condition considering failure surface as a planner rupture surface and develop a methodology using modified pseudo-dynamic approaches of slope considering log-spiral failure mechanism based on the visco-elastic behavior of homogeneous $c - \phi$ soil slope. Using the limit equilibrium principle in which the failure wedge of the slope is divided into a number of vertical slices.

The horizontal acceleration can be obtained by

$$a_h(z,t) = \frac{k_h g}{\left(C_s^2 + S_s^2\right)}\begin{bmatrix} (C_s C_{sz} + S_s S_{sz})\cos(\omega_s t) \\ +(S_s C_{sz} - C_s S_{sz})\sin(\omega_s t) \end{bmatrix} \tag{13}$$

The vertical acceleration can be expressed as

$$a_v(z,t) = \frac{k_v g}{C_s^2 + S_s^2}\begin{bmatrix} (C_s C_{sz} + S_s S_{sz})\cos(\omega_p t) + \\ (S_s C_{sz} - C_s S_{sz})\sin(\omega_p t) \end{bmatrix} \tag{14}$$

The expression for factor of safety (FOS) of the slope is given by,

$$\text{FOS} = \frac{M_{\text{res}}}{M_h + M_v + M_t} \tag{15}$$

Hazari et al. (2017) introduced a vertical slice method to evaluate the stability of soil slope considering the pattern of failure surface as log-spiral using pseudo-static method. The pseudo-static seismic forces in both horizontal and vertical directions are considered with the various parameters of soil parameters, viz. slope angle, soil friction angle, etc.

The factor of safety (FOS) is given by

$$\text{FOS} = \frac{M_{\text{res}}}{M_h + M_v + M_t} \tag{16}$$

where M_{res}, M_h, M_v and M_t are the resisting moment, horizontal driving moment, vertical driving moment and moment due to tangential, respectively. Also, x_i and x_j denote the distance of c.g. for ith and jth slice, respectively.

3 Results

Tables 1, 2, 3 and 4 represent the results obtained from different approaches to seismic analysis considering different failure mechanisms. From the global optimization curve, the minimum factor of safety is taken as the optimized value of factor of safety (FOS).

Table 1 Pseudo-static factor of safety at $c/\gamma H = 0.05$ and $q/\gamma H = 0$ (pseudo-static: log-spiral)

ϕ	$k_v = 0.0$			$k_v = k_h/2$			$k_v = k_h$		
	$k_h = 0.1$								
	β								
	20°	25°	30°	20°	25°	30°	20°	25°	30°
20°	0.95	0.94	0.93	0.91	0.90	0.89	0.87	0.87	0.86
30°	1.07	1.04	1.02	1.02	1.00	0.98	0.98	0.96	0.94
40°	1.22	1.19	1.15	1.18	1.14	1.11	1.14	1.10	1.07

Table 2 Factor of safety at pseudo-dynamic condition for $q/\gamma H = 0$ (pseudo-dynamic: circular)

ϕ	$k_v = 0.0$			$k_v = k_h/2$			$k_v = k_h$		
	β								
	20°	25°	30°	20°	25°	30°	20°	25°	30°
35°	1.2735	1.2282	0.9974	1.2558	1.2095	0.9364	1.2225	1.1736	0.8825
40°	1.3769	1.3408	1.1646	1.3561	1.3307	1.0933	1.3170	1.2970	1.0303
45°	1.6124	1.5439	1.3582	1.5881	1.5323	1.2751	1.5424	1.4896	1.2016

Table 3 Results of pseudo-dynamic factor of safety at $c/\gamma H = 0.05$ and $q/\gamma H = 0$ (pseudo-dynamic: log-spiral)

ϕ	$k_v = 0$			$k_v = k_h/2$			$k_v = k_h$		
	$k_h = 0.1$								
	β								
	20°	35°	50°	20°	35°	50°	20°	35°	50°
20°	1.0244	0.8292	0.7764	0.9810	0.8046	0.6701	0.9446	0.7824	0.6398
30°	1.2173	1.0910	0.8355	1.1735	1.0308	0.7853	1.1328	0.9888	0.7546
40°	1.3802	1.2265	0.9513	1.3409	1.1728	0.8910	1.2972	1.1421	0.8361

Table 4 Factor of safety at dynamic condition $\zeta = 20\%$, $c/\gamma H = 0.05$ and $q/\gamma H = 0$ (modified pseudo-dynamic: log spiral)

ϕ	$k_v = 0.0$			$k_v = k_h/2$			$k_v = k_h$		
	$k_h = 0.1$								
	β			β			β		
	20°	35°	50°	20°	35°	50°	20°	35°	50°
20°	1.1218	0.8526	0.6720	1.0914	0.8361	0.6485	1.0840	0.8144	0.6082
30°	1.3438	1.1516	0.9198	1.3397	1.0560	0.8685	1.3219	0.9746	0.8200
40°	1.4722	1.3210	1.1564	1.4688	1.3003	1.0923	1.4481	1.2605	1.0325

4 Comparative Study

Table 5 shows the comparison of factor of safety of slope obtained by different methods of seismic analyses considering different failure mechanisms. It is observed that factor of safety obtained by the seismic analysis of different methods considering log-spiral failure surface is lower than those obtained considering circular rupture surface. It is also seen that damping has a great effect on factor of safety of slope. For an instance, at $k_h = 0.1$, $\phi = 40°$, $\beta = 20°$ and $k_v = k_h/2$ Hazari et al. (2018b) found factor of safety 21 and 2% less than Chanda et al. (2017a) and (b) as Hazari et al. (2018b) considered log-spiral failure surface, whereas Chanda et al. (2017a) and (b) considered planer and circular rupture surface, respectively. Similarly, for a pseudo-dynamic case, factor of safety obtained considering log-spiral failure surface gives a lower value than those obtained assuming circular failure surface. From Table 5, it is observed that Hazari et al. (2018a) found considerable value through numerical analyses by keeping all the parameters as used in the analytical solution.

5 Conclusion

The study involves a planer, circular and log-spiral failure surfaces of homogeneous soil slope made up of $c - \phi$ soil that was used to determine factor of safety by different methods of seismic analyses. Results were obtained from pseudo-static, pseudo-dynamic and modified pseudo-dynamic methods with respective mathematical models considering different failure surface suggested by different authors and presented in tabular form for the comparison. Values obtained from the different methods also compared with the value obtained by numerical analyses suggested by the authors. It is observed that results obtained considering a log-spiral failure surface give the lower value of factor of safety than those obtained considering planer and circular failure surfaces.

Table 5 Comparison of factor of safety for different methods considering different rupture surfaces

k_h	k_v	β	ϕ	Chanda et al. (2015) (pseudo-dynamic: circular rupture surface)	Hazari et al. (2020) (pseudo-dynamic: circular rupture surface)	Hazari et al. (2018a) (pseudo-dynamic: log-spiral rupture surface)	Chanda et al. (2017a) (modified pseudo-dynamic: planer rupture surface)	Chanda et al. (2017b) (modified pseudo-dynamic: circular rupture surface)	Hazari et al. (2018b) (modified pseudo-dynamic: log-spiral rupture surface)	Hazari et al. (2018a) numerical results
0.1	$k_h/2$	20°	20°	1.01	0.99	0.98	0.95	0.76	0.75	-
			30°	1.29	1.19	1.17	1.18	0.98	0.97	-
			40°	1.50	1.35	1.34	1.34	1.16	1.10	1.22
		35°	20°	0.85	0.86	0.80	0.72	0.66	0.62	-
			30°	1.05	1.04	1.03	0.91	0.79	0.78	-
			40°	1.19	1.19	1.17	1.10	0.98	0.95	1.07
		50°	20°	0.68	0.71	0.67	0.67	0.58	0.52	-
			30°	0.81	0.82	0.78	0.84	0.73	0.70	-
			40°	0.89	0.95	0.89	0.99	0.86	0.81	< 1

References

Bellezza I (2014) A new pseudo-dynamic approach for seismic active soil thrust. Geotech Geol Eng 32(2):561–576

Bellezza I (2015) Seismic active earth pressure on walls using a new pseudo-dynamic approach. Geotech Geol Eng 33(4):795–812

Bishop AW (1955) The use of slip circle in the stability analysis of earth slopes. Geotechnique 5(1):7–17

Chanda N, Ghosh S, Pal M (2017a) Analysis of slope using modified pseudo-dynamic method, J. Geotech. Engrg., Taylor & Francis, 13(6):548–559

Chanda N, Ghosh S, Pal M (2017b) Analysis of slope using modified pseudo-dynamic method, J Geotech Eng 12(4):337–346 (Taylor & Francis)

Chanda N, Ghosh S, Pal M (2015) Pseudo-dynamic analysis of slope considering circular rupture surface. J Geotech Eng 10(3):288–296 (Taylor & Francis)

Choudhury D, Nimbalkar S (2005) Seismic passive resistance by pseudo-dynamic method. Geotechnique 55(9):699–702

Choudhury D, Basu S, Bray J (2007) Behavior of slopes under static and seismic conditions by limit equilibrium method. GSP 161 Embankments, Dam and slopes

Choudhury D, Nimbalkar SS (2006) Pseudo-dynamic approach of seismic active earth pressure behind retaining wall. Geotech Geol Eng 24(5):1103–1113

Choudhury D, Nimbalkar S (2007) Seismic rotational displacement of gravity walls by pseudo-dynamic method: passive case. Soil Dynam Earthquake En 27(3):242–249

Chowdhury RN (1978) Slope analysis. Elsevier, New York

Duncan JM (1996) State of the art: limit equilibrium and finite element analysis of slopes. J Geotech Eng 122(7):577–596

Felleunius W (1936) Calculation of stability of earth dams. In: Proc. 2nd Congr. Large Dams, Washington, 4, pp 445–462

Ghosh P (2007) Seismic passive pressure behind a non-vertical retaining wall using pseudo-dynamic method. Geotech Geol Eng 25(6):693–703

Ghosh S, Sharma RP (2010) Pseudo-dynamic active response of non-vertical retaining wall supporting c-ϕ backfill. Geotech Geol Eng (Springer) 28(5):633–641

Ghosh S, Sharma RP (2012) Pseudo-dynamic evaluation of passive response on the back of a retaining wall supporting c-ϕ backfill. Int J Geomech Geoeng 7(2):115–121

Griffiths DV, Lane PA (1999) Slope stability analysis by finite elements. Geotechnique 49(3):387–403

Hazari S, Ghosh S, Sharma RP (2020) Swedish circle method for pseudo-dynamic analysis of slope. Geotech Geol Eng. https://doi.org/10.1007/s10706-019-01170-y

Hazari S, Ghosh S, Sharma RP (2017) Pseudo-static analysis of slope considering log spiral failure mechanism. In: Proceedings of Indian Geotechnical Conference, IIT Guwahati

Hazari S, Ghosh S, Sharma RP (2018a) Seismic stability of slope: analytical and numerical solution (Communicated)

Hazari S, Ghosh S, Sharma RP (2018b) Seismic analysis of slope considering log-spiral failure mechanism with numerical validation (Communicated)

Janbu N (1954) Applications of composite slip surfaces for stability analysis In: Proc. of European Conf. on the Stability of Earth Slopes, Stockholm, Sweden, 3, pp 43–49

Janbu N (1973) Slope stability computations. In: Hirschfeld RC, Poulos S (eds). Wiley, New York, pp 47–86

Koppula SD (1984) Pseudo-static analysis of clay slopes subjected to earthquakes. Geotechnique 34(1):71–79

Leshchinsky D, San KC (1994) Pseudostatic seismic stability of slopes: design charts. J Geotech Eng ASCE 120(9):1514–1532

Matsui T, San KC (1992) Finite element slope stability analysis by shear strength reduction technique. Soils Found. 32(1):59–70

Morgenstern NR, Price VE (1965) The analyses of the stability of general slip surfaces. Geotechnique 15(1):79–93
Nimbalkar S, Choudhury D (2008) Seismic design of retaining wall considering wall–soil inertia for active case. Int. J. Geotech. Engng 2(4):319–328
Pain A, Choudhury D, Bhattacharyya SK (2015) Seismic stability of retaining wall–soil sliding interaction using modified pseudo-dynamic method. Geotechnique 5(1):56–61
Pain A, Choudhury D, Bhattacharyya SK (2016) Seismic uplift capacity of horizontal strip anchors using a modified pseudo-dynamic approach. Int J Geomech 16(1):1–12
Pasternack SC, Gao S (1988) Numerical methods in the stability analysis of slopes. Comput Struct 30(3):573–579
PLAXIS 2D 8.2 (2004) [Computer software]. Delft, Netherlands, Plaxis
Saha A, Ghosh S (2014a) Pseudo-dynamic analysis for bearing capacity of foundation resting on c–ϕ soil. J Geotech Eng 9(4):379–387
Saha A, Ghosh S (2014b) Pseudo-dynamic bearing capacity of shallow strip footing resting on c-ϕ soil considering composite failure surface: bearing capacity analysis using pseudo-dynamic method. J Geotech Eng 6(2):12–34
Saha A, Ghosh S (2015) Pseudo-dynamic bearing capacity of shallow strip footing resting on c-ϕ soil considering composite failure surface: bearing capacity analysis using pseudo-dynamic method. J Geotech Eng 6(2):12–34, 9 Jun 2020. https://doi.org/10.4018/IJGEE.2015070102
Sarma SK (1979) Stability analysis of embankments and slopes. J Geotech Eng 105(12):1511–1524
Spencer E (1967) A method of analysis of the stability of embankments assuming parallel interslice forces. Geotechnique 17(1):11–26
Spencer E (1973) Thrust line criterion in embankment stability analysis. Geotechnique 23(1):85–100
Steedman RS, Zeng X (1990) The influence of phase on the calculation of pseudo-static earth pressure on a retaining wall. Geotechnique 40(1):103–112
Terzaghi K (1950) Mechanisms of land slides. Engineering geology (Berkeley) volume, Geological Society of America
Zhu DY, Lee CF, Jiang HD (2003) Generalised framework of limit equilibrium methods for slope stability analysis. Geotechnique 53(4):377–395

Landslide Susceptibility Analysis and Mapping in Sastha Valley of Periyar River Basin

P. Preji and Biju Longhinos

Abstract Earth attracts objects toward its surface through the gravitational force and keeps everything in place. In the case of a slope, this gravitational force may cause the materials to move along it causing what is known as landslides. This study mainly focuses on analyzing the landslide susceptibility in Sastha valley of Periyar river basin, Idukki district, Kerala. The study area shows some paleo-landslide movements indicated by sparse tea plantations and sharp meandering of the river flowing through the valley. This motivated to analyze the stability of the slopes through an extensive field investigation in the area. The stability analysis is carried out under both steady and transient states using SLOPE/W and SEEP/W software. A factor of safety map of the study area is also prepared from the GIS-TISSA model. Finally, a landslide susceptibility map of the area is generated by leveraging the results from both field study and GIS model.

Keywords Landslides · Slope stability · River basin · GIS model · Field study

1 Introduction

Earth attracts objects toward its surface through the gravitational force. This constant vector quantity keeps everything in place. In the case of a slope, gravity becomes a curse under certain circumstances and leads to hazards to people, animals, and properties. The gravitational force on a slope has two components, one which is perpendicular to slope and one which is parallel to the slope. In certain conditions, the perpendicular component may decrease and the parallel component increase to overcome the frictional force that resists the motion of particles. This leads to the

P. Preji (✉) · B. Longhinos
College of Engineering, Trivandrum, Kerala, India
e-mail: preji1994@gmail.com

B. Longhinos
e-mail: longhinos2cet@gmail.com

M. Latha Gali and P. Raghuveer Rao (eds.), *Geohazards*, Lecture Notes in Civil Engineering 86, https://doi.org/10.1007/978-981-15-6233-4_3

movement of materials forming the slope. Such instability of the slope occurs when soil, rock, regolith, and vegetation moves downward or outward under the force of gravity is called the landslide. The conditions that trigger landslides can be geological, morphological, physical, and due to human activities.

Because of the rapid growth of urbanization and population, the incidents of reporting landslides are increasing day by day. Earth's natural environment comprising the topography, forests, grasslands, and habitats of native flora and fauna are getting affected by it. Moreover, it has a serious impact on human life ranging from property damage, injury, and to even death. On account of the serious impact of these hazards on human life and property, many researchers began to investigate the geotechnical aspects of landslides. The selection of suitable remedial measures to control the movement of earth slopes and to secure their safety represents the most important geotechnical engineering activity. Landslide investigations and forecasting have got greater attention in recent years due to the unrecoverable damages caused by it.

1.1 Scenario in Kerala

Kerala is a land of beauty and wonders blessed with natural resources. Due to the geographical location, weather conditions, population, and unsustainable land use the state is prone to several natural and human-induced disasters. Among them, the landslides especially debris flows (*urul pottal* in local vernacular) are the most common. Studies reveal that in Kerala, landslides occur in the high land region that contains steep slopes with over-saturated soil due to prolonged rainfall and population growth. Hence, about 1500 km^2 area in Western Ghats of Kerala is vulnerable to the landslide (Kerala State Disaster Management Authority 2012). As a result, more than 257 human lives were lost since 2010.

The occurrence of landslides is frequent in Idukki district of Kerala, particularly, along road cuttings and hill slopes causing damage to life and property. The geographical characteristics of the district (e.g., dense forests, steep hills, deep valley, and heavy rainfall) made 60% of its area prone to landslides. Also, several projects such as tourism development, hill highway projects, and residential projects are going to deploy in Idukki. The waving topological features emphasize the need for careful analysis and sustainable land use in the Idukki district to handle the landslides. Many researchers investigated these problems to map the vulnerable area and assessed mitigation strategies that protect people, property, and various resources. However, inventing better methods for landslide susceptibility mapping and hazard assessment are required not only to cope with the development process and scientific cultivation but also to keep everything (including natural and manmade resources) safe.

1.2 Study Area

Study area known as Sastha valley forms a part of the Western Ghats in Vandiperiyar panchayat of Idukki district in Kerala. It is located at latitude 9.5266° N & 9.5283° N and longitude 77.0757° N & 77.1092° N. This valley is an important catchment area of Periyar River, and on the South, it is surrounded by Sabarimala hill ranges. An important road connecting to Sathram pilgrimage point of Sabarimala devotees is passing through this valley. Landslides are frequent at many places in this road. The panchayat is planning for future development of this area to attract more tourist and Sabarimala devotees. One of the peculiarities of this area is that it shows some signs of paleo-landslides, i.e., the landslides that were occurred in some geological past. This is indicated by the changing course of the stream flowing in the valley at many places.

1.3 Objectives of the Study

The objective of this study is to conduct an extensive investigation of the landslide on the selected area, the Sastha valley, to prepare a landslide susceptibility map for the area. To achieve this goal, this paper follows a structured methodology that comprises software-based and GIS-based slope stability analysis. The software-based stability analysis is conducted based on the field investigation in the selected locations of the Sastha valley. Selection of these locations depends on the observed paleo-landslide symptoms. Since the selected area is prone to heavy rainfall, one of the key ideas behind the software-based analysis is to use the historical rainfall data to understand the behavior of the slope under rainfall and other climatic conditions. Integrating the result from software analysis and GIS method, this study prepares a landslide zonation map that shows vulnerable locations in the area.

2 Literature Review

Landslide is one of the major natural hazards on the earth. According to the study conducted by Ubechu and Okeke (2017), several attributes such as the rate at which material move, type of material, and the nature of movement along with the topology/morphology of the area determine the type of landslide. Namdar and Yahaya (2014) approach the landslide as a geo-hazard induced by geological, geotechnical, and geomorphological features of the area. The erosion process that occurs due to rainfall and less vegetation changes the geomorphology of a site. Thus, soil mineralogy is an important factor associated with landslides. Soil bio-engineering measures are a kind of landslide mitigation technique (Bhattacharjee 2016; Raut and Gudmestad 2018). These techniques utilize the living materials to strengthen the soil and rocks by leveraging the abilities inherent in plants (e.g., spreading shoot and root system for biomass development).

There are several pieces of works that analyzes landslides quantitatively and qualitatively. These works can be classified into two categories: (1) approaches for risk analysis, mitigation, and monitoring and (2) approaches for susceptibility analysis and hazard zonation mapping. Isaza-Restrepo et al. (2016) present a quantitative risk assessment method, especially the geotechnical risk for natural hazards. This method includes the probabilistic analysis of hazards and vulnerabilities and the numerical representation of the risk. Weather conditions are one of the important factors that trigger landslides. So the analysis of slope stability under the climate change is crucial for landslide risk management. Almeida et al. (2017) proposed a physical model that deal with uncertainties in landslides under climate changes, especially under rainfall. Also, researchers use GIS method to analyze the landslide to identify better mitigation strategies (Pereira et al. 2017; Sitányiová et al. 2015). Developing an early warning system is an important step in landslide risk management. Naidu et al. (2017) developed a simple, cost-effective early warning system by correlating rainfall threshold with the factor of safety (FS) for landslides in Amboori, Kerala.

Apart from the normal landslides, paleo-landslides are continuous events subjected to the climate changes such as heavy rainfall, earthquake, and geomorphological factors. Hence, slope stability analysis for landslide susceptibility assessment requires careful analysis of the soil properties under the influence of paleo-landslide. Some researchers explored this area for analyzing slope stability (Husein et al. 2010; Hammond et al. 2009; Gonzalez-Gallego et al. 2008; Francioni et al. 2018).

Furthermore, several pieces of works investigated the landslides in Kerala. A devasting landslide which took place at Muvattupuzha, Ernakulam, Kerala on June 27, 2015 was studied by Ajin et al. (2016). From the study, the authors concluded that the root cause of the disaster was manmade one due to the unscientific cutting of the hill. They also inferred that the existing slope is not at all stable implies by the low factor of safety. Abraham et al. (2013) prepared the landslide hazard zonation maps in and around the Thodupuzha–Idukki–Munnar road in Idukki district, Kerala in their study. They adopted a geospatial approach which includes preparing various thematic maps on a GIS platform using remote sensing and field data. Sajinkumar et al. (2017) studied a headward retreat landslide that occurred behind the Government college Munnar. The main objective is to assess the landslide affected area, to identify the causative factors, and to suggest suitable measures to minimize the landslide hazard. The study involved mapping the area using the total station, determination of geotechnical properties of soil, vertical electrical soundings to infer soil thickness, analyzing the stability, deriving the rainfall threshold based on rainfall data. Also, Jishnu and Ajith (2015) used remote sensing and GIS to generate the landslide hazard zonation maps of Wayanad district of Kerala.

In summary, the landslides are one of the major natural hazards that have to be detected before it happens. Most of the existing works focus on the analysis of the landslide areas to take appropriate mitigation measures. There are few works that try to identify the locations vulnerable to landslides. These methods generate landslide zonation maps for the area under investigation, thereby generates early warning systems and proper safety measures. Developing such methods either uses field-based or GIS-based investigations. Inspired by these methods, this study aims to generate a landslide zonation map with the result taken from the field study and GIS-based mapping. As the Western Ghats of Kerala is found to be one of the major landslide-prone areas, the investigation extends to one of the important locations called Sastha valley in Kerala.

3 Methodology

Because of the topographical peculiarity and trace of paleo-slides this study follows a hybrid methodology to generate a landslide susceptibility map for Sastha valley. Figure 1 explains the flow of the methodology. The work starts with the field investigation which consists of reconnaissance survey, identifying critical locations and collecting undisturbed soil samples at both fields and saturated conditions from the locations identified. The samples obtained from the field investigations are then subjected to laboratory experiments to determine geotechnical parameters. A software-based stability analysis is conducted on the locations under steady and transient conditions. The results from software are validated using an explicit equation. A GIS-based model is also employed to analyze the factor of safety of slopes in the study area. Finally, the landslide zonation map of the study area is prepared from both field investigation and GIS model.

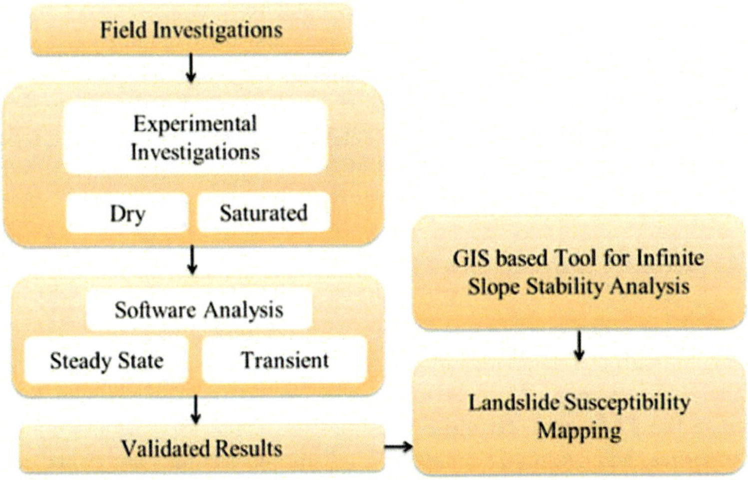

Fig. 1 Workflow

3.1 *Field Investigations*

Field investigations form an integral part of landslide susceptibility analysis and mapping. It is mainly grouped into three stages, namely mapping of the area, geological and geotechnical investigations. The objective of these investigations is to collect data for evaluating the stability of the slope, determining the conditions under which failure may occur and ascertaining remedial measures that are rational and feasible in the study area.

3.1.1 Mapping of the Area

The area under the study should be mapped in detail. Field maps are prepared that gives the plan of the affected area and the typical cross-section which can be used for analysis. General observations are also made covering the conditions of the slope, covering the aspects such as the extent and nature of vegetation cover, surface runoff characteristics, and presence of springs. Erosion of the toe and tension cracks of the crown area is ought to be observed in detail. The toposheet of the area is also studied as a part of the field investigation.

3.1.2 Geological Investigations

Geological investigations comprise of preparing a plan of the landslide area incorporating the geological data. Structural geological features such as bedding planes, joint planes, faults, folds, shear zones, and unconformity are studied in the field in detail and plotted.

The rock types in the slide area are identified and their qualities are assessed. The presence of any soft pockets of beds or interlayers is carefully observed during the field study. Geomorphological features such as elevated and depressed zones, break in slope, erosional and depositional zones, and mass movement are carefully observed and marked for further investigation.

3.1.3 Geotechnical Investigations

Geotechnical investigations are carried out with the objective of determining the nature and strength characteristics of the material comprising the slope. Disturbed and undisturbed samples of soil or rock are collected from the critical locations identified in the study area. Disturbed samples are made use for determining the index properties, and grain size analysis. Undisturbed samples are collected from open pits or boreholes using an appropriate type of sampling tubes. Good quality undisturbed samples are a basic requirement for reliable evaluation of shear strength parameters. The depth and seasonal fluctuations of the water table also form an important component of the data required for landslide investigation.

3.2 Experimental Investigations

The experimental investigations mainly comprise of determination of density as well as shear strength properties of soil samples collected. The density of soil is weight per unit volume of the soil. It is obtained by simply taking the ratio of weight to volume of the sample. Field density of the sample is the density at in situ water content. Saturated density is the density at which sample is fully saturated with water. The shear strength property of the soil is determined by conducting direct shear tests. The test is performed at both field and saturated conditions.

3.3 Software Analysis

Stability of slopes is determined under both steady-state and transient conditions using software analysis. SLOPE/W is used for analyzing the stability of slopes under steady conditions. A combination of both SLOPE/W and SEEP/W is used for transient analysis.

3.3.1 Steady-State Analysis Using SLOPE/W

Generally, there are five steps to define a SLOPE/W stability problem. The first step in evaluating the stability of a terrain in SLOPE/W is to outline the geometry of the slope. Geometry is defined based on the concepts of regions and points. After defining the geometry, the next step is to assign the strength properties of the material forming the slope. The most commonly adopted one is the Mohr–Coulomb method in which parameters such as c and Φ are used to describe the material properties. In order to establish a critical mode of failures, trail slip surface extents are to be demarcated. This forms the third step in the stability analysis. The position and shape of the slip surfaces can be controlled. The next step is to properly define the pore water pressure regime which forms a vital part of a slope stability analysis. For an effective stress analysis, realistic pore pressure conditions are required to be defined. Defining the surcharge loads that are supposed to act on the slope forms the last step in the analysis. The surcharge loads can be point loads and uniformly distributed loads.

3.3.2 Transient Analysis Using SEEP/W

The first step in the transient analysis is to create the geometry of the slope under the steady-state condition in SEEP/W. Then, the required boundary conditions are applied to geometry. After that, a transient seepage analysis is started with an initial steady state as parent one. The corresponding rainfall data is then applied as the step function in the water flux boundary condition. The analysis is then carried out to determine the pore water pressure distribution. Then, a SLOPE/W analysis is carried out with this transient as the parent to determine the FS under transient conditions.

3.4 *Software Validation*

This module describes the validation of SLOPE/W software. Several pieces of work used Taylor chart to find the variation of the FS with respect to slope angle and stability number. By analyzing Taylor charts an explicit equation has been proposed, which has proven accurate results. According to this equation, the FS of soil can be represented as,

$$\mathrm{FS} = \frac{\tan(\varphi)}{\tan(\varphi_m)} \tag{1}$$

$$\varphi_m = \frac{-b - \sqrt{b^2 - 4ac}}{2a} \tag{2}$$

$$a = 5.9466 \times 10^{-6} \tag{3}$$

$$b = -0.00807 + 3.41 \times 10^{-5}\beta - \lambda\pi/180 \tag{4}$$

$$c = 0.042186 + 0.004905\beta - 6.44 \times 10^{-5}\beta^2 + 4.07 \times 10^{-7}\beta^3 \tag{5}$$

$$\lambda = \frac{c}{\gamma H \tan(\varphi)} \tag{6}$$

where φ_m = mobilized friction angle in degrees, β = slope angle in degrees, c = cohesion in kPa, γ = unit weight of soil in kN/m^3, H = height of slope in meters, and Φ = internal angle of friction in degrees. The explicit equation is only applicable for saturated soil conditions and it can be adapted to complete submergence, complete sudden drawdown, steady seepage, and zero boundaries neutral force.

3.5 GIS-TISSA Model

It is a GIS-based tool for infinite slope stability analysis (TISSA). It is a newly created computer program that calculates FS with continuous rasters as input. The input rasters are:

- Digital elevation model (DEM) of the study area.
- Normalized difference vegetation index (NDVI) obtained from Landsat 8 imagery. NDVI indicates root cohesive strength and tree surcharge.
- Soil map of the area obtained from National Bureau of Soil Survey (NBSS)

The tool processes these input rasters to prepare the slope map and hill shade map and finally creates the zonation map of the area.

3.6 Landslide Zonation Map Generation

Field investigations and GIS-based approaches are two independent directions to find out the stability of the slopes. Field study uses experimental and analytical procedures to find out the FS of the soil. One of the limitations of this method is it can identify vulnerabilities of selected locations rather than giving global information about the study area. On the other hand, the GIS-based approach gives the global view of the study area based on topography. However, these approaches cannot give fine-grained geotechnical aspects of the locations. The map generation module takes the advantage of both field study and GIS-based approach to generate a new zonation map.

4 Results and Discussions

4.1 Field Investigations

An extensive field investigation was carried out on the study area by conducting reconnaissance survey with some geological and geotechnical insights. The presence of paleo-landslides was clearly observed in the area. It was indicated by the meandering of the river that flows through the valley. Based on the investigations 15 critical locations were identified in the area. The Google Earth view of these locations is shown in Fig. 2. Undisturbed core samples were collected both at the field and saturated conditions from the locations using a PVC pipe of diameter 6 cm and of height 10 cm.

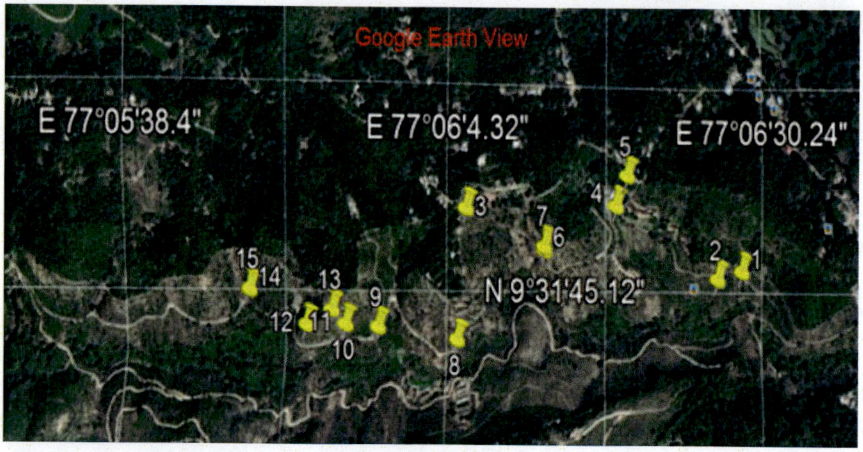

Fig. 2 Location map of the area and samples collected

4.2 Experimental Results

The samples collected from the field were taken to the laboratory for carrying out experimental investigations. The main properties determined were density and shear strength parameters of the soil. Direct shear tests were performed to determine the shear characteristics and density were determined by simply taking the ratio of the weight of the sample by its volume. The test results for the sample at field conditions are reported in Table 1. From the results, it can be inferred that the samples are showing a high value of cohesion at field condition. The test results for the sample at saturated condition are given in Table 2. The results indicate that in the saturated condition the cohesion value of the soil is decreased drastically. It is inferred that the presence of water reduced the inter-particle forces between the particles and thus reduced the cohesion between them.

4.3 Steady-State Stability Analysis Results

The stability of slopes of critical locations was analyzed after the experimental investigations. A steady-state analysis is carried out on SLOPE/W with properties of the soil that do not change with time. The corresponding results for the field and saturated condition are given in Table 3. The results show that the slopes identified have a good FS under field conditions. Also, under the saturated conditions, the FS reduces significantly for all the locations. From the analysis results, around 10 locations were identified as critical.

Table 1 Direct shear test results of sample locations under field condition

Sample location	Field density (g/cc)	Cohesion c (kPa)	Angle of internal friction Φ
1	1.497	16.7	26
2	1.544	18.0	25
3	1.424	16.0	27
4	1.428	16.0	27
5	1.516	17.1	24
6	1.543	14.3	23
7	1.420	13.4	25
8	1.522	12.6	27
9	1.423	15.3	26
10	1.480	14.0	25
11	1.342	15.2	27
12	1.437	16.0	27
13	1.405	17.1	24
14	1.419	11.3	23
15	1.302	10.4	25

Table 2 Direct shear test results of sample locations under saturated condition

Sample location	Saturated density (g/cc)	Water content (%)	Cohesion (kPa)	Φ (degrees)
1	1.692	13	5	27
2	1.837	19	3	30
3	1.623	14	6	23
4	1.656	16	1	30
5	1.819	20	4	34
6	1.713	11	3	34
7	1.733	22	5	33
8	1.705	12	4	28
9	1.636	15	2	24
10	1.672	13	8	20
11	1.624	21	6	22
12	1.696	18	6	26
13	1.686	20	1	30
14	1.618	14	4	26
15	1.615	24	2	30

Table 3 FS of selected locations under field and saturated conditions

Sample location	FS (field condition)	FS (saturated condition)
1	2.423	0.929
2	2.826	1.57
3	1.536	0.765
4	1.942	0.897
5	2.698	1.312
6	1.875	1.006
7	1.726	1.052
8	1.968	1.003
9	2.233	0.871
10	2.455	0.901
11	2.505	0.81
12	1.635	0.955
13	2.126	0.901
14	1.346	0.497
15	1.236	0.409

4.4 *Software Validation Result*

The stability analysis results obtained from SLOPE/W were validated using an explicit equation. Since the equation is complex, a Python code is used for computing the result using the equation. The results show a close agreement between equation and SLOPE/W and thus validated. The corresponding validation results are shown in Table 4.

4.5 *Transient Stability Analysis Results*

The steady-state stability analysis is carried out based on the assumption that soil properties are not changing with time. But in reality, the soil is subjected to varying climatic conditions. So a transient slope stability analysis is carried out incorporating the effect of rainfall in the study area. SEEP/W software is used for this purpose.

4.5.1 Characterization of Rainfall Data

The past rainfall data recorded for the study area shows a decreasing trend in the rainfall intensity for past few years. A good rainfall data was recorded for the year 2014. The rainfall data for the year 2014 for months from May to December was

Table 4 Software validation result with the values of parameters of the explicit equation

Sample location	Soil properties			Equation parameters					Factor of safety	
	C' (kPa)	Field density (kN/m^3)	Φ (degrees)	λ	*a*	*b*	*c*	ø m	Calculated	Analyzed
1	5	11.850	27	0.0695	0.0000594466	−0.00757751	0.177311	30.8811	0.852	0.929
2	12	12.070	33	0.1284	0.0000594466	−0.00860706	0.177311	24.8738	1.400	1.570
3	6	13.095	23	0.0393	0.0000594466	−0.00636872	0.209577	27.9529	0.800	0.765
4	1	14.250	30	0.0145	0.0000594466	−0.00695855	0.161394	31.8715	0.929	0.897
5	12	14.260	34	0.0766	0.0000594466	−0.00753178	0.184865	33.2935	1.027	1.312
6	3	15.373	34	0.0167	0.0000594466	−0.00631554	0.192558	32.6867	1.051	1.006
7	5	12.110	33	0.0533	0.0000594466	−0.00746641	0.169588	29.7696	1.135	1.052
8	8	13.050	28	0.0953	0.0000594466	−0.00785868	0.184865	30.6126	0.899	1.003
9	5	15.120	24	0.0623	0.0000594466	−0.00745274	0.177311	31.9169	0.715	0.871
10	12	14.370	25	0.1033	0.0000594466	−0.00799904	0.184865	29.6398	0.819	0.901
11	6	16.130	22	0.0773	0.0000594466	−0.0077134	0.177311	29.8583	0.704	0.810
12	12	13.190	26	0.1306	0.0000594466	−0.0084743	0.184865	26.8852	0.962	0.955
13	1	12.000	30	0.0172	0.0000594466	−0.00700626	0.161394	31.4029	0.946	0.901
14	4	8.205	26	0.0466	0.0000594466	−0.00666699	0.20069338	40.8579	0.564	0.497
15	2	14.936	30	0.0162	0.0000594466	−0.00647795	0.18486563	42.6064	0.628	0.409

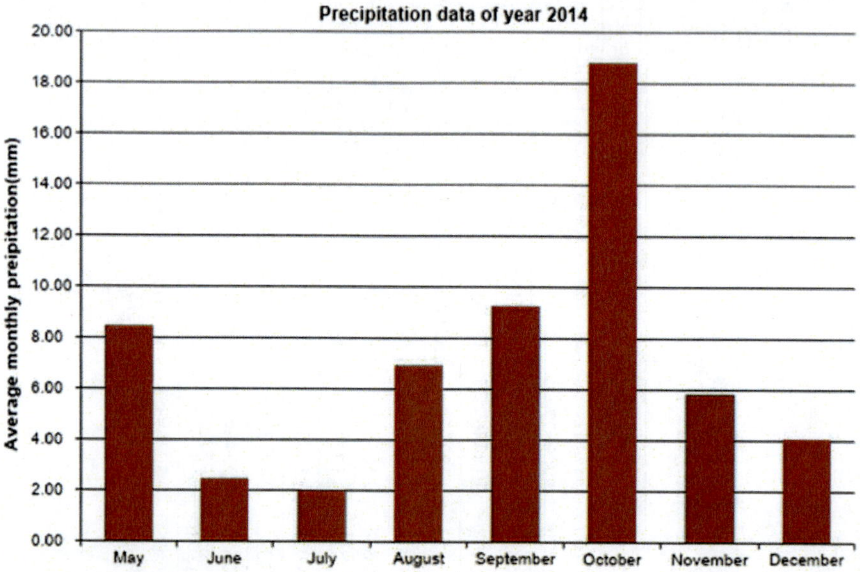

Fig. 3 Rainfall data 2014

collected. The average monthly rainfall for all the months is then calculated and shown in Fig. 3.

4.5.2 Variation of Pore Water Pressure

The transient analysis was carried out for locations 2, 5, 14, and 15. The variation of pore water pressure (PWP) for each location with average monthly rainfall is shown in Fig. 4. The results show that pore water pressure was high after a considerable rainfall has occurred. From the figure, it is clear that PWP is negative in the month of May for all the locations after that it increases gradually and becomes maximum in the month of July where average monthly rainfall is less. This trend gives a perception that PWP increases not at the time of rainfall but a period followed by the rainfall. The PWP then shows a gradual increase from the month of August to October. After that, it shows a steady increase owing to the large rainfall occurred in the month of October.

4.5.3 Variation of FS

The variation of the FS with average monthly rainfall for the four locations obtained from the transient analysis is shown in Fig. 5. From the results, it can be inferred that FS decreases in the month of June and July and then gradually increases during August and then shows a gradual decrease during September and October. After

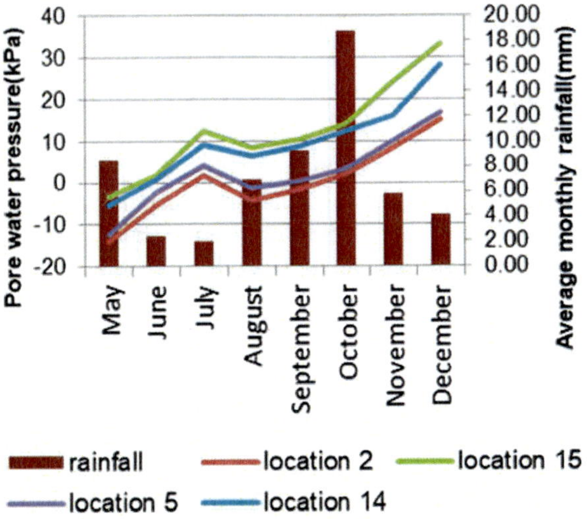

Fig. 4 Variation of pore water pressure subjected to climatic conditions

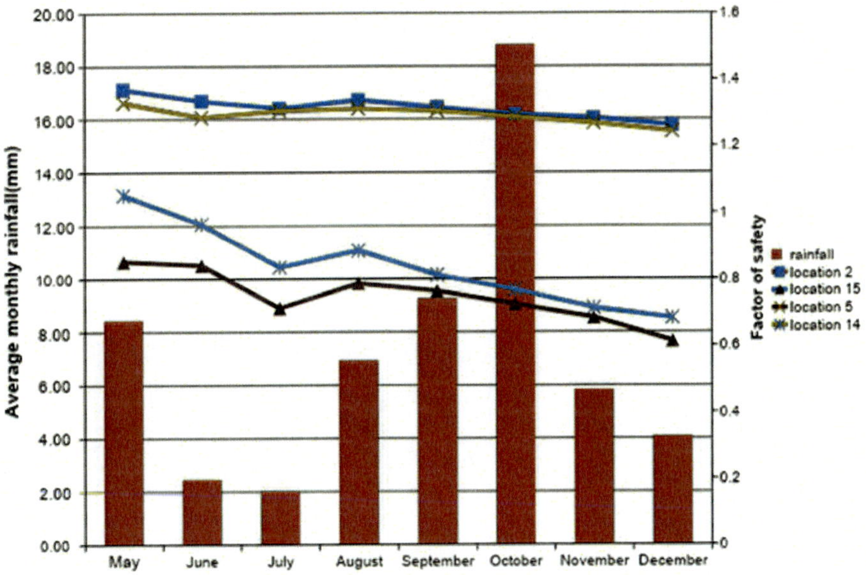

Fig. 5 Variation of FS under climatic conditions

that, it shows a steady decrease toward December. It is thus inferred that FS varies inversely with the pore water pressure. The FS for the locations 14 and 15 becomes very less during the month of December.

4.6 GIS-TISSA Model

GIS-TISSA model is a newly created ArcGIS tool for calculating the FS with continuous raster. The main working procedure of this model begins from taking digital elevation map (raster form) of the area as input followed by preparing hill shade map and finally computing the FS of the soil. The slope map of the study area is created in GIS using the digital elevation model (DEM) of the study area as shown in Fig. 6. Based on this, a FS map of the study area is prepared as shown in Fig. 7.

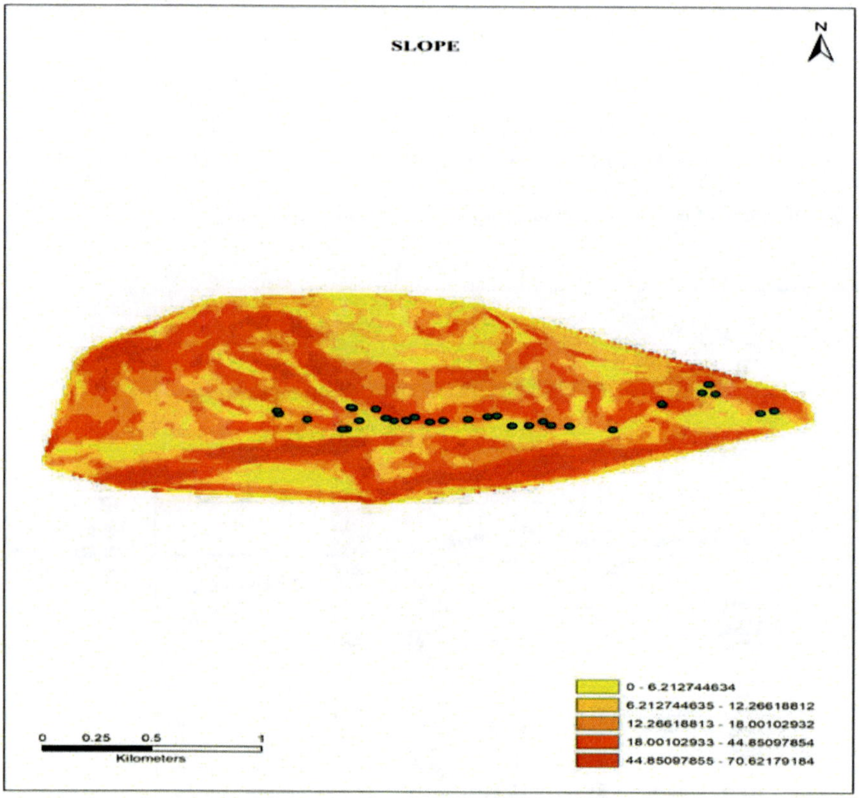

Fig. 6 Slope map generated from GIS-TISSA model

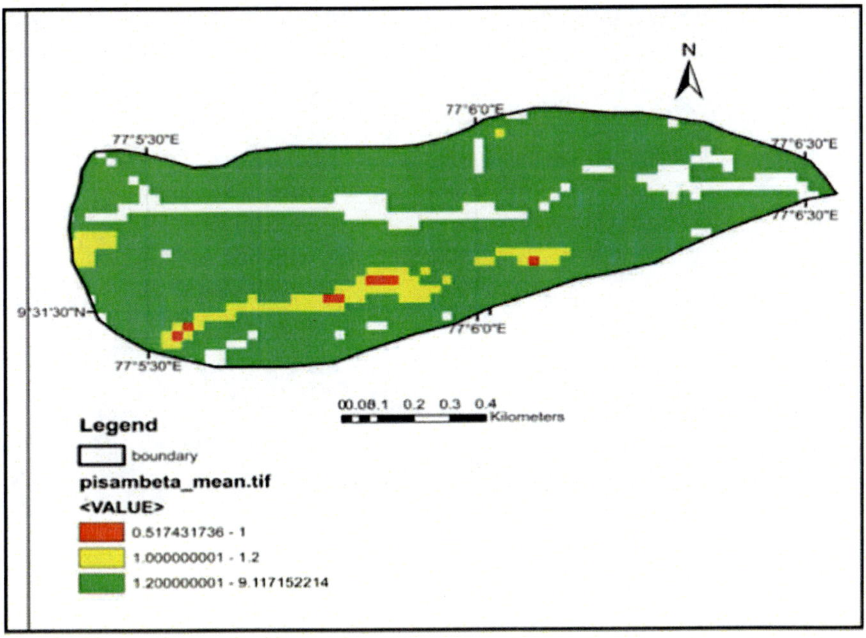

Fig. 7 Landslide zonation map from GIS-based method

4.7 *Landslide Zonation Map*

Based on the field investigation and GIS model, a landslide zonation map of the study area is prepared and is shown in Fig. 8. The map contains ten critical locations selected from field investigation and four critical locations selected from the GIS-based approach for the study area.

5 Conclusion

Analyzing and preparing landslide zonation map is one of the important directions toward avoiding the impacts of landslides on human life and properties. Since Sastha valley is an important location for landslide investigation, the landslide susceptibility of this area was analyzed in this study. Among the 15 location, 10 locations were selected and found to be critical, having the factor of safety less than 1 after the stability analysis. The factor of safety reduces drastically in saturated condition compared with field conditions. This emphasize that the prolonged rainfall and other climatic conditions may lead to the degradation of the stability of the slopes. By analyzing the impact of rainfall, it was found that antecedent rainfall has more influence on pore water pressure compared with daily rainfall. Also, factor of safety shows an inverse relation with pore water pressure. It has its lowest value after the peak

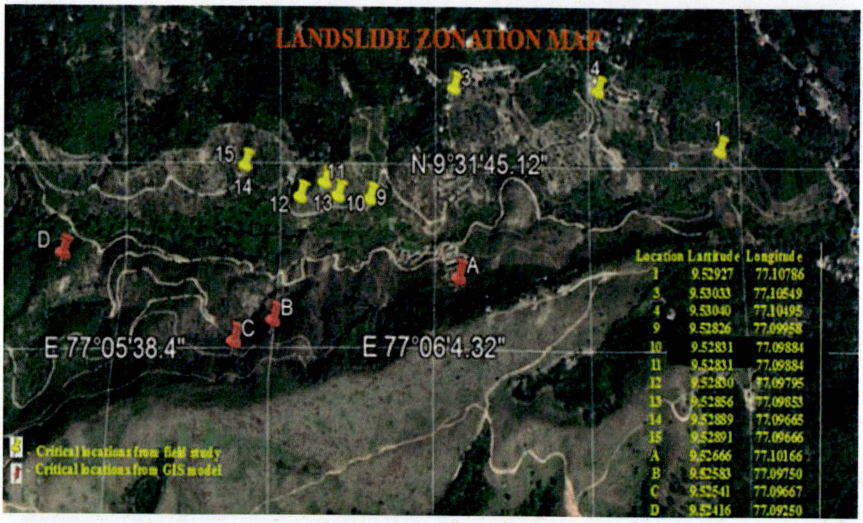

Fig. 8 Revised landslide susceptibility map for the Sastha valley

rainfall. Moreover, a relatively high factor of safety results was obtained from the GIS-TISSA model. These observations conclude that unsustainable land use in this area will lead to landslides causing heavy damages. Thus, the zonation map based on the field investigation and GIS method can be used to identify the locations that are vulnerable to landslide.

In the future, paleo-landslides movements are to be closely monitored. It can give a good insight into landslide mechanism occurring in the valley. The stratigraphy of soil and its properties is the other direction of investigation. It can give a better idea of failure mechanism and nature of slip surface of the landslides.

References

Abraham PB, Shaji E (2013) Landslide hazard zonation in and around Thodupuzha-Idukki-Munnar road, Idukki district, Kerala: a geospatial approach. J Geol Soc India 82(6):649–656

Ajin MM, Geethumol ST, Gino B, Shaharban P, Anniamma C (2016) A case study on landslide: a geotechnical investigation. Int Res J Eng Technol 3(4):2507–2512

Almeida S, Holcombe EA, Pianosi F, Wagener T (2017) Dealing with deep uncertainties in landslide modelling for disaster risk reduction under climate change. Nat Hazards Earth Syst Sci 17(2):225–241

Bhattacharjee J (2016) Landslides mitigation using various techniques. Int J Appl Eng Res 11(8):5772–5776. ISSN: 0973–4562

Francioni M, Stead D, Clague JJ, Westin A (2018) Identification and analysis of large paleo-landslides at Mount Burnaby British Columbia. Environ Eng Geosci 24(2):221–235

Gonzalez-Gallego J, Robles JM, García de la Oliva JL, de Santayana FP (2008) Stabilization of a large paleo-landslide reactivated because of the works to install a new ski lift in Formigal skiing resort

Hammond CM, Meier D, Beckstrand D (2009) Paleo-landslides in the tyee formation and highway construction, Central Oregon Coast Range. GSA Field Guides, 15, pp 481–494
Husein S, Sudarno I, Pramumijoyo S, Karnawati D (2010) Paleostress analysis to interpret the landslide mechanism: a case study in Parangtritis, Yogyakarta. J Appl Geol 2(2)
Isaza-Restrepo PA, Carvajal HEM, Montoya CAH (2016) Methodology for quantitative landslide risk analysis in residential projects. Habitat Int 53:403–412
Jishnu ES, Ajith Joseph K (2015) Identification of potential landslide vulnerable zones of Wayanad district. Kerala using Remote Sensing and GIS, Thesis
Kerala State Disaster Management Authority (2012) Kerala State Disaster Management Plan Profile
Naidu S, Sajinkumar KS, Oommen T, Anuja VJ, Samuel RA, Muraleedharan C (2017) Early warning system for shallow landslides using rainfall threshold and slope stability analysis. Geosci Front
Namdar A, Yahaya FM (2014) Effect of natural hazards on types of landslide. Electron J Geotech Eng (EJGE) 19:1519–1632
Pereira S, Garcia RA, Zêzere JL, Oliveira SC, Silva M (2017) Landslide quantitative risk analysis of buildings at the municipal scale based on a rainfall triggering scenario. Geomat Nat Hazards Risk 8(2):624–648
Raut R, Gudmestad OT (2018) Use of bioengineering techniques to prevent landslides in Nepal for hydropower development. Int J Des Nat Ecodynam 12(4):418–427
Sajinkumar KS, Asokakumar MR, Sajeev R, Venkatraman NV (2017) A potential headward retreat landslide site at Munnar, Kerala. J Geol Soc India 89(2):183–191
Sitányiová D, Vondráčková T, Stopka O, Myslivečková M, Muzik J (2015) GIS based methodology for the geotechnical evaluation of landslide areas. Proc Earth Planetary Sci 15:389–394
Ubechu BO, Okeke OC (2017) Landslide: causes, effects and control. Int J Curr Multi Stud 3(3):647–663

Investigations and Mitigation Measures of Landslide Affected Areas in Hill Roads of East Africa—Case Study on Projects of Ethiopia

Avik Kumar Mandal, S. Sailesh, and Pradyot Biswas

Abstract Under the road development programme of Ethiopian Roads Authority (ERA) in Ethiopia, many roads spreading all over the country were upgraded from the existing gravel road to the widened asphalt road. Due to the rugged and mountainous terrain condition, natural and man-made slope failures such as landslides usually occur in almost all the hilly and mountainous terrain in Ethiopia. The main cause for the occurrences of failures at different locations along project roads which were widened either by cutting the hill slope and/or filling the valley slope was high rainfall in the area during the month of June to September. In this paper, the case study on the investigations of landslide failures along hill slope, valley slope, slope of bridge approaches, slope of hillside on upstream and valley side on downstream of culverts along project roads is described. The analysis of landslides and its different methods of mitigation measures, namely flattening of slope with or without provision of intermediate berms, provision of lateral support by gravity retaining wall at toe of hill slope or at road edge on valley slope side including proper slope protection and drainage measures adopted in the rehabilitation works along different project roads in Ethiopia, are presented here.

Keywords Hill and valley slope · Bridge approach and culvert upstream–downstream · Landslide analysis and its mitigation measures

A. K. Mandal (✉) · S. Sailesh · P. Biswas
LEA Associates South Asia Pvt. Ltd, New Delhi, India
e-mail: avik@lasaindia.com

S. Sailesh
e-mail: sailesh@lasaindia.com

P. Biswas
e-mail: pradyotbiswas@lasaindia.com

M. Latha Gali and P. Raghuveer Rao (eds.), *Geohazards*, Lecture Notes in Civil Engineering 86, https://doi.org/10.1007/978-981-15-6233-4_4

1 Introduction

The roads under the administration of Ethiopian Roads Authority (ERA) represent the major trunk road system connecting capital city of Addis Ababa with the centres of Ethiopia's different regions. The government, in recognition of the crucial role that the road sector plays, set out a Road Sector Development Program (RSDP) to speed up the improvement and expansion of the road network. Under this road development programme, many roads spreading all over the country were upgraded from the existing narrow gravel road to the widened asphalt road.

The ground condition of Ethiopia is predominantly rolling and mountainous. The mountain slope of the country is steep, and slope height from valley bottom to top ridge is also quite high. The project roads considered in this paper are mostly traversing through such hilly area of high rainfall with varying climatic condition. There were several occurrences of slope failures, i.e. landslips or landslides at different locations along the project roads which were widened either by cutting the hill slope and/or filling the side of valley slope. The main cause of landslide occurrences was high rainfall during the month of June to September in the project area consisting of mostly erosive and fragile geological condition. In this paper, the investigation of landslide occurrences and different methods of restoration measures for stabilization of hill and valley slope, bridge approaches, culvert upstream side hill slope and downstream side valley slope, etc., is discussed. The typical details of various rehabilitation works against landslide occurrences are also presented here.

2 Project Details: Location, Topography, Climate and Geology

In this paper, the experiences of three project roads in different parts of Ethiopia, namely Adiremet–Dejena–Dansha in northern part, Kombolcha–Mekane Selam in north-eastern part and Bedele–Metu in western part, are discussed. The brief details of all these project corridor locations and its topography, climate and geology are described in the following paragraphs.

2.1 Locations

The total length of project road "Adiremet–Dejena–Dansha" is around 100 km. The project road is located in northern Ethiopia, starts at Adiremet town passing through Dejena town and finally ends at Dansha town and makes T-junction with Gonder–Humera road. The project road crosses rivers mainly Adishani, Zani, Dema Baria, Maogola, Kulita Masah and Kaza. The project road connects two major link roads, i.e. Gonder–Humera and Gonder–Axum in north of Ethiopia.

The other project road, namely "Kombolcha–Mekane Selam" starts at 17 km north of Kombolcha town and 6 km south of Dessie town and ends at Mekane Selam, after traversing through total length of around 180 km. The notable habitations along the project road were Guguftu, Gimba, Tula-walia, Aksta and Mekane Selam. It is a road of considerable importance, as it provides a direct east–west route from the ports to large parts of northern and eastern Ethiopia.

The project road "Bedele–Metu" is part of Djibouti–Juba–Kampala Road Corridor in Ethiopia. The total length of this project corridor is 50 km, and it is located in the mid-western part of Ethiopia in Oromia Region. Bedele–Metu lies in the ILLU zone between km 61 + 000 and km 111 + 659. It is accessible either via Addis Ababa–Wolliso–Jima–Bedele (about 550 km) or Addis Ababa–Ambo–Nekemte–Bedele (about 500 km).

The typical elevation map of Ethiopia showing the above-said project corridor is shown in Fig. 1.

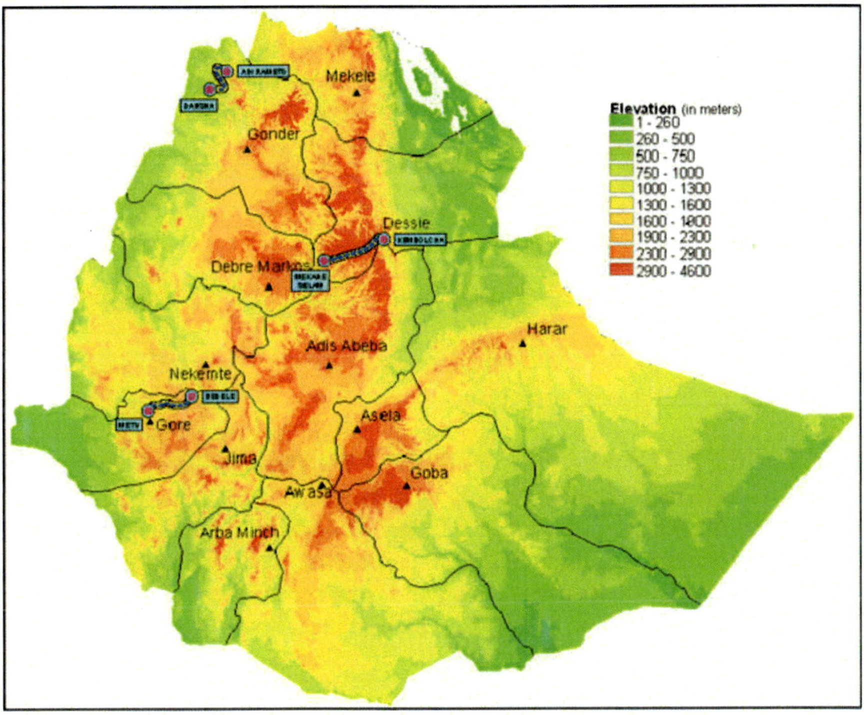

Fig. 1 Elevation map of Ethiopia showing project roads

2.2 Topography

The project corridor "Adiremet–Dejena–Dansha" is characterized by a series of diverse intense and mutually contrasting topographic features ranging from hilly and mountainous to rolling and some intermediate flat lying ground. The road starts from Dansha with an elevation of 768 m and traverses a ridge line up to 18 km. It runs parallel to the Kaza River, crosses the Kaza River at 59 km and then enters escarpments. The elevation at Dejena area is 2178 m, and at Adiremet it is 2065 m.

The project route "Kombolcha–Mekane Selam" largely traverses over mountainous terrain with only few interruptions of rolling and plain terrain. The elevation along the project road varies mostly in range of 2400–3280 m.

The project road, namely "Bedele–Metu", traverses mainly through rolling and mountainous terrain and short escarpment area. The altitude along the project alignment varies from 1400 to 2100 m. Rolling terrain comprises 54%, mountainous terrain is of 37%, and the remaining 9% is escarpment terrain of the total terrain classification along the project road. There are four deep gorges in the whole stretch of the road. The rolling sections are mainly located ahead and past the gorges, and most of the mountainous and escarpment sections are contained in sides of the gorges. Other than in the gorges, some of the escarpment and mountainous sections are found in the ridge tops dominated by rolling terrain but with short stretches. The rolling section ranges from very gentle to steep cross-fall. In addition, ridge top of terrain is also inherent but it is short in stretch.

2.3 Climate

The distinctions in climate characteristics along the project roads are mainly caused by altitude differences and its location where rainfall in the area is highly influenced by the "Intertropical Convergence Zone (ITCZ)". The climate zone experiences low temperature, high rainfall and less evapotranspiration.

As per the "Meteorological Map of Ethiopia" (Ethiopian Meteorological Service 1979), the project area of "Adiremet–Dejena–Dansha" is climatically classified as warm to very warm. Its effective temperature is higher than 29 °C making it very warm and especially uncomfortable. The highest mean seasonal rainfall recorded in this region is 379.9 mm in 2006, and the rainfall season is around four months starting from middle of June and ending in middle of October.

The climate in the project area of "Kombolcha–Mekane Selam" is typical tropical nature of mountainous area with a high diurnal temperature range and rainfall due to aerographic effect, depending on the direction of moisture-laden winds and altitude. The mean annual temperature ranges from 8.5 °C at 3700 m to 24.0 °C at 1200 m indicating the close relation between temperature and altitude. The mean annual rainfall varies between about 600 mm and 1300 mm. The relatively highland areas receive more rainfall than the lowland areas along the project road.

The maximum and minimum temperature of the project area "Bedele–Metu" does not have significant variation through the year. The mean maximum monthly temperature along the project road is around 29 °C. The average value of minimum monthly temperature is 9 °C in Bedele area and 10.3 °C in Metu area. The project area gets maximum rainfall during the months of May through September in the ranges from 232.2 mm to 314.8 mm in Bedele area and from 211.8 mm to 278.9 mm at Metu area. The area gets minimum rainfall during the months of November through February in the ranges from 17.5 mm to 40.5 mm at Bedele area and from 23.4 mm to 54.63 mm at Metu area. Because of the above integrated factors, the project area is densely forested and has got thick overburden of plastic soil, prevailing surface and subsurface water which in turn has been the driving factor for major slope stability problem along the project road.

2.4 Geology

Geological profile of the project road "Adiremet–Dejena–Dansha" is dominated by Cenozoic rocks comprising Tertiary and Quaternary basalt and sediments (sandstones, shales, conglomerates, limestones and quartzites). The project route corridor from Myhanse to Tekeze River to Dansha town is dominated by this group of rocks. Along the Tekeze River, early Tertiary sediments are exposed and ascending to the highland towards Dansha town; early to mid-Tertiary, flood basalts with rare basic tuffs are exposed. The hilly to mountainous terrain along the project road is covered by massive to widely jointed and moderate to strongly fractured basalt and sedimentary rock that sometimes covered with thin light brown or dark-to-dark brown silt clay soils. The low-lying terrain is mostly covered by dark-to-dark brown silt clay soils of alluvial origin. The side slopes are often made up of weathered and strongly fractured basalt that are susceptible to movement when wetting and the existing slope is steep.

The geological condition of the project road "Kombolcha–Mekane Selam" is similar to the project corridor "Adiremet–Dejena–Dansha". The deep underneath weathered basalt rock is overlain by variable thickness of brownish grey expansive soil mixed with little gravel and pieces of boulder along the slope of hill and valley along this project road.

The rocks cropping out in the project area of "Bedele–Metu" are the reflection of regional geology, which comprise dominantly Mekonen basalt with some rhyolite and volcanic ashes. The basalt is affected by different degrees of weathering (both physical and chemical) which can be useful for the road construction. Sedimentary rocks in and outside the project area are identified by horizontal beddings and the presence of coal deposits in the vicinity areas; for example, YAYO Coal deposits an indication of sedimentary rocks. There are also trachyte, trachy–basalt, ignimbrite, rhyolite, welded tuff and other volcanic products. Furthermore, transported soils (colluviums) and residual soil mainly moisted reddish brown, dark grey to black clay or black cotton soil, etc., are frequently seen along the project road. Weathered material such as natural gravel could be useful for sub-base, capping, embankment and replacement, whereas the fresh variety is useful as crushed aggregate for base and surface course.

3 Occurrences of Erosion, Landslides and Its Locations

Occurrence of erosion from slope is the natural process of removing soil particles by external agents such as wind or water. In the Ethiopian wet climate, this involves rainfall which is responsible for the removal of surface layers, resulting in gullies of about 10–60 cm depth. However, over time the rills and gullies deepen further and these cause slopes to overstep, thus triggering the slope instability.

Gravitational forces are always acting on a soil mass or rock beneath a slope. As long as the shear strength of soil mass is equal to or greater than the gravitational forces, the forces are in balance, the mass is in equilibrium and movement does not occur. An imbalance of forces results in slope failure, and movement in the form of creep, falls, slides, avalanches or flows does takes place when driving forces exceed the resisting forces.

Slope failure is a major issue for the road safety and stability in Ethiopia. Such slope failures in Ethiopia are predominant in warm, humid climates and occur during years of unusually heavy precipitation or during periods of heavy concentration of rainfall or during wet years. The slip or slide of slope mass was happened mostly due to significant ingress of rainwater run-off within slope material and subsequent generation of high pore water pressure inside slope mass which resulted in reduction in shear strength of slope material against rotational/transitional mode of failure.

However, prediction of landslip failure is often uncertain and slope stability is an important element depending upon slope geometry, inherent soil strength and groundwater or pore water pressure characteristics.

The occurrences of erosion and subsequent landslips were observed at different locations along the project roads in Ethiopia. These are described under the following sections.

3.1 *At Approaches of Bridges*

There were occurrences of erosion and subsequent slips of slope mass in bridge approaches consisting of mostly less cohesive granular material in Ethiopia. The erosion cum landslip was happened in a number of approach slopes of bridges along project corridors, namely "Adiremet–Dejena–Dansha" and "Kombolcha–Mekane Selam".

Typical photograph of landslips at bridge approach locations along the project road in Ethiopia is shown in Fig. 2.

Fig. 2 Landslip from bridge approach slope along "Adiremet–Dejena–Dansha"

Fig. 3 Landslide from hill slope of rocky nature along "Adiremet–Dejena–Dansha"

3.2 *At Hill Slope of Rocky Material*

The occurrences of slides from hill slope consisting of mostly fragmented large to moderate size of rocks, boulder, gravel, etc., were observed at many locations along the project roads of Ethiopia.

Typical photograph of landslides at hill slope of rocky nature along the project roads in Ethiopia is shown in Fig. 3.

3.3 *At Hill Slope Having Other Than Rocky Material*

Erosion and subsequent slip/slide/collapse of hill slope consisting of mostly brownish and greyish low-to-moderate expansive soil mixed with fragmented rocks, gravel and boulder were observed at several places along the project corridors of Ethiopia.

Typical photographs of landslides at hill slope of non-rocky nature in Ethiopia are shown in Figs. 4 and 5.

Fig. 4 Landslide from hill slope along "Kombolcha–Mekane Selam"

Fig. 5 Landslide from hill slope along "Adiremet–Dejena–Dansha"

3.4 At u/s and d/s Sides, Approaches of Pipe Culverts

At some pile culvert locations along the project roads in Ethiopia, there were occurrences of erosion and subsequent slip/slide of u/s side hill slope and d/s side valley slope consisting of mostly granular material of less cohesive granular material, namely boulder, gravel, etc., mixed with or without brownish/greyish soil and collapse of d/s head wall of pipe culvert. There was huge flow of rainwater run-off discharge mostly over the road surface at culvert location due to no drainage of water through the clogged inlet on u/s side.

Typical photograph of landslip of u/s side hill slope and d/s side valley slope at pipe culvert locations along the project roads in Ethiopia is shown in Fig. 6.

3.5 At u/s and d/s Sides, Approaches of Slab Culverts

The occurrences of erosion and slip of u/s side hill slope and d/s side valley slope consisting of mostly fragmented rocks, boulder, gravel, etc., mixed with or without

Fig. 6 Landslip of d/s side valley slope at pipe culvert location along "Kombolcha–Mekane Selam"

Fig. 7 Landslip at slab culvert location along "Adiremet–Dejena–Dansha"

brownish/greyish soil and also collapse of wing wall supported approaches of slab culvert were observed along the project roads in Ethiopia.

Typical photograph of occurrence at slab culvert location in Ethiopia is shown in Fig. 7.

3.6 *At Valley Side Slope*

There were major erosion and subsequent formation of rain cuts and gullies along the valley side slope of the project corridors of Ethiopia. The occurrences of vertical settlement and longitudinal cracking followed by lateral sliding/slip of valley side slope consisting of mostly less cohesive granular material, namely boulder and gravel, were observed along the project roads.

Typical photographs of erosion cum landslides of valley side slope along the project corridors in Ethiopia are shown in Figs. 8 and 9.

Fig. 8 Landslide of valley side slope along "Adiremet–Dejena–Dansha"

Fig. 9 Landslide of valley side slope along "Kombolcha–Mekane Selam"

4 Causes of Landslide Occurrences

In general, the factors responsible for occurrence of landslides are slope topography, lithology and its structural features, conditions producing a change, namely excavations, seismic events and variation in groundwater levels, etc.

The following are the main causes which make the slope unstable and trigger the movement, i.e. slip or slide of slope mass.

4.1 Geological Factors

The nature of geological formations plays an important role in determining the stability of hill and valley slope. The composition and type of rocks, orientation of bedding planes and joints, the presence of faults and folds play a significant role in assessing the stability of slope.

In all the project corridors of Ethiopia described here, the dip of rocky material in hill slope is such that bedding planes inclined towards the face of the slope and

Fig. 10 Hill slope excavation along "Adiremet–Dejena–Dansha"

the slope comprised frequent joints in the mass. Generally, the bedding planes of the rocky material in hill slope of Ethiopia are filled with gouge material through which the critical failure surfaces occur.

4.2 *Change in Slope Gradient*

An increase in slope gradient due to the previous rockfall and excavation produces a change in the internal stress of rock mass, and equilibrium conditions of slopes are disturbed by the increase of shear stress.

Along the said project roads in Ethiopia, the stability of hill slope got disturbed due to increase in slope gradient by previous several rockfalls and also with the excavation of the natural slope for making the roads. Figure 10 shows the increase in slope gradient due to excavation of natural slope for accommodating the proposed width of the roads.

4.3 *Change in Surcharge*

The change in surcharge condition over the slope either may be due to natural causes like rain, accumulation of previous landslide materials etc., or due to construction activities like placing of fills, spoil heaps, weight of any kind of building increase the shear stress and also increase the pore water pressure in colluvium soil which finally result in decreased shear strength of the slope mass.

In Ethiopia, the hill and valley slopes become unstable due to change in overburden or surcharge condition mostly by the ingress of rainwater, deposition of previous landslide materials or spoil heaps from construction. The change in surcharge condition due to construction activities is shown in Fig. 11.

Fig. 11 Construction of hill road along "Adiremet–Dejena–Dansha"

4.4 Shocks and Vibrations

Shocks and vibrations due to blasting, earthquake or by any construction machinery affect the equilibrium of slope on account of the temporary changes of stress that are caused by oscillations of different frequencies. Thus, shock or vibration due to either earthquake or any other construction activities like blasting increases shear stress and also decreases the shear resistance which finally results in slide of slope mass.

In Ethiopia along the said project corridors, the occurrences of landslides were observed at hill and valley slopes due to sudden shocks and vibration as stated above.

4.5 Change in Water Contents

Rainfall is the chief factor for causing the change in water contents of hill slope mass. The rainwater penetrates through the joints and triggers landslides of the slope mass which was already disturbed by excavations/blasting. Due to the ingress of rainwater inside the slope mass, the pore water pressure and seepage force are increased. Rainwater infiltration in slope also increases mass unit weight and promotes surface erosion. Increase in water content of hill slope mass decreases its shearing resistance consequently resulting in failure of slope.

In the same way, the hill slope along the project roads of Ethiopia becomes unstable due to increase in water content by the ingress of rainwater through the colluvium soil, jointed and highly weathered rock mass.

4.6 Change in Weathering

The agents like rain and fluctuation of temperature do the changes in mechanical weathering of slope material. As a result of the change in weathering, softening

Fig. 12 Change in weathering condition of slope material along project roads in Ethiopia

of colluvium soil and also physical disintegration of rock occur. This results in deterioration of the strength characteristics of the slope material and finally increases the instability of slope.

The stability of hilly roads of Ethiopia becomes critical due to change in weathering caused by rain, seepage of water and also significant temperature variations. Figure 12 shows the change in weathering condition of slope material along roads in Ethiopia.

4.7 Role of Development Activities

The development activities in hilly areas by slope cutting for roads, etc., and the denudation of hill due to clearing of forests do change the equilibrium state and also reduce the internal inherent strength which finally results in occurrence of landslides.

Similarly, there were occurrences of landslides due to the increased rate of development activities in Ethiopia.

4.8 Disturbances in Path of Streams and Falls

Construction activities in hilly areas like blasting, excavation and also the previous landslide cause disturbances in the natural path/channel of water flow and change the water course. The blocked and diverted flows of water wash out the soluble cementing substances, weaken the inter-granular bonds and reduce the mechanical strength of bed material along the slope. The confined water of the natural channel exerts an upward pressure on overlying bed which results in slides of mass from the bed over slope.

The above-said situation does exist along all the project roads in Ethiopia which affect the stability of hill slope.

4.9 *Role of Vegetation*

The development activities and previous landslide cause loss of vegetation. Roots of trees maintain the stability of slopes by their mechanical effects to the ground. The trees also contribute to the drying of slope by absorbing a part of the groundwater through its root which in result enhances the stability of hill slope.

In Ethiopia, loss of vegetation due to the various development activities of road widening and new construction activities causes numbers of landslides from hill and valley slope.

5 Material Investigations

To know the engineering nature of existing material at different affected locations, namely at hill slope, valley slope, u/s and d/s of pipe and slab culvert and bridge approach locations along the project corridors, a detailed material and geotechnical investigation programme were undertaken in the respective affected locations. The best locally available methods for finding out the strength of slope material were adopted in the projects. The detailed material and geotechnical investigation work comprised hand-operated "Dynamic Cone Penetration Test (DCPT)" in the identified respective sites, collection of material samples from the different affected sites, conducting laboratory tests on the collected soil material samples to know its engineering classification, compaction characteristics, expansive nature and its CBR values.

- Field Probing Test like DCP Test—Nearer to the proposed founding level of retaining structure, say in between 0.50 and 2.00 m below ground level at the proposed location of restoration measure along toe of hill slope and at an appropriate intermediate locations say around 4.00–8.00 m below finished road level along valley slope, the hand-operated DCPT as per the guidelines given in publication of TRL, Overseas Road Note (ORN9) by DFID, London, had been conducted. The penetration record obtained through hand-operated DCPT was converted to the equivalent SPT, i.e. N value based on correlation in between DCPT value in mm/blow and SPT (N) value in blows/300 mm as given in Table 3-3 of Transport Research Laboratory (TRL) Overseas Road Note (ORN9). The following typical photograph shows the DCPT conducted at the affected locations.
- Laboratory Investigation—At the proposed locations of restoration measure by retaining wall either at toe of hill or at intermediate position of valley for its foundation including its backfill soil and also at proposed locations of restoration measures mostly by flattening of hill slope and also at the proposed locations of restoration measures by rebuilding of valley slope, the following laboratory tests had been conducted over the collected material samples:

i. Grain size analysis
ii. Atterberg's limit tests (LL, PL, PI)
iii. Free swell index (FSI)
iv. Modified proctor compaction test
v. CBR test including CBR swell.

The typical range of different engineering parameters as determined from the field and laboratory investigations of the collected material samples along the road, namely "Adiremet–Dejena–Dansha", is given in Table 1.

Similarly, the summary of test results of collected material samples along the affected locations of project corridor, namely "Kombolcha–Mekane Selam", is given in Table 2.

The shear strength parameters, i.e. cohesion (c) and angle of internal friction (φ), were determined from the co-relation based on unified soil class as given in the book, namely "Stability Analysis of Earth Slopes" by "Yang H. Huang". Additionally, by using the available basic engineering parameters of the collected material samples as tested and by referring other reputed literature, namely "Foundation Analysis and Design" by Bowles (1997), etc., the shear strength parameters of hill or valley slope and its foundation material had been finalized. The typical range of undrained cohesion and angle of shearing resistance of the materials were found as 1.5–4.00 t/m^2 and 22°–32°, respectively.

Table 1 Summary of test results of material samples along "Adiremet–Dejena–Dansha" project road

Material type and its colour		Brownish/dark brownish/greyish soil mixed with gravel
Soil classification		Clayey/silty gravelly sand **or** gravelly clayey/silty sand **or** gravelly sandy silt/clay **or** gravelly sandy/silty clay
Recorded DCPT values (mm/blows)		0.46–1.38
Correlated SPT values (blows/300 mm)		70–beyond 75
Grain size/sieve analysis (% passing through sieve)	2.00 mm	97.00–67.80
	0.425 mm	81.80–33.80
	0.075 mm	72.20–13.80
Atterberg's limits (%)	LL	65.90–42.80
	PL	49.60–18.20
	PI	23.80–15.10
FSI (%)		5.00–50.00
MDD (gm/cm^3)		1.82–1.56
OMC (%)		21.90–15.40
CBR (%)	2.54 mm	45.60–1.20
	5.08 mm	52.50–1.40
CBR swell (%)		0.80–2.92

Table 2 Summary of test results of material samples along project road "Kombolcha–Mekane Selam"

Material type and its colour		Black cotton soil or yellowish/greyish weathered and disintegrated rock with some gravel or yellowish/brownish grey gravelly soil
Soil classification		Sandy/gravelly silty clay or silty sandy clay or clayey gravelly sand or clayey sandy gravel
Recorded DCPT values (mm/blows)		2–36
Correlated SPT values (blows/300 mm)		68–6
Grain size/sieve analysis (% passing through sieve)	2.00 mm	92.70–11.38
	0.425 mm	84.10–7.11
	0.075 mm	75.90–3.11
Atterberg's limits (%)	LL	72.70–33.00
	PL	37.70–14.60
	PI	36.60–4.50
FSI (%)		12.50–54.60
MDD (gm/cm^3)		1.15–2.06
OMC (%)		53.20–11.50
CBR (%)	2.54 mm	66.00–0.60
	5.08 mm	84.00–0.60
CBR swell (%)		1.10–9.00

6 Details of Restoration Measures

6.1 Objective

The main objectives of providing suitable restoration measures at different affected locations along the project corridor are as follows:

i. To arrest any erosion, slips and slides from hill and valley slope
ii. To prevent any erosion, slips and slides from approaches of bridges
iii. To keep the culvert in proper working condition by keeping the inlet clear and protecting from any future erosion and subsequent slip/slide of backfill slope on both u/s and d/s sides, and u/s hill and d/s valley slopes along the approaches of culverts, protecting head wall of culvert on both u/s and d/s sides from any breakage
iv. To keep the road and pavement condition proper by not allowing continuous flow of rainwater over the road and pavement surface during monsoon
v. To keep the roadside drains in clear and good condition by preventing any fall of debris due to slips or slides from hillside to drain.

Basic Considerations. The basic considerations of finalization of restoration measures were to provide best possible option based on the geotechnical and geological analysis and proper engineering judgement. It was required to adopt the mitigation measures in such a way that the different involved items for doing the rehabilitation works are within the items of project BOQ, available construction techniques, plants and materials and also within the provisions of project contract. The durability of the different restoration measures and its compatibility with the existing natural condition were also kept in consideration while finalizing the rehabilitation works. The total cost and time required for doing the mitigation measures against landslides were also kept within possible minimum.

Philosophy of Restoration Measures. The combination of restoration measures was provided to counterbalance the occurrence of landslide. The following two broad measures were generally adopted in combinations to prevent the failure of slope along the project roads in Ethiopia:

- Reduction of driving forces which tend to cause movement of slope mass
- Increase in resisting force of the slope material.

The reduction of driving forces of slope mass could be achieved by providing adequate drainage, reducing head height of slope, flattening of slope and providing intermediate benching/berms. The increase in resisting forces was obtained by providing any kind of retaining wall like gabion/sausage wall and/or stone masonry wall at toe of hill like toe wall or at intermediate position like breast wall, etc.

Drainage. Provisions of proper surface and subsurface drainage measures are one of the best restoration works. The draining out of water from slope mass reduces the weight of the mass tending to slide, also reduces the pore water pressures and seepage forces and does increase shear strengths of material along the sliding plane. As a result of this, the resisting force against the slope failure increases. The drainage measures for the hill slope can be provided by the following ways:

- Providing catch water drain
- Doing improvement in roadside drain
- Augmenting the capacity in road cross drainage works
- Channelizing the hillside drain
- Providing the box-type stepped chutes of stone masonry
- Doing stone pitching (to prevent surface erosion)
- Including the adequate subsurface drainage like drainage blanket, horizontal or inclined drain inside slope, etc., as required based on the site condition
- Adopting the bioengineering works along the hill or valley slopes for surface protection.

Reduction in Head Weigh. It aims at unloading or taking away a relatively large quantity of material from the head of slope which is prone to slip or slide, thereby reducing the activating force to the hill or valley slope.

Slope Flattening and Benching/Berm. Flattening of slope reduces the driving forces on a potential or existing slide. The stable angle of slope of hill or valley varies depending upon its height and material condition. Benching in slope increases the stability of slope by dividing it into segments of smaller ones. Benching in slopes serves as catch area for falling rocks and also provides space for accommodating the intermediate catch water drain.

Provision of Gravity-Type Retaining Wall. The gravity-type retaining wall can be provided in the form of gabion/sausage walls, stone masonry retaining wall, etc. These retaining walls exert sufficient dead weight near the toe of the unstable slope to prevent the movement. So by this, an additional resisting moment is provided which finally increases factor of safety (FOS) against slope failure.

Types of Restoration Measures for different locations. Based on the basic considerations and philosophy of restoration works as stated in previous sections, the details of mitigation measures against erosion cum landslip or landslide occurrences were finalized for different locations along the project roads in Ethiopia. All these various types of restoration measures for the different locations along the project corridors are described below.

For approaches of bridge. The restoration works for the approaches of bridges were done by the following ways:

- Extending stone masonry (SM) retaining wall type like return wall/wing wall along bridge approaches,
- Rebuilding/reconstruction of backfill with minimum side slope of 1 V:1.5H behind SM wall and also approach slope at the end of SM return wall/wing wall (if required),
- Providing cement-grouted stone pitching over existing or rebuilt backfill slopes and/or embankment fill slope along the approaches for bridges.

Figure 13 shows the typical slope protection works done at bridge approach locations in Ethiopia.

For hill slope of mostly rocky material. The mitigation measures for hill slope having mostly rocky material consisted of the following activities:

- Removal of loose material from hill slope
- Clearing of fallen debris from road surface including hard shoulder and also from existing roadside drain

Fig. 13 Rehabilitation works done at bridge approaches along project roads in Ethiopia

Fig. 14 Mitigation measures at hill slope of rocky nature along project roads in Ethiopia

- Providing gravity-type stone masonry retaining wall at toe of hill (if required) and rebuilding of backfill slope of 1 V:2H behind retaining wall with selected granular soil
- Making new stone masonry lined drain (if required) at toe of hill and/or at road edge on valley side depending upon the site condition.

The above-stated rehabilitation measures for hill slope of rocky material are shown in Fig. 14.

For hill slope having other than Rocky material. The restoration measures for hill slope having other than rocky material comprised the following activities:

- Removal of loose material from hill slope
- Clearing of fallen debris from road surface including hard shoulder and also from existing roadside drain
- Reshaping/flattening of hill slope not steeper than 1 V:1.5H (if required) with 2.00-m-wide intermediate berm after every 4.00 m interval of slope height and with minimum top 0.50-m thickness cover soil over the hill slope and grassing over that cover soil
- Providing gravity-type stone masonry retaining wall at toe of hill (if required) and rebuilding of backfill slope of 1 V:2H behind retaining wall with selected granular soil
- Making the provision of stone masonry lined drain at toe of hill in one or (if any) both sides
- Incorporating stone masonry lined drain (if required) at road edge on valley side depending upon site condition.

Figures 15 and 16 show the typical restoration measures adopted for hill slope in Ethiopia.

For u/s and d/s sides, approaches of pipe culverts. The restoration works for the u/s hillside, d/s valley side and approaches of pipe culverts were undertaken in the following ways:

- Clearing of existing pipes or reconstruction (if required) with increased nos. of pipes in culvert

Fig. 15 Hill slope done at project roads in Ethiopia

Fig. 16 Provision of stone masonry retaining wall at toe of hill slope in Ethiopia

- Removal of fallen debris/loose material from existing inlet and its catch pit of pipe culvert on u/s or provision of proper inlet and its catch pit on u/s of pipe culvert
- Providing check dam or chute in stepped manner (minimum 10–15 m before from u/s head wall) along u/s side hill slope
- Extending the existing stone masonry head wall or reconstruction (if required) of new extended stone masonry head wall in the form of retaining wall along approaches on d/s and/or (if required) u/s side of pipe culvert
- Rebuilding of backfill slope behind u/s and/or d/s head wall with suitable soil
- Making cement-grouted stone pitching over backfill slope of both u/s and d/s head walls of pipe culvert and also over the backfill slope of extended stone masonry retaining wall at the approaches of pipe culvert extended till up to finished road level (FRL)
- Providing proper d/s cascade or wide chute protection in stepped manner for adequate length (minimum 10–15 m away from d/s head wall along valley slope)
- Making of stone masonry lined drain either only u/s side along toe of hill or both at u/s and d/s sides by the side of road for the required length of stretch discharging into culvert u/s or d/s side.

Fig. 17 SM head wall cum retaining wall done at pipe culvert locations in Ethiopia

Fig. 18 D/s side valley slope protection works done at pipe culverts in Ethiopia

The above-said rehabilitation measures for pipe culvert locations are shown in Figs. 17 and 18.

For u/s and d/s sides, approaches of slab culverts. The restoration measures for the u/s hillside, d/s valley side and approaches of slab culverts consisted of the following activities:

- Clearing of fallen debris/loose material from existing inlet and its catch pit of slab culvert on u/s or provision (if required) of proper inlet and its catch pit on u/s of slab culvert
- Making provision (if required) of check dam in stepped manner (minimum 10–15 m before from u/s head wall) along u/s side hill slope
- Extending the existing stone masonry return wall or reconstruction (if required) of new extended stone masonry return wall in the form of retaining wall along approaches of u/s of slab culvert
- Similarly extension of existing stone masonry wing wall or reconstruction (if required) of new extended stone masonry wing wall in the form of retaining wall along approaches of d/s of slab culvert
- Rebuilding of backfill slope behind u/s and/or d/s return or wing walls with suitable soil

- Providing cement-grouted stone pitching over existing or rebuilt backfill slope of both u/s and d/s of slab culvert and also over the backfill slope of extended stone masonry retaining wall, namely return wall and/or wing wall at the approaches of slab culvert extended till up to finished road level (FRL)
- Including proper d/s cascade protection in stepped manner for adequate length (minimum 10–15 m away from d/s head wall along valley slope)
- Providing stone masonry lined drain either only u/s side along toe of hill or both at u/s and d/s sides by the side of road for the required length of stretch discharging into culvert u/s or d/s side.

For valley side slope of road. The mitigation measures for valley side slope comprised the following activities:

- Rebuilding of valley side slope with stone masonry retaining wall (if required) at road edge on valley side and having stone masonry parapet at top of retaining wall extended over finished road top level
- Remaking of backfill for stone masonry retaining wall (if required) with suitable material or reconstruction of normal valley slope embankment fill with side slope not steeper than 1 V: 1.5H using suitable material in layered compacted method
- Reconstructing the front fill slope (if required) of stone masonry retaining wall starting from the designed/specified exposed height below the FRL and for minimum specified valley slope depth below the base of stone masonry wall with suitable soil in proper manner
- Providing cement-grouted stone masonry pitching over front fill slope (if required) of stone masonry retaining wall starting from the designed/specified exposed height below the FRL and for minimum specified valley slope depth below the base of stone masonry wall in proper manner
- Providing (if required) minimum top 0.50-m thickness cover soil over the valley slope and grassing over that cover soil over the rebuilt normal valley slope
- Making provision of single or double pipe culverts (if required) with proper inlet, catch pit, stone masonry head walls both u/s and d/s sides and extension of d/s side head wall in the form of stone masonry retaining wall along the valley slope on road edge and cascade or chute protection along d/s valley slope
- Incorporating stone masonry lined drain (if required) at toe of hillside as well as at road edge on valley side depending upon site condition.

Figure 19 shows the typical restoration measures adopted for valley slope in Ethiopia.

Fig. 19 Typical photograph of restoration measures of valley slope along project roads in Ethiopia

7 Conclusions

The case studies on occurrences of landslides of three project corridors in Ethiopia are presented in this paper. The different locations of occurrences and its probable causes are described here. The basic considerations and philosophy of mitigation measures against occurrences of erosion cum slips and slides are covered. The various component restoration measures and its combinations as adopted at different locations of occurrences along the project roads are also included here. The rehabilitation measures as implemented in the referred projects against the occurrences of landslides are found safe, durable and economic ones.

Acknowledgements The authors thank senior managements, design and construction supervision team of engineers of M/s. LEA Associates South Asia Pvt. Ltd. (LASA) for support and encouragement in preparation for the paper.

References

BIS 6403 (1981) Code of practice for determination of bearing capacity of shallow foundations.
Bowles JE (1997) Foundation analysis and design. The McGraw-Hill Companies, Inc
BS 8002: Code of practice for Earth Retaining Structures, London, United Kingdom
BS 8004: Code of Practice for Foundations, LONDON, United Kingdom
ERA Guidelines and Technical Specifications for Road Works, Addis Ababa, Ethiopia
Ethiopian Meteorological Service (1979) Meteorological Map of Ethiopia, Addis Ababa
Huang YH (1983) Stability analysis of earth slopes. Van Nostrand Reinhold Company, New York, USA
IRC:75–1979, Guidelines for the design of high embankments
IRC:SP:48–1998, Hill Road Manual
Site Investigation Manual (2002) Ethiopian Roads Authority. Addis Ababa, Ethiopia
TRL Overseas Road Note 9 (2000) A design manual for small bridges. Transport Research Laboratory, Crowthorne, Berkshire, United Kingdom

Stability Assessment of Cut Slopes Along Shivpuri–Kaudiyala Road (NH-58), Uttarakhand Himalayas

Neeraj, Koushik Pandit, and Shantanu Sarkar

Abstract In the present study, slope stability assessment of selected cut slopes on a road stretch of approximately 25 km length along the National Highway (NH-58) between Shivpuri and Kaudiyala in the Lesser Himalaya of Uttarakhand has been carried out. This NH-58 connects Rishikesh with Badrinath and Kedarnath, two important Hindu shrines in Uttarakhand and also the border areas of the country. Hence, it is obvious that this highway is important from the pilgrimage, tourism and defense point of view. The present study includes characterization of rock mass, computation of slope mass classification ratings and comparison of different rock mass classification systems (such as RMR, GSI, SMR and CSMR) for 25 numbers of rock slopes along the selected road stretch. Also, kinematic analyses have been performed for all these slopes to understand the mode of slope failure. Further, stability analyses of a few significant road cut slopes have been carried out using a finite element tool.

Keywords Slope stability · Lesser Himalayas · Rock mass classifications · Kinematic analysis · Finite element method

1 Introduction

Road networks and highways play an indispensable role in the Garhwal Himalayan terrains for mass conveyance, tourism and its socioeconomic development. Along highways, stability of slopes is of major concern for ensuring a safe transportation solution. Slopes may fail by natural or anthropogenic agents that lead to disruption in traffic flow, loss of properties and lives and environmental degradation by soil erosion and loss of plantation. One of the common and often neglected areas of concern is

Neeraj
Department of Geology, Kurukshetra University, Kurukshetra, India

K. Pandit · S. Sarkar (✉)
Geotechnical Engineering Division, CSIR-Central Building Research Institute, Roorkee - 247667, India
e-mail: shantanu_cbri@yahoo.co.in

M. Latha Gali and P. Raghuveer Rao (eds.), *Geohazards*, Lecture Notes in Civil Engineering 86, https://doi.org/10.1007/978-981-15-6233-4_5

the systematic geological and geotechnical investigation before excavation of a road cut. Unscientific design of a road cut may weaken toe of a natural slope, which may fail in rainy season even due to slightest of rainfall or in the wake of an earthquake event. Thus, slopes may become highly vulnerable and a threat to local commuters.

According to researchers (Dudeja et al. 2017; Siddique et al. 2017), landslides generally and recurrently occur in the Garhwal Himalayan terrains due to the presence of geologically weak rocks, high rainfall intensity during the monsoon season, tectonically induced seismicity and occasionally due to anthropogenic activities like blasting of tunnels for hydropower projects, unplanned excavation or sideway cuttings for road widening (Sati et al. 2011) and deforestation. In recent times, frequency of landslides in the Himalayan region has raised due to heightened infrastructure development activities like road widening and hydropower projects (Singh et al. 2010).

The present study endeavors to perform stability assessment of selected road cut slopes along Shivpuri (after Rishikesh)–Kaudiyala (before Devprayag) route of NH-58 in the Uttarakhand Himalayas. This is the busiest route in this region as every year a huge number of pilgrims access this route for traveling to the Char Dham shrines. This route is also having strategic importance as it connects to the international borders of the country. Another reason to choose this stretch as study area is that large-scale slope failures and associated problems are very common in this region. All of these need immediate stability evaluation and effective control measures to reduce the related vulnerabilities. For investigating the vulnerable slopes in this region, field survey has been carried out, and pertinent data required for different rock mass classification systems like RMR, GSI, SMR and CSMR have been collected. These data were then used to evaluate stability of the screened road cut slopes in this region. Kinematic analysis technique was also used to identify types of slope failures and failure direction. Two-dimensional stability analysis software based on finite element code was also employed to evaluate stability condition of two critical slopes. From all these analysis, it has been observed that nine slopes are stable, four slopes are having potential of wedge failure, and rests of the twelve slopes are susceptible to planar failure. These results indicate that the techniques employed in this study corroborate each other in assessing the slope stability which is a prerequisite for designing appropriate remedial measures to prevent slope failures and minimizing damage consequences. Overall, this study gives an insight into the landslide dynamics pertinent to slope failures in the lesser Himalayan region through kinematic and finite element analysis and presents a quantitative comparison between the available and popular slope mass classification methods.

2 Study Area

The present study area is a part of Garhwal Himalaya along the national highway NH-58 from Shivpuri to Kaudiyala in Tehri Garhwal district of Uttarakhand state. This area is located between latitude 30°4′9.9″ to 30°8′15.2″ and longitude 78°24′29.7″

Fig. 1 Stretch for all the studied slopes along NH-58 (from Google map)

to 78°30′59″. The highway is situated along right bank of the Ganga River. This is an important route which connects Indo-Gangetic plain to hilly areas of Uttarakhand. This road stretch has been selected primarily because of the frequency and extent of damage which take place every year due to landslides. The road stretch on which the critical slopes were identified and studied is shown on a Google map in Fig. 1.

Geologically, the present study area falls under a part of Garhwal Lesser Himalaya which lies between the Greater Himalaya and Siwalik Range and bounded by Main Central Thrust (MCT) and Main Boundary Thrust (MBT), respectively (Heim and Gansser 1939). The Lesser Himalayan belt is one of the most structurally complex litho-tectonic units of Himalaya. The slopes investigated in this study include meta-sedimentary rock and meta-volcanic rock (Frank et al. 1995). During the field survey, the rock types encountered in the studied slopes are predominantly sandstone, but shale, limestone, slate and phyllite are also present to some extent.

The climatic conditions in the Lesser Himalaya are also very diverse and extreme in terms of amount of rainfall and fall in temperature during winter. The rainy season follows the monsoon cycles which begins in this region from June to August. The winter season is very cold particularly from December to March. Throughout the year, the temperature varies from −3 to 30 °C, and maximum rainfall is about 250 cm (Lawrence 1977).

3 Slope Stability Assessment from Rock Mass Classification Systems

In this study, the rock slopes were selected in such a way that they have varied geological and slope instability conditions. The data collected from field include details of the slope (its location, height and orientation), rock type, joint set numbers, joint spacing, joint condition (roughness, weathering, infilling), joint alteration, joint orientation, rock mass structure, hydrological condition and compressive strength evaluation from Schmidt hammer's rebound number.

Fig. 2 Field photographs of some studied slopes

These field data were collected to determine GSI, RMR, SMR and CSMR of the studied slopes. Also, slope profiles at two locations have been mapped from the coordinates obtained by a laser distance meter instrument. This geometry of slopes is a basic input to perform stability analysis by finite element method. Field photographs of some of the studied slopes are shown in Fig. 2.

Once all the field data have been collected and compiled, different rock mass and slope mass classification systems were used to evaluate stability of the studied slopes. In the present study, the taken rock mass rating systems are: Rock Mass Rating (RMR), Geological Strength Index (GSI), Slope Mass Rating (SMR) and Continuous Slope Mass Rating (CSMR).

RMR is calculated according to Bieniawski's method (1989) by adding rating values for five parameters: (i) strength of intact rock, (ii) rock quality designation (RQD) value (measured or estimated), (iii) spacing of discontinuities, (iv) condition of discontinuities and (v) water inflow through discontinuities. According to RMR_{basic} system, five classes are defined as very poor rock (Class V: 0–20), poor rock (Class IV: 20–40), fair rock (Class III: 40–60), good rock (Class II: 60–80) and very good rock (80–100).

In this study, the parameters of discontinuities of the selected slopes have been found to vary considerably. Joint spacing and opening also show a wide range of values. The presence or absence of filling was noted along with type of filling those discontinuities. RQD value was estimated from its empirical relationship with the number of joints per cubic meter or volumetric joint count (Jv). The hydrological conditions of the slopes were also observed. In most of the cases, the slopes were found to be dry. Along with these data, attitude of discontinuities and slope angles were also measured. For evaluating the compressive strength of rock mass (UCS), rebound numbers were taken from each studied slopes which showed large variation according to different rock type. UCS is calculated with the help of the following correlation from (Saptono et al. 2013):

$$\mathrm{UCS} = 0.308R^{1.327} \quad (1)$$

where R = Rebound hammer value from N-type hammer.

The GSI system is another rock mass classification technique proposed by Hoek and Brown (1997). This rock mass classification system which is an easy method for field application is primarily based on the blockiness of the rock mass and the surface conditions of discontinuities. Later, there have been many modifications of the GSI, and a quantitative method to estimate more precise values was given by Sonmez and Ulusay (2002) which provided a modified chart. This modified quantitative rock mass classification is based on the structure rating (SR) and surface condition rating (SCR). SR is based on volumetric joint count (Jv), while SCR depends on the roughness, weathering and infilling in joints. The values of Jv and SR are given in the chart to obtain the rating for SR. The modified chart represents five rock mass categories ranging between 5 and 100. The relevant correlations are as follows:

$$\mathrm{SR} = -17.5\ \ln(J_v) + 79.8 \quad (2)$$

$$\mathrm{SCR} = R_r + R_w + R_i \quad (3)$$

To determine GSI, structure of rock mass and surface condition of the discontinuities has been studied in detail in the field. For surface condition rating, data on degree of rock mass weathering, roughness and infilling material were collected. Surface roughness data show a wide range from very rough to slickensides. Weathering was found to be varied from low to decompose state.

In Slope Mass Rating (SMR) approach (Romana 1985), stability of rock slope is evaluated from Bieniawski's Rock Mass Rating (RMR) (Bieniawski 1989) by adding adjustment factor of the joint-slope relationship and adding a factor depending on method of excavation (Romana et al. 2003).

$$\mathrm{SMR} = \mathrm{RMR}_{\mathrm{basic}} + (F_1 \cdot F_2 \cdot F_3) + F_4 \quad (4)$$

where $\mathrm{RMR}_{\mathrm{basic}}$ is evaluated according to Bienaiwski (1989, 1979) by adding the rating of five parameters. Here, F_1, F_2, and F_3 are adjustment factors related to joint orientation with respect to slope orientation, and F_4 is the correction factor for method of excavation.

One of the recent popular techniques, Continuous Slope Mass Rating (CSMR) is also utilized in this study for slope stability assessment of rock slopes. This rating system was proposed by Tomas et al. (2007) in order to get more precise evaluation of stability grades. It has been noticed that SMR value slightly deviates from real-field condition. This may occur during computation of adjustment factors when the exact value lies at border of predefined class interval (Umrao et al. 2011). However, consideration of continuous functions in CSMR has suppressed this problem and made this method a robust technique while dealing with assessment of slope stability.

To compute CSMR, the adjustment factors ratings (F_1, F_2 and F_3) were modified, while other parameters and method are similar to SMR. The modified adjustment factors are defined as

$$F_1 = \frac{16}{25} - \frac{3}{500}\arctan\left\{\frac{1}{10}|A| - 17\right\} \quad (5)$$

where A is $|\alpha_j - \alpha_s|$ is for planar failure, $|\alpha_j - \alpha_{s\text{-}} - 1|$ for toppling failure and $|\alpha i\text{-}\alpha s|$ for wedge failure.

$$F_2 = \frac{9}{16} - \frac{1}{195}\arctan\left\{\frac{17}{100}B - 5\right\} \quad (6)$$

where B is β_j for planar failure, β_i for wedge failure, while F_2 remains 1 for toppling mode of failure.

$$F_3 = -30 + \frac{1}{3}\arctan C \quad (7)$$

where C is $(\beta_j - \beta_s)$ for planar failure, and $(\beta_j - \beta)$ for wedge failure. Also,

$$F_4 = -13 - \frac{1}{7}\arctan(C - 120) \quad (8)$$

where C is $(\beta_j - \beta_s)$ for toppling failure.

In all these above equations, α_s is the dip direction of slope, α_j is the dip direction of joint, β_s is the dip amount of slope, β_j is the dip amount of joint, α_i is the dip direction of line formed by the intersection of two discontinuity, and β_i is the amount of plunge of line formed by the intersection of two discontinuity. The value of arctan is in degrees.

The stability classes of SMR values, rock mass description, stability and probability of failure given by Romana (1985) are also applicable for CSMR classification Table 1.

In the present study, SMR and CSMR values were calculated from several rock mass and slope mass parameters based on visual inspection and laboratory experimentation. The result illustrates that 1 slope is stable, 7 slopes are partially stable, 8 slopes are unstable, and 9 slopes fall under completely unstable condition. The results of RMR, SMR and CSMR and their inferred stability grade are shown in Table 2. The stable, partially stable, unstable and completely stable slopes as evaluated from the rock mass classification systems have been shown in Fig. 3 with their respective locations on a Google map excerpt.

Table 1 Various stability classes as per SMR values (Romana 1985)

Class No.	V	IV	III	II	I
SMR value	0–20	21–40	41–60	61–80	81–100
Rock mass	Very bad	Bad	Normal	Good	Very good
Stability	Completely unstable	Unstable	Partially stable	Stable	Completely stable
Failure	Big planar or soil-like or circular	Planar or big wedge failure	Planar along some joint and many wedge failure	Some block failure	None
Probability of failure	0.9	0.6	0.4	0.2	0

From the present study, it has been observed that SMR values obtained for different slopes slightly deviate from computed CSMR values. This difference in values is due to the consideration of discrete and continuous functions in SMR and CSMR, respectively. Hence, it may be apprehended that the CSMR values are much closer to real prevailing field conditions.

The GSI values for the 25 slopes obtained in the present study were compared with the RMR values. It has been already established that RMR can be predicted from GSI and the relation between GSI and RMR_{89} after (Hoek and Brown 1997) is as follows (Singh and Goel 1999):

$$GSI = RMR_{89} - 5 \text{ (for GSI} > 18) \tag{9}$$

While studying a few road cut slopes in Garhwal Himalaya (Sarkar et al. 2012; Sarkar et al. 2012) also found:

$$GSI = RMR_{89} - 9 \tag{10}$$

For this study, the GSI and RMR values have been plotted, and a new correlation has been established (Fig. 4):

$$GSI = RMR - 6 \tag{11}$$

Further, SMR and CSMR values have been also plotted in Fig. 5, and a correlation has been found:

$$CSMR = 0.8\ SMR + 3.8 \tag{12}$$

Table 2 Comparison of RMR_{basic}, SMR and CSMR values for the studied slopes along with their stability grades (SG)

Slope	RMR_{basic}	SMR	SG	CSMR	SG
1	64	24	US	19	CUS
2	64	56	PS	47	PS
3	55	47.5	PS	49	PS
4	53	13	CUS	24	US
5	38	13	CUS	2	CUS
6	59	9	CUS	38	US
7	54	64	S	64	S
8	45	20	US	17.4	CUS
9	63	65.5	S	55	PS
10	51	44.2	PS	43	PS
11	55	23	US	19	CUS
12	53	13	CUS	16	CUS
13	62	47	PS	47	PS
14	62	54	PS	46	PS
15	48	15.5	CUS	7	CUS
16	45	4	CUS	7	CUS
17	48	39	US	30	US
18	43	15.3	CUS	7.5	CUS
19	38	4	CUS	4	CUS
20	58	18	CUS	22.6	US
21	62	37	US	24	US
22	64	31	US	23.5	US
23	60	26	US	30	US
24	30	18	CUS	21	US
25	58	58	PS	52	PS

Here, *US* unstable, *CUS* completely unstable, *PS* partially stable and *S* stable

4 Kinematic Analysis of Slopes

Kinematic analysis has been performed to understand mode of slope failure. Kinematic analysis evaluates the potential failure mechanisms (e.g., planar, wedge or toppling failures) by analyzing the discontinuity-slope relation. Discontinuities considered in the study are primarily bedding, foliation and joint planes. A plane failure is likely to occur when a discontinuity dips in the same direction (within 20°) as the slope face, at an angle gentler than the slope angle but greater than the friction angle along the failure plane. A wedge failure may occur when the intersection line of two discontinuities forming the wedge-shaped block plunges in the same direction

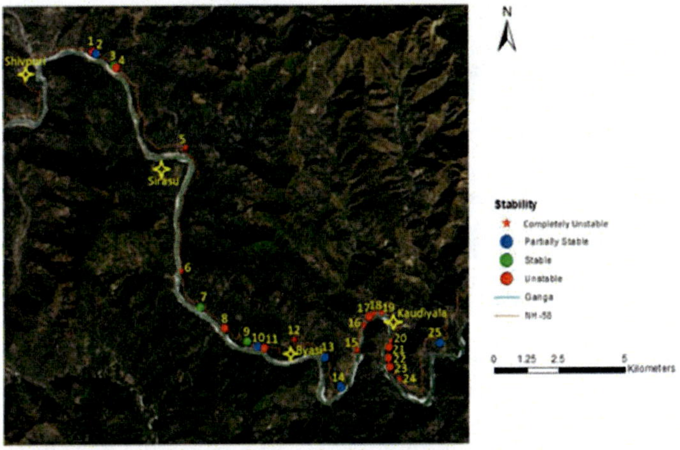

Fig. 3 Locations and evaluated stability of studied slopes along NH-58

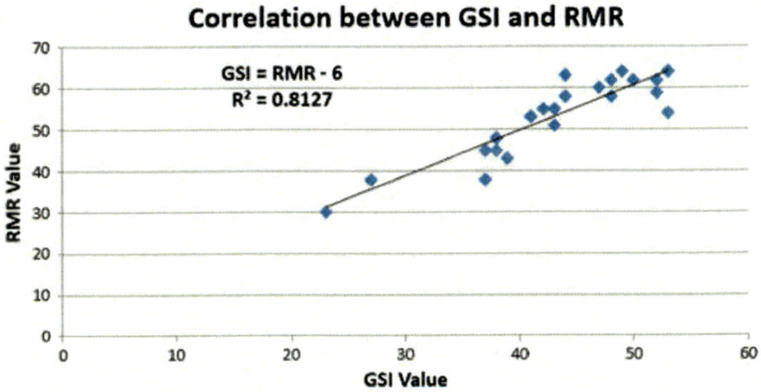

Fig. 4 Correlation between computed GSI and RMR values for the studied slopes

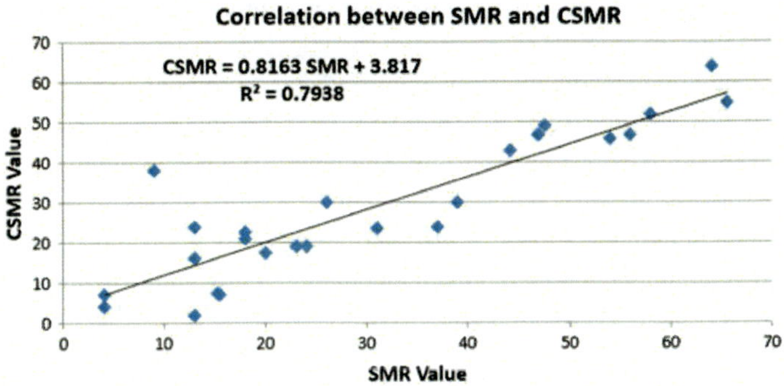

Fig. 5 Correlation between computed SMR and CSMR values for the studied slopes

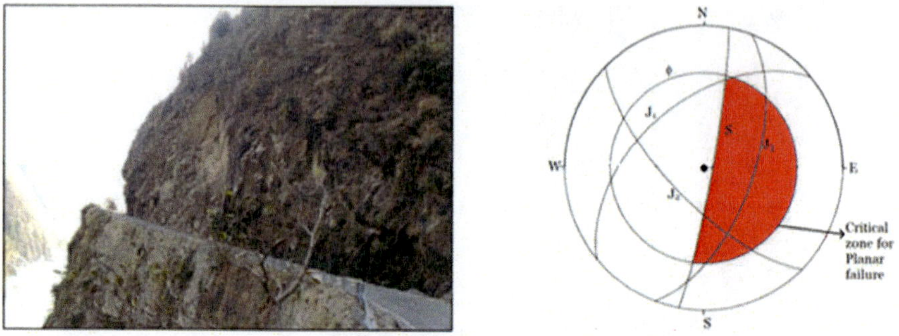

Fig. 6 Kinematic analysis of slope S-16 with site photograph and stereo-plot

as the slope face and the plunge angle are less than the slope angle but greater than the friction angle along the planes of failure (Hoek and Bray 1981). Out of all 25 slopes studied in this work, four slopes (S-4, S-9, S-21, S-25) are predicted to undergo wedge failure by typical double plane sliding and occasional small planar failure, whereas twelve slopes (S-1, S-5, S-6, S-8, S-11, S-12, S-15, S-16, S-17, S-18, S-19 and S-20) are predicted to undergo planar failure. The rest of the slopes are found to be stable due to favorable discontinuity orientations and high shear strength. A typical kinematic analysis of slope S-16 with site photograph and stereo-plot has been shown in Fig. 6.

5 Finite Element Analysis of Slopes

It has been observed that numerical modeling is a unique approach for slope stability analysis as it encompasses slope geometry, layering of strata, different material properties (elastic and plastic state) and different constitutive models which result into a reliable quantitative term, factor of safety of the slope. For determining factor of safety of slopes, in this study, a FEM program RS2 9.0 has been utilized by incorporating engineering and physical properties of slope material determined in laboratory and field. The code, RS2 9.0, is a 2D finite element program for slope stability analysis in which finite element slope stability analysis can be done using the shear strength reduction (SSR) method. It also provides user defined Mohr–Coulomb or Hoek–Brown strength parameters.

In this study, slope stability analysis of two selected critical slopes (S-2 and S-4) has been carried out. By definition, the factor of safety (FoS) of a slope is the "ratio of actual soil shear strength to the minimum shear strength required to prevent failure" or the factor by which soil shear strength must be reduced to bring a slope to the verge of failure (Duncan 1999). In the SSR finite element technique, elasto-plastic strength is assumed for slope materials. The shear strength values of slope materials are progressively reduced until the collapse occurs.

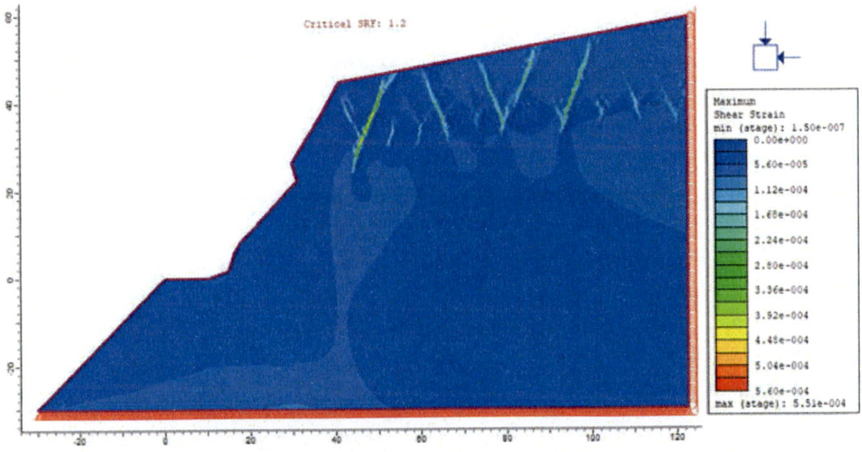

Fig. 7 Contour of the maximum shear strain for S-2

For Mohr–Coulomb material, shear strength reduced by a factor (of safety) F which can be determined from the equation:

$$\frac{T}{F} = \frac{C'}{F} + \frac{\tan\varphi}{F} \tag{13}$$

This equation can be rewritten as

$$\frac{T}{F} = C^{*} + \tan\varphi^{*} \tag{14}$$

In this case,

$$C^{*} = \frac{C'}{F} \text{ and } \varphi^{*} = \arctan\left(\frac{\tan\varphi'}{F}\right) \tag{15}$$

In this present study, factor of safety values for the chosen slopes have been calculated by using the generalized Hoek–Brown (GHB) criterion (Hoek et al. 2002). The obtained FoS values show that one slope is partially stable (FoS = 1.2) and another one is unstable (FoS = 0.49). The interpretation of results for both the slopes is shown in Figs. 7 and 8.

6 Conclusions and Recommendations

In the present study, 25 rock slopes (S-1 to S-25) along the stretch from Shivpuri to Kaudiyala in Uttarakhand Himalaya were selected for slope stability assessment. For all these slopes, the values or ratings of GSI, RMR_{basic}, SMR and CSMR were

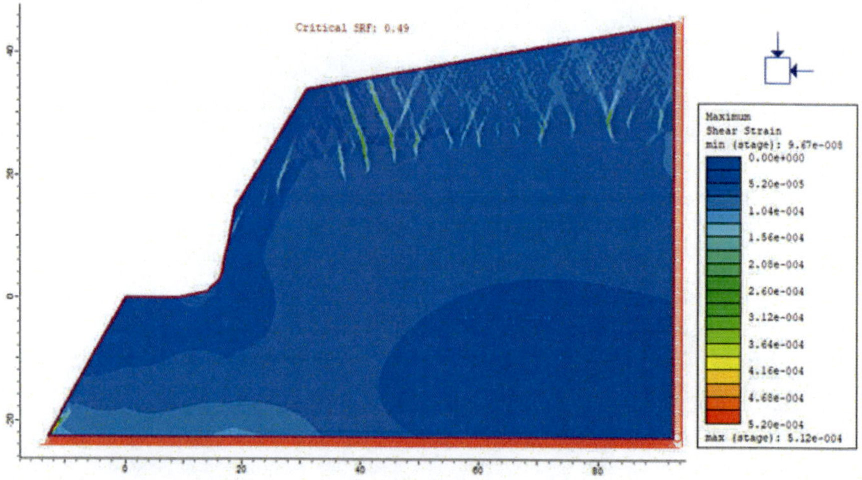

Fig. 8 Contour of the maximum shear strain for S-4

evaluated, and kinematic analysis was carried out. From these analyses, unstable slopes have been identified. These slopes need immediate remedial measures such as reinforcement by rock bolts and anchors as they are rock slopes and have layers of weak weathered rocks on top of hard rock stratum. Based on the kinematic analysis, it has been concluded that out of all 25 locations, four slopes are predicted to undergo wedge failure and occasional small planar failure, whereas the fifteen slopes are predicted to undergo planar failure. A table was constructed showing comparative values of all the ratings obtained from different rock mass classification systems. Further, correlations between GSI and RMR and SMR and CSMR have been established from the compiled pool of field data. However, this kind of correlation study needs more data to establish strong relationships with more confidence level. Also, stability analysis of two slopes has been performed by finite element analysis, the results from which substantiate the findings about the stability grade of those slopes by rock mass classification systems Table 2.

The present study will help in understanding the slope stability of the road cut slopes. The study shows the importance of such investigation before constructing the hill roads. Such study is also important to design adequate remedial measures which can improve the stability of slopes. The results of this study give deeper insights to landslide dynamics by a unique culmination of different slope mass classification systems, kinematic analysis and finite element analysis. The results from this study indicate that all the techniques corroborate with each other in assessing the slope stability if field and laboratory data have been critically obtained and analyzed. Finally, the results obtained here in this study will help in proper planning and design of road cut slopes in the hilly terrain, especially in the ongoing Char Dham Yatra project in Uttarakhand Himalayas.

Acknowledgements The authors would like to acknowledge Director, CSIR–CBRI, Roorkee, for giving his kind permission to communicate this research work for publication.

References

Bieniawski ZT (1979) The geomechanical classification in rock engineering applications. in: Proceedings of the 4th international congress rock mechanics, Montreux, Balkema, Rotterdam, vol 2, pp 41–48

Bieniawski ZT (1989) Engineering rock mass classifications: a complete manual for engineers and geologists in mining, civil, and petroleum engineering. Wiley International, pp 40–47

Dudeja D, Bhatt SP, Biyani AK (2017) Stability assessment of slide zones in Lesser Himalayan part of Yamunotri pilgrimage route, Uttarakhand. India Environ Earth Sci 76(1):2–18

Duncan JM (1999) State of the art: limit equilibrium and finite-element analysis of slopes. J Geotech Eng (ASCE) 122(7).

Frank W, Grasemann B, Guntli P, Miller C (1995) Geological map of the Kishwar Chamba-Kulu region (NW Himalayas India). Jahrb Geol Bundesanst 138:299–308

Heim AA, Gansser A (1939) Central Himalaya: geological observations of the Swiss Expedition. 1936, India, Hindustan Publishing, Delhi

Hoek E, Bray J (1981) Rock slope engineering. Stephen Austin and Sons Limited Publishers, Hertford

Hoek E, Brown ET (1997) Practical estimates of rock mass strength. Int J Rock Mech Min Sci 34(8):1165–1186

Hoek E, Carranza-Torres C, Corkum B (2002) Hoek-Brown criterion—2002 edition. In: Hammah R, Bawden W, Curran J, Telesnicki M (eds) Mining and tunneling innovation and opportunity, proceedings of the 5th North American rock mechanics symposium and 17th tunnelling association of Canada conference. Toronto, Canada. University of Toronto, Toronto, pp 267–73

Lawrence FJ (1977) Geology of the area around Uttarkhasi, Garhwal Himalaya, Uttar Pradesh. MTech Thesis, IIT Roorkee

Romana M (1985) New adjustment ratings for application of Bieniawski classification to slopes. In: Proceedings of international symposium on "The role of rock mechanics", Zacatecas, pp 49–53

Romana M, Serón JB, Montalar E (2003) SMR geomechanics classification: application, experience and validation. In: Merwe JN (ed) Proceedings of the 10th congress of the International Society for rock mechanics, ISRM 2003–Technology roadmap for rock mechanics, South African Institute of Mining and Metallurgy, pp 1–4

Saptono S, Kramadibrata S, Sulistianto B (2013) Using the Schmidt hammer on rock mass characteristic in sedimentary rock at Tutupan coal mine. Procedia Earth Planet Sci 6:390–395

Sarkar S, Kanungo DP, Kumar S (2012) Rock mass classification and slope stability assessment of road cut slope in Garhwal Himalaya, India. Geotech Geol Eng 30:827–840

Sati SP, Sunderiyal YP, Rana N, Dangwal S (2011) Recent landslides in Uttarakhand: nature's furry or human folly. Curr Sci 100(11):1617–1620

Siddique T, Pradhan SP, Vishal V, Mondal A, Singh TN (2017) Stability assessment of Himalayan road cut slopes along National Highway 58. India. Environ Earth Sci. 76:759

Singh B, Goel R (1999) Rock mass classifications—a practical approach in civil engineering, 1st edn. Elsevier Science

Singh TN, Verma AK, Sarkar K (2010) Static and dynamic analysis of a landslide. Geomaterial Nat Hazards Risk 1(4):323–338

Sonmez H, Ulusay R (2002) A discussion on the Hoek-Brown failure criterion and suggested modification to the criterion verified by slope stability case studies. Yerbilimleri (Earth Sci) 26:77–79

Tomás R, Marchal JD, Serón JB (2007) Modification of slope mass rating (SMR) by continuous functions. Int J Rock Mech Min Sci 44(7):1062–1069

Umrao RK, Singh R, Ahmad M, Singh TN (2011) Stability analysis of cut slopes using continuous slope mass rating and kinematic analysis in Rudraprayag district, Uttarakhand. Geomaterials 1:79–87

Multivariate Adaptive Regression Spline Based Reliability Analysis of Stability of Durgawati Earthen Dam

V. Kumar, P. Samui, A. Burman, and N. Himanshu

Abstract In the present paper, multivariate adaptive regression splines (MARS) method is used to carry out reliability analysis of Durgawati earthen dam. The steady and transient state seepage is considered while calculating reliability index (β). The earthen dam is an important water retaining structure and serves the catchment area of Durgawati River in Bihar, India. The factor of safety (FOS) of the earthen dam against slope failure under steady and transient state is performed using GeoStudio-07 software. In order to account for the variability of material parameters over the body of the dam, a data set of material parameters [i.e., cohesion (c), angle of internal friction (φ) and soil unit weight (γ)] are prepared, and MARS is used to calculate reliability index (β) for different realizations over the dam geometry. The reliability index (β) thus estimated shows that the dam is stable and safe from slope failure point of view under different seepage loadings.

Keywords Slope failure · Bishop's method · Durgawati earthen dam · Multivariate adaptive regression splines (MARS) · Steady seepage · Transient seepage

1 Introduction

Performing reliability analysis is important to consider the variability of geotechnical properties of soil in nature for any structure. For slope stability analysis, reliability analysis can be important information about the effect stemming from model uncertainty also. In such situations, correct and reliable assessment of factor of safety (FOS) of slopes becomes a very difficult proposition. A probabilistic framework of analysis is better suitable in these cases so that the analyses can also inform about the chances of failure. Such analyses also give an estimate about the risks involved in the process. Many investigators (El-Ramly et al. 2002; Liang et al. 1999) have earlier studied reliability analysis of slopes with probabilistic representation of related input parameters. Many others performed reliability-based slope analyses and helped to

V. Kumar (✉) · P. Samui · A. Burman · N. Himanshu
Department of Civil Engineering, NIT Patna, Patna, Bihar 800005, India
e-mail: vinaykumarnitp@gmail.com

M. Latha Gali and P. Raghuveer Rao (eds.), *Geohazards*, Lecture Notes in Civil Engineering 86, https://doi.org/10.1007/978-981-15-6233-4_6

popularize such practices (Salgado and Kim 2014; Roh and Hong 2009; Christian et al. 1994). The spatial variability of geotechnical parameters is usually taken care of during reliability analyses of structures like earth slopes (Li et al. 2015; Griffiths et al. 2009; Griffiths and Fenton 2004).

The first-order second moment (FOSM) method estimates the effect of variabilities of related parameters leading to reliable analysis of FOS values as a function of input soil parameters such as cohesion (c), soil unit weight (γ) and soil friction angle (φ). However, there are few limitations of this method. Linearizing model's output function centred about the mean of the input parameters to describe the stochastic/probabilistic behaviour of the model output over the solution domain is not a very easy task (Melching and Yoon 1996). Durgawati earthen dam, which is situated in Kaimur district of the state of Bihar is a very important water retaining structure. Reliability analyses of this structure are performed accounting for the variability of geotechnical properties of soil over the body of the dam and presented in this paper. Steady and transient state seepage conditions are considered in the analyses. MARS is used for the purpose of regression analysis. MARS (Friedman 1990) is a multivariate non-parametric regression technique. MARS is an accurate and fast artificial intelligence (AI)-based algorithm. It automatically searches for the nonlinear input/output relationship inside any large database. The results present in this paper show that MARS is very capable to predict the FOS of the slope. A nonlinear predictive relationship made up of basis functions in terms of c, ϕ and γ is developed which can predict the FOS of the dam under steady and transient seepage state with high degree of accuracy. Bishop's method based on limit equilibrium technique is used for determination of FOS of the dam. The determination of FOS against slope failure under transient state seepage condition is carried out to simulate the drawdown behaviour of the adjacent reservoir on upstream portion of the dam. GeoStudio-07 software is used for this purpose. The FOS values of the earthen dam are calculated for different seepage conditions for many realizations with varying material parameters (c, φ, γ). After that, the structure under consideration is subjected to reliability analysis, and reliability index (β) is determined. Himanshu and Burman (2017) developed a slope stability model of the earthen dam situated on Durgawati river using GeoStudio software for seepage and stability analysis. The same model is also used for reliability analyses performed here.

2 Methodology

2.1 Multivariate Adaptive Regression Splines (MARS)

Multivariate adaptive regression splines (MARS) was initially formulated by Friedman (1990). It is a nonlinear, a nonparametric regression method. Outputs are determined based on a large number of input parameters representing different realizations of the study domain using MARS. MARS generally performs additive, recursive, spline and recursive partitioning regression. MARS generates the output in the form of basis functions with the help of "forward" and "backward" algorithms with very good adaptive capability. The output generated by MARS is usually highly accurate compared to other methods. MARS uses an equation of following nature for performing the task of nonparametric regression:

$$y_i = f(x_{i1}, x_{i2}, \ldots, x_{ik}) + \varepsilon_i \quad (1)$$

Here, $f(x_{i1}, x_{i2}, \ldots, x_{ik})$ is the regression function and ε_i is the estimate of error in performing the task. Ideally, $f(x_{i1}, x_{i2}, \ldots, x_{ik})$ should be chosen as a smooth, continuous function (Fox 2002). MARS develops the model for predicting output y in the form:

$$y = C_0 + \sum_{m=1}^{M} C_m B_m(x) \quad (2)$$

where C_0 is a constant, $B_m(x)$ is basis function, x is the input variable and C_m is the coefficient of $B_m(x)$. In this paper, the input parameters are soil density (γ), friction angle (φ) and cohesion (c). The FOS value of the slope is produced as output in the form $y =$ FOS and $\{x\} = \{\gamma, \phi, c\}^T$. The spline function is made up of two truncated left and right sided functions as shown in Eqs. 3a and 3b, respectively.

$$b_q^-(x - t) = [-(x - t)]_+^q = \begin{cases} (t - x)^q & \text{if } x < t \\ 0 & \text{otherwise} \end{cases} \quad (3a)$$

$$b_q^+(x - t) = [+(x - t)]_+^q = \begin{cases} (x - t)^q & \text{if } x > t \\ 0 & \text{otherwise} \end{cases} \quad (3b)$$

Here $b_q^+(x - t)$ and $b_q^-(x - t)$ are the spline functions with t as positions of the knots. In general, multivariate adaptive regression splines (MARS) model is based on the following three steps:

Construction.
Pruning.
Selection of optimum MARS.

First phase, i.e., construction phase, uses basis functions to define "Eq. 2". Generalized cross validation (GCV) is performed to validate the choice of the functions $B_m(x)$. GCV represents the residual sum of squares. In this procedure, a penalty is embedded to represent the complexity associated with the model with sole purpose of avoiding too many basis functions. The GCV value is calculated from the following "Eq. 4":

$$\mathrm{GCV}\,(M) = \left(\frac{1}{n}\right)\frac{\sum_{m=1}^{M}(y_i - \hat{y}_i)^2}{\left[1 - \frac{C(M)}{n}\right]^2} \tag{4}$$

where

n	number of data objects
y_i	response for i th object
$\hat{y}_i$	predicted response of i th object
$C(M)$	penalty factor.

Equation 5 is used to calculate the penalty factor $C(M)$:

$$C(M) = M + dM \tag{5}$$

Here, d is a cost penalty factor which optimizes the functions $B_m(x)$. For many choices of function $B_m(x)$, data over fitting may pose difficulties. Pruning is carried out to discard few of these over fitted functions. After finishing all necessary steps, the final optimum MARS is selected.

2.2 *Reliability Analysis.*

Following steps are involved in carrying out reliability analysis of stability of slopes of Durgawati earthen dam under steady and transient state seepage:

- To represent the variability of the soil properties within the study domain, 100 data sets are generated by maintaining certain upper and lower bounds using Eq. 6

$$y = \mathrm{Round}(\mathrm{Rand}(), 2) \times \alpha_l + |\alpha_u - \alpha_l| \tag{6}$$

  where α_l = lower limit and α_u = upper limit.
- The SEEP-W and SLOPE-W modules of GeoStudio-2007 software are used to perform seepage and stability analyses of Durgawati dam. Limit-equilibrium-based Bishop's method is selected for determining the FOS of the slope of the dam.
- After normalizing the input parameter, data set is divided into two groups which are known as testing and training data set. MARS takes input of these data sets which also designates different realization of the embankment geometry with varying permeability values.
- A dimensionless parameter called coefficient of variation (COV) represent a dispersion of the data sets. It is expressed as:

$$\mathrm{COV} = \frac{\sigma_x}{\mu_x} \times 100\% \tag{7}$$

Table 1 Suggested COV values

Parameter	Coefficient of variation (%)	References
ρ or γ	3–10	Lee et. al. (1983), Harr (1987), Beacher and Christian (2003)
φ - sand	10	Harr (1987), Lee et. al. (1983)
φ- clay	10–50	Lee et. al. (1983), Beacher and Christian (2003)
φ- mine tailings	5–20	Beacher and Christian (2003)
c_u or q_u	40	Harr (1987), Lee et. al. (1983)

where σ_x = standard deviation and μ_x = mean, the COV used to describe the variation of geotechnical soil properties, i.e., cohesion (c), angle of internal friction (ϕ) and unit weight of soil (γ). The values of COV used in the paper are presented in Table 1. The ranges presented in Table 1 have been chosen from the previous works of other investigators.

Reliability index is used to express the degree of uncertainty in the calculated FOS, and it is presented as (Malkawi et al. 2000)

$$\beta = \frac{E(F) - 1.0}{\sigma(F)} \tag{8}$$

where

β Reliability index.
$E(F)$ The expected value of the FOS.
$\sigma(F)$ Standard deviation.

2.3 *Model Analysis*

In the current work, MARS is used for reliability analysis of earthen dam against slope failure. It is important to check the degree of match between predicted and actual data. Few parameters such as root mean square error (RMSE), weighted mean absolute percentage error (WMAPE), maximum determination coefficient value (R^2), adjusted determination coefficient (Adj. R^2), variance account factor (VAF), performance index (PI) and Nash–Sutcliffe coefficient (NS) are useful for checking the fit between actual FOS value and the predicted FOS value by MARS (Gokceoglu and Zorlu 2004; Nayak et al.2005; Wang et al. 2009).

3 Results and Discussions

3.1 Description of Durgawati Earthen Dam

Durgawati dam construction began in 1975 and finished in 2015. The target irrigation area of the dam is 17,267-ha in Rohtas and Kaimur district. The dam is 46.30 m high and 1615.4 m long. Two sections CH 21.0 at RD 640.09 m and CH 22.0 at RD 670.57 m are chosen for the purpose of reliability analysis. Himanshu and Burman (2017) reported a case study on Durgawati dam considering stability under steady and transient state seepage loading conditions. Table 2 shows the material properties of the dam which are also considered as mean values for analysis purpose. Figure 1 shows the geometric model of the dam. The geotechnical properties of the soil, i.e., cohesion, friction angle and unit weight, are determined in NIT Patna geotechnical laboratory from the design report of Durgawati dam.

The input data are cohesion (c), angle of internal friction (φ) and unit weight of soil (γ) for clay core, shell and foundation portion of the dam body. The material properties considered for filter and toe drain portion are angle of internal friction (φ) and unit weight (γ) only. These parameters for different parts of the dam are identified as c_{core}, φ_{core}, γ_{core}, c_{shell}, φ_{shell}, γ_{shell}, c_{fdn}, φ_{fdn}, γ_{fdn}, φ_{filter}, γ_{filter}.and φ_{toe}, γ_{toe}. The FOS values of the dam are determined for different combinations of these parameters for cases 1–8 outlined below. GeoStudio-07 is used for this purpose. The considered cases are:

Case 1 CH 21—Steady state seepage, downstream side.
Case 2 CH 21—Steady state seepage, upstream side.
Case 3 CH 21—Transient state seepage for 57 days in downstream side for drawdown simulation.
Case 4 CH 21—Transient state seepage for 57 days in upstream side for drawdown simulation.
Case 5 CH 22—Steady state seepage, downstream side.
Case 6 CH 22—Steady state seepage, upstream side.
Case 7 CH 22—Transient state seepage for 57 days in downstream side for drawdown simulation.

Table 2 Properties of different component of Durgawati earthen embankment dam

Soil designation	Effective cohesion (c)	Angle of internal friction (φ)	Unit weight (γ) (KN/m^3)
Embankment core	20	24.5	21.07
Embankment shell	20	24.5	21.07
Foundation	10	30	20
Filter	0	24.5	16.5
Toe drain	0	24.5	16.5

Case 8 CH 22—Transient state seepage for 57 days in upstream side for drawdown simulation.

For developing the MARS model, training and testing data sets are built. Training data use 70% of the entire data, whereas 30% are used as testing data set.

3.2 *Results of Analysis from MARS*

The results of reliability analysis of Durgawati dam using MARS as a regression tool is performed and presented in this section. Furthermore, the reliability index β is determined for steady state and transient state over the period of 57 days to simulate drawdown scenario for section CH 21.0 and CH 22.0, respectively. Basis functions obtained during construction phase are sent for pruning, and the final MARS model is developed. Suitable basis functions are derived in the constructive phase for developing MARS model. For all the eight cases listed above, the expressions of basis functions are presented in Table 3.

Here Y = FOS of the dam, C_m = coefficient of basis function and $B_m(x)$ is the basis function with x as the input parameters. Figures 2, 3, 4 and 5 show the performance of testing and training data set of the earthen dam for steady and transient state seepage condition at CH 21.0 and CH 22.0, respectively. The match between

Table 3 Expressions of the MARS

Section 1: CH 21.00

S. No.	cases	Equations of MARS
1	CH 21-Downstream side, steady state condition	$Y = 0.465 + \sum_{m=1}^{22} C_m B_m(x)$
2	CH 21-Upstream side, steady state condition	$Y = 0.315 + \sum_{m=1}^{24} C_m B_m(x)$
3	CH 21-Downstream side, transient state condition	$Y = 0.476 + \sum_{m=1}^{15} C_m B_m(x)$
4	CH 21-Upstream side, transient state condition	$Y = 0.388 + \sum_{m=1}^{18} C_m B_m(x)$
5	CH 22-Downstream side, steady state condition	$Y = 0.483 + \sum_{m=1}^{21} C_m B_m(x)$
6	CH 22-Upstream side, steady state condition	$Y = 0.700 + \sum_{m=1}^{6} C_m B_m(x)$
7	CH 22-Downstream side, transient state condition	$Y = 0.439 + \sum_{m=1}^{11} C_m B_m(x)$
8	CH 22-Upstream side, transient state condition	$Y = 0.614 + \sum_{m=1}^{9} C_m B_m(x)$

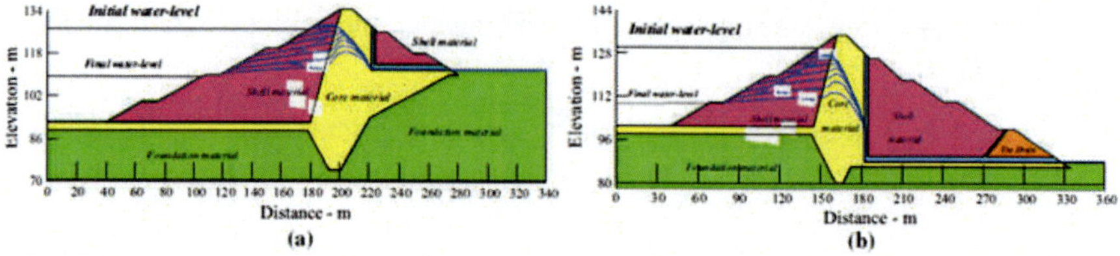

Fig. 1 Durgawati earthen dam at section **a** CH 21.00 and **b** CH 22.00 (Himanshu and Burman, 2017)

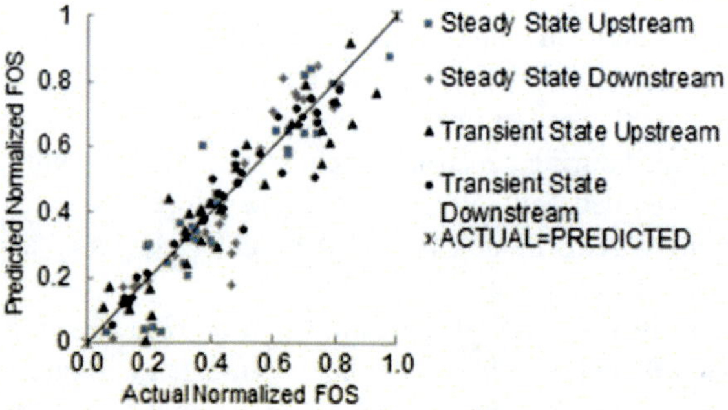

Fig. 2 Performance of testing data set at section CH 21.0

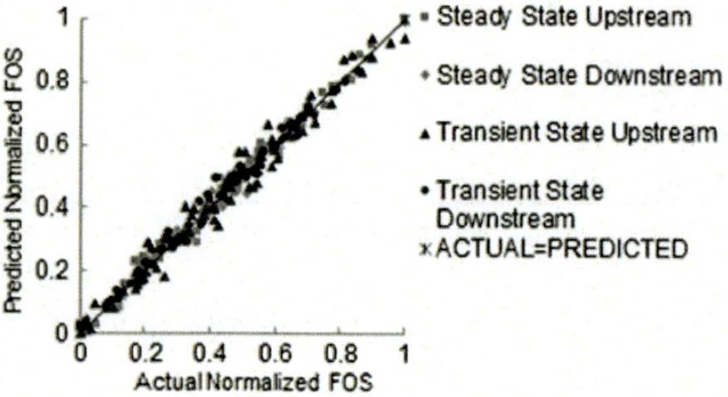

Fig. 3 Performance of training data set at section CH 21.0

predicted FOS by MARS and the actual FOS obtained by using GeoStudio-07 software shows that overtraining of data has been avoided and the MARS model is satisfactory.

3.3 Performance of MARS Model

The parameters such as RMSE, VAF, R^2, Adj. R^2, PI, WMAPE are identified and investigated to check the performance of MARS model in predicting the FOS value of the earthen embankment. The input parameters are cohesion, friction angle and unit weight. The results are presented in Tables 4 and 5.

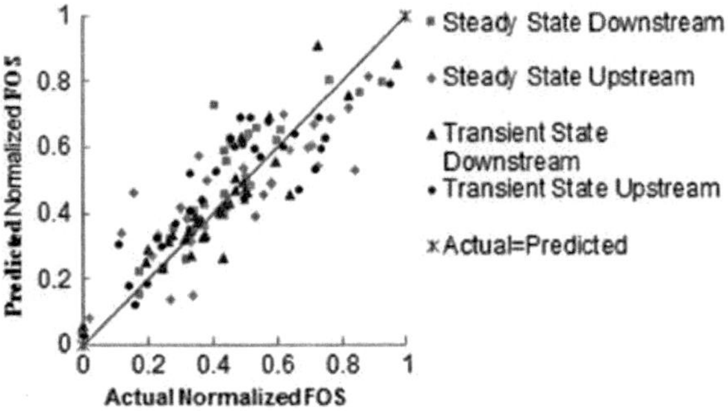

Fig. 4 Performance of testing data set at section CH 22.0

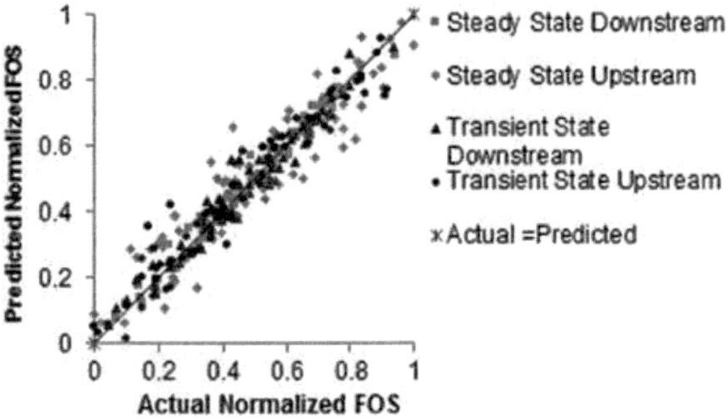

Fig. 5 Performance of training data set at section CH 22.0

Table 4 Model analysis for section CH 21.0

	Section CH 21: Training observed			
	Steady state		Transient state	
	Downstream	Upstream	Downstream	Upstream
Model	MARS	MARS	MARS	MARS
WMAPE	0.017	0.020	0.0189	0.033
RMSE	0.048	0.078	0.054	0.083
VAF	99.20	98.82	99.00	97.20
R^2	0.992	0.988	0.990	0.972
Adj. R^2	0.990	0.985	0.987	0.965
NS	0.992	0.988	0.990	0.972
PI	1.933	1.894	1.923	1.854
Section CH 21 Testing observed				
WMAPE	0.07	0.084	0.050	0.082
RMSE	0.227	0.313	0.169	0.204
VAF	83.52	81.15	91.13	83.68
R^2	0.873	0.848	0.910	0.837
Adj. R^2	0.771	0.725	0.837	0.705
NS	0.833	0.799	0.908	0.826
PI	1.379	1.223	1.579	1.337

Table 5 Model analysis for section CH 22.0

	Section CH 22: Training observed			
	Steady state		Transient state	
	Downstream	Upstream	Downstream	Upstream
Model	MARS	MARS	MARS	MARS
WMAPE	0.026	0.078	0.035	0.044
RMSE	0.067	0.276	0.093	0.119
VAF	97.90	86.14	96.06	94.14
R^2	0.978	0.86	0.960	0.937
Adj. R^2	0.973	0.826	0.951	0.922
NS	0.980	0.861	0.960	0.941
PI	1.885	1.411	1.818	1.744
Section CH 22: Testing observed				
WMAPE	0.070	0.117	0.068	0.095
RMSE	0.193	0.404	0.18	0.225
VAF	80.94	63.81	83.31	74.27
R^2	0.82	0.636	0.835	0.741
Adj. R^2	0.671	0.341	0.701	0.531
NS	0.795	0.637	0.833	0.727
PI	1.287	0.574	1.355	1.048

From Tables 4 and 5, it is observed that there are differences in the estimated performance parameters for training and testing part.

The GeoStudio-07 modules SEEP-W and SLOPE-W are utilized for seepage and stability analysis of the dam. The seepage and slope failure analysis of the earthen dam is performed using SEEP-W and SLOPE-W modules of GeoStudio-2007 software. In this study, Bishop's method is used to determine the minimum FOS. Two hypothetical scenarios are presented to study the reliability index and uncertainty of obtained values of FOS for slope analysis of Durgawati earthen dam using the MARS model. Different COV values as given in Tables 6 and 7 are considered, and the analyses

Table 6 Coefficient of variation of first scenario

Soil designation	Coefficient of variation of effective cohesion (%)	Coefficient of variation of angle of internal friction (%)	Coefficient of variation of unit weight (%)
Embankment core	40	40	10
Embankment shell	40	40	10
Foundation	40	40	10
Filter	40	10	10
Toe drain	40	20	10

Table 7 Coefficient of variation for second scenario)

Soil designation	Coefficient of variation of effective cohesion (%)	Coefficient of variation of angle of internal friction (%)	Coefficient of variation of unit weight (%)
Embankment core	35	35	5
Embankment shell	35	35	5
Foundation	35	35	5
Filter	35	5	5
Toe drain	35	15	5

are rerun. The comparative values of reliability indexes are shown in Figs. 6, 7 for CH 21.0 and in Figs. 8 and 9 for CH 22.0. The results show that the reliability indexes obtained in these situations for both training and testing data are favourably comparable.

The MARS model's performance for steady state condition in both upstream and downstream is very good. The performance of MARS in transient state upstream sections is found to be superior to the actual value at CH 21.0 as observed in Fig. 6. In steady state, downstream and upstream side as well as transient state downstream and upstream side is better as found in Fig. 7 for CH 21.0. The performance of actual model in steady state downstream and upstream side, transition state downstream and upstream is found better than FOSM based MARS at CH 22.0 for first scenario depicted in Fig. 8. The performance of actual model in steady state downstream,

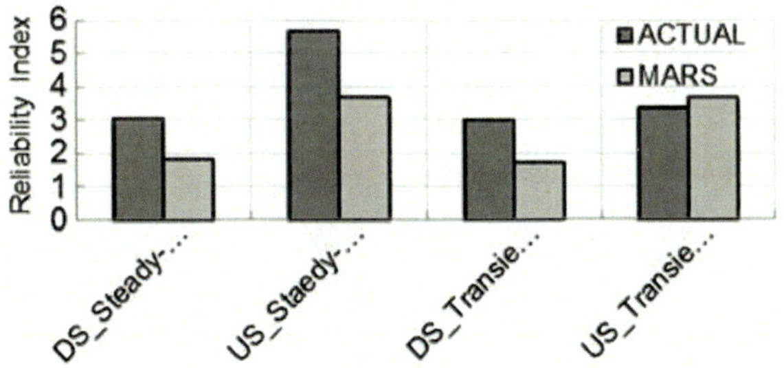

Fig. 6 Reliability index for section at CH 21.00

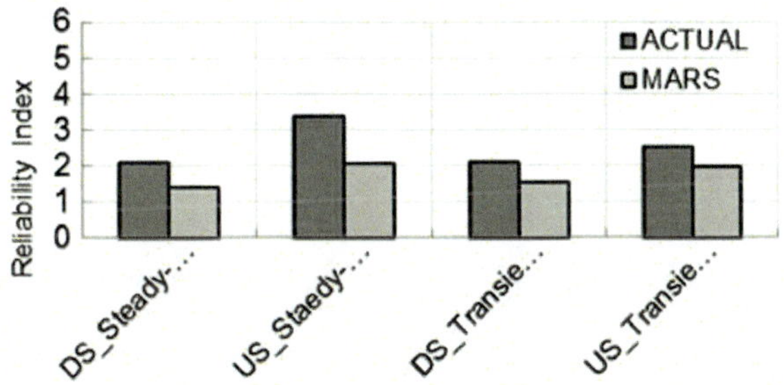

Fig. 7 Reliability index for section at CH 21.0

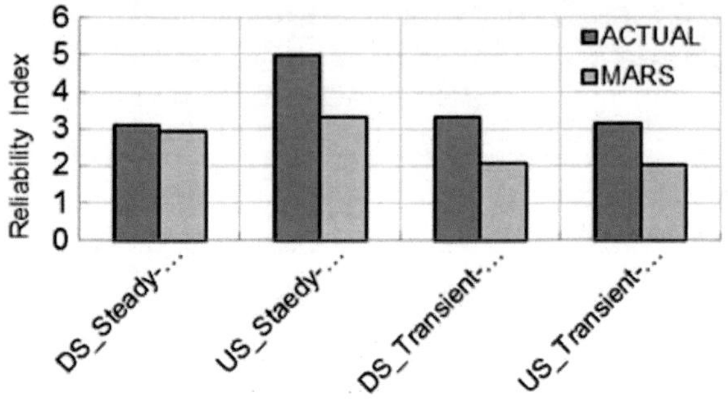

Fig. 8 Reliability index for section at CH 22.00

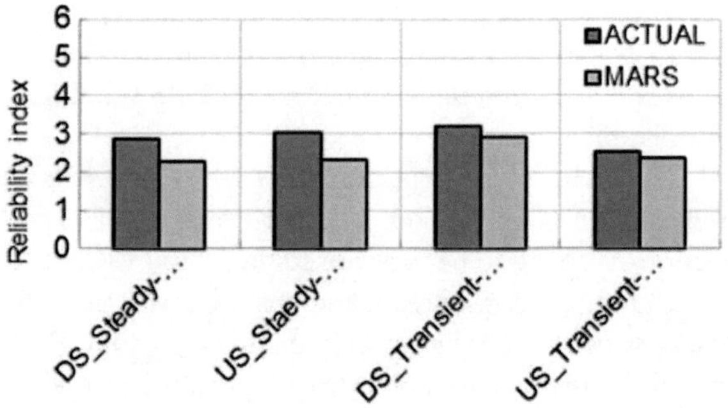

Fig. 9 Reliability index for section at CH 22.00

steady state upstream, transient state downstream and transient state upstream is better than FOSM based MARS at CH 22.0 for second scenario in Fig. 9.

4 Conclusions

MARS is applied successfully for determination of FOS of Durgawati earthen dam against slope failure for steady and transient state seepage conditions at section 2CH 21.0 and CH 22.0, respectively. The relevant methodology is presented in detail. It is found that MARS based FOSM model is able to satisfactorily predict the FOS value of the dam. Equations in the form of basis functions are developed by using MARS for this purpose which again shown excellent performance.

References

Baecher GB, Christian JT (2003) Reliability and statistics in geotechnical engineering. Wiley, West Sussex

Christian JT, Ladd CC, Baecher GB (1994) Reliability applied to slope stability analysis. J Geotech Eng 120(12):2180–2207

El-Ramly H, Morgenstern NR, Cruden DM (2002) Probabilistic slope stability analysis for practice. Can Geotech J 39:665–683

Friedman JH (1990) Multivariate adaptive regression splines. Ann Statist 19:1–67

Fox J (2002) Nonparametric regression. Appendix to an R and S-PLUS companion to applied regression. Sage, Thousand Oaks, CA

Griffiths DV, Fenton GA (2004) Probabilistic slope stability analysis by finite elements. J Geotech Geoenviron 130(5):507–518

Griffiths DV, Huang J, Fenton GA (2009) Influence of spatial variability on slope reliability using 2-D random fields. J Geotech Geoenviron 135(10):1367–1378

Gokceoglu C, Zorlu K (2004) A fuzzy model to predict the uniaxial compressive strength and the modulus of elasticity of a problematic rock. Eng Appl Artif Intell 17:61–72

Harr ME (1987) Reliability-based design in civil engineering. McGraw-Hill, New York

Himanshu N, Burman A (2017) Seepage and stability analysis of Durgawati earthen dam" A case study. Indian Geotech J, 1–20

Li DQ, Jiang SH, Cao ZJ, Zhou W, Zhou CB, Zhang LM (2015) A multiple response surface method for slope reliability analysis considering spatial variability of soil properties. Eng Geol 187:60–72

Lee IK, White W, Ingles OG (1983) Soil variability. Geotechnical Engineering, Pitman, Chapter 2

Liang RY, Nusier OK, Malkawi AH (1999) A reliability based approach for evaluating the slope stability of embankment dams. Eng Geol 54:271–285

Malkawi AIH, Hassan WF, Abdulla FA (2000) Uncertainty and reliability analysis applied to slope stability. Struct Saf 22:161–187

Melching CS, Yoon CG (1996) Key sources of uncertainty in QUAL2E model of Passaic River. J Water Resour Plan Manag 122(2):105–113

Nayak PC, Sudheer KP, Rangan DM, Ramasastri KS (2005) Short-term flood forecasting with a model. Water Resour Res 41:W04004. https://doi.org/10.1029/2004WR003562.

Roh GH, Hong HP (2009) Calibration of information-sensitive partial factors for assessing earth slopes. J Geotech Eng 4(3):93–102

Salgado R, Kim D (2014) Reliability analysis of load and resistance factor design of slopes. J Geotech Geoenviron 140(1):57–73

Wang CW, Chau KW, Cheng CT, Qiu L (2009) A comparison of performance of several artificial intelligence methods for forecasting monthly discharge time series. J Hydrol 374:294–306

Landslides Hazard Mapping Using High-Resolution Satellite Data

Saloni Jain, Rakesh Khosa, and A. K. Gosain

Abstract Landslides are one of the severe natural hazards induced by heavy rainfall, deforestation, slope failure and urban expansion. It can lead to significant loss of life and property in hilly and gully regions. Field studies that identify and map landslides are expensive and time-consuming as it includes the cost of the survey, travelling, workforce, and instrument. Although progression in technology and availability of high-resolution remote sensing data has now made it possible to identify landslides (satellite images and aerial photographs), accessibility to high-resolution satellite data is still an expensive and tedious procedure. Several studies have conducted in a GIS environment to map landslide zones, but the resolution of the open-source data is commonly coarse (30 m), which adds to the uncertainty of the outcome. In this study, application of the appropriate rule set with object-based image analysis (OBIA) technique has been used to identify landslides zones, through a combination of spectral, textural and geometrical properties of imagery and topographic data. It overcomes the shortcomings induced by pixel-based classification. For the current study, High spatial resolution data such as Google Earth imagery and CartoDEM (30 m) has been used. This approach shows an excellent prospect for quick and near-to-actual assessment of landslides zones which are generally induced by extreme rainfall events in the hilly regions of India. The methodology used has the potential to facilitate more reliable disaster management strategies. This study shows the potential of open-source data and emerging technology in the field of landslide assessment.

Keywords Landslides · Object-based image analysis · Google Earth

1 Introduction

Landslides count among the most exacting of natural hazards because of its potential to cause loss to human life as well as socio-economic disruption. Landslides occur over the area with sharp changes in relief and generally triggered by a range

S. Jain (✉) · R. Khosa · A. K. Gosain
Department of Civil Engineering, Indian Institute of Technology (IIT), Delhi, New Delhi, India
e-mail: cez178078@iitd.ac.in

M. Latha Gali and P. Raghuveer Rao (eds.), *Geohazards*, Lecture Notes in Civil Engineering 86, https://doi.org/10.1007/978-981-15-6233-4_7

of processes such as rainfall, earthquakes. Due to burst in the human population, anthropological activities are bound to continue to expand into landslide-prone environments; hence, the recognition of the scope and the magnitude of the hazard has increased (Cruden and Varnes 1996; De Blasio 2011; Woebbecke et al. 1995).

It is necessary to predict and map the possible landslide regions in order to mitigate the disaster. Researchers over the decades have presented different ideas about landslide mapping such as landslide inventory mapping is a representation of the spatial distribution of the landslides at a predefined cartographical scale (Malamud et al. 2004; Mayr et al. 2016); landside mapping can be categorised by displacement of vegetation together with unconsolidated material in surveying methods in mountain region (Motohka et al. 2010; Selby 1993).

Geospatial technologies are proven to be "New tool to an Old Problem" in the context of landslides. It can play an essential role in this field such as mapping of past or active slope failures, identifying the possible sites for a landslide, landslide zonation and predicting its time of occurrence. Geospatial technologies provide abilities to study the interrelations and interactions between the various disciplines to reveal the underlying fundamental processes fixed in the images from a distance (Platt and Rapoza 2008). The underline assumption in mapping landslide is the occurrence of this event leaves visible marks on the territory. Recently, detached landslide is easy to recognise on the satellite imagery due to their colour difference from the surroundings as they appear lighter in tone (Tsai et al. 2010; Santurri et al. 2010). However, as clock shifts, it gets harder to identify the boundary of slides due to fuzziness brought by environmental and physical processes, so optical remote sensing promises mapping of only shallow and recent landslides. Due to progressive enhancement in the resolution of satellite imagery, researches have inclined their attention towards the mapping of the landslide by using remote sensing the techniques over the conventional methods. However, the field observations are still the main bases of soil mapping as it captures the three-dimensional properties of Pedon (the smallest unit of soil). Currently, no satellite data ensures this level of detailed information. Advantages and limitations of conventional methods are exhaustively presented in previous reviews (Guzzetti et al. 2012; Weidner 2008).

The development in satellite optical imagery opened up new possibilities in the practice of visual interpretation for landslide investigation. Remote sensing can play an essential role in landslides inventory studies (Weidner 2008), especially in inaccessible and remote regions. The new remote sensing data could provide equivalent results even in areas where landslides have left faded marks only (Fiorucci et al. 2011). Due to the high price of data acquisition, these high-resolution images are generally utilised in a specific small region and are rarely applied in large ones (Guo et al. 2016).

Google Earth (GE) offers a welcome solution to the above-discussed issues; it provides open, high spatial resolution images suitable for landslide mapping. However, Google Earth images are restricted to a three-band colour code (R, G and B), which is expected to inferior the classification performance due to its poor spectral resolution (Zhan et al. 2005). The potential for the classification of spatial characteristics by Google Maps has been underestimated (Drăgut and Eisank 2010). By analysing the tone, texture and geometric features in a GE image, experts can recognise eroded surfaces with high assurance.

Image processing and analysis for marking landslides is a necessary process and wildly used in the field of disaster studies in order to mark the extent of the disaster. Over some time, several methods of feature extraction have been proposed. It includes manual and automatic extraction, later can be further classified as pixel-based and object-based image analysis. Pixel-based image classification classifies based on the brightness value of pixel while the object-based image analysis (OBIA) partition land-cover parcel into image objects and classified by expert rules. Image segmentation algorithms such as the multiresolution segmentation can be applied to create image objects, which serve as the basis for the classification process. A probable solution to the difficulties associated with pixel-based classification could be the need to function at the spatial scale of the things of concern themselves, moderately than depending on the extent of image pixels (Rau et al. 2011); hence, the methods of object-based image analysis (OBIA) for generating and updating geographical information are becoming more critical. Despite the advantages of OBIA, the visual interpretation of ortho-photos or pixel-based classification approaches are still the predominant methods used for mapping landslides. Due to the particular properties of landslides (e.g. shape) and the enhanced resolution of available imagery, pixel-based classification techniques tend to result in noteworthy commission and omission errors (Casagli et al. 2016). The outcomes of OBIA are more sensitive compared to the pixel-based classification process, which generally produces the pepper–salt effect (Sharma 2017; Scaioni et al. 2014). Apart from the mentioned problems, total reliability on brightness value may reduce the accuracy as two pixels with same spatial reflectance might be entirely different types of objects/ features (e.g. building and roads) or two pixels with very different reflectance may be part of the same object type (e.g. different rooftop materials of buildings).

Arithmetic operations with image bands can highlight specific object classes. Vegetation indices are often used to categorise vegetation and separate it from other categories (e.g. bare earth) in satellite data. The Excess Green Vegetation Index (ExG) is one of the most comprehensive indices if only bands in the visible range are presented (Zhan et al. 2005). It uses "panchromatic" brightness layer than brightness value (R, G and B). It was calculated for the multispectral images by dividing the sum of the three spectral bands by the number of bands. For the classification of landslides, this panchromatic layer was valuable exposure of bare ground because landslides appear brighter as compared to their immediate surroundings on the imagery (Hölbling et al. 2015; Mulders 1987).

In this paper, a methodology has been proposed to map shallow, eroded areas with a high level of detail and accuracy using Google Earth imageries and CartoDEM. An object-based image analysis approach is used to classify eroded surfaces from the surroundings, and also a new technique of accuracy assessment is applied which judges the location as well as shapes and size of the extracted objects with reference objects.

Fig. 1 Study area

2 Study Area and Data

The Kalsi is a beautiful village at the intersection of Yamuna and Tons rivers (Fig. 1). The 79.28 hectare study area is located approximately 49 kms (NE) from Dehradun city, India. It comprises pastoral hill country on moderately indurated Tertiary sandstone and mudstone, with relief in the order of 780 m above sea level. Most of the area falls in the zone of heavy rainfall—as a consequence, rain-triggered shallow landslide erosion is frequent incidence.

Hindustan Times published a report on 13 July 2017, 20:20 IST (Fig. 1) which focuses on the landslides incidences in this Kalsi region of Uttarakhand, especially following heavy monsoon rains and discussed in brief about Kalsi often hold up traffic for several days in the hilly areas. An image of the recent landslide is taken from the Google Earth of the area where the landslides have blocked the path. This portion affected by landslides of Kalsi district has been chosen as a study area. High spatial resolution Google Earth imagery is used, which consists of the spatial resolution 0.5 m. For extraction of drainage pattern and topographical attributes such as slope and elevation, CartoDEM has been used. CartoDEM is a freely available data provided by National Remote Sensing Centre, Hyderabad. The spatial resolution of data is 30 m.

3 Methods

The proposed OBIA methodology for landslide mapping is shown in Fig. 2. The detail of the developed algorithms and how to perform accuracy assessment is described in the following sections.

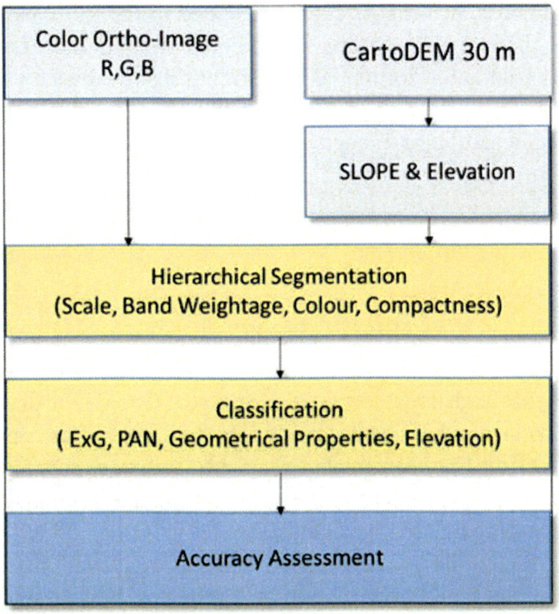

Fig. 2 Methodology framework for landslide mapping

3.1 Segmentation

Image segmentation which is an integral part of OBIA is a process of extraction of the similar objects that subdivide the image into dissected regions (Campbell 2011). The multiresolution segmentation is an optimisation technique which locally minimises the average heterogeneity of image objects for a particular resolution and maximises their respective homogeneity. The segmentation process used in the study is based on region merging algorithm, which begins with one pixel and consecutively merging neighbour pixels, based on the threshold of chosen scale, spectral and shape parameters. A larger-scale parameter will result in more significant-sized image objects. On the other hand, selecting a smaller scale will cause over-segmentation and small objects (Laliberte et al. 2012). Although there are tools available for scale parameter estimation, it is difficult to find a suitable value of the scale parameter without performing the trial-and-error test (Dragut and Blascke 2006).

Other parameters such as colour homogeneity are based on the standard deviation of the spectral values. The shape homogeneity depends on the variation of a compact (or smooth) shape. Homogeneity criteria can be customised by weighting shape and compactness criteria. The shape and colour criterion can be given up to the value of 0.9. This ratio determines the degree shape which influences the segmentation compared to colour—for example, a shape weighting of 0.6 results in a colour weighting of 0.4. In the same way, the value the compactness gives it a relative weighting against smoothness.

Different weights can be assigned to the R, G and B bands by their importance in the mapping. In this study, equal weights have been assigned to all the bands. By trial-and-error method, different sets of values for scale, colour and compactness are run on the images. By observing the outcomes, final sets of parameters for the segmentation have been chosen as scale = 36, shape = 0.8 and compactness = 0.3. The image has a very high spatial (resolution 0.5 m) while a low spectral resolution (3 bands) so more weight is assigned to the shape than colour.

3.2 Visual Interpretation of Landslides

Visual landslide interpretation from the Google Earth was carried out. Each landslide was then subdivided visually into "old landslides" and "new landslides", and also, an effort has been made to separate sediment sources (scars) from areas receiving sediment (debris tails). The ultimate aim of the mapping was to identify all landslides, including old and new landslides.

Regarding spatial location, the feature is well distributed and are covering areas where it could be landslides or in the regions that are landslides, but they have not been mapped due to errors or misinterpretation.

3.3 Classification

Automated thresholding for the classification with spectral features is implemented to the characteristics of each scene individually to classify the imageries.

Excess Green Vegetation Index (ExG)

Excess Green Vegetation Index (ExG; (Mayr et al. 2016; Yu and Gong 2011)) is used to classify eroded surface from the surroundings. Google Earth image consists of the red (*R*), green (*G*) and blue (*B*) colour bands which are used in the estimation of ExG. The mathematical expression of ExG is as follows in Eq. 1.

$$\text{ExG} = 2g - b - r \tag{1}$$

where

$$b = \frac{B}{(B+G+R)} \quad g = \frac{G}{(B+G+R)} \quad r = \frac{R}{(B+G+R)}$$

The data was acquired on 22 October 2017, so according to the phenological cycle of the grass and trees, their colours appear as dark green, while landslide areas appear as shades of brown. By observing the equation, it can understand that more weight is assigned to a green colour, so the object reflects more green band will have higher ExG value and vice versa. As the eroded surface has low reflectivity in the green band, the extracted ExG value is less than zero. These indices work better when there is an enhanced separability between the grass and eroded surface.

Green Red Vegetation Indices (GRVI)

This index is beneficial to extract vegetation from Google Earth images. As these images do not contain NIR band (vegetation tends to have a high reflectance in NIR bands), these indices use only red and green bands for extraction of vegetation (Rau et al. 2011). The mathematical expression for GRVI is as follows in Eq. 2.

$$\text{GRVI} = \frac{\text{Green} - \text{Red}}{\text{Green} + \text{Red}} \tag{2}$$

Its value ranges from -1 to $+1$, which is useful for differentiating forest area from the sparse vegetation as forest area has a higher GRVI value than sparse vegetation. As in the image, sparse vegetation appears brownish; it is very tedious to differentiate sparse vegetation from the eroded surface. Nevertheless, GRVI has a different value for each class as green vegetation (GRVI $>$ 0), soils (GRVI $<$ 0) and water/snow (GRVI close to 0) (Motohka et al. 2010). They are used to remove the sparse vegetation regions from the eroded surface regions.

Panchromatic (PAN)

An arithmetic average was calculated for the multispectral images by dividing the sum of the three spectral bands by three, which is termed as "Panchromatic". For the classification of landslides, this panchromatic layer was significant since landslide regions appear lighter in tone than their immediate surroundings on the photographs due to the removal of existing features which causes the exposure of bare ground (Hölbling et al. 2012). PAN was used to identify the landslides which were left due to positing ExG value. By referring to imagery, recent landslides appear as a bright object which shows that they have higher PAN value.

Other Characteristics

Object-based image analysis has various advantageous over pixel-based classification; it allows the user to work with other properties such as geometric and texture than spectral properties of images. Digital elevation model is used as an ancillary data for landslide detection as landslide generally occurs at steep slopes, so the integration of digital elevation models (DEM) and its derivatives (e.g. slope, drainage pattern) add to the accuracy of the extraction.

Geometrical properties such as length/width ratio are proven to be very useful in the extraction of road feature. In the image, in some places it is difficult to distinguish between the road and eroded surface as they both have high reflectance properties which lead to the misclassification of the way into the eroded surfaces, but road is a linear feature, so its L/B ratio is higher than the other classes. By training the segmentation algorithm according to the mentioned properties, roads can be classified differently from the eroded surfaces.

3.4 Accuracy Assessment

The assessment was carried out by a few parameters of the discrepancy methods. The assessment has done by the reference training object which was digitised manually by visual interpretation of imagery. This method focused on the area parameters of both segmented and reference objects. The area-based method assesses the accuracy of objects based on both location and geometry. Several parameters have been used in the study: (1) the relative area of an overlapped region to a reference object (RA_{or}), (2) the relative area of an overlapped region to a segmented object (RA_{os}), (3) the quality rate (qr), (4) the SimSize, given as below (Taylor et al. 2015). Each factor judges the efficiency of segmented objected concerning the reference object. The mathematical expression of each element is shown in Eqs. 3, 4, 5 and 6, respectively.

$$RA_{or}\% = \frac{1}{n}\sum_{i=1}^{n}\frac{A_o(i)}{A_r} \times 100\% \tag{3}$$

$$RA_{os}\% = \frac{1}{n}\sum_{i=1}^{n}\frac{A_o(i)}{A_s(i)} \times 100\% \tag{4}$$

$$qr = \frac{1}{n}\sum_{i=1}^{n}1 - \frac{A_o(i)}{A_u(i)} \tag{5}$$

$$\text{SimSize} = \frac{1}{n}\sum_{i=1}^{n}\frac{\min\left(A_r, Aa_{\ S(i)}\right)}{\max\left(A_r, A_{S(i)}\right)} \tag{6}$$

where n is the number of segmented objects created by an algorithm. A_r is the area of the reference object, A_s is the area of the ith segmented object, $A_o(i)$ is the area of the ith overlapped region associated with the reference object, and $A_u(i)$ is the area of the union between the references object and the ith segmented object. RA_{or} and RA_{or} assess the accuracy by measuring the overlay region between the reference and segmented objects. When objects are well-segmented, the overlapping area will be more so both RA_{or} and RA_{os} values will be close to 100. The quality rate parameters focus on both the geometry and location of the segmented objects. qr parameter (Weidner 2008) ranges between 0 and 1. The values close to zero indicate a perfect match while values close to one indicate an over- or under-segmentation. The SimSize (Zhan et al. 2005) measures the similarity regarding the size of the segmented object. It ranges between 0 and 1, with one being ideal. It assesses how accurately segmented objects have preserved the shape of the feature. The more same object will have a higher value (close to 1) and vice versa.

4 Results and Discussions

The outcome of the study is divided into two portions, as follows: landslide mapping and accuracy assessment.

4.1 Landslides Mapping and Factors Affecting

A necessary input to perform classification is objects. Objects are the group of the pixel that shares similar characteristics according to a prespecified threshold. Segmentation is performed to convert pixels into objects. Segments produced for a scene by seeded region growing are shown in Fig. 3. The parameters used for creating segments are scale, shape and compactness. By applying hit-and-trial method, parameter scale = 36, shape = 0.9 and compactness = 0.1 have given a better outcome though segmented objects were smaller as compared to the purposes of interest (eroded areas). All the three bands were given an equal weighting.

The classification of the image objects was executed by using membership functions, based on fuzzy logic theory combined with user-defined rules. Rule sets for classification are as follows: ExG has been used to extract eroded surface (possible landslides). For the eroded surface, objects have been trained such that all purposes which are having ExG value less than zero and the slope value higher than 25 are classified as the eroded surface. Some objects that belong to road class are misclassified as landslides due to the similarity in reflection with the eroded surface. That pixel can be removed by considering geometrical properties like length/width ratio.

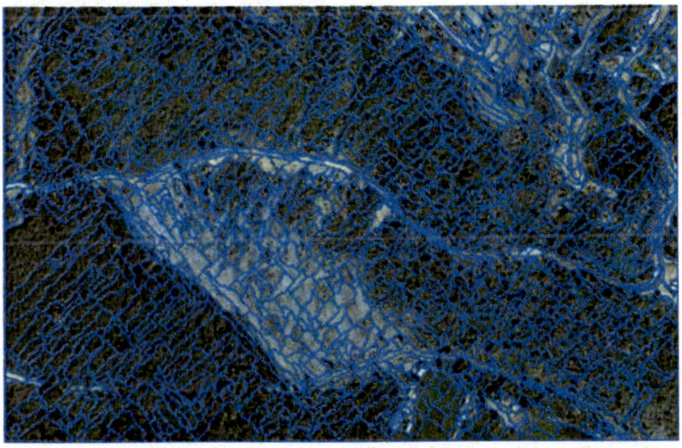

Fig. 3 Results of the segmentation

As the road is a linear feature, objects were trained such that all object which is having a length/width ratio higher than 3.5 should be classified as roads. To extract and discriminate both forest and sparse vegetation classes, GRVI indices are used. As forest will have high GRVI value (greater than 0.4) than sparse vegetation (0.1–0.2) and for rest, the average GRVI value was approximately 0.01–0.03. To detect freshly detached landslide, PAN is used as they tend to have a higher reflection, but they consist of positive ExG value due to low red band reflection. So all the objects which are having PAN value higher than 175 are classified as a freshly detached landslide, and they were later on merged with the landslide class. The detailed imagery is shown in Fig. 4b. As built-up is in a regular shape, so by using the geometrical properties of shape index, they are classified. A detached portion of land is represented as a yellow colour. A total of 13 landslides were observed from the imagery. While keeping the previous year imagery (Fig. 4a), there was a minimal trace of landslides. These figures help in understanding the magnitude of the disaster.

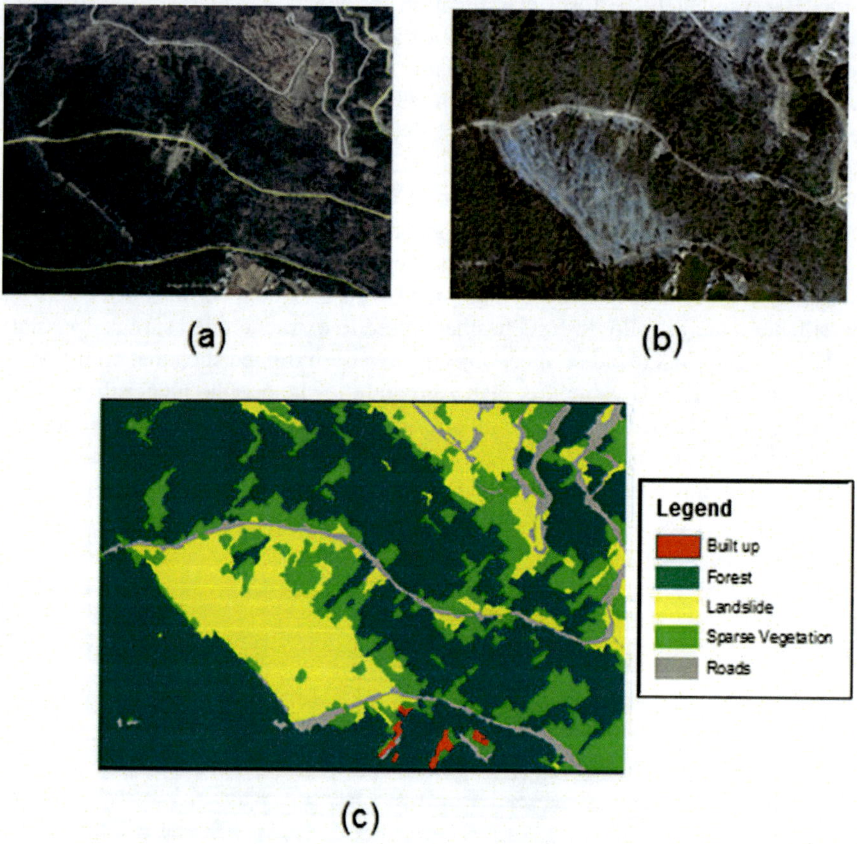

Fig. 4 **a** and **b** Sample image (May, 2016 and October, 2017) and **c** classified image

Fig. 5 Topography and drainage pattern of the study area

To gather more information regarding the occurrence of the landslide, few features have been obtained. By using the watershed delineation tool of ArcSWAT, drainage patterns are extracted, shown in Fig. 5. A drainage pattern is defined as topographical features from which a stream gets runoff, through flow, and groundwater flow which can be divided by topographic barriers called a watershed. By observing the drainage pattern keenly (Fig. 5), it has appeared that most of the reaches pass through the landslide area, which might indicate that more runoff passed through it during heavy rainfall of July 2017 which may further lead to detachment of surface. Also, slopes greater than 25° are more prone to a landslide as most of the detached area met this criterion.

One more observation is reflected from the imagery that landslide detachment occurs near to the road network. That confirms the frown upon that "anthropogenic activities are causing landslides". In the previous year image also, there were traces of landslides near the road network that is later on translated to greater disaster due to heavy rainfall, though there can be many other factors such as geological and hydrological behind this landslide other than above-discussed reasons.

4.2 Accuracy Assessment

Most of the classified scencs contain one central eroded area and sometimes a couple of small ones. To assess the consequences of the automated classification, they are compared to a manual classification of the nine scenes based on the same Google Earth scene. By observing Fig. 6, it can be seen that most of the landslides are mapped (location-wise) as compared to the reference, but the shape and size of the objects are slightly different. Overall accuracy is 96.99% if the conventional method of landslides assesses accuracy.

When comparing the total area of mapped landslides between the semi-automated object-based mapping and manual mapping, only minor differences are detected (Fig. 7). A slight trend towards overestimating the landslide area with OBIA compared to manual mapping can be recognised.

Fig. 6 Extracted and reference landslides

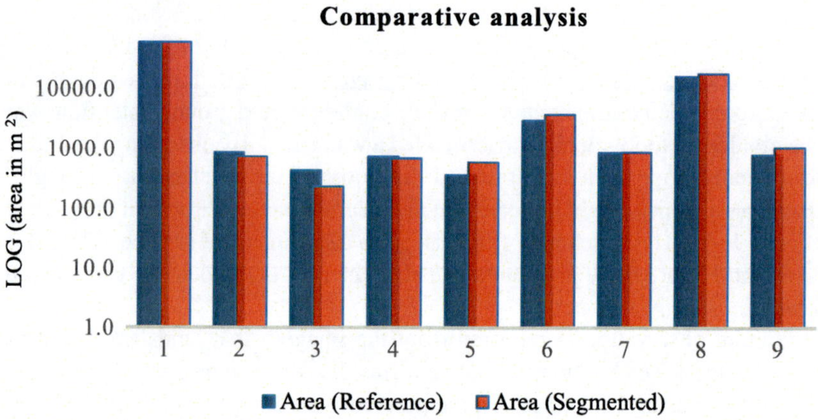

Fig. 7 Comparative analysis between the reference and extracted possible landslides

Accuracy assessment is performed on the area-based method included the various parameters into consideration such as (1) the relative area of an overlapped region to a reference object (RA_{or}), (2) the corresponding area of an overlapped region to a segmented object (RA_{os}), (3) the quality rate (qr) and (4) the SimSize. The outcomes reflect a different scenario. A total of nine samples are taken for the analysis. For parameters RA_{or} and $RA_{os,}$ extracted objects indicate a low accuracy of 79.77 and 75.44%, respectively.

The reason could be understood by observing Fig. 6 as extracted objects are very irregular in shape and continuity as compared to reference objects that lead to a smaller the overlay region which lowers the accuracy of these parameters. The qr parameter value lies between 0 and 1. The values close to zero indicate a perfect match, i.e. overlay and union areas are equal, while values close to one indicate an over- or under-segmentation, i.e. overlay region is tiny as compared to the union because object lies away from each other. The value obtained is 0.29, which indicates a functional similarity between overlay and union regions. The SimSize processes

Table 1 Accuracy assessment of Landslide area extraction

S. No	RA(or) %	RA (os) %	qr	SimSize
1.00	86	87	0.36	0.99
2.00	64	76	0.33	0.85
3.00	89	84	0.23	0.53
4.00	71	64	0.31	0.95
5.00	88	72	0.35	0.60
6.00	80	84	0.32	0.80
7.00	65	64	0.29	0.99
8.00	81	73	0.22	0.89
9.00	76	75	0.24	0.75
Overall	79.77	75.44	0.29	0.82

the similarity regarding the size of the *i*th segmented object and ranges between 0 and 1, with 1, being ideal parameters considering both geometric and spatial properties of the feature into consideration. SimSize value is 0.82, which indicates the geometrical similarities of objects are quite high.

The manual approach demonstrations gains for delineating single landslides or splitting up multiple landslides into smaller landslide regions. This is a challenge in OBIA since objects produced through segmentation rarely correspond to single landslides due to scale issues which correspond to over- or under-segmentation. Under-segmentation occurs when two or more segments may represent a single object, and over-segmentation occurs when a single segment may contain several objects, respectively (Clinton et al. 2008). Advanced split and merge algorithms could be used to refine the delineation of image objects, or manual editing can be done, but it is a tedious and time-consuming process. However, the creation of "meaningful" objects about a particular context or aim can be very complicated (Blaschke et al. 2014). Thus, instead of associating the absolute number of mapped landslides, the overlapping area was used for calculating the mapping accuracy in this study. Although results from manual mapping performed by local experts are often the only reference available, they cannot constitute an utterly accurate reference as their generation depends on various factors. Above-mentioned factors have to be well-thought-out when interpreting accuracy values (Table 1).

5 Uncertainty in Outcome

Remote sensing methods work on the basic principle of optics such as reflection, refraction and absorption of radiation. How any object will respond to a specific wavelength could be understood by its spectral reflectance curve. The reflected radiation is further converted to a brightness value and is a necessary input to automatic extraction of features. If the reflection curve of soil is seen, then few observation can be noted: (1) higher reflection for longer wavelength (Till 2.2 μm) and (2) absorption band (1.4, 1.9 and 2.2 μm) (Fabre et al. 2015). It implies that all the signification outcome lies in the longer wavelength. Now coming to the study, Google Earth imagery is used for feature extraction that comprises only visible bands, which indicates loss of valuable information as for soil reflection increases steadily towards longer wavelength in the visible region, but so does construction material such as concrete and asbestos. Due to this problem, a significant amount of misclassification can be observed between roads, built-up and possible landslides. It was the main reason behind the inclusion of barren land into a potential landslide. This caused uncertainty in the outcome.

As ancillary information, digital elevation model is used to map possible landslides. However, CartoDEM has a spatial resolution of 30 m which is larger than the spatial resolution of Google Earth 0.46 m. It dilutes the information for each pixel. The significant difference in spatial resolution of both layers makes the outcome from overlay operation unreliable. This error can be reduced to a great extent by using high-resolution stereo pair data such as IKONOS, Cartosat, but they carry a limitation of cost and availability within.

Object-based image analysis was used for the automatic extraction of various land uses/land covers; it does not promise to replicate the real scenario; it just helps in portraying a crude representation of reality. Main drawbacks of visual interpretation are the uncertainty of outputs, the subjectivity and the strict dependence on human expertise. The reference layer is prepared through manual mapping, which is generally accurate within the limits of image quality, is a very time-consuming process and also its level of accuracy depends on the efficacy of the user. This has to be considered when reading accuracy values. Reference layer (manually digitised) cannot be viewed as a real layer as landslide depends on the various geological, anthropogenic and hydrological factors which cannot be identified through visual interpretation. In this study, all of the elements which trigger the landslides are not extracted. So field survey is must to testify the outcomes.

6 Conclusion

The OBIA has been proven as a promising method for landslides identification and classification. An advantage of the OBIA approach, compared to a pixel-based classification, is that it copes with the salt–pepper noise of the high-resolution data, and also it considers spectral properties as well as shape and texture of the objects. In order to increase accuracy, more attention should be given to the classification rules and the type of data in the analysis. The lack of the DEM with a spatial resolution equivalent to the Google Earth and the lack of a more detailed lithological and geological map have brought many uncertainties in the classification process.

In accuracy assessment, the developed landslide detection algorithm achieved 79.77, 77.54, 29 and 82% for parameters RA_{or}, RA_{os}, qr and SimSize, respectively. Location-based accuracy parameter, also known as overall accuracy, reflects 96.7% accuracy. The massive difference between the outcomes is due to consideration of different factors in the analysis. Area-based accuracy methods give more reliable information as it captures both geometry and location into account, while the later focuses on the location only.

The study demonstrates that landslide mapping can be done by using open-source data with considerable accuracy.

References

Blaschke T, Hay GJ, Kelly M, Lang S, Hofmann P, Adding E (2014) Geographic object-based image analysis—towards a new paradigm. ISPRS J Photogram Remote Sens 87:180–191

Campbell JB, Wynne RH (2011) Introduction to remote sensing, 5th edn. The Guilford Press, New York, 667 pp

Casagli N, Cigna F, Bianchini S, Hölbling D, Füreder P, Righini G, Del Conte S, Friedl B, Schneiderbauer S, Iasio C, Vlckoh J, Greif V, Proske H, Granica K, Falco S, Lozzi S, Mora O, Arnaud A, Novali F, Bianchi M (2016) Landslide mapping and monitoring by using radar and optical remote sensing: examples from the EC-FP7 project SAFER. Remote Sens Appl Soc Environ 4:92–108

Clinton N, Holt A, Yan L, Gong P (2008) An accuracy assessment measure for object-based image segmentation. Int Arch Photogramm Remote Sens Spatial Inf Sci XXXVII(Part B4)

Cruden DM, Varnes DJ (1996) Landslide types and processes, Special Report, Transportation Research Board. Natl Acad Sci 247:36–75

De Blasio FV (2011) Introduction to the physics of landslides. Springer, Dordrecht, Heidelberg, London, New York

Dragut L, Blascke T (2006) Automated classification of landform elements using object-based image analysis. Geomorphology 81:330–344

Drăgut L, Eisank C (2010) ESP: a tool to estimate scale parameters for multi-resolution image segmentation of remotely sensed data. Int J Geogr Inf Sci 24(6):859–871

Fabre S, Briottet X, Lesaignoux A (2015) Estimation of soil moisture content from the spectral reflectance of bare soils in the 0.4–2.5 μm domain. Sensors 15:3262–3281

Fiorucci F, Cardinali M, Carlà R, Rossi M, Mondini AC, Santurri L, Ardizzone F, Guzzetti F (2011) Seasonal landslide mapping and estimation of landslide mobilization rates using aerial and satellite images. Geomorphology 129:59–70

Guo Z, Shao X, Xu Y, Miyazaki H, Ohira W, Shibasaki R (2016) Identification of village building via Google Earth images and supervised machine learning methods. Remote Sens 8:271

Guzzetti F, Mondini A, Mondini A, Cardinali M, Fiorucci F, Santangelo M, Chang K (2010) Landslide inventory maps: new tools for an old problem. Earth-Sci Rev 112:42–66

Hölbling D, Füreder P, Antolini F, Cigna F, Casagli N, Lang S (2012) A semi-automated object-based approach for landslide detection validated by persistent scatterer interferometry measures and landslide inventories. Remote Sens 4:1310–1336

Hölbling D, Friedl B, Eisank C (2015) An object-based approach for semi-automated landslide change detection and attribution of changes to landslide classes in northern Taiwan. Earth Sci Inf 8(2):327–335

Laliberte AS, Browning DM, Rango A (2012) A comparison of three feature selection methods for object-based classification of sub-decimeter resolution Ultracam-L imagery. Int J Appl Earth Obs 15:70–78

Malamud BD, Turcotte DL, Guzzetti F, Reichenbach P (2004) Landslide inventories and their statistical properties. Earth Surf Process Landforms 29:687–711

Mayr A, Rutzinger M, Bremer M, Geitner C (2016) Mapping eroded areas on mountain grassland with terrestrial photogrammetry and object-based image analysis. In: ISPRS annals of the photogrammetry, remote sensing and spatial information sciences, Volume III-5, XXIII ISPRS Congress, Prague, Czech Republic, vol 5, pp 137–144

Motohka T, Nasahara KN, Oguma H, Tsuchida S (2010) Applicability of green-red vegetation index for remote sensing of vegetation phenology. Remote Sens 2:2369–2387

Mulders MA (1987) Remote sensing in soil sciences. Elsevier Science Publishers B.V, The Netherlands

Platt RV, Rapoza L (2008) An evaluation of an the object-oriented paradigm for land use/land cover classification. Prof Geogr 60:87–100

Rau JY, Jhan JP, Lob CF, Linb YS (2011) Landslide mapping using imagery acquired by a fixed-wing UAV. In: International archives of the photogrammetry, remote sensing and spatial information sciences, Volume XXXVIII-1/C22, ISPRS Zurich Workshop, Zurich, Switzerland, pp 195–200

Santurri L, Carlà R, Fiorucci F, Aiazzi B, Baronti S, Cardinali M, Mondini A (2010) Assessment of very high-resolution satellite data fusion techniques for landslide recognition. Int Arch Photogramm Remote Sens Spat Inf Sci 38(B7):492–496

Scaioni M, Longoni L, Valentina Melillo V, Monica Papini M (2014) Remote sensing for landslide investigations: an overview of recent achievements and perspectives. Remote Sens, 6, 1-x manuscripts. https://doi.org/10.3390/rs60x000x

Selby MJ (1993) Hillslope materials and processes. Oxford University Press, Oxford

Sharma N (2017) Frequent landslides in Uttarakhand worries scientists. Hindustan Times, July 17

Taylor P, Montaghi A, Larsen R, Greve MH (2015) Accuracy assessment measures for image segmentation goodness of the Land Parcel Identification System (LPIS) in Denmark. Remote Sens Lett 4(10):946–955

Tsai F, Hwang JH, Chen LC, Lin TH (2010) Post-disaster assessment of landslides in southern Taiwan after 2009 Typhoon Morakot using remote sensing and spatial analysis. Nat Hazard Earth Syst 10:2179–2190

Weidner U (2008) Contribution to the assessment of segmentation quality for remote sensing applications. In: Proceedings of the 21st congress for the International Society for Photogrammetry and Remote Sensing, 3–11 July, Beijing, China

Woebbecke DM, Meyer GE, Von Bargen K, Mortensen DA (1995) Colour indices for weed identification under various soil, residue, and lighting conditions. Trans ASAE 38(1):259–269

Yu L, Gong P (2011) Google Earth as a virtual globe tool for earth science applications at the global scale: progress and perspectives. Int J Remote Sens 33:366–3986

Zhan Q, Molenaar M, Tempfli K, Shi W (2005) Quality assessment for geospatial objects derived from remotely sensed data. Int J Remote Sens 26(14):2953–2974

Analysis of Rainfall-Induced Slope Failure Using Monte Carlo Simulations: A Case Study

A. S. S. Raghuram and B. Munwar Basha

Abstract Climatic changes influence the depth of slip surface drastically. Slope failures particularly, failures caused by rainfall, slip surface lie above the water table (i.e., in unsaturated zone). In this zone, suction develops, which increases the slope stability in dry conditions. During rainfall, water infiltrates into the soil pores due to which suction stress decreases significantly, and thereby, stability of slope decreases. Relation between soil suction and soil water content can be represented using soil water characteristic curve (SWCC). The measured SWCC, hydraulic conductivity (k_s) of the soil are inherently variable. Conventional approaches may render unrealistic safety of the slope sometimes, as they do not account for variability. Therefore, it is highly appropriate to develop a reliability-based design (RBD) for stability analysis of rainfall-induced slopes. In this study, an efficient reliability analysis using Monte Carlo Simulations (MCS) is presented for rainfall-induced slope failures. A case study on shallow slope failures near Seattle, Washington, which are of sandy colluvium deposits on steep coastal hill slopes, is presented.

Keywords Unsaturated soil · Shear strength · Rainfall induced slopes · Suction stress · Monte Carlo simulations

1 Introduction

The depth of slip surface is significantly influenced by the climatic changes. Rainfall-induced slope failure problem is a global problem. They are quite common in tropical and subtropical climates and in other parts of the world (Zhang et al. 2016). In case of rainfall triggered slope failures, slip surface lies in the unsaturated zone. Soil water characteristics curve (SWCC) governs the behavior of unsaturated soil. SWCC

A. S. S. Raghuram · B. M. Basha (✉)
Department of Civil Engineering, Indian Institute of Technology Hyderabad, Kandi, Sangareddy 502285, India
e-mail: basha@ce.iith.ac.in

A. S. S. Raghuram
e-mail: raghuram.ammavajjala@gmail.com

M. Latha Gali and P. Raghuveer Rao (eds.), *Geohazards*, Lecture Notes in Civil Engineering 86, https://doi.org/10.1007/978-981-15-6233-4_8

relates the soil water content to the soil suction stress (Fredlund et al. 2012). Limit equilibrium (LE) methods have been used for most of the slope stability analysis in which suction stress is not quantified. In LE method, the most common assumption made is that the pore water pressure is positive or zero along the failure plane. However, in reality under rainfall conditions, pore pressure within the slope could be highly variable, and it could be negative as well. In dry conditions, suction stress increases and thereby increases the slope stability. However, LE methods do not account for suction stress and rainfall characteristics such as hydraulic conductivity and rainfall infiltration.

Therefore, it is appropriate to modify the shear strength equation, in order to accommodate suction stress and rainfall characteristics for estimating the slope stability of rainfall-induced slope failures. Alonso (1976) reported that the shear strength parameters, suction stress and hydraulic conductivity in a slope are highly variable. Due to the uncertainties associated with rainfall-induced slopes, it is not logical to apply the same LE approach. Therefore, it is highly necessary to develop a probabilistic method, which provides a quantitative and systematic tool to address the influence of these uncertainties. In recent decades, successful contributions to slope reliability analysis have been achieved. Previous studies (Li and Lumb 1987; Griffiths and Fenton 2004; Low 2005; Zhang et al. 2011; Li et al. 2015; Jiang et al. 2015) indicate that Monte Carlo Simulations (MCS) is the most efficient and popular approach, as it provides unbiased reliability results.

This paper aims to propose a theoretical framework for the reliability analysis using MCS of the rainfall-induced slope failure. A shallow slope failure at Seattle area near Washington, USA, is presented as a case study to examine the reliability indices. A performance function for calculating reliability index is formulated against sliding failure of infinite slope. In addition, the results are presented in the form of design charts from which the designer can choose the reliability index value required to ensure the stability of a particular slope.

2 Past Studies on Reliability Analysis of Rainfall-Induced Slopes

A number of parametric analysis have been conducted to examine the influence of soil permeability and various rainfall events on the slope stability (Ng and Shi 1998). The authors found that the matric suction has reduced due to rainwater infiltration. Lu and Griffiths (2004) developed suction stress equation based on the SWCC and soil permeability characteristic curve. Babu and Murthy (2005) performed the sensitivity analysis and presented the reliability index for unsaturated soil slopes. Lu and Godt (2008) presented a generalized framework for the stability of infinite slopes. Authors developed the framework under steady-state unsaturated seepage conditions. In addition, the authors presented a case study of shallow slope failures of sandy colluvium near Seattle, Washington.

Oh and Vanapalli (2010) considered the rainfall infiltration and found the stability analysis for compacted soil slopes. Authors have performed the slope stability analysis for different scenarios including short-term and long-term stability. Oh and Lu (2015) presented two case studies of rainfall triggered slope failures. Authors expanded finite element methods and the traditional LE for unsaturated conditions. Chiu et al. (2012) presented the reliability of the soil SWCC using Bayesian framework. Zhang et al. (2016) reported that the fitting parameter 'α' (inverse of parameter 'a') and 'n' are the uncertain parameters and are normally distributed. Raghuram and Basha (2017) have performed the stability of rainfall triggered slope failure in the light of reliability. The authors concluded that the suction stress variation with respect to the distance between slip surface and water table depends up on the rainfall intensity and SWCC fitting parameters.

Table 1 presents the statistical values of the hydraulic conductivity reported in the literature. From Table 1, it can be observed that the coefficient of variation (COV) of k_s varies significantly. The reported values of COV of k_s range from 50 to 240%, which shows high degree of variability exist. Phoon et al. (2010) presented the mean and COV values of SWCC fitting parameters 'α' and 'n' and are shown in Figs. 1, 2, 3 and 4. It can be observed from Figs. 1 and 3 that the minimum and maximum COVs of fitting parameters 'α' and 'n' vary from 81 to 152% and 9 to 19%, respectively. The above discussion warrants that the variability associated with fitting parameters of SWCC cannot be ignored and should be given due consideration while estimating the stability of unsaturated infinite slopes.

Table 1 Statistics of k_s available in the literature

Author	k_s	
	Mean (m/s)	COV (%)
Zhang et al. (2014b)	1.59×10^{-6}	86.9
Sahis et al. (2014)	2.50×10^{-7}	90–240
Zhang et al. (2014a)	8.00×10^{-6}	50

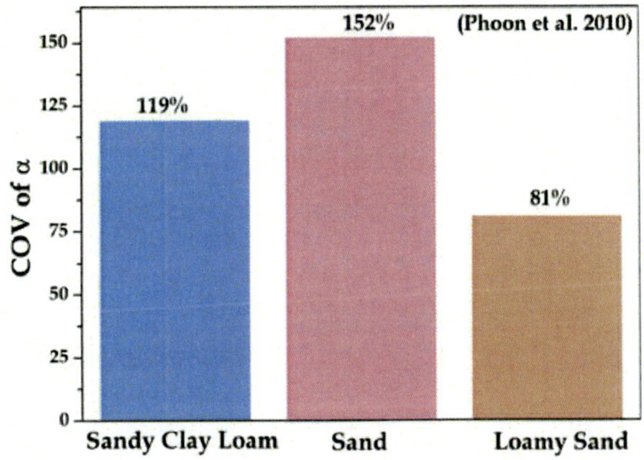

Fig. 1 COV of SWCC fitting parameter α for three different soils

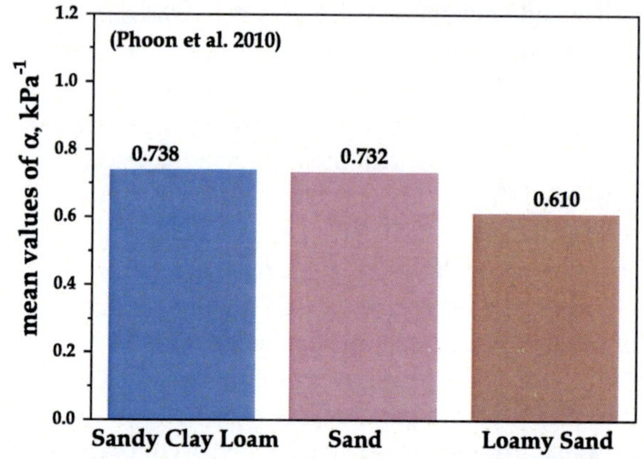

Fig. 2 Mean values of SWCC fitting parameter '*α*' for three different soils

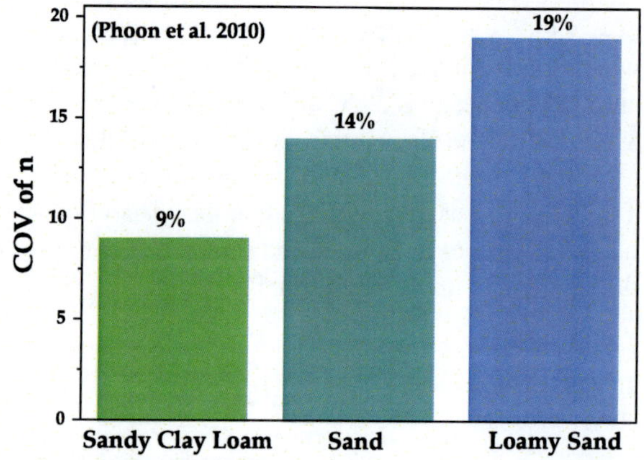

Fig. 3 COV of SWCC fitting parameter '*n*' for three different soils

Fig. 4 Mean values of SWCC fitting parameter '*n*' for three different soils

3 Methodology

The geometry and properties of the failed slope near Seattle area, Washington, USA, are shown in Fig. 5. In Fig. 5, H_w and H represent the depths of the water table and slip surface. The depth of water table is fixed at 5 m from the ground level. In general, it is assumed that the friction angle is constant throughout the slope as per classical infinite slope theories. However, in the field, this assumption is rarely valid. This assumption is not valid, particularly, when the formation of hill slopes is due to compaction, mechanical transport and weathering (Lu and Godt 2008). Therefore, a constant value of friction angle cannot be assumed. The shear strength for shallow soil (<2 m) is mainly due to inter-particle friction. Lu and Godt (2008) have given Eq. 1 to calculate the variation in friction angles with respect to the weathered zone.

$$\phi' = \phi'_o + \frac{\Delta\phi}{1 + \frac{Z_w}{H}} \tag{1}$$

where ϕ'_o = frictional angle at the ground surface, $\Delta\phi$ = range of frictional angle variation within the weathered zone, Z_w = depth of weathered zone which is 0.5 m in the present study. The Eq. 1 is used to take care of the variation of friction angle with respect to the weathered zone. Bishop (1959) presented the unsaturated shear strength (τ_{us}) and can be expressed as shown in Eq. 2.

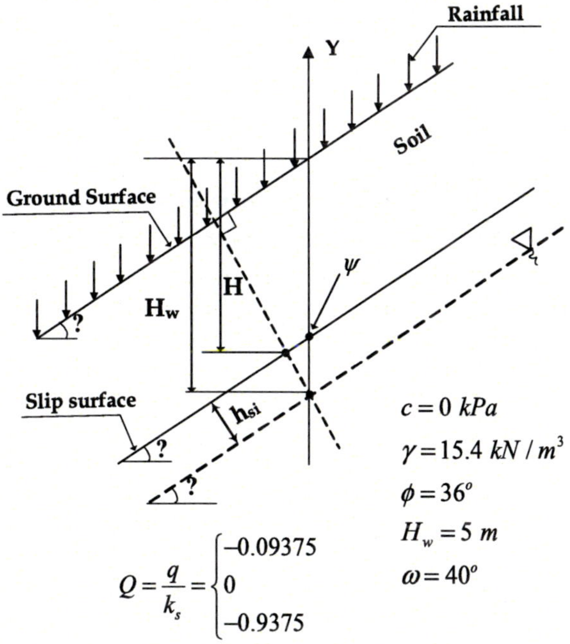

Fig. 5 Geometry and properties of failed slope

$$\tau_{\mathrm{us}} = c' + (\sigma_{\mathrm{f}} - u_{\mathrm{a}}) \tan \phi' + \chi (u_{\mathrm{a}} - u_{\mathrm{w}}) \tag{2}$$

where c' = effective cohesion, σ_{f} = total stress, u_{a} = pore air pressure which is considered as '0' in the current study, ϕ' = effective friction, $\chi (u_{\mathrm{a}} - u_{\mathrm{w}})$ = increase in cohesion due to suction stress.

Lu and Griffiths (2004) developed the suction stress equation by considering rainfall characteristics as presented in Eqs. 3 and 4 considering the rainfall infiltration. The same equation is considered in the current study for evaluating the reliability indices. Lu and Griffiths (2004) expressed χ and $(u_{\mathrm{a}} - u_{\mathrm{w}})$ as

$$\chi = \left[\frac{1}{1 + \left\{ -\ln\left[\left(1 + \frac{q}{k_{\mathrm{sat}}}\right) \mathrm{e}^{-\alpha \gamma_{\mathrm{w}} h_{\mathrm{si}}} - \frac{q}{k_{\mathrm{sat}}} \right] \right\}^{n}} \right]^{\left(1 - \frac{1}{n}\right)} \tag{3}$$

$$(u_{\mathrm{a}} - u_{\mathrm{w}}) = \frac{-1}{\alpha} \ln\left[\left(1 + \frac{q}{k_{\mathrm{sat}}}\right) \mathrm{e}^{-\alpha \gamma_{\mathrm{w}} h_{\mathrm{si}}} - \frac{q}{k_{\mathrm{sat}}} \right] \tag{4}$$

where $(u_{\mathrm{a}} - u_{\mathrm{w}})$ = the suction stress, α = inverse of air entry value, n = fitting parameter of SWCC, k_{s} = hydraulic conductivity, q = rainfall infiltration, γ_{w} = unit weight of water. Seattle area is prone to thousands of slope failures as Seattle area mainly consists of colluvial deposits, where shallow landslides are very common as they often triggered by rainfalls (Lu and Godt 2008).

To facilitate analysis of the above model and to generalize the equation for all soil types and infiltration rates, the suction stress equation can be rearranged as follows:

$$U = \frac{-\ln\left[(1 + Q)\mathrm{e}^{-Z} - Q\right]}{\left[1 + \left(-\ln\left[(1 + Q)\mathrm{e}^{-Z} - Q\right]\right)^{n}\right]^{\frac{(n-1)}{n}}} \tag{5}$$

where $U = \alpha \chi (u_{\mathrm{a}} - u_{\mathrm{w}})$ dimensionless suction stress.
$Z = (\alpha \gamma H)\left(\frac{\gamma_{\mathrm{w}} h_{\mathrm{si}}}{\gamma H}\right)$ dimensionless depth coordinate.
$Q = \frac{q}{k_{\mathrm{s}}}$ dimensionless infiltration.
For the FS under unsaturated state, the effective stress (Eq. 2) can be substituted into the classical infinite slope stability, and the FS equation can be shown as Eq. 6.

$$F = \frac{\tan \phi'(Z_{\mathrm{w}})}{\tan \omega} + \frac{2c'}{\gamma H \sin(2\omega)} + \frac{\chi (u_{\mathrm{a}} - u_{\mathrm{w}})}{\gamma H} (\tan \omega + \cot \omega) \tan \phi'(Z_{\mathrm{w}}) \tag{6}$$

The right hand side of Eq. 6 consists of three terms. They are due to internal frictional resistance, soil cohesion and suction stress.

Table 2 Statistics of soil properties, SWCC parameters and rainfall characteristics

Random variable	Statics		
	Mean	COV (%)	Distribution
Q	−0.09375, 0, −0.9375	20	Normal
ϕ'	36°	20	Normal
α	14	20	Normal
n	2.12	20	Normal
γ	15.4	5	Normal

4 Reliability Analysis Using MCS

The reliability analysis of rainfall-induced slope is carried out by performing MCS. The COVs of each design variable considered in the present study are shown in Table 2. It is quite important to find the number of simulations required for obtaining results which are of reasonable accuracy. In the present study, a total of 10^6 simulations are performed. In the probabilistic framework, the statistics for the output results can be represented either by reliability index (β) or probability of failure (p_f). The probability of failure (p_f) is given as

$$p_f = \frac{\sum_{i=1}^{NS} I(\tau_{us} < \tau)}{NS} \tag{7}$$

where NS = total number of simulations, τ_{us} = unsaturated shear strength mobilized and τ = shear stress.

Indicator function, $I(g(x) < 0)$, decides the total unsuccessful simulations. Where, $g(x)$ is the performance function. For a particular random sample, $I(\tau_{us} < \tau)$ takes the value of 1 if $\tau_{us} < \tau$ condition occurs. Else, it takes 0 if $\tau_{us} \geq \tau$ condition occurs. Finally, the reliability indices can be obtained using Eq. 8.

$$\beta = \Phi^{-1}(1 - p_f) \tag{8}$$

where Φ = standard normal distribution.

5 Results and Discussions

Two rainfall intensities along with zero rainfall intensity are considered for the reliability analysis in order to evaluate the influence of rainfall intensity on the reliability index against sliding failure of infinite slope. Seattle-Tacoma, WA airport reported the rainfall intensities, namely the monthly maximum intensity (q) of -1.5×10^{-7} m/s (i.e., November 2006) and daily maximum intensity (q) of -1.5×10^{-6} m/s (i.e., on 20 October 2003) and are considered in the present study to find the reliability indices. The suction stress and factor of safety variations with respect to the depth of water table are presented by Raghuram and Basha (2017). The authors have concluded that, in the absence of rainfall (i.e., $q = 0$ m/s), the suction stress is 1.37 kPa when water table depth is at 0.3 m. The authors also concluded that the factor of safety in case of maximum monthly rainfall scenario is greater than the maximum daily rainfall case. This is due to the decrease in suction stress in sands due to steady-state infiltration. The factor of safety increases nonlinearly as the distance between the slip surface and water table increases.

The results presented in Fig. 6 show the change in reliability indices with respect to the various suction heads (distance between slip surface and water table). It can be observed that when the distance between slip surface and water table is 4 m, the reliability index in case of small infiltration rate ($Q = -0.09375$) is greater than no rainfall condition ($Q = 0$). This is due to the suction stress variation with respect to the distance between slip surface and water table. Table 3 represents the expected performance level, and the corresponding reliability indices recommended by US Army Corps of Engineers (1999).

An interesting observation that can be made from Fig. 6 is that the reliability indices are less than 1.0 at the distances above water table for maximum monthly rainfall and maximum daily rainfall conditions. However, as per US Army Corps of Engineers (1999), the expected performance of the considered slope is hazardous. Therefore, the results presented based on reliability analysis using MCS clearly indicate the fact that the variability associated with the fitting parameters of SWCC and rainfall characteristics has to be given due consideration to evaluate the actual performance of the rainfall triggered slope failures.

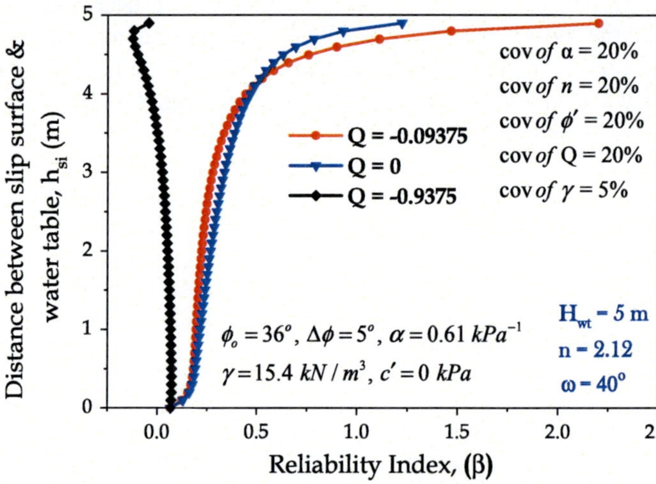

Fig. 6 Influence of distance between water table and slip surface on reliability index (β) for three different rainfall infiltration conditions

Table 3 Expected performance level and corresponding reliability indices (U. S. Army Corps of Engineers 1999)

Expected performance level	Reliability index (β)
High	5.0
Good	4.0
Above average	3.0
Below average	2.5
Poor	2.0
Unsatisfactory	1.5
Hazardous	1.0

5.1 Influence of SWCC Fitting Parameter α on Reliability Index

To understand the influence of SWCC fitting parameter α on the reliability index, a study is performed by varying $\alpha\gamma H$ from 5 to 20 for different slip surface positions as shown in Fig. 7a–d. Four different slip surface positions were considered in the present study (0.4 m, 0.7 m, 1 m and 1.3 m from the ground surface). Figure 7a–d shows that the reliability index significantly decreases with an increase in the fitting parameter α and the slope angle. As mentioned earlier, α is the fitting parameter related to inverse of AEV. Therefore, for a given water content, matric suction decreases, as the α increases, which in turn decreases the reliability index and stability of the slope. Moreover, as the slope angle increases, the reliability index and stability of slope decrease.

For an illustration, for a given slope angle of 45°, the reliability indices are 2.31, 2.02, 1.90, 1.84 for corresponding $\alpha\gamma H$ values of 5, 10, 15 and 20, respectively, when the slip surface is located at 0.7 m below the ground level. Similarly, the influence of the slip surface position with respect to the water table is shown in Fig. 7a–d. It can be noted from Fig. 7a–d that the reliability indices decrease with increase in the depth of slip surface. This is due to the fact that, when depth of slip surface increases for a given depth of water table (5 m in the present study), suction head decreases, which further decreases the stability of the slope. As an example, for a given slope angle of 45° and $\alpha\gamma H = 5$, the reliability indices are 3.34, 2.31, 1.76 and 1.45, respectively, for four different slip surface positions at 0.4 m, 0.7 m, 1.0 m and 1.3 m, respectively, below the ground surface.

5.2 Effect of SWCC Fitting Parameter n on Reliability Index

To understand the influence of SWCC fitting parameter *n* on the reliability index, a study is performed by varying fitting parameter '*n*' from 2 to 3.5 for different slip surface positions as shown in Fig. 8a–d. Four different slip surface positions are considered in the present study (0.4 m, 0.7 m, 1 m and 1.3 m from the ground surface). Figure 8a–d shows that the reliability index significantly decreases with an increase in the fitting parameter n and the slope angle. As mentioned earlier, n is the fitting parameter related to the SWCC of the slope. For a given water content, matric suction decreases, with an increase in n value, which further decreases the stability

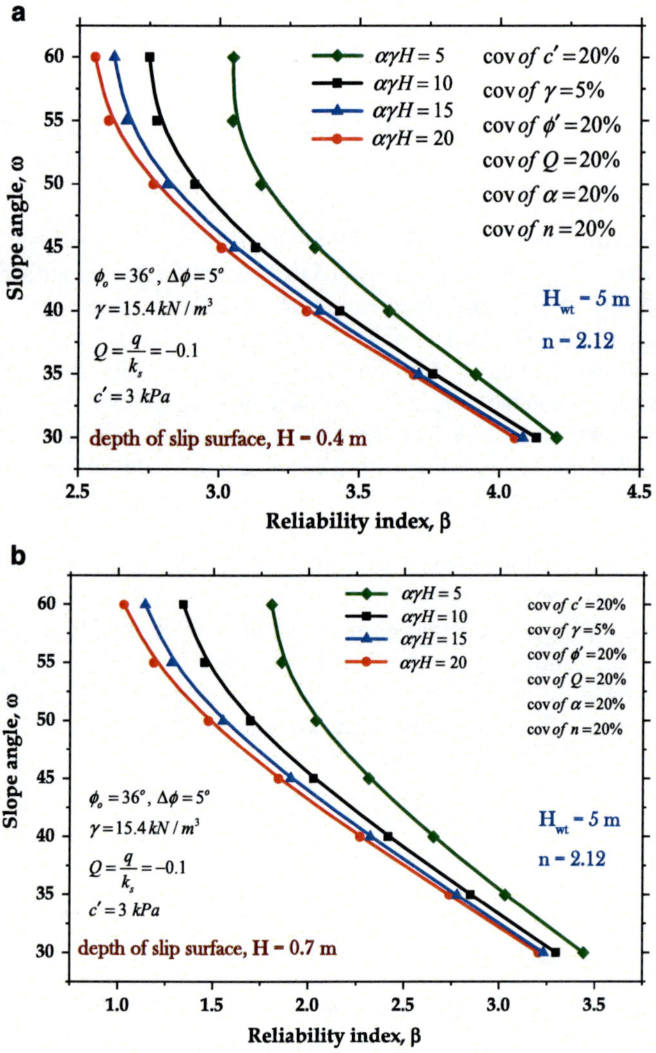

Fig. 7 **a** Influence of SWCC fitting parameter α on reliability index (β) for slip surface depth of 0.4 m. **b** Effect of SWCC fitting parameter α on reliability index (β) for slip surface depth of 0.7 m. **c.** Effect of SWCC fitting parameter α on reliability index (β) for slip surface depth of 1.0 m. **d** Influence of SWCC fitting parameter α on reliability index (β) for slip surface depth of 1.3 m

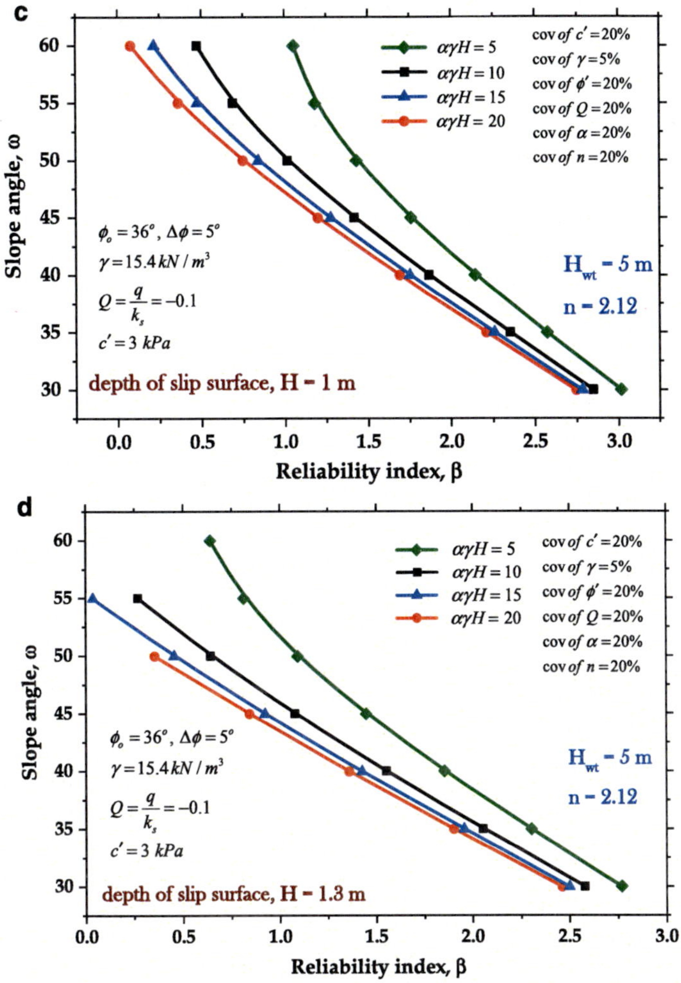

Fig. 7 (continued)

of slope. Moreover, as the slope angle increases, the reliability index and stability of slope decrease.

For example, for a given slope angle of 60°, the reliability index values are 2.67, 2.57, 2.49, 2.44 for corresponding n values of 2, 2.5, 3 and 3.5, respectively, when the slip surface is located at 0.4 m from the ground level. Similarly, the influence of the slip surface position with respect to the water table is shown in Fig. 8a–d. It can be noted from Fig. 8a–d that the reliability indices decrease with increase in the depth of slip surface. As an example, for a given slope angle of 30° and $n = 2.5$, the reliability indices are 4.06, 3.21, 2.75 and 2.47, respectively, for four different slip surface positions at 0.4 m, 0.7 m, 1.0 m and 1.3 m, respectively, from the ground surface.

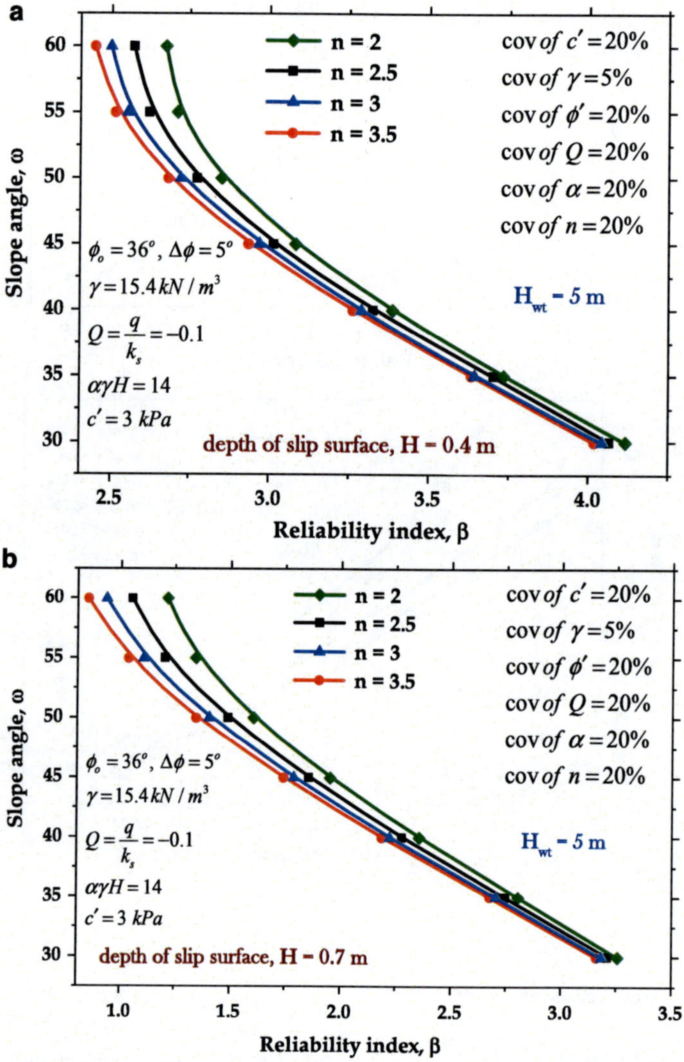

Fig. 8. **a** Effect of SWCC fitting parameter n on reliability index (β) for slip surface depth of 0.4 m. **b** Influence of SWCC fitting parameter n on reliability index (β) for slip surface depth of 0.7 m. **c** Influence of SWCC fitting parameter n on reliability index (β) for slip surface depth of 1.0 m. **d** Effect of SWCC fitting parameter n on reliability index (β) for slip surface depth of 1.3 m

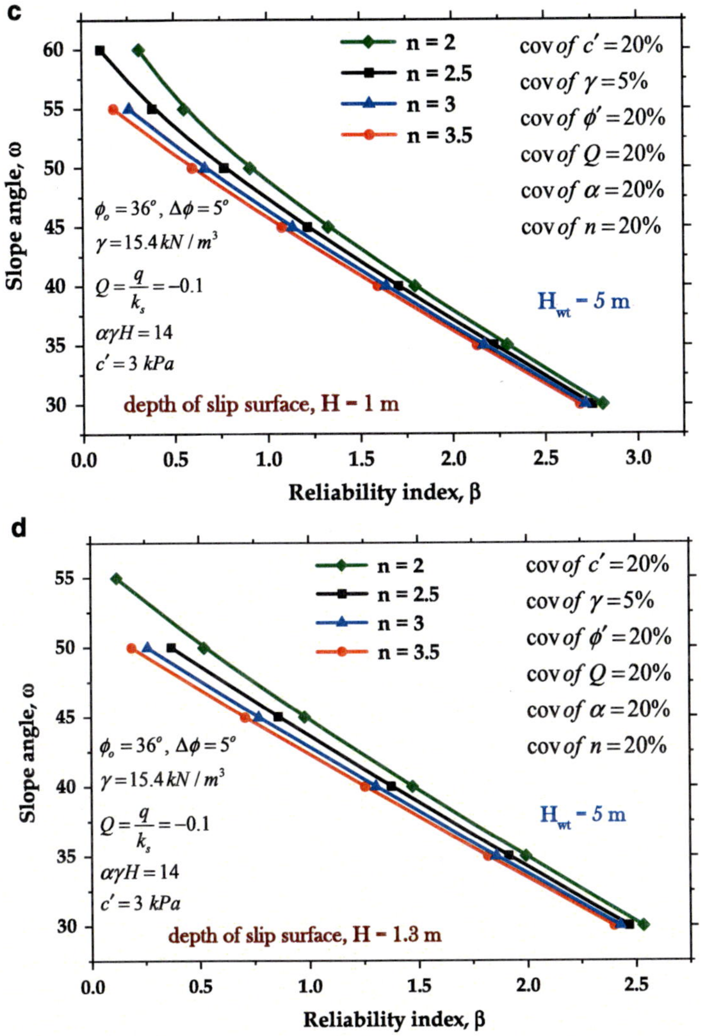

Fig. 8. (continued)

5.3 *Influence of Rainfall Infiltration q on Reliability Index*

To find the effect of rainfall intensity (q) on the reliability index, a study is performed by varying $Q = q/k_s$ from -1 to -0.25 for various slip surface positions as shown in Fig. 9a–d. Figure 9a–d shows that the reliability index significantly increases with decrease in the rainfall infiltration (q). One important and interesting observation is that, when rainfall infiltration is almost equal to the hydraulic conductivity, the reliability index is very less. This is due to the fact that suction stress disappears when (qk_s) (Raghuram and Basha 2017).

For an illustration, for a given slope angle of 50°, the reliability indices are 2.58, 2.66, 2.75, 2.81 for corresponding Q values of −1, −0.75, −0.5 and −0.25, respectively, when the slip surface is located at 0.4 m from the ground level. Similarly, the influence of the slip surface position with respect to the water table is shown in Fig. 9a–d. It is obvious to note from Fig. 9a–d that the reliability indices decrease with increase in the depth of slip surface. This is due to the decrease in suction

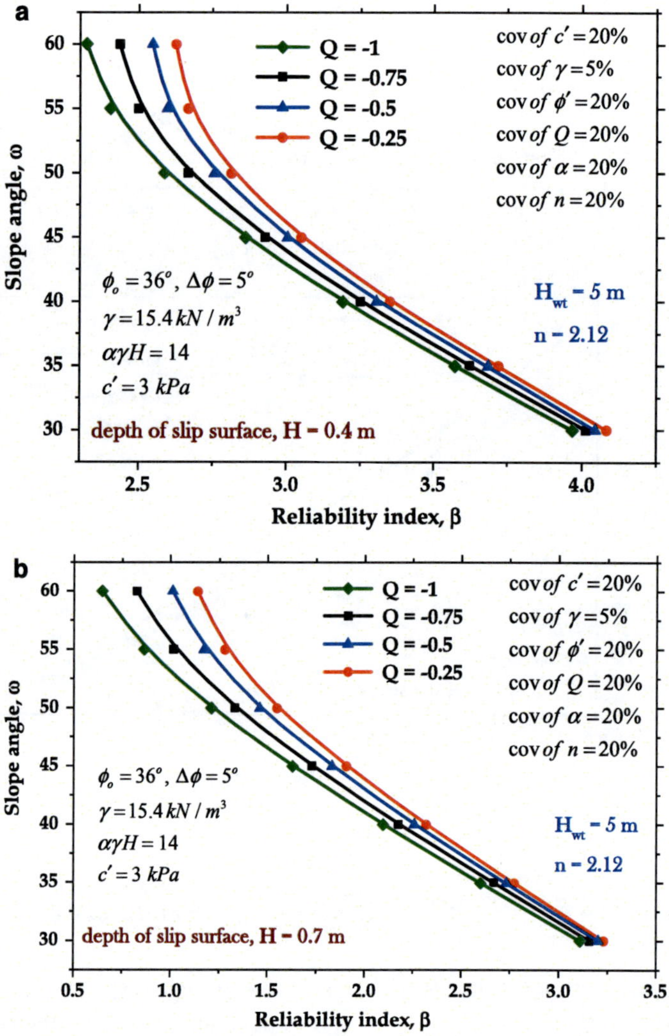

Fig. 9 **a** Influence of rainfall infiltration and hydraulic conductivity (Q) on reliability index (β) for slip surface depth of 0.4 m. **b** Influence of rainfall infiltration and hydraulic conductivity (Q) on reliability index (β) for slip surface depth of 0.7 m. **c** Effect of rainfall infiltration and hydraulic conductivity (Q) on reliability index (β) for slip surface depth of 1.0 m. **d** Influence of rainfall infiltration and hydraulic conductivity (Q) on reliability index (β) for slip surface depth of 1.3 m

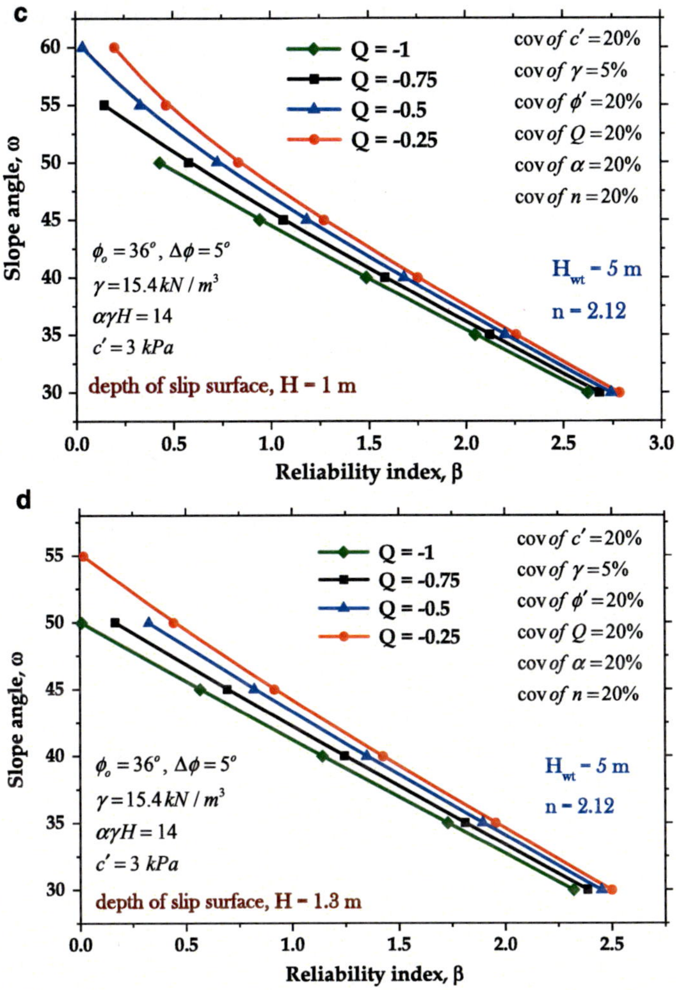

Fig. 9 (continued)

head, which further decreases the stability of the slope. As an example, for a given slope angle of 45° and $Q = -1$, the reliability indices are 2.86, 1.63, 0.94 and 0.57, respectively, for four different slip surface positions at 0.4 m, 0.7 m, 1.0 m and 1.3 m, respectively, from the ground surface.

6 Conclusions

The key advantage of the MCS method is its simplicity, which offers a tool for accommodating the variability associated with the SWCC fitting parameters and rainfall characteristics. The methodology for the reliability analysis using MCS of rainfall triggered slope failures is presented in the paper. Shallow slope failure of sandy colluvium soil near Seattle, Washington, is discussed in the light of reliability analysis and is presented as a case study. The following are the major conclusions:

1. The study highlighted the importance of considering the variability associated with fitting parameters of SWCC and rainfall characteristics.
2. The reliability analysis illustrates that the expected performance of sandy colluvium soil slope near Seattle, Washington, USA, is hazardous for all the three rainfall intensities considered. In addition, the magnitude of the reliability index is greatly influenced by the magnitude of the rainfall intensity.
3. The SWCC fitting parameter α significantly influences the stability of the slope. The reliability index has reduced from 2.31 to 1.84 when $\alpha\gamma H$ has increased from 5 to 20 for a given slope angle of 45°.
4. For a given slope angle of 60°, the reliability index has reduced from 2.67 to 2.44 when n increases from 2 to 3.5.
5. One of the interesting observations from the study is that, when the rainfall intensity is equal to the hydraulic conductivity, the reliability index is very less. This is due to the loss of suction stress.
6. The required reliability value can be chosen from the design charts to ensure the stability of the slope, which accounts for the variability associated with parameters mentioned above.

The present study concludes that it is highly appropriate to consider the variability associated with the fitting parameters of SWCC, rainfall intensity and hydraulic conductivity of the soil in order to assess the actual performance of slope.

References

Alonso EE (1976) Risk analysis of slopes and its application to slopes in Canadian sensitive clays. Geotechnique 26(3):453–472

Babu GLS, Murthy DSN (2005) Reliability analysis of unsaturated soil slopes. J Geotech Geoenviron Eng 131(11):1423–1428

Bishop AW (1959) The principle of effective stress. Lecture delivered in Oslo, Norway, 1955. Technisk Ukeblad 106(39):859–863

Chiu CF, Yan WM, Yuen KV (2012) Reliability analysis of soil–water characteristics curve and its application to slope stability analysis. Eng Geol 135–136:83–91

Fredlund DG, Rahardjo H, Fredlund MD (2012) Unsaturated soil mechanics in engineering practice. Wiley, Hoboken

Griffiths DV, Fenton GA (2004) Probabilistic slope stability analysis by finite elements. J Geotech Geoenviron Eng 130(5):507–518

Jiang SH, Li DQ, Cao ZJ, Zhou CB, Phoon KK (2015) Efficient system reliability analysis of slope stability in spatially variable soils using Monte Carlo simulation. J Geotech Geoenviron Eng 141(2):04014096

Li KS, Lumb P (1987) Probabilistic design of slopes. Can Geotech J 24(4):520–535

Li DQ, Jiang SH, Cao ZJ, Zhou W, Zhou CB, Zhang LM (2015) A multiple response surface method for slope reliability analysis considering spatial variability of soil properties. Eng Geol 187:60–72

Low BK (2005) Reliability-based design applied to retaining walls. Geotechnique 55(1):63–75
Lu N, Godt J (2008) Infinite slope stability under steady unsaturated seepage conditions. Water Resour Res 44
Lu N, Griffiths DV (2004) Suction stress profiles in unsaturated soils. J Geotech Geoenviron Eng 130(10):1063–1076
Ng CWW, Shi Q (1998) A numerical investigation of the stability of unsaturated soil slopes subjected to transient seepage. Comput Geotech 22(1):1–28
Oh S, Lu N (2015) Slope stability analysis under unsaturated conditions: case studies of rainfall-induced failure of cut slopes. Eng Geol 184:96–103
Oh WT, Vanapalli SK (2010) Influence of rain infiltration on the stability of compacted soil slopes. Comput Geotech 37:649–657
Phoon KK, Santoso A, Quek ST (2010) Probabilistic analysis of soil water characteristic curves. J Geotech Geoenviron Eng 136(3):445–455
Raghuram ASS, Basha BM (2017) Reliability analysis of rainfall induced slope failure at Seattle area near Washington. In: Indian geotechnical conference, IIT, Guwahati
Sahis MK, Bhattacharya G, Chowdhury R (2014) Reliability analysis of rainfall induced slope instability of unsaturated soil slopes. In: Khalili N, Russell A, Khoshghalb A (eds) Unsaturated soils: research & applications, pp 1287–1293
U.S. Army Corps of Engineers (1999) Risk-based analysis in geotechnical engineering for support of planning studies, engineering and design, Rep. No. 20314-1000. Department of Army, Washington, DC
Zhang J, Zhang LM, Tang WH (2011) New methods for system reliability analysis of soil slopes. Can Geotech J 48(7):1138–1148
Zhang J, Huang HW, Zhang LM, Zhu HH, Shi B (2014a) Probabilistic prediction of rainfall-induced slope failure using a mechanics-based model. Eng Geol 168:129–140
Zhang LL, Zeng J, Cheng Y (2014b) Model uncertainty of unsaturated hydraulic properties and effects on slope reliability. In: Vulnerability, uncertainty, and risk, pp 2330–2339
Zhang L, Li J, Zhang J, Zhu H (2016) Rainfall-induced soil slope failure: stability analysis and probabilistic assessment. CRC Press, Taylor and Francis Group, UK

Probabilistic Assessment of Paglajhora Landslide Using SLOPE/W

Abhirup Dikshit and Neelima Satyam

Abstract This paper reports stability analysis of the dynamic and recurrent Paglajhora landslide which is situated in the upper subpart of the Shiva Khola basin in the Kurseong region of Darjeeling Himalayas in the state of West Bengal, India. The height of the affected slope is approximately 1540 m, having two porches with fluctuating slope variations. The lower portion frequently fails and causes blockage of roads damaging the NH-55 highway and cutting the Kalimpong region from other parts of the country. The problem of slope failure is further adverse due to various geological features, steep slopes, rugged topography and intense monsoonal rainfall. The study area is divided into two major lithotectonic units, the Higher Himalayan Crystalline Sequence (HHCS) and the Lesser Himalayan Sequence (LHS), separated by major ductile shear zone known as the Main Central Thrust. Parametric analysis of the slope with variation in slope height and slope angle has been performed using SLOPE/W software. The resulting change in the factor of safety is calculated. The analysis would help in understanding the possible failure scenarios of the landslide area, and effective measures could be taken to reduce the risk.

Keywords Landslide · Paglajhora · Probabilistic analysis · SLOPE/W

A. Dikshit (✉) · N. Satyam
Discipline of Civil Engineering, Indian Institute of Technology Indore, Indore, Madhya Pradesh 452020, India
e-mail: abhirupdikshit@gmail.com

M. Latha Gali and P. Raghuveer Rao (eds.), *Geohazards*, Lecture Notes in Civil Engineering 86, https://doi.org/10.1007/978-981-15-6233-4_9

1 Introduction

1.1 A Subsection Sample

Landslides are extensive and uncertain phenomena that affect all continents, creating a severe threat to the population (Petley 2012). The Indian Himalayan region has been immensely affected by the increase in the frequency of landslide occurrences (Dikshit and Satyam 2018). There are two national highways (NH-31 A, NH-55) across Darjeeling region, of which the Paglajhora region is situated in NH-55. The study area is selected due to the importance of the NH-55 which links the hill and plains which are often blocked by several landslides causing serious damage. The present study determines the factor of safety with the variation of slope and elevation along with spatial variability of soil properties. Such study would help in understanding the critical sections of the Paglajhora landslide.

The slope stability analysis is usually carried out to assess the safe and economic design of natural or man-made slopes. Slope stability can be defined as the resistance of an inclined surface to failure by either sliding or collapsing. The choice of analysis technique can be attributed to site conditions and the possible failure mechanism.

The stability of slopes has generally been analysed utilising deterministic analysis method. In the deterministic analysis, the mean value of various geotechnical parameters (unit weight, cohesion, internal friction) is used. However, slope stability issues are characterised by multiple uncertainties like geological anomalies, inherent anisotropy in rock mass properties, variable environmental conditions and human roles. In some cases, slope failure may occur for a factor of safety greater than unity due to mean value used as input parameters for analysis. The probabilistic analysis provides a range of geotechnical parameter which expresses the uncertainties involved in slope stability analysis (Singh et al. 2013).

2 Study Area

The study region is situated in the upper subpart of Shiva Khola basin located in the Kurseong division of Darjeeling Himalayas. The area stretches from 26° 52′ N to 26° 57′ N and 88° 18′ E to 88° 19′ E. The maximum elevation of the landslide is 1540 m with its toe at the height of 780 m. The region intersects NH-55 at two distinct elevations, at 1185–1190 and 1230–1335 m affecting the highway for about 3 km (Sengupta et al. 2010). Geologically, the region comprises gneiss, mica-schist and granulitic rocks including mylonitised granite with subparallel thrust, phyllite, silvery-mica-chlorite-schist, grey sericite (Mondal and Maiti 2013). The rocks are traversed by quartz and quartzo-felspathic which are highly metamorphosed. The presence of clay minerals induces seepage leading to slope instability (Mondal and Maiti 2013). A variety of slope-forming material characterises the Paglajhora region. In the Paglajhora slide area, moderate-to-highly steeper slopes are occupied either by exposed bedrock with or without thin slope material (Fig. 1). The entire region is a bulb shaped with the broader upper part towards the north and a narrow toe towards the south-east.

The increase in frequency and risk associated with landslides in seismologically active Darjeeling Himalayan region poses a dire need to implement the probabilistic methodology. Himalayan rocks, young and dynamic, and its load deformation behaviour pose a serious problem to the stability of slopes. The situation is worsened due to the heterogeneous morphometry, complex geology and several

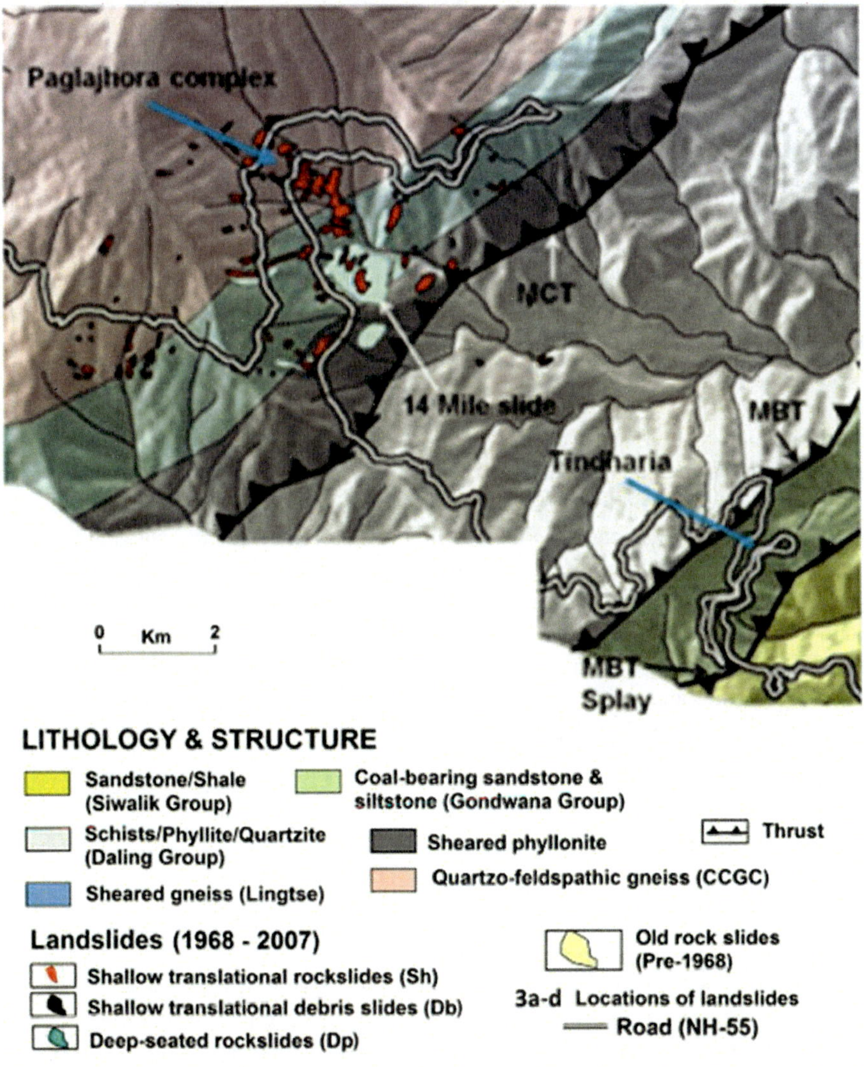

Fig. 1 Lithology of the study region and past landslides in the study area (*Source* Geological Survey of India)

Fig. 2 Landslide damages in Paglajhora region (*Source* Save The Hills)

local factors (Korup et al. 2006). The high-intensity rainfall further aggravated the problem of stability due to high porosity and low shear strength of soil (Korup et al. 2006). The pore water pressure and mountain rivulets also play a significant role in slide initiation. The apex of the sliding zone is predominated with organic matter which encourages high water holding capacity and volume expansion along with deforestation resulting in erosion.

The Paglajhora region has severely affected the NH-55 (the road leading to Darjeeling town from Siliguri) in between Gayabari–Mahanadi (lower Paglajhora) and Mahanadi–Giddapahar (upper Paglajhora) sectors for a cumulative length of ~3 km. The Darjeeling Himalayan Toy Train (World Heritage Site of UNESCO) tract is also aligned along the road bench of NH-55 in this sector, which passes through this landslide complex. Since 1968, Paglajhora has had ten landslide events affecting NH-55, and disrupting communication line between Siliguri and Darjeeling was utterly interrupted, from days to months. Paglajhora sinking zone faced massive slope failures in 1998, 2002, 2005 and 2011, which indicates that the occurrence of landslides in the region is ongoing and the increasing slides is a tremendous threat to upslope settlements and the Hill Cart Road (connecting Siliguri and Darjeeling town) (Sengupta et al. 2010) (Fig. 2).

3 Probabilistic Assessment

The analysis has been conducted using GeoStudio software. SLOPE/W includes a general comprehensive algorithm for probabilistic analyses. The profile of the study region was generated on Google Earth and later provided as an input for the software. The soil parameters were collected from literature and applied in the input model. The geotechnical parameters were assigned a probability distribution, and a Monte Carlo scheme is then used to compute a probability distribution of the resulting safety factors. In a Monte Carlo simulation, discrete values of random variables are generated which are consistent with their probability distribution, and a limit state function is determined for every generated set. The process is repeated multiple times to evaluate the probability of failure.

It is not possible to distinguish the accurate effect of these safety factors on a safety level. The determination of factor of safety using the probabilistic approach is more precise compared to the deterministic approach. The use of deterministic analysis is constrained due to the uncertainties in soil properties and environmental conditions (Alonso 1976). Thus, the probabilistic analysis produces a direct estimate of the distribution of either the factor of safety or critical height associated with a design or analysis situation. However, the probabilistic analysis does not affect deterministic solution, and the probabilistic analysis is performed on the critical surface only. The software computes the factor of safety of all slip surfaces first and determines the critical slip surface with mean parameters as if no probabilistic analysis is chosen. The input value of a parameter represents the mean value, and the variability of the parameter is assumed to be normally distributed with a known standard deviation.

4 Results

As the slope failure surface approaches the crown, the factor of safety decreases resulting in a high probability of slope failure. The range of elevation was selected between 1100 and 1350 m as this is the section where it intersects with NH-55. A similar analysis was conducted for slope angle, and the results indicate that factor of safety decreases with an increase in slope angle. The spatial variability of soil was taken in consideration for the analysis, and the probabilistic density function of geotechnical parameters (Cohesion, Unit weight and Internal Friction) is depicted in Figs. 3, 4 and 5, respectively.

The variables of the geotechnical parameters depicted in Figs. 3, 4 and 5 are used as an input parameter. The final results shown in Figs. 6 and 7 are the probability distribution determined using Monte Carlo simulation. There are several approaches for probabilistic evaluation like Monte Carlo, point estimate method. However, in this study, Monte Carlo has been conducted due to its simplicity and can be performed without the extensive knowledge of mathematical and statistical expertise.

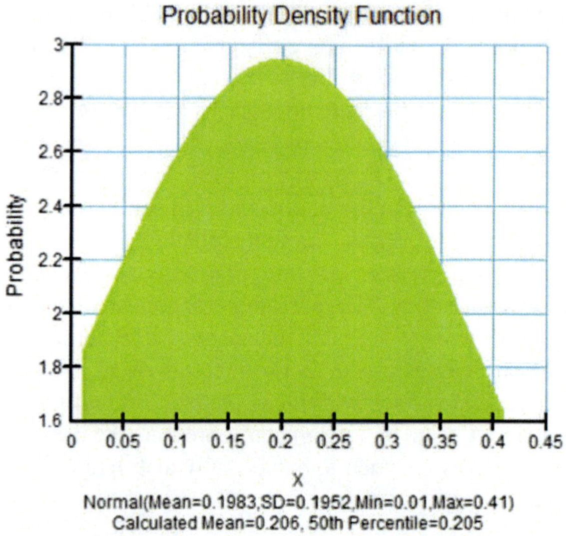

Fig. 3 Probabilistic density function of cohesion

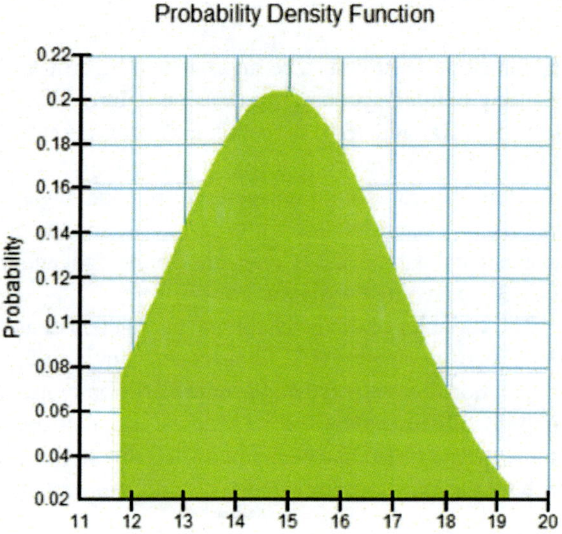

Fig. 4 Probability density function of unit weight

The results were determined using the distribution of input parameters instead of a single parameter value. After that, the factor of safety was determined using probabilistic analysis, and the results have been depicted in Figs. 6 and 7. The results provide a probabilistic distribution of factor of safety concerning slope angle and elevation.

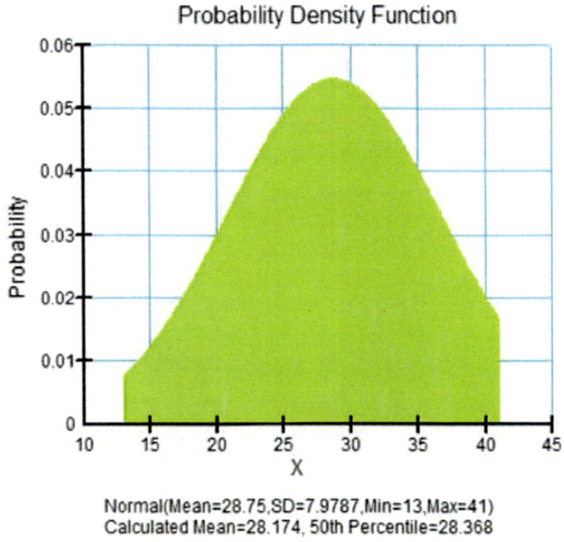

Fig. 5 Probability density function of internal friction

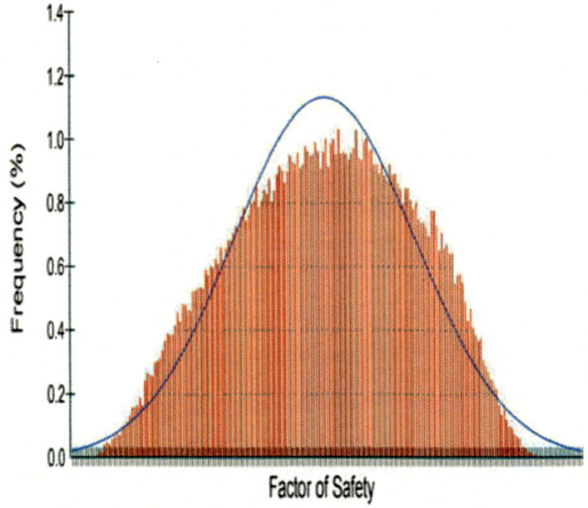

Fig. 6 Variation of the factor of safety with slope angle

5 Conclusion

Landslide is one of the prominent geohazards in the Himalayan region which leads to loss of life and property. This study is on Paglajhora landslide zone which has a history of landslides and affects the NH-55 which connects the hills to the cities. A probabilistic analysis provides a more accurate estimate of the distribution of either the factor of safety or critical height associated with the analysis. In this study, a probabilistic analysis was conducted considering the spatial variability of soil by collecting various soil sample results from published literature, and the factor of

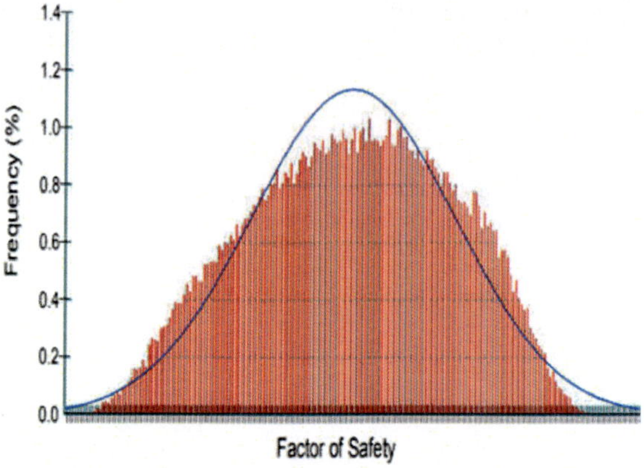

Fig. 7 Variation of the factor of safety with slope height

safety was determined for variation of slope and elevation. Monte Carlo simulation was used to determine the factor of safety by incorporating the input parameters, and a probability distribution was assigned. Such analysis is critical compared to deterministic approach as the variation of soil properties leads to a more realistic determination of factor of safety. This analysis concludes that slope stability problems should be dealt with using a probability distribution of the slope characteristics which can be confirmed with the results obtained in this paper. The results were determined for elevation where NH-55 intersects the landslide zone, and it depicts the probabilistic distribution of the factor of safety in variation with the slope as well as height. The results implicate that region is highly vulnerable to slope failures and leads to disruption of traffic on the national highway. Such analysis can also help in determining the probability of slope failure.

References

Alonso EE (1976) Risk analysis of slope and its application to slopes in Canadian sensitive clays. Geotechnique 26(3):453–547

Dikshit A, Satyam DN (2018) Estimation of rainfall thresholds for landslide occurrences in Kalimpong, India. Innovative Infrastruct Solutions 3:24

Korup O, Strom AL, Weidinger JT (2006) Fluvial response to large rock-slope failures—examples from the Himalayas, the Tien Shan, and the New Zealand Southern Alps. Geomorphology 78:3–21

Mondal S, Maiti R (2013) Integrating the analytical hierarchy process (AHP) and the frequency ratio (FR) model in landslide susceptibility mapping of Shiv-khola watershed, Darjeeling Himalaya. Int J Disaster Risk Sci 4:200

Petley D (2012) Global patterns of loss of life from landslides. Geology 40(10):927–930

Sengupta A, Gupta S, Anbarasu K (2010) Landslide investigations and mitigation in Eastern Himalayan region. J Indian Roads Congr 71(2):133–142

Singh R, Umrao RK, Singh TN (2013) Probabilistic analysis of slope in Amiyan landslide area, Uttarakhand. Geomatics Nat Hazards Risk 4(1):13–29

Stability Analysis of Hill Slopes in The Nilgiris, Tamil Nadu

Praveen Kumar and Krishnamurthy Premalatha

Abstract Landslide is a sudden slide of rock, debris and earth down the slope. The Nilgiris have always been under the threat of landslides in monsoon season. The Nilgiris is a hilly district with an area of 2500 km^2 located in the North-Western part of Tamil Nadu. Rainfall is the major triggering factor of landslides in The Nilgiris (Chandrasekaran in Assessment of damages induced by recent landslides in Ooty, Tamil Nadu, India. IIT Bombay, Mumbai, India, pp 687–688, 2010). In this study, two sites have been selected where landslides occurred during monsoon in November 2009. The soil samples were collected from the sites of landslides. A detailed investigation was carried out on the soil samples and index properties; chemical composition and shear strength parameters were drawn. The critical surface for the slope was drawn graphically using Fellenius (1936) method, and the same was incorporated into the software through co-ordinates. Many studies have been carried out using numerical analysis for The Nilgiris area but in this study, realistic site conditions and parameters have been incorporated. The analysis was carried using wetting depths of 2 m, 3 m, 4 m and 5 m at the edge of the slopes and shear strength parameters in GEO 5 and PLAXIS 2D software. The analysis result shows that the infiltration of water into the slope results in the reduction of shear strength parameter of the soil which leads to the failure of the slope. The results of Geo 5 analysis and PLAXIS 2D analysis were compared, and the conclusions were drawn from it.

Keywords Landslide · Monsoon · Infiltration · Wetting depth

P. Kumar (✉) · K. Premalatha
Department of Civil Engineering, Anna University, Chennai, Tamil Nadu 600025, India
e-mail: praveen04kumar08@yahoo.com

K. Premalatha
e-mail: kvpremalatha@yahoo.com

M. Latha Gali and P. Raghuveer Rao (eds.), *Geohazards*, Lecture Notes in Civil Engineering 86, https://doi.org/10.1007/978-981-15-6233-4_10

1 Introduction

The stability of slopes has always been under severe threat in many parts of The Nilgiris. The landslides in The Nilgiris has always caused a great loss to human life and its economy. The Nilgiris is a hilly district with an area of 2500 km^2 and is located in the North-Western part of Tamil Nadu. Slope failures are very common geotechnical phenomenon in many countries like Hong Kong, Singapore and India. Many peaks in The Nilgiris have steep rocky escarpments with or without soil cover around it. Rainfall is the major triggering factor of landslides and debris flow in The Nilgiris. The other factors include cutting of toe of slopes for the construction of roads and houses. The mechanism behind the landslide is that on a very steep slope with loose type of soil when receives a high-intensity rainfall of longer duration causes the saturation of the loose soil media resulting in the reduction of the shear strength. Many studies had been carried out using numerical methods but in this study, realistic site condition and parameters have been considered. The soil samples from top and slide mass were collected. The basic index properties, chemical composition and shear strength parameters of the soil samples were found by laboratory.

1.1 Slope Stability

The stability of slope is of great importance to civil, geotechnical and mining engineers worldwide for a wide variety of engineering projects, such as cuts, dams, embankments and natural slope stabilization. Slope stability analysis is required because failure of natural slopes and man-made slopes has resulted in destruction and many deaths in The Nilgiris in the past.

2 Study Area

The Nilgiris district is located in the state of Tamil Nadu in India. The word 'Nil' in its name means Blue and 'Giri' means Mountains; hence, it is also known as the Blue Mountains. The Nilgiris district is surrounded by the state of Karnataka in northern part, Erode in eastern part, Coimbatore on the southern part and the state of Kerala on the western part. The topographical features of The Nilgiris district include rugged hilly terrain and steep covered with thick vegetation cover, with hills altitude varying in the range of 850–2637 m. The highest peak of The Nilgiris is Doddabetta Mountain (2637 m). According to the IS code [IS 1893–Part I (2016)], The Nilgiris district is located in the seismic zone III. The plateau, nearly co-terminus with The Nilgiris district, is drained by hundreds of streams. The major ones are the Bhavani, Moyar, Pandiyar, Punnapuzha, Katteri, Coonoor, Kulacombe, Kundah, Pykara, Kallar, Burliar, Gathada Halla and Billithada Halla Rivers. The catchment for the drainage towards the north is larger in area compared to those of the easterly and south-east drainages. The drainage is radial in many places due to the dominant high points. The landslide locations selected for the present study are Nadduhatti slide (Location 1, 11° 24.7′ N 76° 49.5′ E) and Achanakal slide (Location 2, 11° 22.8′ N 76° 42.2′ E).

2.1 *Geomorphology*

The Nilgiris hills rise from the higher elevation of the Coimbatore plateau, sloping steeply and gradually merging with the Western Ghats in the north-west and south-west. The famous hills are Ooty hills, Doddabetta, Kodaibetta and Devabetta. Among these, Doddabetta is the highest peak in the region. The drainage pattern of the region is usually radial and dendritic due to dominant high peaks. Most part of The Nilgiris district is deeply weathered and in some places reaches 40 m thickness (Rajakumar et al. 2007). The region comprises Archaean metamorphic rocks like charnockite, biotite gneiss, magnetite quartzite and pyroxene granulite.

2.2 *Rainfall*

The Nilgiris district receives rain during both rainy seasons, south-west monsoon and north-east monsoon. The parts of Gudalur, Pandalaur and Kundah taluks receive rainfall from the south-west monsoon. The parts of Udhagamandalam taluk, Coonoor and Kotagiri receive rainfall from the north-east monsoon. The average precipitation in The Nilgiris region varies in the range of between 1500 and 3000 mm (IMD 2011). Last 5 years rainfall data of The Nilgiris is shown in Fig. 1.

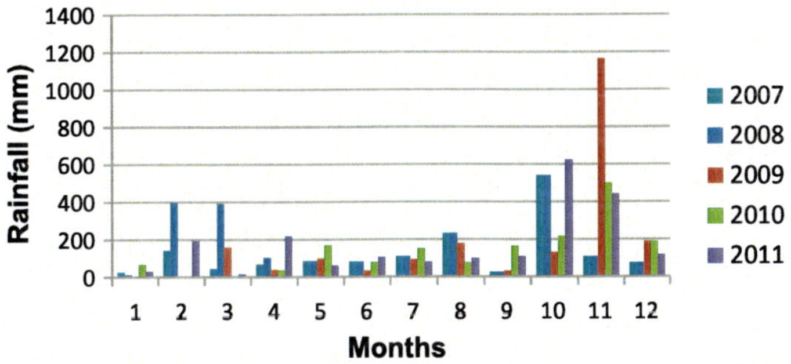

Fig. 1 Five years rainfall data of The Nilgiris

2.3 *Infiltration and Wetting Depth*

Infiltration is the process of water entry into soil from rainfall. Soil water percolation is the process of water flow from one point to another point within the soil. In the rainy season, due to infiltration of rainwater on the ground surface, the top layer of soil gets saturated. The depth up to which the soil gets saturated due to percolation of rainwater is known as wetting front depth (Lumb 1975).

3 Materials and Methods

For this study, two sites, namely Achanakal and Nadduhatti, were selected and detailed site investigation was done by site visit. The soil samples were collected from top of the hills with the help of sampling tube from both the sites. The soil samples were also collected from the slide mass of soil at the foot of the hills. The samples were transported in protective condition so that the soil does not lose its natural properties. The field density of the soil samples was determined at the site. Detailed experimental investigation was done on the samples in the laboratory. Photographic view of the site and detailed information about the sites were collected from locals of the area. All the landslides occurring in these areas were due to heavy rainfall during monsoon seasons. Using the information provided by the locals and photographs that were taken during the site visit, the slope layout was plotted in stepped form to replicate the roads and the houses constructed after cutting of the slopes roughly with an angle of elevation of 60° (Figs. 2 and 3).

The shear strength parameters were found by triaxial test. The infiltration rate was calculated using the Horton's equation, and wetting depth was found using Lumb's (1975) equation.

3.1 *Laboratory Investigations on Soil Samples*

Detailed experimental investigations were carried out on soil samples from the two locations where the landslides occurred in 2009. Specific gravity and dry and wet sieve analysis were carried out on the soil samples based on ASTM standards. The soil samples had no clay contents so liquid limit and plastic limit tests were not carried out. The value of k (coefficient of permeability) was arrived by falling head permeability test based on IS 2720 (17) which was 1.61×10^{-3} cm/s for Anchakal site and 1.55×10^{-3} cm/s for Nadduhatti site. The soil samples of the two locations have high percentage of fine contents 52% for Nadduhatti and 54% for Achanakal. The fines have very low plasticity limit and no liquid limit. Both locations have low value of permeability. The soils were classified as per unified soil classification system.

Fig. 2 Hill slope of Nadduhatti site

Fig. 3 Hill slope of Achanakal site

The soil types are silty sand (SM) at Nadduhatti and sandy silt (ML) at Achanakal. The properties of the soil samples are given in Table 1.

Table 1 Properties of soil samples of two locations

Properties	Nadduhatti	Achanakal
Specific gravity	2.36	2.48
Water content	20	24
Unit weight (kN/m^3)	18	17.6
Sand	44	56
Silt	51	36
Clay	2.6	4.4
Soil classification	ML	SM

4 Stability Analysis of Slope

A number of trials were carried out using manual marking of the slip surface in the software. The critical slip surface was drawn by Fellenius method graphically. The co-ordinates of the critical failure surface drawn were entered textually in the software. The critical failure surface was converted into circular form, and analysis was carried out. The factor of safety in LEM method depends upon the direction of force acting on each slice in the assumed slope. In FEM approach, the factor of safety emerges naturally from analysis without any particular form of mechanism. The factor of safety depends on shear strength parameters of the soil. Many numbers of trials were carried out using different depth of wetting and above experimental results. The wetting depth considered was 2 m, 3 m, 4 m and 5 m for both the sites. The optimization consists of finding the circular slip surface with smallest stability. The optimized analysis results have been given importance in this study because the optimized analysis gives the failure surface with the minimum value of factor of safety. The soil parameter for slope stability analysis is given in Table 2.

4.1 GEO 5 Software

The GEO 5 is a modern tool for geotechnical engineers to solve various geotechnical problems. The package of this software includes the stability analysis, excavation

Table 2 Soil parameters for slope stability

Tests	Nadduhatti	Achanakal
Undrained shear strength (kN/m^2)	316	196
Undrained shear strength (kN/m^2) (In wetting zone)	152	96
Poisson's ratio	0.3	0.3
Undrained deformation modulus (kN/m^2)	30,000	20,000
Undrained deformation modulus (kN/m^2) (in wetting)	15,000	10,000

design, deep foundation, settlement analysis, tunnels and shaft, walls and gabions and shallow foundations. The GEO 5 software includes five methods of analysis, namely Bishop (1955), Janbu (1973), Spencer (1967), Fellenius (1936) and Morgenstern and Price (1965) Methods. The GEO 5 software consists of two types of analysis, namely all methods and optimized method.

4.2 PLAXIS 2D

PLAXIS 2D is a finite element program used for the simulation and analysis of deformation and stability of different geotechnical applications. Results of the analysis are highly sensitive to mesh size and the number of nodes per element. The instability of a slope can be based on three different criteria of SRM.

1. Non-convergence of the solution
2. Extension of the plastic zone from toe to the top of the slope or equivalent shear strain along the potential slip surface
3. Rapid increase in nodal displacement on the slope surface.

4.3 PLAXIS 2D Analysis Results

4.4 GEO 5 Analysis Results

5 Results and Discussion

The analysis was done for both sites considering the height of the slope to be 80 m. The analysis was also carried out for the slopes without the role of infiltration in it. For the analysis, water table was not taken into the consideration. The results of all analysis are given in Tables 3 and 4 and Figs. 4, 5 and 6.

The infiltration rate was calculated based on Horton's (1939) equation. From this, it is observed that the increase in rainfall duration decreases the infiltration rate. The wetting front depth was calculated based on Lumb's (1975) equation. In GEO 5, Nadduhatti site 30° slope is safe for even maximum wetting front depth of 5 m. The above slopes are safe for maximum of 24 h duration of rainfall. Slope of 45° and 60° are stable only for 3 m wetting front depth. The rainfall duration for 3 m wetting front depth is 18 h. For Anchakal site, 30° slope is not safe for even minimum wetting front depth of 2 m. The slopes of this sites are very much prone to slides even for lesser duration of rainfall due to lower values of shear strength parameters.

In PLAXIS 2D, Nadduhatti site 30° slopes are safe upto wetting front depth of 3 m. From the rainfall duration point of view, the slopes are safe in factor of safety but they undergo large displacement which makes them unsafe. The slopes of Nadduhatti site could withstand the rainfall duration of upto 10 h safely with the slope angle of 30°.

Table 3 Slope stability analysis results of Geo 5 of angle 30°, 45° and 60°

Type of analysis	FOS Achanakal			FOS Nadduhatti		
All methods analysis	30°	45°	60°	30°	45°	60°
Without wetting depth	1.60	1.36	1.14	2.60	2.23	1.68
With 2 m wetting depth	1.67	1.36	1.13	2.58	2.16	1.62
With 3 m wetting depth	1.56	1.29	1.08	2.49	2.1	1.58
With 4 m wetting depth	1.29	1.15	F	2.29	1.84	1.53
With 5 m wetting depth	1.18	1.10	F	2.0	1.75	1.44
Optimized analysis						
Without wetting depth	1.14	1.08	1.06	1.67	1.63	1.58
With 2 m wetting depth	1.12	1.10	1.07	1.66	1.60	1.58
With 3 m wetting depth	1.08	1.06	1.02	1.64	1.57	1.54
With 4 m wetting depth	F	F	F	1.60	1.44	1.42
With 5 m wetting depth	F	F	F	1.58	1.39	1.36

Table 4 Slope stability analysis results of PLAXIS 2D of angles 30°, 45° and 60°

Type of analysis	Nadduhatti			Achanakal		
	30°	45°	60°	30°	45°	60°
Without wetting depth	2.8	2.4	2.1	2.7	2.4	2.1
With 2 m wetting depth	3.2	3.1	1.8	1.2	2.1	1.5
With 3 m wetting depth	3.0	2.3	1.7	1.8	1.2	F
With 4 m wetting depth	F	F	F	F	F	F
With 5 m wetting depth	F	F	F	F	F	F

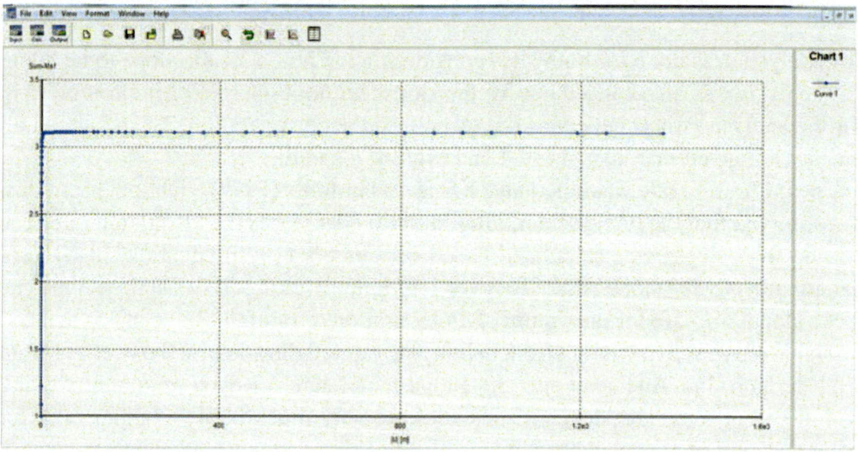

Fig. 4 FOS analyses result with 45° and 2 m wetting for Nadduhatti site

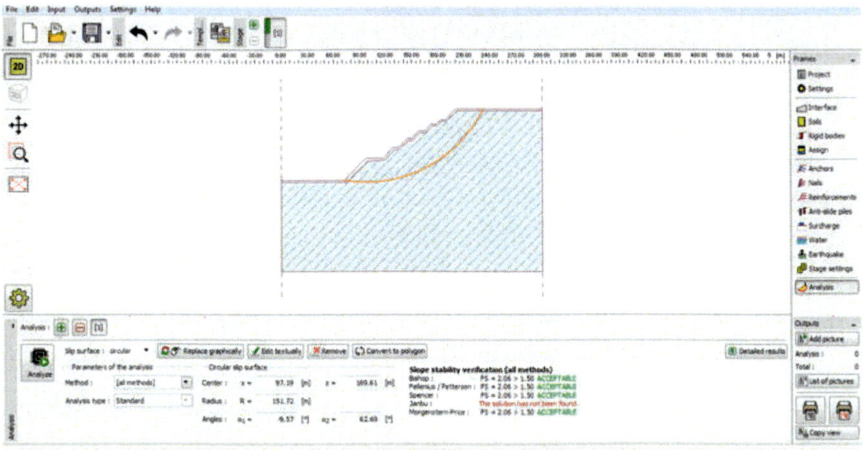

Fig. 5 All methods analysis with 45° and 2 m wetting for Nadduhatti site

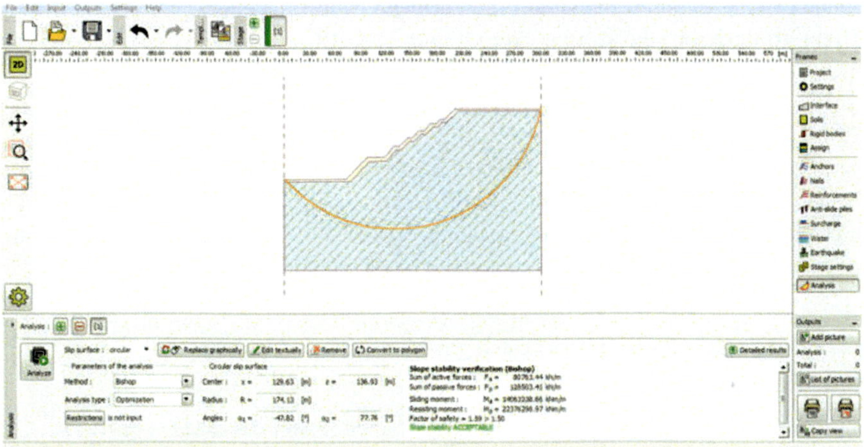

Fig. 6 Optimized analyses with 45° and 2 m wetting for Nadduhatti site

For Anchakal site, the slopes are very much prone to failures during rainy seasons. The slopes with even 30° angles are not safe, and it undergoes large displacement at the toe of the slope. The slopes of Anchakal sites are unstable due to lower shear strength parameters of the site.

Table 5 Safe slopes for maximum duration of rainfall for Achanakal site

Duration of rainfall (h)	Wetting front depth (m)	Slope angles in degree
30	5	30
18	3	45
18	3	60

Table 6 Safe slopes for maximum duration of rainfall for Achanakal site

Duration of rainfall (h)	Wetting front depth (m)	Slope angles in degree
12	2	30

6 Conclusions

Increase in the duration of rainfall increases the wetting front depth. For 23 h of continuous rainfall during the year 2009, the wetting front depth is 3.7 m. Further increase in the duration of rainfall increases the wetting front depth; i.e. for 144 h of rainfall duration, the wetting front depth is 23.18 m which caused many landslides. The loose soil with fair permeability is bound to fail for longer durations of rainfall. The major triggering factor of landslides in both the area is due to ponding of rainwater, since infiltration rate decreases with increase in duration of rainfall. Based on the LEM and FEM method of analysis, the recommended safe slope for Nadduhatti site and Achanakal site for different duration of rainfall is given in Tables 5 and 6, respectively. To minimize the damage from the slides, proper monitoring of infiltration and ponding during monsoon seasons is required.

References

Bishop AW (1955) The use of the slip circle in the stability analysis of slopes. Geotechnique 5(1):7–17

Blazka P, Fischer Z (2014) Moisture, water holding, drying and wetting in forest soils open. J Soil Sci 4(5):11. Article ID: 46264

Chandrasekaran SS (2010) Assessment of damages induced by recent landslides in Ooty, Tamil Nadu, India. In: Indian geotechnical conference, IIT Bombay, Mumbai, India, no 2, pp 687–688

Chatra AS, Dodagoudar R, Maji VB (2017) Numerical modeling of rainfall effects on the stability of soil slopes. Int J Geotech Eng 13(5):425–437

Fellenius W (1936) Calculation of stability of earth dam. In Transactions 2nd Congress Large Dams, Washington, DC, Vol 4, pp 445–462

Horton RE (1939) Analysis of runoff-plat experiments with varying infiltration-capacity. Trans Am Geophys Union 20(4):693–711

IMD (2011) Nilgiri district rainfall for last five years, India Meteorological Department. http://www.imd.gov.in/section/hydro/distrainfall/webrain/tamilnadu/nilgiri. Accessed 04 Oct 2017

Janbu N (1973) In: Hirschfeld RC, Poulos SJ (eds) Slope stability computations. Embankment Dam Engineering-Casagrande, pp 47–86

Lumb P (1975) Slope failures in Hong Kong. Q J Eng Geol 8(1):31–65
Morgenstern NR, Price VE (1965) The Analysis of the Stability of General Slip Surfaces. Géotechnique 15(1):79–93
Rajakumar P, Sanjeevi S, Jayaseelan S, Isakkipandian G, Edwin E, Balaji P, Ehanthalingam G (2007) Landslide susceptibility mapping in a hilly terrain using remote sensing and GIS. J Indian Soc Remote Sens 35(1):31–42
Spencer E (1967) A method of analysis of the stability of embankments assuming parallel inter-slice forces. Geotechnique 17(1):11–26

Effect of Joint Orientation on the Seismic Stability of Rock Slope with Transmission Tower

Mohammad Zaid, Abdullah Talib, and Md. Rehan Sadique

Abstract In India, the Himalayan region lies in the earthquake zone V. The occurrence of landslide hazard is common in the area. Therefore, it renders a necessity to analyze the stability of the slope before designing a slope or a structure near or on it. This paper includes the qualitative stability analysis of a slope located at Karcham Wangtoo in Wangtu, Kinnaur, Himachal Pradesh, India. The qualitative sense lies in the fact that the least stable condition is found by calculating the total maximum deformation and the maximum stresses. The factor of safety was not exclusively calculated. The effect of joint set orientation was analyzed for the stability of slope having static and dynamic loading. The static load includes the gravity load of rockmass, footing and transmission tower. The dynamic load was applied using the acceleration-time history of the Koyna earthquake of magnitude 6.4 on the Richter scale. The analysis was done using the Abaqus/CAE 6.13 software package. It was found that in static condition, 30° of angle of orientation of joint with slope are the critical angle for the instability. For the seismic loading, the critical angle of orientation of joint is found to be 45°. As the distance of footing from the edge of the slope was increased, the stability of slope increases.

Keywords Slope stability · Qualitative analysis · Joint orientation · Seismic loading · Footing

M. Zaid (✉) · Md. R. Sadique
Department of Civil Engineering,
Aligarh Muslim University, Aligarh, Uttar Pradesh 202002, India
e-mail: mohammadzaid1@zhcet.ac.in

A. Talib
Department of Civil Engineering, IISc, Bangalore, Karnataka 560012, India

M. Latha Gali and P. Raghuveer Rao (eds.), *Geohazards*, Lecture Notes in Civil Engineering 86, https://doi.org/10.1007/978-981-15-6233-4_11

1 Introduction

The stability analysis of slope is important importance as the failure of the slope can cause loss of life, vegetation, property and several other losses. In comparison to slopes having no or less construction activity and the one having buildings, transmission towers, etc. are more susceptible to a landslide.

The study of slope stability is a very interesting topic among researchers. Lian-Heng Zhao et al. have studied the stability of slopes having cracks under earthquake loading condition (Zhao et al. 2016). They have concluded that with an increase in the angle of friction increases failure of slope subjected to earthquake load, and seismic forces have more effect than cracks on gentle slopes.

A. Azhari and U. Ozbay have created a database to understand the effect of different external and internal factors on slope failure under earthquake loading condition (Azhari and Ozbay 2017). They have concluded that the presence of water in the region where an earthquake event has taken place increases the chances of slope failure due to earthquake loading. L. Lu et al. studied the presence of water on the stability of slope during an earthquake event (Lu 2015). They have obtained cumulative plastic deformation and the critical value of deformation to analyze the stability of the slope. They concluded that the presence of water reduces the stability of the slope and increases the probability of slope failure even at earthquake having lower energy.

W. R. Azzam has studied the effect of earthquake loading on the skirted foundation which is near a slope of sand (Azzam 2015). He has utilized finite element method of numerical analysis. He concluded that skirted foundation reduces the deformation of slope thus stabilizing it against earthquake.

Qingchao Lv et al. carried out the stability analysis of rock slope, which is induced by an earthquake, based on back analysis of shear strength parameters (Lv et al. 2017). They have shown in their study that complex geological conditions and lack of detailed survey information make it difficult to analyze the stability of rock slope.

Wenqi Du et al. have studied the effect of the variability of property of slope on its stability (Liu 2017). They have concluded that to design safe structure on a slope, the variability of parameters such as properties of slope material like effective friction angle, cohesion etc. has to be taken into account.

Ozer Cinicioglu and Anil Erkli have developed design charts considering the influence of all the parameters affecting the undrained bearing capacity (Cinicioglu and Erkli 2018). They concluded that bearing capacity must be checked for stability before going for the shallow foundation on slopes and the different combination of geometry, property and load leads to different types of slope failure.

Due to heavy rainfall and earthquake, there are increased chances of a landslide in this region. Different landslides that took place in the region of study are Thangi slide, Kadra Dhang slide, Pangi slide, Powari slide, Barua slide, Uri rockfall, Nathpa landslide, Shoulding Khad slide, Jhakri landslide. On November 24, 2015, a massive

Table 1 Seismic events within 200 km of the study area

Date of occurrence	Epicenter		Magnitude on richter scale
	Latitude (°N)	Longitude (°E)	
05.03.1842	30	78	6.5
16.06.1902	31	79	6.0
13.06.1906	31	79	6.0
28.02.1908	32	77	7.0
20.10.1937	31.1	78	6.0
12.05.1939	32.5	78	6.3
22.06.1945	32.8	76.9	6.5
10.06.1947	32.6	75.9	6.0
27.06.1955	32.5	78.5	6.0
12.04.1963	32	78.79	6.0
19.01.1975	32.35	78.76	6.8
20.10.1991	30.75	78.86	6.6

Source Seismic Parameters for Nathpa-Jhakri Hydroelectric Project Site, H.P. Project No. 575, Department of Earthquake Engineering, University of Roorkee, 1993

landslide on national highway-5 in Kinnaur had taken place affecting people and causing loss of lakhs of rupees. These reasons have made authors took up this region to study and analyze the slope having transmission tower for stability under earthquake loading.

This paper focuses on the stability of a slope in Wangtu which is at 1823 m elevation from mean sea level, near Karcham Wangtoo powerhouse alongside the Satluj River. The region lies in seismic zone V which is highly susceptible to earthquakes. There are several earthquake events noted in this region some of them are shown in Table 1. The authors have carried out the analysis of jointed rock slope. The angle of orientation of joint with slope has been varied and the location of the footing of transmission tower has been changed to analyze the stability of the slope. As the region of the study lies in seismic zone V; hence, the Koyna earthquake's data of acceleration-time history has been adopted for the analysis of rock slope. Qualitative method of analysis has been used to study the stability and failure of slope. Total maximum deformation at the edge of the slope has been taken under consideration to compare different cases for stability.

2 Numerical Model

The geometry of the slope is shown in Fig. 1. The slope has a 75° dip from the horizontal. The dimension of the footing of the transmission tower was taken as 4.5

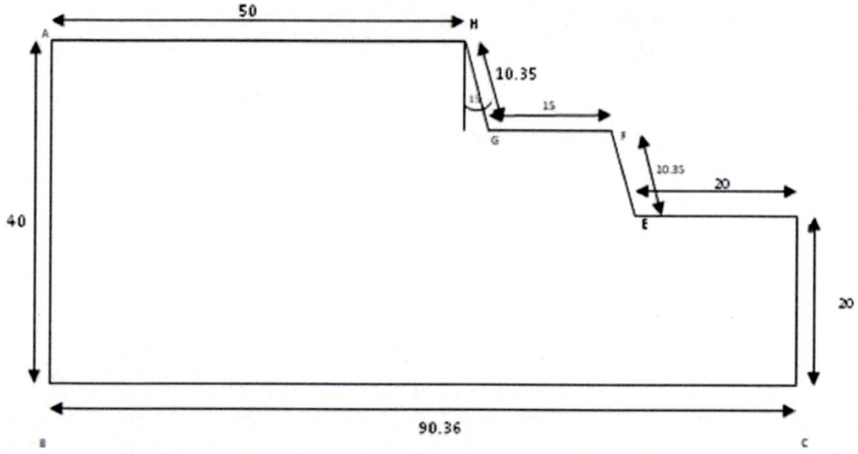

Fig. 1 The geometry of the Karcham Wangtoo slope

× 4.5 m ($d = 4.5$ m). The orientation of the joint was considered initially as 0° and was changed up to 90° clockwise from the slope. The footing of the transmission tower was considered initially at the edge of the slope and the position of transmission tower was varied between 0d (0 m) and 5d; here; the distance between the edge of the slope and near the edge of footing was considered (Zaid et al. 2018).

The elasto-plastic 2D plane strain model was considered for the 2D analysis and the Mohr–Coulomb theory was applied for the failure of rockmass in the Abaqus/standard 6.13 (Abaqus 2013). The property of the rockmass and concrete footing is given in Table 2 (Gahoi et al. 2017).

The analysis was carried out in two steps, at first, the static analysis was done for 1 s and the earthquake was applied for which acceleration-time history is shown in Fig. 2.

The interaction of the coefficient of friction 0.85 was applied as general contact to the model and normal hard as contact property. The interaction property of coefficient of friction between the footing and rockmass was applied as 0.55 (Naqvi et al. 2017).

Table 2 Property of numerical model

	Property	Values
Basalt	Mass density	2910 kg/m^3
	Young's modulus	46.5 GPa
	Poisson ratio	0.186
	Friction angle	63.38
	Cohesion	26.25 MPa
Concrete	Mass density	2400 kg/m^3
	Young's modulus	25 GPa
	Poisson ratio	0.15

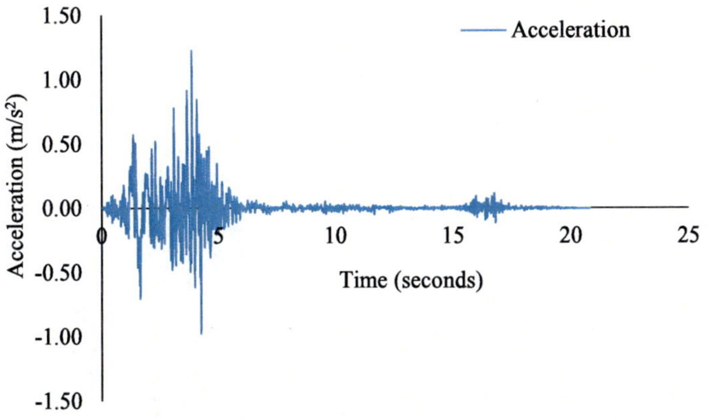

Fig. 2 Acceleration-time history of Koyna earthquake

The load of 28,430 N/m was applied on the footing as the weight of the upper structure of transmission tower. The gravity was applied to the whole model as −1 component in the load. The absorbing boundary condition was applied to the model for the seismic analysis (Fig. 3).

The elements for the static analysis were meshed as CPE3 a three-node plane strain triangle element type. The model was meshed as an element type of CPE4R. A four-node bilinear plane strain quadrilateral, reduced integration, hourglass control and the absorbing boundary condition were meshed as CINPE4 element type (Naqvi et al. 2017) (Fig. 4).

3 Analysis

The numerical analysis was done using Abaqus/standard and explicit method of modeling. The paper has two parts, the first step has static analysis and the second step has dynamic implicit analysis.

In the first step, the stability of jointed rockmass has been studied and the results were plotted.

In the second step of the analysis, an earthquake loading was applied to the base of the model. The acceleration-time history of the Koyna earthquake was applied (Fig. 2). The results are obtained for total maximum deformation at the edge of the slope to study the stability of the slope.

The analysis of the model followed the different steps as mentioned below:

1. The geometry of the model to be analyzed was created in this step by specifying the dimensions as shown in Fig. 1.
2. Creating a directory of property to be applied to the model and application of the property to the footing and the rockmass.
3. Assembling of footing and rockmass.

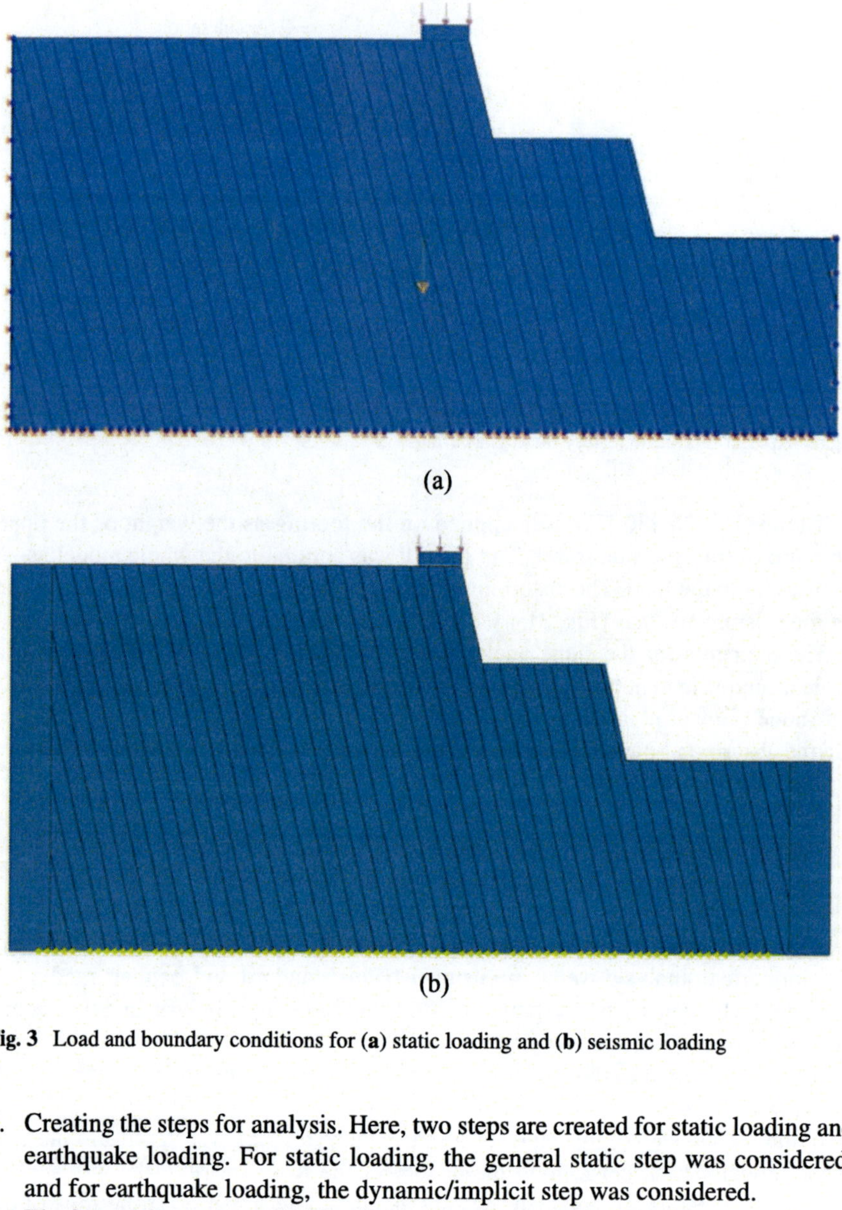

Fig. 3 Load and boundary conditions for (a) static loading and (b) seismic loading

4. Creating the steps for analysis. Here, two steps are created for static loading and earthquake loading. For static loading, the general static step was considered, and for earthquake loading, the dynamic/implicit step was considered.
5. The interaction was applied for footing, rockmass and joints.
6. Loads and boundary conditions were applied. The earthquake was applied at the base of the model in the x-direction (Naqvi et al. 2017).
7. Meshing was done and absorbing boundary condition was created.
8. The final analysis job was created and the results were extracted.

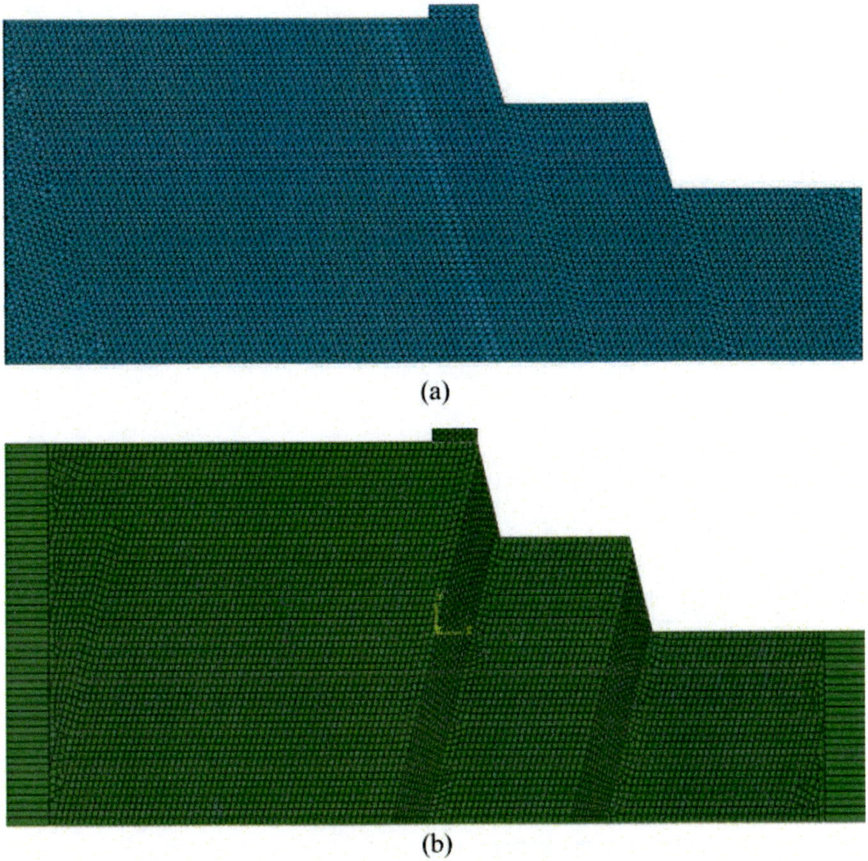

Fig. 4 The meshing of the model in the (**a**) static analysis and (**b**) seismic analysis

4 Results

This paper is composed of two parts of analysis; first, the static loading results are presented. In static analysis, the self load of footing and the load from the transmission tower, which was been calculated as 28,430 N/m, has been considered. The total maximum deformation on the edge slope due to the fixed location of transmission footing and varying orientation of the joint set is shown in Fig. 5.

Figure 6 shows the graph of total maximum deformation with varying angle of orientation of joint and for different location of footing ranging from 0 m to 5d.

Figure 7 has been plotted for thc study of stress concentration at the edge of the slope for different cases of static analysis.

The above results are for the static step of the analysis. In the second step, the seismic analysis of the slope was carried out using the earthquake model represented earlier.

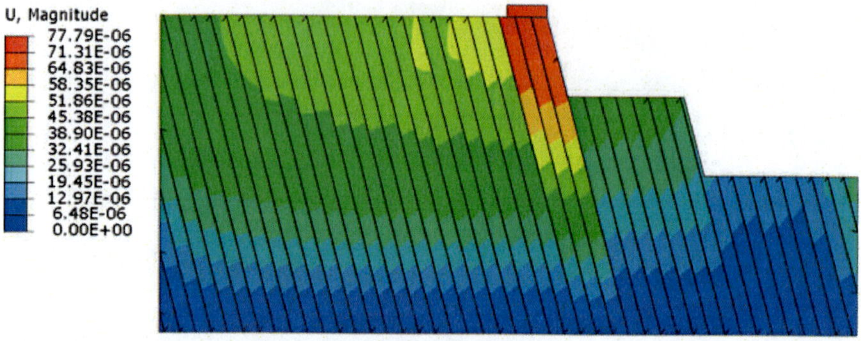

(a). 0d and 0-degree Joint Orientation with Slope

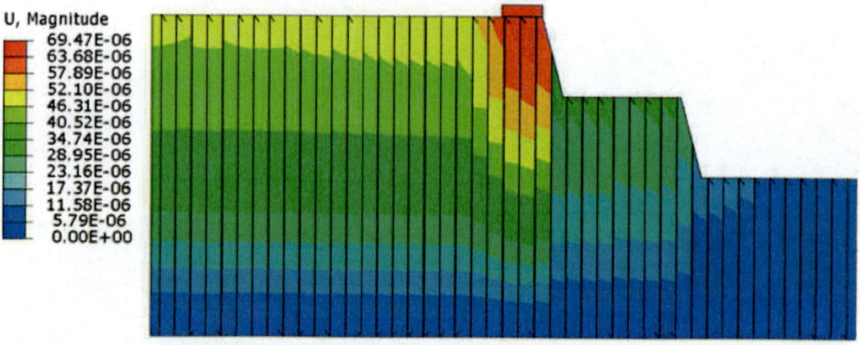

(b). 0d and 15-degree Joint Orientation with Slope

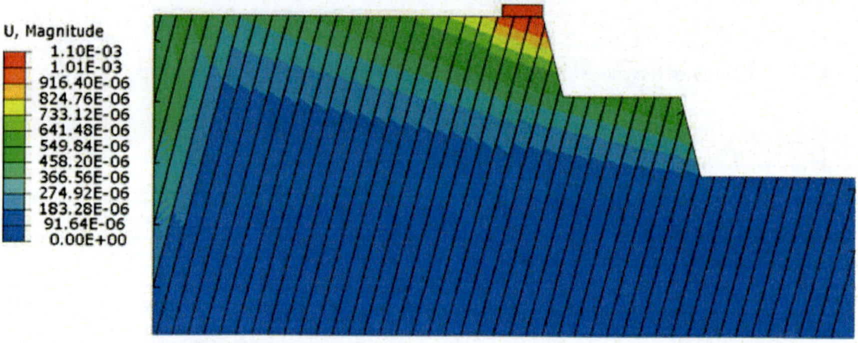

(c). 0d and 30-degree Joint Orientation with Slope

Fig. 5 Deformation contours for different orientation of joint having footing of transmission tower at the edge of slope (in meter)

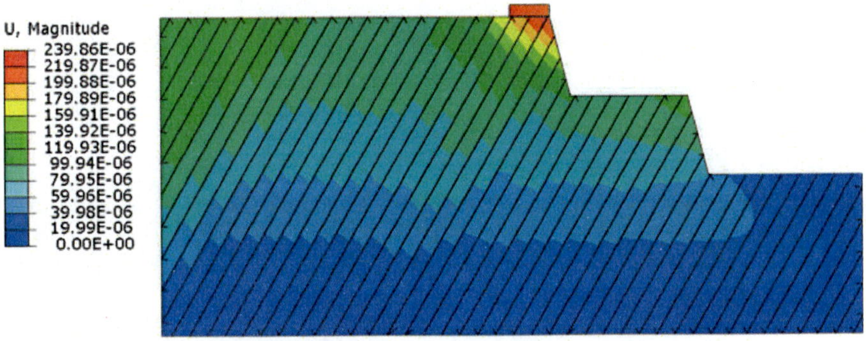

(d). 0d and 45-degree Joint Orientation with Slope

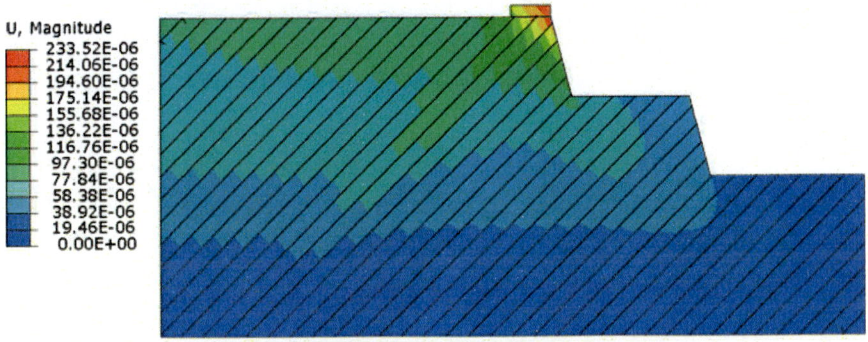

(e). 0d and 60-degree Joint Orientation with Slope

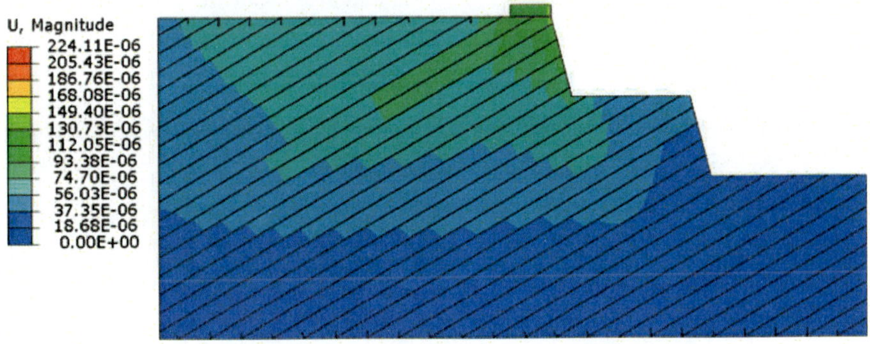

(f). 0d and 75-degree Joint Orientation with Slope

Fig. 5 (continued)

(g). 0d and 90-degree Joint Orientation with Slope

Fig. 5 (continued)

Fig. 6 Graph of deformation versus orientation of joint for static loading

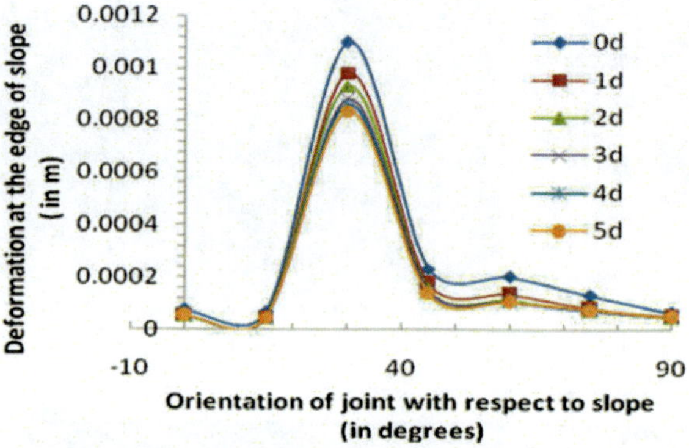

Fig. 7 Graph of stress at the edge of the slope for varying position of the footing and change in orientation of the joint in case of static loading

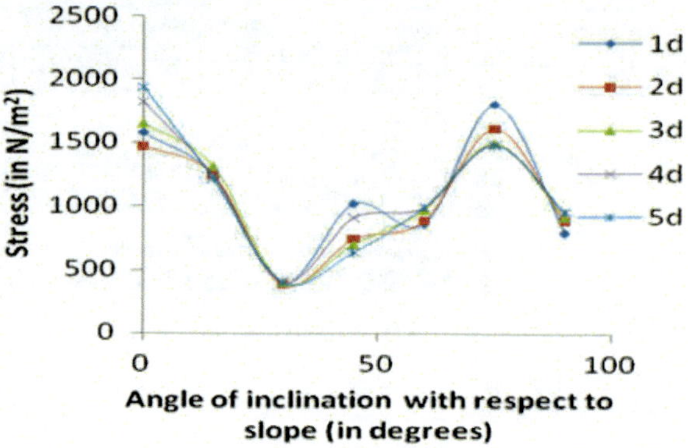

Figure 8 shows the deformation for different locations of the footing of transmission tower with changing the angle of orientation of the joint set.

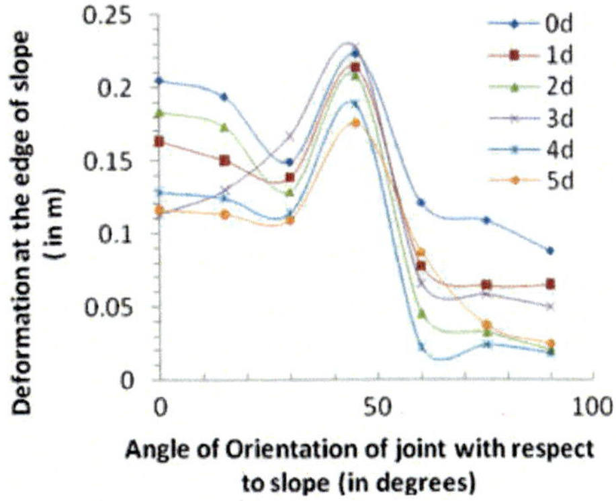

Fig. 8 Graph of deformation versus angle of orientation of joint for different location of footing in case of seismic loading

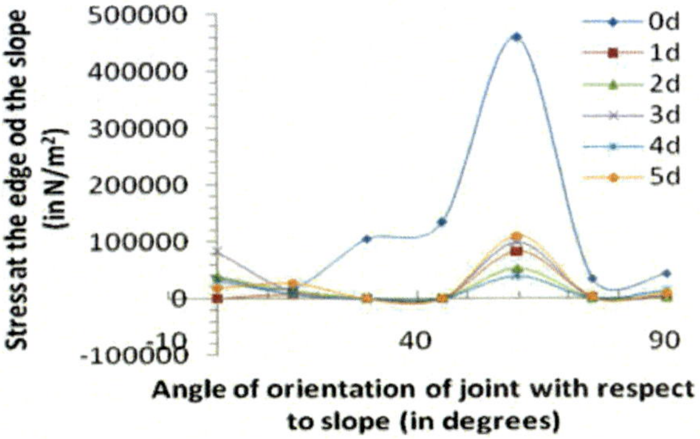

Fig. 9 Graph of stress at the edge of slope versus angle of orientation of joint for varying location of footing in case of seismic loading

Figure 9 shows the plots for the stress at the edge of the slope for both the static and earthquake loading. It shows the variation of the stress at the edge of the slope for different angles of orientation of joint set and varying location of the footing of the transmission tower.

5 Discussion

The paper showed the variation of deformation and stresses at the edge of the slope in both static as well as seismic loading cases.

In static loading case, from Fig. 6, it appears that for the angle of orientation of joint having 30° with slope, maximum deformation has been noted for each change in location of footing. In Fig. 7, at 45° of orientation of joint, minimum stress can be noted at the edge of the slope for static loading.

In seismic loading case, Fig. 8 shows that at 45° of angle of orientation of joint with slope, the slope has maximum deformation at the edge. In Fig. 9, which shows a plot between the stress and the angle of orientation for the varying location of the footing of transmission tower shows that at 60° of angle of joint orientation, maximum stress can be noted for each location of footing.

6 Conclusion

The study of the effect of joint orientation on the stability of rock slope subjected to earthquake loading has been carried out by using finite element software Abaqus/Standard 6.13. The joint orientation parallel to the slope which is having 75° of dip with the horizontal said to be 0° angle of orientation. The orientation of joint with the slope was varied up to 90°. The Koyna earthquake time history has been utilized for the seismic loading of the model. The conclusions drawn from this study are as follows:

- The distance between the near edge of the footing of the transmission tower with the edge of the slope must be more than the width of the footing for achieving the least unstable conditions.
- It can be concluded that if joint orientation is between 30° and 60° with slope, then the occurrence of the least stable condition can be expected. As the results show that for this range of joint orientation, either the deformation or stress is maximum for static and dynamic cases.
- The finite element method of analysis has proved as a time-saving, effective, economical method of analyzing typical problems related to earthquake loading.
- Analysis of stability of slope by observing the total maximum deformation has proved to be an effective method of stability analysis. This method is although a qualitative method to analyze the stability, but as it gives the results faster as compared to methods based on the calculation of factor of safety of degree of freedom. Therefore, this method can be used for the initial prediction of the stability of the slope for its further analysis and design.
- For the future study of this topic, joint spacing can be varied, multiple joint sets can be used, fault zone and shear zone can also be incorporated, and different loads other than transmission tower can also be used.

References

Abaqus V (2013) 6.13 documentation. Dassault Systemes Simulia Corporation

Azhari A, Ozbay U (2017) Investigating the effect of earthquakes on open pit mine slopes. Int J Rock Mech Min Sci 100:218–228. Available at: https://dx.doi.org/10.1016/j.ijrmms.2017.10.005.

Azzam WR (2015) Finite element analysis of skirted foundation adjacent to sand slope under earthquake loading. HBRC J 11(2):231–239. Available at: https://linkinghub.elsevier.com/retrieve/pii/S1687404814000327

Cinicioglu O, Erkli A (2018) Seismic bearing capacity of surficial foundations on sloping cohesive ground. Soil Dynam Earthq Eng 111:53–64. Available at: https://doi.org/10.1016/j.soildyn.2018.04.027

Gahoi A et al (2017) Numerical analysis of the tunnels subjected to impact loading. In: INDOROCK. Indorock2017, New Delhi

Liu Y et al (2017) Probabilistic stability analyses of undrained slopes by 3D random fields and finite element analysis. Geosci Front. Available at: https://linkinghub.elsevier.com/retrieve/pii/S1674987117301597.

Lu L et al (2015) Stability analysis of slopes with ground water during earthquakes. Eng Geol 193:288–296

Lv Q, Liu Y, Yang Q (2017) Stability analysis of earthquake-induced rock slope based on back analysis of shear strength parameters of rock mass. Eng Geol 228:39–49. Available at: https://dx.doi.org/10.1016/j.enggeo.2017.07.007

Naqvi MW et al (2017) Dynamic analysis of rock tunnels considering joint dip angle: a finite element approach. In: 13th international conference on vibration problems. Indian Institute of Technology Guwahati, India

Zaid M, Talib A, Sadique MR (2018) Stability analysis of rock slope having transmission tower. IJRECE 6(2)

Zhao LH et al (2016) Stability analysis of seismic slopes with cracks. Comput Geotech 77:77–90. Available at: https://dx.doi.org/10.1016/j.compgeo.2016.04.007.

Landslide Hazard and Risk Assessment Along NH-108 in Parts of Lesser Himalaya, Uttarkashi, Using Weighted Overlay Method

Pankaj Kumar, Anupam Mital, P. K. Champati Ray, and Shovan L. Chattoraj

Abstract The present study is focused on the landslide incidences which are very frequent in the Himalayan region. Economy in most of the Himalayan region thrives more on tourism; hence, it is essential to take proper attention in these areas. The objectives of the present study include generation of a landslide inventory, damage assessment, hazard mapping and subsequently risk analysis. The aim is to identify the causative factors responsible for landslide occurrences in these areas and then prepare susceptibility and vulnerability zonation maps. The study area is a 46 km long stretch of NH-108 covering 184 km^2 of the area in Uttarkashi district of Uttarakhand, India. The entire analysis is carried out using weighted overlay approach in ArcGIS software. The susceptibility map of the region is prepared and categorized from low to very high susceptible zones. Accordingly vulnerability and risk map of the region are also prepared. The entities considered in risk assessment include agricultural land, roads and places of human settlements. The validation of the analyses results is carried out through field investigations and satellite imageries.

Keywords Landslide · Remote sensing and GIS · Weighted overlay method · Risk assessment

1 Introduction

Since inception of the Himalaya natural hazards (flash flood, cloud burst, lake burst, forest fire, landslide, etc.) has been threatening human lives and ecosystem, landslide is one of the major geohazards causing huge losses in terms of human life, culture, economy, infrastructure, public utilities and other resources which can have long-term impact on the society and to the nation. For carrying out any study related to landslide, a thorough understanding of the phenomenon needs to be understood.

P. Kumar (✉) · A. Mital
Civil Engineering Department, NIT Kurukshetra, Kurukshetra, Haryana, India
e-mail: sampankaj23@gmail.com

P. K. C. Ray · S. L. Chattoraj
Geosciences and Geohazard Department, IIRS, Dehradun, Uttarakhand, India

M. Latha Gali and P. Raghuveer Rao (eds.), *Geohazards*, Lecture Notes in Civil Engineering 86, https://doi.org/10.1007/978-981-15-6233-4_12

Landslide is a dynamic phenomenon occurring every year in rugged mountainous terrains. Landslide is defined as the movement of masses of rocks, debris or earth down a slope (Cruden 1991). The landslide phrase is formed from two different words, the initial one depicting the type of material, i.e. rock, debris, earth, etc., and the latter one depicting the type of movement like fall, flow, slide, etc.

Landslides are caused due to instability of slopes and are generally triggered by natural causes (precipitation, earthquakes, aftershocks, etc.) and human interferences with nature (by unscientific road cutting, dumping of debris along roadsides, land use for construction projects, deforestation, illegal mining operations, etc.) rendering hill slopes unstable. In addition to these, forest fires also cause soil erosion and induce landslides due to loss of vegetation cover.

The causes that increase the shear stress acting on the slope and/or degree of cohesion in the slope materials result in the movement of slope causing landslide. These conditions can arise due to erosion, excavation, slope steepening, overburden, earthquakes and increase in pore water pressure. According to Terzaghi (1950), landslides can result from external causes which cause an increase in shear stress (rapid drawdown, water table fluctuation, unloading slope toe, etc.) and internal causes which decrease shear resistance (erosion, weathering, etc.).

2 Study Area

The Western Himalayan region is well known for frequent landslide incidences especially during rainy seasons because of its fragile landscape. The Garhwal Himalaya region has a history of landslides. Some very well-known landslides witnessed by Garhwal Himalaya are co-seismic in nature, like 1991 Uttarkashi earthquake caused numerous massive landslides, especially on a 42-km road section between Bhatwari and Uttarkashi (Jain et al. 1992). Besides this, in the year 1998, Malpa landslide in Uttarkashi killed 300 people including 60 pilgrims of Kailash-Manasarovar Yatra. However, the most horrible and devastating one was the Kedarnath tragedy of June 2013 caused due to cloud burst induced heavy rainfall. The numerous landslides that occurred blocked several arterial roads and national highways disrupting movement of people and goods for weeks, thus disconnecting the region from rest of the state.

The study area is a 2.5-km buffer area in Uttarkashi district along the NH-108 (35 km) and -94 (8 km) in the Bhagirathi River valley. This area spans around 184.3 km^2 and extends from 30° 36′ to 30° 47′ North latitude and 78° 17′ to 78° 30′ East longitude, falling in the survey of India topographic sheet numbers 53J01, 02, 05 and 06, covering Chinyalisour, Dharasu, Nakuri, Matli, Uttarkashi, Gangori and Siror as major places. The location map of study region is shown in Fig. 1.

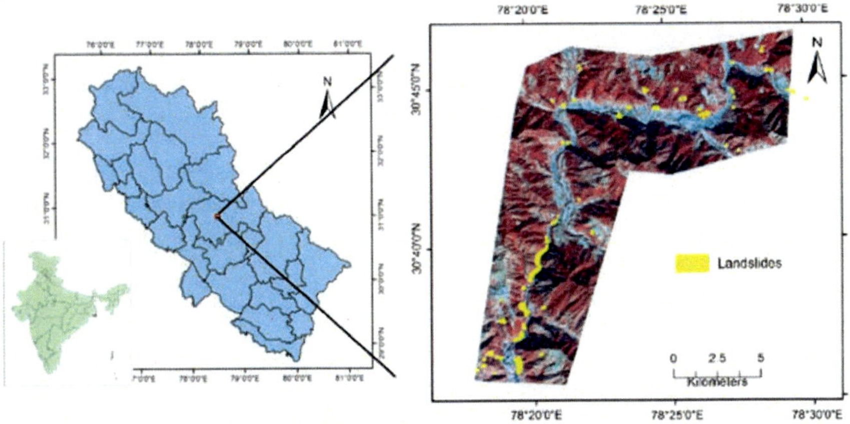

Fig.1 Location map of NH-108

3 Geology and Geomorphology

The rock formations of this area belong to Lesser Himalaya and some from Higher Himalaya sequence and comprise Berinag Formation (consisting of massive, coarse grained to pebbly and usually 'sericitic quartz arenite' of white, pale purple and green colour with metamorphosed amygdaloidal vesicular basalts and tuffites), Damta Formation (consisting of deep purple, violet and pink fine-grained cross-bedded quartzite interbedded with ripple marked and mudcracks bearing purple slates), Chandpur Formation (consisting of dark grey, green, maroon and purple green coloured phyllites, slates and shales), Nagthat Formation (consisting of fine to coarse-grained quartzarenite with subordinate amount of purple to grey sandstone, siltstone–shale and conglomerates) and central crystallines (Valdiya 1983) as shown in Fig. 2. Geomorphologically, the area is divided into six major units, i.e. the high denudational mountain and the river valleys. The general geomorphic features of the study region include highly dissected hills and valleys, moderately dissected hills and valleys, active flood plain, water bodies, mass wasting and piedmont alluvium plain.

4 Hazard and Risk Assessment

The hazard is the probability a particular landslide which occurs within a given time and space (Varnes 1984). The risk assessment study involves analysis of hazard, its consequences and determining mitigation measures based on level of risk to various elements (settlements, roads and agricultural lands). The assessment results are represented in the form of susceptibility, vulnerability and risk maps. The susceptibility

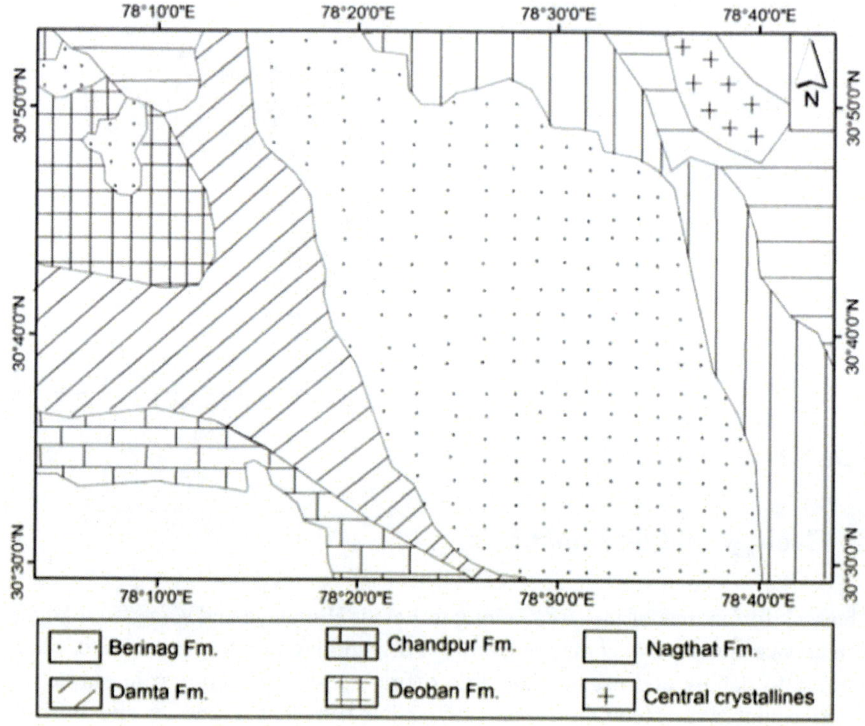

Fig. 2 Geological map of study region (After Valdiya 1983)

of an area is the degree to which it is influenced by soil/rock movements or an estimate of where mass movements can occur. Vulnerability here refers to inability of elements to withstand the effects from a landslide event. Risk is presented by Varnes (1984) as "the expected number of lives lost, persons injured, damage to property and disruption of economic activity due to a particular damaging phenomenon for a given area and reference period".

For the preparation of landslide susceptibility map, various thematic layers (slope, aspect, geology, distance to linear features faults/lineaments, land use/land cover, soil, vegetation, erosion, geomorphology, drainage density and landslide locations) are generated in Arc Map 10.3. Subsequently, all the vector layers are converted into raster layers as shown in Figs. 3, 4, 5, 6, 7, 8, 9, 10, 11 and 12. Details about the thematic layers prepared for both susceptibility as well as vulnerability map are explained in Table 1.

After thematic layer preparation, weighted overlay technique is applied for the generation of susceptibility and vulnerability maps, in which each layer is given some weightage and each class of all layers is also given ranks. The distribution of weights amongst different thematic layers on which landslide occurrence is dependent is

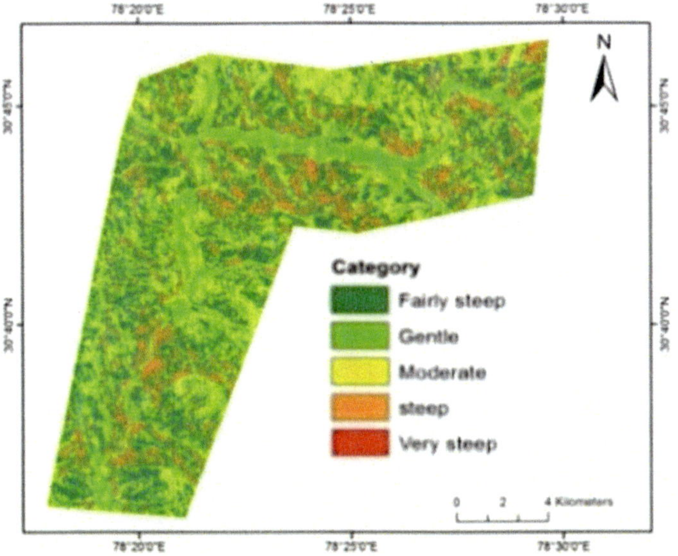

Fig. 3 Slope map

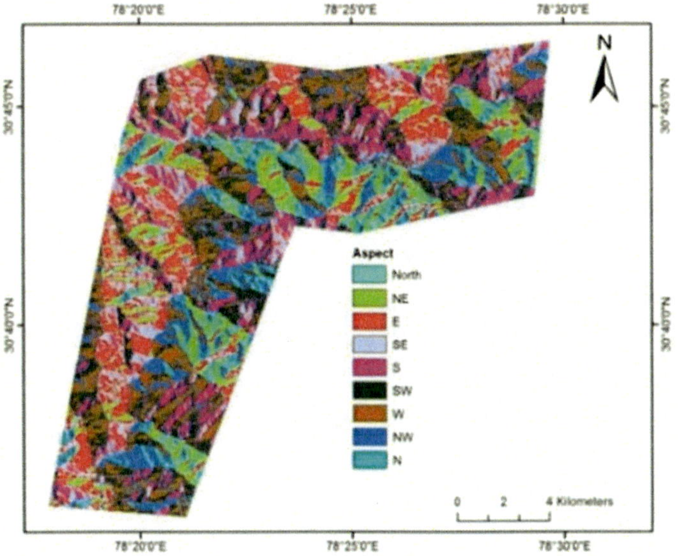

Fig. 4 Aspect map

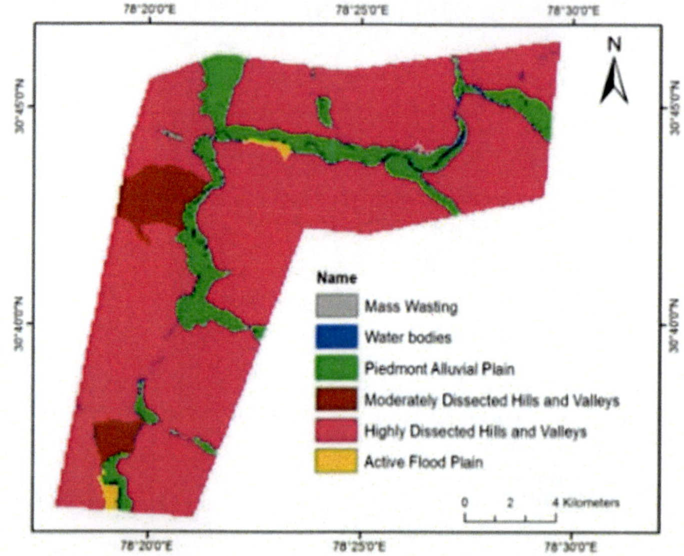

Fig. 5 Geomorphology map

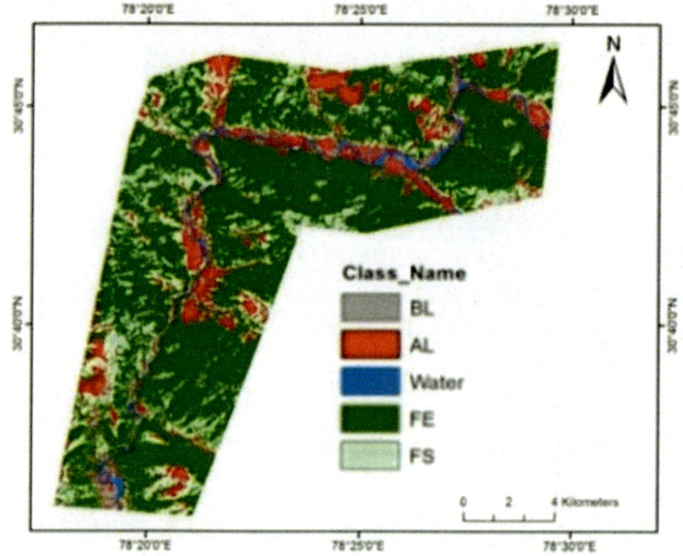

Fig. 6 LULC map

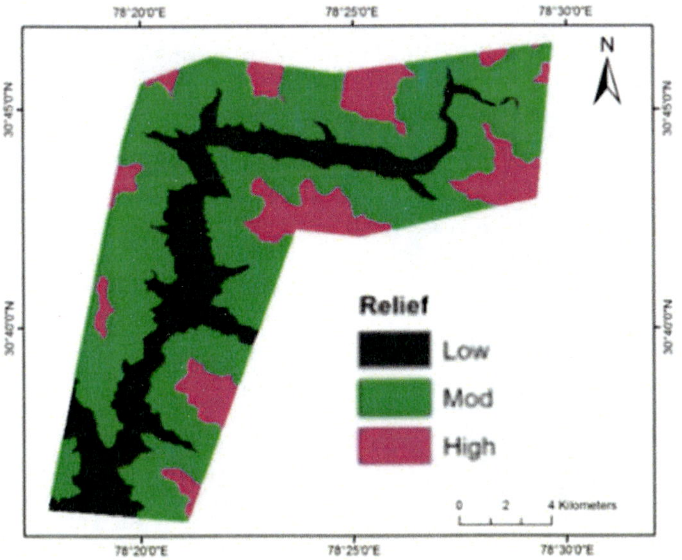

Fig. 7 Relief map

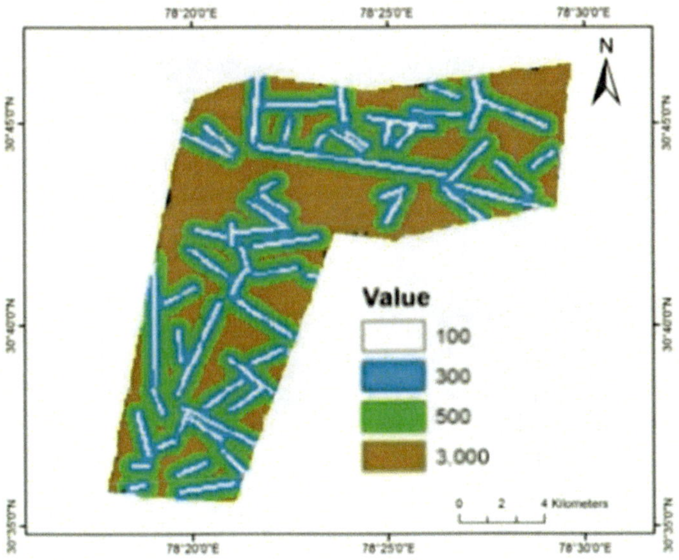

Fig. 8 Lineament map

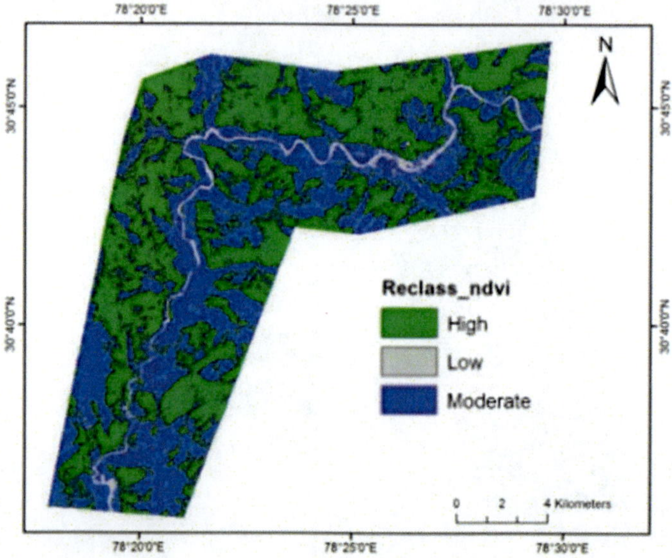

Fig. 9 Vegetation map

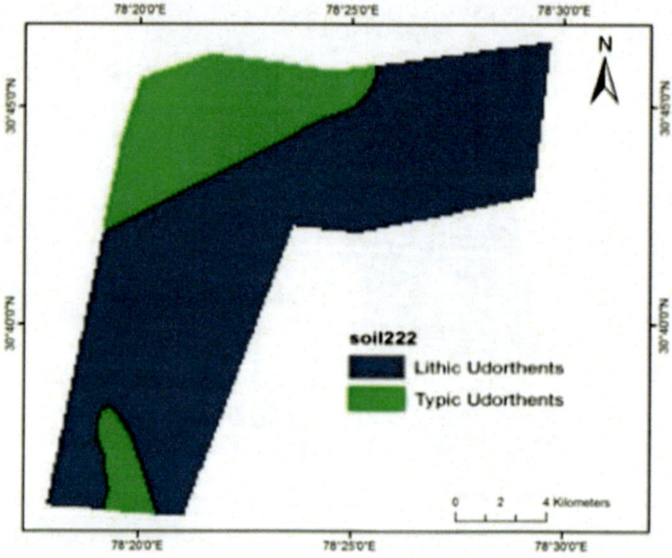

Fig. 10 Soil map

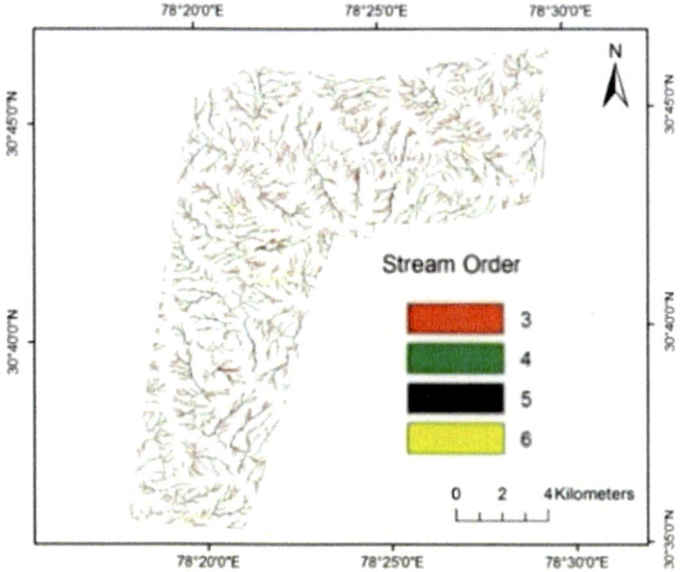

Fig. 11 Drainage density map

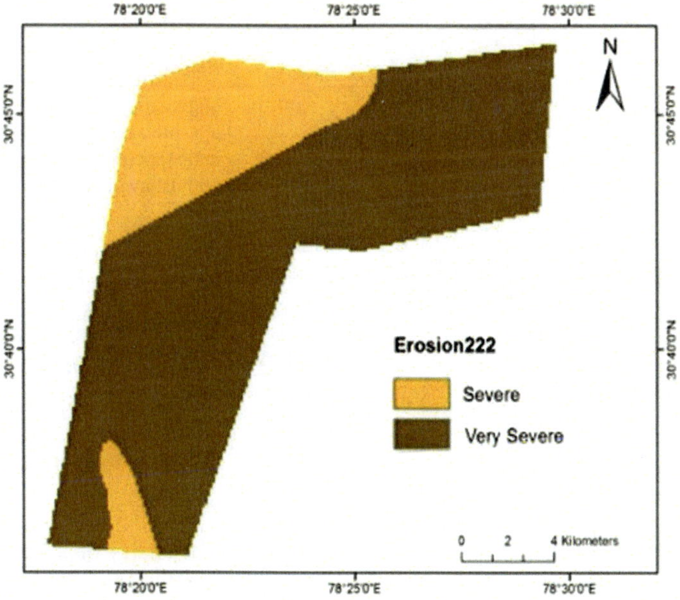

Fig. 12 Erosion map

Table 1 Thematic layers and their source of generation

S. No.	Thematic layers	Source
Landslide inventory		
1	Landslide location map	IRS-P6 (LISS-4) image
Susceptibility mapping		
2	Slope	Generated form SRTM DEM (30 m)
3	Aspect	Generated form SRTM DEM (30 m)
4	Geology	Geological map (Valdiya 1983)
5	Structural features (Fault and lineaments)	Image enhancement (edge detection) Landsat-8 imagery
6	Drainage density	Hydrologic processing using Cartosat-1 DEM (30 m)
7	Land use/land cover	Digitization based on visual interpretation over LISS-IV MS image
8	Landslide locations	Digitized over LISS-IV imagery
9	Soil and erosion	Soil map of NBSS-LUP
10	Vegetation type	Supervised classification (Max likelihood) (Validated with Bhuvan portal and field input)
11	Geomorphology	Digitization based on visual interpretation of LISS-IV multispectral image, validated with Bhuvan portal and field input
Risk mapping		
12	Settlement	Digitization based on visual interpretation of LISS-IV multispectral image and SOI toposheets, cross checked with Google Earth images
13	Road	
14	Agricultural field	Digitization based on visual interpretation over LISS-IV multi-spectral image

shown in Table 2. The weightage is given based on the relationship observed amongst the causative factors for landslide occurrence as observed in field and also the previous literature by assigning a bivariate landslide susceptibility index (W_i) according to Eq. 1 for each causative factor on the basis of aerial densities (Yin and Yan 1988; Conforti et al. 2014):

$$S = \frac{\sum W_i S_{ij}}{\sum W_i} \quad (1)$$

Table 2 Distribution of weights to thematic layers for weighted overlay technique

Raster	% influence	Field	Scale
Geology	13	Volcanics of Garhwal	2
		Berinag formation	6
		Damta formation	9
Slope	17	0–15 Gentle	1
		15–25 Moderate	4
		25–40 Fair steep	8
		40–60 Steep	9
		>60 Very steep	3
Drainage density	7	Low	2
		Moderate	4
		High	7
Geomorphology	12	Mass wasting	5
		Water bodies	1
		Piedmont alluvial plain	3
		Mod. dissected hills	6
		Highly dissected hills	7
		Active flood plain	4
NDVI	9	Low	9
		Moderate	5
		High	3
Land use	10	Built-up	2
		Agricultural land	2
		Water bodies	1
		Forest-evergreen	3
		Barren land/scrub	8
Aspect	5	North	1
		Northeast	3
		East	5
		Southeast	6
		South	9
		Southwest	6
		West	5
		Northwest	2
Relative relief	4	Low	1

(continued)

Table 2 (continued)

Raster	% influence	Field	Scale
		Mod	3
		High	6
Soil erosion	5	Severe	3
		Very severe	7
Lineament buffer distance	12	100	7
		300	4
		500	2
		>500	1

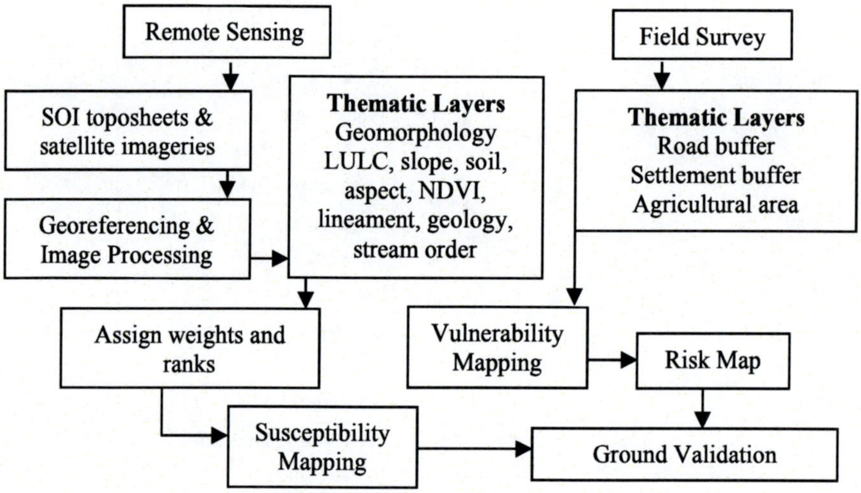

Fig. 13 Flowchart depicting methodology of the study

where W_i is weight of ith factor map, S_{ij} is the ith spatial class weight of jth factor map and S is the spatial unit value in output map. The methodology adopted to obtain the susceptibility, vulnerability and risk maps using weighted overlay method is shown in Fig. 13.

5 Results

The results of the study are presented in the form of landslide susceptibility, vulnerability and risk maps as discussed below.

5.1 Landslide Susceptibility Analysis

The landslide susceptibility is prepared through weighted overlay method. The landslide susceptibility map is classified into five classes, viz. very less, less, moderate, high and very high susceptible zones. The total area is 184.3 km^2, and out of the total area, 3.53% shows very less susceptibility, 11.81% shows less susceptibility, 31.96% lies under moderate susceptible zone and 49.87% lies under high susceptible zone and 2.83% is very high susceptible zone as shown in Fig. 14 and Table 3. The validation of result is done through landslides observed during field and also from the LISS-IV satellite image on which more than 90 landslides, as shown in Fig. 1, were digitized.

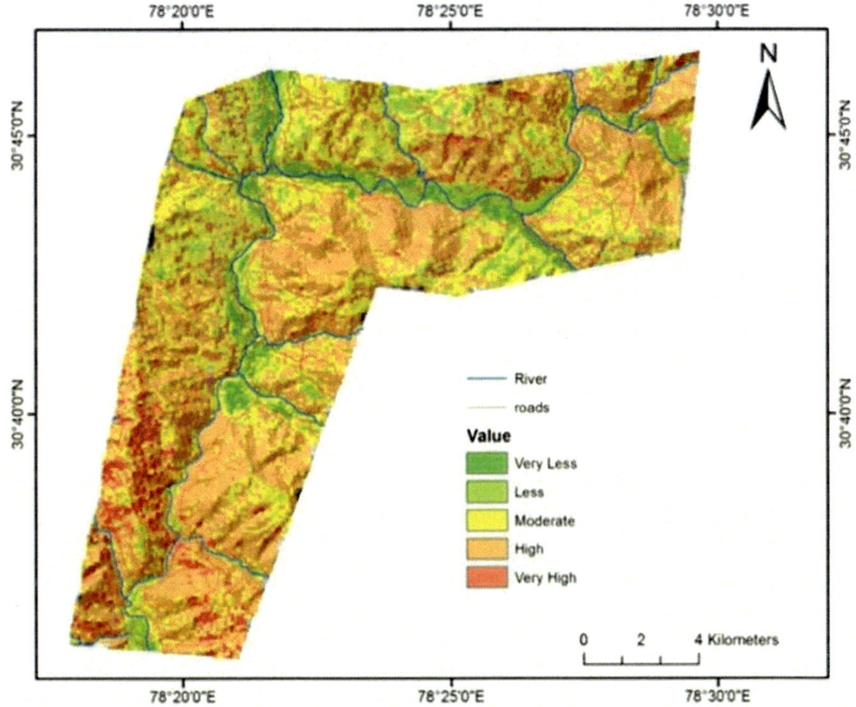

Fig. 14 Landslide susceptibility map

Table 3 Statistics from landslide susceptibility analysis

Susceptibility class	Area (km^2)	%age area	Cumulative area (%)
Very less	06.50	03.53	03.53
Less	21.77	11.81	15.34
Moderate	58.90	31.96	47.30
High	91.92	49.87	97.17
Very high	05.21	02.83	100

5.2 Landslide Vulnerability Analysis

The landslide vulnerability map is also prepared through weighted overlay method considering the three thematic layers, i.e. road, built-up and agricultural land, and assigning equal weightages to the thematic layers. The vulnerability map is classified into three classes, viz. low, moderate and high vulnerable zones. The total area is 184.3 km^2, and out of the total area, 59.29% lies in low, 33.17% lies in moderate and 7.54% lies in high vulnerable zones as shown in Fig. 15 and Table 4.

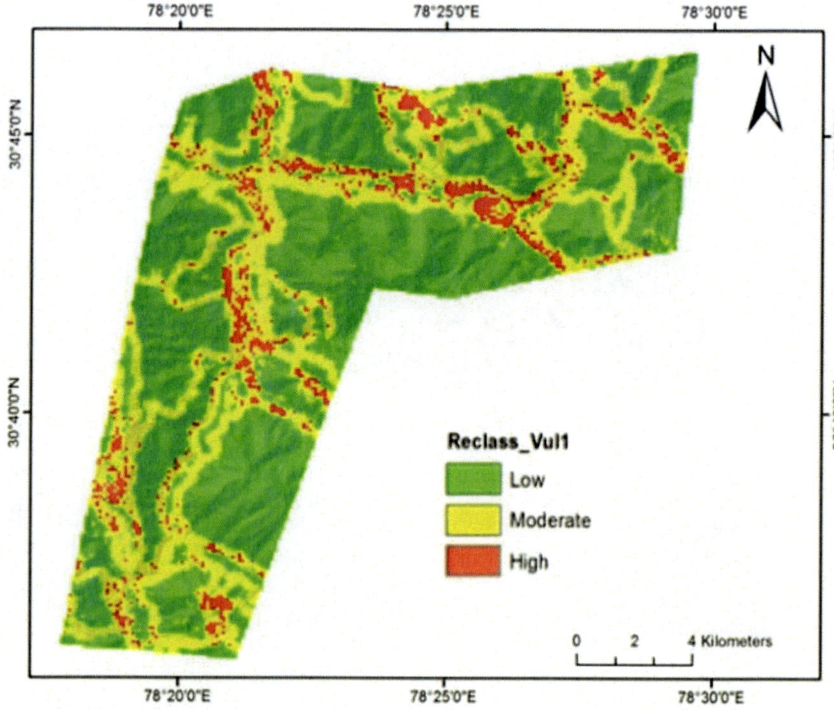

Fig. 15 Landslide vulnerability map

Table 4 Statistics from landslide vulnerability analysis

Vulnerability class	Area (km^2)	%age area	Cumulative Area (%)
Low	109.28	59.29	59.29
Moderate	61.13	33.17	92.46
High	13.89	07.54	100

5.3 *Landslide Risk Analysis*

Risk is the product of susceptibility and vulnerability maps, i.e. risk map is derived by crossing the susceptibility and vulnerability map. The risk map as shown in Fig. 16 is divided into five class, viz. very low-, low-, moderate-, high- and very high-risk zones. Out of 184.3 km^2 of study area, 22.95% lies in very low-risk zone, 37.83% lies in low-risk zone, 18.01% lies in moderate-risk zone, 16.75% lies in

Fig. 16 Validation of risk assessment results from field observations. **a**, **b**, **c**, **d**, **e** and **f** shows the risk due to landslide to tunnel area, tourist lodge, bridge, 60-m road stretch, agricultural land and settlement concentrations, respectively

Table 5 Statistics from landslide risk analysis

Risk class	Area (km^2)	%age area	Cumulative area (%)
Very low	42.30	22.95	22.95
Low	69.72	37.83	60.78
Moderate	33.19	18.01	78.79
High	30.88	16.75	95.55
Very high	08.21	04.45	100

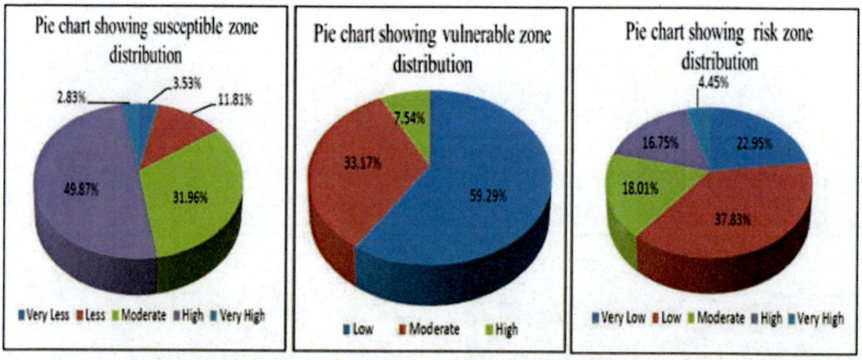

Fig.17 Pie chart distribution of susceptible, vulnerable and risk zones

high and only 8.21% is under very high-risk zone. The results of risk analysis are validated from field observations as well as from LISS-IV satellite images. The risk statistics based on pixel densities are shown in Table 5. The pie chart distribution of landslide susceptible zones, vulnerable zones and risk zones is shown in Fig. 17. Overlay analysis of risk elements reveals that 14.43% of agriculture land is under very low-risk, 13.63% is under low-risk, 44.53% lies in moderate-risk zone, 18.35% is under high-risk and 9.07% lies in very high-risk zone. In case of road features, it is observed that 2.77% is under low-risk, 41.71% is under moderate-risk, 42.93% is under high-risk and 12.59% of the road area lies in very high-risk zone. With respect to settlement, it is observed that 7.24%, 5.17%, 37.21%, 31.78% and 18.60% of settlements are in very low-, low-, moderate-, high- and very high-risk zones, respectively. The percentage distribution of agricultural land, roads and settlements at risk is well represented in the form of pie chart in Fig. 18.

6 Conclusions

Remote sensing and geographic information system have been very useful in understanding the hazard zonation for the Chinyalisour to Siror region along NH-108, through satellite data of Landsat-8, SRTM DEM and LISS-IV. The high-resolution

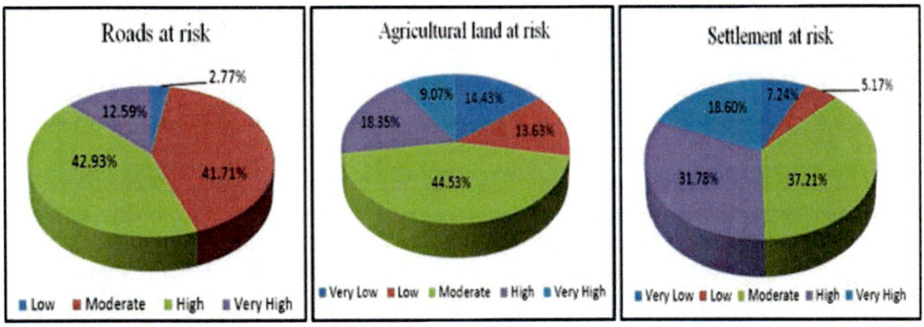

Fig. 18 Pie chart distribution of roads, agricultural land and settlements at risk

image of LISS-IV is useful in identifying the landslides and validating the landslide susceptibility zonation results. On the basis of weightages and calculation, it is observed that slope, geology and structural features (faults and lineaments) are most influencing factors in the study region. The landslide risk assessment reveals that out of 184.3 km^2 of study area, 22.95% lies in very low-risk zone, 37.83% lies in low-risk zone, 18.01% lies in moderate-risk zone, 16.75% lies in high and only 8.21% is under very high-risk zone. The study also infers that 27.42% of agriculture land, 55.52% of the road area and 49.38% of settlements are lying in high and very high-risk zones, respectively. From field observations, it is observed that the susceptibility of region is due to steepening of slopes for road widening works, absence of proper drainage networks along man-modified slopes and toe erosion along the banks of river and roads during high precipitation and flood events. During rainy days, water percolates down the slope material, reduces its shear strength and weakens the intermolecular bond resulting in landslide. It is suggested that adequate provisions for controlling drainage by providing paved catch drains, culverts, retaining wall, toe protection walls, gabion walls, etc., with adequate weep holes, just after the cutting and excavation of slopes, should be done. After construction works, locally growing plants should be grown to increase the root tensile strength of soil as it binds the soil particles firmly. Benching of slopes and use of asphalt mulch, jute/coir fibre netting, shotcreting, rock anchoring, etc., are also useful to stabilize the slopes.

Acknowledgements This paper has immensely benefited from contributions and consultations with all colleagues in Geosciences and Geohazards and other departments of Indian Institute of Remote Sensing. Organizational support and overall guidance provided by Dr. A. Senthil Kumar, Director, IIRS and Dr. Sarnam Singh, Dean (Academics) are also duly acknowledged. SLC and PKC are thankful to Indian Space Research Organization, Department of Space, Government of India, for the financial support provided in TDP project.

References

Conforti M, Rago V, Muto F, Versace P (2014) Preliminary analysis of spatial landslide risk along the highway in Calabria (Northern Italy). In: 17th joint geomorphological meeting, Liege, pp 50–51

Cruden DM (1991) A simple definition of a landslide. Bull Int Assoc Eng Geol 43:27–29

Jain SK, Singh RP, Gupta VK, Nagar A et al (1992) Garhwal earthquake of October 20, 1991. EERI special report. EERI Newslett 26(2):1–4

Terzaghi K (1950) Mechanism of landslides. In: Engineering geology, Berkley Volume. The Geological Society of America, pp 83–123

Valdiya KS (1983) Lesser Himalayan geology: crucial problems and controversies. Curr Sci 52(18):839–857

Varnes DJ (1984) Landslide hazard zonation: a review of principles and practice. UNESCO, Daramtiere, Paris, p 61

Yin KL, Yan TZ (1988) Statistical prediction models for slope instability of metamorphosed rocks. In: Bonnard C (eds) Landslides, proceedings of the fifth international symposium on landslides, vol 2, Balkema, Rotterdam, pp 1269–1272

Seismic Response of Gravity Retaining Wall

Monica Joseph and Subhadeep Banerjee

Abstract The behavior of retaining wall during an earthquake is a topic of major concern nowadays due to its grave consequences. IS 1893 adopts the force-based pseudo-static method for the design of retaining walls under seismic condition, which incorporates the additional seismic force into the static Coulomb's theory. The lateral earth pressures derived from the M–O method are only true in its range of assumptions. In this paper, a numerical study based on the displacement-based approach is conducted for analyzing the seismic response of gravity retaining wall using the finite difference software FLAC-2D. A nonlinear time-history analysis of gravity retaining wall is conducted with actual earthquake data including the recent earthquakes in India. The validation of the numerical model is done by comparing the results with the already published literature. The results are presented in graphs and can be adopted for further studies.

Keywords FLAC-2D · Gravity retaining wall · Displacement-based design

1 Introduction

Retaining walls have always been an important topic for civil engineers due to its inevitable uses and structural challenges during normal and seismic events. A number of retaining wall failures have been reported during the post-earthquake reconnaissance. The structural response of retaining wall is a complex topic due to the soil structure interaction and uncertainties in the determination of actual earthquake data and soil properties. The frequent earthquake events followed by the failure have increased the demand on the seismic design of retaining walls. The classic earth pressure theory of Coulomb and Rankine (Coulomb 1773; Rankine 1857) is widely used to determine the lateral earth pressure on the retaining wall. The static analysis of retaining

M. Joseph (✉) · S. Banerjee
Indian Institute of Technology Madras, Chennai, Tamil Nadu 600036, India
e-mail: monicajoseph93@gmail.com

S. Banerjee
e-mail: subhadeep@iitm.ac.in

M. Latha Gali and P. Raghuveer Rao (eds.), *Geohazards*, Lecture Notes in Civil Engineering 86, https://doi.org/10.1007/978-981-15-6233-4_13

wall for various ground conditions has been carried out using these theories. Seismic design of retaining wall is important in the earthquake prone zones. The seismic design methods of retaining wall can be mainly classified as force-based method, which also called as Mononobe–Okabe (Mononobe and Matsuo 1929; Okabe 1924) method and displacement-based method. The former method is the direct extension of classic Coulomb's method. In the Mononobe and Okabe method, the dynamic forces is converted into static inertial forces by incorporating two coefficients called as seismic horizontal and vertical coefficients into it. Even though the method is widely adopted for finding the active and passive forces acting on the wall, the method holds back many limitations. The simplicity of Mononobe–Okabe method made it as the initial estimate for determining the dynamic earth pressure. The method is valid only for continuous granular backfill and the cases where the water table fluctuations are not encountered. Since the method has many assumptions to simplify it, M–O method has been evaluated by many researchers. Seed and Whiteman (1970) have confirmed that the active earth pressure using M–O method shows good agreement with actual cases. But the passive earth pressure has to be renewed to match with the actual cases. Many researchers have contributed their results for modifying the M–O method. Assumption of a planar rupture plane in M–O method is not practically accepted, and the result of that shows that it underestimates active earth pressure and overestimates passive earth pressure. A design-based on M–O method will end up in uneconomical sometimes, unsafe design. After considering all the limitations of M–O method, a displacement-based method can in many ways give better, safe, and economical result. The static factor of safety required to maintain no displacement during an earthquake is so high that it will results in a highly uneconomical design. Allowing some displacement during an earthquake by maintaining its stability can reduce the design requirements, and the design will be economical.

Newmark (1965) developed a sliding block model to determine the permanent displacement of an earth dam during a seismic event. He developed charts of standardized displacement for normalized earthquake. Later, this model is modified by many researchers to determine the seismic displacement of gravity retaining wall, settlement of embankment. Richards and Elms (1979) used the same sliding block model and modified the normalized displacement curve for determining the maximum displacement of a gravity retaining wall during an earthquake. A standard method of design of gravity retaining wall based on a permissible displacement is suggested. According to this method, the designer can choose an allowable displacement and uses this displacement to compute a design acceleration coefficient, and thereby, this acceleration coefficient can be used to compute the wall dimensions. In the sliding block theory of Newmark, he assumed a constant value for yield ground acceleration for constant soil strength properties. They conducted small-scale shaking table tests and found out that the yield ground acceleration decreases for consecutive shaking input pulses for peak friction angle. They also found out that the yield acceleration remains constant for residual peak friction angle (Φ_{res}). Richards and Elms (1979) proposed a method for designing retaining wall under seismic prone area by choosing an allowable displacement. According to this method, the designer can choose an allowable displacement. This displacement is further used to determine the design seismic coefficient, and this coefficient is used to find the retaining wall mass required. The obtained wall dimensions are cross-checked for overturning, allowing the structure to only slide during motion. Wartman et al. (2005) performed a scaled recorded earthquake data from Hyogoken-Nambu earthquake of 1995 and reported that the use of peak strength of soil properties resulted in 80% overestimation of results, and in other hand, use of residual strength properties of soil resulted in 60% underestimation of seismic displacement. They suggested a more reasonable approach for using the soil properties for determining the actual values.

Byrne and Salgado (1981) presented a SDOF model which allows both the earthquake-induced forces and displacements to be computed, considering weight of the wall and flexibility and strength of backfill and foundation soil. Prakash et al. (1981) proposed a sliding block model to estimate base translation of a retaining wall using a bilinear wall force–displacement relation, which is characterized by a yield–displacement and a force–displacement slope on the active and passive sides.

Nadim and Whitman (1983) computed the amount of permanent tilting and sliding of a gravity retaining wall due to earthquake loading. Limiting accelerations, where the wall begins to tilt or slide, were evaluated by considering the dynamic equilibrium of the wall. Siddharthan et al. (1990) presented an analytical model for predicting the seismic displacements of rigid retaining walls, supporting sand backfill. The response is calculated in terms of sliding and tilting in a coupled manner based on the model proposed by Nadim and Whitman (1983). By performing a parametric study on the position of center of rotation along wall base, they conclude that when center of rotation is away from the wall toe, a coupled mode of deformation occurs, and total horizontal movement can be as much as 78% higher than those given by sliding considerations alone.

In the present study, the response of a gravity retaining wall is undertaken for different earthquake conditions. A gravity retaining wall is modelled in the finite difference software, FLAC 7.0. The validation of the model is done by comparing the results with the existing literature.

1.1 Description of the Numerical Model

A finite difference software FLAC-2D is used for modelling the gravity retaining wall. A trapezoidal gravity retaining wall of 0.5 m top width, 3 m bottom width, and height 6 m is modelled as shown in Fig. 1. The model is dimensioned such a way to avoid the boundary effects. A uniform square grid is provided for the generated mesh, allowing a maximum frequency of $f_{\max} = V_s/[(8/10)\Delta_l]$. The model is resting on a foundation soil of 13 m depth overlying a rock of 7 m depth. The backfill soil is having the same soil properties as of foundation soil. The mesh generated for the numerical model is shown in Fig. 2.

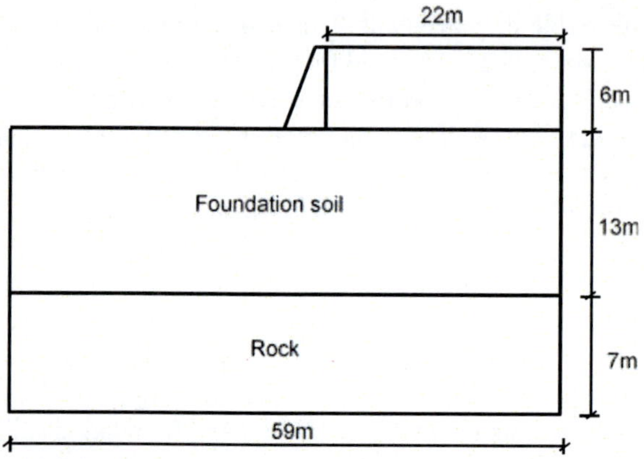

Fig. 1 Sketch of soil layers and retaining wall

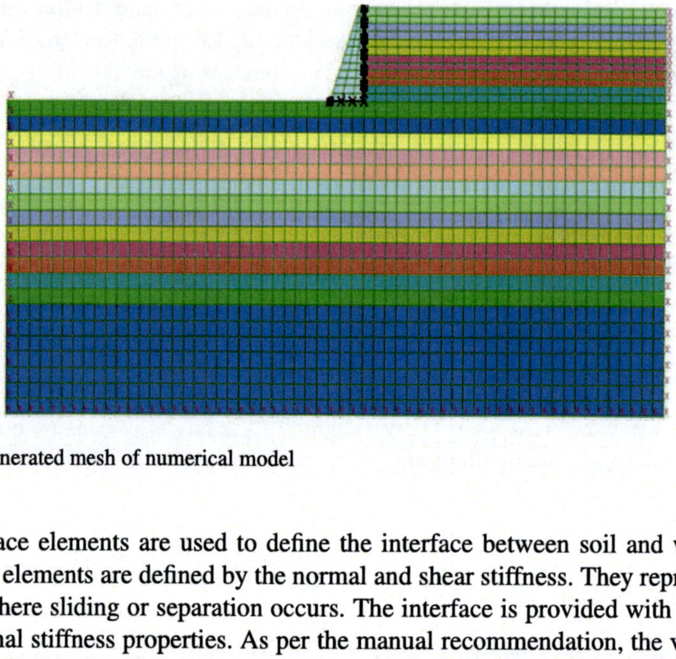

Fig. 2 Generated mesh of numerical model

Interface elements are used to define the interface between soil and wall. The interface elements are defined by the normal and shear stiffness. They represent the planes where sliding or separation occurs. The interface is provided with the shear and normal stiffness properties. As per the manual recommendation, the values for these properties k_s and k_n are set to ten times the equivalent stiffness of the stiffest neighboring zone. The apparent stiffness (expressed in stress-per-distance units) of a zone in the normal direction is given as $k_n = 10 * \max\{[K + 4/3G]/\Delta Z_{\min}\}$. The normal and shear stiffness has given same values for analysis.

The analysis is done in two stages. In the first stage, the static equilibrium of model under gravity loads in considered. The unbalanced force is brought to nearly zero before the dynamic analysis is carried out. The free-field boundaries are provided at both the vertical sides to minimize the reflection of seismic waves back to the model. The free-field model consists of a one-dimensional "column" of unit width,

Table 1 Material properties (from Deyanova et al. 2016)

Soil type	Backfill	Wall	Base
Void ratio	0.45	–	–
Mass density (kg/m^3)	2000	2400	2400
Friction angle (°)	37.5	–	–
Dilation angle (°)	5	–	–
Shear modulus (MPa)	$7000\frac{(2.17-e)^2}{(1+e)}(\sigma_0)^{0.5}$	2.2e+9	2e9
Cohesion (MPa)	0	–	–
Poisson's ratio	0.3	0.2	0.29
Model	Nonlinear hystertic + Mohr–Coulomb	Linear elastic	

simulating the behavior of the extended medium (FLAC user's manual). The plane waves propagating upward suffer no distortion at the boundary because the free-field grid supplies conditions that are identical to those in an infinite model. The bottom of the model is provided with quiet boundary which is based on the use of independent dashpots in the normal and shear directions at the model boundaries.

1.2 Soil Types and Constitutive Models

The response of the retaining wall is very much influenced by the type and properties of the soil condition on the backfill and foundation region. For the present study, a dense cohesionless soil is assigned at the backfill. The same soil is provided at the foundation level also. The soil properties change with effective mean stress at different levels. The change in properties with increase in confining pressure is accounted based on Ishihara (1996). The foundation and the backfill soil is modelled as nonlinear media with shear degradation rule according to Darendali (2001) and with Mohr–Coulomb failure criterion. The base and retaining wall is modelled as linear elastic materials. The soil properties and constitutive model types are mentioned in Table 1.

1.3 Selection of Earthquake Data

The real records of earthquakes are used for the present analysis. The validation of model is done based on Deyanova et al. (2016). The earthquake data used for the analysis are spectrum compatible records. The raw earthquake data are corrected for baseline correction and filtered for removing the unwanted noise. These data are then matched with the target spectrum based on Eurocode 8. The spectrum is matched such a way that the difference between the average and target spectra should not exceed

Table 2 Earthquake data used for validation

No	Magnitude (Mw)	PGA (g)	Scaling factor	PGA (g)		Predominant frequency
A1	7.3	0.373	1.1	0.412	0.16	
A2	6.9	0.442	0.403	0.178	0.38	
A3	6.6	0.141	2	0.282	0.1	

A1—Tabas, Iran (1978), A2—Loma Prieta, California, U.S. (1989), A3—Kyushu, Japan (2005)

more than 10% between the period 0.15 and 2 s. The earthquake data used for the analysis is mentioned in Table 2. Since quiet boundary is provided at the bottom, the input motion is given as stress wave propagating upward. The acceleration time history is converted to stress time history as per the manual provisions. The vertical component of the accelerogram is neglected, as the self-weight of the gravity retaining wall is considered to be enough to prevent vertical motion of the retaining wall. The accelerogram from two horizontal components are considered for the analysis.

2 Static Analysis of Retaining Wall

Rankine's and Coulomb's (Coulomb 1773; Rankine 1857) earth pressure theories are the widely adopted methods for static analysis of retaining wall. Coulomb's theory is developed for a general geometry retaining wall. The theory considers the wall friction (δ) for a retaining wall with rough vertical face and a cohesionless backfill. He considered the static equilibrium of the soil wedge close to the wall which is moving with the wall during its active and passive state. Rankine developed a much simple theory for a retaining wall with smooth vertical back wall and a cohesionless horizontal backfill. He ignored the wall friction making it more conservative. The present numerical analysis is compared with the classic Rankine's and Coulomb's theory in Fig. 3.

Fig. 3 Comparison of earth pressure

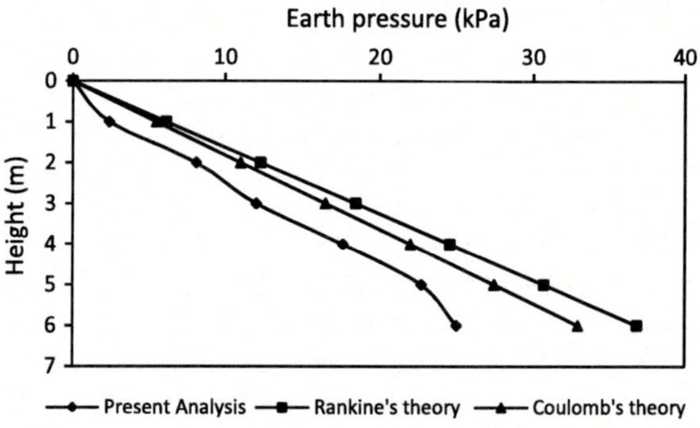

Analyzing the figure, we can understand that the results from Rankine's theory is comparatively higher than Coulomb's theory as the reduction since the wall friction considered in the Coulomb's theory reduces the lateral earth pressure to some extent. The lateral earth pressure obtained in the present study is lesser than both the theories. Since the classic theories are on the conservative side, design based on numerical study will be more economical. The results from numerical study also show a nearly triangular distribution of earth pressure on the wall.

The horizontal displacement contours after the static analysis is shown in Fig. 4. The displacements are more toward the wall. When the wall moves away from backfill, corresponding displacement of the wall will be toward left, which is given a negative sign convention. The vertical displacement contour under the static analysis is shown in Fig. 5.

The downward settlement is given negative sign as per the sign convention. Vertical displacements are observed above 7 m from the bottom of the model, below which, a rigid rock base is provided. The horizontal and vertical effective stress contours are shown in Figs. 6 and 7. The effective vertical and horizontal stress vari-

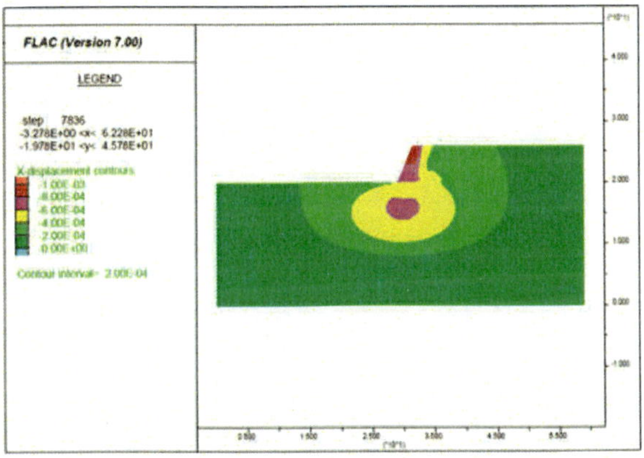

Fig. 4 Horizontal displacement contour after stage 1

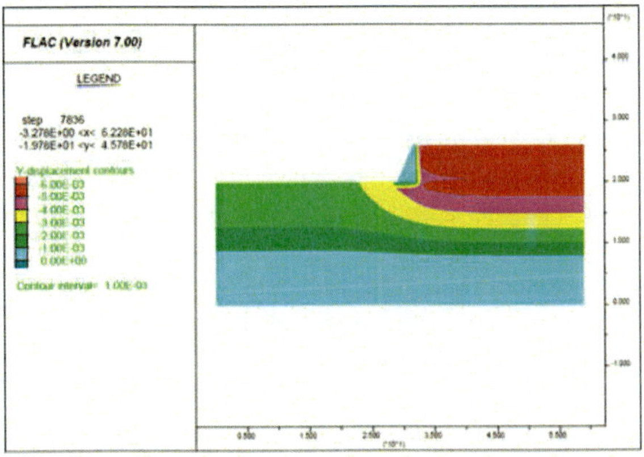

Fig. 5 Vertical displacement contour after stage 1

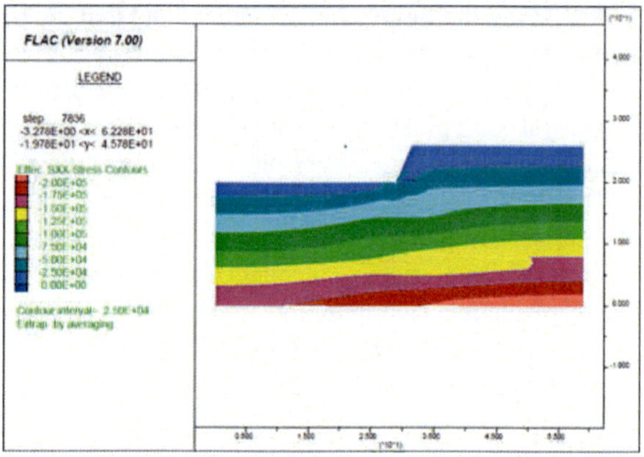

Fig. 6 Effective stress contour after stage 1

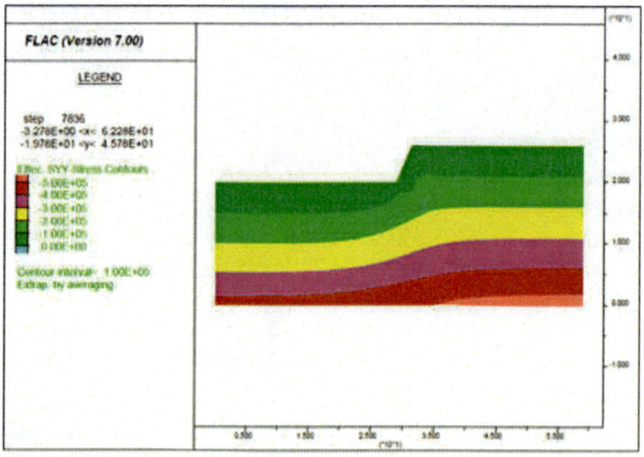

Fig. 7 Effective vertical stress contour under static load

ation with respect to depth is shown clearly in the figure. The effective vertical stress increases vertically downward with increasing overburden pressure.

3 Validation of the Model

The numerical model developed in FLAC 7.0 is validated with the existing literature by comparing the results. The earthquake data mentioned in Table 2 are given as input, and the results as acceleration time histories, displacement time histories, and effective stresses at various points are obtained. The horizontal and vertical displacements along the wall height and behind the wall are also measured.

The tilt of the retaining wall under different earthquakes is measured, and the results are compared with Deyanova et al. (2016). The results are shown in Fig. 8. Tilt is calculated as the difference in horizontal displacement of the wall top and bottom divided by the height of the retaining wall.

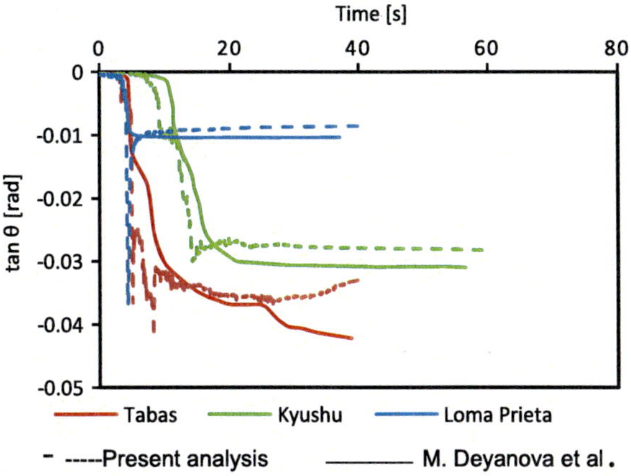

Fig. 8 Comparison of tilt of the retaining wall from FLAC 7.0 with the Deyanova et al. (2016)

4 Dynamic Analysis

The spectrum-matched real earthquake data are converted into stress history, as the quiet boundary applied at the bottom boundary will not be efficient if acceleration data is given as input. The dynamic analysis is started only after assuring static equilibrium of the model. The input stress wave is applied at the bottom of the model. The stress histories, acceleration histories, and displacement histories are determined at different levels of model. The acceleration time history of Tabas earthquake for 20 s is shown in Fig. 9. The high peaks of acceleration are visible from 5 to 10 s. The acceleration data are filtered and corrected for base line correction.

The effective stress time history along the wall height for Tabas earthquakes is shown in Fig. 10. The effective stress peaks are observed during the peak acceleration durations (between 5 and 10 s). The horizontal displacement contour is shown in Fig. 11. The displacement is more toward the wall and reduces away from the wall. The displacement is not much significant below 10 m depth.

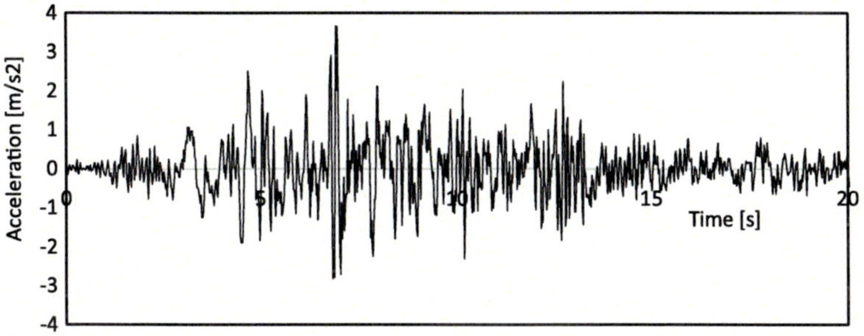

Fig. 9 Acceleration time history for Tabas earthquake (1978) for 20 s

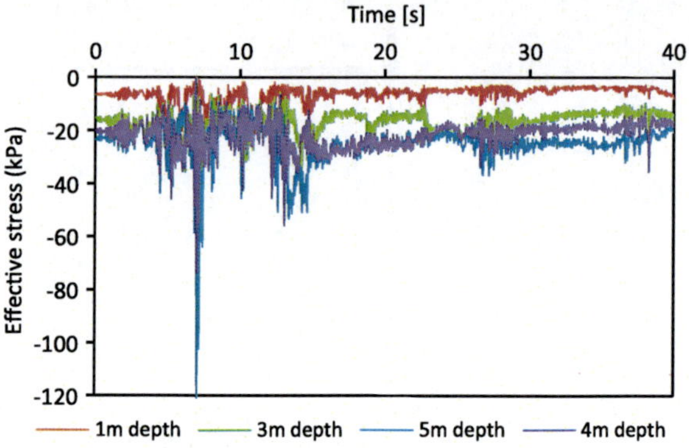

Fig. 10 Effective stress time history for Tabas earthquake along the wall height

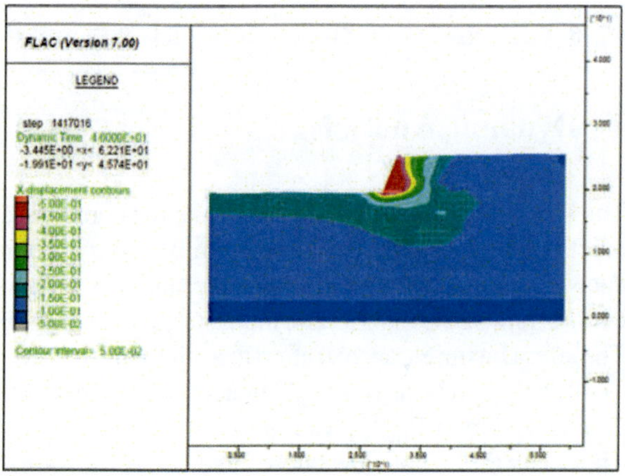

Fig. 11 Horizontal displacement contour after subjected to Tabas earthquake

The effective lateral pressure acting on the wall is very important for the seismic design of retaining wall. Since water table is not considered in the analysis, there is no problem of pore pressure development in the backfill. The total and effective stresses are equal in this case. The tilt of the wall during the earthquake is very less that the chance that it can cause rotation of wall with respect to the base is small. The response of retaining wall under an earthquake is also influenced by the soil properties of backfill and foundation. Since dense soil is assigned to foundation and backfill, the settlements in the soil are not very significant.

The response spectra of an earthquake can be directly used for the seismic design of retaining wall. Knowing the natural frequency of the retaining wall, it is easy to determine the peak response of the structure. The response spectra at the bottom of the retaining wall are shown in Fig. 12. The maximum peak is found to be at 0.14 s. The response spectra for the same earthquake at passive and active side of the retaining wall are also shown in Figs. 13 and 14.

The response spectra at the top of the wall are shown in Fig. 15. Analyzing the response spectra, it is easily visible that the responses are more at the active side of the

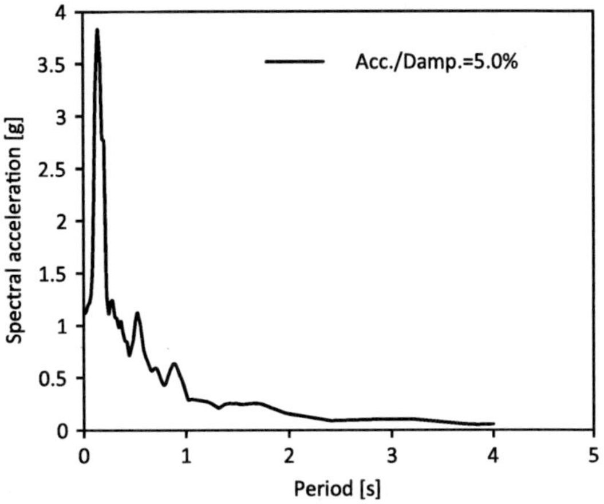

Fig. 12 Response spectra at the bottom of the wall under Tabas earthquake

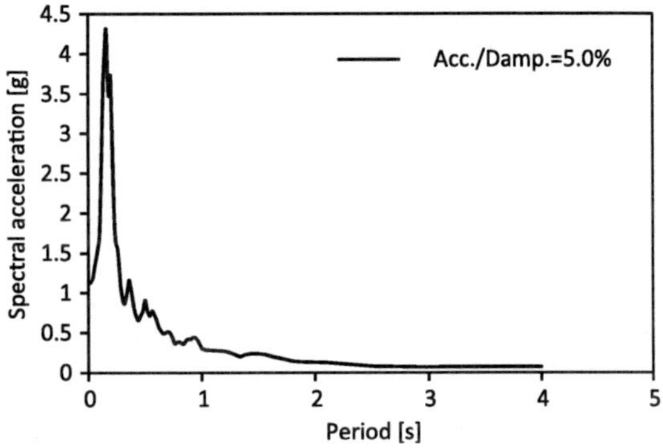

Fig. 13 Response spectra at the active side of the wall under Tabas earthquake

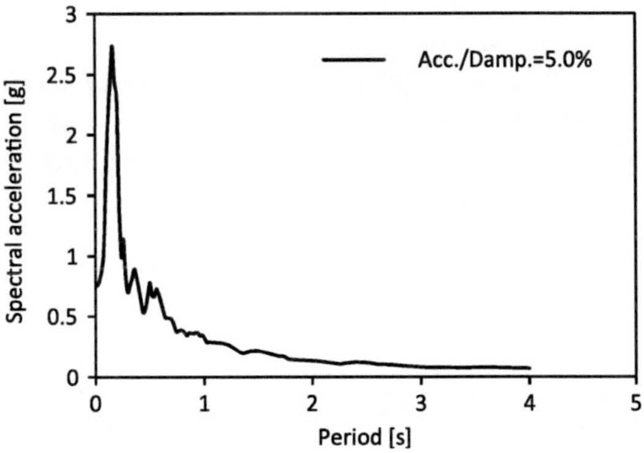

Fig. 14 Response spectra at the passive side of the wall under Tabas earthquake

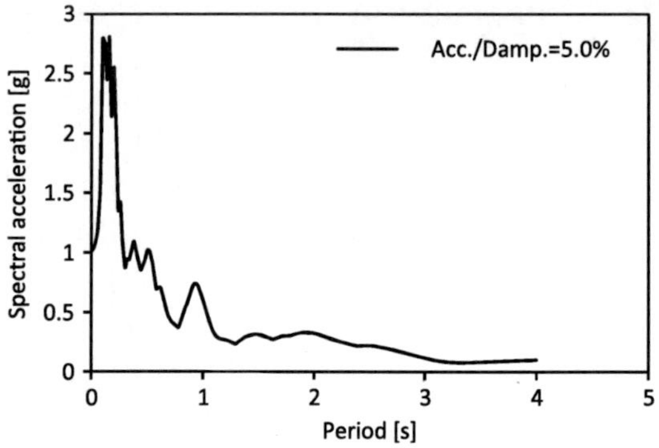

Fig. 15 Response spectra at the top of the wall under Tabas earthquake

wall than the passive side. The lateral earth pressure will correspondingly increase at the active side, leading the wall to fail. The failure of the wall is considered to be at a wall movement of 10% of the wall height. The condition can be changed based on the importance of structures. In that case, 5% of wall height is also adopted.

5 Conclusion

A numerical model of gravity retaining wall is developed in FLAC 7.0, and validation of the model is done with an existing journal. The static and dynamic analysis is carried out in the gravity retaining wall model, and results are shown in figures. The comparison of static analysis results with the classic Coulomb's and Rankine's theory shows that the results of classic theories are very much on the conservative side and relying the results from numerical study which will result in a more economical design of retaining wall. Under the seismic analysis, the response of retaining wall under Tabas earthquake is analyzed. The output in the form of stress contours and response spectra is given in Figures. The response spectra from the figures can be used for the design for safety against peak responses.

References

Byrne PM, Salgado F (1981) Seismic response of retaining. In: Proceedings of international conference on recent advances in geotechnical earthquake engineering and soil dynamics. Scholars Mine, Missouri, pp 727–732

Coulomb CA (1773) Sur une application des regles de maximis et minimis a quelques problemes de statique relatifs a l'architecture. Memoires de Mathematique et de Physique Par Divers Savants

Darendeli BM (2001) Development of a new family of normalize modulus reduction and material damping curves. Ph.D. thesis, University of Texas at Austin, Austin, USA

Deyanova M, Carlo GL, Mario M (2016) Displacement-based parametric study on the seismic response of gravity earth-retaining walls. J Soil Dyn Earthq Eng 80:210–224

FLAC. Fast Lagrangian analysis of continua. User's manual
Ishihara K (1996) Soil behaviour in earthquake geotechnics. Oxford University Press, New York
Mononobe N, Matsuo H (1929) On the determination of earth pressure during earthquake. In: Proceedings of the world engineering conference, vol 9. World Engineering Congress, Japan, pp 177–185
Nadim F, Whitman R (1983) Seismically induced movement of retaining walls. J Geotech Eng 109:915–931
Newmark NM (1965) Effects of earthquakes on dams and embankments. Geotechnique 15:139–160
Okabe S (1924) General theory on earth pressure and seismic stability of retaining wall and dam. J Jpn Soc Civ Eng 10(6):1277–1323
Prakash S, Puri VK, Khandoker JU (1981) Displacement analysis of rigid retaining walls in rocking. In: Proceedings of international conference on recent advances in geotechnical earthquake engineering and soil dynamics. Scholars Mine, Missouri, pp 1021–1036
Rankine WJM (1857) On the stability of loose earth. Philos Trans R Soc Lond 147(1):9–27. https://doi.org/10.1098/rstl.1857.0003
Richards R Jr, Elms DG (1979) Seismic behavior of gravity retaining walls. J Geotech Eng Div ASCE 105:449–464
Seed HB, Whitman RV (1970) Design of retaining structures for dynamic loads. In: Proceedings of specialty conference on lateral stresses in ground and design of earth retaining structures. ASCE, Ithaca, pp 103–145
Siddharthan R, Ara S, Anderson JG (1990) Seismic displacements of rigid retaining walls. In: Proceedings of 4th U.S. national conference on earthquake engineering. EERI, El Cerrito, California, pp 673–682
Wartman J, Seed RB, Bray JD (2005) Shaking table modeling of seismically induced deformations in slopes. J Geotech Geoenviron Eng 131(5):610–622

Some Studies on the Pseudostatic Analyses of Water Retention Type Tailings Dams

Pankaj Kumar and B. V. S. Viswanadham

Abstract Tailings dams have been widely constructed to store the milled and mined out residual waste or tailings from the mining industry. These residues could be hazardous and/or toxic in nature; therefore, it is pertinent to design these storage structures in such a way that they are stable on the wake of slope failures due to accidents arising from natural and anthropogenic activities. In this paper, an attempt has been made to study the seepage and stability behaviour of water retention type tailings dams of different heights under static and seismic conditions. To this effect, the effect of horizontal drain on the seepage and stability behaviour of tailings dams was investigated using Soil Vision software. Seepage analyses were carried out using SVFLUX, and pseudostatic slope stability analyses were performed using SVSLOPE software for various combinations of horizontal and vertical seismic accelerations. It was found that the use of horizontal drain ensured the depletion of phreatic surface and improved the stability of the dam in the downstream side under both static and seismic conditions. A comparison has been made for obtaining the most suitable configuration for the construction.

Keywords Tailings dams · Pseudostatic analyses · Seepage analysis · Slope stability analysis

1 Introduction

A tailings dam is typically an earth-fill embankment dam used to store by-products of mining operations after separating the ore from gangue. Millions of tons of non-radioactive residue from uranium mining activities are produced per year by various industrial units distributed across India. These mining by-products are unstable and do not form part of the aerobic ecological system so it can damage environment. Therefore, the deposition and storage of these residual materials from the mining operations are done by constructing dams, embankments and dykes. The main job

P. Kumar · B. V. S. Viswanadham (✉)
Department of Civil Engineering, Indian Institute of Technology Bombay, Mumbai, Maharashtra 400076, India
e-mail: viswam@civil.iitb.ac.in

M. Latha Gali and P. Raghuveer Rao (eds.), *Geohazards*, Lecture Notes in Civil Engineering 86, https://doi.org/10.1007/978-981-15-6233-4_14

of a tailings impoundment dam is to contain the residue/tailings from mining and milling operations and to recycle the water from top of the impoundment after settling of solids via a reclamation barge back into the mill.

The tailings embankment which is a false representation of dam with a water reservoir at upstream is however designed as a water retention type dam for slope stability and seepage requirements (Cowherd et al. 1993). Tailings placed within the embankment consolidate very slowly and there always remains a possibility for development of high pore water pressure. This can cause reduction of effective stress and consequently the shear strength of the tailings. Due to this, the slope of tailings embankment becomes unstable and is prone to failure. In terms of mechanical instability, tailings dams are reported as very high-risk structures. Statistics indicate that tailings dams have failed majorly due to slope instability triggered by natural hazards/rainstorms (Rico et al. 2008). In case of breaching, landslide like situation prevails, and the saturated and viscous tailings then start flowing in the direction of breach. Other major causes of tailings dam failure include overtopping, piping failure due to erosion, weak foundations, liquefaction and excessive pore water pressure.

The behaviour of such tailings dams under seismic conditions is mostly studied through finite element-based numerical methods, full-scale model tests in the field at normal gravity and small-scale model tests in geotechnical centrifuge at high gravity. However, field tests are expensive; difficult to conduct within a confined space, duration and environment; and repeatability of full-scale tests is also a problem. Whereas, these issues do not arise while using finite element-based numerical techniques with appropriate choice of material model and boundary conditions. Physical modelling techniques such as centrifuge model-based studies have also emerged as a major tool in the past few decades to understand in detail the deformation behaviour and stability aspects of similar structures like retaining walls, embankments, levees and hill slopes.

Since the 1920s, inertial instability concept is being used for assessment of stability of earthen dams and embankments during earthquakes. Like earthen dams, seismic stability of a tailings dam can be carried out using pseudostatic approach. "Many researchers (Byrne et al. 1984; Castro et al. 1985; Akhlaghi and Neishapouri 2007; Chakraborty and Choudhury 2009, 2010, 2013) have carried out seismic slope stability analyses of tailings dams using different FEM-based and FDM-based software". In the recent past "Reyes and Parra (2014) carried out 2D and 3D slope stability analyses of a tailings mine waste facility using Soil Vision". "Akhlaghi and Nikkar (2014) carried out seismic slope stability analyses of 24.4 m high Upper San Fernando Dam and 24 m high Kitayama Dam using Plaxis-2D". "Sitharam and Hegde (2017) performed seismic analyses for rock-fill tailings dam using FLAC^{2D}".

This paper addresses the seepage and slope stability behaviour of typical 7.2 and 9.6 m high water retention type tailings dams (WRTDs). The analyses are carried out using SVOFFICE 5 (2018) software considering both static and pseudostatic conditions. The main aim of the present study is to ascertain the stability behaviour of the WRTDs configurations during seismic events.

2 Model Geometry and Materials Used

The seepage and slope stability analyses of water retention type tailings dams under seismic condition are carried out for two different geometrical configurations, i.e. dam A and dam B, both having upstream and downstream side slopes of $1.5H{:}1V$. The geometrical model of the tailings dams considered for the present study is shown in Fig. 1 and various geometrical parameters for both the dams are presented in Table 1. The dam sections are provided with horizontal drains of thickness 450, 600, 750 and 1000 mm. The length (L) of horizontal drains is considered with respect to height (H) of the dam as $H/2$, $2H/3$, $0.75H$, H, $1.25H$, $1.5H$ and $1.75H$. The properties of materials used for different sections of the water retention type tailings dams are presented in Table 2.

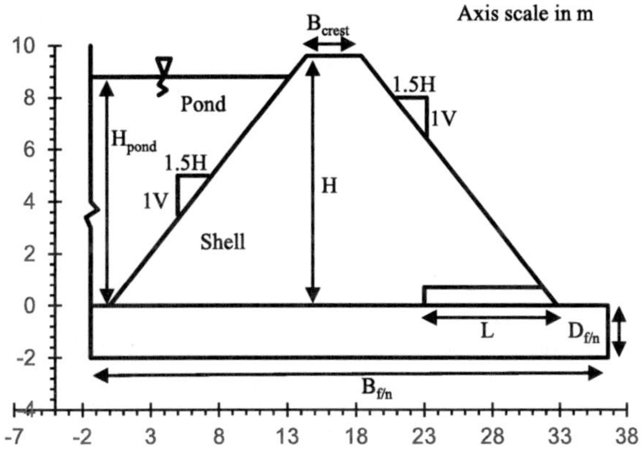

Fig. 1 Typical cross section of a water retention type tailings dam

Table 1 Geometrical parameters of water retention type tailings dams

Parameters	Dam A	Dam B
Dam height, H (m)	7.2	9.6
Side slopes ($H{:}V$)	1.5:1	1.5:1
Dam base width, B_{base} (m)	24.6	32.8
Dam crest width, B_{crest} (m)	3	4
Pond height, H_{pond} (m)	6.6	8.8
Foundation width, $B_{f/n}$ (m)	28.5	38
Foundation depth, $D_{f/n}$ (m)	1.5	2

Table 2 Properties of materials used in WRTDs

Parameters	Properties of material at different sections		
	Shell/foundation	Tailings pond	Drains
Density (kN/m^3)	18.30	19	20
Cohesion (kPa)	31	14.7	0
Friction angle (φ)	28°	12°	32°
Saturated VMC	0.3	0.25	0.3
Permeability (m/s)	1×10^{-8}	1×10^{-8}	1×10^{-4}
Material model	MC	MC	MC

3 Methodology and Modelling

Stability analysis of water retention type tailings dams can be accomplished by various limit equilibrium methods for determining the critical slip surface and corresponding factor of safety (FS). The methodology adopted in the present study for analysing tailings dams is described in subsequent sections.

3.1 Seepage Analysis

A 2D steady-state seepage analysis is conducted using the SVFLUX-GE module of SVOFFICE 5 geotechnical software. The governing partial differential equation for steady-state seepage (assuming vapour flow is negligible) assuming saturated soil conditions is given by Eq. 1, where k_w is the hydraulic conductivity function.

$$\frac{\partial}{\partial x}\left[k_x^w \frac{\partial h}{\partial x}\right] + \frac{\partial}{\partial y}\left[k_y^w \frac{\partial h}{\partial y}\right] = 0 \tag{1}$$

In the present study, respective hydraulic heads at strategic locations are assigned constant values to serve as necessary seepage boundary conditions, as shown in Fig. 2. The finite element solver used by SVOFFICE 5 automatically generates mesh in the 2D model and implements mesh refinement.

3.2 Static Slope Stability Analyses

The seepage analysis results obtained at various stages of analysis using SVFLUX are incorporated into the limit equilibrium-based software SVSLOPE for computing the static slope stability of water retention type tailings dams. For this case, Bishop's (1955) and generalized limit equilibrium (GLE) methods are used for computing the

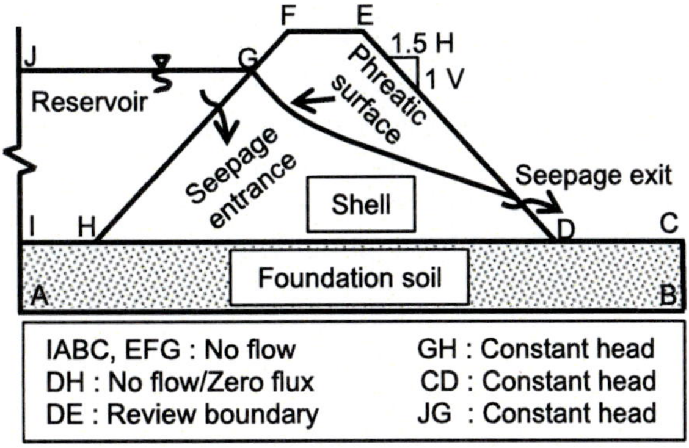

Fig. 2 Boundary conditions used for seepage analysis

factor of safety by considering circular slip surfaces. The GLE method is broadly outlined in Fredlund and Krahn (1977) and Fredlund et al. (1981) and is based on the calculation of factor of safety equations using moments and horizontal equilibrium as given by Eq. 2.

$$\text{FS} = \frac{\text{Available shear strength}}{\text{Mobilised shear stress}} \tag{2}$$

3.3 *Pseudostatic Slope Stability Analyses*

A factor of safety (FS) against slope failure is computed using static limit equilibrium procedure in which horizontal and/or vertical inertia forces (destabilizing effect of earthquake) are applied to the potential sliding mass. In actual analysis, inertia forces F_h and F_v acting through centroid of the sliding mass as computed using Eqs. 3 and 4, are applied which acts out of the slope direction. In this method, the cyclic earthquake motion is replaced with constant horizontal acceleration (a_h) and/or vertical seismic accelerations (a_v) expressed as ($k_h \times g$) and ($k_v \times g$), respectively, where k_h and k_v are horizontal and vertical seismic coefficients. The FS is given by Eq. 5.

$$F_h = ma_h = \frac{Wa_h}{g} = k_h W \tag{3}$$

$$F_v = ma_v = \frac{Wa_v}{g} = k_v W \tag{4}$$

$$\text{FS} = \frac{\text{Resisting forces}}{\text{Driving forces}} = \frac{\text{cl} + [(W - F_v)\cos\beta - F_h \sin\beta]\tan\phi}{(W - F_v)\sin\beta + F_h\cos\beta} \tag{5}$$

where W is the weight of the soil failure wedge, and β is the inclination of failure plane with the horizontal. The stability of the sliding mass depends critically on the selection of appropriate seismic coefficients. The value of k_h can also be determined in laboratory using physical models testing with tilting table set-up (Ohishi et al. 1995) as given by Eq. 6, where α is the tilting angle of the base of the slope from horizontal. For stability analysis, the values of k_h for various design seismic zones are obtained from Indian design standard (IS 1893 (Part-1) 2016) and corresponding α values can be used for laboratory investigations or vice versa.

$$k_h = \tan \alpha \tag{6}$$

3.4 *Model Analysis*

In this paper, the static and pseudostatic slope stability analyses have been presented for water retention type tailings dam configurations. The aim of this study is to fulfil three main objectives. The first objective is to select an appropriate seepage control measure (horizontal drain) with suitable dimensions (length, thickness, etc.) for both dam configurations. The second objective is to find out static and pseudostatic FS against slope failure for WRTDs under steady-state seepage condition. Third objective is to compare the FS values against slope failure for both the dams. In view of this, the WRTDs have been analysed for the following cases:

Case-1 Static slope stability analysis for downstream slope when entire reservoir is filled with water.

Case-2 Pseudostatic slope stability analysis for downstream slope when entire reservoir is filled with water.

Case-3 Static slope stability analysis for downstream slope when entire reservoir is filled with tailings slurry.

Case-4 Pseudostatic slope stability analysis for downstream slope when entire reservoir is filled with tailings slurry.

4 Results of Analysis and Discussions

This section presents seepage and slope stability analysis results obtained to fulfil various objectives considering different cases as mentioned earlier.

4.1 Selection of Appropriate Seepage Control Measure

In order to obtain appropriate type and dimensions of seepage control measures for 7.2 and 9.6 m high WRTDs, seepage and slope stability analyses were carried out considering case-1.

Seepage analysis for selection of drain parameter. A series of seepage tests is carried out for both the WRTDs with different combinations of thickness and length of horizontal drain. Figures 3, 4, 5, 7, 8 and 9 show obtained phreatic surfaces for a particular thickness (say 300 mm or 600 mm or 1000 mm) and different lengths (0.5*H*, 0.67*H*, 0.75*H*, *H*, 1.25*H*, 1.5*H* and 1.75*H*) of horizontal drain for 7.2 and 9.6 m high WRTDs, respectively. It can be clearly observed that with the provision of horizontal drain, phreatic surface does not intersect the downstream slope of the dam and lies within the dam body. As the length of the horizontal drain increases from 0.5*H* to 1.75*H*, the phreatic surface also depletes and confines itself well within the body of the dam. Similar pattern in depletion of phreatic surface is also noticed

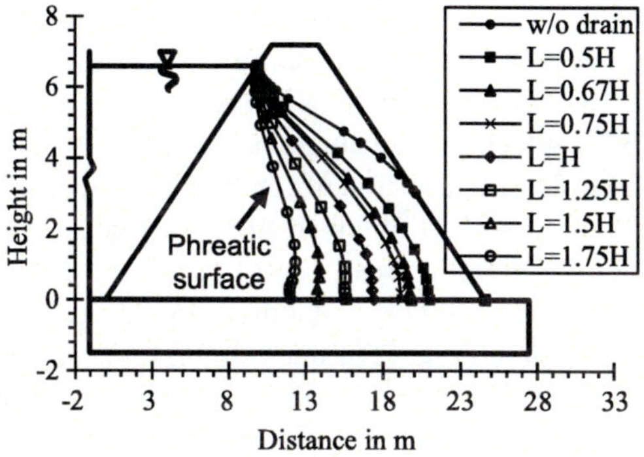

Fig. 3 Variation of phreatic surfaces with length for 7.2 m high WRTD having 300 mm thick horizontal drain

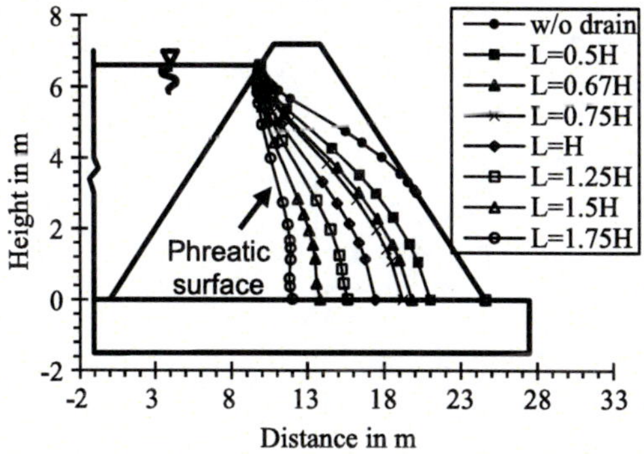

Fig. 4 Variation of phreatic surfaces with length for 7.2 m high WRTD having 600 mm thick horizontal drain

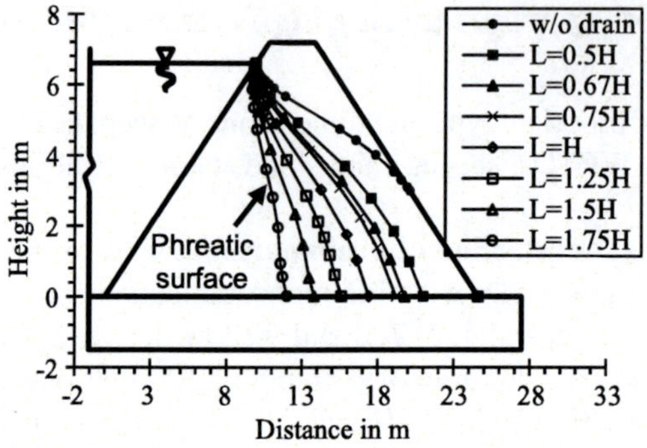

Fig. 5 Variation of phreatic surfaces for 7.2 m high WRTD having 1000 mm thick horizontal drain

with the variation of thickness of horizontal drain for both the dams. Figures 6 and 10 show comparison of phreatic surfaces for different thicknesses (300 mm, 600 mm and 1000 mm) and different lengths (0.5*H*, 0.75*H*, and *H*) of horizontal drain for 7.2 and 9.6 m high WRTDs, respectively. For a particular length of horizontal drain, very small difference in orientation of phreatic surfaces can be observed with different thicknesses for both the configurations of WRTDs.

Stability analysis for selection of drain parameter. Static slope stability analysis is carried out for both the dams with different combinations of thicknesses and lengths of horizontal drain. Phreatic surface obtained from seepage analysis is used as input for coupled analysis of WRTDs. Figures 11 and 12 show the distribution of factor of safety with different *L*/*H* ratios of horizontal drain for 7.2 and 9.6 m high WRTDs, respectively. It can be observed that FS values slightly decreased with the increase in L/H ratio from 0.50 to 0.75 and then increased to become constant for *L*/*H* ratio of greater than equal to 1, for horizontal drains of thicknesses 450, 600, 750 and 1000 mm for 7.2 m high dam. In the case for 9.6 m high dam, the FS values increased with the increase in *L*/*H* ratio from 0.50 to 1 and became constant for *L*/*H* ratio of

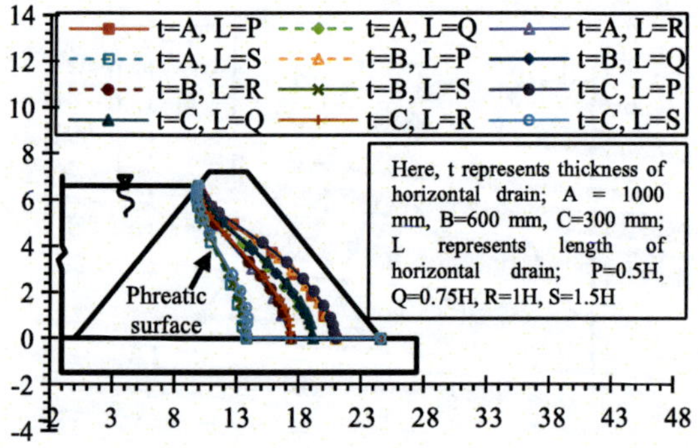

Fig. 6 Comparison in phreatic surfaces for 7.2 m high WRTD for various thicknesses of horizontal drain

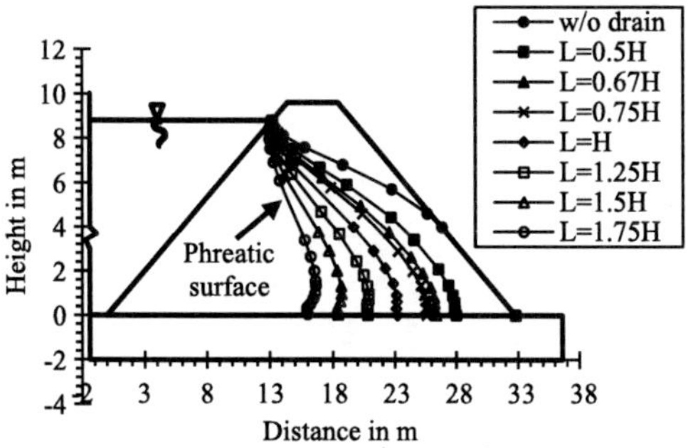

Fig. 7 Variation of phreatic surfaces with length for 9.6 m high WRTD having 300 mm thick horizontal drain

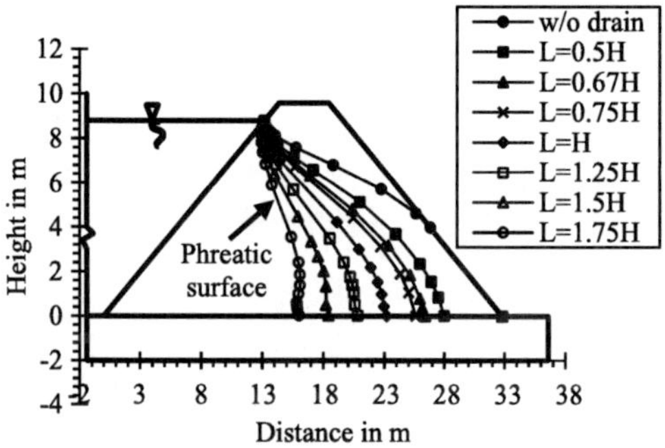

Fig. 8 Variation of phreatic surfaces with length for 9.6 m high WRTD having 600 mm thick horizontal drain

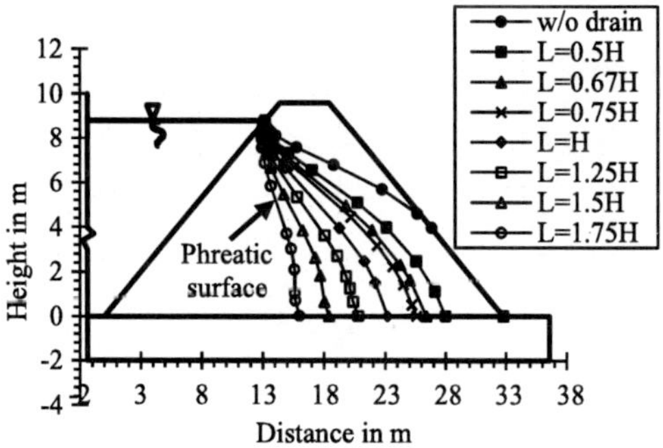

Fig. 9 Variation of phreatic surfaces for 9.6 m high WRTD having 1000 mm thick horizontal drain

greater than equal to 1.25, for horizontal drains of thicknesses 450 and 600 mm, whereas for 750 and 1000 mm thickness of horizontal drain in 9.6 m dam, the FS values first increased up to *L*/*H* ratio of 0.75, then decreased and became constant

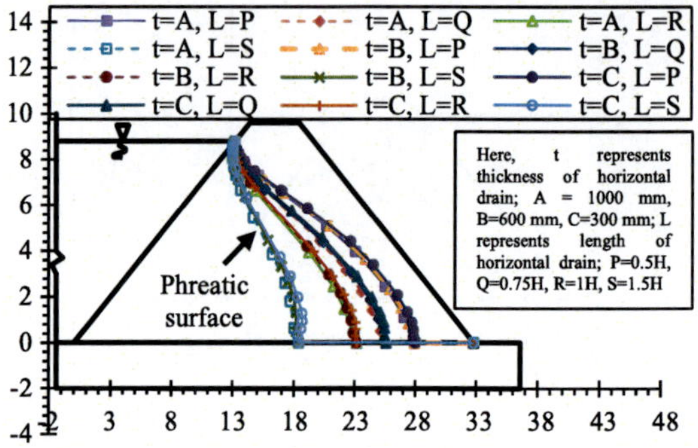

Fig. 10 Comparison in phreatic surface for 9.6 m high WRTD for various thickness of horizontal drain

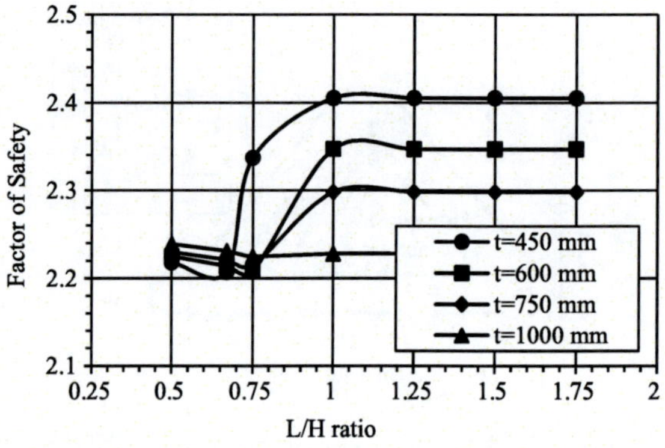

Fig. 11 FS values (Bishop's method) for different *L*/*H* ratio of horizontal drain for 7.2 m high WRTD

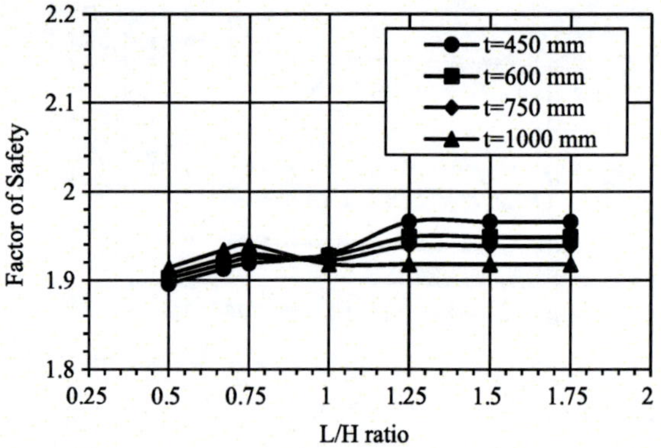

Fig. 12 FS values (Bishop's method) for different *L*/*H* ratio of horizontal drain for 9.6 m high WRTD

for *L/H* ratio of greater than equal to 1.25. This is quite a different behaviour of FS values with *L/H* ratio in comparison to results obtained for 7.2 m high dam. But it can be identified that beyond *L/H* ratio of 1, the variation in FS values is observed to be either very small or nearly constant for both the dams. That is, stability of small dams (less than 10 m high) from slope failure point remain unaffected from an increase in the length of horizontal drains beyond height of the dams. It is also observed that horizontal drains of thickness 450 and 600 mm provide greater FS. Based on these results, further analysis is carried out with horizontal drain of thickness 600 mm and *L/H* ratio of 0.75.

4.2 *Static and Pseudostatic Analysis*

Static and pseudostatic slope stability analysis is carried out for 7.2 and 9.6 m high WRTDs having 600 mm thick horizontal drain and *L/H* ratio of 0.75 for all the cases mentioned above.

Seepage analysis. Seepage analysis is carried out for both WRTDs with upstream reservoir filled with water only (case 1 and 2) and tailings slurry only (case 3 and 4). Figures 4 and 8 shown earlier depicts phreatic surfaces for 7.2 and 9.6 m high WRTDs having 600 mm thick horizontal drain and *L/H* ratio 0.75, respectively, for cases 1 and 2. Figures 13 and 15 show variation of pore water pressure for 7.2 and 9.6 m high WRTDs having 600 mm thick horizontal drain and *L/H* ratio 0.75, respectively, for cases 3 and 4. Figures 14 and 16 show total head contours for 7.2 and 9.6 m high WRTDs having 600 mm thick horizontal drain and *L/H* ratio 0.75, respectively, for cases 3 and 4. It can be observed that phreatic surfaces for case 3 and 4 are well within the dam body at higher elevation as compared to phreatic surfaces for case 1 and 2. Also, phreatic surfaces for case 3 and 4 for 9.6 m high WRTDs are well inside the dam body as compared to phreatic surfaces for case 3 and 4 for 7.2 m high WRTDs. Corresponding stability conditions are discussed in subsequent section.

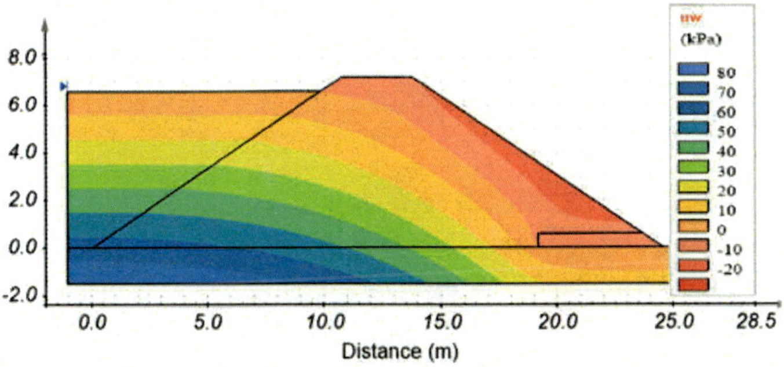

Fig. 13 Variation of pore water pressure for 7.2 m high WRTD modelled using SVFlux

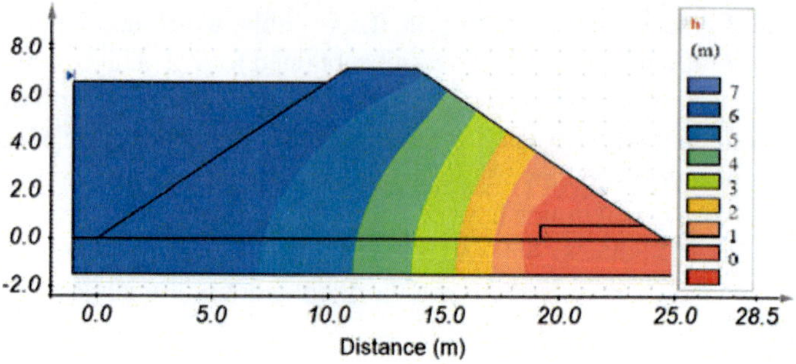

Fig. 14 Total head contours for 7.2 m high WRTD modelled using SVFlux

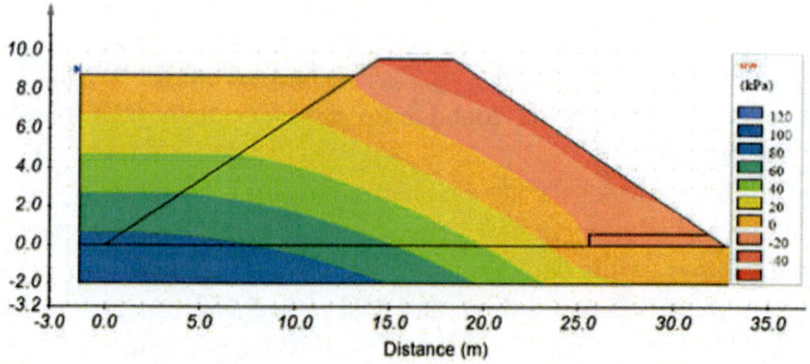

Fig. 15 Variation of pore water pressure for 9.6 m high WRTD modelled using SVFlux

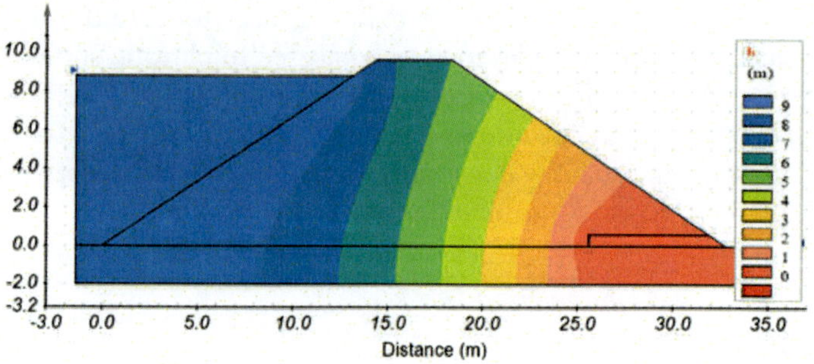

Fig. 16 Total head contours for 9.6 m high WRTD modelled using SVFlux

Stability analysis. Figures 17 and 18 present results of stability analysis in terms of variation of factor of safety (FS) values for static and pseudostatic slope stability analysis of 7.2 and 9.6 m high WRTDs for case 1 and 2, respectively. Figures 19 and 20 present results of stability analysis in terms of variation of FS values for static and pseudostatic slope stability analysis of 7.2 m high and 9.6 m high WRTDs for case 3 and 4, respectively. The pseudostatic FS values are obtained for various k_h values in combination with $k_v = k_h/2$ and/or $k_v = 0$. It can be noted from Figs. 17,

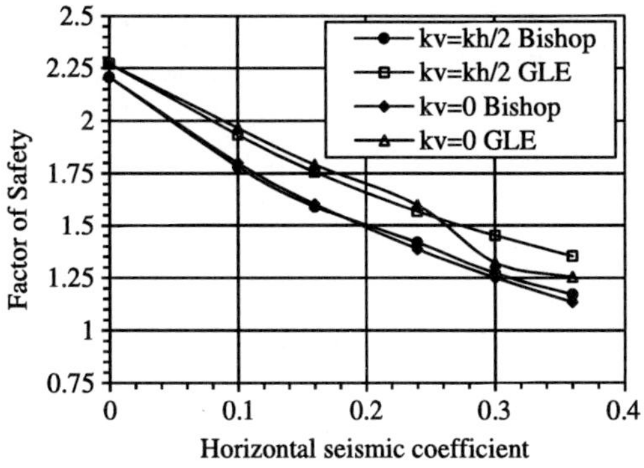

Fig. 17 Results of stability analysis of 7.2 m high WRTD for case 1 and 2 using SVSLOPE

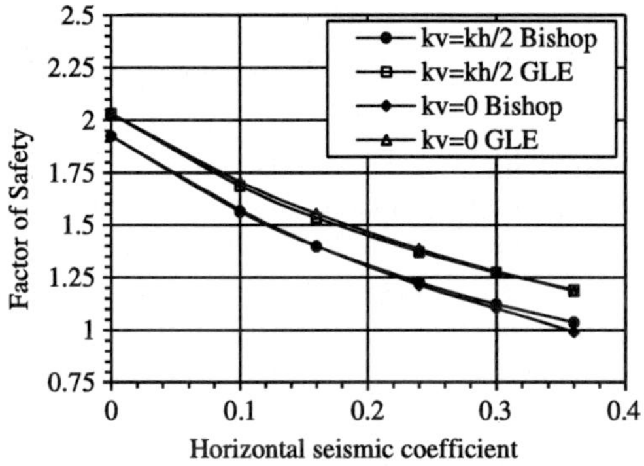

Fig. 18 Results of stability analysis of 9.6 m high WRTD for case 1 and 2 using SVSLOPE

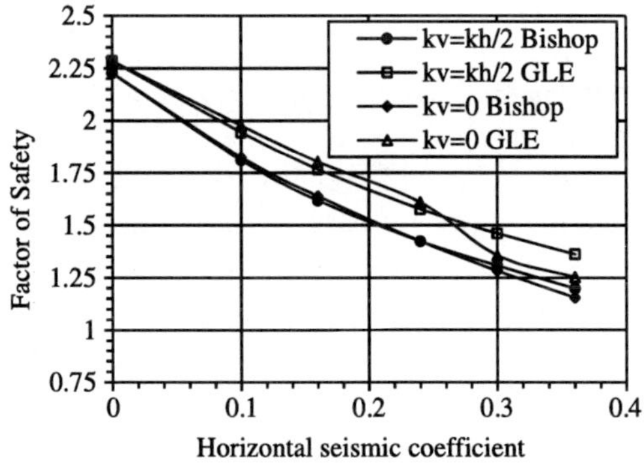

Fig. 19 Results of stability analysis of 7.2 m high WRTD for case 3 and 4 using SVSLOPE

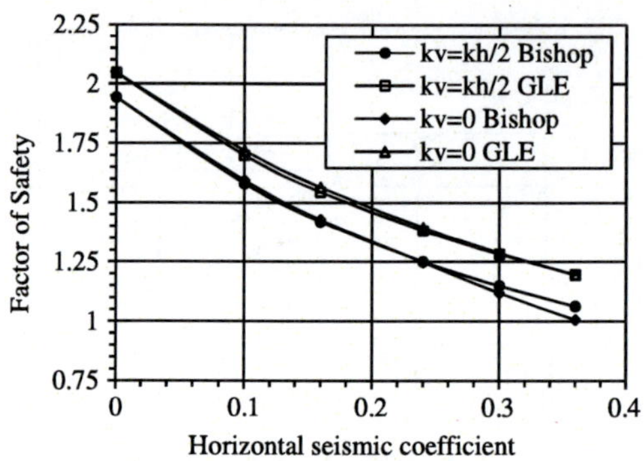

Fig. 20 Results of stability analysis of 9.6 m high WRTD for case 3 and 4 using SVSLOPE

18, 19 and 20 that static FS values for 7.2 and 9.6 m high WRTDs increase in cases 3 and 4 as compared to cases 1 and 2, respectively. It can also be observed that the FS values keep on decreasing with the increase in design horizontal seismic coefficient (k_h) for both the dam configurations. Also, FS values obtained from Bishop's method being smaller/lower seems more optimistic for design as compared to GLE method. For a 7.2 m high WRTD when the reservoir is filled with water only, minimum FS value obtained for static case is 2.208 and for one of the seismic case is 1.135 (for $k_h = 0.36$ and $k_v = 0$), whereas when the reservoir is filled with tailings slurry only, minimum FS value obtained for static case is 2.227 and for seismic case is 1.155 (for $k_h = 0.36$ and $k_v = 0$). For a 9.6 m high WRTD when the reservoir is filled with water only, minimum FS value obtained for static case is 1.925 and for seismic case is 0.99 (for $k_h = 0.36$ and $k_v = 0$), whereas when the reservoir is filled with tailings slurry only, minimum FS value obtained for static case is 1.943 and for seismic case is 1.007 (for $k_h = 0.36$ and $k_v = 0$).

The FS values are obviously higher in case 3 and 4 due to additional stability provided by the tailings material, which also increases as the tailings material self-consolidates with time.

Figures 21 and 22 show static and pseudostatic ($k_h = 0.16$ and $k_v = 0.08$) FS values of 7.2 m high tailings dam for cases 3 and 4, respectively. Figures 23 and 24 show static and pseudostatic ($k_h = 0.16$ and $k_v = 0.08$) FS values of 9.6 m high tailings dam for cases 3 and 4, respectively.

5 Conclusions

Based on analysis and interpretation of results, following conclusions can be drawn.

1. As the length of the horizontal drain increases from 0.5H to 1.75H, the phreatic surface also depletes and confines itself well within the body of the dam. For a particular length of horizontal drain, very small difference in orientation

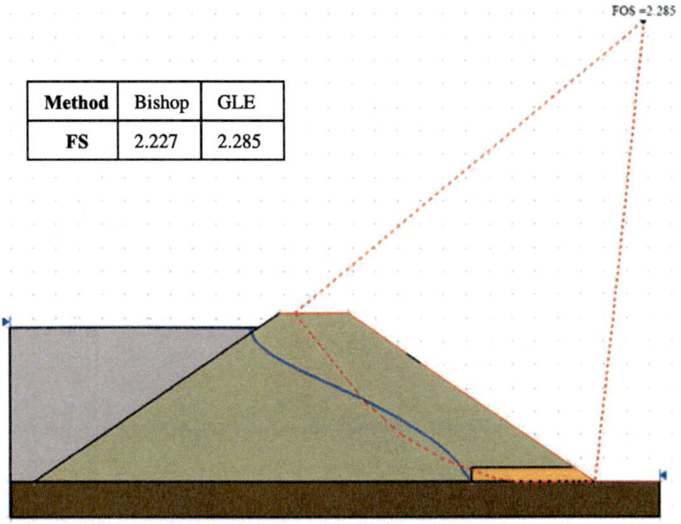

Method	Bishop	GLE
FS	2.227	2.285

Fig. 21 Static FS value for 7.2 m high WRTD modelled using SVSLOPE

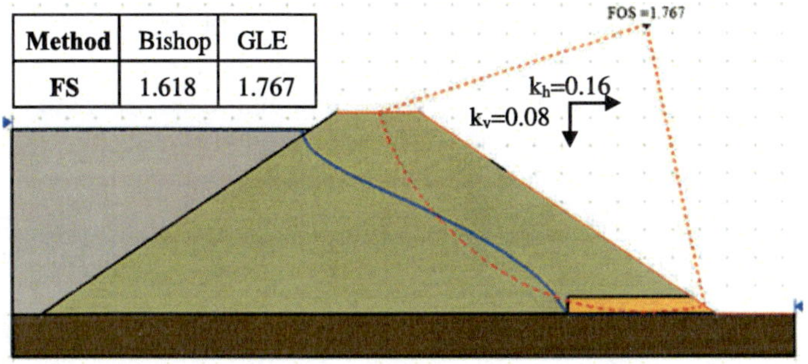

Method	Bishop	GLE
FS	1.618	1.767

Fig. 22 Pseudostatic FS value for 7.2 m high WRTD modelled using SVSLOPE

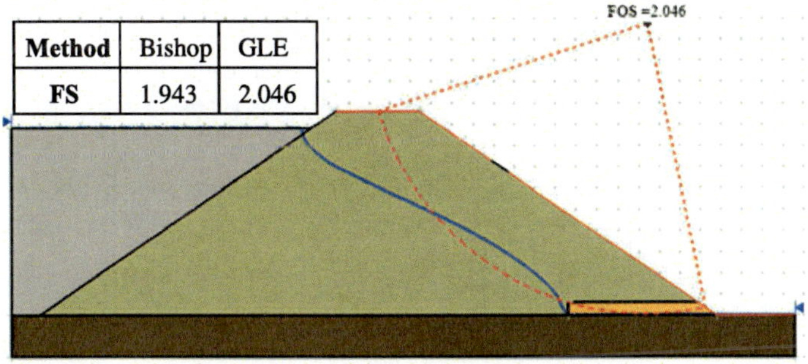

Method	Bishop	GLE
FS	1.943	2.046

Fig. 23 Static FS value for 9.6 m high WRTD modelled using SVSLOPE

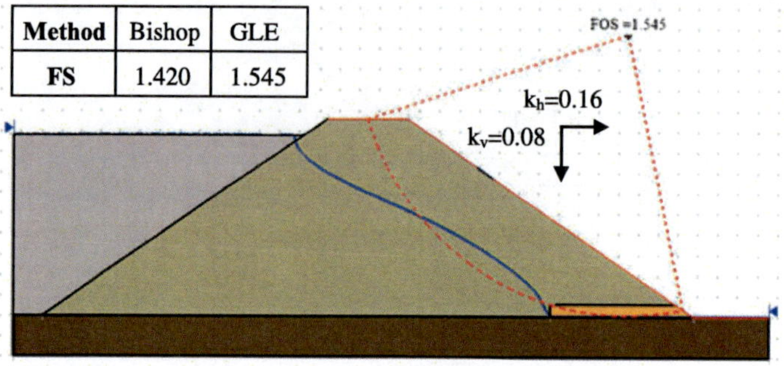

Fig. 24 Pseudostatic FS value for 9.6 m high WRTD modelled using SVSLOPE

of phreatic surfaces can be observed with different thicknesses for both the configurations of WRTDs.

2. Beyond *L*/*H* ratio of 1, the variation in FS values was observed to be either very small or nearly constant for both the dam configurations. That is, stability of small dams (7–10 m height) from slope failure point remains unaffected from increase in the length of horizontal drains beyond height of the dams.
3. The horizontal drains having thicknesses 450 and 600 mm provide greater FS compared to others.
4. In the present case, for WRTDs of height 7–10 m, it is advisable to adopt 450–600 mm thick horizontal drain and *L*/*H* ratio in range of 0.67–1.
5. It was observed that FS values keep on decreasing with the increase in design horizontal seismic coefficient for both the dam configurations. Also, FS values obtained from Bishop's method being lower are more optimistic for design considerations in comparison to GLE method of slope stability.
6. From static and pseudostatic slope stability analysis, it can be concluded that both the WRTDs configurations are safe against slope failure except the case of 9.6 m high dam at $k_h = 0.36$ and $k_v = 0$ where FS value drops below 1.

Acknowledgements The authors would like to express their sincere gratitude to Atomic Energy Regulatory Board (AERB), Anushakti Nagar, Mumbai, Government of India, and Dean R&D, Industrial Research and Consultancy Centre (IRCC), IIT Bombay, for extending financial assistance and funding the research work.

References

Akhlaghi T, Neishapouri M (2007) A study on the pseudostatic analysis of the upper San Fernando dam using FE simulations and observed deformations during the 1971 earthquake. In: Fourth international conference on earthquake geotechnical engineering, pp 1–12

Akhlaghi T, Nikkar A (2014) Evaluation of the pseudostatic analyses of earth dams using FE simulation and observed earthquake-induced deformations: case studies of Upper San Fernando and Kitayama dams. Sci World J 2014:1–12. https://doi.org/10.1155/2014/585462

Bishop AW (1955) The use of the slip circle in the stability analysis of slopes. Géotechnique 5(1):7–17

Byrne PM, Morris DV, Caldwell JA (1984) Seismic stability of a tailings impoundment on soft clayey silt deposits. In: Eighth world conference on earthquake engineering, vol 3, San Francisco, pp 381–388

Castro G, Poulos SJ, Leathers FD (1985) Re-examination of slide of lower San Fernando dam. J Geotech Eng 111(9):1093–1107

Chakraborty D, Choudhury D (2009) Investigation of the behavior of tailings earthen dam under seismic conditions. Am J Eng Appl Sci 2(3):559–564

Chakraborty D, Choudhury D (2010) Seismic slope stability analysis of tailings earthen dam using TALREN 4. In: Indian geotechnical conference, GEOtrendz, pp 187–190

Chakraborty D, Choudhury D (2013) Pseudostatic and pseudodynamic stability analysis of tailings dam under seismic conditions. Proc Nat Acad Sci Sect A Phys Sci 83(1):63–71

Cowherd DC, Miller KC, Perlea VG (1993) Seepage through mine tailings dams. In: Proceedings of third international conference on case histories in geotechnical engineering, St. Louis, Missouri, pp 463–468

Fredlund DG, Krahn J (1977) Comparison of slope stability methods of analysis. Can Geotech J 14(3):429–439

Fredlund DG, Krahn J, Pufahl DE (1981) The relationship between limit equilibrium and slope stability methods. In: International conference on soil mechanics and foundation engineering, vol 3, pp 409–416

IS 1893 (Part-1) (2016) Guidelines for criteria for earthquake resistant design of structures, sixth revision. Bureau of Indian Standard (BIS), New Delhi

Ohishi K, Saitoh K, Katagiri M, Katakami N, Terashi M (1995) Comparison of embankment behaviors in static tilting and shaking table tests. In: Ishihara (ed) Earthquake geotechnical engineering, Balkema, Rotterdam, ISBN 905410578X, pp 1075–1080

Reyes A, Parra D (2014) 3D slope stability analysis by using limit equilibrium method analysis of a mine waste dump. Proc Tailings Mine Waste 2014:127–139

Rico M, Benito G, Salgueiro AR, Herrero AD, Pereira HG (2008) Reported tailings dam failures: a review of the European incidents in the worldwide context. J Hazard Mater 152(2):846–852

Sitharam TG, Hegde A (2017) Stability analysis of rock-fill tailing dam: an Indian case study. Int J Geotech Eng 11(4):332–342. https://doi.org/10.1080/19386362.2016.1221574

SVOFFICE 5 (2018) User's manual. The SoilVision Systems Ltd., Sasktoon, Saskatchewan, Canada

Behavior of Reinforced Retaining Wall Against Railway Embankment Using Midas Under Static and Seismic Loading

Shilpa S. Vadavadagi and Sowmiya Chawla

Abstract Retaining walls are commonly used in the design of roadways and railways. Basically retaining walls are meant to carry lateral loads. Although many works from the literature suggest vivid methods to analyze and design these retaining structures, still there is enough scope to explore the problems faced in the analysis of retaining structures. In the present study, a retaining wall is considered to be supporting railway embankment. Backfill consists separate layers like natural soil, sub-ballast, and ballast. Load of sleepers and rails is also accounted. Load of rails is simulated using two point loads. Soil is assumed to be linearly elastic and is of Mohr–Coulomb's characteristics. Geogrid is considered as reinforcing material in retaining wall. Midas is used to do FEM modeling of the problem. The wall is analyzed under static and seismic loading conditions. A detailed sensitivity analysis is carried out to understand the displacements of wall.

Keywords Retaining wall · Railway embankment · Static and dynamic loading

1 Introduction

Mechanically stabilized earth (MSE) wall is stabilized internally by reinforcing material. It has a facing angle of 70–900. It is a composite structure of backfill, reinforcement facing panel and foundation. It has wide application because of its ease of construction with less time and site preparation, cost of effective and pleasing aesthetics. Another advantage of being more popular compared to conventional retaining wall is technical feasibleness. Mechanically stabilized earth (MSE) retaining structures are widely used because of their wide variety of applications in geotechnical structures. Geosynthetics have four different functions such as reinforcement, drainage, separator, and filter. Apart from these, performance of the MSE

S. S. Vadavadagi (✉) · S. Chawla
Indian Institute of Technology (Indian School of Mines) Dhanbad, Dhanbad, India
e-mail: svshilpas@gmail.com

S. Chawla
e-mail: sowmiya@iitism.ac.in

M. Latha Gali and P. Raghuveer Rao (eds.), *Geohazards*, Lecture Notes in Civil Engineering 86, https://doi.org/10.1007/978-981-15-6233-4_15

wall also depends on the behavior of the backfill soil. Backfill of MSE wall can be either borrowed from the different sites which have good shear strength or the locally available soil can be used. For the successful design and performance of MSE wall, earthen materials which will be used as backfill material and their interaction with manufactured components of the MSE wall are of critical importance for its performance. So, it is much necessary to use the backfill material which has high quality with its good durability and good reinforcement interaction with reinforcing material; well-graded granular soil will serve the purpose of good and durable backfill by Hossain et al. (2011).

Another proof of importance of backfill selection was done by Soong and Koerner (1999) where twenty case histories were reported on poor performance of marginal backfill which resulted in reinforced wall failure. According to them, positive pore pressures were found to be developed when fine-grained soils were used. That also led to serviceability problems. According to Elias et al. (1996), backfill should be of free drainage material and it should not be having organic material and deleterious material. Costa et al. (2016) conducted centrifuge evaluation for the behavior of geotextile-reinforced soil walls and found that walls showed were affected through stability by showing their time-dependent behavior when subjected to constant acceleration.

Reinforcement materials also have influence over the performance of MSE wall in terms of their length and spacing. Lesser the spacing of the reinforcement layers lesser the deformation of the wall.

2 Problem Definition

A numerical study has been carried for the retaining wall supporting railway track system. In the present study, retaining wall is supporting backfill on which rails, sleepers, ballast, and sub-ballast layers exist. Reinforcement layers of geogrid in the backfill are also been reinforced to length of 6.6 m. Beyond the reinforced zone of backfill, unreinforced zone is also supported by retaining wall below which foundation is laid.

Height of the retaining wall is considered to be 4.8 m as it was modeled by Yarivand et al. (2017). Train load considered is 150 kN. Train load has been simulated on the top of edges of both the rails as static loads at the midnode generated on the top of the rail. Static loads here considered on the node of each rail was 150 kN. This load has been calculated assuming a single boogie load of twenty tones. This boogie load has been multiplied with the dynamic factor of 1.5. The obtained value was divided by two in order to distribute the value for two rails. The point load to be considered on each rail depends upon the weight or load of each bogie.

Table 1 Material properties of soil

Material properties	Retaining fill	Foundation
Young's modulus (MPA)	150	200
Poisson's ratio (μ)	0.3	0.3
Density (γ)	18	22
Friction angle (ϕ)	9	25
Cohesion (c)	25	30

Table 2 Material properties of railway track

Heading level	Rail	Sleeper	Ballast	Sub-ballast
Young's modulus (MPA)	2100	10,349	207	138
Poisson's ratio (μ)	0.33	0.37	0.37	0.33

Depth of the foundation has been extended to 10 m below the ground level with its width 32 m at the bottom. Properties assigned for backfill and foundation soil used has been shown in Table 1. Properties of rail, sleepers, ballast, and sub-ballast used by Sowmiya (2013) are shown in Table 2.

3 Methodology

Analysis has been done in the software Midas/GTS/NX. A 2D mesh for the model is shown in Fig. 1. Meshing has been done by size control in which edges are selected. In size control type of mesh number of divisions' method has been used. If the edge

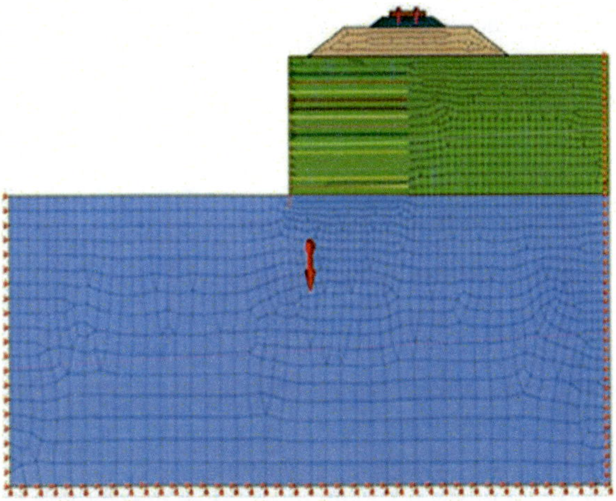

Fig. 1 Meshed diagram of MSE wall

selected is shorter, then a number of divisions for nodes to be generated will be small. Vice versa if the selected edge is longer. So, the number of divisions considered for rail is four whereas for geogrid is 30. Likewise, the number of divisions of nodes has been varied for each edge of the model. Properties of all the materials that used boundary conditions are autogenerated.

For mesh generation of geogrid and diaphragm wall, the extract tool in the Midas has been generated. Using the tool extract, a geogrid is selected and then material property of the geogrid is assigned. Self-weight of the model was considered in *Y*-direction. Boundary condition for the model was autogenerated. Skinner and Rowe (2005) analyzed retaining wall supporting bridge abutment with approach roads loading. Train load considered is 150 kN. Train load consideration has been explained in problem definition section. Geogrid length is considered to be 0.7 times height of the retaining wall in the first case. In the second case, length of geogrid considered is 6.65 m. Young's modulus and Poison's ratio of geogrid are 1000 mPa and 0.4, respectively, whereas for wall it is 13,800 mPa and 0.2, respectively.

In the analysis case of the model, slope stability with strength reduction method (SRM) has been selected as a part of solution type. In the first case of analysis, the analysis name is analysis without reinforcement. In this case, rails, sleeper, ballast, sub-ballast, backfill, and foundation are activated to the active mesh set except geogrids and wall. Second case of analysis name is analysis without geogrid in which analysis has been done without reinforcing geogrids, and in this particular case, rails, sleeper, ballast, sub-ballast, wall, backfill, and foundation are activated to the active mesh set except geogrids. Geogrid is deselected to active mesh set. And the analysis is performed. Third analysis case name is analysis with reinforcement. In the analysis case, geogrids are also activated to active mesh set and the analysis has been performed.

The analysis for the above cases has been carried out for single-tier and two-tier retaining wall. For single-tier retaining wall, geogrid lengths are varied. Initially, geogrid length considered is 0.7 times the height of the retaining wall ($0.7 * H$). Geogrid length is then considered to be 6.65 m. This length is selected in such a way that the other end of the geogrid will end at the midway of the railway track system. For two-tier retaining wall also, length variation done is the same as done for single-tier retaining wall.

4 Results and Discussions

4.1 Single-Tier Retaining Wall

4.1.1 Considering Geogrid Length Equal to 0.7*H*

For the analysis of one tied retaining wall, geogrid length considered is 0.7 times the height of the retaining wall. When the analysis is run failure, envelop is found

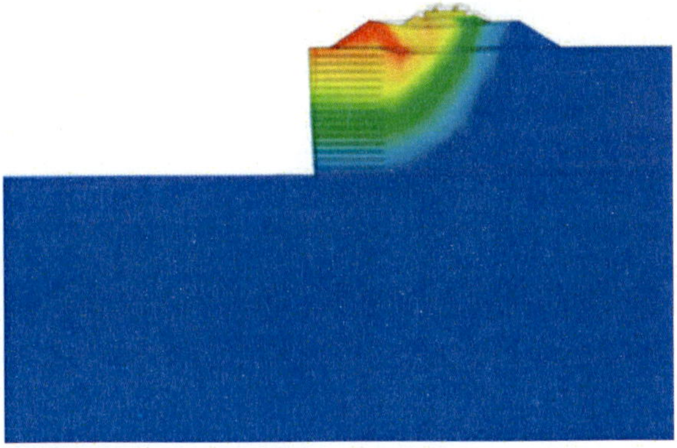

Fig. 2 Failure envelop passing within reinforced zone

to pass within reinforced zone as shown in Fig. 2. Factor of safety is found to be more for the analysis case, analysis with wall and geogrid than analysis with wall and analysis case without reinforcement.

4.1.2 Considering Geogrid Length Equal to 6.65 m

In this case, when the geogrid length is increased to 6.65 m, factor of safety is increased than the previous case of consideration of geogrid length. Failure envelop is found to increase and it is passing beyond reinforced zone as shown in Fig. 3.

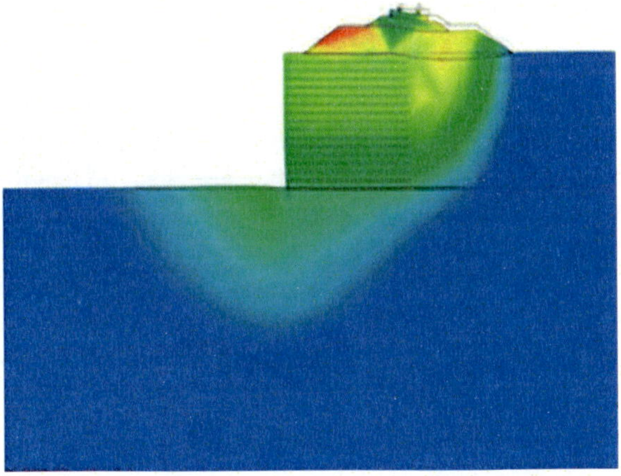

Fig. 3 Failure zone passing beyond the reinforced zone for single-tier retaining wall

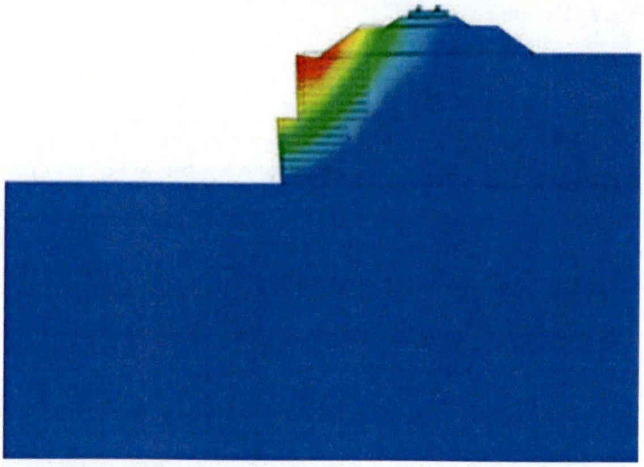

Fig. 4 Failure zone passing within geogrid-reinforced zone

4.2 *Two-Tier Retaining Wall*

4.2.1 Considering Geogrid Length Equal to 0.7*H*

For the same height of the retaining wall supporting railway track system, a single-tier retaining wall is modeled as two-tier retaining wall. In this case, factor of safety is found to be more for the first analysis case, analysis without reinforcement when compared to one-tier retaining wall with the same first analysis case (Fig. 4).

Factor of safety for remaining two analysis case did not very much. Again when geogrid length used is 0.7 times the height of the retaining wall, it is found that failure envelop is passing within reinforced zone.

4.2.2 Considering Geogrid Length Equal to 6.65 m

In the two-tier retaining wall case also, geogrid length is increased to 6.65 m. Factor of safety is found to be increased for analysis compared to full length retaining wall with the same geogrid length. Factor of safety for the analysis case of analysis with wall and geogrid for two-tier retaining wall with the geogrid length equal to 0.7H is compared. It is found more factor of safety is from 3.6 to 5.1. Because of an increase in geogrid length, failure envelop is passing beyond the reinforced zone as shown in Fig. 5.

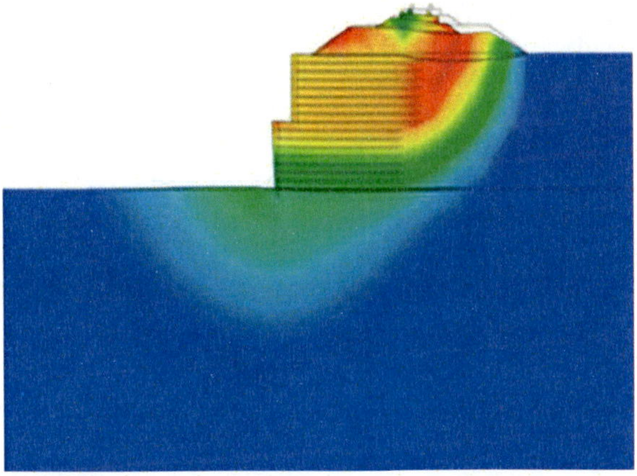

Fig. 5 Failure zone passing beyond reinforced zone for two-tier retaining wall

4.3 Stress Distribution

4.3.1 One-Tier Retaining Wall

Stress distribution for geogrid along its length has been plotted. It is shown in Fig. 6. Stress distribution for first and last geogrid has been plotted. From the figure, it is been found that first geogrid at the wall side has shown higher stresses. Then, it has decreased to zero stress for a distance of 2 m to around 4.5 m. Furthermore, stress increased linearly up to its other end at 6.65 m to a stress increment of 10 kN/m^2.

For the last geogrid, stress at the wall side is low around 2 kN/m^2. It decreased to 0 kN/m^2 at a geogrid length of 1 kN/m^2. Further stresses are increased linearly up to geogrid length of 4 m and then increased till the other end of the geogrid length.

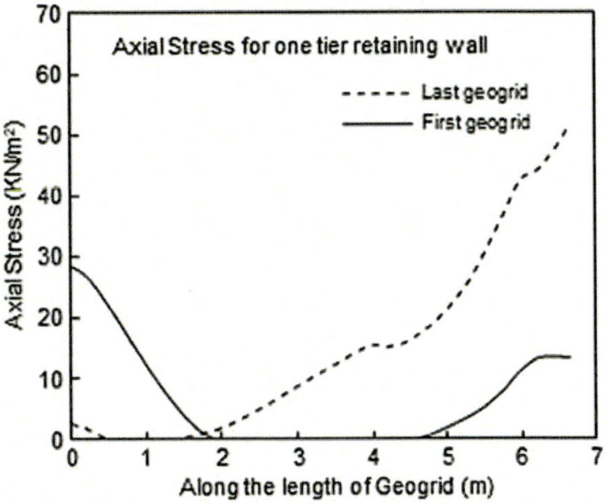

Fig. 6 Axial stress distribution for one-tier wall for first and last geogrid

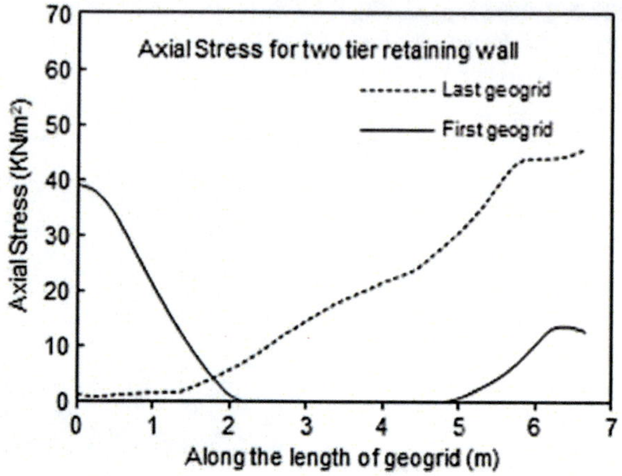

Fig. 7 Axial stress distribution for geogrid in two-tier retaining wall

4.3.2 Two-Tier Retaining Wall

For two-tier retaining wall also, variation in axial stress for first and last geogrid has been analyzed. In the last geogrid, stress distribution at the wall end is less as compared to its other end length which has higher stresses. It has shown a mere linear increase in axial stresses as shown in Fig. 7, whereas first geogrid is considered first from the bottom of the retaining fill soil which will be usually laid at the time of construction. So, the first geogrid has shown more axial stresses at the wall end, showing the axial stress value around 39 kN/m^2. Axial stress decreased linearly up to 2 m geogrid length and up to 4.8 along the length of geogrid the axial value has remained zero. Again the value increased linearly up to its other end.

4.3.3 Along the Length of Wall for One-Tier Retaining Wall

Axial stress distribution for wall also has been plotted. It can be seen from Fig. 8 that at the bottom end of the wall, stress taken by wall is more hence stresses taken by the last geogrid in Fig. 6 are lesser.

It can also be observed from the above figure that at the top end of the wall axial stresses are found to decrease linearly. So, it can be well compared that stresses taken by last geogrid in one-tier retaining wall are more.

For all the cases of analysis, the horizontal deformation of wall is found to be more for unreinforced condition. And the deformation is comparatively less for reinforced model for one-tier and two-tier retaining wall when compared to unreinforced condition of the retaining wall.

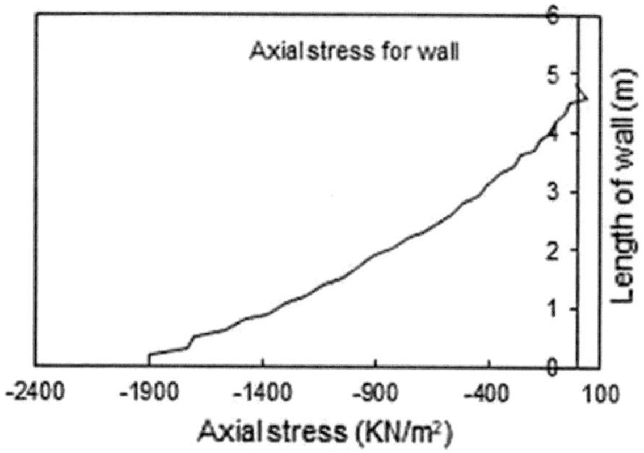

Fig. 8 Axial stress distribution for retaining wall

References

Costa CML, Zornberg JG, de Souza Bueno B, Costa YDJ (2016) Centrifuge evaluation of the time-dependent behavior of geotextile-reinforced soil walls. Geotext Geomembr 44(2):188–200

Elias V, Barry PE, Christopher R (1996) Mechanically stabilized earth walls and reinforced soil slopes design and construction guidelines: participants manual. FHWA Demonstration Project 82, Ground Improvement. Federal Highway Administration

Hossain MS, Kibria G, Khan MS, Hossain J, Taufiq T (2011) Effects of backfill soil on excessive movement of MSE wall. J Perform Constr Facil 26(6):793–802

Skinner GD, Rowe RK (2005) Design and behaviour of a geosynthetic reinforced retaining wall and bridge abutment on a yielding foundation. Geotext Geomembr 23(3):234–260

Soong T, Koerner RM (1999) Geosynthetic reinforced and geocomposite drained retaining walls utilizing low permeability backfill soils. GRI report 24

Yarivand A, Behnia C, Bakhtiyari S, Ghalandarzadeh A (2017) Performance of geosynthetic reinforced soil bridge abutments with modular block facing under fire scenarios. Comput Geotech 85:28–40

Analysis of Retaining Wall in Static and Seismic Condition with Inclusion of Geofoam Using Plaxis 2D

Pankajkumar Yadav, D. K. Singh, P. P. Dahale, and A. H. Padade

Abstract This paper presents the study carried out to analyze the static and seismic behavior of the retaining wall and backfill soil. The lateral earth pressure on retaining wall plays an important role to ensure safety of such structure. The dynamic lateral earth pressure on retaining wall during earthquake condition is always greater than static lateral earth pressure and can induce large destabilizing force. Plaxis 2D software accomplishes the analysis of retaining wall in static and seismic conditions. The retaining wall with backfill material was analyzed with inclusion of compressive material EPS geofoam, which was placed at the interface of retaining wall and backfill material as an absorber to decrease lateral earth pressures on retaining wall. Parametric study of retaining wall has been done by changing various densities of EPS geofoam and thickness. This study shows that inclusion of EPS geofoam reduces the pressure on retaining wall in static as well as seismic condition.

Keywords Retaining wall · Seismic analysis · Geofoam · FEM

P. Yadav (✉)
Department of Civil Engineering, GHRAET, Nagpur, India
e-mail: pankajyadav1806@icloud.com; yadavpankaj1806@gmail.com

D. K. Singh
Department of Civil Engineering, National Fire Service College, Nagpur, India
e-mail: dhananjay.singh86@gmail.com

P. P. Dahale
Department of Civil Engineering, Shri Ramdeobaba College of Engineering & Management, Nagpur, India
e-mail: dahale.p.prasad@gmail.com

A. H. Padade
Department of Civil Engineering, Visvesvaraya National Institute of Technology, Nagpur, India
e-mail: amit51085@gmail.com

M. Latha Gali and P. Raghuveer Rao (eds.), *Geohazards*, Lecture Notes in Civil Engineering 86, https://doi.org/10.1007/978-981-15-6233-4_16

1 Introduction

The heavy soil mass is supported by retaining walls. In the field of geotechnical engineering, retaining structures are used in highways, railways, tunnels, dams, basement of buildings, etc. The soil at higher elevation would tend to move down without any structural support, and it exerts pressure on the structure. The pressure exerted on structure is called as lateral earth pressure. The classification of earth pressure is a function of absolute and relative movement of the backfill soil and retaining structure. If the movement of wall is away from the backfill then the pressure on wall is known as active earth pressure and if the movement of wall is towards the backfill then the pressure on wall is known as passive earth pressure, both pressures can be computed by the methods given by Coulomb (1776) and Rankine (1857) in static condition.

In earthquake-prone area, earthquake can induce large destabilizing force in retaining wall and backfill soil, and seismically induced force has greater influence on lateral earth pressure. Mononobe-Okabe (1929) developed a method to evaluate magnitude of dynamic earth pressure based on pseudo-static approach, and Seed-Whiteman (1970) suggested dynamic earth pressure can be divided into static part and dynamic part, the static part acting one-third of height of wall and the dynamic component of earth pressure acting at 0.6H from base, where '*H*' is height of retaining wall.

Wood (1973) considered the backfill is uniform and elastic; in this case, the dynamic thrust acts at 0.63H. Pseudo-static approach neglected the time effect of dynamic force, dynamic amplification and damping. Steedman and Zeng (1990) considered pseudo-dynamic approach to calculate dynamic earth pressure. A new approach has been carried out by Richards-Elms (1979) based on pseudo-static displacement approach. They derived an equation to evaluate displacement of rigid retaining wall during earthquake. Whitman and Liao (1985) identified several modeling errors that result from the simplifying assumptions of Richards-Elms procedure of evaluating displacement of retaining wall during earthquake, and they found that displacement was lognormal distributed (probabilistic). Variability of ground motions, uncertainty of soil properties and friction angle 'δ', combining all these sources of uncertainty the permanent displacement can be characterized by lognormally distributed variable.

In the context to reduce the pressure on retaining structures, the geofoam (lightweight elastic material) can be used at interface of retaining wall and backfill material. Horvath (1997) found that the geofoam can reduce lateral earth pressure to even less those active conditions. The uniformity and compressibility of geofoam play an important role in the active and passive state of retaining wall. The total lateral earth pressure on retaining wall would decrease because some amount of pressure will dissipate to compress the geofoam. Bathurst et al. (2007) investigated the performance of seismic geofoam buffers by carrying out physical shaking table tests on 1-m-high non-yielding rigid wall with granular backfill and found a maximum dynamic force reduction up to 31%.

2 Material Properties

2.1 Material Properties of Sand

The backfill material is considered as cohesionless soil (sand). The specific gravity of sand was 2.8, the void ratio at loose state (e_{max}) was 0.82, and dense state (e_{min}) was 0.48. The mechanical properties of sand were computed in laboratory by bender element test. This laboratory test is based on wave propagation through soil sample. It consists of source element and receiver element arranged in triaxial cell base. When high-frequency electrical pulse is applied to source element, it produces a stress wave that travels through the specimen toward the receiver element, and it generates a voltage pulse which is measured by receiver element. Shear waves and primary wave will generate due to high-frequency electrical pulse, which is applied to source element by producing a stress wave. The piezo-ceramic bender element is an electromechanical transducer, which is capable of converting mechanical energy to electrical energy on very small strain level.

Shear wave and primary wave velocity determination in sands is an important parameter for analyzing and predicting safety of various structures located on it. The small strain shear modulus of soil is a fundamental parameter used in various kinds of geotechnical analysis, especially in earthquake geotechnical engineering and soil dynamics.

For this test, the sample length was taken of height 12 cm and the diameter was 6.0 cm. These dimensions were selected to obtain the best results from bender element tests because a slenderness ratio is two or greater than two give good results (Camacho-Tauta 2012). In this test, the P-wave and S-wave velocities (Figs. 1 and 2) are computed by dividing the length of the sample by the time taken by waves to move from the source end to the receiver end. Table 1 shows results of this test at different densities of sand.

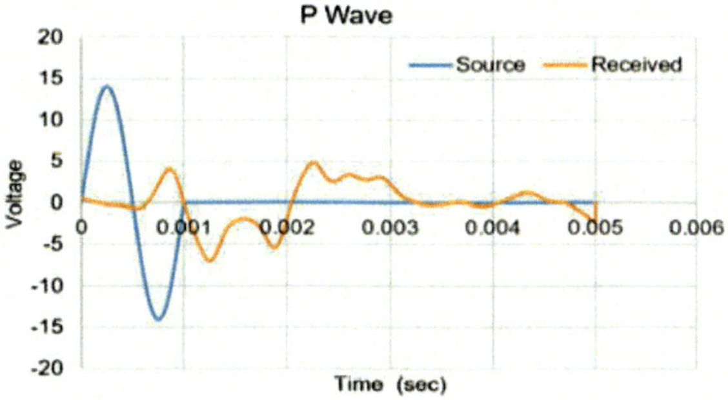

Fig. 1 Bender element test result for P-wave of sand at 16.5 kN/m^3 density

Fig. 2 Bender element test result for S-wave of sand at 16.5 kN/m^3 density

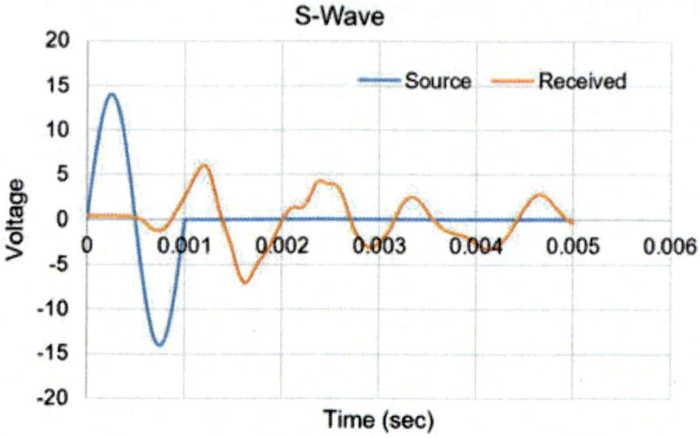

Table 1 Summary of test results of bender element test on sand

Density (kN/m^3)	Wave type	Velocity (m/s)	υ	*E* (kN/m^2)
15.5	P-wave	153	0.29	27,453
	S-wave	83		
16.5	P-wave	179	0.30	38,973
	S-wave	95		
17.5	P-wave	194	0.30	48,923
	S-wave	104		

E = Modulus of elasticity, υ = Poisson's ratio

2.2 *Properties of Geofoam (Extruded Polystyrene)*

In textile term, the geofoam is called as expended polystyrene (EPS). The geofoam is a super-light material which is available in the form of blocks. As per ASTM D 4439, the density varies from 11 to 40 kg/m^3. It is very much compatible with conventional construction materials such as concrete and steel. Geofoam is used as compressible inclusion at the interface of the backfill and the retaining wall.

The hysteresis loss of energy in the foam was calculated according to ASTM D 3574–17 to determine damping of energy (Fig. 3). The compression force displacement (CFD) procedure was followed. For the hysteresis loss and for used EPS foam, it is 25%. The behavior of EPS geofoam under compression test is affected much due to its density (Fig. 4). Higher the density, higher is the compressive strength. In initial linear response of curve, the stress–strain behavior of EPS geofoam is almost linear up to 1.5% of strain level and reaches about 80–85% of total compressive strength. Initial tangent modulus (E_{i}) is an important parameter of EPS geofoam which characterizes the stiffness. After linear curve, the nonlinear stress–strain curve is called as yielding. This yielding zone is extended between a strain range of 1.5 and 5%. In zone 3 beyond the yielding, compressive stress increases marginally with increase in strain with linear variation. In the present study, the specimens were tested for 15% of strain. Therefore, the zone of work hardening is limited between these ranges of

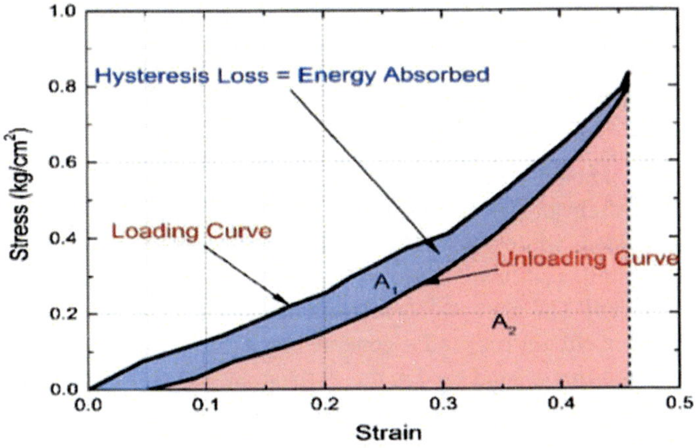

Fig. 3 Hysteresis loss in geofoam

Fig. 4 Stress–strain curve of geofoam

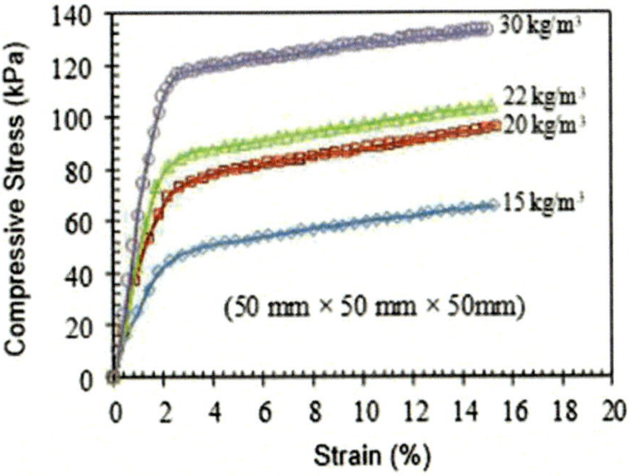

strain. A cube of size 50 mm × 50 mm × 50 mm was taken for compression test of different densities 15, 20, 22 and 30 kg/m^3. In this paper, the geofoam of density 15, 20 and 22 kg/m^3 was taken for analysis and Table 2 shows the modulus properties of respective geofoams.

Table 2 Properties of (EPS) geofoam

Properties	Unit weight (kg/m^3)	E (kN/m^2)	υ
Geofoam1	15	2400	0.086
Geofoam2	20	4000	0.114
Geofoam3	22	5000	0.170

E = Modulus of elasticity, υ = Poisson's ratio

3 Methodology

3.1 Analysis of Retaining Wall in Static Condition

A retaining wall of height 6 m retains sand, modeled in Plaxis 2D. It is finite element software; plane strain modeling is suitable for analysis of retaining wall. Bottom of geometry is fixed, while vertical movement is allowed. The material modeling for backfill sand in this software is Mohr–Coulomb model, which required modulus properties of backfill and strength parameters like angle of shearing resistance and cohesion; however, strength of sand is governed by angle of shearing resistance only. The material properties of sand used in this software was evaluated in laboratory and is shown in Table 3. Plaxis works well for cohesionless soil, to avoid numerical instability, while using Plaxis for cohesionless soil instead of taking the value of cohesion zero, it was taken 1 kN/m^2.

The plate element is used in this software to represent retaining wall of thickness 0.85 m. The retaining walls are made up of concrete. The modulus of elastic (E) for concrete (cement and aggregates) can go up to 50GPa (ACI 318-08). It can be also calculated according to grade of concrete as per IS-456 2000 ($E = 5000\sqrt{f_{ck}}$), where f_{ck} is characteristic compressive strength of concrete. Table 4 represents the modulus and stiffness properties of plate (retaining wall).

The static analysis of retaining wall includes only effect of gravity loads of backfill and retaining wall. The geometry of retaining wall with backfill is shown in Fig. 1, and height of retaining wall plate is 6 m and 2 m embedded in foundation soil with 0.85 m thickness (Fig. 5). Backfill material was 12 m extended behind the wall to show proper failure pattern in backfill.

Table 3 Properties of sand for Plaxis 2D

Properties	Unit	Value
Unit weight	kN/m^3	16.5
Young's modulus	kN/m^2	40,000
Poisson's ratio	–	0.3
Cohesion	kN/m^2	1
Friction angle	Degrees	32°

Table 4 Material properties of plate

Properties	Unit	Value
Axial stiffness (EA)	kN/m	4.25×10^6
Flexural rigidity (EI)	kNm^2/m	2.56×10^5
Poisson's ratio (υ)	–	0.15
Weight per area	kN/m/m	20.4

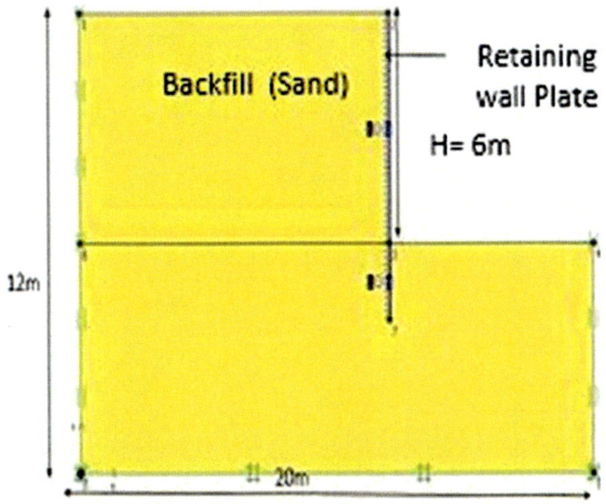

Fig. 5 Geometric model of retaining wall and backfill in Plaxis 2D for static condition

The first step of analysis was to generate the mesh formation and calculate initial stress at rest, and then plastic analysis was done. After completion of analysis, the output contains deformed mesh (Fig. 6) and stress variation in geometry of model.

Plaxis 2D gave maximum pressure 29.72 kN/m^2 for retaining wall of height 6 m. In static condition, according to Coulomb's method, the maximum pressure (stress) at bottom of retaining wall ($K_a \gamma H$) equals to 28.22 kN/m^2 where K_a is 0.285 and Rankine's method gave maximum pressure (stress) at bottom of retaining wall ($K_a \gamma H$) equal to 30.42 kN/m^2 where K_a is coefficient of active earth pressure equal to 0.307. The value of K_a obtained by Plaxis was found to be 0.293. From Fig. 7, it is clear that the static lateral earth pressure on retaining wall was found apparently same, which was evaluated by Coulomb's method, Rankine's method and Plaxis 2D.

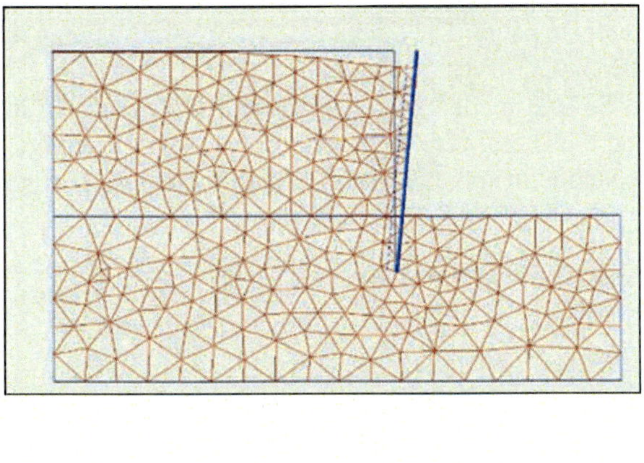

Fig. 6 Deformed mesh of retaining wall and backfill after completion of static analysis

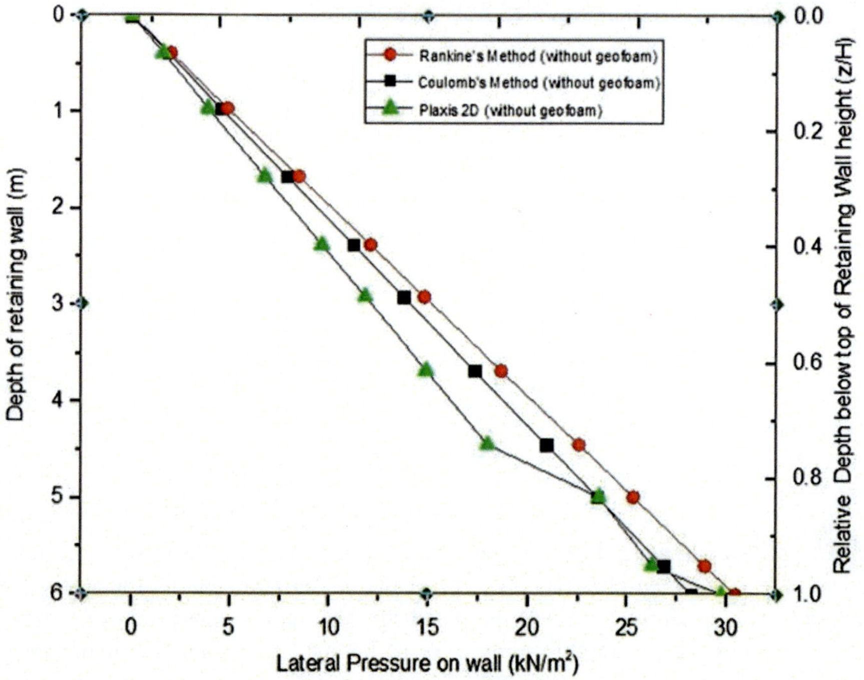

Fig. 7 Lateral earth pressure on wall by different methods

3.2 *Inclusion of Geofoam in Retaining Wall*

The thick panels of geofoam of various thicknesses were provided at interface of retaining wall and backfill which act as absorber and reduce the pressure on retaining wall due to compressibility. Table 5 shows the various combinations of thickness with different densities of geofoam been taken for analysis. 't' is the thickness of geofoam panel. The thickness-to-height of retaining wall ratio was taken in range of 0.085–0.33 (Figs. 8, 9 and 10).

Several models of retaining wall with geofoam inclusion of different densities and thickness were analyzed respectively. From Fig. 11, the average of percentage reduction in static pressure on wall by inclusion of geofoam1 thickness of 0.5 m is

Table 5 Various thicknesses of geofoam

Thickness (t)-to-height (H) ratio	Thickness Geofoam1 (m)	Thickness Geofoam2(m)	Thickness Geofoam3(m)
$(t/H) = 0.0845$	0.5	0.5	0.5
$(t/H) = 0.167$	1	1	1
$(t/H) = 0.335$	2	2	2

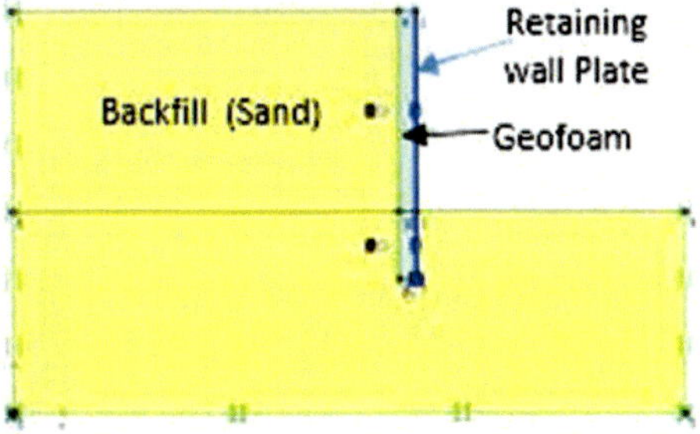

Fig. 8 Retaining wall with geofoam inclusion at interface of wall and backfill in static condition

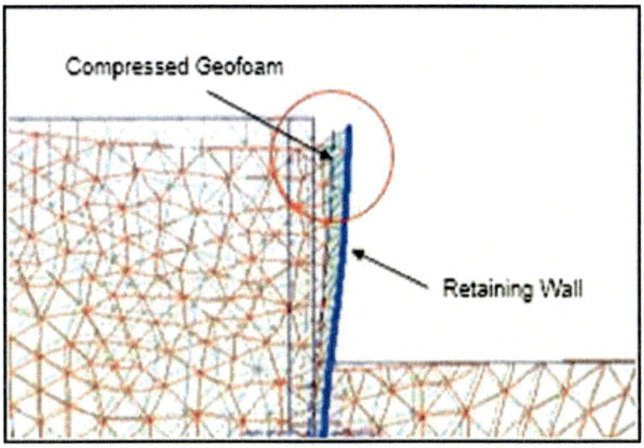

Fig. 9 Deformed mesh showing compressed geofoam

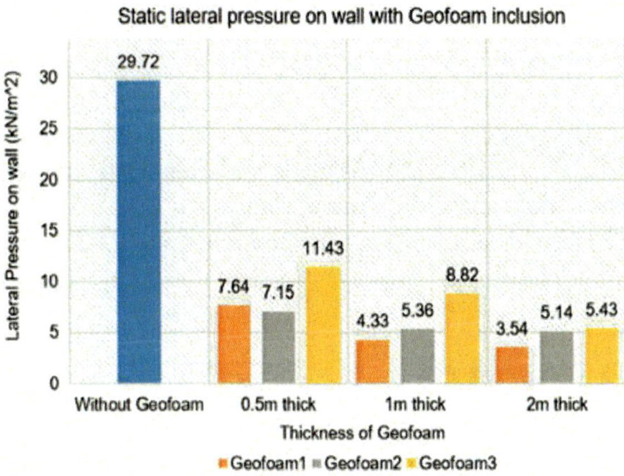

Fig. 10 Static lateral earth pressure with geofoam of different densities and thickness

70.79%, by inclusion of geofoam2 of thickness 1 m is 79.23% and by inclusion of geofoam3 of thickness 2 m is 84.17%.

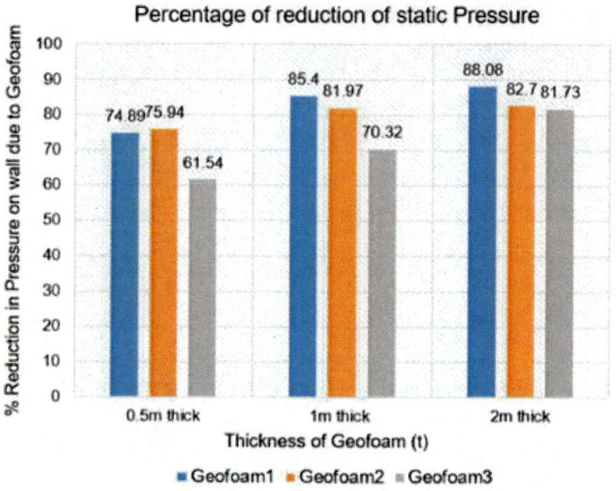

Fig. 11 Percentage of reduction in static lateral earth pressure with geofoam inclusion of different densities and thickness

3.3 Seismic Analysis of Retaining Wall

Soil and structures are often subjected not only to static loads but also to dynamic loads. If the loads are powerful, as in earthquakes, they may cause severe damages. With this software's dynamic analysis module, we can analyze the effects of vibrations in the soil. The earthquake is modeled by imposing a prescribed displacement at the bottom boundary. At the far vertical boundaries, absorbent boundary conditions are applied to absorb outgoing waves. For plane strain models, the standard absorbent boundaries are generated at the left-hand, the right-hand and the bottom boundary. The absorbent boundaries reduce the box effect while analysis. A real accelerogram of earthquake in standard SMC format (Strong Motion CD-ROM) given as input to horizontal prescribed displacement to the bottom boundary is shown in Fig. 13. The maximum peak ground acceleration in this accelerogram is 0.25 g. The seismic analysis is completed in two stages: first the plastic analysis and then dynamic analysis. The time interval of dynamic analysis is 10 s (Fig. 12).

The mechanical properties obtained by bender element test corresponds to very small strain (10^{-6} to 10^{-5}%), the shear modulus of soil at such strain rate is a

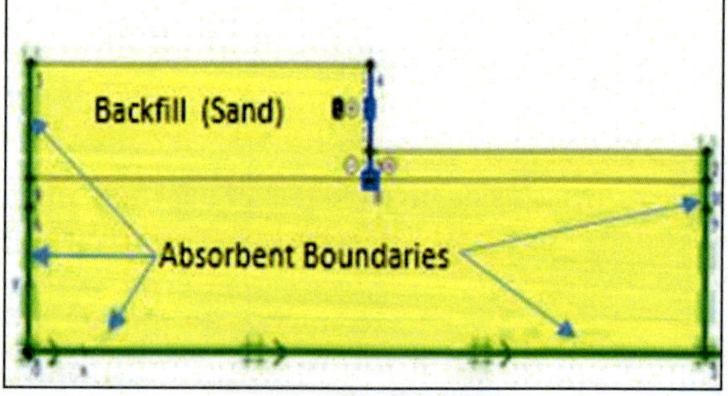

Fig. 12 Geometry of retaining wall without geofoam in seismic condition

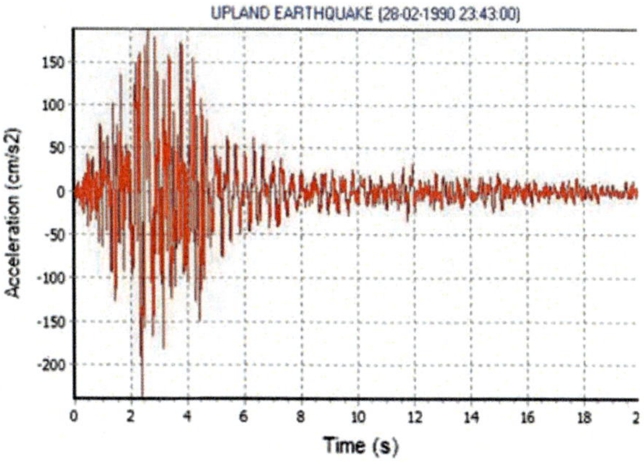

Fig. 13 Input acceleration history of actual earthquake of 0.25 g horizontal acceleration (in SMC format)

fundamental parameter used in various kinds of geotechnical analysis especially in earthquake geotechnical engineering and soil dynamics. The parameters of backfill material, geofoams and retaining wall plate are same as used in static condition.

From Fig. 14, it is clear that the maximum input horizontal acceleration is 0.25 g gives an amplification of acceleration in backfill. The acceleration at the top of retaining wall is 0.38 g, while input at bottom is 0.25 g, and the amplification factor is 1.52 times of input horizontal acceleration.

Permanent displacement in retaining wall during earthquake can be evaluated by method of Richards-Elms (1979) and Whitman-Liao (1985), but these methods are only applicable to gravity retaining wall (Fig. 15).

Mononobe-Okabe (1929) has developed a method to calculate dynamic earth pressure during earthquake based on pseudo-static approach that is popularly known

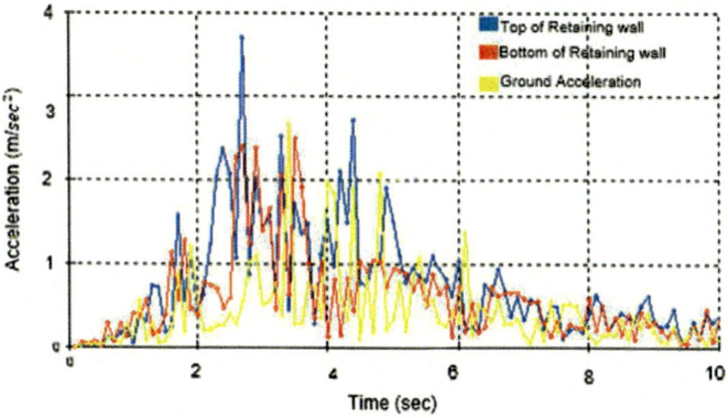

Fig. 14 Acceleration spectrum at top and bottom of retaining wall and ground level

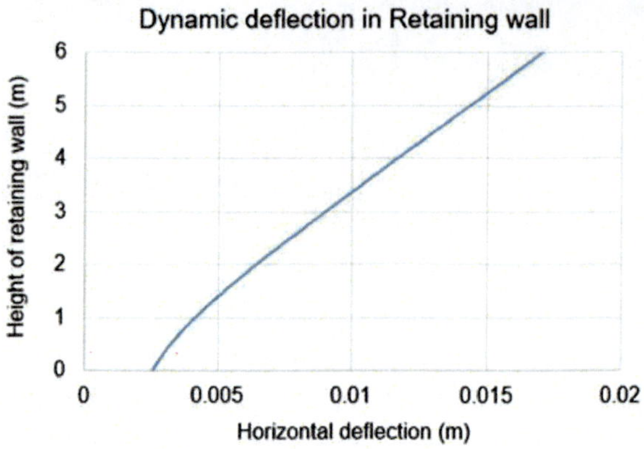

Fig. 15 Permanent deflection in retaining wall

as M–O method. This method is extension of static coulomb's wedge theory to pseudo-static condition. The dynamic earth pressure can be calculated by Eq. 1.

Dynamic active earth pressure (P_{AE})

$$P_{\mathrm{AE}} = \frac{1}{2} K_{\mathrm{AE}} \gamma H^2 (1 - k_v) \tag{1}$$

The coefficient of dynamic active earth pressure (K_{AE}) by Mononobe-Okabe method can be calculated by Eq. 2.

$$K_{\mathrm{AE}} = \frac{\cos^2(\varnothing - \theta - \varphi)}{\cos\varphi \cos^2\theta \cos(\delta + \theta + \varphi)\left[1 + \sqrt{\frac{\sin(\delta+\varnothing)\sin(\varnothing-\beta-\varphi)}{\cos(\delta+\theta+\varphi)\cos(\beta-\theta)}}\right]^2} \tag{2}$$

Seed-Whiteman has derived the formula for coefficient of dynamic lateral earth pressure (K_{AE}) based on the course of experiments shown in Eq. 3.

$$K_{\mathrm{AE}} = k_A + 3/4 k_h \tag{3}$$

K_A coefficient of active earth pressure
K_{AE} coefficient of dynamic active earth pressure
k_h coefficient of horizontal acceleration
k_v coefficient of vertical acceleration
P_{AE} dynamic active earth pressure
$\varnothing$ angle of shear resistance
θ batter angle of retaining wall
φ $\tan^{-1}\left(\frac{k_h}{1-k_v}\right)$
β angle of inclination of backfill
δ angle of friction between wall and backfill

The coefficient of dynamic earth pressure (K_{AE}) by Mononobe-Okabe method is 0.502, and by Seed-Whiteman method it is 0.456. The dynamic earth pressure is found linear, but Plaxis 2D shows time-dependent dynamic lateral earth pressure shown in Fig. 16. M–O method gives maximum value of dynamic lateral earth pressure which would be safe for designing retaining wall.

Further, the dynamic analysis of retaining wall with geofoam inclusion of different densities and thickness was done and Fig. 17 shows the results.

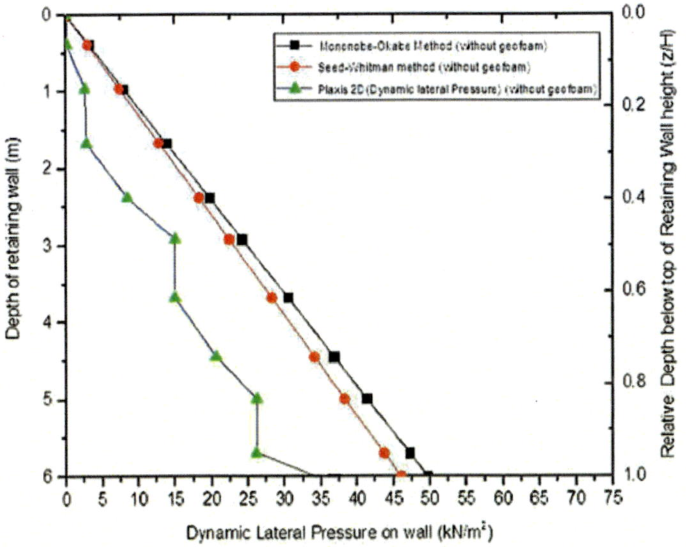

Fig. 16 Dynamic earth pressure by different methods

Fig. 17 Dynamic earth pressure on retaining wall with geofoam of different densities and thickness

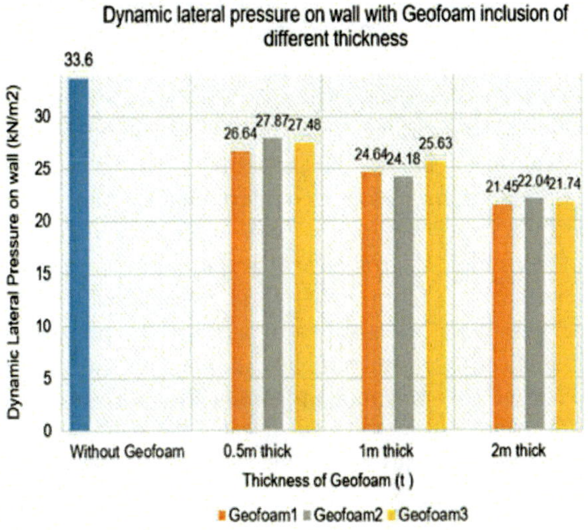

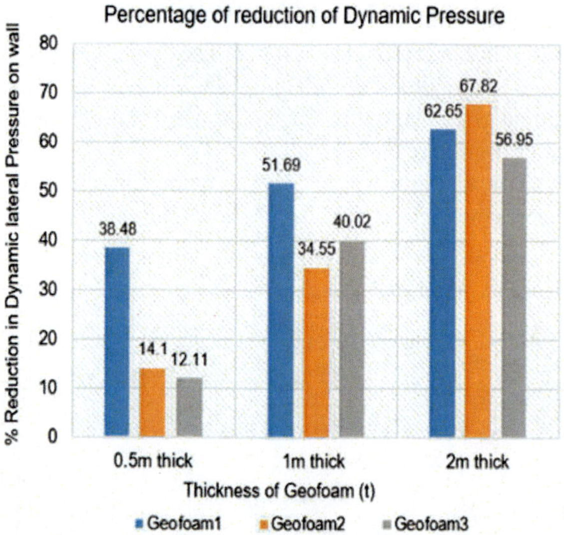

Fig. 18 Percentage of dynamic earth pressure reduction by geofoam of different densities and thickness

From Fig. 18, the average of percentage reduction in dynamic lateral pressure by inclusion of geofoam1 of thickness 0.5 m is 21.58%, geofoam2 of thickness 1 m is 4.208% and geofoam3 of thickness of 2 m is 62.57%.

3.4 Inclusion of Geofoam in Cantilever Retaining Wall

A cantilever retaining wall of 6-m height, 3-m base slab and embedded depth of retaining wall 1.65 m was also modeled in Plaxis 2D (Fig. 19). The thickness of plate is 0.85 m, first the model was analyzed without geofoam and deformed mesh for this condition is shown in Fig. 20, the maximum displacement at top of cantilever retaining wall without geofoam was 47.5 mm (Fig. 21).

The mean shading of horizontal stress in backfill (Fig. 22) shows few plastic failure lines that can be also interpreted as triangular failure pattern taking place in backfill.

From Fig. 23, it is clear that the failure zone can be traced by dotted line that represents soil failure wedge and presents point of application of all stress (active and passive). The lateral earth pressure is represented in Fig. 24 which gives point of application of pressure. The moment on wall (stem) will be higher as compared to normal generalization of lateral earth pressure.

In cantilever retaining wall, base slab is provided, and the soil above base slab increases the stability of wall. To keep this point in mind, a minimum thickness (0.5 m) of geofoam is applied at interface of wall and backfill as shown in Fig. 25

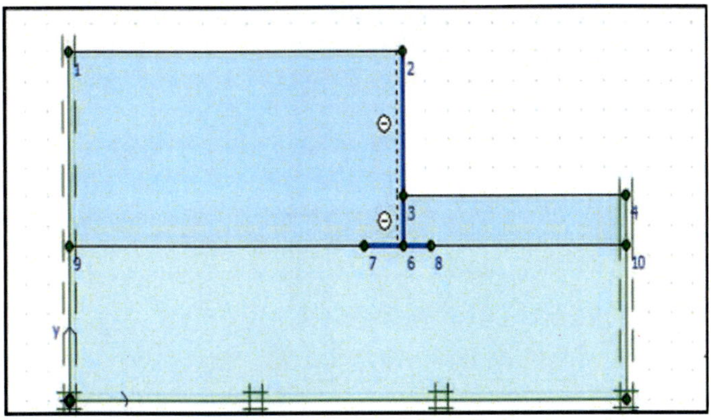

Fig. 19 Geometry of cantilever retaining wall in Plaxis 2D

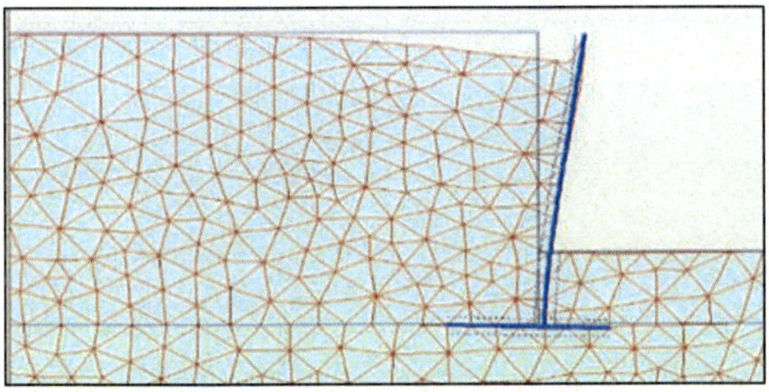

Fig. 20 Deformed mesh of cantilever retaining wall system

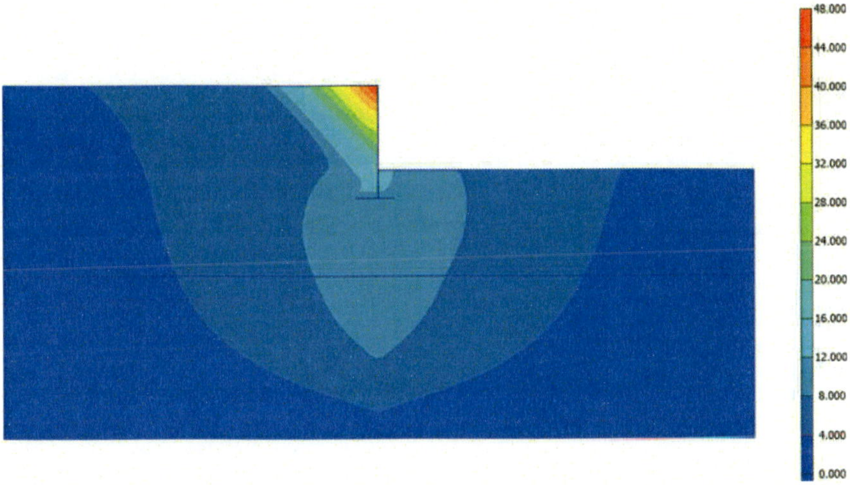

Fig. 21 Horizontal displacement (mm) in cantilever retaining wall system without geofoam (mean shading)

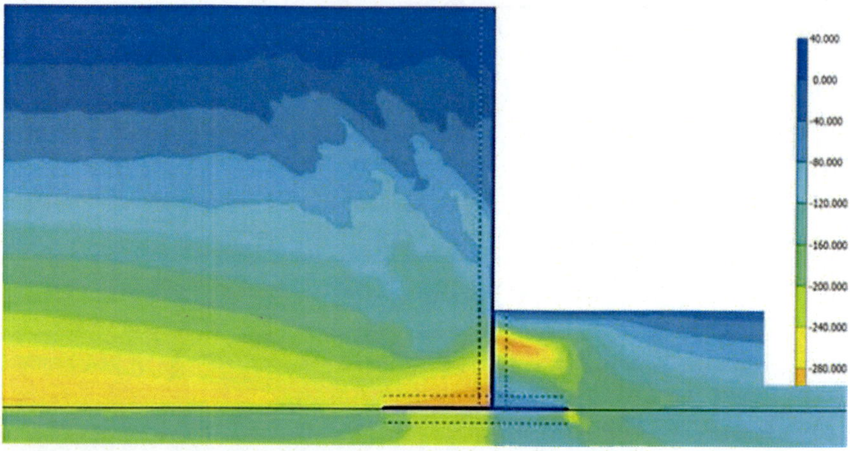

Fig. 22 Mean shading of horizontal stress (KN/m^2) in backfill soil at active state of wall

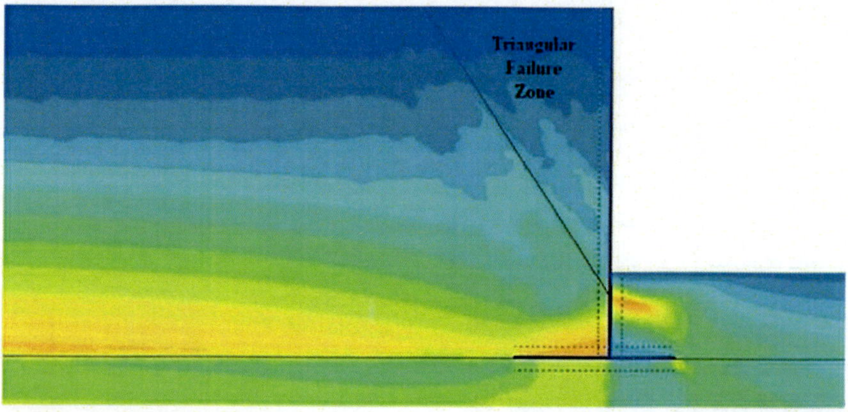

Fig. 23 Interpretation from stressed soil body

Fig. 24 Point of application of earth pressure on cantilever retaining wall

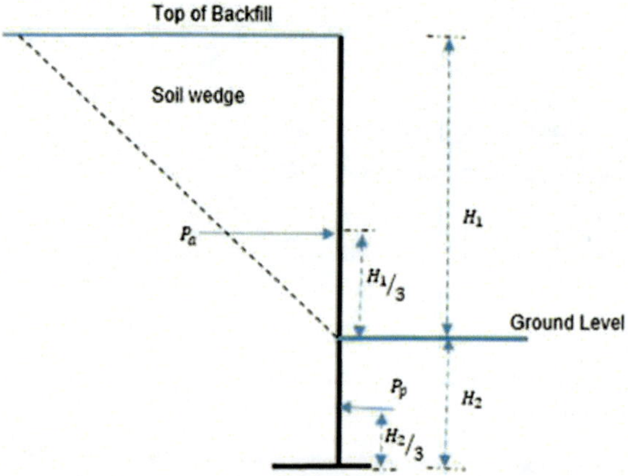

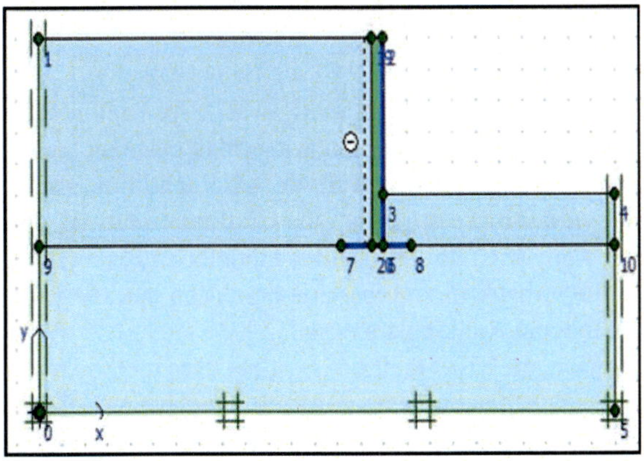

Fig. 25 Inclusion of geofoam (green portion) in cantilever retaining wall at interface

and analysis was done which reduces the pressure on wall as well as increases stability by reducing horizontal movement.

The analysis of cantilever retaining wall with inclusion of geofoam reduces the outward movement of wall, and it gets reduced (Fig. 26).

Fig. 26 Mean shading Geofoam inclusion reduces lateral movement (mm) of cantilever retaining wall

4 Conclusion

The static and seismic behaviors of retaining wall have been studied in this paper. Coulomb's method and Rankine's method are used to evaluate the lateral earth pressure on retaining wall for static condition. In static condition, the Rankine method gives greater value of earth pressure than the Coulomb method, which may be safe to design retaining wall. The retaining wall is modeled in Plaxis 2D to evaluate earth pressure on wall due to backfill soil in static condition and the value found similar to Coulomb's method and Rankine's method.

In seismic condition, the M–O method is widely used method to estimate dynamic lateral earth pressure. At the conclusion of such intensive modeling of the retaining wall with geofoam as compressible inclusion at interface of backfill and retaining wall in Plaxis 2D, many observations can be made and many questions have been clarified. The major objective of this work was to reduce the pressure on retaining wall. The EPS geofoam in static condition, average reduction of the lateral earth pressure on retaining wall is up to 78.06%, and the average reduction of the dynamic lateral earth pressure on retaining wall is up to 42.07%. As the thickness of geofoam increases, the reduction in lateral pressure on retaining wall increases.

References

Bathurst RJ, Zarnani S, Gaskin A, Gaskin A (2007) Shaking table testing of geofoam seismic buffers. Soil Dynam Earthquake Eng 27(4):324–332. https://doi.org/10.1016/j.soildyn.2006.08.003

Camacho-TautaJ.,Álvarez-Jiménez JD, Reyes-Ortiz OJ (2012) A procedure to calibrate and perform the bender element test. DYNA 79:10–18. Electrónico, pp 2346–2183. ISSN impreso 0012–7353

Coulomb CA (1776) Essai sur une application des regles des Maximis et Minimis a quelquels problemes de Statique relatifs, a l'Architecture. Memoirs Academie Royal Pres. Division Sav. 7, Paris, France (in French)

Horvath JS (2010) Lateral pressure reduction on earth-retaining structures using geofoams: correcting some misunderstandings. In proceedings of Earth Retention Conference 3, Bellevue, Washington, ASCE GSP-208, pp 862–869. https://doi.org/10.1061/9780784411285

Mononobe N, Matsuo H (1929) On the determination of earth pressures during earth-quakes. In Proceedings of the World Engineering Congress, Tokyo, Japan, 9

Rankine W (1857) On the stability of loose earth. Philos Trans Royal Soc London 147

Richards RJ, Elms D (1979) Seismic behaviour of gravity retaining walls. J Geotech Eng Div ASCE 105(GT4):449–464

Seed HB, Whitman RV (1970) Design of earth retaining structures for dynamic loads. In Proceedings of the ASCE Specialty Conference: Lateral Stresses in the Ground and Design of Earth Retaining Structures, pp 103–147

Steedman RS, Zeng X (1990) The influence of phase on the calculation of pseudo-static earth pressure on a retaining wall. Geotechnique 40(1):103–112

Wood JH (1973) Earthquake-induced soil pressures on structures, report EERL 73-05, California Institute of Technology

Whitman RV, Liao S (1985) Seismic design of gravity retaining walls. US Army Corps of Engineers, Vicksburg, Misc. Paper GL-8501

Static and Seismic Analysis of Twin Metro Underground Tunnels

Manendra Singh, M. N. Viladkar, and N. K. Samadhiya

Abstract In any metropolitan city, underground structures are key elements of a mass rapid transit system. In situations where metro underground construction is to be undertaken in poor soil strata, it is not possible to excavate a single large diameter tunnel for accommodating the two-way traffic of trains. It becomes therefore essential to have two parallel tunnels aligned either horizontally or vertically. In this paper, an attempt has been made to predict the influence of construction of a second tunnel, aligned either horizontally or vertically, after the construction of the existing first tunnel. The analysis has been carried out for both static and seismic conditions by varying the pillar width between the tunnels. Elasto-plastic finite element analysis has been carried out through Plaxis 2D software for the 1999 Chamoli earthquake of lower Himalaya. The analysis brought forth the fact that vertical stresses at critical points and the forces in RC liners of both horizontally and vertically aligned twin tunnels increase during the earthquake for a pillar width equal to half the diameter of tunnels. These vertical stresses and forces in RC liners have been found to reduce with increase in pillar width in case of vertically aligned twin tunnels, whereas a reverse situation arises in case of horizontally aligned twin tunnels.

Keywords Seismic · Tunnels · Metro

M. Singh
Department of Civil Engineering, National Institute of Technology Hamirpur, Hamirpur 177005, India
e-mail: manendra@nith.ac.in

M. N. Viladkar (✉) · N. K. Samadhiya
Department of Civil Engineering,
Indian Institute of Technology Roorkee, Roorkee 247667, India
e-mail: mnviladkar50@gmail.com

N. K. Samadhiya
e-mail: nksamfce@iitr.ac.in

M. Latha Gali and P. Raghuveer Rao (eds.), *Geohazards*, Lecture Notes in Civil Engineering 86, https://doi.org/10.1007/978-981-15-6233-4_17

1 Introduction

In any metropolitan city, metro underground tunnels and underground rail stations form key components of a Mass Rapid Transit (MRT) system. With the growth in population and the limitation of space in a metro city, there is a growing need to enlarge these transportation networks. In areas which lie in seismic zones, there is always the vulnerability of such underground facilities to seismic loading and it is a very sensitive issue. A large earthquake not only can cause the potential loss of human life but can also damage many other infrastructural facilities. In turn, it can result in considerable economic losses, particularly if the time required to restore the functionality of the network is large. This loss can be reduced if the possible risk and the associated damage can be reduced. For these reasons, it is essential to understand as to how metro underground tunnels suffer damage during earthquakes and how to enhance the service efficiency. A single tunnel provided for both the up and down tracks in a metro underground network usually leads to a very large diameter of the tunnel. Construction of such a large size tunnel may not be always feasible depending of course upon the site-specific conditions. It is therefore usual to provide another tunnel aligned in parallel either horizontally or vertically. In this paper, the influence of construction of a second tunnel after the construction of the existing first tunnel has been investigated by varying the pillar width (clear distance) between the tunnels. Both static and seismic analyses of twin tunnels have been performed. Many authors such as Soliman et al. (1993), Kawata and Ohtsuka (1993), Saitoh et al. (1994), Perri (1995), Yamaguchi et al. (1998), Shahrour and Mroueh (1997), Hu et al. (2003), Chu et al. (2007), Chehade et al. (2008), Chen et al. (2009a, b), Afifipour et al. (2011), Li et al. (2012), Hussein et al. (2012) and Comodromos et al. (2014) have studied the static behavior of twin tunnels. Many authors, namely Kumari et al. (2012, 2013, 2014), Motaal et al. (2013), Sahoo et al. (2013), Shaalan et al. (2014), Azadi et al. (2014), Rahim et al. (2015) and Shirinabadi (2016), also studied the dynamic behavior of twin tunnels.

2 Horizontally Aligned Twin Tunnels

2.1 Twin Tunnel Geometry

As a part of the current investigation, it was decided to consider the behavior of twin tunnels of DMRC which were excavated through alluvium deposits, generally known as Delhi silt. The geometric parameters of twin tunnel section are defined in Table 1. For the present study, the soil mass was treated as homogeneous.

Properties of the soil medium surrounding the tunnel are presented in Table 2. These properties remain unaltered with depth of soil strata. No water table was

Table 1 Geometric details of tunnel (Yadav 2005)

Properties	Values
Diameter of both tunnels, *D*	6.0 m
Overburden depth, *H*	12 m
Support system	Segmental reinforced concrete (RC) liners
Thickness RC liners	0.28 m
Clear distance between tunnels	0.5 D–5.0 D
Elastic modulus of RC liners, E_c	3.16×10^7 kPa
Poisson's ratio of concrete	0.15

Table 2 Properties of soil medium (Yadav 2005)

Properties	Values
Unit weight, γ_{bulk}	18 kN/m^3
Saturated unit weight, γ_{sat}	20 kN/m^3
Cohesion, *c*	0
Friction angle, φ	35°
Dilatational angle, ψ	5°
Poisson's ratio	0.25
Elastic modulus, *E*	25 MPa

encountered during tunnel excavation. Complete geometry of physical model of soil–tunnel system is shown in Fig. 1 along with the locations of various critical points where the response was monitored.

2.2 *Numerical Modeling*

Two-dimensional plane strain finite element analysis has been carried out for DMRC metro tunnels using Plaxis 2D software. The extent of the model in the two directions, after carrying out sensitivity analysis, has been taken as 180 m × 60 m. Six-noded triangular elements were considered for modeling of soil domain. Segmental RC liners of tunnel were simulated using plate bending elements. The stress–strain behavior of soil was considered as elasto-plastic which follows the Mohr–Coulomb yield criterion. For RC liners, elastic behavior was considered. No-slip (perfect bond between soil and liners) condition has been assumed between the tunnels and the surrounding soil medium. Damping in soil and RC liners of tunnel were taken as 15 and 2%, respectively. Response spectra compatible time history for 1999 Chamoli earthquake of lower Himalaya, as shown in Fig. 2, was preferred for the analysis. For static response, nodes along vertical boundaries of finite element mesh were

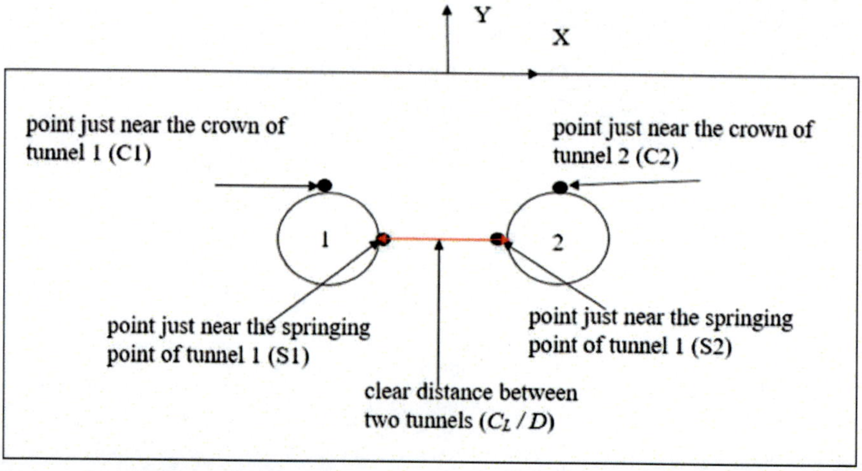

Fig. 1 Layout of geometry of twin tunnels

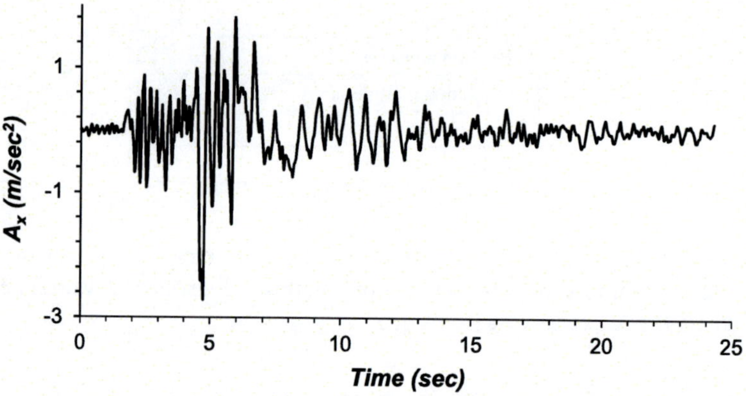

Fig. 2 Response spectra compatible time history of 1999 Chamoli earthquake

restrained in X (horizontal) direction and were free to move in Y (vertical) direction, whereas the bottom boundary was fixed in both directions. For dynamic analysis, viscous absorbent boundary, proposed earlier by Lysmer and Kulhemeyer (1969), was used to represent the displacement condition along both vertical boundaries. Coefficients of absorbent boundary, C_1 and C_2, have been taken as 1 and 0.25, respectively.

2.3 Stages of Calculations

Stage 1: Only initial stresses are generated using K0-procedure. Both the tunnels and other loadings are kept inactive in this phase.
Stage 2: Soil inside tunnel 1 has been removed, and RC liner was applied as the permanent support system. Tunnel 2 is kept inactive in this phase.
Stage 3: Contraction of 3% was applied for simulating the volume loss during the construction of tunnel 1. This completes the construction of tunnel 1. Tunnel 2 is still kept inactive.
Stage: 4: Repeat step 2 for tunnel 2.
Stage 5: Repeat step 3 for tunnel 2. Static analysis of the whole system was first carried out, and the resulting stresses were stored as initial stresses in the whole system. This defines the state of stress surrounding the tunnel before the occurrence of an earthquake.
Stage 6: For dynamic analysis, the boundaries are treated as viscous absorbent boundaries and corresponding constants *a, b* are defined. The time history of earthquake loading is therefore applied along the base of the model, and the seismic analysis is carried out.

2.4 Results and Discussion

Vertical stress concentration around the tunnels and the forces mobilized in RC liners during static and dynamic analysis are presented here. For studying the effect on vertical stress concentration during static and dynamic analysis, four critical points, namely C1 (point just near the crown of tunnel 1), S1 (point just near the springing point of tunnel 1), C2 (point just near the crown of tunnel 2) and S2 (point just near the springing point of tunnel 2), are selected, as shown in Fig. 1.

2.4.1 Static Analysis

Firstly, tunnel 1 has been constructed. Stresses were redistributed after the construction of tunnel 1, and these resulting stresses were stored as initial stresses in the whole system prior to the construction of tunnel.

The aim is to predict the effect of construction of tunnel 2 on tunnel 1 and the surrounding soil. Figure 3 shows the incremental vertical stress after the construction of tunnel 2 for different values of the pillar widths where σ_{s2} is the vertical stress after the construction of tunnel 2 and σ_{s1} is vertical stress after the construction of tunnel 1. It can be seen from Fig. 3 that vertical stress (σ_{s2}) at both critical points of tunnel 1, namely C1 and S1, increases by about 1.7 times σ_{s1} for a pillar width of 0.5 D. It can as well be noticed that vertical stress at points C1 and S1 of tunnel 1 decreases when the pillar width increases from 0.5 D to 3.0 D. Thereafter, it remains constant even if the pillar width is increased to 5.0 times the diameter.

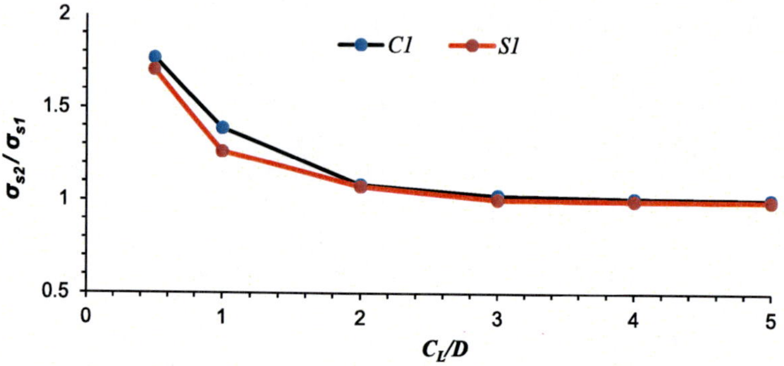

Fig. 3 Increment in vertical stress at different points after construction of tunnel 2

Figure 4 shows the horizontal displacement at ground surface after the construction of tunnels for pillar width of 0.5 D. Maximum horizontal displacement after the construction of first tunnel is of the order of 10 mm. It can be also seen that the maximum horizontal displacement increases to 19 mm after the construction of the second tunnel.

Values of maximum horizontal displacement at the ground surface have been summarized in Table 3 for different pillar widths. It can be noticed that horizontal

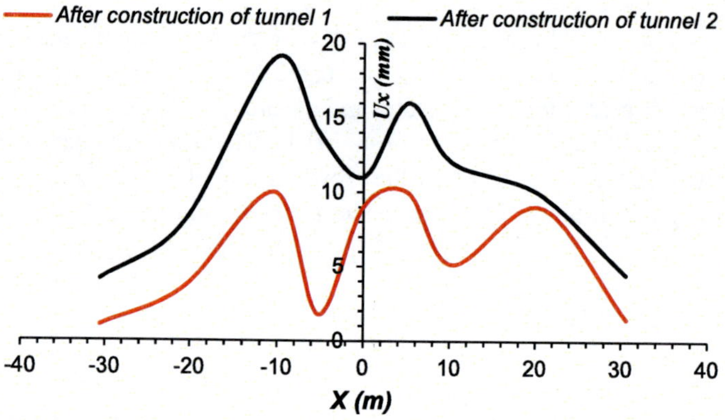

Fig. 4 Horizontal displacement at ground surface after construction of both tunnels ($C_L = 0.5$ D)

Table 3 Displacements at ground surface (0, 0) after construction of tunnel 2

Pillar width, C_L	U_x (mm)	U_y (mm)
0.5	19	44
1.0	16	41
3.0	10	23
5.0	7.5	22

displacement at ground surface reduces with increase in pillar width between the tunnels. Maximum values of horizontal displacement are of order of 19, 16, 10 and 7.5 for pillar widths of 0.5 D, 1.0 D, 3.0 D and 5.0 D, respectively.

The values of vertical displacement at the ground surface (ground subsidence), which occurs after the construction of both tunnels, are presented in Fig. 5. The maximum vertical displacement after the construction of first tunnel was about 25 mm. It can also be seen from Fig. 5 that maximum vertical displacement increases up to 44 mm after the construction of the second tunnel. These values of maximum vertical displacement at ground surface decrease with increase in pillar width as shown in Table 3.

Forces in RC liners after the construction of tunnel 1 are taken as initial condition prior to construction of tunnel 2. Then, percent increase in forces in RC liners of tunnel 1, after the construction of tunnel 2, is presented in Fig. 6.

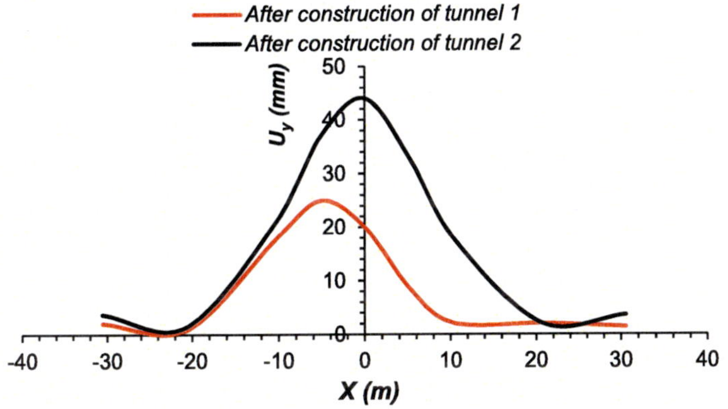

Fig. 5 Vertical displacement at ground surface after construction of both tunnels ($C_L = 0.5$ D)

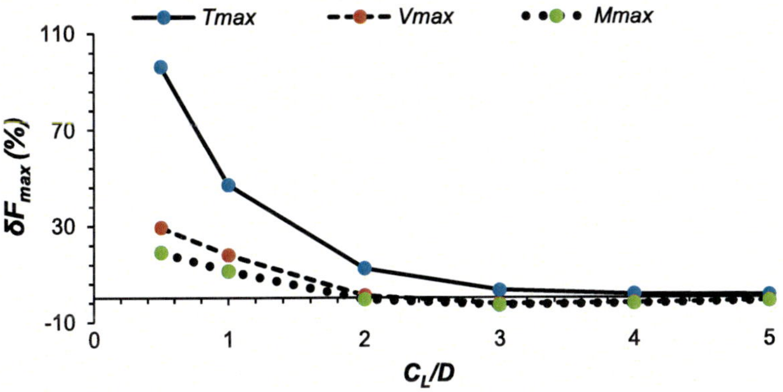

Fig. 6 Percent increase in forces in RC liners of tunnel 1 due to construction of tunnel 2

The maximum forces in RC liners are denoted as axial force, Tmax, shear force, Vmax, and the bending moment, Mmax, respectively. The incremental values of these forces are plotted in Fig. 6 with respect to pillar width. It shows that axial force, shear force and bending moment increase significantly when the pillar width reduces to half the diameter of tunnel. The corresponding increment in axial force, shear force and bending moment is of order of about 96, 29 and 19%, respectively. Forces in RC liners of tunnel 1 were found to reduce significantly as the pillar width between the tunnels increases from half to 3 times the diameter of tunnel. Thereafter, the forces attain almost a constant value.

2.4.2 Seismic Analysis

Stresses are redistributed after the construction of the second tunnel, and these resulting stresses were stored as initial stresses in the whole system prior to applying the earthquake loading. Figure 7 shows the increase in vertical stress at different points of both the tunnels during the earthquake. In Fig. 7, σ_{s2} is the vertical stress at different points of both tunnels after the construction of tunnel 2, and σ_{eqm} is the maximum vertical stress during the earthquake. It is seen that vertical stress during earthquake increases at all points with increasing pillar width and that it is maximum at the crown points of both tunnels.

Then, percentage increment in forces in RC liners of tunnel 1 and tunnel 2 during the earthquake was estimated and their variation with respect to size of pillar width is presented in Figs. 8 and 9, respectively.

Both the figures suggest that increment in shear force and bending moment in the liners during the earthquake is practically independent of the pillar width. The increment in axial force in the liners of tunnel 1 increases significantly from about 28% for pillar width equal to 0.5D to almost 141% for pillar width equal to 3D and subsequently remains practically constant (Fig. 8). Figure 9 shows similar trend during earthquake for shear force and bending moment in the liners of tunnel 2.

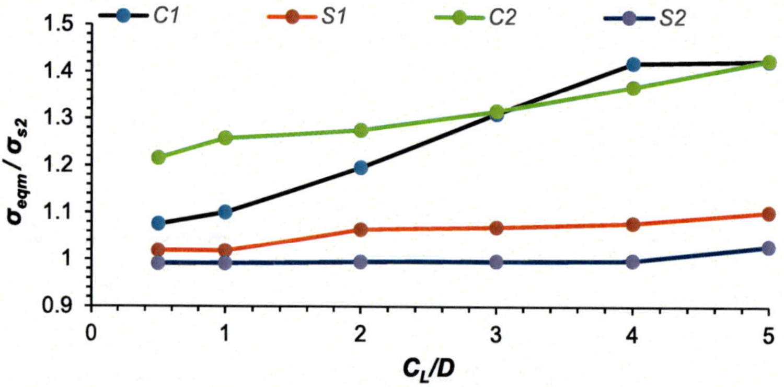

Fig. 7 Increase in vertical stress at different points during the earthquake

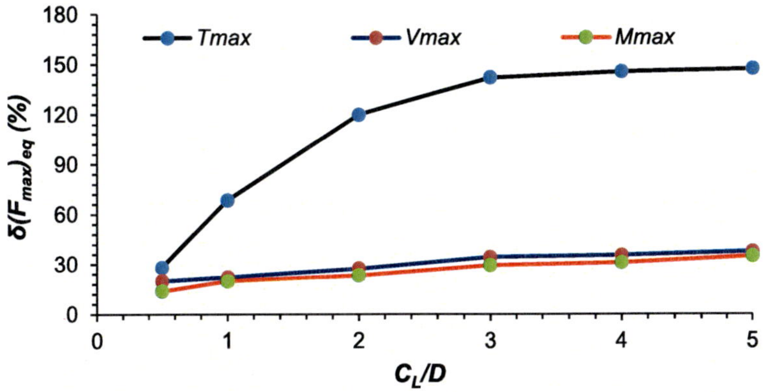

Fig. 8 Percent increase of forces in RC liners of tunnel 1 during earthquake

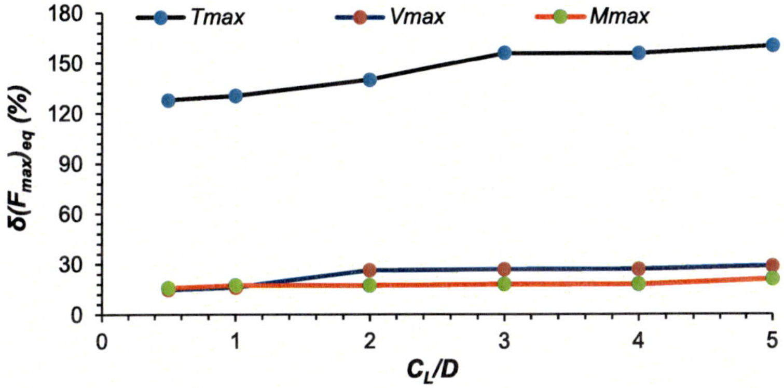

Fig. 9 Percent increase of forces in RC liners of tunnel 2 during earthquake

However, incremental axial force in liners increases from 127% for a pillar width of 0.5 D to about 151% for a pillar width of 3D and thereafter it remains constant.

3 Vertically Aligned Twin Tunnels

In this section, the case of vertically aligned twin tunnels is considered for analysis. The layout of twin tunnels is shown in Fig. 10. It is treated that lower tunnel 2 is first constructed and then the upper tunnel 1 is constructed. The depth of overburden above the upper tunnel 1 is two times its diameter. All other properties were kept the same as in case of horizontal twin tunnels, except the position of the tunnels. For studying the vertical stress concentration and forces in RC liners during static and seismic analysis and also for studying the influence of varying pillar width, six

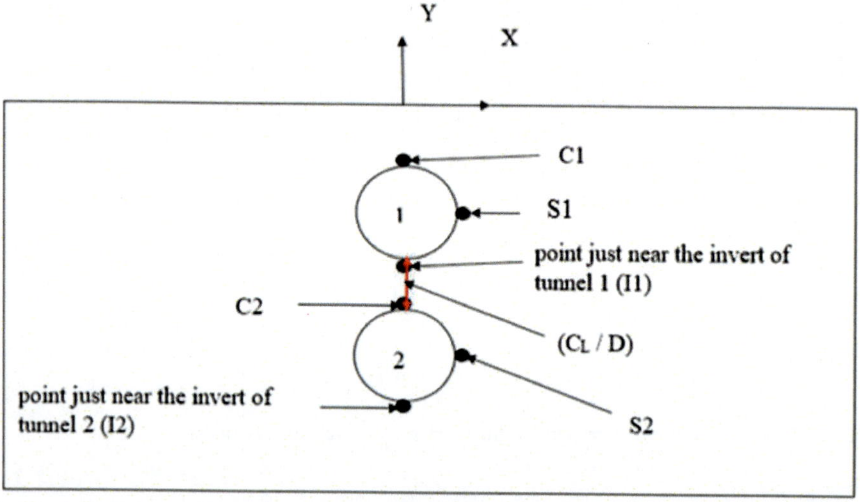

Fig. 10 Layout of geometry of vertical twin tunnels (not to scale)

critical points, namely C1 (point just at the crown of tunnel 1), I1 (point at the invert of tunnel 1), S1 (point just at the springing point of tunnel 1), C2 (point at the crown of tunnel 2), I2 (point at the invert of tunnel 2) and S2 (point at the springing point of tunnel 2), have been chosen, as shown in Fig. 10. Results of static and seismic analysis are presented in the following subsections.

3.1 Static Analysis

Figure 11 shows the change in vertical stress concentration at critical points of lower tunnel 2 after the construction of upper tunnel 1 for different pillar widths. σ_{s2} is the vertical stress after the construction of lower tunnel, whereas σ_{s1} is the vertical stress after the construction of upper tunnel. It can be seen from Fig. 11 that vertical stress concentration factor (σ_{s1}) at all critical points of lower tunnel 2, namely C2, I2 and S2, was found to be 0.52, 0.88 and 0.87 times that of σ_{s2}, respectively, for a pillar width of 0.5 D. This is due to the fact that soil mass surrounding tunnel 2 has entered into plastic state. The minimum concentration factor is at the periphery of the tunnel only. This vertical stress concentration around lower tunnel was found to reduce with increasing pillar width between the tunnels till finally at a pillar width of five times the diameter, the stress concentration factor at all points, C2, I2 and S2, attains a value of unity indicating an all-round elastic state in soil mass.

Figure 12 shows the horizontal displacement at ground surface after the construction of the both tunnels for a pillar width of 0.5 times the diameter of tunnel. Maximum horizontal displacement after the construction of lower tunnel 2 is of order

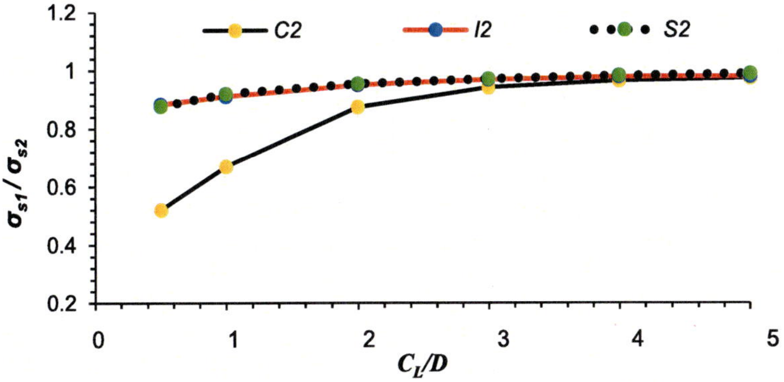

Fig. 11 Change in vertical stress at different points around lower tunnel (2) after construction of upper tunnel (1)

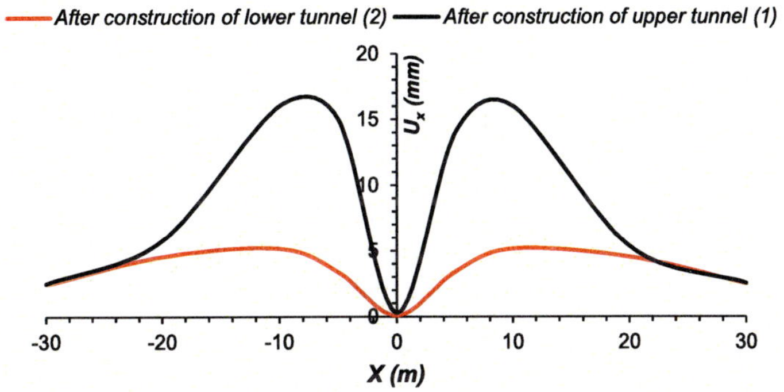

Fig. 12 Horizontal displacement at ground surface after the construction of both tunnels (C_L = 0.5 D)

of 5.21 mm. From Fig. 12, it is also clear that maximum horizontal displacement increases to more than 16 mm after the construction of the upper tunnel 1.

The effect of increasing the pillar width on horizontal displacement at ground surface has been shown in Fig. 13. It can be noticed that horizontal displacement at ground surface reduces with increasing pillar width between the tunnels due to reduction in stress concentration around the tunnels. Maximum values of horizontal displacement are, respectively, of the order of 16, 15, 13 and 12 mm corresponding to pillar widths of 0.5 D, 1.0 D, 2.0 D and 5.0 D.

The profile of vertical displacement, which occurs at ground surface after the construction of tunnels, is shown in Fig. 14 for a pillar width of 0.5 D. It has been found that maximum vertical displacement after the construction of lower tunnel 2 was maximum above the tunnel with a value of 14 mm. This maximum vertical displacement was found to increase up to 39 mm after the construction of the upper tunnel 1.

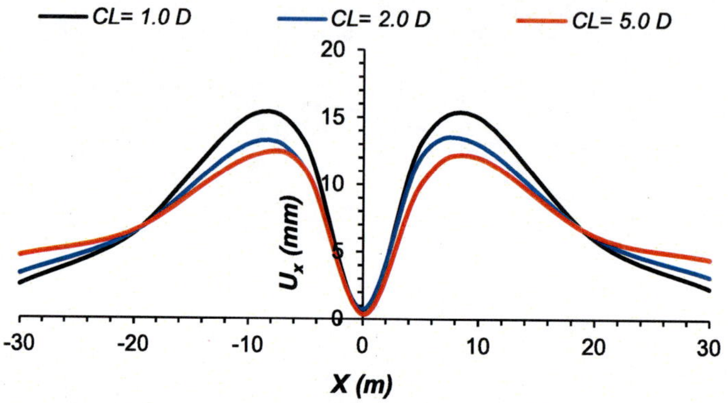

Fig. 13 Horizontal displacement at ground surface for different pillar widths after construction of both tunnels

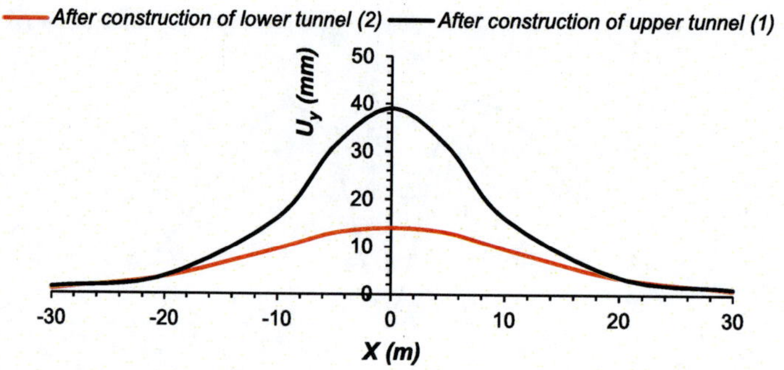

Fig. 14 Vertical displacement profile at ground surface after the construction of both tunnels ($C_L = 0.5\ D$)

Maximum vertical displacement at ground surface was found to be almost similar for all values of pillar width except for pillar width of 0.5 D as is obvious in Fig. 15.

Then, percentage changes in forces in RC liners of lower tunnel 2, due to construction of upper tunnel 1, are presented in Fig. 16. Negative sign along y-axis of Fig. 16 suggests that forces in RC liners of lower tunnel actually reduce significantly after the construction of the upper tunnel.

It can be seen that axial force, shear force and bending moment in lower tunnel 2 reduce significantly after the construction of upper tunnel 1, by about 19.94, 46.87 and 53.12%, respectively, for a pillar width of 0.5 times the diameter of tunnel. This percentage reduction further reduces significantly with increasing pillar width from 0.5 times to 3 times the diameter of tunnel, and thereafter becomes almost constant at a pillar width of about 5.0 times of diameter of tunnel.

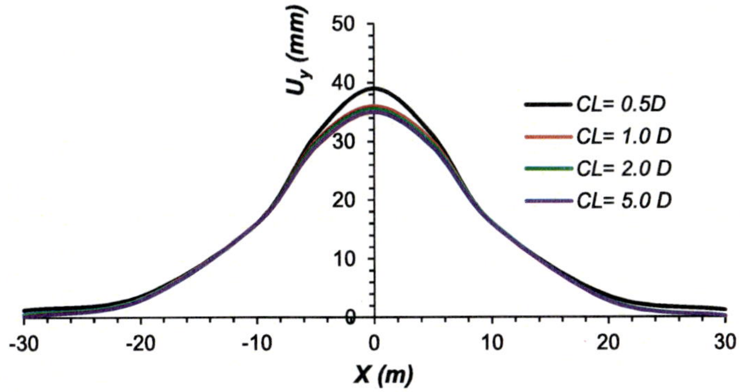

Fig. 15 Vertical displacement profile at ground surface after construction of both tunnels

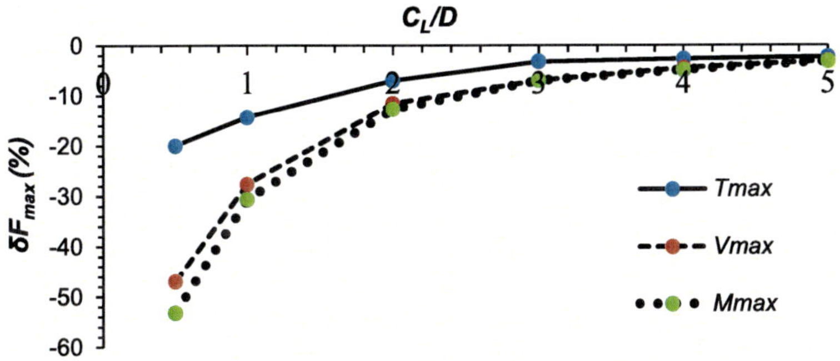

Fig. 16 Percentage change in forces of liners of lower tunnel 2 after construction of upper tunnel 1

3.2 *Seismic Analysis*

Seismic soil–structure interaction analysis of vertically aligned twin tunnels was also carried out. Figures 17 and 18 show the vertical stress concentration after the Chamoli earthquake for lower tunnel 2 and upper tunnel 1, respectively.

From Figs. 17 and 18, it can be noticed that vertical stress at crown and invert of both the tunnels increase slightly after the earthquake for a pillar width of 0.5 D, and these stresses reduce with increasing pillar width between the tunnels. But the vertical stress at springing points of both the tunnels was found to be almost similar after the earthquake for different values of the pillar width.

Then, percentage increment in liner forces for both the tunnels during the Chamoli earthquake was estimated and these values are presented in Figs. 19 and 20, respectively, for the two tunnels. It can be seen that the forces in the RC liners of both the tunnels increase significantly during the earthquake for pillar width of 0.5 D and these forces then reduce with increasing pillar width between the two tunnels.

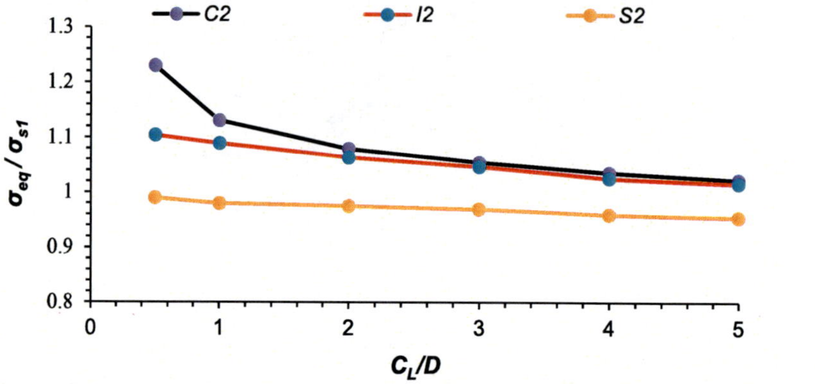

Fig. 17 Change in vertical stress at different points of lower tunnel (2) due to Chamoli earthquake

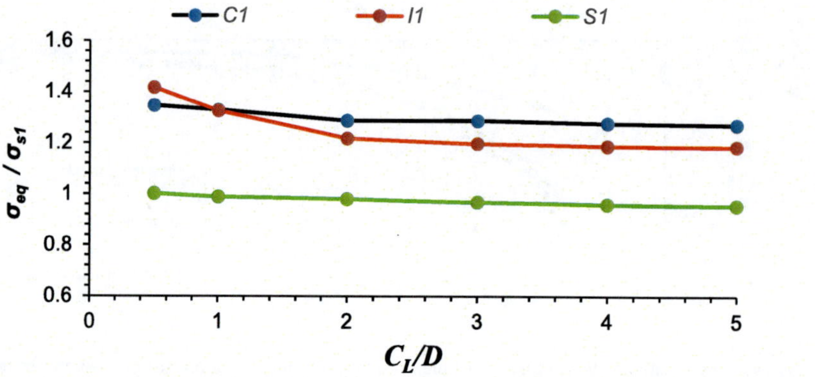

Fig. 18 Increment in vertical stress at different points of upper tunnel (1) due to Chamoli earthquake

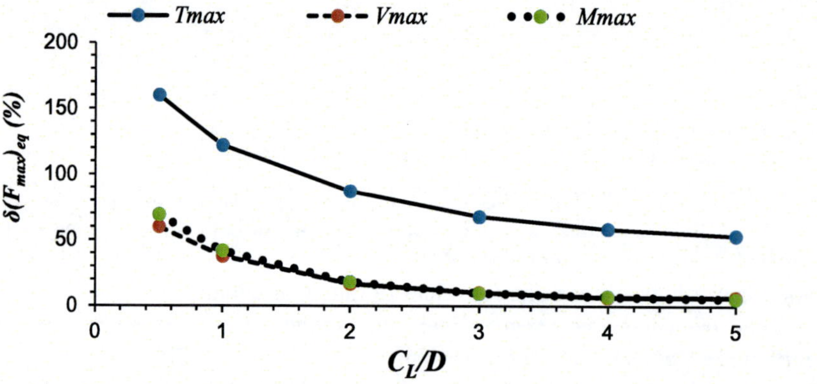

Fig. 19 Percentage increment in liner forces of upper tunnel during Chamoli earthquake

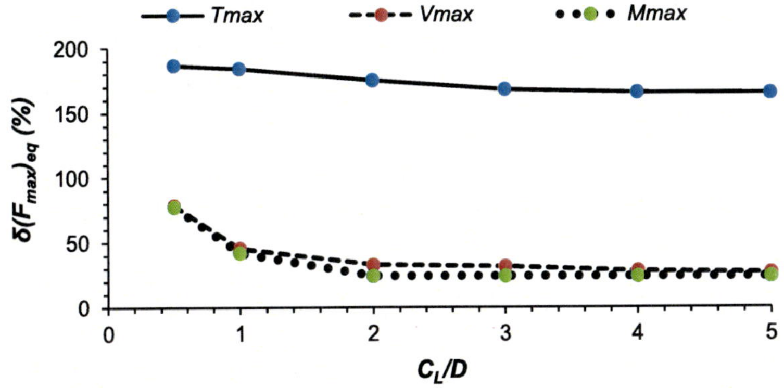

Fig. 20 Percentage increment in liner forces of lower tunnel during earthquake

4 Concluding Remarks

I. Static analysis of horizontally aligned twin tunnels:

   (a) After the construction of tunnel 2, vertical stress (σ_{s2}) at both critical points of tunnel 1, namely crown and springing point, increases by about 1.7 times that of σ_{s1} for a pillar width of 0.5 D. It decreases when pillar width reduces from 0.5 D to 3.0 D, and thereafter it becomes constant at a pillar width of five times the diameter.

   (b) Axial force, shear force and bending moment in RC liners increase significantly after the construction of tunnel 2, for a pillar width of 0.5 D. Liner forces in tunnel 1 reduce after the construction of tunnel 2 and with the increasing pillar width between the twin tunnels.

II. Seismic analysis of horizontally aligned twin tunnels:

   (a) Vertical stresses at critical points of both the tunnels increase during the earthquake, and with increasing pillar width, the earthquake causes an increase in vertical stress in both the tunnels.

   (b) Forces in RC liners of both the tunnels increase significantly, and these values increase with increasing pillar width between the twin tunnels.

III. Static analysis of vertically aligned twin tunnels: Vertical stress around the lower tunnel and liner forces in lower tunnel reduce significantly after the construction of upper tunnel and with increasing pillar width between the tunnels.

IV. Seismic analysis of vertically aligned twin tunnels: Vertical stresses at critical points and liner forces in both the tunnels increase during the earthquake for a pillar width of 0.5 D. However, this increment in vertical stress and liner forces shows a decreasing trend with increasing pillar width.

References

Afifipour M, Sharifzadeh M, Shahriar K, Jamshidi H (2011) Interaction of twin tunnels and shallow foundation at Zand underpass, Shiraz metro Iran. Tunn Undergr Space Technol 26(2):356–363

Azadi M, Kalhor M (2014) Study of the effect of seismic behavior of twin tunnels position on each other. Int J Civil Environ Struct Constr Architect Eng 8(6):625–627

Chehade FH, Shahrour I (2008) Numerical analysis of the interaction between twin-tunnels: influence of the relative position and construction procedure. Tunn Undergr Space Technol 23(2):210–214

Chen SL, Lee SC, Gui MW (2009) Effects of rock pillar width on the excavation behavior of parallel Tunnels. Tunn Undergr Space Technol 24(2):148–154

Chen SL, Li GW, Gui MW (2009) Effects of overburden, rock strength and pillar width on the safety of a three-parallel-hole tunnel. J Zhejiang Univ Sci 10(11):1581–1588

Chu BL, Hsu SC, Chang YL, Lin YS (2007) Mechanical behavior of a twin-tunnel in multi-layered Formations. Tunn Undergr Space Technol 22(3):351–362

Comodromos EM, Papadopoulou MC, Konstantinidis GK (2014) Numerical assessment of subsidence and adjacent building movements induced by TBM and EPB tunnelling. J Geotech Geoenviron Eng 140(11):04014061–1–12

Hu ZF, Yue ZQ, Tham LG (2003) Design and construction of a deep excavation in soft soils adjacent to Shanghai metro tunnels. Can Geotech J 40(5):933–948

Hussein MFM, Shaban M, Talebinejad A (2012) Relationship between twin tunnels distance and surface subsidence in soft ground of Tabriz Metro–Iran. In: Coal operators conference, The University of Wollongong, Australia, Feb 16–17

Kawata T, Ohtsuka M (1993) Observational construction of large scaled twin road tunnels with minimum interval. In: Reith JL (ed), Infrastructures Souterraines de Transports. Balkema, Rotterdam

Kumari SDA, Vipin KS, Sitharam TG (2012) Seismic response of twin tunnels in weathered rocks. Geo-Congress, Oakland, California, March 25–29, pp 3268–3274

Kumari SDA, Vipin KS, Sitharam TG (2013) Twin tunnels under seismic conditions. In: Indorock, Fourth indian rock conference, Solan, Himachal Pradesh, India, pp 550–558

Kumari SDA, Vipin KS, Sitharam TG (2014) Effect of shape of twin tunnels during seismic Loading. J Rock Mech Tunn Technol 20(1):49–59

Li XG, Yuan DJ (2012) Response of a double-decked metro tunnel to shield driving of twin closely under-crossing tunnels. J Tunn Underground Space Technol 28:18–30

Lysmer J, Kuhlemeyer RL (1969) Finite dynamic model for infinite media. J Eng Mech Div 95:859–878

Motaal MAA, Nahhas FMEI, Khiry AT (2013) Mutual seismic interaction between tunnels and the surrounding granular soil. HBRC J 10(3):265–278

Perri G (1995) Analysis of the effects of the new twin-tunnels excavation very close to a big diameter tunnel of Caracas Subway. In: Salam A (ed) Tunnelling and ground conditions. Balkema, Rotterdam, pp 523–530

Rahim HH, Abd Enieb M, Khalil AA, Ahmed ASH (2015) Seismic analysis of urban tunnel systems for the greater Cairo metro line No. 4. Electron J Geotech Eng 209(10):4207–4222

Sahoo JP, Kumar J (2013) Stability of long unsupported twin circular tunnels in soils. Tunn Undergr Space Technol 38:326–335

Saitoh A, Gomi K, Shiraishi T (1994) Influence forecast and field measurement of a tunnel excavation rossing right above existing tunnels. In: Salam A (ed) Tunnelling and Ground Conditions. Balkema, Rotterdam, pp 83–90

Shaalan OA, Salem TN, Eman A, Shamy EI, Mansour RM (2014) Dynamic analysis of two adjacent tunnels. Int J Eng Innov Technol (IJEIT) 4(4):145–152

Shahrour I, Mroueh H (1997) Three-dimensional nonlinear analysis of a closely twin tunnels. In: Sixth international symposium on numerical models in geomechanics (NUMOG VI), Montreal, Quebec, Canada, July 2–4, 2, pp 481–487

Shirinabadi R, Moosavi E (2016) Twin tunnel behavior under static and dynamic loads of Shiraz metro. Iran J Mining Sci 52(3):461–472
Soliman E, Duddeck H, Ahrens H (1993) Two and three-dimensional analysis of closely spaced double-tube tunnels. Tunn Undergr Space Technol 8(1):13–18
Yadav HR (2005) Geotechnical Evaluation of Delhi metro tunnels. Ph.D. thesis, Department of Civil Engineering, IIT Delhi, India
Yamaguchi I, Yamazaki I, Kiritani K (1998) Study of ground-tunnel interactions of four shield tunnels driven in close proximity, in relation to design and constructions of parallel shield tunnels. Tunn Undergr Space Technol 13(3):289–304

Soil-Structure Interaction Analysis of a Raft Foundation Supporting RC Chimney

Shilpa Dixit, Srinivasa Phanikanth Vedula, and Srinivas Kakaraparthi

Abstract Foundations supporting superstructure need to be designed for the reactions due to wind/seismic loads in addition to dead load, imposed load of the superstructures. In case of tall structures like chimneys, the base moments will be high and hence need to be checked for both maximum and minimum bearing pressures and ensure that the safe bearing pressure is not exceeded and also the possible uplift if any is limited to the guidelines suggested by the codes. In the present study, a foundation raft supporting 60 m high RCC ventilation stack is analyzed for various load combinations. The effects of soil-structure interaction have been considered in the detailed finite element analysis carried out. Soil-structure interaction (SSI) effects in both static and dynamic case with and without embedment effects are considered. In the foundation design, uncertainties in SSI are also accounted. The results obtained for various aspects mentioned above have been discussed.

Keywords Soil-structure interaction · Ventilation stack · Uncertainty analysis · Raft foundation

1 Introduction

Foundations supported on tall structures transfer large bending moments to the foundation raft and hence need to be checked for the maximum bearing pressure and ensure that the same do not exceed the allowable bearing pressure. Also, such structures need to be checked against possible uplift of the raft due to very high bending moments transferred from superstructure. In the present study, behavior of a foundation raft supporting a 60 m high ventilation stack is studied in detail. Soil-structure interaction (SSI) effects are considered both under static as well as dynamic conditions. The SSI effects for static loading are considered using Bowles (1996). Other formulations such as Vesic (1961) are also widely used by designers. Wolf (1985)

S. Dixit · S. P. Vedula (✉) · S. Kakaraparthi
Bhabha Atomic Research Center, Mumbai, Maharashtra, India
e-mail: vphanikanth@gmail.com

M. Latha Gali and P. Raghuveer Rao (eds.), *Geohazards*, Lecture Notes in Civil Engineering 86, https://doi.org/10.1007/978-981-15-6233-4_18

describes in detail about the two basic methods involved in the consideration of soil-structure interaction (SSI) problems. These are the direct method and substructure method. In the direct method, the entire structure foundation soil system is modeled and analyzed in a single step. The method however requires the consideration of nonlinear soil properties, and in case of dynamic problems, it requires consideration of modulus degradation and damping curves. Seed and Idriss (1970) have proposed set of curves for cohesionless soils, for estimation of modulus reduction and damping ratios. For clayey soils, modulus reduction and damping curves are proposed by Vucetic and Dobry (1991). Another approach is substructure approach which requires replacing the soil by equivalent springs. For dynamic SSI analysis, impedance function approach suggested by ASCE 4-98 is used in the present study. The spring constant is (surface) considered in all three translational modes, i.e., one in vertical and two in horizontal direction. Beside the consideration of surface springs, embedment effects are also considered in the analysis, and the effects of embedment are discussed.

A detailed seismic SSI analysis of the ventilation stack was carried out by Jaya et al. (2009), and the authors concluded that the response of the stack for hard rock ($Vs_{30} > 1500$ m/se and rock ($760 < Vs_{30} <= 1500$ m/s) conditions is similar to the fixed base condition, and hence, fixed base analysis is adequate for the stack-like structures on similar sites. The authors further concluded that the effect of SSI appears to be beneficial for the stack structure founded on soft rock (average shear wave velocity lower than 750 m/s). Hence, in the present study, fixed base condition is assumed for superstructure analysis of ventilation stack.

The basic input parameters considered are presented in Table 1. The foundation raft proposed in the present study is solid circular in geometry and has a diameter of 17.5 m with raft thickness of 1.8 m. The concrete grade is M30, and hence, Young's modulus is considered as 2.73e07 kN/m^2. The allowable bearing pressure based

Table 1 Basic input parameters

Parameter	Input value
Radius of foundation raft	8.75 m
Foundation thickness	1.80 m
Axial load transferred to foundations raft from superstructure	1.1193 e04 kN
Bending moment transferred at the foundation raft from superstructure due to wind loads	1.01881e05 kNm
Maximum moment transferred at the foundation raft due to seismic loads	8.7907e04 kNm
Grade of concrete	M30
Young's modulus of concrete	2.73 e07 kN/m^2
Allowable bearing pressure	400 kN/m^2
Static subgrade modulus	40,000 kN/m^3
Shear wave velocity(V_s30)	1060 m/s
Poisson's ratio	0.30

on geotechnical investigations conducted for the proposed structure is obtained as 400 kN/m^2. The subgrade modulus based on Bowles (1996) thus is evaluated as 4.0e04 kN/m^3. The axial load considered at the foundation base is 1.1193e04 kN which is obtained from a separate superstructure analysis of the RC chimney, the details of which are not in the scope of the work.

2 Transfer of Loads

The chimney is analyzed for wind loads and also for seismic loads. For wind loads, both along wind loads and across wind loads are considered in the analysis as per IS 4998:2015. FE analysis of the chimney is performed considering fixed base assumption. For seismic loads, site-specific response spectrum is considered for evaluating the earthquake forces. The obtained reactions of the superstructure analysis of the chimney are transferred to the foundation raft at the center of the raft, and these are converted to equivalent vertical loads (Fig. 1) for FE analysis along the periphery of the stack shell, i.e., nodes along the periphery of stack shell radius at $r = 3.0$ m as described below.

The vertical load is distributed equally along the peripheral nodes of the shell location of the chimney, i.e., at radius $r = 3.0$ m for the axial load reaction of the superstructure. The moment obtained (in a given direction say X or Y) is assumed to vary based on cosine distribution as shown in Eq. 1, [Manohar (1985)] and converted to equivalent vertical loads (forming a couple) along the periphery of stack shell location, i.e., at $r = 3.0$ m.

For the obtained superstructure reactions in terms of external load P, and bending moment Mx and My, the distribution of equivalent vertical load (Pi, see Fig. 2) at any node i is obtained as per Eq. 2. The obtained loads at each node are given in Table 2 for both wind loads as well as seismic loads.

Let $n =$ no. of nodes, $P =$ external load, $M_x = M$ along x, and $M_y = M$ along y, and also let R_x and R_y as the reaction forces for M_x and M_y, respectively, and vary as $R\cos\theta$.

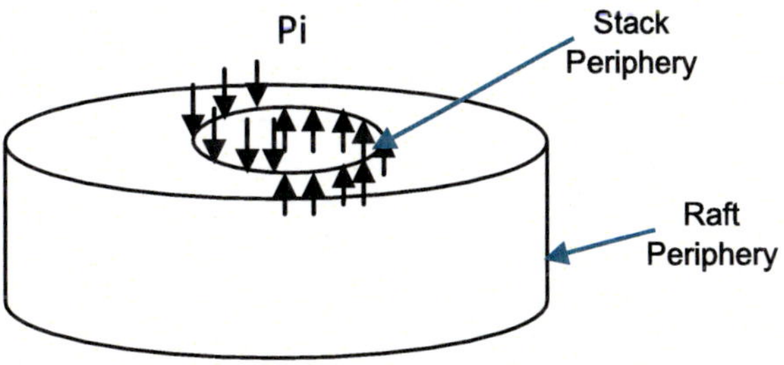

Fig. 1 Moments at center converted to equivalent vertical load applied on stack peripheral nodes based on cosine distribution of bending moment [Manohar (1985)]

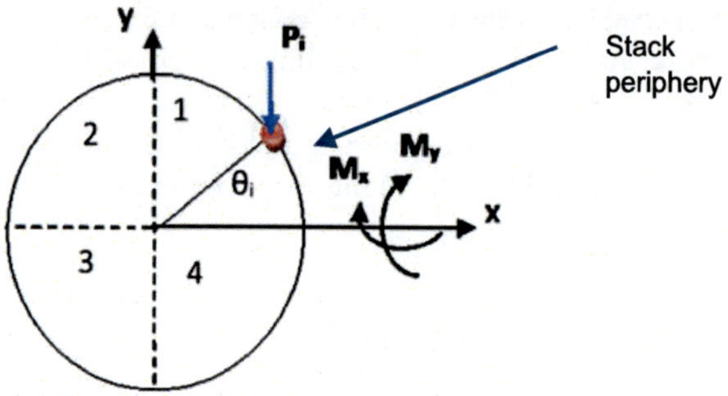

Fig. 2 Application of vertical loads on stack periphery (at $r = 3.0$ m)

Table 2 Equivalent vertical load at $r = 3.0$ m

Angle (degrees)	Wind load (WL), kN	Seismic load (EL), kN
0	3296.45	2908.29
15	3200.02	2825.09
30	2917.30	2581.14
45	2467.55	2193.08
60	1881.43	1687.35
75	1198.88	1098.42
90	466.42	466.42
105	−266.05	−165.59
120	−948.60	−754.52
135	−1534.72	−1260.25
150	−1984.46	−1648.31
165	−2267.18	−1892.25
180	−2363.61	−1975.46
195	−2267.18	−1892.25
210	−1984.46	−1648.31
225	−1534.72	−1260.25
240	−948.60	−754.52
255	−266.05	−165.59
270	466.42	466.42
285	1198.88	1098.42
300	1881.43	1687.35
315	2467.55	2193.08
330	2917.30	2581.14
345	3200.02	2825.09

$$M_x = 4R_x r \sum_{\theta_i=0^\circ}^{90^\circ} \cos^2\theta_i \quad (1)$$

Similarly, M_y is related to R_y as above and by replacing $\cos\theta$ with $\sin\theta$. R_x and R_y are obtained from above expression (Eq. 1), and P_i is obtained at various angles on stack periphery using Eq. 2 as given below.

$$P_i = \frac{P}{n} \pm R_x \cos\theta_i \pm R_y \sin\theta_i \quad (2)$$

3 Finite Element Modeling of the Foundation Raft

The raft is modeled using three-noded shell elements at the center of the raft up to a radial distance of $r = 0.5$ m, and beyond $r = 0.5$ m radius, four-noded shell elements are considered. There are total 24 segments along any circumferential section of the raft, and radially, the element spacing is considered as 0.5 m except at stack shell periphery (from inner radius = 2.7 m to outer radius = 3.3 m, i.e., for the thickness of 600 mm of stack shell at bottom). Additional grids are provided at these sections to accommodate stack shell location where loading from superstructure will be imposed on the raft nodes. Also, the last circumferential section along radius is 0.25 m thick. Thus, the total no. of shell elements of the FE model are: 480, and the no. of the nodes are 481. Equivalent vertical loads at the location of stack peripheral nodes are applied as per Table 2. Commercial finite element software *SAFE* is used for the present study. The diameter of foundation raft is considered as 17.5 m which is obtained from the consideration of keeping the bearing pressures within the allowable limits. The FE model considered is shown in Fig. 3. Soil is replaced by equivalent springs both for static and dynamic loading. For static analysis, the modulus of subgrade is considered as proposed by Bowles (1996), and for dynamic analysis, impedance approach as given in *ASCE 4-98* is adopted. The translational springs considered for the present analysis is given in Table 3. These are evaluated based on shear wave velocity (V_{s30}) of 1060 m/s and Poisson's ratio of 0.30, respectively. Thus, the total no. of spring

Table 3 Spring stiffness based on ASCE 4-98

S. No.	Motion	Impedance function	Spring stiffness, kN/mm
1	Horizontal	$k_x = 32(1-\mu)GR/(7-8\mu)$	99,786.38
2	Vertical	$k_z = 4GR/(1-\mu)$	122,556.7

elements in vertical as well as two horizontal directions is 481 in each direction. The obtained spring stiffness of the raft is distributed to each node in proportion to the contributory area over the total area and is as shown in Eq. (3).

$$k_i = k_{\text{raft}} * A_i / A_{\text{raft}} \quad (3)$$

where k_{raft} is the spring stiffness of the foundation, A_{raft} is the area of the raft, and parameters k_i and A_i are the stiffness of the spring at node i and influence area of the node under consideration, respectively, for any given direction, i.e., x, y, z.

Load application: In the present study, the stack shell thickness at the base of the raft is 600 mm, and hence, the foundation raft actually experiences pressure load than concentrated loads, as the given concentrated load along the periphery of the raft tends to distribute as pressure. Hence, an attempt is made to observe the response of the foundation raft by applying concentrated load as well as pressure load, and the obtained results are discussed in detail in the following sections. The pressure load application at stack shell peripheral element locations is shown in Figs. 4 and 5.

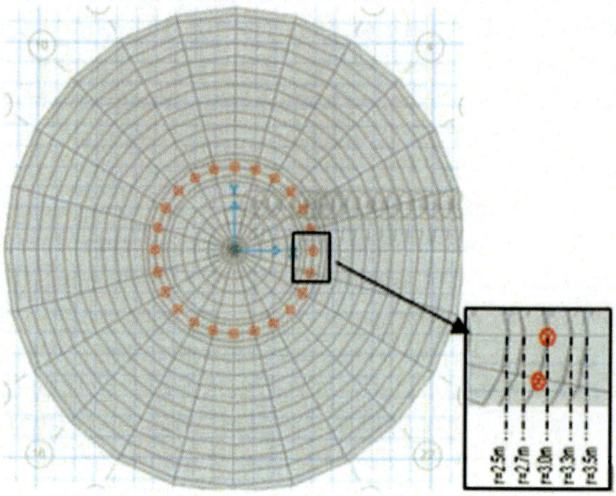

Fig. 3 FE model considered for the present study no. of shell elements: 480, nodes: 481, translational springs in x-, y-, and z-direction are 481 each

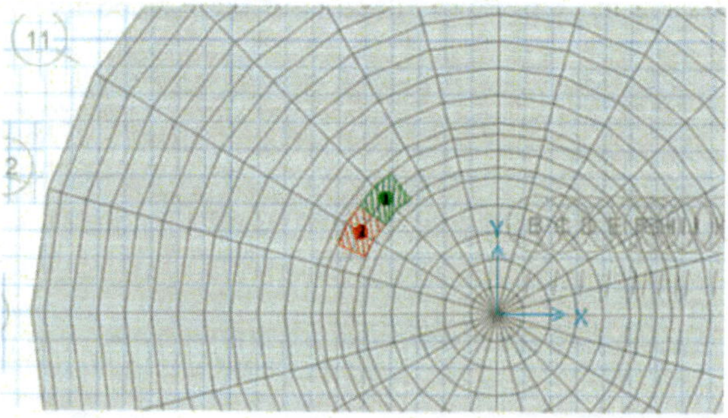

Fig. 4 Distribution of uniform pressure around given concentrated loads

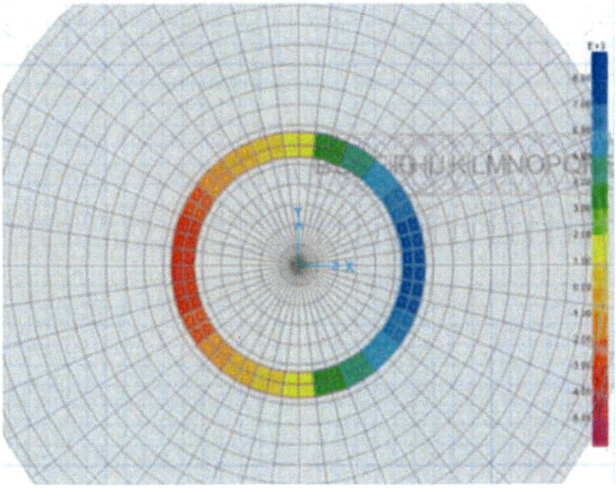

Fig. 5 uniform pressure loads at the location of stack shell periphery ($r = 3$ m)

4 Validation of FE Analysis Results

The detailed finite element analysis is carried out both for static case considering the subgrade modulus evaluated based on Bowles (1996) for wind loads. For seismic loads, spring stiffness in all translational directions is evaluated using *ASCE 4-98*, as presented in Table 3.

The obtained FE analysis results in terms of radial bending moments, tangential moments, and bearing pressures are also verified with the analytical solutions available in the literature. Expressions for obtaining the radial and tangential bending moments for axial load and external moment at the foundation raft location for solid circular raft are obtained from Manohar (1985).

The method described in Manohar (1985) consists of evaluating bending moments in the raft for (a) axial loading due to chimney dead weight, etc., (b) external moment at the foundation raft due to superstructure moments of the chimney (viz wind load/earthquake load), and (c) bending moment due to backfill load if any. The total bending moment will be obtained by principle of superposition to obtain the total response. For the present study, backfill effects are ignored.

5 Results and Discussion

The radial and tangential bending moments obtained from FE analysis results are compared with the solutions available in the literature as discussed above for the input parameters presented in Table 1 considering wind load case, and the obtained results based on both these methods are presented in Table 4. The results are also plotted in Fig. 6. for radial and tangential bending moments vs. radius of the raft. From these results, it can be seen that excellent matching is obtained from the present FE analysis when compared with the analytical formulations available. Also, it is

Table 4 Comparison of bending moments obtained from finite element analysis vs. analytical results (wind loads)

r (m)	$f = r/a$	Analytical method [Manohar (1985)] (kNm/m)		Finite element method (kNm/m)	
		M_r	M_t	M_r	M_t
0	0.00	729.3	729.3	253.5	750.7
0.5	0.06	1286.9	987.3	1173.0	844.7
1	0.11	1850.8	1250.3	1720.6	1174.0
1.5	0.17	2422.9	1518.4	2319.2	1457.1
2	0.23	3004.9	1791.5	2930.5	1744.5
2.5	0.29	3598.6	2069.6	3560.1	2043.2
2.7	0.31	3839.8	2182.2	3758.2	2150.2
3	0.34	4205.8	2321.9	4168.6	2341.0
3.3	0.38	3420.2	2265.6	3267.1	2237.4
3.5	0.40	2991.2	2201.1	2846.7	2152.4
4	0.46	2147.6	1997.1	1918.6	1905.6
4.5	0.51	1528.0	1775.1	1305.9	1660.2
5	0.57	1058.2	1560.4	857.1	1432.6
5.5	0.63	697.6	1363.1	523.7	1229.7
6	0.69	422.5	1186.7	277.9	1053.0
6.5	0.74	218.4	1032.4	102.4	902.0
7	0.80	76.1	900.2	−13.5	775.8
7.5	0.86	−9.9	789.7	−77.4	673.1
8	0.91	−43.4	700.5	−94.1	592.5
8.5	0.97	−26.6	632.1	−67.1	532.9
8.75	1.00	0.3	605.6	−21.9	513.1

where M_r and M_t are radial and tangential bending moments in the raft, respectively, and a = radius of raft

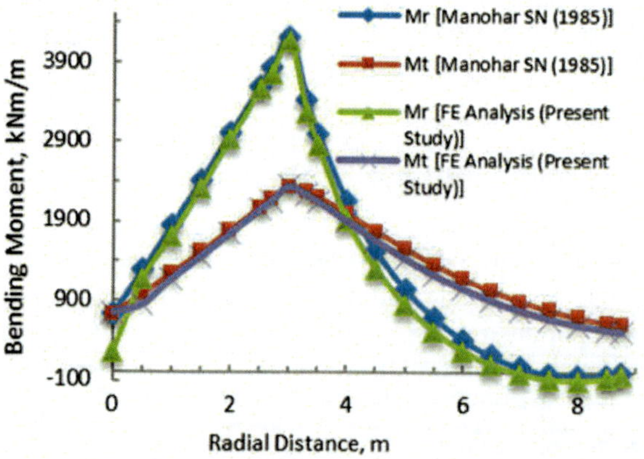

Fig. 6 Comparison of bending moments in the raft for wind loads: FE analysis versus Manohar (1985)

Table 5 Comparison of soil bearing pressures for wind loads: analytical method vs. finite element analysis results

Bearing pressure (kN/m^2)	Analytical method [Manohar (1985)]	Finite element analysis
Maximum	391.77	339.76
Minimum	4.51	5.76

observed that the maximum radial and tangential moments occur at the location of the stack shell, i.e., at $r = 3.0$ m radius.

The bearing pressure obtained using both these methods are also compared and are presented in Table 5 for static case considering wind loads, and again, it can be seen that there is reasonable agreement in results from the FE analysis results when compared with the analytical methods available in the literature.

5.1 *Response Due to Wind Loads and Earthquake Loads*

The maximum bending moments (M_r, M_t) due to wind loads are obtained with subgrade modulus evaluated as per Bowles (1996). The maximum bending moments are also evaluated for earthquake loads with spring stiffness using impedance approach as per ASCE 4-98 and considering embedment effects. The moments are evaluated for concentrated loads (CL) at the stack shell location ($r = -3.0$ m), and all these results are presented in Fig. 7a. It can be seen that the design is governed by the wind load case. The higher bending moment for wind loading case may be due to lesser soil stiffness under static loads when compared to earthquake loads where higher spring stiffness resulted in lesser bending moments. Figure 7b shows the minimum radial and tangential bending moments for wind loads and earthquake loads at $r = -3.0$ m location for both concentrated loads and pressure loads. Figure 7c shows the downward displacement for both static (wind loads) and dynamic loading (earthquake loads) considering embedment effects, and it is observed that the displacements are higher for wind loads as the spring stiffness considered in this case is as per Bowles (1996) and is lower compared to dynamic spring stiffness and hence resulted in higher displacements. Also, it is observed that the location of maximum displacement for wind load analysis is at the edge of the foundation, while in case of earthquake loading the maximum displacement is observed to be at the location of stack shell. This may be due to high spring stiffness and higher downward loads acting at the stack shell which has resulted in local deformation.

Also, superstructure reactions from the stack shell for earthquake loads considering impedance approach are imposed by dispersing the concentrated loads due to stack shell reactions as equivalent pressure over the area of the stack shell, i.e., over annular space between $r = 2.7$ m and $r = 3.3$ m. It can be seen that the application of pressure loading gives lower bending moments which will result in economic sections (raft thickness). The maximum bending moment M_r is reduced by 14% for application of pressure load compared to concentrated load, and the results are also

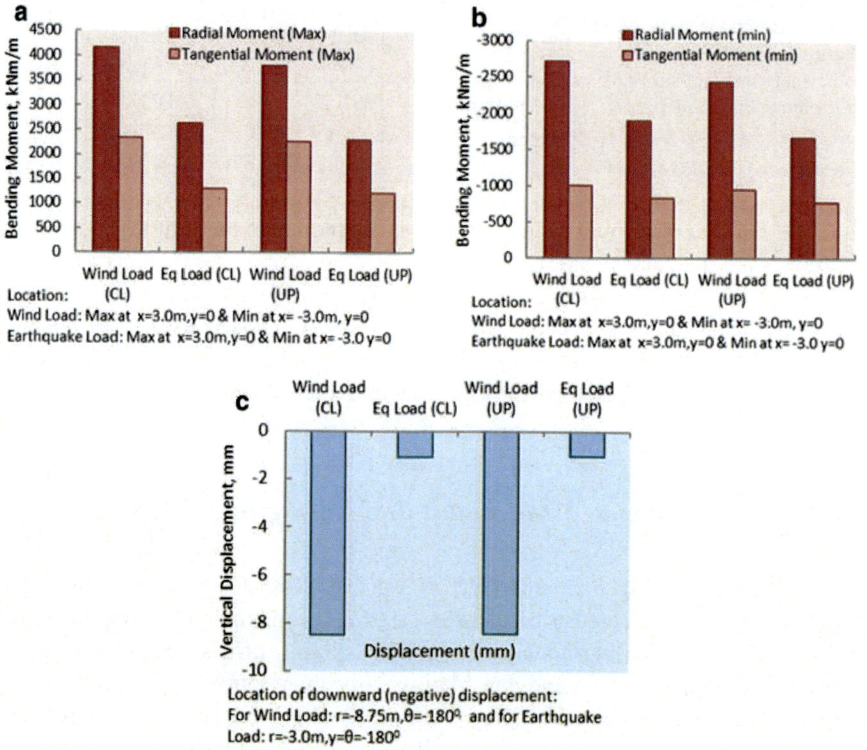

Fig. 7 **a** Maximum radial and tangential bending moments (at the stack shell location, r = 3.0 m) considering wind loads (static) and earthquake loads with embedment. **b** Minimum radial and tangential bending moments (at the stack shell location, $r = -3.0$ m) considering wind loads (static) and earthquake loads with embedment. **c** Negative (downward) displacements (at the stack shell location, $r = -3.0$ m , and $r = 3.0$ m) considering wind loads (static) and earthquake loads with embedment of springs (dynamic case)

shown in Fig. 8a, whereas M_t is reduced by 6% for pressure loading. Similarly, the minimum bending moments are also plotted for both concentrated loads and pressure loading, and the results are shown in Fig. 8b. It is observed that the bending moments M_r are reduced by 13% for pressure loading and tangential moment is reduced by 7% for pressure loading when compared with concentrated loads.

5.2 *Uncertainties in Soil-Structure Interaction*

As can be seen from the impedance approach, the soil stiffness is derived from shear wave velocities which is conducted at few locations of the given site, the solution obtained from these results needs to be looked into with respect to the

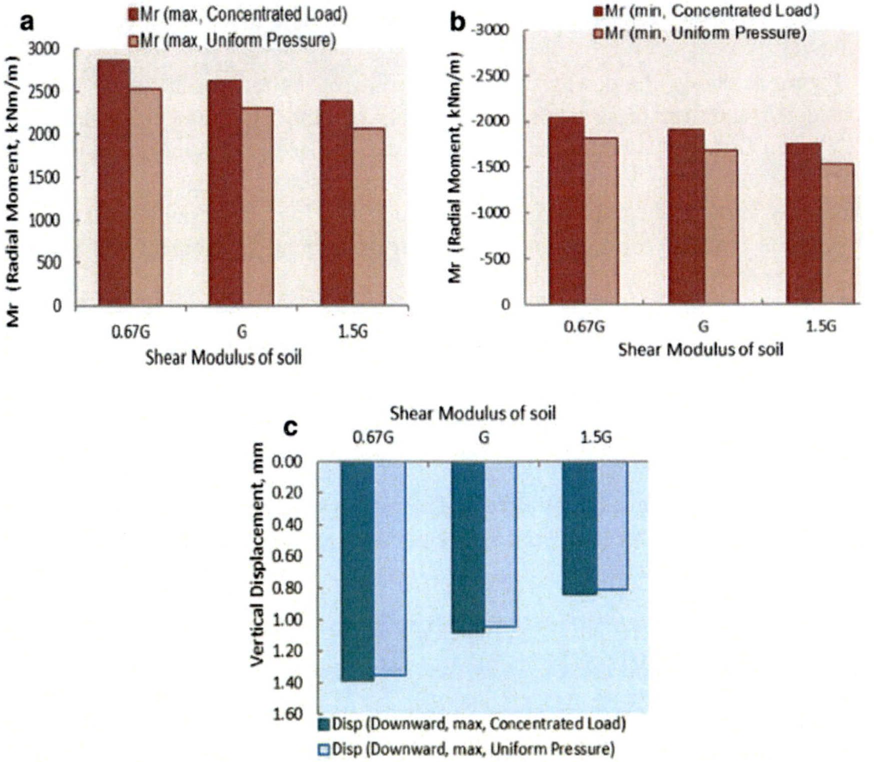

Fig. 8 **a** Maximum radial bending moments in the raft ($r = 3.0$ m). **b** Minimum radial bending moments (Min) in the raft ($r = - -3.0$ m). **c** Downward maximum displacement (mm), $r = 3.0$ m. Here, CL: Concentrated load and UP: Uniform pressure loads

possible uncertainties associated more so when the soil layer thickness tend to vary horizontally. In the present study, the uncertainties are considered as suggested in ASCE 4-98 by varying the shear modulus from 0.67G to 1.5G, respectively. The obtained response in terms of bending moment, displacement, bearing pressure, etc., is examined, and the results are discussed here. The maximum bending moments (+ve) have been plotted in Fig. 8a considering shear modulii of 0.67G, 1.0G, and 1.5G (i.e., considering $C_v = 0.5$), respectively, for both concentrated shell loads as well as pressure loads at shell location of the stack. It can be seen that as the soil stiffness is high, the bending moments are reduced, and the reduction in bending moment is 9.17% for point loads and 10.3% for uniform pressure loads, when G is increased to 1.5G. So also, the increase in bending moment is observed as 8.8% for point loads and is 9.9% for uniform pressure loads when G is reduced to 0.67G.

Similarly, the effect of variation of shear modulus on the minimum bending moments is also plotted and is shown in Fig. 8b. In this case, it is observed that the bending moment is reduced by 8.4% for point loads, and 9.4% for uniform pressure loads, when G is increased to 1.5G, whereas the increase in bending moment is

7% for point loads, while it is 7.8% for uniform pressure loads when G is reduced to 0.67G.

Figure 8c shows the downward displacements of the raft for the shear modulii considered, and from these results, it can be seen that the displacement is reduced by 22% when G is increased to 1.5G, and also, the displacement is increased by 29% when G is reduced to 0.67G. Finally, the obtained results considering uncertainties with and without embedment effects are presented in Table 6 for point loads as well as pressure loads in terms of radial and tangential bending moment and vertical displacements.

5.3 Effects of Embedment

The behavior of foundation raft is examined for the surface spring stiffness and also by considering the embedment effects. The results obtained are presented in this section, and the same are discussed. The embedded spring stiffness as suggested by ASCE 4-98 is as given below:

$$k_{ze} = \eta_z k_z \tag{4}$$

where

k_z = vertical spring stiffness = 4GR/(1 − μ),
$\eta_x = \eta_y$ = (1 + 3.36H/R), and η_z = (1 + 0.99H/R)

The results obtained by considering surface springs as well as embedded spring stiffness are presented in Fig. 9a. It is observed that by considering embedment effects, the radial moment is reduced by 3.27% for 0.67G, and the moment is reduced by 3.8% for 1.0G, whereas the reduction is about 4.3% when shear modulus is considered as 1.5G. Also, variation of minimum radial moments (i.e., at $r = -3.0$ m) considering with and without embedment effects was plotted in Fig. 9b. Again, it is observed that by considering embedment effects, the radial moment is reduced by 4.54% for 0.67G, and the moment is reduced by 5.21% for 1.0G, whereas the reduction is about 5.85% when shear modulus is considered as 1.5G.

The above results show the importance of consideration of uncertainties in the soil properties for safety-related structures to ensure the structure performs well during earthquakes. The maximum downward displacements are plotted in Fig. 9c considering the embedment effects and also considering surface spring stiffness. It can be seen that consideration of embedment effects reduced the displacements by 10.6% for 0.67G, whereas the reduction is by about 10.5% for 1.0G, and in case of 1.5G, the reduction is observed to be about 10.5%. In this study, the results are based on ht. of embedment considering 1.8 m thk.

Table 6 Effect of uncertainties in SSI with and without embedment effects

Case		Displacement (CL)		Radial moment (CL)		Tangential moment (CL)		Displacement (UP)		Radial moment (UP		Tangential moment (UP)	
		mm		kNm/m		kNm/m		mm		kNm/m		kNm/m	
		Max	Min	Max	Min	Max	Min	Max	Min	Max	Min	Max	Min
Static	–	−0.45	−7.88	3696.5	−2243	2111	−783	−0.48	−7.83	3359	−2011	2030	−735
Dynamic spring without embedment ($G = 0.67G$)	0.67G – E	0.05	−1.55	2953	−2088	1516	−886	0.04	−1.51	2623	−1860	1440	−836
Dynamic spring without embedment ($G = G$)	G – E	0.16	−1.20	2729	−1970	1359	−855	0.14	−1.17	2402	−1743	1284	−804
Dynamic spring without embedment ($G = 1.5G$)	1.5G – E	0.21	−0.94	2492	−1820	1198	−801	0.19	−0.91	2167	−1596	1126	−750
Dynamic spring with embedment ($G = 0.67G$)	0.67G + E	0.10	−1.39	2857	−2040	1447	−876	0.10	−1.35	2528	−1813	1372	−825

(continued)

Table 6 (continued)

Case		Displacement (CL)		Radial moment (CL)		Tangential moment (CL)		Displacement (UP)		Radial moment (UP		Tangential moment (UP)	
		mm		kNm/m		kNm/m		mm		kNm/m		kNm/m	
		Max	Min	Max	Min	Max	Min	Max	Min	Max	Min	Max	Min
Dynamic spring with embedment ($G = G$)	G + E	0.18	−1.08	2626	−1907	1288	−834	0.17	−1.05	2299	−1682	1215	−783
Dynamic spring with embedment ($G = 1.5G$)	1.5G + E	0.22	−0.84	2385	−1747	1128	−771	0.20	−0.81	2062	−1524	1057	−720

Here, maximum bending moments are obtained at stack shell periphery ($r = 3.0$ m, $r = -3.0$ m) for both wind and seismic loading; For displacements, the location for seismic loads is $r = 3.0$ m, −3.0 m (stack periphery), and for wind loads, the locations is at the edge of the raft, $r = 8.75$ m, −8.75 m. E represents embedment effects, $-E$ stands for raft without embedment, while $+E$ stands for raft with embedment

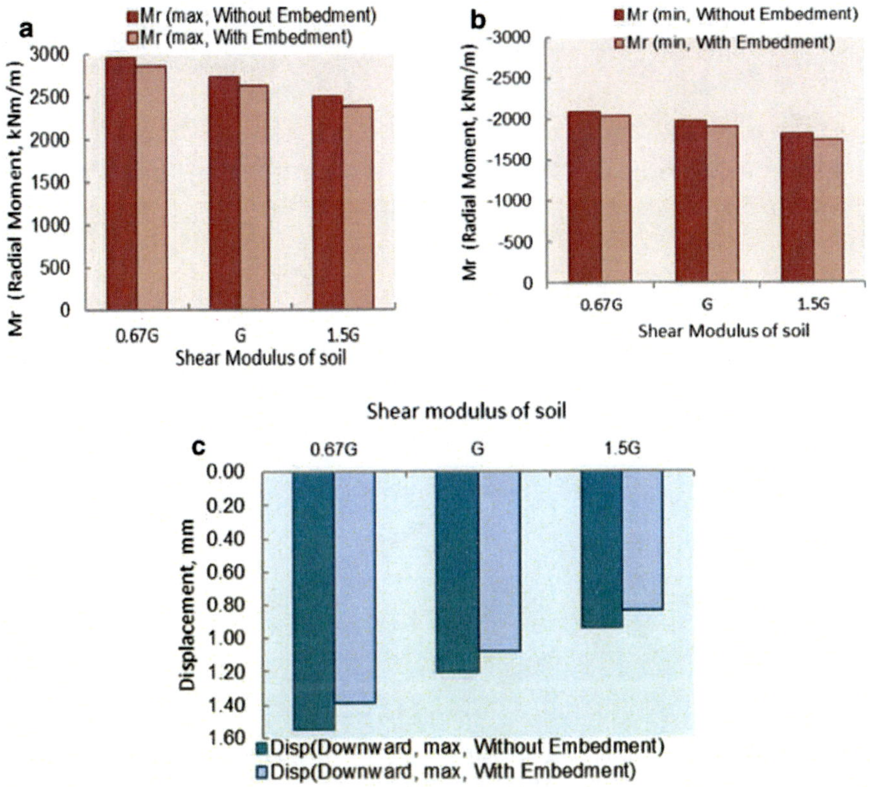

Fig. 9 **a** Maximum radial bending moment in the raft at $r = 3.0$ m with and without embedment effects. **b** Minimum radial bending moment in the raft at -3.0 m) with and without embedment effects. **c** Maximum downward displacement in the raft (at $r = 3.0$ m) with and without embedment effects

6 Summary

Figure 10a shows the variation of maximum (+ve) radial bending moment in the raft by considering superstructure reactions of the stack shell in terms of concentrated loads along the stack periphery and also the superstructure reactions applied in terms of pressure loads over the annular space of the stack shell location (i.e., from $r = 2.7$ m to $r = 3.3$ m) for various shear modulii in an attempt to examine the effects of uncertainties in foundation response. It is observed that the moments are reduced by 11.2% when pressure loads are applied over the foundation as against application of concentrated loads on the stack shell periphery for $G = 0.67$G. It is also observed that the reduction is same for all values of G and hence do not vary with shear modulus. Again, the minimum radial moments (−ve) are plotted in Fig. 10b for both concentrated loads as well as pressure loads (at $r = -3.0$ m) for various shear modulii, and it can be seen that the bending moments are reduced by 10.9% for

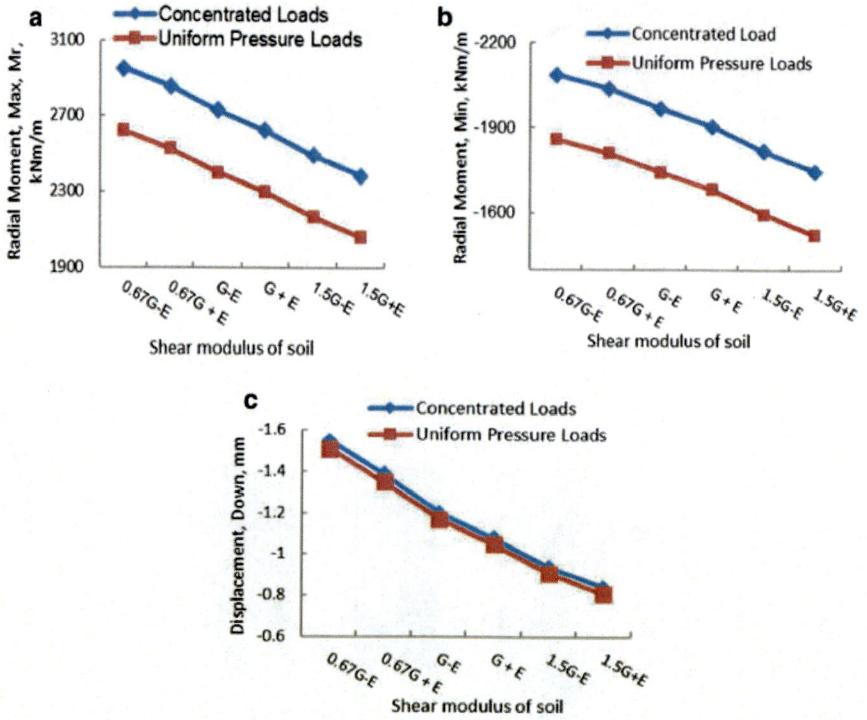

Fig. 10 **a** Variation of maximum radial bending moment in raft ($r = 3.0$ m). **b** Variation of minimum radial bending moment in raft ($r = 3.0$ m). **c** Variation of maximum downward displacements in the raft ($r = 3$ m)

pressure loads when compared to concentrated loads. Figure 10c shows the downward displacements for concentrated loads and pressure loads. It is observed that the applied load in terms of concentrated load versus pressure load has little influence on displacements; however, the response is highly influenced by the shear modulus of the soil and the variation of displacement with shear modulus is found to be linear. Figure 11a shows the displacement contour, for application of concentrated loads on periphery of stack shell with surface springs. It is found that maximum downward displacement is 1.2 mm. When the embedment effects are considered, the maximum downward displacement is found to be 1.08 mm, and the displacement contour with embedment effects for concentrated load application is shown in Fig. 11b. The displacement contour without embedment effects was also observed for pressure loading and is shown in Fig. 11c from which it can be seen that the maximum downward displacement is 1.172 mm. When the embedment effects are considered with the pressure loading application on the stack shell periphery, the maximum downward displacement is observed as 1.047 mm which is shown in Fig. 11d.

The effect of embedment on the radial and tangential bending moments of the foundation raft is also examined in the present study for application of concentrated loads as well as pressure loads applied on the stack periphery. Figure 12a shows the

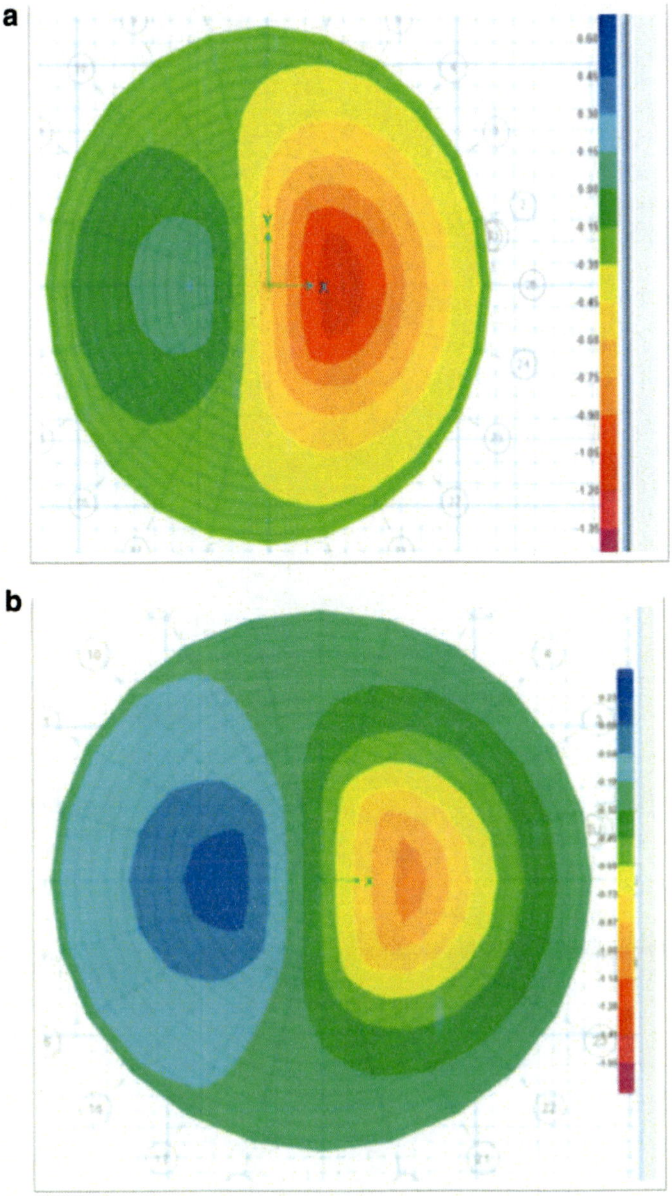

Fig. 11 **a** Displacement contour for concentrated loads without embedment effects, with $G = 1.0G$, $z\text{max} = 0.156$ mm, $z\text{min} = -1.2$ mm. **b** Displacement contour for concentrated loads with embedment effects, $G = 1.0G$; $z\text{max} = 0.1844$ mm, $z\text{min} = -1.08$ mm. **c** Displacement contour for pressure loads without embedment effects $G = 1.0G$, $z\text{max} = 0.1417$ mm, $z\text{min} = -1.172$ mm. **d** Displacement contour for pressure loads with embedment effects, $G = 1.0G$, $z\text{max} = 0.168$ mm, $z\text{min} = -1.047$ mm

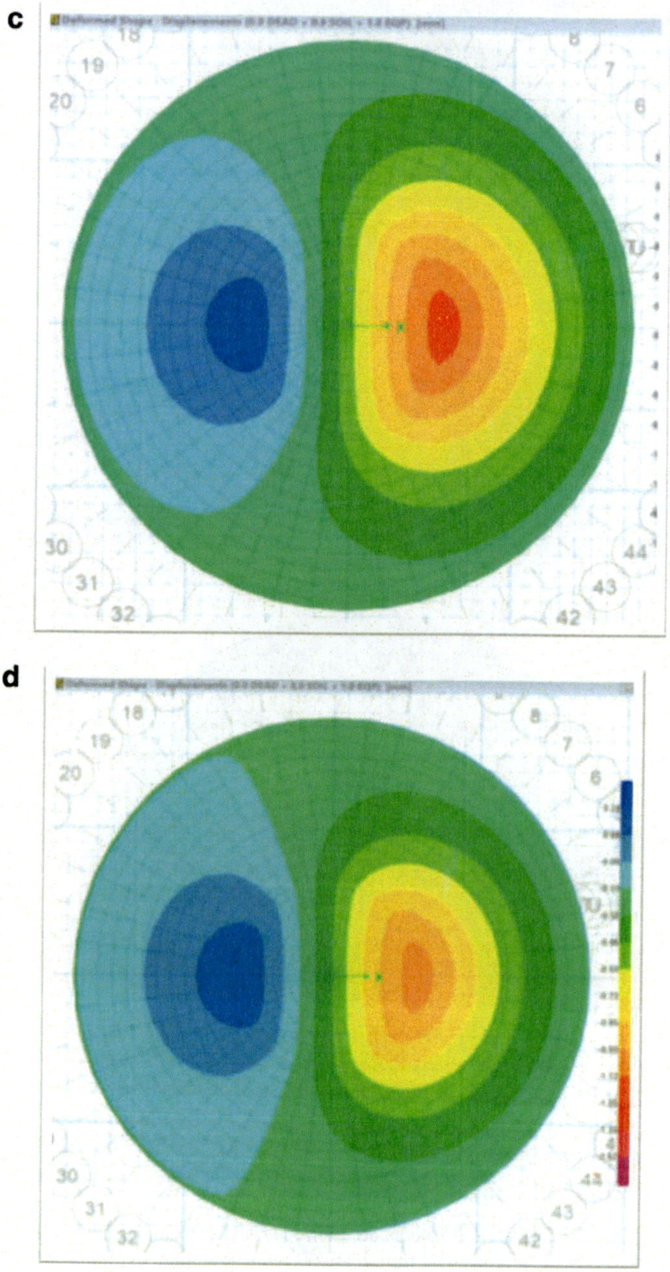

Fig. 11 (continued)

a

b

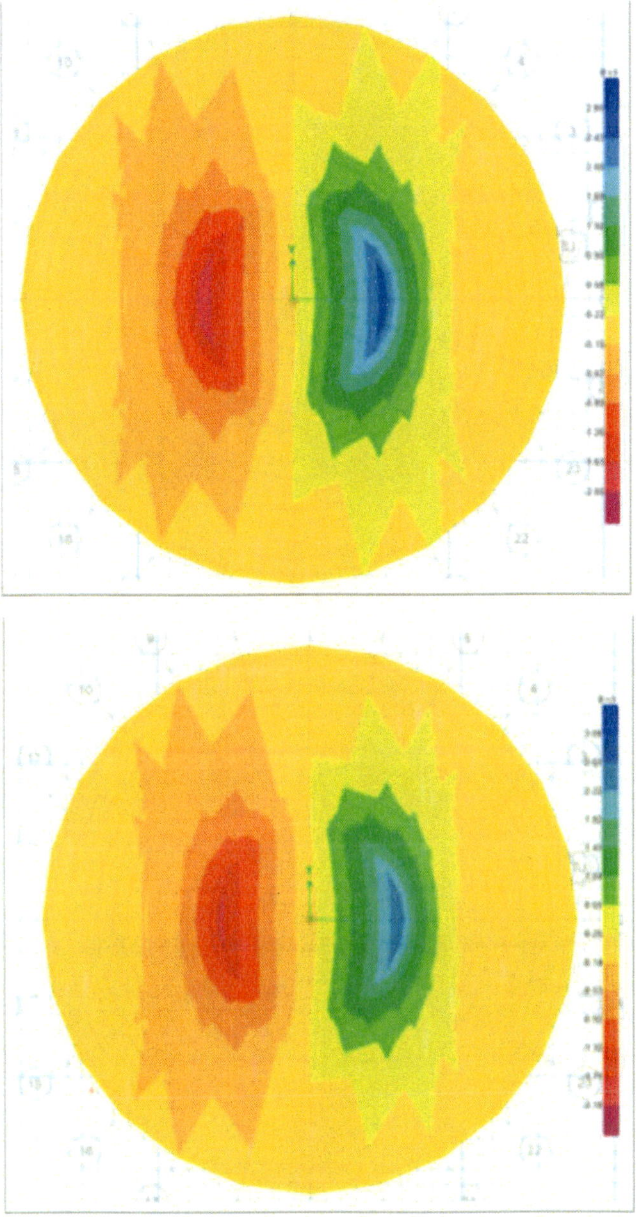

Fig. 12 **a** Radial bending moment in the raft for concentrated loads/without embedment/$G = G$, $M_r$max = 2729.26 kNm/m, $M_r$min = −1969.498 kNm/m . **b** Radial bending moment in the raft for concentrated loads with embedment effects, $G = 1.0$G, $M_r$max = 2625.38 kNm/m, $M_r$min = −1907 kNm/m. **c** Radial bending moment in the raft for Pressure Loads without Embedment effects G=1.0G; $M_r$max=2401.87kNm/m; $M_r$min=−1743.39kNm/m. **d** Radial bending moment in the raft for Pressure Loads/With Embedment G=G, $M_r$max=2299.34kNm/m, $M_r$min=−1681.52 kNm/m

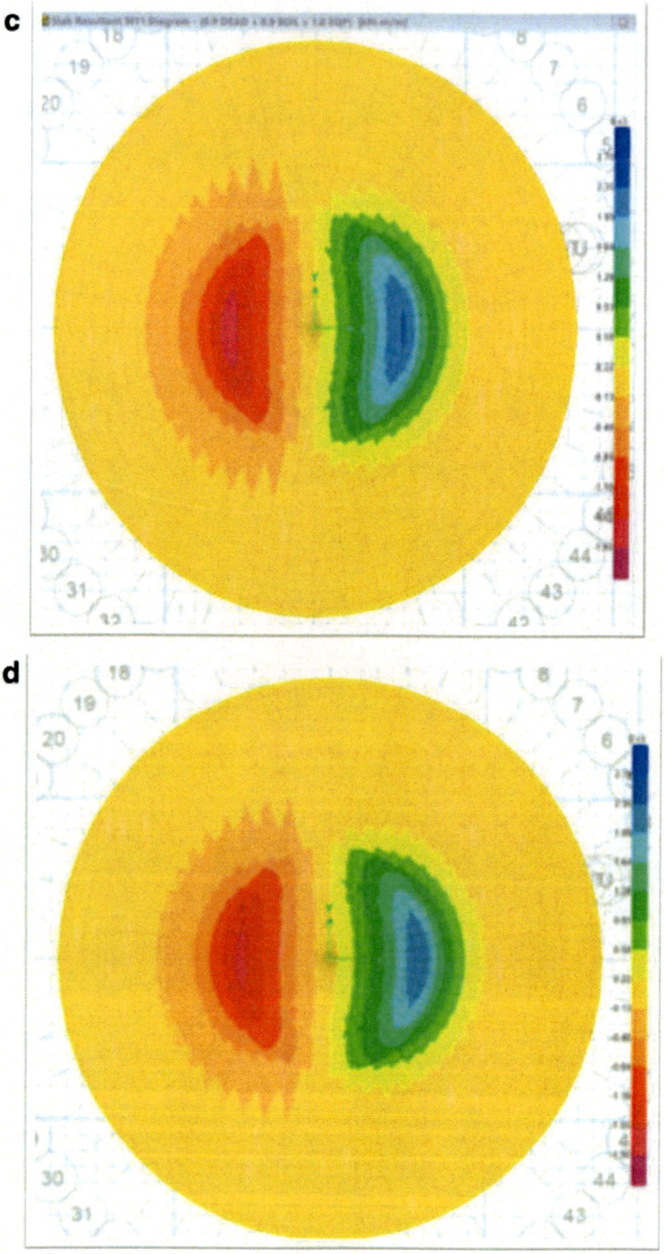

Fig. 12 (continued)

radial moments for application of concentrated loads with surface spring stiffness (without considering embedment effects). The maximum radial moment is found to be 2729.3 kNm/m which is observed at $r = 3.0$ m and $\theta = 00$, whereas the minimum radial moment is found to be −1969.50 kNm/m which is noted at $r = -3.0$ m and $\theta = 1800$. When the embedment effects are considered, the maximum radial moment was reduced to 2625.38 kNm/m, and hence, the moment is reduced by 3.81%.

Similarly, the min. (−ve) radial bending moment was found to be −1907 kNm/m, and hence, the min. radial bending moment is reduced by about 3.18%, and these results are shown in Fig. 12b. When the pressure loading is applied on the stack periphery, the maximum radial moment without embedment effects is found to be 2401.87 kNm/m, and hence, the radial moment is reduced by about 12%, and these results are shown in Fig. 12c. Figure 12d shows the radial moments when pressure loading is applied considering embedment effects, and it is observed that the maximum radial moment is 2299.34 kNm/m which shows that the reduction in bending moment is about 4.27% when embedment effects are considered. Similarly, the effects of embedment for both concentrated loads and pressure loads with and without embedment effects are studied and are presented in Fig. 13a–d) for tangential bending moments. Again, it is seen that the consideration of embedment effects reduces the bending moments. Also, the consideration of pressure loading further reduces the bending moments.

7 Conclusions

The study considers detailed FE analysis of a foundation raft by considering the spring stiffness both for static and dynamic conditions. The FE analysis results are validated with the analytical solutions available in the literature and an excellent match is observed for the radial as well as tangential bending moments in the raft. Maximum and minimum radial, tangential bending moments are found to be at the stack shell location ($r = +.0$ m, $r = -3.0$ m). The bearing pressures obtained with the FE analysis performed are also compared with the analytical solution results and are found to be in reasonable agreement. Also, it is observed that the maximum bearing pressure is within the allowable bearing pressure of the study region. The spring stiffness is considered with and without embedment effects. The effect of embedment and uncertainties in soil-structure interaction is also considered in the present study. It is observed that when the foundation stiffness is reduced to 0.67G, the raft bending moments at stack peripheral location increase when compared to 1.0G and hence governs the designs. Also, the radial and tangential bending moment for wind loads is higher and is observed to be the governing load case for the foundation design. The study shows the importance of consideration of uncertainties in SSI in the dynamic analysis of foundations.

a

b

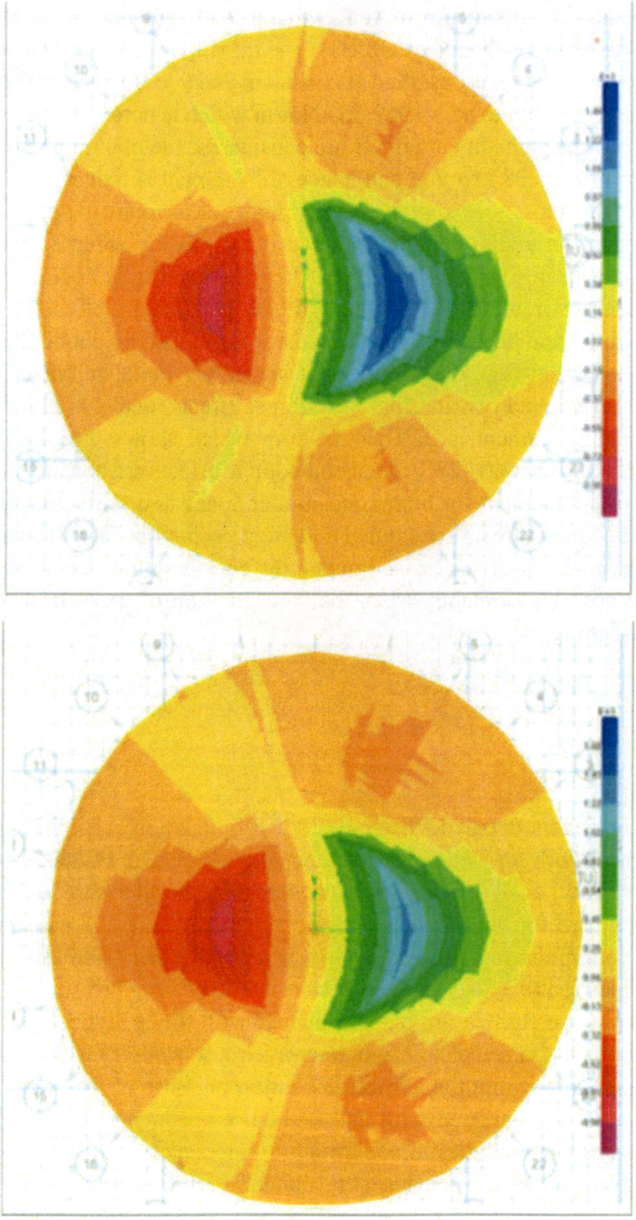

Fig. 13a **a** Tangential bending moment in the raft for concentrated loads/without embedment/$G = G$, $M_t\max = 1358.5$ kNm/m, $M_t\min = -855.3$ kNm/m. **b** Tangential bending moment in the raft for concentrated loads/with embedment/$G = G$, $M_t\max = 1287.78$ kNm/m, $M_t\min = -833.92$ kNm/m. **c** Tangential bending moment in the raft for pressure loads/without embedment/$G = G$, $M_t\max = 1284.445$ kNm/m, $M_t\min = -804.405$ kNm/m. **d** Tangential bending moment in the raft for pressure loads/with embedment/$G = G$ $M_t\max = 1215.0$ kNm/m, $M_t\min = -783.0$ kNm/m

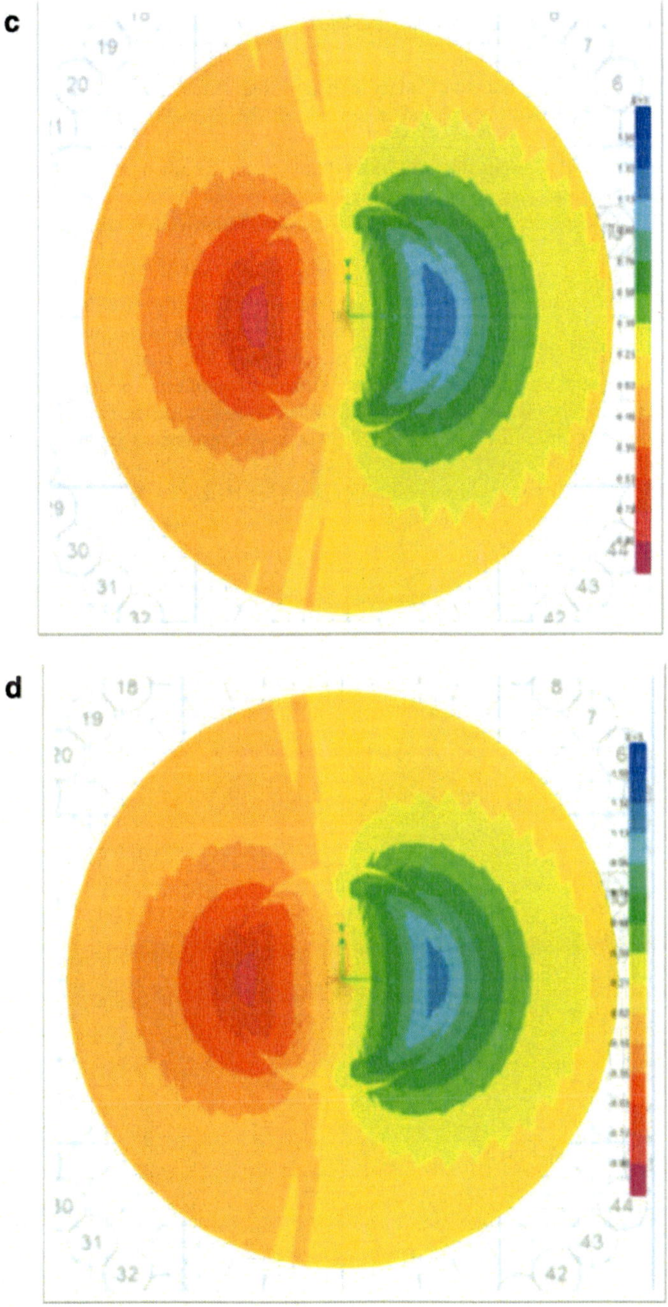

Fig. 13a (continued)

References

ASCE 4-98 (1998) Seismic analysis of safety related nuclear structures and commentary

Bowles JE (1996) Foundation analysis and design, 5th edn. McGraw-Hill Publishing Company Ltd. New York

Jaya V, Dodagoudar GR, Boominathan A (2009) (2009) Seismic soil-structure interaction analysis of ventilation stack structure. Indian Geotech J 39(1):116–134

Manohar SN (1985) Tall chimneys, design & construction. Tata McGraw hill publishing company Ltd. New Delhi

SAFE v12.3.1 (2010). A commercial finite element Software for RC foundation analysis, design and detailing. Computers and Structures Inc., (CSi) Structural and Earthquake Engineering Software

Seed HB, Idriss IM (1970) Soil moduli and damping factors for dynamic response analyses. Report EERC, pp 70–10. Earthquake Engineering Research Centre, University Of California, Berkeley

Vesic AB (1961) Beams on elastic subgrade and Winkler's hypothesis. In: Proceedings of the 5th international conference on soil mechanics and foundation engineering, Paris, pp. 845–850

Vucetic M, Dobry R (1991) Effect of soil plasticity on cyclic response. J Geotech Eng Div ASCE 117(1):89–107

Wolf JP (1985) Dynamic soil-structure interaction. Prentice Hall, New York, p 1985

Evolution in Liquefaction Strength of Ganga River Sand Due to Intrusion of Non-plastic Silt

Nishant Nilay and Pradipta Chakrabortty

Abstract Previous studies claim that the presence of higher proportion of plastic clay or silt particles in sandy deposits enhances the liquefaction strength of soil. However, the behavior of sand mixed with non-plastic silt is still a part of the enduring discussion among geotechnical researchers. The present study is carried out to determine the liquefaction strength of Ganga sand containing different percentages of non-plastic silt ranging from 0%, 5%, 10%, 20%, 30%, and 100%. Undrained strain-controlled cyclic triaxial tests have been carried out at 0.5 Hz frequency, 0.65% axial strain level, and confining pressure of 150 kPa. Considering generation of excess pore water pressure (EPWP) value of 0.95 as the criteria for liquefaction, the result suggests that the rate of EPWP generation has decreased initially upto 10% silt content, and thereafter, at 30% silt content the rate has increased tremendously which eventually got reduced at soil sample having 100% silt.

Keywords Liquefaction · Non-plastic silt · Strain-controlled test · Cyclic triaxial

1 Introduction

Liquefaction is a perilous phenomenon that results in substantial loss of shear strength of soil. The loss of shear strength and stiffness of soil occurs due to the generation of positive pore water pressure which in turn reduces the effective stress of soil deposit under static or cyclic loading condition. Being region-specific phenomenon, liquefaction is dependent on various parameters that vary from one place to another like relative density, soil type, water table depth, and topography. It has generally been observed that loose saturated cohesionless soil deposits like clean sand or sandy

N. Nilay (✉) · P. Chakrabortty
Indian Institute of Technology Patna, Bihta 801103, India
e-mail: nishantnilay.iitp@gmail.com

P. Chakrabortty
e-mail: pradipt@iitp.ac.in

M. Latha Gali and P. Raghuveer Rao (eds.), *Geohazards*, Lecture Notes in Civil Engineering 86, https://doi.org/10.1007/978-981-15-6233-4_19

silt are highly susceptible to liquefaction; however, higher density and cohesiveness make the soil deposits intact.

The presence of non-plastic silt is very common in alluvial deposit of the Indo-Gangetic flood plain. In past, regions in and around Indo-Gangetic plain has witnessed many devastating earthquakes such as Bihar–Nepal (1833), Kangra (1905), Bihar (1934), Kinnaur (1975), Nepal (1988), and Nepal (2015). The ejaculated sand boils on the ground surface during many of these earthquakes have shown the presence of sand and silt mixed soil. Numerous past studies on the effect of fines on seismic response of sand have yielded that the presence of higher proportion of plastic fines increases the liquefaction resistance of sand, e.g., (Kishida 1970; Zhou 1981; Andrews and Martin 2000). However, the behavior of sand in the presence of non-plastic silt under monotonic/cyclic loading condition is still a part of enduring discussion because of various contradicting outcomes. Some studies (Chang et al. 1982; Kuerbis et al. 1988; Amini and Qi 2000; Muley et al. 2015) reported increase in liquefaction resistance with the increase in fine content while some others (Singh 1994; Polito and Martin 2001; Xenaki and Athanasopoulos 2003; Dash and Sitharam 2009; Belkhatir et al. 2010; Karim and Alam 2014, 2017) observed that liquefaction potential increased till certain limit with the increase in fine content.

Several occurrences of liquefaction phenomena have been observed in and around the Gangetic plane near Patna in the past which has resulted in the devastation of life as well as property. This area lies in the seismic zones III and IV (BIS 2016). The alluvial deposit found around this area consists of mainly fine sands mixed with fines (clay and silt). In an attempt to study, the effect of non-plastic fines on seismic response of loose saturated Ganga sand, a series of experimental investigation has been carried out using strain-controlled undrained cyclic triaxial test on isotropically consolidated soil specimens at confining pressure of 150 kPa and 0.5 Hz loading frequency. The main objective of this study is to inspect the cyclic behavior possessed by Ganga sand in the presence of different percentages of non-plastic silt (0%, 5%, 10%, 20%, 30%, and 100%).

2 Material Used

The sand used in this study was collected from the bank of river Ganga near Patna, India, and the silt was obtained after sieving the Gangetic sand using 75 micron IS sieve. Various preliminary tests have been carried out on sand and silt to ascertain their index and engineering properties. Particle size distribution of soil helps in its classification based on gradation. In the present study, the gradation of soil was achieved using dry sieve analysis test and hydrometer test. Following the procedure recommended by IS code (BIS 1985b), the particle size distribution curve for Ganga sand and Silt has been obtained and is shown in Fig. 1.

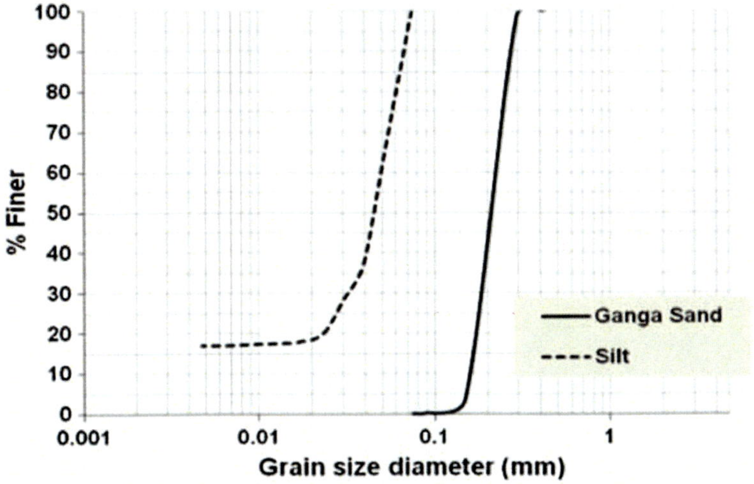

Fig. 1 Particle size distribution curve for Ganga sand and silt

Table 1 Specific gravity of soil specimens

Samples	'G'
100% sand + 0% silt	2.67
95% sand + 5% silt	2.69
90% sand + 10% silt	2.71
80% sand + 20% silt	2.71
70% sand + 30% silt	2.68
0% sand + 100% silt	2.68

Specific gravity (G_s) for each soil sample was calculated according to the recommended procedure by IS code (BIS 1980). The G_s values for Ganga sand with different silt percentage are given in Table 1. It is evident from Table 1 that for sand, specific gravity ranges from 2.67 to 2.71. The maximum and minimum dry density of soil is not unique as it is very much dependent on the method used for its determination, e.g., (Lade et al. 1998). The IS code (BIS 1983a) recommends the calculation of maximum dry density of cohesionless soil using vibratory table. However, the maximum dry density achieved using a heavy compaction test yielded little bit higher values than that obtained using vibratory table. Hence in the present study, maximum dry density has been achieved using heavy compaction test according to the recommendations of IS code (BIS 1983b). For achieving minimum dry density of soil samples, ASTM method C (2016) has been used in this study. The maximum and minimum dry density values of Ganga river sand intruded with different silt percentage is given in Table 2. Using the relationship between dry density and void ratio, maximum and minimum void ratio of all soil samples was obtained and is shown in Fig. 2.

Table 2 Maximum and minimum dry density of soil sample

Samples	Ganga sand	
	$\rho_{d,\ max}$ (gm/cc)	$\rho_{d,\ min}$ (gm/cc)
100% sand + 0% silt	1.628	1.274
95% sand + 5% silt	1.613	1.286
90% sand + 10% silt	1.622	1.395
80% sand + 20% silt	1.698	1.389
70% sand + 30% silt	1.720	1.431
0% sand + 100% silt	1.530	1.121

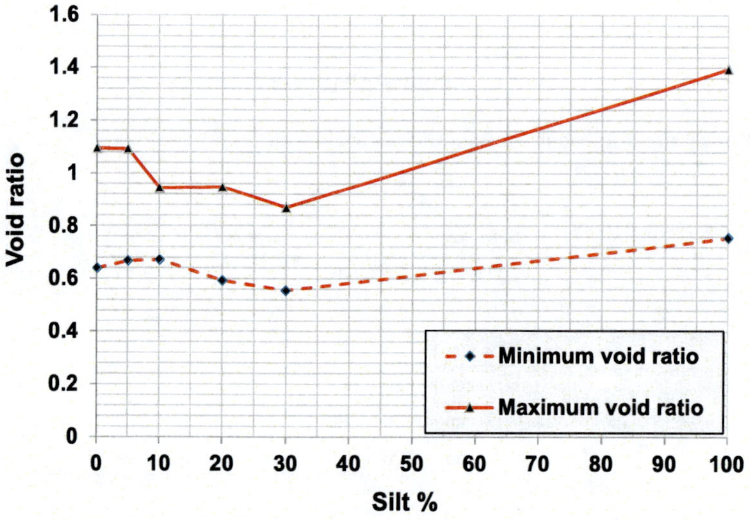

Fig. 2 Maximum and minimum void ratio of Ganga sand at different silt percentage

Liquid limit of silt was determined using cone penetration method IS: 2720 (part-5) (BIS 1985a). The test results show that the liquid limit value for the silt that is being used is less than 2%. After many attempts for getting plastic limit value of silt, it is inferred that due very low or no plasticity in silt it was not possible to mold it into a thread of 3.2 mm.

Shape and size of sand and silt particles were analyzed using the images obtained at 20X magnification in optical microscope. Figure 3 clearly delineates that the particles of Ganga sand are smaller than 425-μm size, and they are rounded to sub-rounded in shape. It is evident from the magnified image for silt particles that more proportion of particles are having size greater than approximately 45 micron, and the same can also be inferred from the particle size distribution curve shown in Fig. 1.

The X-ray diffraction (XRD) analysis is based on the diffraction pattern obtained after the bending of light from the geometrical shadow of the obstacles. In the present

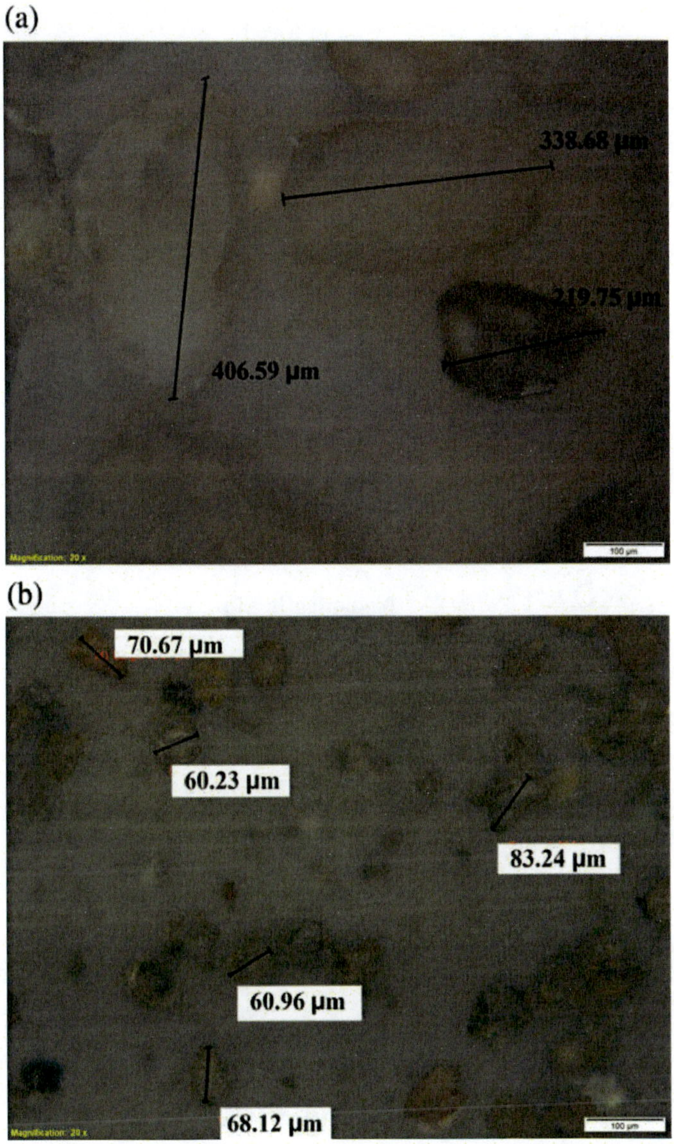

Fig. 3 Magnified images from optical microscope of **a** Ganga sand particles. **b** Silt particles

study, XRD pattern of Ganga sand and silt has been obtained using the X-ray diffractometer. The X-ray of energy copper k-α which corresponds to the X-ray wavelength of 1.540 Å was passed through the soil samples and using the detector the number of X-rays observed at each angle 2θ. The intensity is recorded as counts per second, and the plot between intensity sand 2θ for different soil samples is shown in Fig. 4. For qualitative analysis, XRD plots were compared with the XRD plot of soil from

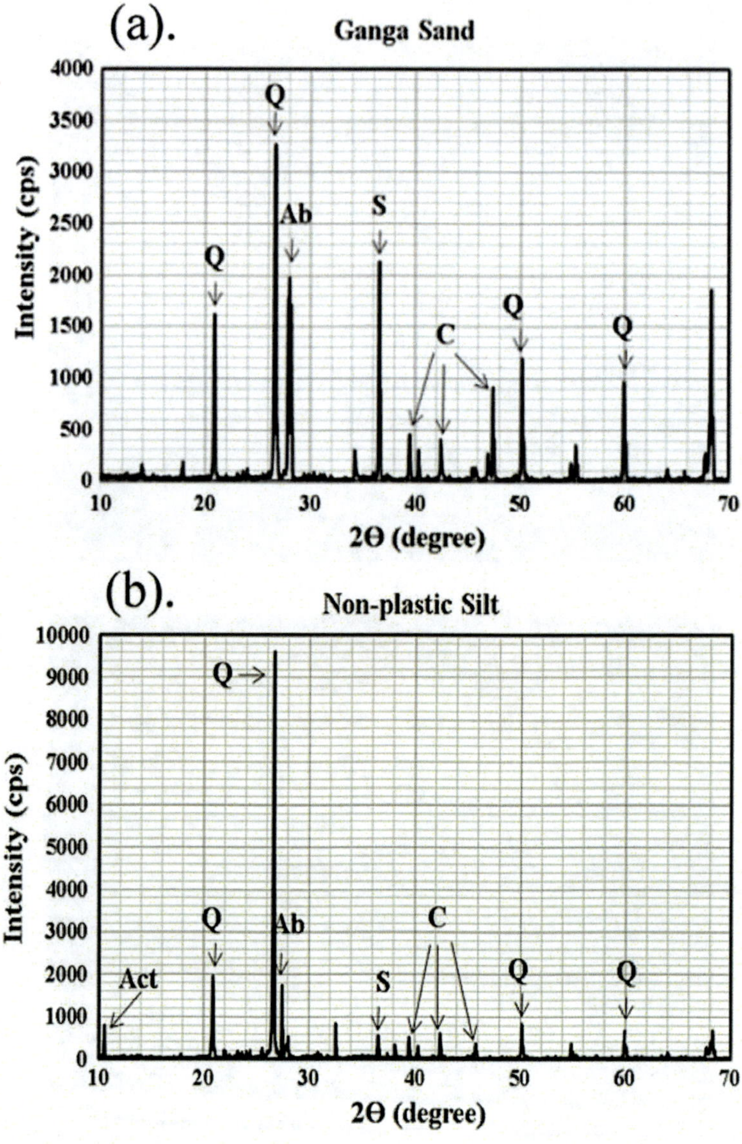

Fig. 4 XRD plot between 2θ (degree) and intensity (count per second) of **a** Ganga sand. **b** Ganga silt (Q—quartz, C—calcite, S—smectite, Ab—plagioclase, and Act—actinolite)

Sossego mine reported by Li et al. (2014). Based on the comparison with the reference XRD plots, the presence of different minerals in soil samples has been detected. The relative amount of minerals based on relative intensities of diffraction peaks cannot be guessed directly as it requires further studies and is out of the scope of present study.

3 Experimental Procedure

The seismic response of Ganga sand intruded with different percentage by weight of non-plastic silt has been studied using strain-controlled cyclic triaxial test in accordance with the approved guidelines of ASTM (1996). Soil sample preparation for cyclic triaxial test is considered as the most important step of the experiment. As few parameters that are chosen as a variable or a constant like density, relative density, and void ratio can only be controlled at this stage of experiment. In the present study, the soil sample is prepared in a split mold of 38 mm diameter and 75 mm height using dry air pluviation method in such a way that the relative density is maintained in-between 10 and 20%. The method involves the dropping of soil particles through funnel from a certain height such that the mass of soil deposited inside the mold yields the desired relative density. After depositing the soil inside the mold, the filter paper and porous stone were kept on top of it and then the sample cap having screwed piston rod was placed over it. Once the cell chamber was attached, it was completely filled with water to apply the confining pressure in later stages of experiment. At the end of this stage, the suction pressure of 10 kPa was passed through the soil specimen for 30 min in order to remove the entrapped air in the voids. The desirable saturation of soil specimens was achieved by passing de-aired water through backpressure line. The difference between confining and backpressure value was kept constant at 20 kPa during the saturation stage. With every increment of confining pressure by 20 kPa, the saturation value was calculated.

Once the desirable saturation of greater than 80% was achieved, the soil specimens were isotropically consolidated to an effecting confining stress of 150 kPa. At the time of consolidation, the volume change was measured through the volume of water leaving the system through backpressure line and getting collected into volume change indicator. The soil samples were allowed for consolidation till the volume change has become constant. After attaining constant volume change, all the soil specimens were loaded with cyclic loading with a frequency of 0.5 Hz. The steps followed from the sample preparation and placing stage to the initiation of saturation phase in sequential order are shown in Fig. 5. The relative densities and void ratios reported in this study from now onwards are post-consolidation relative densities and void ratios. The pre- and post-consolidation relative density of Ganga sand at different silt percentage is shown in Fig. 6.

4 Results and Discussion

The results of typical cyclic triaxial test performed on clean Ganga fine sand specimen prepared at post-consolidation relative density of 10–25% and loaded at axial strain level of 0.65% are presented in Fig. 7. It is to be noted that the occurrence of initial liquefaction in the present study has been considered when the development of excess pore water pressure becomes equal to 90% of initial confining pressure.

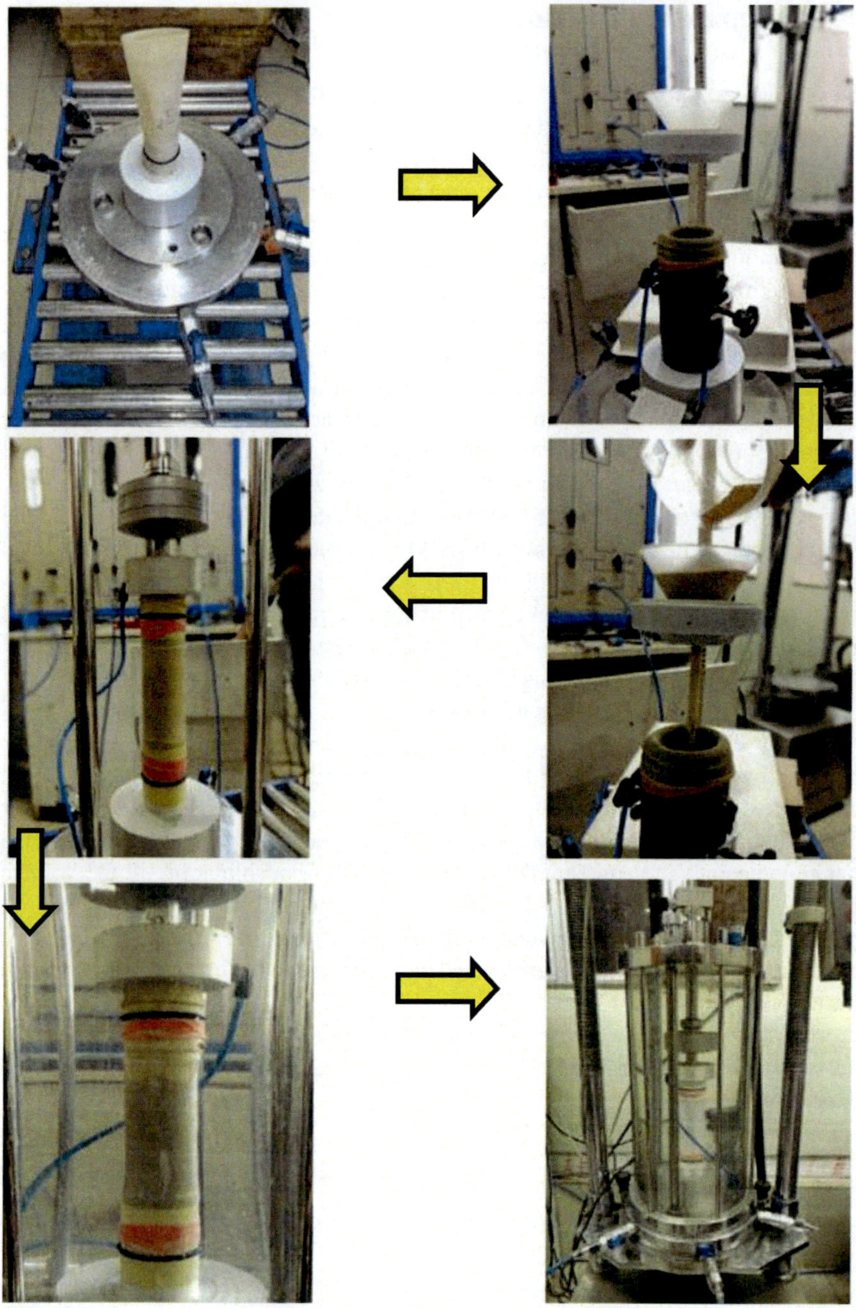

Fig. 5 Steps followed from sample preparation stage to the saturation stage in sequential order

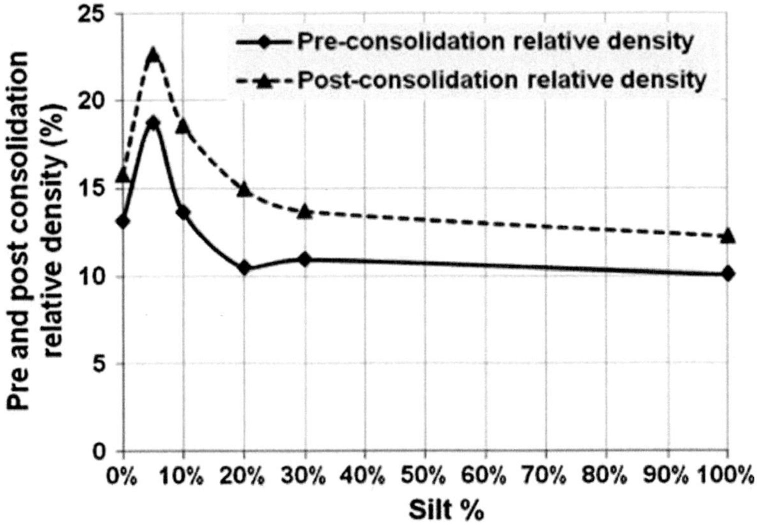

Fig. 6 Pre- and post-consolidation relative density of Ganga sand specimens at different percentages of silt

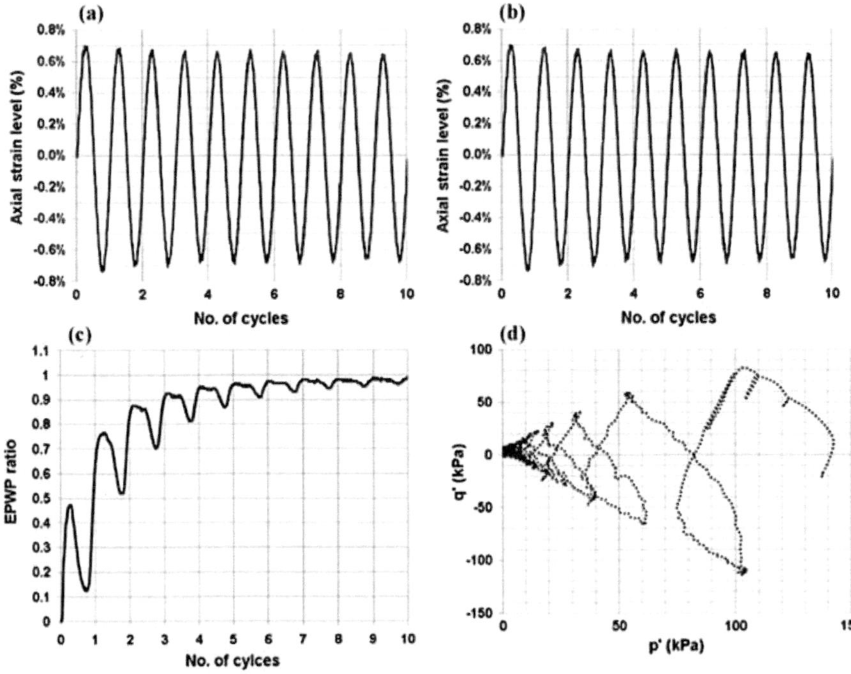

Fig. 7 Typical cyclic triaxial test results of clean Ganga sand. **a** Axial strain level (%) versus number of cycles of loading. **b** Deviator stress versus number of cycles of loading. **c** EPWP ratio versus number of cycles of loading. **d** Mean effective stress (p') versus mean deviatoric stress (q')

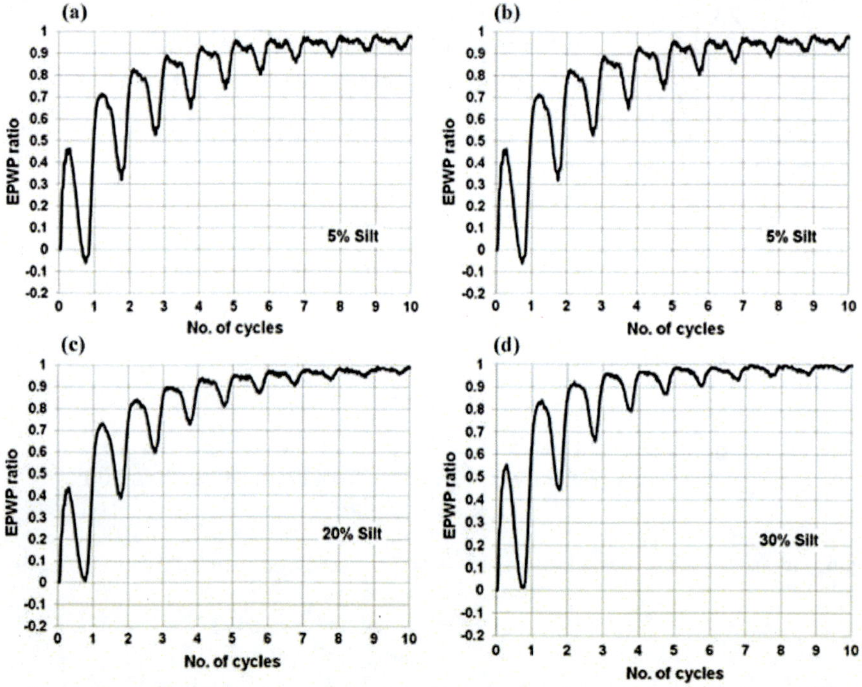

Fig. 8 EPWP ratio versus number of cycles of loading for various sand specimens with **a** 5% silt, **b** 10% silt, **c** 20% silt, **d** 30% silt

From Fig. 7, it is evident that clean Ganga sand specimen is liquefying in third loading cycle. It is beyond doubt that the deviatoric stress will reduce continuously with increase in loading cycles, and the same can be observed in Fig. 7b where after attaining maximum deviatoric stress of around 90 kPa, the stress value decreased to around 20 kPa in third loading cycle. Figure 8 delineates the number of cycles required for liquefaction $[N_L(R_{u,\max} = 0.9)]$ for soil specimens of Ganga sand at different silt percentage. It is evident from Figs. 7c and 8 that with addition of silt in Ganga sand, the liquefaction potential is decreasing. It can be inferred from Fig. 7c that the number of loading cycles for the occurrence of liquefaction in Ganga sand started decreasing when silt content exceeded 10% silt content by weight. It is interesting to note that the pore water pressure has diminished from its initial value in first cycle for those soil specimens in which liquefaction resistance has increased; however, for all the other sand specimens, the EPWP ratio remains always positive. This peculiar behavior can be justified with the fact that the soil specimen prepared with 5% and 10% silt has relatively higher relative density than the others and the same can be observed in Fig. 6. Hence, more dilation has occurred in first loading cycles which caused the reduction in pore water pressure generation from its initial value.

The soil specimen with 100% silt has been found to get liquefied within third cycle only. Being completely non-plastic fines and less permeable, the dissipation of pore water pressure was difficult, and thus, it was expected that liquefaction will occur early. The typical cyclic triaxial test results of soil specimen with 100% silt are shown in Fig. 9. It is evident from the test results that there is not much difference in the maximum deviatoric stress in extension and compression obtained in clean Ganga sand and soil specimen with 100% silt.

Table 3 represents the number of loading cycles required by each soil specimen for the occurrence of liquefaction. It is perceptible from all the test results presented here that the number of cycles required for occurrence of liquefaction (i.e., EPWP ratio becomes equal to 0.9) in all the Ganga sand specimens and 100% silt are more or less same as all the soil samples are liquefying within 5 loading cycles only at axial strain level of 0.65%.

Polito and Martin (2001) have performed stress-controlled test on Monterey and Yatesville sand specimens with different percentages of silt. The cyclic triaxial test result obtained for Yatesville sand when investigated at 30% relative density illustrates that there will not much change in cyclic resistance due to intrusion of silt till limiting fine content (LFC) value. However, after LFC the cyclic resistance has

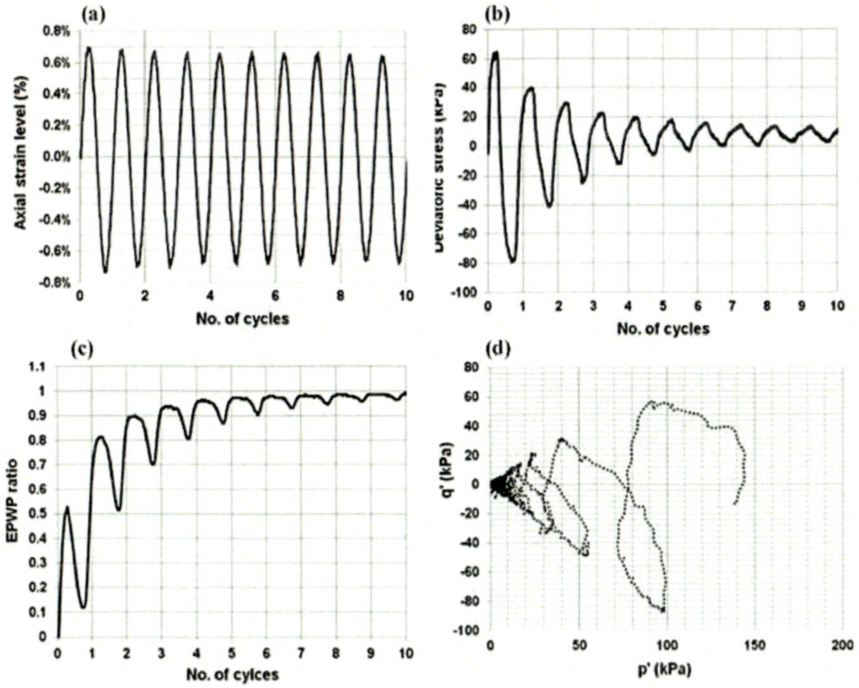

Fig. 9 Typical cyclic triaxial test results of soil specimen with 100% silt. **a** Axial strain level (%) versus number of cycles of loading. **b** Deviator stress versus number of cycles of loading. **c** EPWP ratio versus number of cycles of loading. **d** Mean effective stress (p') versus mean deviatoric stress (q')

Table 3 Number of loading cycles required by Ganga sand for the occurrence of liquefaction at confining pressure of 150 kPa and axial strain of 0.65%

Samples	No. of cycles for $R_{u,\max} = 0.9$
100% sand + 0% silt	3
95% sand + 5% silt	4
90% sand + 10% silt	5
80% sand + 20% silt	4
70% sand + 30% silt	2
0% sand + 100% silt	3

decreased with increase in silt content. It is very interesting to note that the mean grain diameter of Yatesville sand and Ganga sand is more or less same, i.e., 0.18 mm and 0.2 mm, respectively. Hence, the behavior of sand based on fabrics can be expected to be similar for both of these sands. Muley et al. (2015) has inspected the cyclic behavior of Solani River sand in the presence of silt at axial strain level of 0.75%. The result obtained in this study gives very good affirmation to the findings of Muley et al. (2015).

The variation of cyclic stress ratio (CSR) at each loading cycles obtained for different soil specimens is shown in Fig. 10. It is quite evident that the soil specimen with 10% silt is having higher CSR value which continues to degrade with increase in loading cycle numbers. It is clearly visible that for all the soil specimens the CSR value becomes lower than 0.1 after eleventh loading cycles; however for most of the soil samples, the slope of curve has started changing after fifth loading cycles.

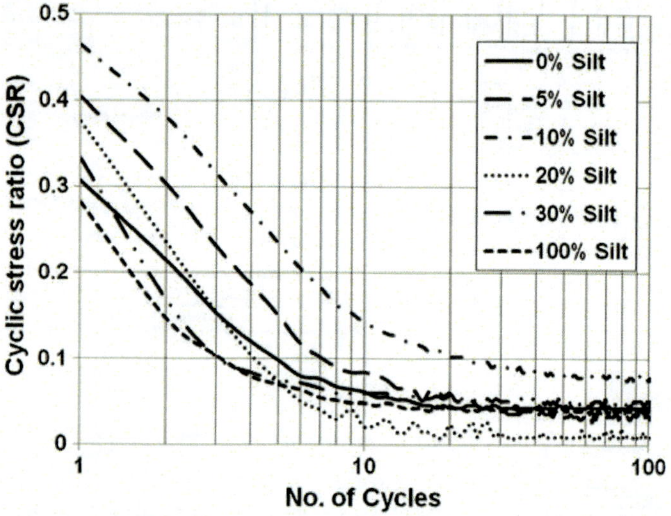

Fig. 10 Maximum cyclic stress ratio value obtained in each loading cycle for different soil specimens

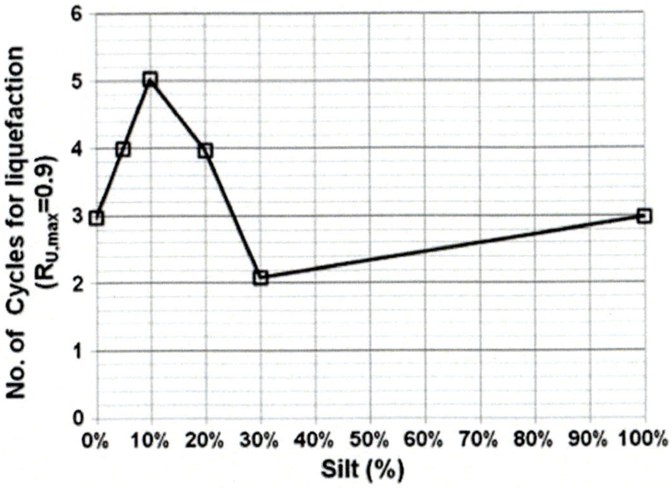

Fig. 11 Number of loading cycles required for liquefaction versus silt percentage

Figure 11 shows the variation of loading cycles required to cause liquefaction for soil specimens with different silt content. It is clearly noticeable that the liquefaction potential has reduced till 10% silt content in Ganga sand. However, beyond LFC (29%) there is no major change in liquefaction potential has been observed. Further study is required for establishing this for the silt percentages between 30 and 100%. As discussed earlier that all the soil specimens are liquefying within five loading cycles only, and the same can be shown in Fig. 11. This behavior of Ganga sand mixed with silt under cyclic loading can be explained using the concept discussed by Finn et al. (1994). The liquefaction resistance will increase till there is some room within the voids of sand to contain the silt particles. Hence, it can be inferred that if the sand particles are smaller in size or silt particle are larger in size than the size of voids in sand matrix, behavior of sand in the presence of silt may not get altered in cyclic or monotonic loading condition.

5 Conclusion

The present study is carried out for inspecting the behavior of Ganga sand and silt mix soil at axial strain level of 0.65%. It has been concluded that the rate of pore water pressure generation has decreased initially upto 10% silt content, and thereafter at 30% silt content, the rate has increased which eventually reduced at soil sample having 100% silt. The investigation also concludes that under cyclic loading condition used here, there will not be much effect on Ganga sand if it is present with silt having mean grain diameter (D_{50}) of 0.045 mm. The results clearly illustrate that the consistency in the behavior of sand mixed with different percentages of silt under cyclic loading condition.

However, on the basis of literature review it has been found that the study on seismic behavior of sand–silt mix deposit is a multi-dimensional problem and is dependent on various parameters and the methodology followed in the study, e.g., Sample preparation technique, test type (stress- or strain-controlled), and value of $R_{u,\max}$ (0.9, 0.95, or 1) chosen to define the liquefaction occurrence. The outcome of the study may change if any part of the methodology or any variable changes.

References

American Society for Tests and Materials (ASTM) (1996) Standard test method for load controlled cyclic triaxial strength of soil. ASTM D 5311–92 (Re-approved 1996). ASTM, West Conshohoken

American Society for Tests and Materials (ASTM) (2016) Method of test for determination of minimum index density and unit weight of soil. ASTM D 4254–16. ASTM, West Conshohoken

Amini F, Qi GZ (2000) Liquefaction testing of stratified silty sands. J Geotech Geoenviron Eng 126(3):208–217

Andrews DC, Martin GR (2000) Criteria for liquefaction of silty soils. In: Proceedings of the 12th world conference on earthquake engineering. New Zealand Society for Earthquake Engineering, Upper Hutt, New Zealand

Belkhatir M, Arab A, Della N, Missoum H, Schanz T (2010) Liquefaction resistance of Chlef river silty sand: effect of low plastic fines and other parameters. Acta Polytech Hung 7(2):119–137

BIS (1980) IS: 2720 (Part 3)-1980, Method of test for determination of specific gravity of soil. Bureau of Indian standards, Manak Bhawan, 9 Bahadur Shah Zafar Marg, New Delhi-12, India. (Reaffirmed 2002)

BIS (1983a) IS: 2720 (Part 14)-1983, Method of test for determination of density index of soil. Bureau of Indian standards, Manak Bhawan, 9 Bahadur Shah Zafar Marg, New Delhi-12, India. (Reaffirmed 2006)

BIS (1983b) IS: 2720 (Part 8)-1983, Method of test for determination of water content-dry density relation using heavy compaction. Bureau of Indian standards, Manak Bhawan, 9 Bahadur Shah Zafar Marg, New Delhi-12, India (Reaffirmed 2006)

BIS (1985a) IS: 2720 (Part 5)-1985, Method of test for determination of liquid and plastic limit. Bureau of Indian standards, Manak Bhawan, 9 Bahadur Shah Zafar Marg, New Delhi-12, India. (Reaffirmed 2006)

BIS (1985b) IS: 2720 (Part 4)-1985, Method of test for grain size analysis of soil. Bureau of Indian standards, Manak Bhawan, 9 Bahadur Shah Zafar Marg, New Delhi-12, India. (Reaffirmed 2006)

BIS (2016) IS: 1893 1893 (Part 1)-2016, Criteria for earthquake resistant design of structure, General Provision and Building. Bureau of Indian standards, Manak Bhawan, 9 Bahadur Shah Zafar Marg, New Delhi-12, India

Chang NY, Yeh ST, Kaufman LP (1982). Liquefaction potential of clean and silty sands. In: Proceedings of the 3rd international earthquake microzonation conference, Seattle, USA, vol 2, pp 1017–1032

Dash HK, Sitharam TG (2009) Undrained cyclic pore pressure response of sand–silt mixtures: effect of non-plastic fines and other parameters. Geotech Geol Eng 27(4):501–517

Finn WL, Ledbetter RH, Wu G (1994) Liquefaction in silty soils: design and analysis. Ground failures under seismic conditions, Geotechnical Special Publication No. 44, ASCE, pp 51–76

Karim ME, Alam MJ (2014) Effect of non-plastic silt content on the liquefaction behavior of sand–silt mixture. Soil Dyn Earthq Eng 65:142–150

Karim ME, Alam MJ (2017) Effect of nonplastic silt content on undrained shear strength of sand–silt mixtures. Int J Geo-Eng 8(1):14

Kishida H (1970) Characteristics of liquefaction of level sandy ground during the Tokachioki earthquake. Soils Found 10(2):103–111

Kuerbis RH, Negussey D, Vaid YP (1988) Effect of gradation and fines content on the undrained response of sand. Geotech Spec Publ 21:330–345

Lade PV, Liggio CD, Yamamuro JA (1998) Effects of non-plastic fines on minimum and maximum void ratios of sand. Geotech Test J 21:336–347

Li Q, Ji HB, Qin F, Tang L, Guo XY, Feng JG (2014) Sources and the distribution of heavy metals in the particle size of soil polluted by gold mining upstream of Miyun Reservoir, Beijing: implications for assessing the potential risks. Environ Monit Assess 186(10):6605–6626

Muley P, Maheshwari BK, Paul DK (2015) Liquefaction potential of Roorkee region using field and laboratory tests. Int J Geosynthetics Ground Eng 1(4):37

Polito CP, Martin JR II (2001) Effects of nonplastic fines on the liquefaction resistance of sands. J Geotech Geoenviron Eng 127(5):408–415

Singh S (1994) Liquefaction characteristics of silt. Ground failures under seismic conditions, Geotechnical Special Publication No. 44, ASCE, pp 105–116

Xenaki VC, Athanasopoulos GA (2003) Liquefaction resistance of sand–silt mixtures: an experimental investigation of the effect of fines. Soil Dyn Earthq Eng 23(3):1–12

Zhou SG (1981) Influence of fines on evaluating liquefaction of sand by CPT. In: Proceedings of international Conferences on recent advances in geotechnical earthquake engineering and soil dynamics, 4

Probabilistic Approach of Liquefaction Assessment for Guwahati Based on Shear Wave Velocity Values

K. S. Vipin and S. D. Anitha Kumari

Abstract Guwahati is one of the fastest growing cities in the North East India. It is situated along the banks of river Brahmaputra and is classified into seismic zone V in the seismic zonation map of India. The high seismic hazard in North East India points to the fact that earthquake-induced liquefaction assessment for these regions is necessary. Due to the large number of uncertainties associated with the evaluation of soil liquefaction, probabilistic liquefaction assessment has gained significance. In this study, liquefaction potential is evaluated by following a sequence of steps which include the generation of an earthquake catalogue, identification of the seismic sources, developing probabilistic seismic hazard, and probabilistic assessment of liquefaction potential. The use of probabilistic methods in liquefaction potential evaluation will help in quantifying the uncertainties involved in seismic hazard and soil resistance in a better way. The spatial maps showing the variation of factor of safety against liquefaction for different return periods are developed based on the shear wave velocity values. These results will help in delineating regions with high liquefaction hazard in the city thereby facilitating risk mitigation activities.

Keywords Liquefaction · Shear wave velocity · Probabilistic approach

1 Introduction

Many earthquakes viz. Niigata (1964), Kobe (1985), Loma Prieta (1989), Bhuj (2001), Christchurch (2011), and Tohoku (2011) have caused severe damages especially due to liquefaction of soils. This indicates the significance of better assessment of liquefaction potential of soils. Liquefaction is generally associated with soft and loose soil deposits. The susceptibility of various soil deposits to liquefaction is

K. S. Vipin
Swiss Re, Bangalore, India
e-mail: ks.vipin@gmail.com

S. D. Anitha Kumari (✉)
M.S. Ramaiah University of Applied Sciences, Peenya, Bangalore, India
e-mail: anitha.vipin@gmail.com

M. Latha Gali and P. Raghuveer Rao (eds.), *Geohazards*, Lecture Notes in Civil Engineering 86, https://doi.org/10.1007/978-981-15-6233-4_20

governed by several factors. There are various field and laboratory tests adopted to estimate the liquefaction susceptibility. The commonly adopted field tests are standard penetration test (SPT), cone penetration test (CPT), shear wave velocity test (V_s) and Becker penetration test (BPT) (Youd et al. 2001). Correlations are developed based on these in situ penetration tests which aids in assessing the initiation of liquefaction due to dynamic loads. The availability of large amount of SPT and CPT data has made the correlation of relative density of soil deposits to liquefaction potential easier. Deterministic methods of liquefaction evaluation based on SPT data was reported by Seed et al. (1985). One of the recent probabilistic liquefaction evaluation methods based on SPT and CPT data was formulated by Cetin et al. (2004) and Moss et al. (2006). Though the method suggested by Cetin et al. and Moss et al. gives the probability of liquefaction, both the methods do not consider the uncertainty in earthquake loading.

The recent advancements in obtaining the shear wave velocity (V_S) profiles of soil deposits have stressed the need for V_S-based liquefaction assessment methods. Moreover, V_S profiles are preferred because this technique is less sensitive to the various field problems including the penetration resistance due to the presence of fines. Hence, in this study, the liquefaction potential assessment of Guwahati city is done based on shear wave velocity by adopting a probabilistic liquefaction potential evaluation method.

Most of the liquefaction potential assessment methods developed consider single earthquake magnitude and ground acceleration value. A performance-based liquefaction potential assessment method was suggested by Kramer and Mayfield (2007). In this method, the entire range of ground motion and earthquake magnitude levels is considered for liquefaction potential assessment (the conventional methods use a single magnitude and acceleration values). Performance-based liquefaction potential assessment based on SPT data was done by Vipin et al. (2010) and CPT-based liquefaction potential assessment was done by Vipin and Sitharam (2012). In this study, the method suggested by Kayen et.al. (2013) and Kramer and Mayfield (2007) is used to develop a probabilistic liquefaction potential map of Guwahati city based on V_S values.

2 Study Area

North eastern parts of India are classified into zone V (IS-1893 (2016) due to the high probability of occurrence of seismic events. Guwahati, capital city of Assam, is the largest and busiest urban area among the North eastern states. Liquefaction was observed in various parts of Assam during the 1897 Great Assam Earthquake (Magnitude 8.3) and the Upper Assam Earthquake of 1950 (Magnitude 8.7). Guwahati, located on the banks of Brahmaputra river, is having alluvial soil deposits with low soil density. Due to its close proximity with the river, the water table at most locations is reported to be shallow. The density of the soils in this region is reported to be in between 16.5 and 17.9 kN/m^3 (Das and Saikia 2010). The depth of water table

Fig. 1 Study area

in various parts of Guwahati city varies from 0 to 6 m (Sharma and Rahman 2016). The soft and loose soil deposits along with shallow depth of water table indicate a high probability of liquefaction in the event of an earthquake. Moreover, the city lies close to the unruptured North East Himalayan fault zone. Figure 1 shows the study area of Guwahati region lying within a latitude and longitude of 26°00′ N—26°13′ N and 91°30′ E—91°55′ E respectively. As per IS–1893 (Part I 2016), Guwahati is classified into high risk area falling under zone V. The rapid development of this city has led to an urban sprawl which points to the increased vulnerability and damages to human life and infrastructure.

Various studies on the earthquake hazard of North East India clearly indicate the vulnerability of these regions to liquefaction (Nath et al. 2008; Raghu Kanth et al. 2008; Raghukanth and Dash 2010). Liquefaction potential evaluation of Guwahati city is reported by Ayothiraman et.al. (2012), Sharma and Hazarika (2013), and Sharma and Begum (2018). However, all the studies reported have not considered the uncertainties in earthquake loading fully.

3 Seismic Hazard Assessment of Guwahati

Due to the high probability of earthquakes in the North Eastern region, the assessment of hazard in this region has gained significant attention. Various researchers have done studies on the seismic hazard assessment of Guwahati (Das et al. 2006, 2016). The probabilistic seismic hazard assessment (PSHA) method adopted by Vipin et al. (2009) and Sitharam and Vipin (2011) is used for hazard assessment in this study as well. The earthquake catalogue, seismicity parameters, and the seismic sources are selected from Kolathayar et al. (2012). In the present study, the seismic sources within a distance of 400 km from Guwahati city are considered. The next generation ground motion prediction equation (GMPE) developed by Boore and Atkinson (2008) for active tectonic region is used for the prediction ofground motion values. A probabilistic seismic hazard assessment of Guwahati city is done based on the above-collected data. The PGA values for return periods of 475 (10% probability of exceedance (PE) in 50 years) and 2500 years (2% PE in 50 years) are calculated. The surface level peak ground acceleration values (PGA) were calculated by considering the site amplification based on V_{S30} values.

4 Probabilistic Evaluation of Liquefaction Potential

The liquefaction potential of soils is expressed in terms of the seismic loading on the soil layer and the resistance of the soil layer to liquefaction. Cyclic stress ratio (CSR) indicates the seismic demand (loading) whereas the resistance is indicated by cyclic resistance ratio (CRR). Studies reported by various researchers like Seed and Idriss (1971), Seed et al. (1985), Youd et al. (2001), Liao et al. (1988), Juang and Jiang (2000), Cetin et al. (2004) adopted deterministic and probabilistic methods using SPT data. Some of the important methods to assess the liquefaction susceptibility based on CPT data were suggested by Stark and Oslon (1995), Robertson and Wride (1998), Juang and Jiang (2000), Lai et al. (2006) and Moss et al. (2006). However, in none of these methods, the uncertainties associated with earthquake loading are fully accounted for. This shortcoming can be overcome by doing the liquefaction potential assessment based on the performance-based method suggested by Kramer and Mayfield (2007).

Presently, the most widely used soil profiling is based on shear wave velocity. Kayen et al. (2013) proposed a new method to assess the probability of liquefaction based on shear wave velocity values. Using this method, the probability of liquefaction is obtained based on the values of shear wave velocity, magnitude of earthquake, cyclic stress ratio, and the effective overburden pressure as shown in Eq. 1:

$$P_L = \Phi\left[-\frac{(0.0073V_{S1})^{2.8011} - 1.946\ln \text{CSR} - 2.6168\ln M_w - 0.0099(\ln(\sigma'_{v0}) + 0.0028\text{FC}}{0.4809}\right] \quad (1)$$

where P_L is the probability of liquefaction; Φ is the standard normal cumulative distribution function; V_{S1} is the shear wave velocity at a given depth; CSR is the cyclic stress ratio without magnitude scaling factor; M_w is the moment magnitude of earthquake, σ'_{v0} is the effective vertical pressure at the given depth; and FC is the fineness content in percentage. The value of depth reduction factor, which is used in calculating CSR, is obtained using the method suggested by Cetin et al. (2004). The major drawback in the above method is that the liquefaction probability is evaluated without considering the uncertainties in earthquake loading; instead, it considers a single value of ground acceleration and earthquake magnitude. But the PGA value for a given site obtained in the PSHA is contributed by different magnitudes of earthquakes with varying frequencies. The deaggregated seismic hazard curve for one of the locations at Guwahati is shown in Fig. 2. The contribution of different magnitudes with varying frequencies to the final PGA value at a given site is clearly observed from this figure. The conventional liquefaction potential analysis fails to account for these uncertainties in earthquake loading.

Kramer and Mayfield (2007) suggested a method to incorporate the uncertainties of earthquake loading in the liquefaction potential analysis. In this method, the contributions from all the magnitudes and all the acceleration levels have been considered. Thus, the uncertainty involved in evaluating the earthquake loading for the initiation of liquefaction is taken care off. According to Kramer and Mayfield (2007), the annual probability of exceedance of given factor of safety value is given by Eq. 2.

$$\Lambda_{FS_L^*} = \sum_{j=1}^{N_M} \sum_{i=1}^{N_a} P[FS_L < FS_L^* | a_i, m_j] \Delta\lambda_{a_i}, m_j \tag{2}$$

where $\wedge_{FS_L^*}$ is the annual rate at which factor of safety will be less than FS_L^*; FS_L—factor of safety against liquefaction; FS_L^*—targeted value of factor of safety against liquefaction; N_M—Number of magnitude increments; N_a—number of acceleration increments; $\Delta\lambda_{a_i}, m_j$ is the incremental annual frequency of exceedance for acceleration a_i and magnitude m_j.

Vipin and Anitha (2016) developed an approach to evaluate factor of safety against liquefaction by modifying the method proposed by Kayen et al. (2013). The modified is method is given in Eq. 3.

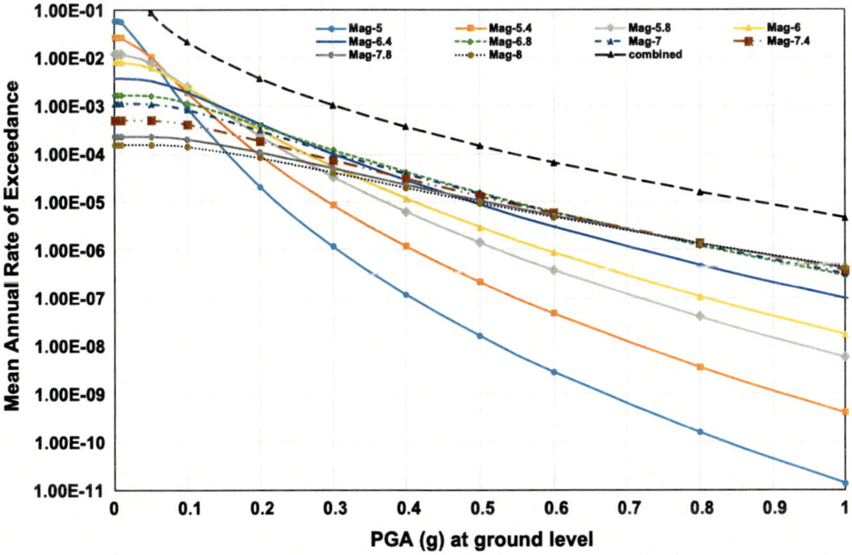

Fig. 2 Deaggregated hazard curve with respect to magnitude

$$P[\mathrm{FS_L} < \mathrm{FS}_L^* | a_i, m_j] = \Phi\left[-\frac{(0.0073V_s)^{2.8011} - 1.946\ln \mathrm{CSR}_{\mathrm{eq},i}\mathrm{FS}_L^* - 2.6168\ln m_j - 0.0099(\ln(\sigma'_{v0}) + 0.0028\mathrm{FC}}{0.4809}\right] \quad (3)$$

In the above equation, Φ is the standard normal cumulative distribution function; V_S is the shear wave velocity at a given depth; σ'_{v0} is the effective vertical stress at the given depth, and FC is the fineness content in percentage. FS_L is the factor of safety against liquefaction, FS_L^* is the targeted value of factor of safety against liquefaction, $\mathrm{CSR}_{\mathrm{eq},i}$ is the cyclic stress ratio without any magnitude scaling factor computed for acceleration value a_i, a_j and m_j(moment magnitude) is obtained from the probabilistic deaggregated seismic hazard curve with respect to magnitude. $\mathrm{CSR}_{\mathrm{eq},i}$, the CSR value calculated for an acceleration a_i, and r_d—stress reduction factor (Cetin and Seed 2004) is given in Eq. 4.

$$\mathrm{CSR}_{\mathrm{eq},i} = 0.65\frac{a_i}{g}\frac{\sigma_{vo}}{\sigma'_{vo}}r_d \quad (4)$$

4.1 *Probabilistic Evaluation of Factor of Safety Against Liquefaction for Guwahati*

The assessment of liquefaction potential includes the quantification of soil strength and the ground acceleration due to earthquakes. This study considers both these aspects explicitly. The ground acceleration is modelled using a probabilistic seismic hazard analysis. Based on the PGA obtained from the PSHA analysis, liquefaction potential is estimated using shear wave velocity values. The V_{S30} profile for the Guwahati city was obtained from the global V_{S30} map provided by USGS. The liquefaction potential was assessed at a depth of 3 m and the depth of water table was assumed as 1 m from the ground surface (Sharma and Rahman 2016). The spatial variation of V_{S30} values for Guwahati is shown in Fig. 3. The V_{S30} values are lower at western part of the city and relatively high along the eastern part of the city. From the V_{S30} values obtained from USGS, the shear wave velocity values (V_S) at 3 m depth were obtained by using the method suggested by Boore et al. (2011).

5 Results and Discussions

The liquefaction potential is expressed as the FS against liquefaction for a given return period at 3 m depth. In the present study, the return periods considered are 10% probability of exceedance in 50 years and 2% probability of exceedance in 50 years. The above probabilities correspond to return period of 475 and 2500 years, respectively. The PSHA analysis is done using linear seismic sources and Fig. 4

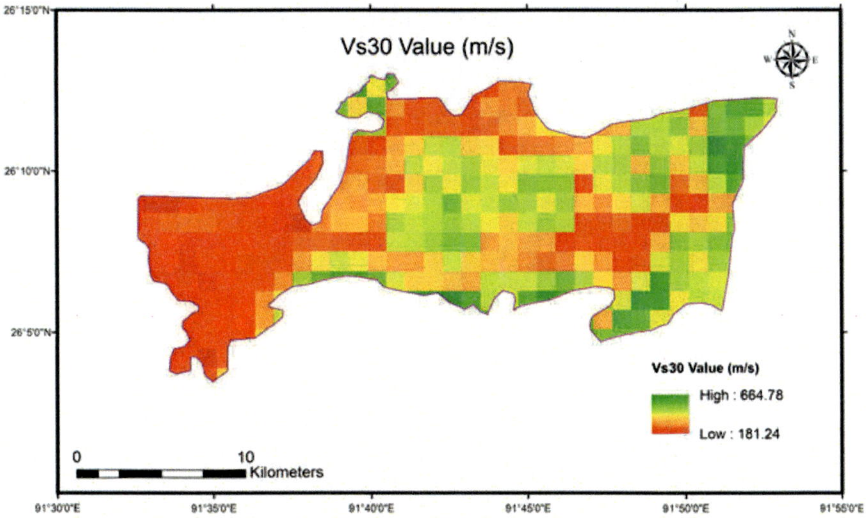

Fig. 3 Spatial variation of V_{S30} values for Guwahati

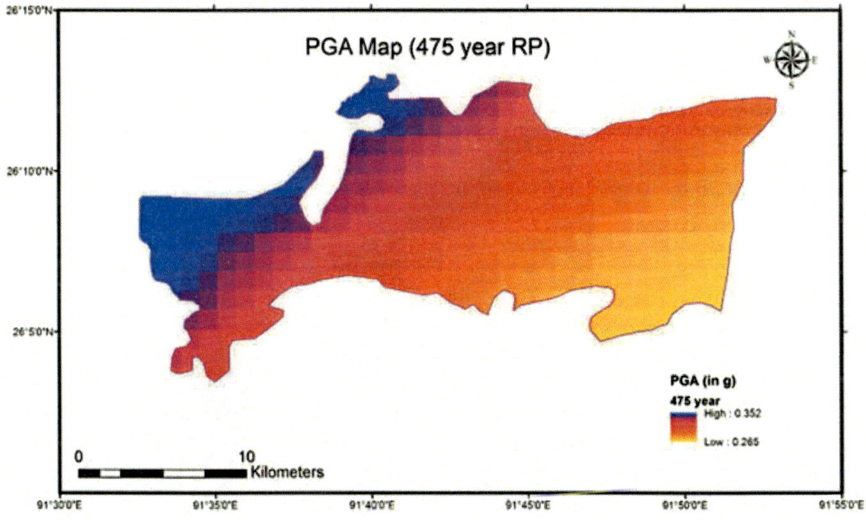

Fig. 4 Peak ground acceleration (in g) at bed rock level for a return period of 475 years

shows the seismic hazard map for a return period of 475 years. The seismic hazard is high for the entire study area and the highest values are observed in the north western part of the city. It can be seen that the PGA values obtained in this study closely match with the design ground acceleration provided by IS-1893 (2016). PGA values were evaluated for a return period of 500 years as well and the spatial variation pattern of PGA values is similar to the one shown in Fig. 3. Using the deaggregated seismic hazard obtained from the PSHA study as an input, the probabilistic liquefaction potential assessment was done. The factor of safety against liquefaction for return periods of 475 and 2500 years are calculated and the same are shown in Figs. 5 and 6, respectively. The general design criteria for non-critical structures are based on 475 year return period. It can be seen that for this return period, except for some regions in the eastern part of the city, majority of the city has a factor of safety against

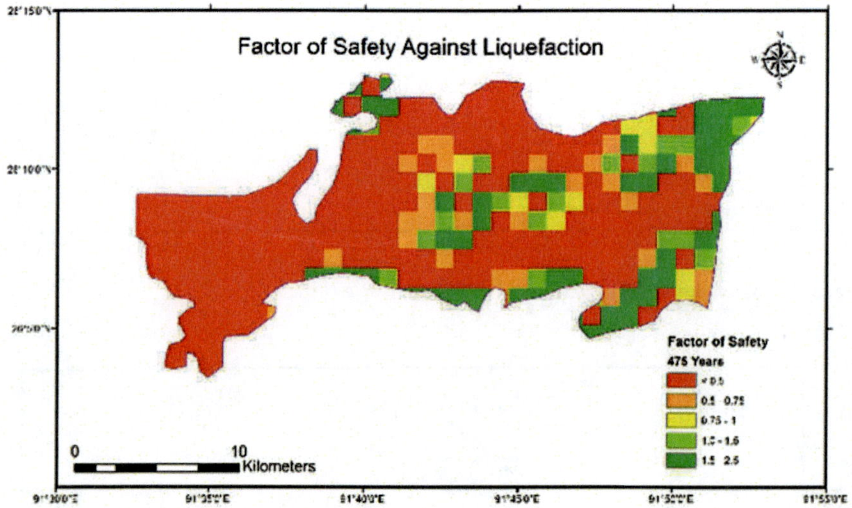

Fig. 5 Factor of safety against liquefaction for a return period of 475 years at 3 m depth

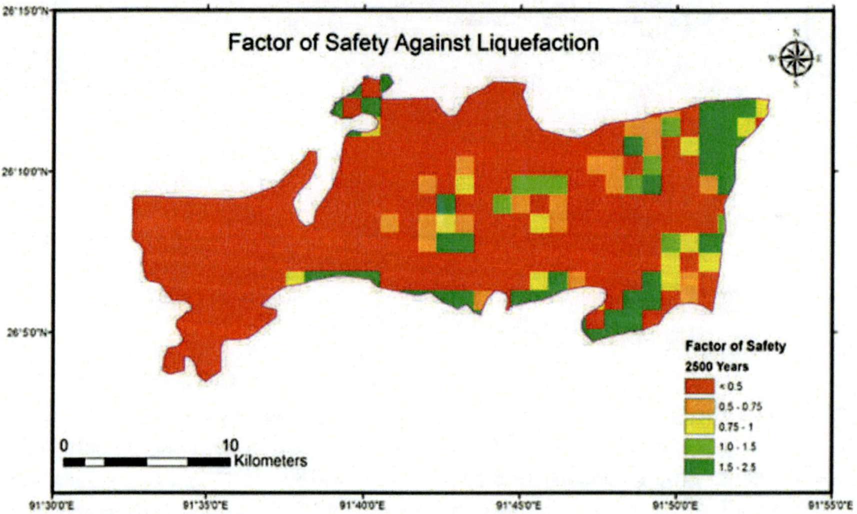

Fig. 6 Factor of safety against liquefaction for a return period of 2500 years at 3 m depth

liquefaction less than 1. These regions include the international airport situated at Guwahati which is the only one in the entire North eastern part of India.

6 Conclusions

The liquefaction potential of Guwahati is assessed as a function of factor of safety against liquefaction for given return periods. The analysis was done using a fully probabilistic methodology which considers all the uncertainties in earthquake loading and soil strength. The main advantage of this study is that the factor of safety against liquefaction can be calculated for any given return period. The study indicates that except the eastern part, majority of the city is susceptible to liquefaction. The reported high seismic activity and proximity to waterbody of the city call for adequate measures for mitigating liquefaction hazard. Even though the study is done based on soil details available on a global scale, it can be used as a first-order map in identifying regions of high liquefaction potential. In-depth studies need to be taken up in these regions for a more precise liquefaction potential assessment.

References

Ayothiraman R, Raghu Kanth STG, Sreelatha S (2012) Evaluation of liquefaction potential of Guwahati: gateway city to Northeastern India. Nat Hazards 63(2):449–460

Boore DM, Atkinson GM (2008) Ground-motion prediction equations for the average horizontal component of PGA, PGV and 5%-damped PSA at spectral periods between 0.01 s and 10.0 s. Earthquake Spectra 24(1):99–138

Boore DM, Thompson EM, Cadet H (2011) Regional correlations of V_{S30} and velocities averaged over depths less than and greater than 30 m. Bull Seismol Soc Am 101(6):3046–3059

Cetin KO et al (2004) Standard Penetration Test-based probabilistic and deterministic assessment of seismic soil liquefaction potential. J Geotech Geoenviron Eng 130(12):1314–1340

Das R, Sharma ML, Wason HR (2016) Probabilistic seismic hazard assessment for NorthEast India Region. Pure Appl Geophys 173(8):2653–2670

Das S, Gupta ID, Gupta VK (2006) A probabilistic seismic hazard analysis of Northeast India. Earthquake Spectra 22(1):1–27

Das UK, Saikia BD (2010) Shear strength of unsaturated residual soils of the hills in Guwahati. In: Proceedings of GEOTrendz—Indian geotechnical conference, pp 679–682, Mumbai

Juang CH, Jiang T (2000) Assessing probabilistic methods for liquefaction potential evaluation. Soil dynamics and liquefaction, Geotechnical Special Publication 107, pp 148–162

Kayen R et al (2013) Shear-wave-velocity based probabilistic and deterministic assessment of seismic soil liquefaction potential. J Geotech Geoenviron Eng 139(3):407–419

Kolathayar S, Sitharam TG, Vipin KS (2012) Deterministic seismic hazard macrozonation of India. J Earth Syst Sci 121(5):1351–1364

Kramer SL, Mayfield RT (2007) Return period of soil liquefaction. J Geotech Geoenviron Eng 133(7):802–813

Moss RE, Seed RB, Kayen RE, Stewart JP, Kiureghian AD (2006) CPT based probabilistic and deterministic assessment of in situ seismic soil liquefaction potential. J Geotech Geoenviron Eng 132(8):1032–1051

Nath SK, Thingbaijam KKS, Raj A (2008) Earthquake hazard in Northeast India—a seismic microzonation approach with typical case studies from Sikkim Himalaya and Guwahati city. J Earth Syst Sci 117(2):809–831

Raghu Kanth STG, Dash SK (2010) Evaluation of seismic soil-liquefaction at Guwahati city. Environ Earth Sci 61(2):355–368

Raghu Kanth STG, Sreelatha S, Dash SK (2018) Ground motion estimation at Guwahati city for a Mw 8.1 earthquake in the Shillong plateau. Tectonophysics 448(1–4):98–114

Seed HB, Idriss IM (1971) Simplified procedure for evaluating soil liquefaction potential. J Soil Mech Found Div, ASCE, 97(8):1249–1274

Sharma B, Begum N (2018) Probabilistic assessment of liquefaction potential of Guwahati City. In: Proceedings of sustainable civil infrastructures: innovative infrastructure technology, GeoMEast, pp 35–45

Sharma B, Hazarika P (2013) Assessment of liquefaction potential of Guwahati city. A case study. Geotech Geol Eng 31(5):1437–1452

Sharma B, Rahman SK (2016) Use of GIS based maps for preliminary assessment of subsoil of Guwahati city. J Geosci Environ Protect 4:106–116

Vipin KS, Sitharam TG (2012) A performance based framework for assessing liquefaction potential based on CPT data. Georisk 6(3):177–187

Vipin KS, Sitharam TG, Anbazhagan P (2010) Probabilistic evaluation of seismic soil liquefaction potential based on SPT data. Nat Hazards 53(3):547–560

Generalized Solution for the Critical Soil Wedge Angle Under Seismic Passive Earth Pressure Condition

Priyam Chatterjee, Bikash Chandra Chattopadhyay, and Joyanta Maity

Abstract In this paper, a generalized solution for the critical soil wedge angle under seismic passive earth pressure condition on the retaining wall is presented with plane failure surface using pseudo-static method. The development of an explicit expression for the critical soil wedge angle is presented in this paper using analytical method. The effects of different parameters like height of wall, slope of backfill, inclination of the back of wall, adhesion between the wall back face and soil backfill, properties of the backfill, and seismic coefficients on the critical soil wedge angle on the chosen wall are studied. The results from this method are compared with standard theoretical methods and show encouraging agreement.

Keywords Earthquake · Critical soil wedge angle · Seismic earth pressure coefficients · Retaining wall · Analytical solution

1 Introduction

Estimation of passive earth pressure is an important topic of research and more so under seismic conditions. This estimation is required for earth retaining structures. In static condition, different theories to compute passive earth pressure are available. But there are few theories available for seismic condition. Among the theories available till date for the estimation of seismic earth pressure, the Mononobe–Okabe method (1926), which is the pioneering work in this field, is commonly used. Although numerous investigators have determined the active earth pressure coefficients under seismic conditions, such as Prakash and Saran (1966), Saran and Prakash (1968), Choudhury and Nimbalkar (2006), Puri and Prakash (2011), John

P. Chatterjee (✉) · B. C. Chattopadhyay · J. Maity
Meghnad Saha Institute of Technology, Kolkata, India
e-mail: pc.bubai@gmail.com

B. C. Chattopadhyay
e-mail: ccbikash@yahoo.com

J. Maity
e-mail: joymaity1975@yahoo.co.in

M. Latha Gali and P. Raghuveer Rao (eds.), *Geohazards*, Lecture Notes in Civil Engineering 86, https://doi.org/10.1007/978-981-15-6233-4_21

et al. (2014), Rahaman and Raychowdhury (2015), Gupta and Chandaluri (2016), very few studies were conducted for seismic passive condition. Some of the studies conducted under seismic passive condition are Morrison and Ebeling (1995), Soubra (2000) and Kumar (2001), Choudhury (2004), Shafiee et al. (2010), Choudhury and Nimbalkar (2005), Subba Rao and Choudhury (2005), Mondal et al. (2011), Shukla (2013).

However, there are very few studies that describe how to determine the critical soil wedge angle and what are the effects of other parameters on the critical soil wedge angle. Therefore, an attempt is made to present the derivation of an explicit analytical expression for the critical soil wedge angle under seismic passive earth pressure condition from c-φ soil backfills for a more generalized field situation of sloping backfill, considering friction and adhesion at the wall-backfill interface along with both the horizontal and vertical seismic inertial forces.

2 Methodology

Using the generalized analytical solution, the influence of various key factors, such as wall inclination α, backfill inclination β, backfill properties c and Ø, friction and adhesion at the wall-backfill interface, horizontal and vertical seismic coefficients k_h and k_v on critical soil wedge angle are studied. Configuration of the retaining wall and force polygon are shown in Figs. 1 and 2, respectively. Solutions are made for Ø varying from 20° to 40°, inclination of backfill from 0° to 10°, and that of back of wall from 0° to 10° for various k_h and k_v and wall friction angle. The angle of

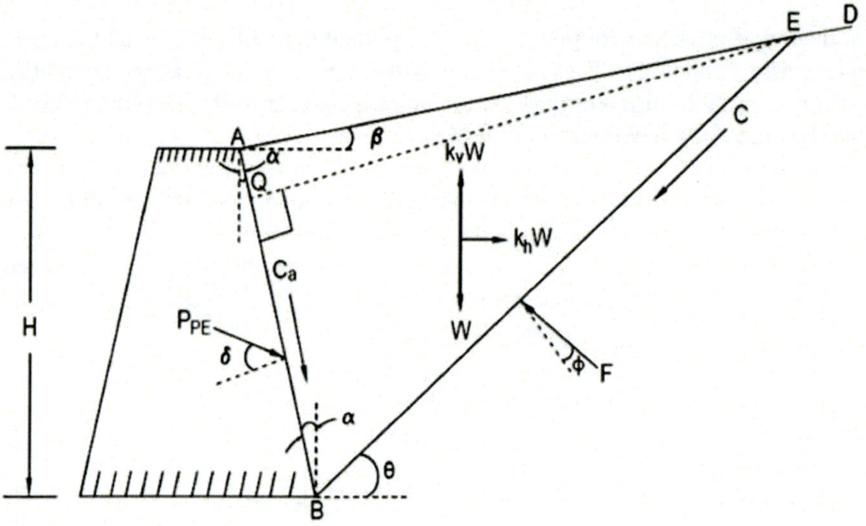

Fig. 1 Configuration of the retaining wall

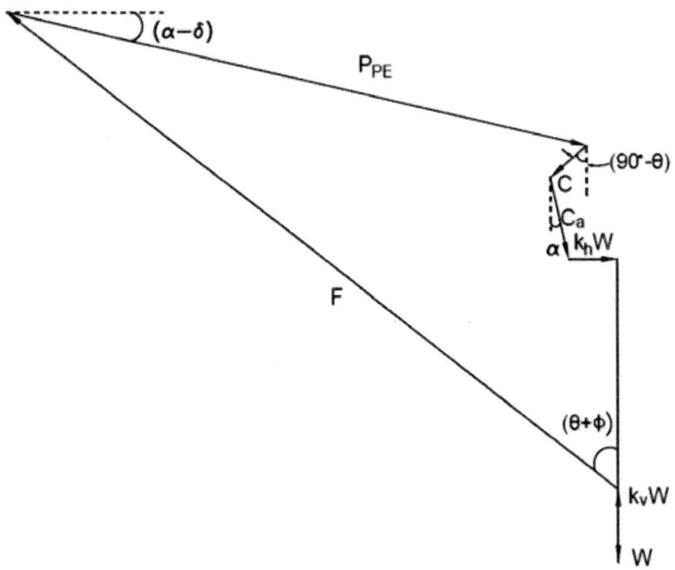

Fig. 2 Force polygon

shearing resistance of cohesion-less soil varies with dry density of the soil. As Ø was varied from 20° to 40°, corresponding values of dry density were selected from standard code as per Swiss Standard SN 670 010B.

3 Analysis

Considering a retaining wall of height *H* with back face *AB*, at an inclination α with vertical is considered (Fig. 1), where the backfill is sloped with horizontal at an inclination β and having unit weight γ and shear strength parameters *c* and φ. Angle of friction with the wall and backfill is δ. Unit adhesion between the soil and the back of the wall is c_a. The forces and force polygon are shown in Fig. 2. Here, for a trial case, it is assumed that the failure plane BE passes through the bottom of the wall and is inclined at an angle θ to the horizontal and cuts the backfill surface at *E*. Weight of the wedge ABE is W and the earthquake force $k_h W$ and $k_v W$ are acting in the shown direction during earthquake, where k_h and k_v are seismic coefficients in horizontal and vertical directions, respectively. Resisting forces along the failure surface *BE*, are *F* and *C*. Where *F* is acting at an angle φ with normal to the surface *BE* and cohesive force *C* (=*cH*, *c* being unit cohesion) acts along *BE*. Resisting adhesive force along the back of the wall *AB* is C_a, where adhesive force C_a (=$c_a \cdot$ AB, c_a being unit adhesion between the wall back face and the soil backfill) is acting along

AB. Total seismic passive earth pressure PPE acts at an angle δ to the normal to the back of the wall *AB*. All the forces shown in the figure are expressed per unit running length of the wall.

The weight of the soil wedge ABE is calculated as

$$W = \frac{\gamma}{2} \cdot H^2 \sec^2 \alpha \frac{\cos(\theta - \alpha)\cos(\alpha - \beta)}{\sin(\theta - \beta)} \tag{1}$$

Cohesion force,

$$C = c \cdot H \sec \alpha \frac{\cos(\alpha - \beta)}{\sin(\theta - \beta)} \tag{2}$$

Adhesion between the back of the wall and soil,

$$C_a = c_a \cdot H \sec \alpha \tag{3}$$

Considering equilibrium of forces in the horizontal direction,

$$P_{\mathrm{PE}} \cos(\alpha - \delta) - C \cos \theta + C_a \sin \alpha + k_h W - F \sin(\theta + \emptyset) = 0 \tag{4}$$

Considering equilibrium of forces in the vertical direction,

$$P_{\mathrm{PE}} \sin(\alpha - \delta) - C \sin \theta - C_a \cos \alpha + F \cos(\theta + \emptyset) = W(1 - k_v) \tag{5}$$

Eliminating F from Eqs. (4) and (5) with the substitution of W, C, and C_a from Eqs. (1), (2), and (3), respectively,

$$P_{PE} = \frac{1}{\cos(\theta + A)}\left[I \cdot \frac{\cos(\theta - \alpha)}{\sin(\theta - \beta)}\{(1 - k_v)\sin(\theta + \varphi) - k_h \cos(\theta + \varphi)\} + \frac{E}{\sin(\theta - \beta)} + G \sin(\theta + D)\right] \tag{6}$$

where

$$A = (\varphi + \delta - \alpha) \tag{7}$$

$$D = (\varphi - \alpha) \tag{8}$$

$$E = c \cdot H \sec \alpha \cos(\alpha - \beta) \cos \varphi \tag{9}$$

$$G = c_a \cdot H \sec \alpha \tag{10}$$

$$I = \frac{\gamma}{2} \cdot H^2 \sec^2 \alpha \cos(\alpha - \beta) \tag{11}$$

For optimum value of P_{PE},
$\frac{\partial P_{\mathrm{PE}}}{\partial \theta} = 0.$
Or,

$$\begin{aligned}
&\Big[I \frac{(1-k_v)}{2\cos\psi} \{ \cos(\alpha+\beta)\cos(A+\psi-\varphi) - \cos(\alpha-\beta)\cos(A+\varphi-\psi)\} \\
&\quad - \frac{G}{2}\sin 2\beta \cos(A-D) + E\sin(A-\beta) \Big] \sin 2\theta \\
&- \Big[I \frac{(1-k_v)}{2\cos\psi} \{ \cos(\alpha-\beta)\sin(A+\varphi-\psi) + \sin(\alpha+\beta)\cos(A+\psi-\varphi)\} \\
&\quad + \frac{G}{2}\cos 2\beta \cos(A-D) + E\cos(A-\beta) \Big] \cos 2\theta \\
&+ \Big[I \frac{(1-k_v)}{2\cos\psi} \sin(A+\alpha+\psi-\beta-\varphi) + \frac{G}{2}\cos(A-D) \Big] = 0
\end{aligned} \tag{12}$$

where

$$\psi = \tan^{-1}\left(\frac{k_h}{1-k_v}\right) \tag{13}$$

ψ is seismic inertia angle.
Equation (12) can be further written as,

$$m_1 \sin 2\theta - m_2 \cos 2\theta + m_3 = 0 \tag{14}$$

where

$$\begin{aligned}
m_1 = I \frac{(1-k_v)}{2\cos\psi} &\{\cos(\alpha+\beta)\cos(A+\psi-\varphi) - \cos(\alpha-\beta)\cos(A+\varphi-\psi)\} \\
&- \frac{G}{2}\sin 2\beta \cos(A-D) + E\sin(A-\beta)
\end{aligned} \tag{15}$$

$$\begin{aligned}
m_2 = I \frac{(1-k_v)}{2\cos\psi} &\{\cos(\alpha-\beta)\sin(A+\varphi-\psi) + \sin(\alpha+\beta)\cos(A+\psi-\varphi)\} \\
&+ \frac{G}{2}\cos 2\beta \cos(A-D) + E\cos(A-\beta)
\end{aligned} \tag{16}$$

$$m_3 = I \frac{(1-k_v)}{2\cos\psi} \sin(A+\alpha+\psi-\beta-\varphi) + \frac{G}{2}\cos(A-D) \tag{17}$$

Equation (14) can be written thus as,

$$(m_2 + m_3)\tan^2\theta + 2m_1\tan\theta + (m_3 - m_2) = 0$$

Or,

$$\tan\theta = \frac{-2m_1 \pm \sqrt{4m_1^2 - 4(m_2 + m_3)(m_3 - m_2)}}{2(m_2 + m_3)}$$

Since θ_{critical} will lie between 0° and 90°,
$\tan\theta$ cannot be negative; therefore, '$+$' or '$-$' should be considered accordingly based on the specific values of m_1, m_2, and m_3 in view of the practical site sit.
Therefore, critical soil wedge angle, θ_{critical} or θ_c can be found as,

$$\theta_{\text{critical}} = \tan^{-1}\left[\frac{-m_1 \pm \sqrt{m_1^2 + m_2^2 - m_3^2}}{m_2 + m_3}\right]. \tag{18}$$

For the real values of θ_{critical}, the expression under the radical sign in Eq. 18 must be positive and the denominator must not be zero, that is,

$$\left(m_1^2 + m_2^2 - m_3^2\right) \geq 0 \text{ and } (m_2 + m_3) \neq 0$$

Equation (18) provides a general expression for the critical soil wedge angle under seismic passive earth pressure condition.

4 Results and Discussions

Critical soil wedge angles under seismic passive earth pressure condition, evaluated from analytical solution, are presented for different range of values of the variables. The values were obtained in a sequential manner in which height of the wall (H) was initially chosen as 4 m. For this height, inclination of the back of wall (α) was chosen as 0°, inclination of the backfill was taken as 0°, and earth pressure was determined for $k_v = 0$ where k_h was varied from 0.05 to 0.2. The process was repeated for other values of k_v, i.e., $k_v = 0.0, 0.05$ and 0.1, respectively. Keeping α constant, β was varied from 5° and 10°. This procedure was thus repeated for $\alpha = 5°$ and 10°. By this process, θ_c for wall height 4 m was evaluated. Similar process was repeated for wall height 6 m.

4.1 Effects of Different Parameters on Critical Soil Wedge Angle

Effects of different parameters on critical soil wedge angle are described sequentially below:

Effect of Angle of Shearing Resistance (Ø) on Critical Soil Wedge Angle. Changes in values of critical soil wedge angle for wall height 4 m and 6 m are presented in Figs. 3 and 4, respectively, for $\alpha = 0°$, $\beta = 0°$, and $\delta = 0°$ against the variation of angle of shearing resistance (Ø) of the backfill.

For both cases, k_v was kept constant and equal to nil and c was kept constant at 2 KN/m^2. It is seen that θ_c decreases with increase in φ value of any chosen value of k_h for any height of wall. However, for higher values of k_h resulting critical wedge angle is lower. For example, for $\varphi = 30°$, when $k_h = 0.05$, θ_c for wall height 4 m is 29.3°, while for $k_h = 0.2$, the value reduces to 26.8°. Similar observations are seen when α and β changes from 5° and 10°.

Fig. 3 Variation of θ_c for wall height 4 m, $\alpha = 0°$, $\delta = 0°$, $\beta = 0°$ for varying Ø value and k_h value

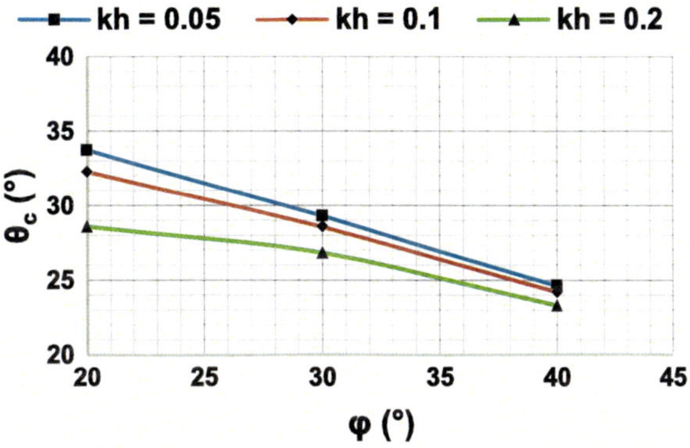

Fig. 4 Variation of θ_c for wall height 6 m, $\alpha = 0°$, $\delta = 0°$, $\beta = 0°$ for varying Ø value and k_h value

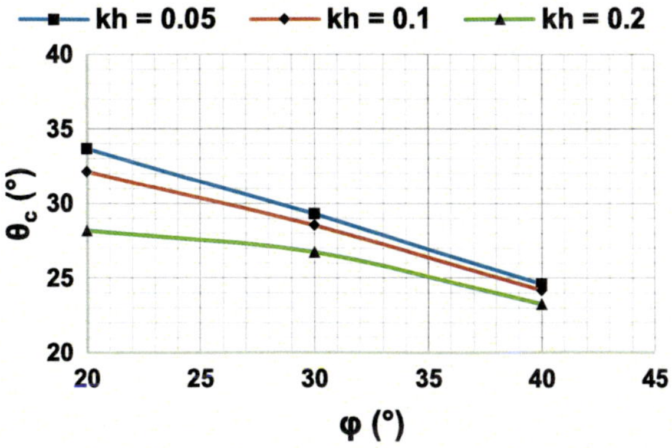

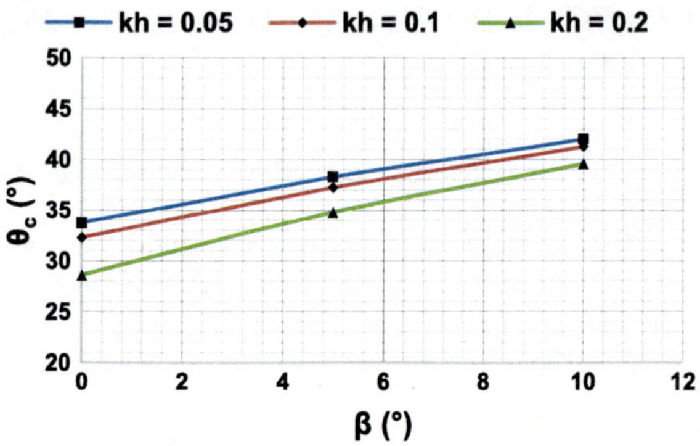

Fig. 5 Variation of θ_c for height = 4 m, $\alpha = 0°$, $\delta = 0°$, $Ø = 20°$

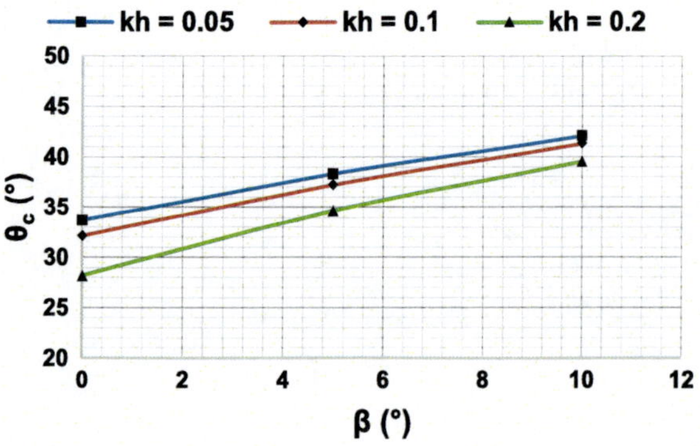

Fig. 6 Variation of θ_c for height = 6 m, $\alpha = 0°$, $\delta = 0°$, $Ø = 20°$

Effect of Inclination of Backfill (β) on Critical Soil Wedge Angle. Changes in value of critical soil wedge angle for wall height 4 m and 6 m are presented in Figs. 5 and 6, respectively, for $\alpha = 0°$, $\delta = 0°$, and $Ø = 20°$ against the variation of angle of inclination (β) of the backfill. It is seen that increase in β results in increase in θ_c. Similarly, as k_h increases, θ_c decreases. Similar observations are seen for height = 6 m and for $Ø = 30°$ and 40°.

Effect of Inclination of Back of Wall (α) on Critical Soil Wedge Angle. Changes in value of critical soil wedge angle for wall height 4 m and 6 m are presented in Figs. 7 and 8, respectively, for $\beta = 0°$, $\delta = 0°$, and $Ø = 20°$ against the variation of inclination of back of wall (α) of the backfill.

It is seen that increase in α results in increase in θ_c. Similarly, as k_h increases, θ_c decreases. Similar observations are seen for height = 6 m and for $Ø = 30°$ and 40°.

Effect of Inclination of Wall Friction (δ) on Critical Soil Wedge Angle. Changes in value of critical soil wedge angle for wall height 4 m and 6 m are presented in Figs. 9 and 10, respectively, for $\beta = 0°$, $\delta = 0°$, and $Ø = 20°$ against the variation of angle of wall friction (δ) of the backfill. It is seen that, as δ increases, θ_c decreases. As k_h increases, θ_c decreases. Similar observations are seen for height = 6 m and for change in Ø from 30° and 40°.

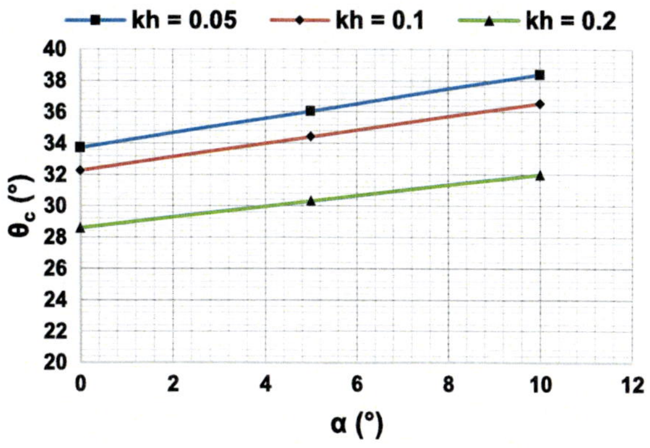

Fig. 7 Variation of θ_c for height = 4 m, $\beta = 0°$, $\delta = 0°$, $Ø = 20°$ for varying α value and k_h value

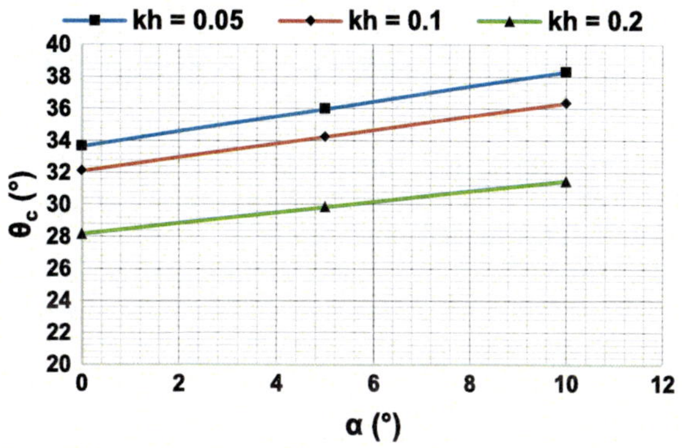

Fig. 8 Variation of θ_c for height = 6 m, $\beta = 0°$, $\delta = 0°$, $Ø = 20°$ for varying α value and k_h value

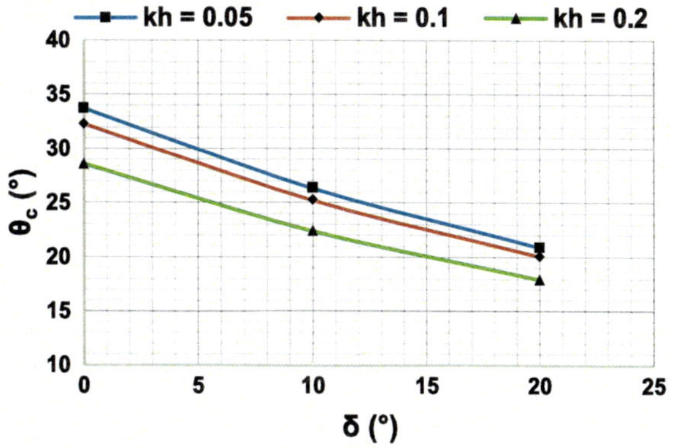

Fig. 9 Variation of θ_c for height = 4 m, $\beta = 0°$, $\delta = 0°$, $Ø = 20°$ for varying δ value and k_h value

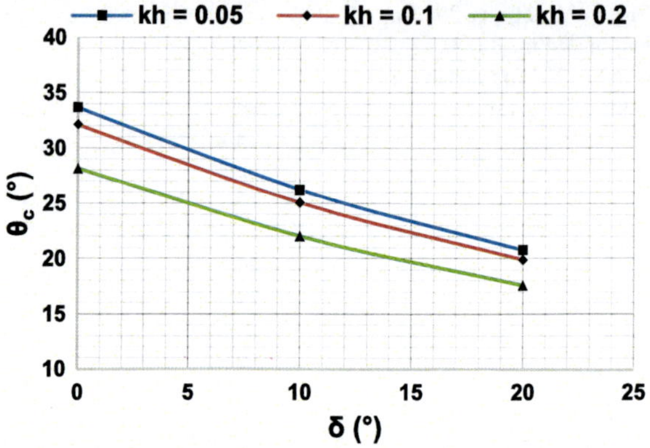

Fig. 10 Variation of θ_c for height = 6 m, $\beta = 0°$, $\delta = 0°$, $Ø = 20°$ for varying δ value and k_h value

5 Comparison of Critical Soil Wedge Angle from Present Study and Mondal et al. Solution (2011)

To check the validity of the results presented in this paper, comparisons are made between the result obtained from present analysis and pseudo-static approach presented by Mandal et al. (2011).

For the purpose of comparison, the critical soil wedge angle obtained from Mandal et al. solution (2011) and present analytical solution for different values of φ is presented in Table 1.

Table 1 Critical soil wedge angle from theoretical solution (Mandal et al. 2011) and present analytical solution for $\alpha = 0°$, $\delta = 2/3\varphi$, $\gamma = 15$ KN/m^2

φ (°)	β (°)	$k_h = k_v$	θ_c	
			Theoretical study (Mandal et al.) (°)	Present study (°)
20	5	0.1	31.957	27.49336
		0.2	35.090	24.41892
	10	0.1	35.157	31.67136
		0.2	37.775	29.72647
30	5	0.1	23.084	20.8679
		0.2	24.968	19.91277
	10	0.1	25.839	24.12098
		0.2	27.378	23.52012
40	5	0.1	14.630	13.67489
		0.2	15.593	13.46051
	10	0.1	17.102	16.45831
		0.2	17.781	16.35664

For graphical comparison, the critical soil wedge angle obtained from Mandal et al. solution (2011) and present analytical solution for different values of φ is shown by Figs. 11, 12, 13, and 14.

From Fig. 11, it is seen that, when $\beta = 5°$ and $k_h = k_v = 0.1$, for $\varphi = 40°$, the result is almost similar but for $\varphi = 20°$ and $\varphi = 30°$, the result from the present study gives slightly lesser values of critical soil wedge angle than the theoretical solution (Mandal et al. 2011).

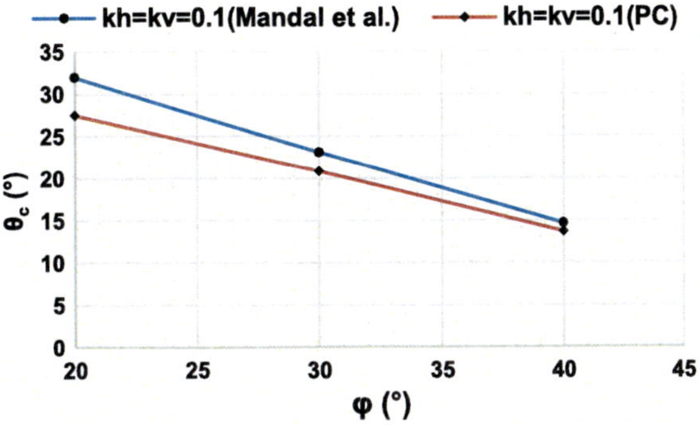

Fig. 11 Variation of θ_c (Theoretical (Mandal et al. 2011) and Present (PC)) with varying φ for $\beta = 5°$, $k_h = k_v = 0.1$

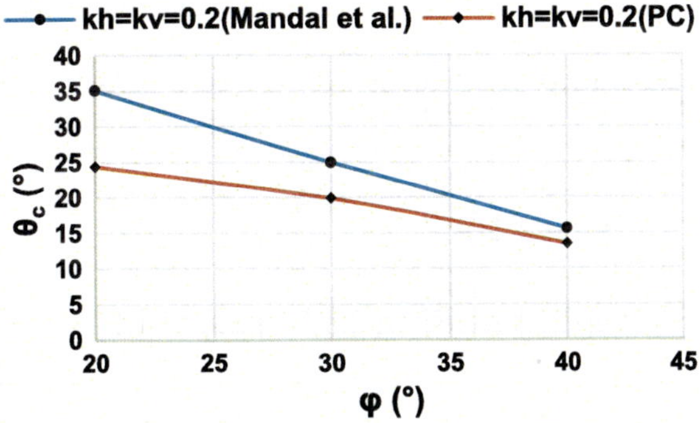

Fig. 12 Variation of θ_c (Theoretical (Mandal et al. 2011) and Present (PC)) with varying φ for $\beta = 5°$, $k_h = k_v = 0.2$

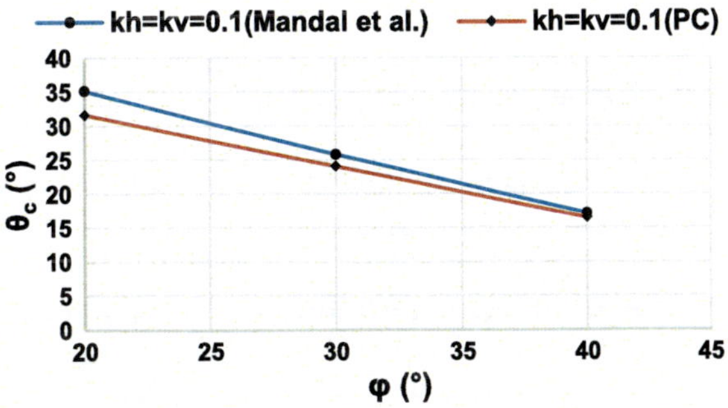

Fig. 13 Variation of θ_c (Theoretical (Mandal et al. 2011) and Present (PC)) with varying φ for $\beta = 10°$, $k_h = k_v = 0.1$

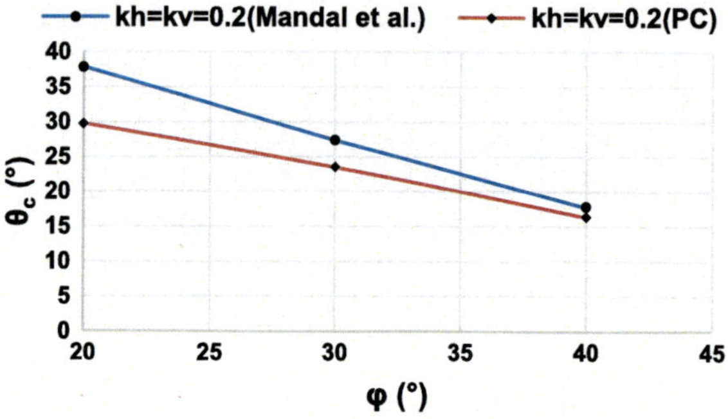

Fig. 14 Variation of θ_c (Theoretical (Mandal et al. 2011) and Present (PC)) with varying φ for $\beta = 10°$, $k_h = k_v = 0.2$

From Fig. 12, it is seen that, when $\beta = 5°$ and $k_h = k_v = 0.2$, for $\varphi = 20°$, the result from present study is much less than the theoretical study (Mandal et al. 2011) and for $\varphi = 30°$ and $\varphi = 40°$, the result from the present study gives slightly lesser values of critical soil wedge angle than the theoretical solution (Mandal et al. 2011).

From Fig. 13, it is seen that, when $\beta = 10°$ and $k_h = k_v = 0.1$, for $\varphi = 40°$, the result is almost similar but for $\varphi = 20°$ and $\varphi = 30°$, the result from the present study gives slightly lesser values of critical soil wedge angle than the theoretical solution (Mandal et al. 2011).

From Fig. 14, it is seen that, when $\beta = 10°$ and $k_h = k_v = 0.2$, for $\varphi = 20°$, the result from present study is much less than the theoretical study (Mandal et al. 2011) and for $\varphi = 30°$ and $\varphi = 40°$, the result from the present study gives slightly lesser values of critical soil wedge angle than the theoretical solution (Mandal et al. 2011).

6 Conclusions

On the basis of this study, the following conclusions can be made-

- An explicit expression [Eq. 18] is derived for the critical soil wedge angle under seismic passive earth pressure condition from the sloping *c*-φ backfills at the back of an inclined rigid retaining walls considering wall–soil friction and adhesion.
- Critical soil wedge angle under seismic passive earth pressure condition is very much dependent on the value of seismic coefficients, inclination of the back of the wall, inclination of backfill, properties of backfill and seismic coefficient due to earthquake.
- For any value of k_h, the critical soil wedge angle decreases with increase in φ value, but for any φ value, critical soil wedge angle decreases with increase of k_h. This is true for any value of α or β or any height of wall.
- When α changes from 0° to any higher value, other parameters remaining same, and critical soil wedge angle always increases.
- Similarly, when the slope of backfill surface, β increases from 0° to a higher value, other parameters remaining same; critical soil wedge angle always increases.

- When angle of wall friction (δ) increases from 0° to a higher value, critical soil wedge angle decreases. This is true for any value of k_h or k_v or height of wall.
- The critical soil wedge angle under seismic passive earth pressure condition from the analytical solution shows a remarkable agreement both qualitatively and quantitatively with results from pseudo-static method presented by Mandal et al. (2011).

References

Choudhury D (2004) Seismic passive resistance at soil-wall interface. In: 13th World conference on earthquake engineering, Vancouver, B.C., Canada, 1–6 Aug 2004, Paper No. 2746

Choudhury D, Nimbalkar S (2005) Seismic passive resistance by pseudo-dynamic method. Ge'otechnique 55(9):699–702

Choudhury D, Nimbalkar S (2006) Pseudo-dynamic approach of seismic active earth pressure behind retaining wall. Geotech Geol Eng 24:1103–1113

Gupta A, Chandaluri VK (2016) Development of design charts for the dynamic active thrust from c-φ soil backfills. In: Indian geotechnical conference, 15–17 December 2016, IIT Madras, Chennai, India

John R, Preethakumari K, Sethi P (2014) Seismic active earth pressure behind retaining wall. Int J Sci Eng Res 5(12)

Kumar J (2001) Seismic passive earth pressure coefficients for sands. Can Geotech J 38:876–881

Mandal T, Jadav R, Bhagat B, Deshmukh VB (2011) Behavior of retaining wall under static and dynamic earth pressure. In: Proceedings of Indian geotechnical conference, 15–17 Dec 2011, Kochi (Paper No. K-321)

Morison EE, Ebeng RM (1995) Limit equilibrium computation of dynamic passive earth pressure. Can Geotech J 32:481–487

Okabe S (1926) General theory of earth pressure. J Jpn Soc Civ Eng 12(1)

Prakash S, Saran S (1966) Static and dynamic earth pressures behind retaining walls. In: Proceedings of the 3rd symposium on earthquake engineering, University of Roorkee, vol 1, pp 277–288

Puri VK, Prakash S (2011) Dynamic earth pressure against retaining walls. In: Proceedings of Indian geotechnical conference 15–17 Dec 2011, Kochi

Rahaman O, Raychowdhury P (2015) Seismic active earth pressure on retaining walls considering soil amplification. In: Indian geotechnical conference, 17-19 Dec 2015, Pune, Maharastra, India

Saran S, Prakash S (1968) Dimensionless parameters for static and dynamic earth pressures behind retaining walls. J Indian Natl Soc Soil Mech Found Eng, July, 295–310

Shafiee AH, Eskandarinajed A, Jahandish M (2010) Seismic passive earth thrust on retaining walls with cohesive backfills using pseudo-dynamic approach. Can Geotech J 28:528–535

Shukla SK (2013) Seismic Passive Earth Pressure from the Sloping c-φ Soil Backfills, Indian Geotechnical Society 2013. Indian Geotech J. https://doi.org/10.1007/s40098-013-0045-7

Soubra AH (2000) Static and seismic passive earth pressure coefficients on rigid retaining structures. Can Geotech J 37:463–478

Subba Rao KS, Choudhury D (2005) Seismic passive earth pressures in soils. J Geotech Geoenviron Eng (ASCE) 131(1):131–135. https://doi.org/10.1061/(ASCE)1090-0241(2005)131:1(131)

SWISS STANDARD SN 670 010B [characteristic Coefficients of Soils, Association of Swiss Road and Traffic Engineers]

Dynamic Slope Stability Analysis of Ash Dykes

P. Hari Prasad, K. Vineeth Reddy, and G. Kalyan Kumar

Abstract Ash, a by-product obtained from the ignition of coal, is predominantly produced from thermal power plants based on coal. Disposal of the ash in land can be minimized by using the ash in various geotechnical, structural, and transportation engineering works. When ash is to be used in these works, properties of the pond ash under dynamic conditions must be ascertained to find its liquefaction potential under cyclic loading. The cyclic loading procedures to determine dynamic properties of ash commonly used in laboratory are simple shear, cyclic triaxial, and resonant column. In the current study, a series of strain-controlled tests were performed at various relative densities and effective confining pressures in cyclic-traxial apparatus to find the liquefaction potential of pond ash in the dyke. Finite element analysis was carried out using the GeoStudio software to find the factor of safety against slope stability. The factor-of-safety values were found to be more than 1.5 under both static and dynamic loading conditions. The bearing capacity of soil at first stage of ash dyke loading and pond ash at higher stages was found to be adequate.

Keywords Cyclic triaxial test · Strain-controlled testing · Pore water pressure ratio · GeoStudio

1 Introduction

In India, since 1920s, wide-scale coal burning has begun for power generation. This resulted in million tons of ash and related by-products. Current worldwide production was estimated around 600 million tonnes all-around the world and its disposal had been a serious environmental problem (Ahmaruzzaman 2010). The ash produced is

P. Hari Prasad
Keller Ground Engineering Pvt. Ltd, Chennai, India

K. Vineeth Reddy
Anurag Group of Institutions, Hyderabad, Telangana, India

G. Kalyan Kumar (✉)
NIT Warangal, Warangal, Telangana, India
e-mail: kalyan@nitw.ac.in

M. Latha Gali and P. Raghuveer Rao (eds.), *Geohazards*, Lecture Notes in Civil Engineering 86, https://doi.org/10.1007/978-981-15-6233-4_22

broadly classified into pond ash and fly ash. Due to rapid urbanization and constructions, availability of sand to use as a fill material is becoming less and is depleting day by day. Hence, a by-product of coal burning, i.e. ash (a non-plastic material) is being used as a replacement of sand. Mostly, loose and uniformly graded sand deposits are prone to liquefaction. Ash being a fine-grained material shows the similar behaviour and also its strength is less than that of sand. Research on liquefaction has begun from past 50 years by several researchers around the world (Casagrande 1965, 1976; Castro 1975; Seed et al., 1975; Finn et al. 1977; Seed 1979; Dobry et al. 1982; Seed et al. 1983; Ishihara 1996; Andrus and Stokoe 2000; Fourie et al. 2001; Olson and Stark 2003; Choudhary et al. 2010; Kokusho 2016). The modest way to mitigate liquefaction hazard for already existing structures is to densify the soil using some common ground improvement methods like deep compaction (dynamic) or vibro-compaction, etc. Other methods such as grouting, underpinning, etc., are being used due to some reasons like in a well-developed area or at confined (Rawlings et al. 2000; Karol 2003; Sawada and Kobayashi 2011).

In laboratories, liquefaction potential is found using the cyclic triaxial, simple shear, shake table, centrifuge tests, etc. (Park and Silver 1975), whereas in field, tests like standard penetration test (SPT) (Cetin et al. 2004), dynamic cone penetration test (DCPT) (Robertson and Campanella 1985), etc., are used (Kramer 1996; Marchetti 1982). Factor of safety against the slope instability is another important parameter to be considered when ash is utilized as a fill material for embankments (Hazen 1918). In the current study, pond ash was collected from Sri Damodaram Sanjeevaiah Thermal Power Station, Nelatur Village, Krishnapatnam, Nellore, Andhra Pradesh. This power plant is a coal-based power plant of Andhra Pradesh Power Development Company Limited (APPDCL). The capacity of proposed thermal power plant is 1600 MW and was in operation from 2014. On the basis of experimental investigation, both representative and undisturbed samples were collected from project site. Samples were found belonging to class F category as per ASTM C 618 and to be non-plastic depending upon classification and other laboratory tests. The project site lies in Zone III according to seismic hazard map of India from IS 1893-2000. The zone factor of study area is 0.16.

In the current study, an attempt was made to study the dynamic slope stability of ash dyke constructed with pond ash using the GeoStudio software. Strain-controlled testing procedure was followed using cyclic triaxial testing apparatus (Dobry et al. 1980).The dynamic soil properties of pond ash were determined under the following conditions: (i) different confining pressures and (ii) different strain rates (ASTM D5311M-13).

2 Methodologys

2.1 *Preliminary Tests*

Geotechnical properties of pond ash were determined according to IS code. The grain size distribution curve and geotechnical properties of pond ash were presented in Fig. 1 and Table 1, respectively. The pond ash was classified as poorly graded sand (SP) as per IS 2720. Minimum and maximum void ratios of pond ash were calculated

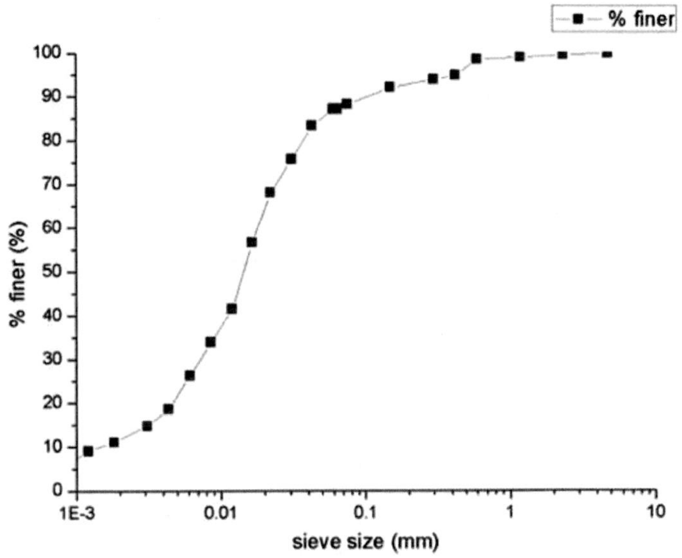

Fig. 1 Grain size distribution curve

Table 1 Geotechnical properties of pond ash

Property	Value
Specific gravity	2.11
e_{max}	1.05
e_{min}	0.68
D_{10}	0.0017 mm
D_{30}	0.0076 mm
D_{60}	0.0179 mm
Uniformity coefficient, C_u	10.53
Coefficient of curvature, C_c	0.95
Soil classification	SP
Cohesion C (kPa)	10
Angle of internal friction, Φ (°)	24
Coefficient of permeability, k (m/s)	3.85×10^{-7}

through relative density experiment. Shear strength parameters were found using direct shear test and were provided in Table 1. Permeability of pond ash was found using variable head permeability test. Properties of the pond ash under dynamic conditions were found by performing the cyclic triaxial tests.

2.2 *Cyclic Triaxial Testing*

The loading combination of the earthquakes is very complex and almost all the energy is conveyed in the form of radially propagating shear waves. This could be approximately reproduced in laboratory with cyclic triaxial experiment, the most commonly used apparatus to find the dynamic behaviour of soil. Widely used methods of sample preparation are (i) Dry pluviation, (ii) Water sedimentation, and (iii) Under compaction (Mulilis et al. 1975; Kumar et al. 2007). In this study, under compaction was adopted for reconstituting the pond ash specimens (Ladd 1978). The mould and set-up used in specimen preparation were shown in Fig. 2.

Specimens of 50 mm diameter and 100 mm length were reconstituted in laboratory using under compaction technique. After reconstituting the sample, it was allowed for saturation which was assumed to be achieved when Skempton's pore pressure parameter, B (=$\Delta u/\Delta\sigma$), reaches 0.95 or higher (Kenan and Ellen 2004). Sample was allowed for consolidation and then cyclic loading tests were performed to get the dynamic soil parameters. The liquefaction potential of pond ash was estimated from the tests (Seed and Idriss 1971; Ishihara 1977; Iwasaki et al. 1984). Bearing capacity was checked at different stages of ash dyke loading.

2.3 *Analytical Study*

GeoStudio is a modelling software for geo-engineers and earth scientists which is capable of rigorous analytical capability with very sophisticated product integration of different software like Slope/W, Seep/W, and Quake/W, etc. In the current study, finite element analysis was carried out using Slope/W of GeoStudio software.

Soil parameters found were used in analytical study to get the slope stability. Slope/W was used for determining the factor of safety of dyke against slope stability in static condition and Quake/W in dynamic condition. The soil profile was modelled to resemble the field condition which was consisting of silty sand, silty clay, sand, and clay. Water table was assumed to 2 m depth below the top layer of dyke.

Fig. 2 Mould and set-up used in specimen preparation

3 Results and Discussions

3.1 Cyclic Triaxial Tests

The results of cyclic triaxial tests carried on specimens subjected to 0.5% strain and 50 loading cycles were presented in Fig. 3. Decrease of shear strength of ash with more number of load cycles was observed. This decrease can be credited to rise in pore water pressure. Degradation of stiffness was observed from flattening of the hysteresis loop (Dobry et al. 1980; Law et al. 1990; Figueroa et al. 1994). Excess pore water pressures became almost equal to confining pressure beyond eight cycles and shear stress offered negligible. The pond ash sample is considered to be liquefied.

Effect of Effective Confining Pressure (Consolidation Stress). Three different consolidation stresses (50, 100, and 150 kPa) were investigated. The change in excess pore water pressure ratios with shear strain at various consolidation stresses for eight

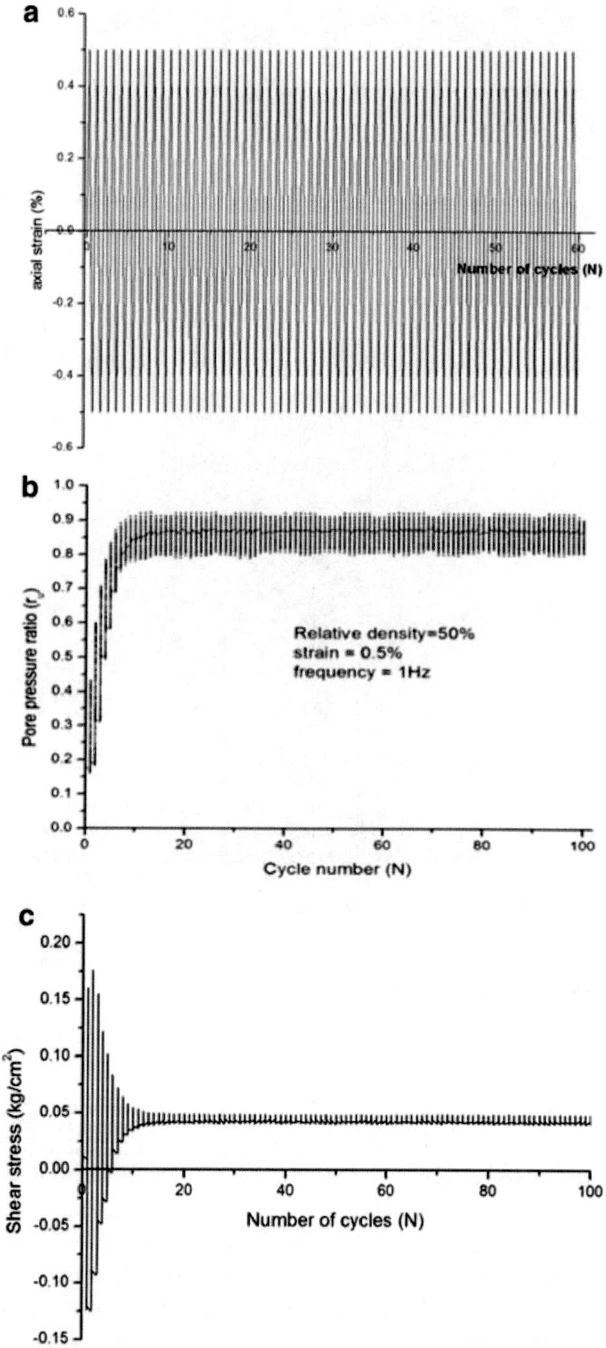

Fig. 3 Strain-controlled test results ($\gamma = 0.5\%, f = 1$ Hz, $D_r = 50\%$ and $\sigma_c{}' = 100$ kPa). **a** Shear strain versus number of load cycles. **b** Pore pressure water ratio versus number of load cycles. **c** Shear stress versus number of load cycles. **d** Shear stress versus shear strain

Fig. 3 (continued)

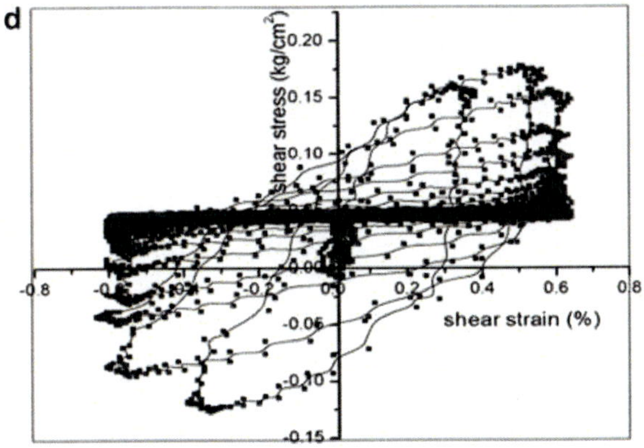

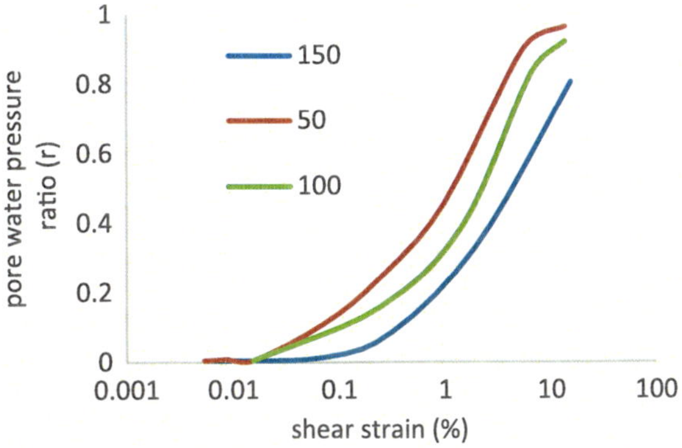

Fig. 4 Change in excess pore water pressure ratio versus shear strain (%) at various confining pressures

loading cycles was presented in Fig. 4. Consolidation stress increase by 50 kPa resulted a deviation of about 8% in excess pore water pressure ratio. At 0.2% shear strain, the pore pressure ratios for specimens under 50, 100, and 150 kPa were 0.45, 0.38, and 0.30 (about 8% difference), respectively.

Effect of Shear Strain and Number of Load Cycles. At lower strains ($\Upsilon_c <$ 0.03%), pore pressure difference between the number of cycles is smaller and with strain level, pore pressures generated also increased (Figs. 5, 6 and 7). The influence region shifted towards higher shear strains with increase in effective confining pressures inferring that the number of loading cycles is more significant on the development of excess pore pressures.

3.2 Slope Stability of Ash Dykes

Slope stability analysis of the ash dyke at the first and third (final) stage of construction was carried out by using GeoStudio software. The results of stability analysis of the

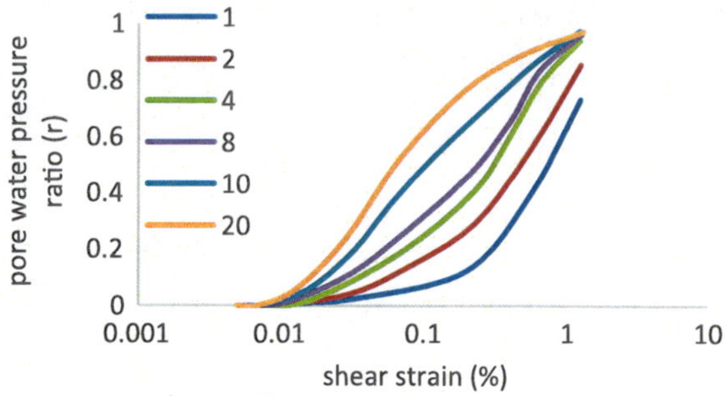

Fig. 5 Change in pore water pressure ratio versus cyclic shear strain at $\sigma_c' = 50$ kPa

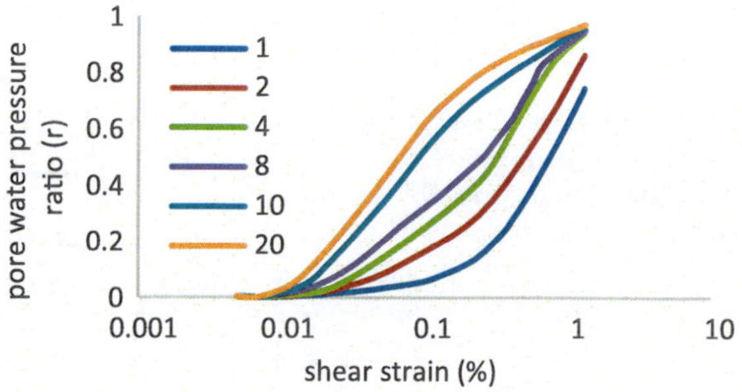

Fig. 6 Change in pore water pressure ratio versus cyclic shear strain at $\sigma_c' = 100$ kPa

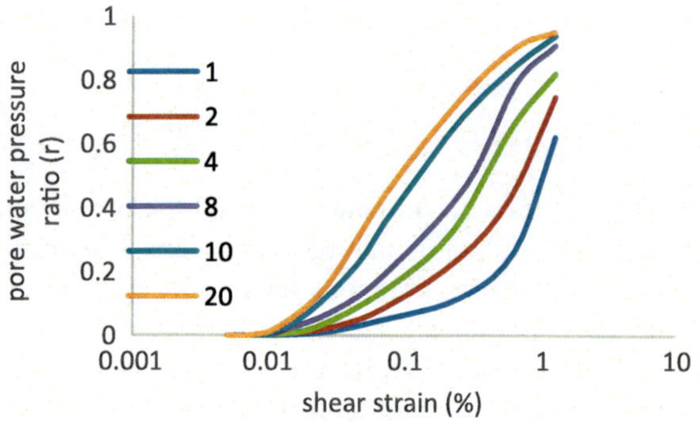

Fig. 7 Change in pore water pressure ratio versus cyclic shear strain at $\sigma_c' = 150$ kPa

ash dyke were presented in Figs. 8 and 9, respectively. The critical factor of safety values were obtained as 2.57 and 1.538 for the first and third stages, respectively, in static condition and in dynamic analysis, factor of safety is 1.53.

3.3 *Bearing Capacity*

Bearing capacity was calculated at ground level and also at the base of third stage of ash dyke loading. The bearing capacity requirement was satisfied with a factor of

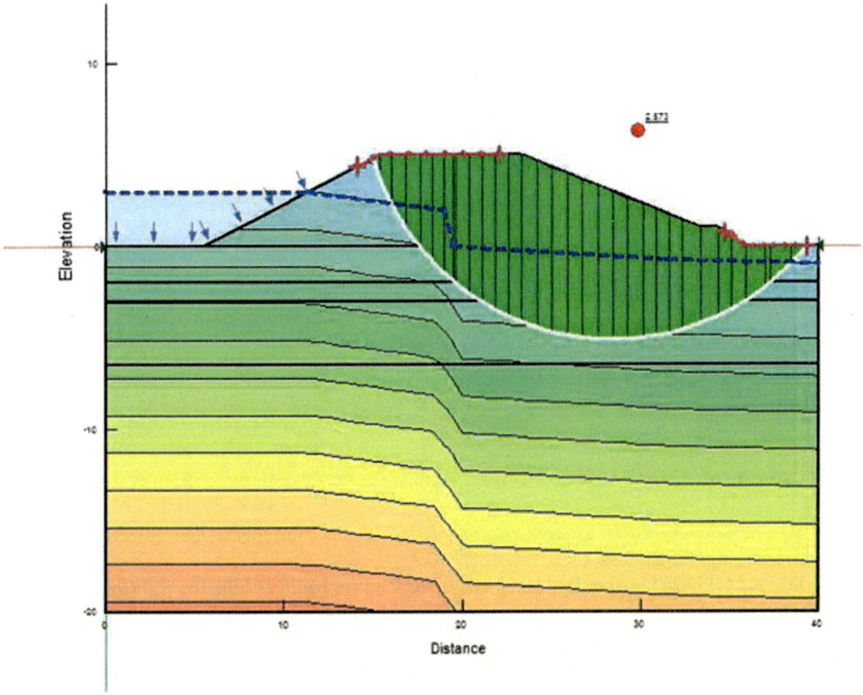

Fig. 8 Static stability analysis for first stage of ash dyke

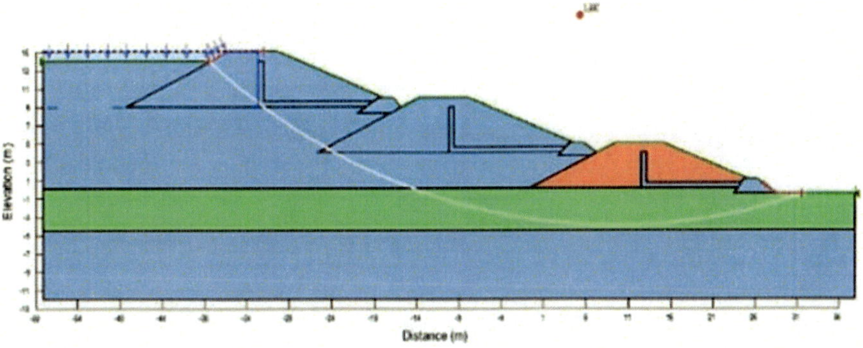

Fig. 9 Static stability analysis for third stage of ash dyke

safety values more than 3 at both ground surface and at the base of third stage with a decreasing tend of bearing capacity. This can be justified with the lesser strength of pond ash compared to the natural ground surface.

4 Conclusions

The following conclusions were drawn from the current work:

- The factors of safety values obtained from the GeoStudio software were greater than 1.5. Hence, pond ash can remain stable with no permanent deformations for the lower magnitudes of earthquakes.
- With increase in effective confining pressure by 50 kPa, approximately 8% deviation in excess pore pressure ratio was noticed.

References

Ahmaruzzaman M (2010) A review on the utilization of fly ash. Prog Energy Combust Sci 36(3):327–363

Andrus RD, Stokoe KH (2000) Liquefaction resistance of soils from shear-wave velocity. J Geotech Geoenviron Eng, ASCE 126(11):1015–1025

ASTM D5311M-13 (2013) Standard test method for load controlled cyclic triaxial strength of soil, pp 1–11

Casagrande A (1976) Liquefaction and cyclic mobility of sands: a critical review. Harvard Soil Mechanics Series 88, Harvard University, Cambridge, MA

Casagrande A (1965) The role of the 'calculated risk' in earthwork and foundation engineering. Soil Mech Found Div ASCE 91(4):1–4

Castro G (1975) Liquefaction and cyclic mobility of sands. J Geotech Eng Div, ASCE 101(6):551–569

Cetin KO et al (2004) SPT-based probabilistic and deterministic assessment of seismic soil liquefaction potential. J Geotech Geoenviron Eng, ASCE 130(12):1314–1340

Choudhary, Maheshwari SS, Kaynia BK, Amir M (2010) Liquefaction resistance of Solani sands under Cyclic loads. In: Indian geotechnical conference—GEOtrendz, 16–18 Dec 2010, pp 115–118

Dobry R, Ladd RS, Yokel FY, Chung RM, Powell D (1982) Prediction of pore water pressure buildup and liquefaction of sands during earthquakes by the cyclic strain method. National Bureau of Standards, Washington, D.C.

Dobry R, Powell DJ, Yokel FY, Ladd RS (1980) Liquefaction potential of saturated sand—the stiffness method. In: Proceedings of the 7th world conference on earthquake engineering, Istanbul, Turkey, pp 25–32

Figueroa JL, Saada AS, Liang L, Dahisaria MN (1994) Evaluation of soil liquefaction by energy principles. J Geotech Eng, ASCE 120(9):1554–1569

Finn WDL, Martin GR, Lee KW (1977) An effective stress model for liquefaction. J Geotech Eng Div 103(6):517–533

Fourie AB, Blight GE, Papageorgiou G (2001) Static liquefaction as a possible explanation for the Merriespruit tailings dam failure. Can Geotech J 38:707–719

Hazen A (1918) A study of the slip in the Calaveras Dam. Eng News Rec 81(26):1158–1164

Ishihara K (1996) Soil behaviour in earthquake geotechnics, 1st edn. Clarendon Press, Oxford, p 350

Ishihara K (1977) Simple Method of Analysis for Liquefaction of Sand Deposits during Earthquakes. Soils Found 17(3):1–17

Iwasaki T, Arakawa T, Tokida K (1984) Simplified procedures for assessing soil liquefaction during earthquakes. Soil Dyn Earthq Eng 3(1):49–58

Karol RH (2003) Chemical grouting and soil stabilization. Marcel Dekker, New York, p 558
Kenan H, Ellen RM (2004) A comparison between in situ and laboratory measurements of pore water pressure generation. In: 13th world conference on earthquake engineering, Vancouver, Canada, 1–6 Aug 2004, p 1220
Kokusho T (2016) Major advances in liquefaction research by laboratory tests compared with in situ behavior. Soil Dyn Earthq Eng 91:3–22
Kramer SL (1996) Geotechnical earthquake engineering. Prentice Hall, New Jersey
Kumar BA, Ghosh A, Ghosh A (2007) Compaction characteristics of pond ash. J Mater Civ Eng 19(4):349–357
Ladd RS (1978) Preparing test specimens using undercompaction. Geotech Test J, GTJODJ 1(1):16–23
Law KT, Cao YL, He GN (1990) energy approach for assessing seismic liquefaction potential. Can Geotech J 27(3):320–329
Marchetti S (1982) Detection of liquefiable sand layers by means of quasi-static penetration tests. In: Proceedings, 2nd European symposium on Penetration testing, Amsterdam, vol 2, pp 458–482
Mulilis JP, Chan CK, Seed HB (1975) The effects of method of sample preparation on the cyclic stress-strain behavior of sands. Report No. EERC 75–18, Earthquake Engineering Research Center, University of California, Berkeley
Olson SM, Stark TD (2003) Yield strength ratio and liquefaction analysis of slopes and embankments. J Geotech Geoenviron Eng 129(8):727–737
Park TK, Silver ML (1975) Dynamic triaxial and simple shear behavior of sand. J Geotech Eng Div, ASCE 101(6):513–529
Rawlings CG, Hellawell EE, Kilkenny WM (2000) Grouting for ground engineering, CIRIA Report C514. Construction Industry Research and Information Association, London
Robertson PK, Campanella RG (1985) Liquefaction potential of sands using the CPT. J Geotech Eng Div, ASCE 111(3):38–40
Sawada R, Kobayashi H (2011) The countermeasure against liquefaction sand with chemical grouting for existing foundation structures. In: Proceedings of the fifth international conference on advances in Earthquake Geotechnical engineering, 5th ICEGE. Santiago, Chile; paper N. TCASA
Seed HB, Idriss IM, Arango I (1983) Evaluation of liquefaction potential using field performance data. J Geotech Eng, ASCE 109(3):458–482
Seed HB, Idriss IM (1971) Simplified procedure for evaluating soil liquefaction potential. J Soil Mech Found Div, ASCE 97(9):1249–1273
Seed HB, Arango I, Chan CK (1975) Evaluation of soil liquefaction potential during earthquakes. Report No. EERC 75-28, Earthquake Engineering Research Center, University of California, Berkeley
Seed HB (1979) Soil liquefaction and cyclic mobility evaluation for level ground during earthquakes. J Geotech Eng Div, ASCE 105(2):201–255

Analysis of Pile Under Seismic Motion Using Pseudo-static Approach

Tanumaya Mitra, Kalyan Kumar Chattopadhyay, and Ambarish Ghosh

Abstract In the present study, pseudo-static approach has been used to assess the behaviour of a single pile subjected to seismic loading by FDM. Imperial Valley earthquake accelerogram has been scaled to 0.10 g, 0.13 g and 0.16 g peak bed rock acceleration (PBRA). These motions have been used to carry out equivalent linear method of ground response analysis (GRA) to get the peak ground acceleration (PGA) and deformation profile of soil column. The effect of shear wave velocity of soil mass, pile diameter, input motion, vertical load at pile head on behaviour of both free head and fixed head piles has been studied using pseudo-static approach. The study shows that the increase in PBRA and vertical load will increase the pile deflection and bending moment, whereas the increase in shear wave velocity of soil mass will cause reduction in pile deformation but an increase in bending moment.

Keywords Finite difference method · Pseudo-static analysis · Piles

1 Introduction

Pile foundations are used to transfer the heavy load of the superstructure to the underlying soil by means of skin friction and end bearing for axially loaded pile and lateral soil resistance may also develop for laterally loaded piles. In some cases, pile may be subjected to combined axial and lateral load, e.g. structures subjected to earthquake, offshore structures like jacked up rig, bridge abutment, etc. Pile foundations continue to fail in major earthquakes all over the world in spite of adopting newly developed analytical procedures in design and construction practices. Each earthquake teaches new lessons to the engineering community and with occurrence of new earthquake, there will be addition of new knowledge. Hence, the behaviour of pile under seismic loading is a major area of research among the investigators.

T. Mitra (✉) · K. K. Chattopadhyay · A. Ghosh
Department of Civil Engineering, Indian Institute of Engineering Science and Technology, Shibpur, Howrah, West Bengal, India
e-mail: tanumayamitra@gmail.com

M. Latha Gali and P. Raghuveer Rao (eds.), *Geohazards*, Lecture Notes in Civil Engineering 86, https://doi.org/10.1007/978-981-15-6233-4_23

Beams on elastic foundation and continuum modelling are the most common procedures for determining the behaviour of pile for both static and dynamic cases. However, Winkler method is still most widely used and preferable method due to its simplicity and less computational time required. In this study, behaviour of single pile subjected to lateral load under seismic loading has been studied using BNWF approach, where the soil surrounding the pile has been replaced by a series of non-linear independent springs and the spring constants represent the stiffness of soil. The spring stiffness can be evaluated using static and cyclic *p-y* curves developed by several researchers. Depending upon the type of problem, static *p-y* curves may be used for pile subjected to static lateral load and cyclic *p-y* curves for dynamically loaded pile. Matlock (1970) and Georgiadis et al. (1992) developed *p-y* relationship for soft clay; Dewaikar and Patil (2006) proposed a hyperbolic model to develop *p-y* curves for clay; Reese et al. (1975) have proposed *p-y* curves for stiff clay. Reese et al. (1974) have developed *p-y* relationship for sand. In this paper, cyclic *p-y* curves proposed by Matlock (1970) have been used. Tabesh and Poulos (2001) developed a simple approximate method to evaluate the behaviour of single pile subjected to seismic loading using pseudo-static approach. Liyanapathirana and Poulos (2005) extended the pseudo-static approach to liquefiable soil. In present analysis, behaviour of single pile subjected to axial load under seismic motion has been studied using pseudo-static approach. In this method, a free field ground response analysis is carried out to get the soil deformation profile to calculate the kinematic load and peak ground acceleration. The vertical load on pile top is accounted for estimating the inertial load. The study reveals that with the increase in PBRA, pile deflection and bending moment increase. This attributes to the fact that increase in PBRA will increase the PGA as well as free field soil deformation, which in turn leads to increase in pile deflection and bending moment. Vertical load also plays an important role on the behaviour of pile under seismic shaking. Increase in vertical load on pile will increase the inertial load as well as pile deflection and bending moment. Increase in shear wave velocity of soil mass will increase the soil stiffness and reduces soil deformation, which causes reduction of kinematic load on pile and as a result, pile deflection decreases. Fixed headed pile will deform less than the free head pile but will attract more bending moment because of increased rigidity.

2 Analysis

The governing differential equation of pile considering axial load and relative displacement between soil and pile at depth z is given by (Architectural Institute of Japan 2001),

$$EI\frac{d^4y}{dz^4} + P\frac{d^2y}{dz^2} + k(y - y_s) = 0 \tag{1}$$

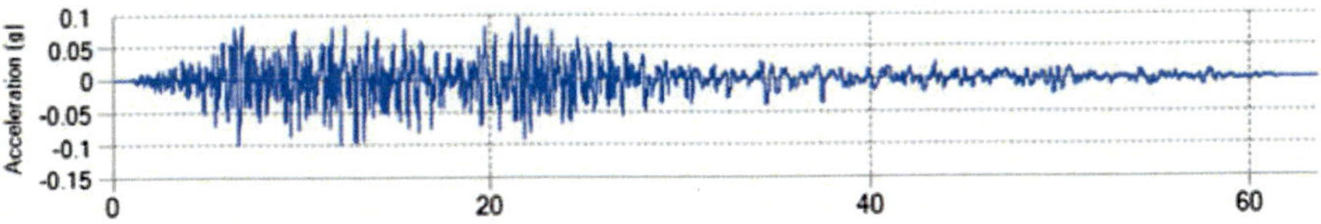

Fig. 1 Imperial Valley earthquake motion scaled to 0.10 g

where

- P is the axial load at the pile head,
- y is the pile displacement at depth z and.
- y_s is the soil displacement at depth z, obtained from ground response analysis.

Equation 1 has been solved using finite difference method (FDM) to get the deformation profile of the pile.

2.1 Steps Followed in Pseudo-static Analysis

The following steps (Liyanapathirana and Poulos 2005) are followed to carry out the pseudo-static analysis.

(i) Free field ground response analysis has been carried out to determine the peak ground acceleration (PGA) and the soil deformation profile.
(ii) The inertial load has been obtained by multiplying the PGA with the vertical load at pile head.
(iii) The deformed soil profile has been applied to the pile by means of non-linear p-y springs.
(iv) The pile deformation has been determined by solving the governing differential equation of pile deformation by applying the inertial load and kinematic load simultaneously.

2.2 Input Motions Used

Imperial Valley earthquake motion has been scaled to 0.10 g, 0.13 g and 0.16 g and these motions have been used as bedrock motions (Figs. 1, 2 and 3).

2.3 Soil Profile

A 10 m thick layer of cohesive soil with bulk unit weight of 18 kN/m^3 was considered in this study. The soil profile was discretized into ten numbers of layers with uniform

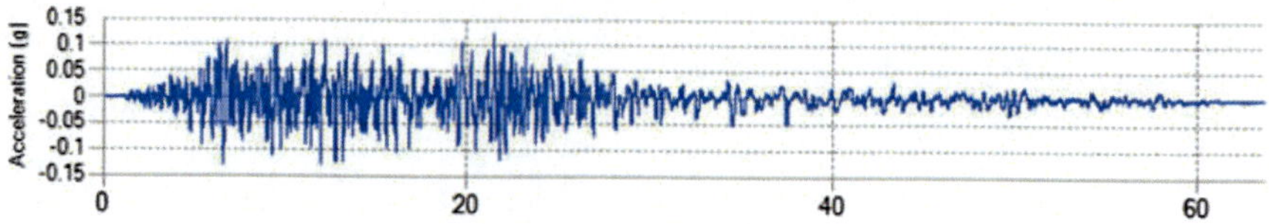

Fig. 2 Imperial Valley earthquake motion scaled to 0.13 g

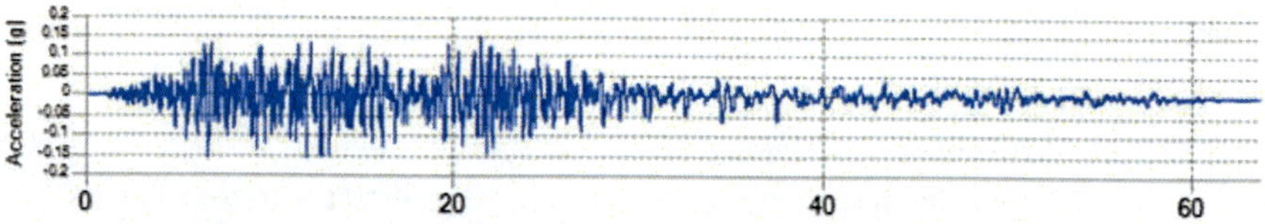

Fig. 3 Imperial Valley earthquake motion scaled to 0.16 g

thickness of 1.0 m. Water table has been considered at the ground surface. Different shear wave velocities (100 m/s, 110 m/s and 120 m/s) were also considered to study the behaviour of piles under different seismic motions.

2.4 *Development of Cyclic* p-y *Curves*

Matlock (1970) has recommended a procedure for constructing *p-y* curves. The *p-y* curves can be constructed with the help of Eq. 2,

$$\frac{p}{p_u} = 0.5\left(\frac{y}{y_{50}}\right)^{1/3} \tag{2}$$

where

- y_{50} is the lateral deflection at one half the ultimate soil resistance, expressed as $2.5\varepsilon_{50}D$; and corresponding lateral strain ε_{50} may be taken as 0.02 for soft clay (Poulos and Davis 1980).
- p is the soil resistance, which starts reducing at $y = 3y_{50}$ to a value of $0.72\ (x/x_r)$ at $y = 15y_{50}$ and beyond.
- x_r is the depth of reduced resistance.
- y is the deformation of soil.
- p_u is the ultimate soil resistance expressed as,

$$p_u = \left[\left(3 + \frac{\gamma' x}{c} + \frac{Jx}{D}\right)cD\right] \text{ or } 9c_u D,$$

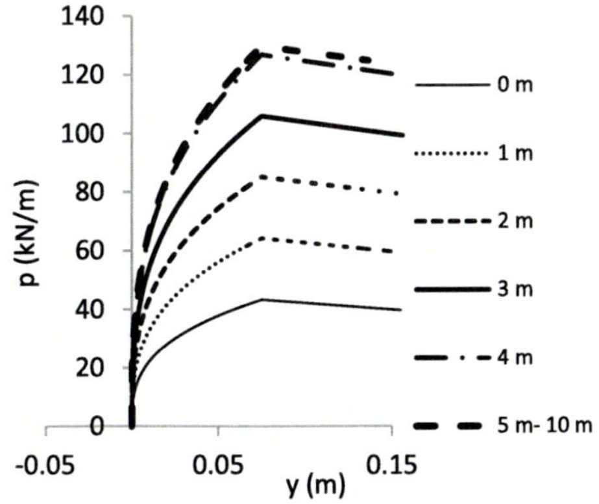

Fig. 4 Generated cyclic *p-y* curves (Matlock 1970)

whichever is less;
In which,

- J is a constant taken as 0.5,
- γ' is the effective unit weight of soil,
- x is the depth,
- c is the cohesion of the soil mass and.
- D is the diameter of the pile.

Cyclic *p-y* curves at different depths have been generated using method suggested by Matlock (1970) as shown in Fig. 4.

2.5 Results

Equivalent linear method of ground response analysis has been adopted in this study to determine the soil deformation profile and acceleration response. Free field soil deformation and maximum horizontal acceleration profile with depth have been shown in Figs. 5 and 6 considering homogenous soil with a typical shear wave velocity of 100 m/s. The motion gradually amplifies as it propagates to the top. A single pile of diameter 0.5 m and length 10.0 m has been considered in this investigation. Figures 7 and 8 illustrate the influence of vertical load on the deflection behaviour of free head and fixed head pile under seismic motion, respectively. The effect of vertical load on bending moment response of free head and fixed head pile has been shown in Fig. 9. Increase in vertical load increases the pile deflection and bending moment for both free head and fixed head pile. The effect of input motion (PBRA) on the pile deflection and bending moment response has been shown in Figs. 10 and 11. Increase in input motion increases pile deflection and bending moment both. The effect of shear wave velocity of soil media on pile deflection and bending moment

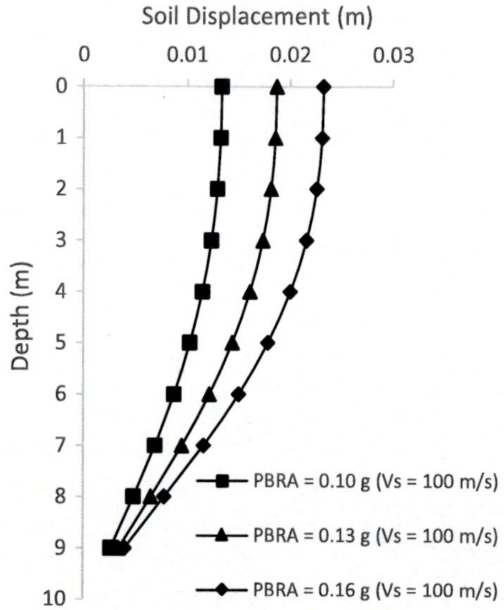

Fig. 5 Soil displacement profile

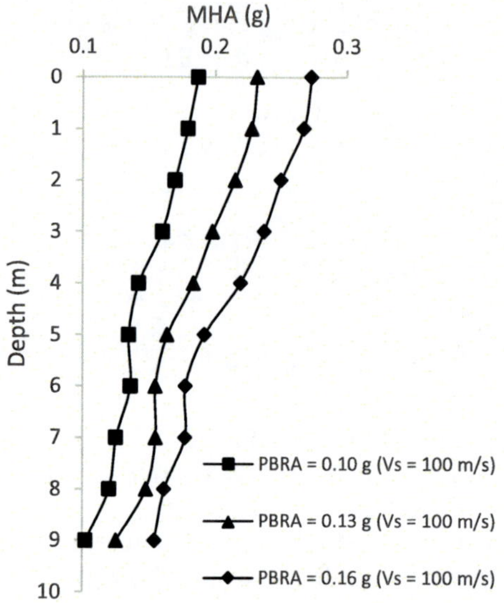

Fig. 6 Maximum horizontal acceleration profile with depth

response is presented in Figs. 12 and 13. Increase in stiffness of soil by enhancing its shear wave velocity will lead to reduction in pile deflection but increases the bending moment of pile.

Fig. 7 Effect of vertical load on free headed pile displacement

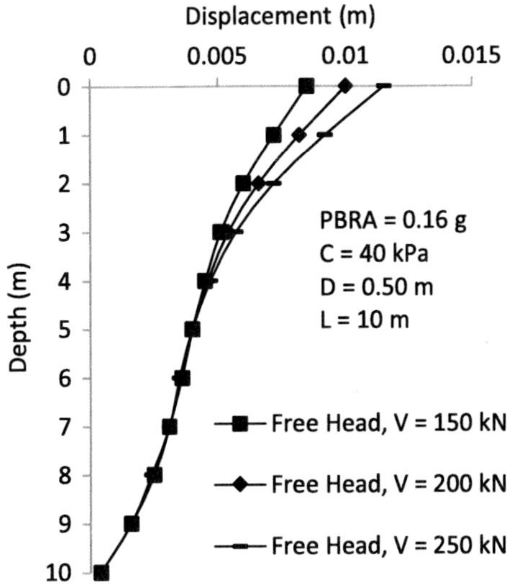

Fig. 8 Effect of vertical load on fixed headed pile displacement

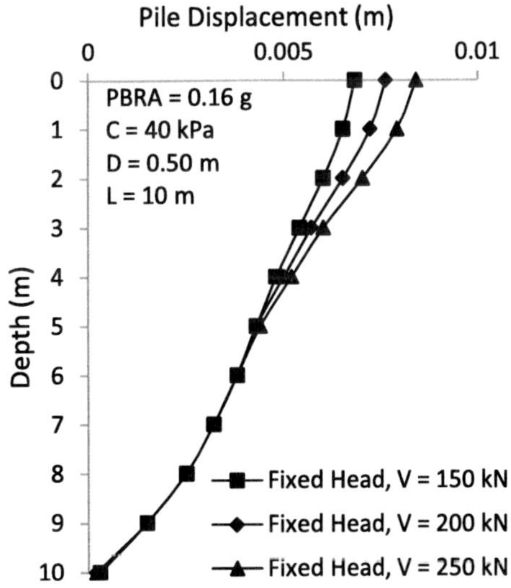

3 Observations

- With the increase in bedrock motion, both the deflection and bending moment of pile increase as the PGA and soil deformation increase.
- Pile deflection is less for fixed head pile, compared to free head pile; but bending moment for fixed head pile is more than free head pile.
- For constant bedrock motion, with the increasing vertical load, i.e. increased inertial load, pile deflection and bending moment increase.

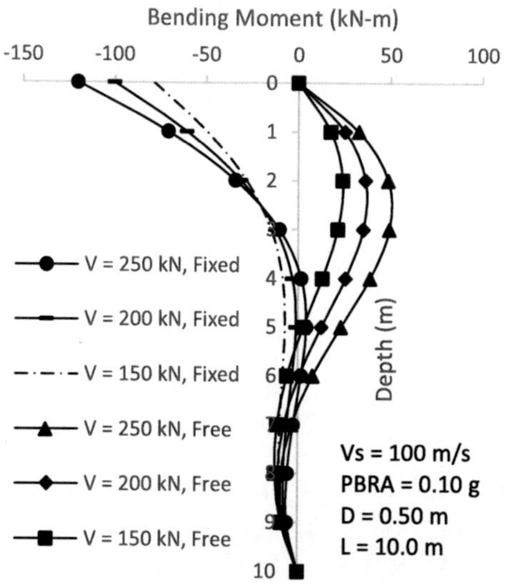

Fig. 9 Effect of vertical load on bending moment of pile

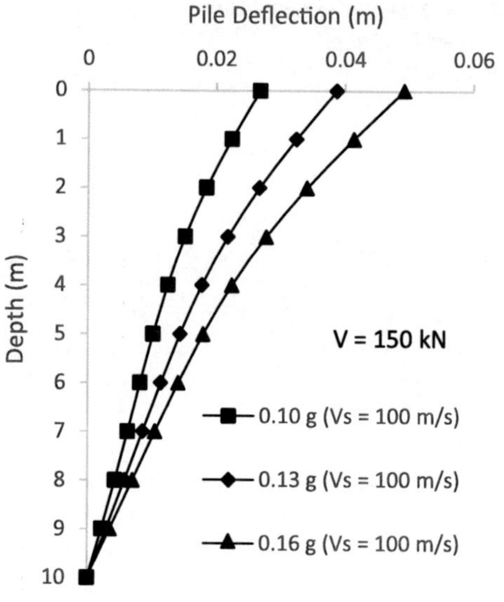

Fig. 10 Effect of input motion on pile displacement response

- With the increase in shear wave velocity of soil, the soil displacement reduces, hence kinematic load imposed on pile will reduce, resulting in reduction in pile displacement but bending moment increases, as the soil stiffness increases with the increasing shear wave velocity of soil mass, pile will behave more rigidly, thereby attracting more bending moment.

Fig. 11 Effect of input motion on bending moment of pile

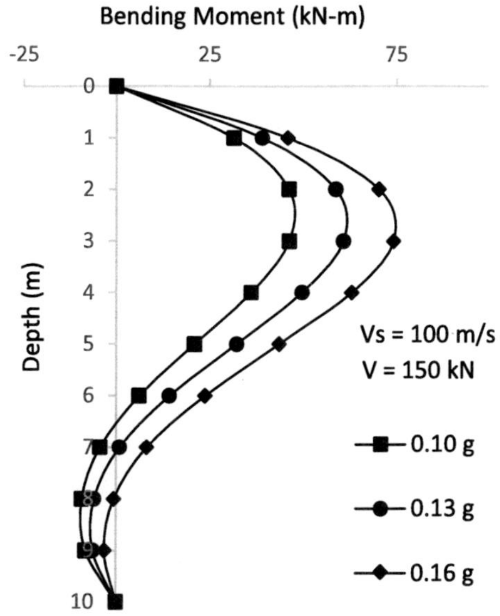

Fig. 12 Effect of shear wave velocity of soil media on pile displacement response

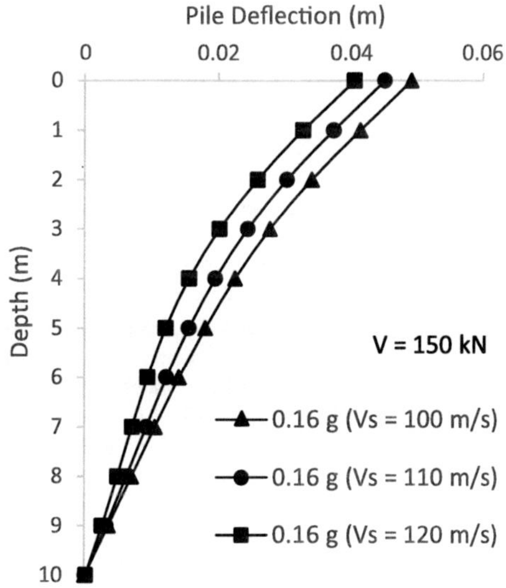

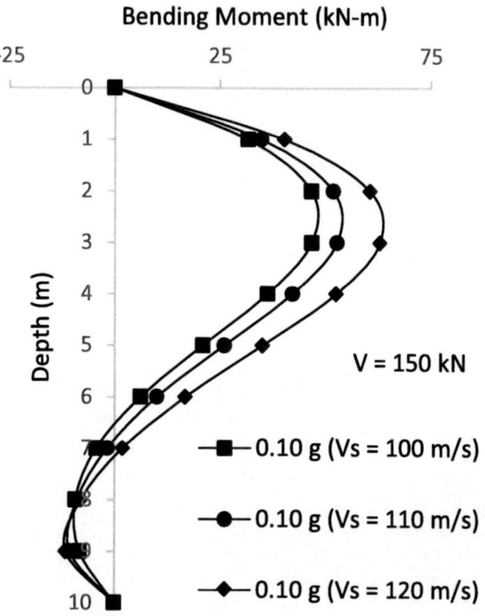

Fig. 13 Effect of shear wave velocity of soil media on bending moment of pile

References

Architectural Institute of Japan (AIJ) (2001) Recommendations for design of building foundations, Japan

Dewaikar DM, Patil PA (2006) A new hyperbolic *p-y* curve model for laterally loaded piles in soft clay. In: Foundation analysis and design: innovative methods, pp 152–158

Georgiadis M, Anagnostopoulos C, Saflekou S (1992) Cyclic lateral loading of piles in soft clay. Geotech Eng 23(1)

Liyanapathirana DS, Poulos HG (2005) Pseudostatic approach for seismic analysis of piles in liquefying soil. J Geotech Geoenviron Eng 131(12):1480–1487

Matlock H (1970) Correlations for design of laterally loaded piles in soft clay. Offshore technology in Civil engineering's hall of fame papers from the early years, pp 77–94.

Poulos HG, Davis EH (1980) Pile foundation analysis and design. Wiley, New York

Reese LC, Cox WR, Koop FD (1974) Field testing and analysis of laterally loaded piles in sand. In: Proceedings of the VI annual offshore technology conference, Houston, Texas, 2(OTC 2080), pp 473–485

Reese LC, Cox WR, Koop FD (1975) Field testing and analysis of laterally loaded piles om stiff clay. In: Offshore technology conference

Tabesh A, Poulos HG (2001) Pseudostatic approach for seismic analysis of single piles. J Geotech Geoenviron Eng 127(9):757–765

Ground Response Analysis and Determination of Liquefaction Potential Index

Tanumaya Mitra, Kalyan Kumar Chattopadhyay, and Ambarish Ghosh

Abstract Four borehole data from 4/1 Moore Avenue, Tollygunge, Kolkata, were collected to carry out ground response and liquefaction analysis. Two scaled input motions with peak bed rock acceleration (PBRA) 0.10 and 0.16 g have been generated from original Imperial Valley Earthquake accelerogram. The peak ground accelerations (PGA) obtained from ground response analysis using DEEPSOIL software have been used to determine the probable depth of liquefiable soil for Mw of 6.8. Ground response analysis shows amplification of PGA from 0.10 to 0.143 g and 0.16 to 0.208 g. Liquefaction analysis result shows that the bore holes analyzed are not susceptible to liquefaction as the liquefaction potential index (LPI) values come out to be 0.0 indicating very low risk for the region.

Keywords Liquefaction · Ground response analysis · Liquefaction potential index

1 Introduction

Soil liquefaction is one of the major problems in Geotechnical Earthquake Engineering. Due to liquefaction, degradation of shear strength and stiffness of loose, saturated, cohesionless soil take place. In some cases, the reduction of shear strength is up to such an extent that failure of soil leads to failure of structures due to the inability of soil to resist the load. Geotechnical engineering community became aware of the disastrous effects of soil liquefaction since the year 1964 when the Great Alaska earthquake ($M_W = 9.2$) occurred, followed by the Niigata earthquake ($M_S = 7.5$) in Japan. Both the earthquakes resulted in the failure of several buildings and bridge foundations along with floatation of buried structures due to soil liquefaction and lateral spreading. Thus, the evaluation of soil liquefaction resistance has become an important aspect of geotechnical engineering research and practice. Simplified procedure of liquefaction analysis was first proposed by Seed and Idriss in 1971. Factor of safety against liquefaction is determined using this approach by taking the

T. Mitra (✉) · K. K. Chattopadhyay · A. Ghosh
Department of Civil Engineering, Indian Institute of Engineering Science and Technology, Shibpur, Howrah, West Bengal, India
e-mail: tanumayamitra@gmail.com

M. Latha Gali and P. Raghuveer Rao (eds.), *Geohazards*, Lecture Notes in Civil Engineering 86, https://doi.org/10.1007/978-981-15-6233-4_24

ratio of cyclic resistance ratio (CRR) to cyclic stress ratio (CSR). CRR represents the capacity of soil to resist liquefaction, and CSR is the ratio of average cyclic shear stress experienced by the soil during earthquake to the effective overburden pressure. Factor of safety for a soil layer can be calculated using several in situ tests such as SPT, CPT, BPT and shear wave velocity test (V_s) test (Youd et al. 2001). In this study, SPT-based liquefaction analysis framework has been used to determine the variation of factor of safety with depth.

Typical soil profile of Kolkata consists of loose fill soil followed by a layer of bluish gray silty clay with varying percentage of semi-decomposed wood up to a depth of about 10 to 15 m. But the soil profile encountered in 4/1 Moore Avenue, Tollygunge, Kolkata region, was totally different from the expected typical soil profile of Kolkata. The top soil consists of soft grayish brown silty clay followed by a layer of medium dense to dense grayish fine silty sand with water table located near ground surface. Ground response analysis has been carried out using equivalent linear method to determine the PGA, and liquefaction analysis has been carried out for this region to determine the liquefaction potential index (LPI).

2 Soil Profile

Soil profile of the site consists of soft, grayish brown silty clay/clayey silt up to a depth of about 3 m, followed by layers of medium dense to very dense grayish or yellowish brown silty sand up to a depth of about 40 m, with water table located at 1.0 m below E.G.L. Four borehole data were collected to carry out the liquefaction analysis and to determine the liquefaction potential index for each borehole. The soil profiles of 4/1 Moore Avenue, Tollygunge, have been shown in Fig. 1a–d, and average soil properties are given in Table 1.

3 Ground Response Analysis

Influence of local geology and soil conditions plays an important role on the ground response while the soil column is subjected to seismic shaking. Ground response analyses are used to predict ground surface motions for developing design response spectra, to determine peak ground acceleration, evaluation of soil liquefaction potential and to determine the earthquake forces that may lead to instability of soil slope as well as retaining structures. Ground response analysis techniques are often categorized based on the dimensionality of the problem and the characteristics of the soil model used in the analysis (Kramer 1996). The problem can be 1D, 2D and 3D and solved by linear, equivalent linear and nonlinear approaches. In case of linear approach, constant values of shear modulus (G) and damping ratio (ξ) for an induced level of small shear strain in each layer are used, whereas for nonlinear approach,

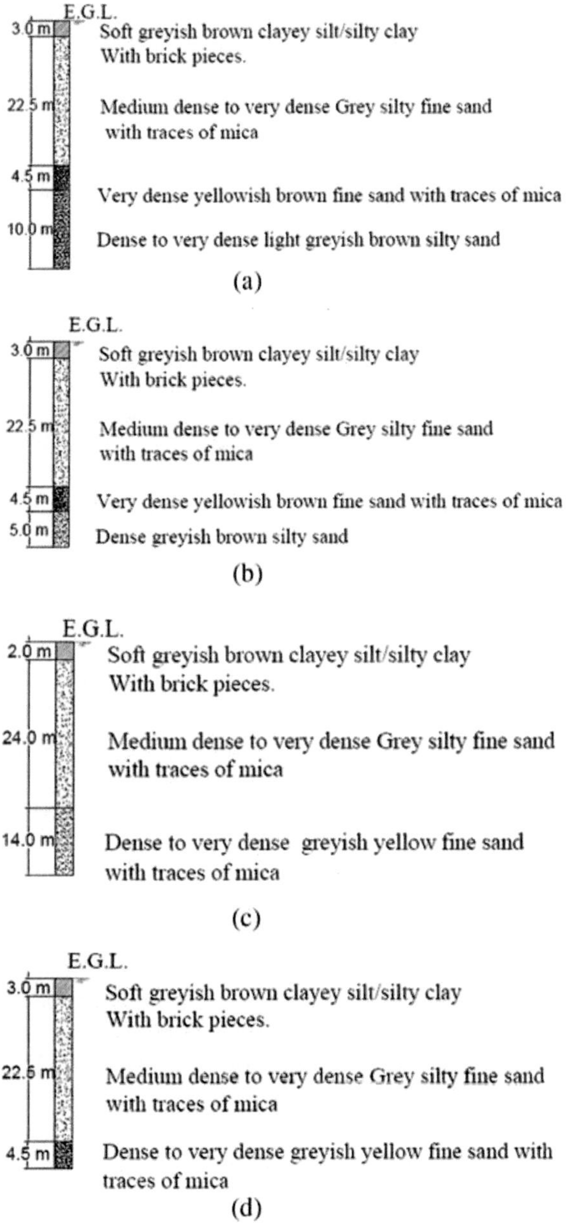

Fig. 1 Soil profile of the site **a** BH 1, **b** BH 2, **c** BH 3, **d** BH 4

Table 1 Average soil properties of 4/1 Moore Avenue, Tollygunge

Soil type	Thickness (m)	γ_b (kN/m^3)	N_{avg}	V_s (m/s)	PI (%)
Clayey silt/silty clay	3.0	17.5	10	154.0	5.0
Greyish silty fine sand	22.5	19.0	29	295.0	0.0
Yellow fine sand	4.5	19.0	45	324.0	0.0

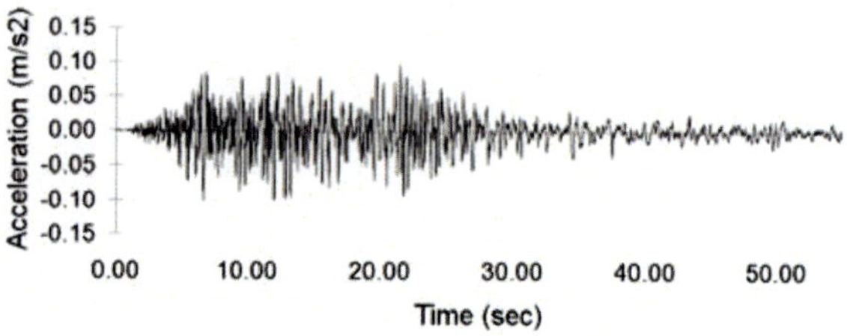

Fig. 2 Scaled motion (PBRA 0.10 g)

cyclic backbone curves are used to simulate the nonlinear response of soil. Equivalent linear approach uses the equivalent linear soil properties by utilizing an iterative procedure. In every iteration, the equivalent linear values of shear modulus and damping ratio are modified corresponding to the effective shear strain in each layer. Even though nonlinear approach simulates the soil response much better than other methods, equivalent linear approach is computationally convenient and provides reasonably accurate results.

3.1 Seismicity of the Region

The most prominent tectonic feature in the Bengal Basin is the NE–SW-trending Eocene Hinge Zone (EHZ) which is 25 km wide and extends to a depth of about 4.5 km below Kolkata (Nath et al. 2014). The EHZ may be taken as the principal contributor of earthquake in and around Kolkata. EHZ is a strike slip fault capable of producing earthquake of maximum estimated magnitude of 6.8 (Nath et al. 2014).

3.2 Selection of Input Motion

Imperial Valley Earthquake accelerogram has been selected as the magnitude of this earthquake is very close to the maximum estimated earthquake for Kolkata. Two input motions have been generated from original Imperial Valley Earthquake accelerogram (Mw = 6.7) by scaling the motion to 0.10 and 0.16 g. These two motions have been used as input motions for carrying out ground response analysis. The scaled input motions have been shown in Figs. 2 and 3. These motions have been applied at a depth of 30.0 m from existing ground level.

3.3 Dynamic Soil Properties

A set of material curves has been defined in DEEPSOIL (Hashash et al. 2015) for defining strain-dependent shear modulus (modulus reduction curves) and damping

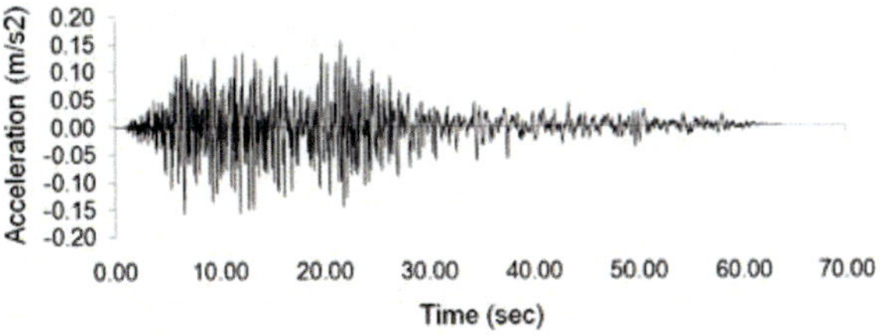

Fig. 3 Scaled motion (PBRA 0.16 g)

ratio for different soils. In the absence of site-specific modulus reduction and damping ratio curves, standard curves proposed by Vucetic and Dobry (1991) (for clay) and Seed and Idriss (1970) (for sand) have been used as the reference curves to simulate the material behavior. Shear wave velocity values of different soil layers have been determined using the empirical equation based on the uncorrected SPT blow counts (Imai and Tonouchi 1982). Water table has been considered at the ground surface.

3.4 Results of Ground Response Analysis

The ground surface motions and PGA values obtained from the ground response analysis have been shown in Figs. 4, 5 and Table 2, respectively.

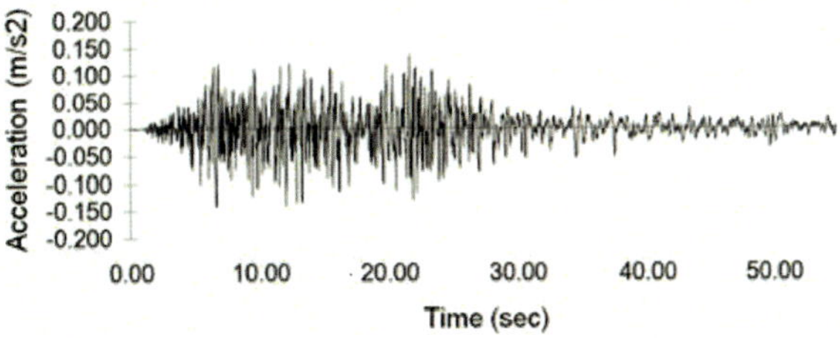

Fig. 4 Ground surface motion (PGA 0.143 g)

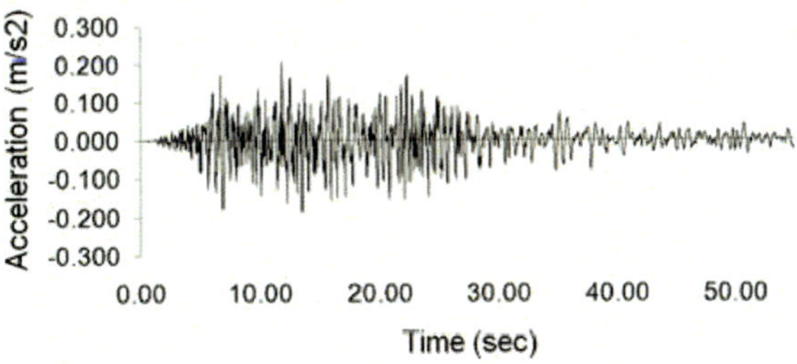

Fig. 5 Ground surface motion (PGA 0.208 g)

Table 2 Results of ground response analysis

Input PBRA (g)	Output PGA (g)
0.10	0.143
0.16	0.208

4 Liquefaction Analysis

Boulanger and Idriss (2006) recommended that, for practical purposes, fine-grained soils can be expected to exhibit clay-like behavior if they have PI $\geq$ 7 and sand-like behavior if PI < 7. For soils that behave more fundamentally like sands (PI < 7), the cyclic strength may be more appropriately estimated within the existing framework of standard penetration test (SPT) and cone penetration test (CPT)-based liquefaction correlations, whereas for PI $\geq$ 7, cyclic strength of soil can be estimated using in situ testing and laboratory testing. As the plasticity index of all the soil layers is below 7, SPT-based liquefaction analysis (Idriss and Boulanger 2014) has been used to determine the liquefaction susceptibility of the site using moment magnitude of earthquake Mw = 6.8. Liquefaction analysis results in the form of factor of safety versus depth have been shown in Fig. 6a–d. Liquefaction potential index (LPI) value (Luna and Frost 1998) for each borehole has been determined. The LPI values have been mentioned in Table 3. Figure 7 shows the variation of factor of safety with depth for the four bore holes considered in the liquefaction analysis.

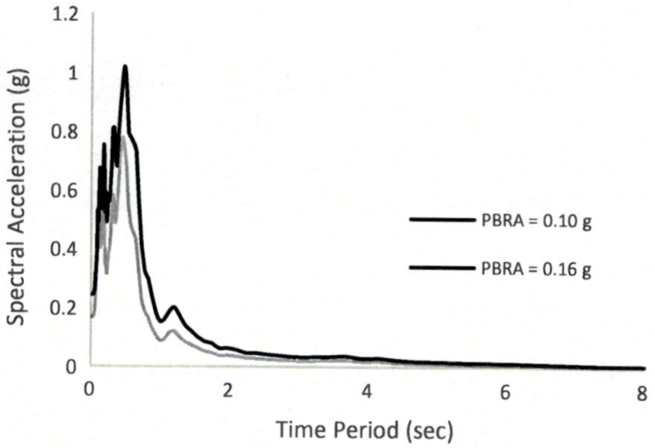

Fig. 6 Site-specific response spectra

Table 3 LPI values obtained from liquefaction analysis

BH No.	LPI	Description
1	0	Non-liquefiable
2	0	
3	0	
4	0	

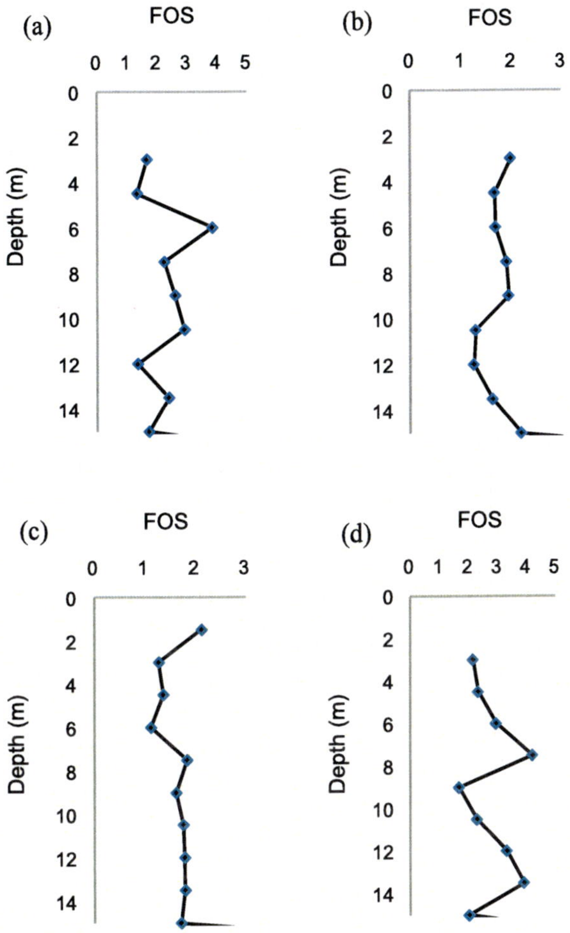

Fig. 7 Liquefaction analysis (factor of safety versus depth) **a** BH 1, **b** BH 2, **c** BH 3, **d** BH 4

5 Conclusion

Ground response analysis and liquefaction analysis have been carried out for four boreholes located at Tollygunge Moore Avenue, Kolkata. From the ground response analysis, it has been observed that the site amplification varies in between 1.30 to 1.43 for 0.10 g and 0.16 g input accelerogram. Liquefaction analysis result shows that the LPI value is 0.0, indicating that the boreholes analyzed are not susceptible to liquefaction. For $(N1)_{60} \geq 30$, granular soil is too dense to liquefy (Youd et al. 2001). And this analysis also revealed that the sand layers analyzed are not susceptible to liquefaction as the N value is about 30. Recent studies have revealed that some locations in Kolkata are susceptible to liquefaction. For those sites, foundation design has to be done considering the depth of liquefied zone, and special foundation design methodology needs to be adopted. The present study revealed that the site is not susceptible to liquefaction. Hence, for each project, site-specific geotechnical investigation along with ground response and liquefaction analysis is important, otherwise it may lead to uneconomical design of foundation if the foundation to be constructed is embedded in non-liquefiable soil or may lead to failure of structure if the liquefied zone is not taken into consideration.

References

Boulanger RW, Idriss IM (2006) Liquefaction susceptibility criteria for silts and clays. J Geotech Geoenviron Eng ASCE 132(11):1413–1426

Hashash YMA, Musgrove MI, Harmon JA, Groholski DR, Phillips CA, Park D (2015) DEEPSOIL 6.1, User manual

Idriss IM, Boulanger RW (2014) CPT and SPT based liquefaction triggering procedure. Centre for Geotechnical Modelling, University of California at Berkeley, Report No. UCD/CGM-14/01

Imai T, Tonouchi K (1982) Correlation of N-value with S-wave velocity and shear modulus. In: Proceedings of the 2nd European symposium of penetration testing, Amsterdam, pp 67–72

Kramer SL (1996) Geotechnical earthquake engineering. Prentice Hall, Upper Saddle River

Luna R, Frost JD (1998) Spatial liquefaction analysis system. J Comput Civ Eng 12(1):48–56

Nath SK, Adhikary MD, Maiti SK, Devaraj N, Srivastava N, Mohapatra LD (2014) Earthquake scenario in West Bengal with emphasis on seismic hazard microzonation of the city of Kolkata India. Nat Hazards Earth Syst Sci 14(2549–2575):2014

Seed HB, Idriss IM (1970) Soil moduli and damping factors for dynamic response analyses. Technical report EERRC-70-10, University of California, Berkeley

Vucetic M, Dobry R (1991) Effect of soil plasticity on cyclic response. J Geotech Eng 117(1):89–107

Youd TL, Idriss IM, Andrus RD (2001) Liquefaction resistance of soils: summary report from the 1996 NCEER and 1998 NCEER/NSF workshops on evaluation of liquefaction resistance of soils

Probabilistic Seismic Hazard Analysis of Vadodara Region

Payal Mehta and T. P. Thaker

Abstract The Vadodara region falls in seismic zone-III as per IS 1893:2016 and surrounded by highly active seismic sources. Seismicity of the Vadodara region has been determined by the evaluation of peak ground acceleration (PGA) model considering probabilistic framework. Probabilistic analysis has been carried out considering the earthquake data for the time interval between 1668 and 2017. The seismic parameter '*b*' has been estimated as 0.815 ± 0.001 by adopting suitable recurrence relation. Four ground motion prediction relationships have been adopted to estimate the hazard of the study region. The PGA model at rock level has been quantified by dividing the study region into the grid size of 1 km × 1 km, which varies from 0.054 to 0.071 g and 0.104 to 0.139 g for 10% and 2% probability of occurrence in 50 years, respectively. The results will be further useful for designing earthquake-resistant structures, risk mitigation and future city planning.

Keywords Peak ground acceleration · Probabilistic seismic hazard analysis · Seismic parameters · Ground motion prediction relationships

1 Introduction

India is facing heavy loss of lives and properties due to unpredictable natural events like earthquake. The loss during any seismic event can be minimized by proper input of ground motion parameters in the design of earthquake-resistant structures. The seismic code of India, IS:1893-2016, gives the broad overview of seismic zonation of the whole country. Many past devastating earthquakes show that the local geology of the region plays an important role during these events. India is the developing country with the new initiation of 'smart city project' taken by the Government of India. Earthquake-resistant design of any structure is the smart solution of the economic development of the country. Ground motion parameters considering geotechnical properties of subsoil and tectonic features of the region are used as an input in the

P. Mehta (✉) · T. P. Thaker
Department of Civil Engineering, School of Technology,
Pandit Deendayal Petroleum University, Gandhinagar, Gujarat 382007, India
e-mail: payalmehta2910@gmail.com

M. Latha Gali and P. Raghuveer Rao (eds.), *Geohazards*, Lecture Notes in Civil Engineering 86, https://doi.org/10.1007/978-981-15-6233-4_25

designing of seismic-resistant structures. Ground motion parameters can be obtained in terms of peak ground acceleration model, which is further useful for the risk mitigation, disaster preparedness, urban rehabilitation and economical development of the region.

In the present paper, peak ground acceleration model at rock level has been developed for the Vadodara region considering the probabilistic approach. In 2001, Bhuj earthquake remarkable damages have been observed in and around the Vadodara City. Few buildings were cracked and damaged in this earthquake. But many fatalities have been reported in the Ahmedabad City which is only 100 km away from the Vadodara City. Vadodara is carrying the high risk of seismicity due to near-field and far-field seismic sources. Vadodara is located near the bank of Vishwamitri River, which spreads the loose sediments near its bank. The city also carries the deposits of the Late Pleistocene fluvial sediments by the Dhadhar River, which originates from the Pavagadh Hill. The general soil stratum of the city is alluvial soil with loose sediments. Hence, the soil stratum of the city is also having the tendency to amplify the ground acceleration at the surface level. Vadodara is to be developed as a 'smart city' of India in the near future. Hence, the detailed hazard analysis is very much crucial for the microzonation study of the region. Seismic catalogue has been developed for the time interval between 1668 and 2017 considering Vadodara City as a centre and covering 350 km radius around it. Seismotectonic model has been generated considering tectonic features and the past earthquake events. Guternberg and Richter relationship has been applied to estimate the seismic parameters. The evaluated seismic parameter '*b*' has been compared with the published literature for the peninsular India. The hazard has been quantified in terms of peak ground acceleration model at the rock level for 10 and 2% probability of occurrence in 50 years, which is further compared with the published literature.

2 Study Region

Vadodara is located at 22.30° N 73.19° E in state of Gujarat covering an area of 235 km^2. Vadodara is the third-largest city of the state with its historical and cultural importance. As per the Indian seismic code IS:1893-2016, Vadodara falls under seismic zone-III. Vishwamitri River is passing through the centre of the city and spreading the loose sediments along its bank. The local soil profile of the city is the alluvial type of soil. Many upcoming constructions under the scheme of 'riverfront project' and 'smart city project' have increased the vulnerability to the city. The city has proximity to some of the seismically active faults with low to moderate seismic activity. Moreover, historical city like Vadodara not only the new constructions need to be designed as a safe construction, but the existing structures and monuments are also having the equal importance. Hence, the ongoing rapid development of the city, existing soft soils strata near the river valley and seismotectonic features of the area demand an in-depth seismic evaluation for the risk mitigation planning of the city. The study region depicts in Fig. 1.

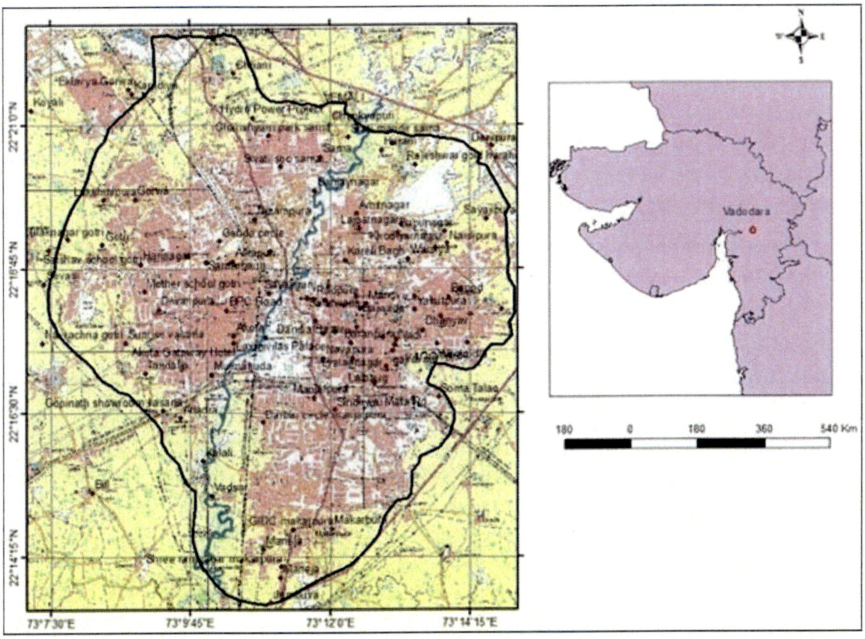

Fig. 1 Study region

3 Geology and Seismotectonic Setup

Geologically, Gujarat State is divided into three regions: Kutch basin, Saurashtra region and Mainland Gujarat. Vadodara City is situated in Mainland Gujarat which consists of two rift zones, namely Cambay rift zone and Narmada rift Zone. The NE part of the mainland is covered with the Proterozoic rocks and Deccan basalt covers the southern part of the mainland (Sinha-Roy et al. 1995). The Deccan traps and alluvium formation are the main features of Vadodara region (Merh 1995). The study region is associated with many near-field and far-field seismic sources, which are having moderate to high seismicity. In the present study, Vadodara City is considered as a control region of tectonic setup covering 350 km radius around it.

The seismic sources have been identified from the Seismotectonic Atlas of India and its environs (SEISAT 2000) and from other published literatures. The seismic catalogue has been compiled from the various sources such as the Institute of Seismological Research (ISR), Gujarat; International Seismological Centre (ISC), UK; Geological Survey of India (GSI); India Meteorological Department (IMD), New Delhi; United States Geological Survey (USGS), National Earthquake Information Centre; Oldham (1883), Chandra (1977), Malik et al. (1999) for the time interval of 1668 to 2017 bounded the study region by latitude from 19° N to 26° N and longitude from 69° E to 77° E. Heterogeneous data has been converted into homogeneous scale of moment magnitude after considering the appropriate relationships

(Mehta et al. 2017a, b). Declustering of the catalogue has been implemented to remove the associate events from the main events. It consists of Gardner and Knopoff (1974), Uhrhammer (1986) and Grunthal algorithms based on temporal and spatial windowing approaches. The conservative results from the Uhrhammer (1986) algorithm have been obtained, and final catalogue of 397 events has been considered in the analysis. All the seismic events of moment magnitude $M_w \geq 1$ have been compiled on the common platform of GIS with seventeen major faults, ten lineaments and seventeen minor faults. The base map has been prepared by superimposing the epicentres of declustered data of past 350 years earthquakes. The major near-field seismic sources are marginal fault, Son Narmada fault, Cambay West fault and Tapti North fault while the far-field sources are Kutch Mainland fault, Island Belt fault and Allah Bund Fault. These near-field sources are carrying the moderate risk of seismicity with the moment magnitude of 5 and above. The far-field sources include Kutch Mainland fault, Island Belt fault and Allah Bund Fault, which are approximately 300 km away from the study region but associated with the high seismicity of moment magnitude 7 and above. Around thirteen earthquakes of moment magnitude $M_w \geq 6$ have been reported in the study area. Figure 2 shows the seismotectonic model of the study region.

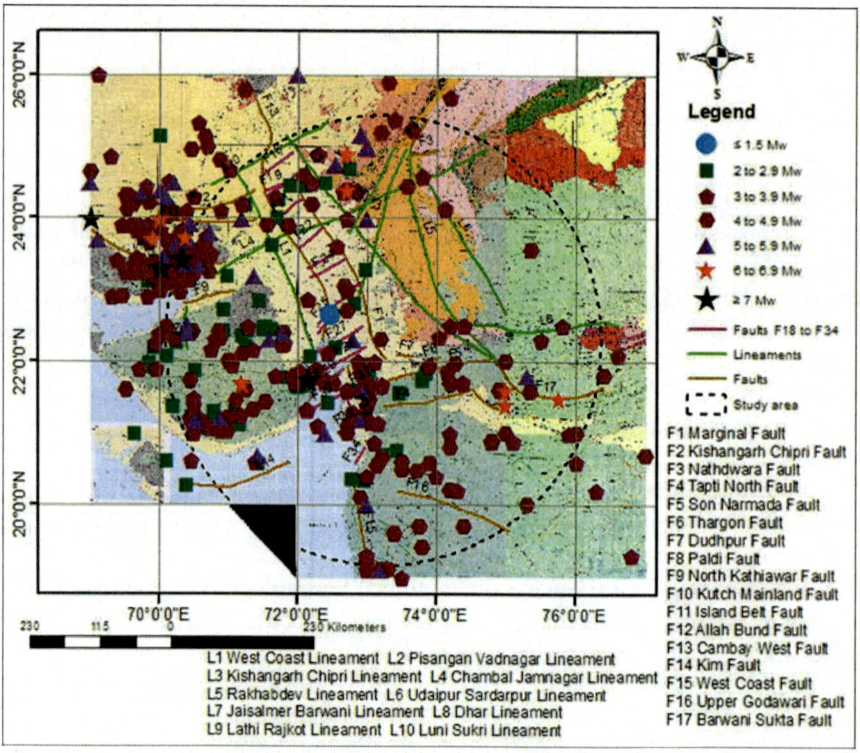

Fig. 2 Seismotectonic model of the study region

In the present paper, the analysis has been carried out considering minimum moment magnitude of 4 and above. Around 210 earthquakes have been reported of moment magnitude 4 and above in the study region, which represents the higher seismicity of the study region.

4 Seismic Parameters

Seismic parameters are used to estimate the seismic activity of the region. They can be quantified by using the simple recurrence law of Gutenberg and Richter (1944), which is expressed as:

$$\mathrm{Log}_{10}(N) = a - b\,M_w \tag{1}$$

where N stands for cumulative number of earthquakes greater than or equal to a particular magnitude M_w per year, parameters 'a' describe the seismic activity (log number of events with $M_w = 0$) and 'b' which is typically close to 1 is a tectonics parameter describing the relative abundance of large to smaller shocks.

As suggested by Wiemer and Wyass (2000), the value of m_0 can be considered from the frequency–magnitude distribution where the data depart from a straight line. In the present study, the value of m_0 has been estimated as '4.0' because the earthquake below this value would not be produced the significant damage to the structures. Stepp (1973) and Cumulative Visual Interpretation (CUVI) (Tinti and Mulargia 1985) methods have been used to check the completeness of the seismic data, which is represented in Figs. 3 and 4.

The frequency–magnitude relationship for the study area is shown in Fig. 5, which suggests that the results obtained from both the methods are in good agreement. Hence, the average value has been adopted in the present study as given below.

$$\mathrm{Log}_{10}(N) = (-0.815 \pm 0.001)M_w + (3.833 \pm 0.002) \tag{2}$$

Table 1 depicts the comparison of 'b' value for PI published in various literatures. It is observed that the 'b' value of the present study is in good agreement with the values of Ram and Rathor (1970), Kaila et al. (1972), Rao and Rao (1984), Seeber et al. (1999), Tripathi (2005), Raghukanth and Iyengar (2006), Jaiswal and Sinha (2006), Anbazhagan et al. (2008), Vipin et al. (2009), NDMA (2010) and Thaker et al. (2012) in PI. Lai et al. (2009) and Menon et al. (2010) show higher values, which may be due to earthquakes of higher magnitude in the study region as compared to Kancheepuram and Tamil Nadu (Thaker et al. 2012).

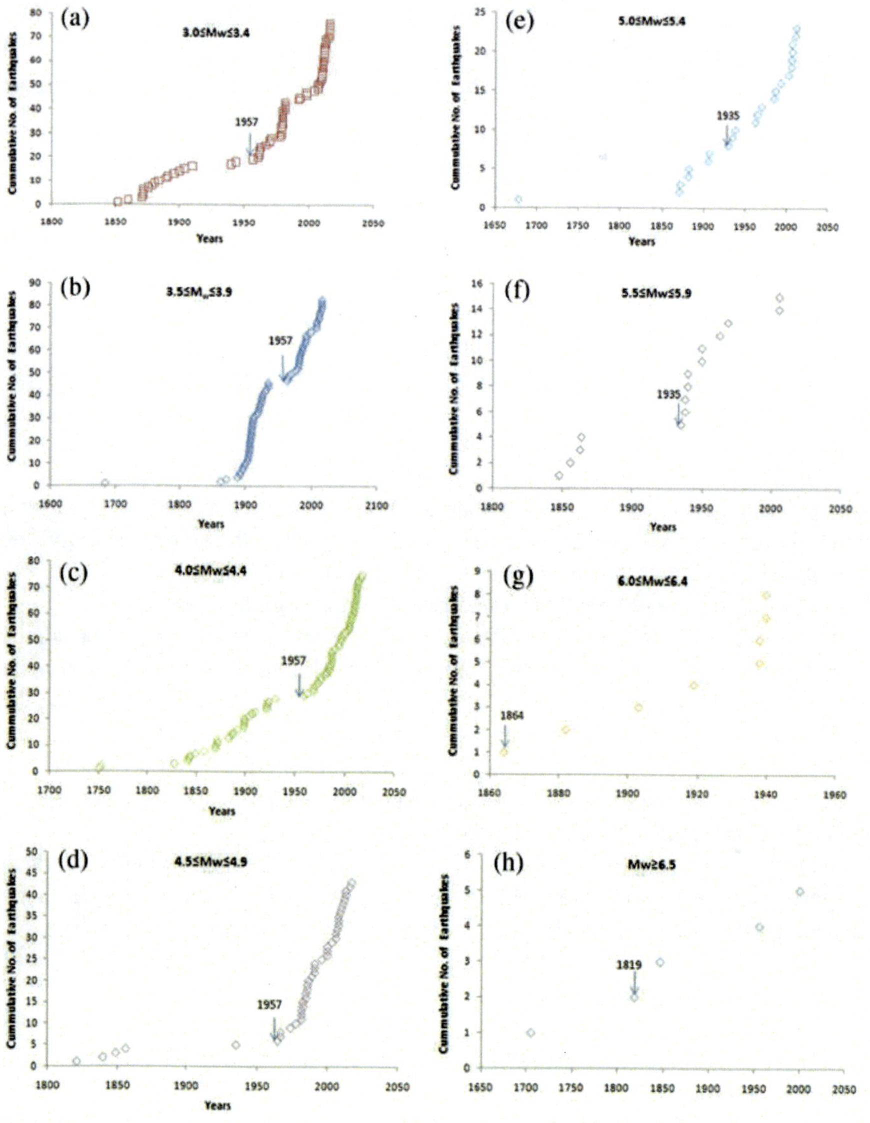

Fig. 3 Completeness period by CUVI method

4.1 Deaggregation

In the present study, the recurrence relation has been developed for the entire region. The approach suggested by Raghukanth and Iyengar (2006) based on heuristic basis invoking the principle of conservation of seismic activity is adopted which is already implemented by Iyengar and Ghosh (2004), Raghukanth and Iyengar (2006), Anbazhagan et al. (2008), Vipin et al. (2009) and Thaker et al. (2012) for PSHA of

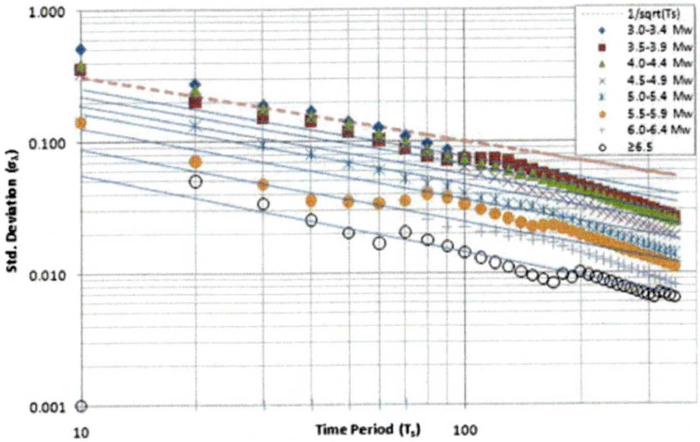

Fig. 4 Completeness period by Stepp method

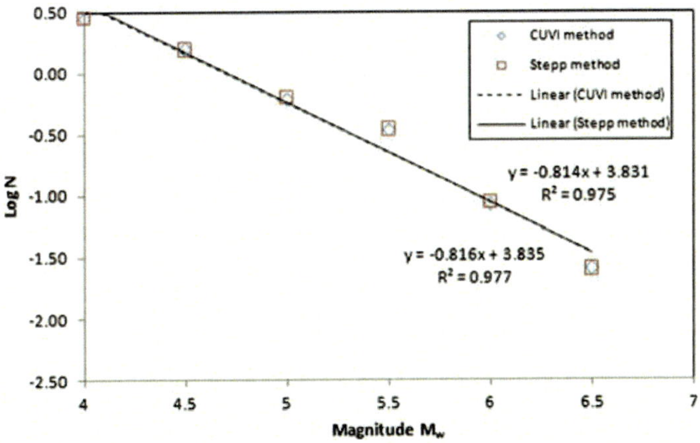

Fig. 5 Frequency–magnitude relationships

Delhi, Mumbai, Bangalore, South India and Surat, respectively. According to this approach, the regional seismicity measured in terms of the number of earthquakes per year with $m \geq m_0$ should be equal to the sum of such events occurring on individual fault. The recurrence relation has been estimated based on two assumptions, i.e. length weighing factor and earthquake weighing factor.

Length factor is associated with the length of individual fault to the total length of all faults in the study region, i.e. $\alpha_i = \frac{L_f}{\Sigma L_i}$, where L_f = Length of the fault. The seismic weighing factor is considered as the ratio of seismic events associated with the individual fault to the total number of the events in the study region, i.e. $\chi_i = \frac{N_i}{N_R}$; where N_i = number of earthquakes close to the seismic source and N_R = total number of earthquakes in the region.

Table 1 Comparison of '*b*' values with values reported by various researchers

Authors	'*b*' Value	Region
Ram and Rathor (1970)	0.81	PI
Kaila et al. (1972)	0.70	PI
Rao and Rao (1984)	0.85	PI
Seeber et al. (1999)	0.89	Maharashtra state
Tripathi (2005)	0.72	Gujarat
Raghukanth and Iyengar (2006)	0.86	Mumbai city
Jaiswal and Sinha (2006)	0.91	PI
Anbazhagan et al. (2008)	0.87 ± 0.03	Bangalore city
Vipin et al. (2009)	0.891 ± 0.07	PI
Lai et al. (2009)	1.26	Kancheepuram, Tamil Nadu
Menon et al. (2010)	1.13	Tamil Nadu
NDMA (2010)	0.87 ± 0.06	Gujarat region
Thaker et al. (2012)	0.89 ± 0.02	Surat city
Shukla and Choudhury (2012)	0.62	Mainland Gujarat
Present study	0.815 ± 0.001	Vadodara city

The recurrence relation for source '*i*' is obtained by averaging both weighting factor and multiplying the regional relation as given below;

$$\nu = N_i(m_0) = 0.5(\alpha_i + \chi_i)N(m_0) \tag{3}$$

In the present study of probabilistic seismic hazard analysis (PSHA), the seismic sources having the PGA above threshold value of 0.01 g have been selected from deterministic analysis (Mehta et al. 2018). The source recurrence relation weighing factors for each source are presented in Table 2. An annual rate of event of magnitude $\geq M_w$ (λ_m) has been estimated by the truncated exponential recurrence model of McGuire and Arabasz (1990), which is given as follows:

$$\lambda_m = \nu \frac{\exp[-\beta(m - m_0)] - \exp[-\beta(m_{\max} - m_0)]}{1 - \exp[-\beta(m_{\max} - m_0)]}; \quad \text{for } m_0 \leq m \leq m_{\max} \tag{4}$$

where m_0 is the threshold magnitude, $\beta = 2.303b$ and N_i (m_0) or ν is the weighing factor for a particular source based on deaggregation. The deaggregation of regional hazard in terms of fault recurrence is presented in Fig. 6. Finally, probability density function f_M (m) has been evaluated for each fault based on the following expression.

$$f_M(m) = P[M < m/m_0 \leq m \leq m_{\max}] = \frac{\beta \exp[-\beta(m - m_0)]}{1 - \exp[-\beta(m_{\max} - m_0)]} \tag{5}$$

Table 2 Source recurrence relation weighing factors

Faults	M_w	L_f (km)	No of eq.	α_i	χ_i	Average weighing factors
Marginal fault (MF)	5.7	290	26	0.118	0.132	0.125
Tapti North fault (TNF)	5.4	400	25	0.113	0.182	0.148
Son Narmada fault (SNF)	6.5	550	15	0.068	0.251	0.159
Paldi fault (PF)	5	15	18	0.081	0.007	0.044
Kutch Mainland fault (KMF)	7.7	170	68	0.308	0.078	0.193
Allah Bund fault (ABF)	7.8	149	18	0.081	0.068	0.075
Cambay West fault (CWF)	7	420	36	0.163	0.192	0.177
Barwani–Sukta fault (BSF)	6.2	198	15	0.068	0.090	0.079

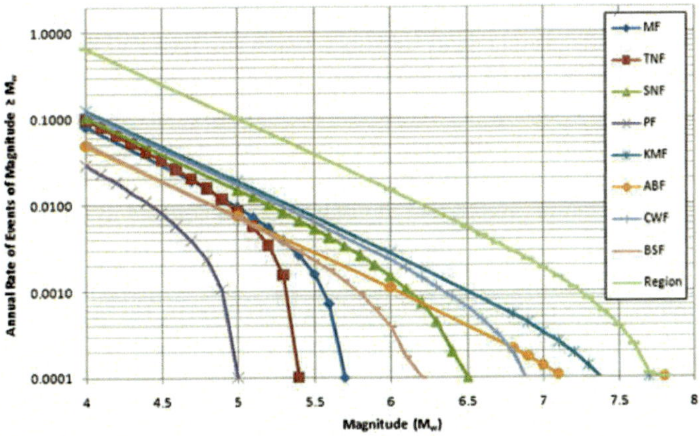

Fig. 6 Deaggregation of regional hazard in terms of fault recurrence

5 Probabilistic Seismic Hazard Analysis

In the present study, classical Cornell (1968) approach has been used to estimate the seismic hazard of the region with probabilistic framework. This approach has been considered by many researchers for the hazard analysis (Iyengar and Ghosh 2004; Raghukanth and Iyengar 2006; Anbazhagan et al. (2008); Vipin et al. 2009; Thaker et al. 2012) and discussed in the published literature. Hence, the detailed procedure for carrying out PSHA will not be repeated here. The detailed analysis has been carried out through Microsoft Excel (Thaker et al. 2012). Assuming that the number of earthquakes occurring on a fault follows a stationary Poisson's process, the probability that the control variable Y_T exceeds level y^*, in terms of a finite time interval T years is given by;

$$P\left[Y > y^*\right] = 1 - \exp\left(-\lambda_y T\right) \tag{6}$$

The mean annual rate of exceedance λ_y is computed at a site for different specified ground motion values y^* in a life period of T, which is given by;

$$\lambda_{y^*} = \sum_{i=1}^{N_S} \sum_{j=1}^{N_M} \sum_{k=1}^{N_R} v_i P\left[Y > y^*/m_j, r_k\right] f_{M_i}\left(m_j\right) f_{R_i}(r_k) \Delta m \Delta r \tag{7}$$

where $m_j = m_o(j-0.5)(m_{\max} - m_o)/N_M$, $r_k = r_{\min} + (k-0.5)(r_{\max} - r_{\min})/N_R$, $\Delta m = (m_{\max} - m_o)/N_M$ and $\Delta r = (r_{\max} - r_{\min})/N_R$.

Vadodara City is divided into the grid size of 1 km × 1 km to estimate the hazard at the centre of each grid point. Four attenuation relationships, i.e. Raghukanth and Iyengar (2007), NDMA (2010), Hwang and Huo (1997) and Toro et al. (2002) have been considered in the present analysis. Raghukanth and Iyengar (2007) and NDMA (2010) relationships are based on the peninsular India. Hwang and Huo (1997) and Toro et al. (2002) relationships are based on ENA. The study region is situated in the peninsular India, and PI is having similar features in terms of seismogenic activities and known seismotectonics with ENA (Schweig et al. 2003). Hence, the average of these four ground motion prediction relations has been adopted in the analysis. The peak ground acceleration (PGA) model at rock level has been developed for 10 and 2% probability of exceedance in 50 and 100 years, which are shown in Figs. 7 and 8. The PGA values range between 0.054–0.071 g and 0.104–0.139 g for 10% and 2% probability of exceedance in 50 years, respectively.

6 Results and Discussions

The present study investigates the seismic hazard of Vadodara region using probabilistic framework. The seismotectonic model has been developed for the study region using ArcGIS 10.3 software considering seventeen major faults and minor faults. The seismic parameter '*b*' is estimated as 0.815 ± 0.001, which is in very good agreement with other published literature for PI. Two regional specific and two ENA-based attenuation relationships have been adopted in the present analysis. The PGA model at rock level has been generated by the PSHA, which ranges from 0.054 to 0.071 g and 0.104 to 0.139 g for 10% and 2% probability of exceedance in 50 years respectively. The maximum credible earthquake (MCE) is related to ground motion with 2475 years return period, while the design basic earthquake (DBE) is related to the 475 years return period (Menon et al. 2010; Thaker et al. 2012). The study region

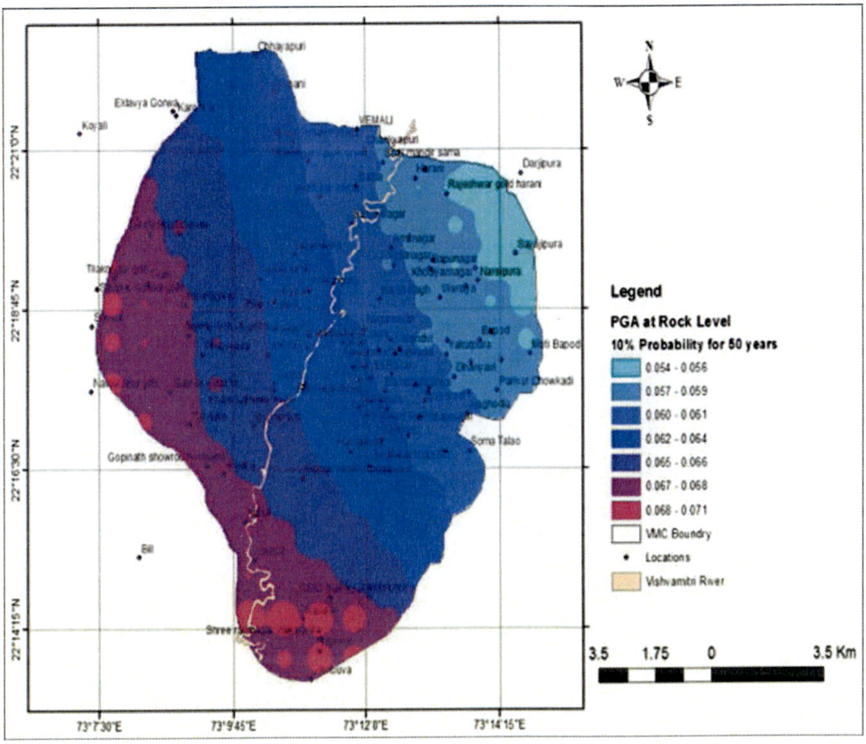

Fig. 7 PGA model at rock level for 10% probability of exceedance in 50 years (475 years return period)

falls under the zone factor III with PGA of 0.16 g (IS:1893-2016), which is likened with 2475 years return period. Hence, the present results are 13% lower as compare to IS code at rock level. Further, IS code does not provide any discrimination for zone factor at rock as well as the soil strata. The literature shows that generally the PGA value increases from 1.5 to 1.8 times at the surface as compared to engineering rock. It shows that the IS code would be underestimated the peak ground acceleration values at soil strata. The soil profile of the study region is loose alluvium type but PGA value at the surface level is not the scope of the present paper. Moreover, the results are also very well matched with the other studies carried out in PI (Thaker et al. 2012; Raghukanth and Iyengar 2006). The results will be further useful for risk mitigation and earthquake-resistant design of structures.

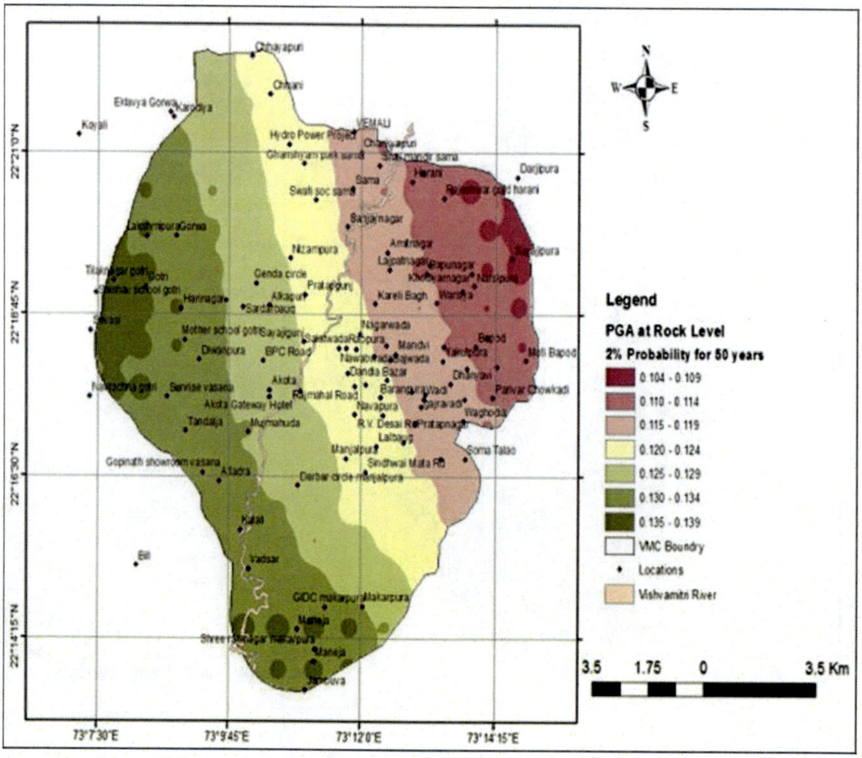

Fig. 8 PGA model at rock level for 2% probability of exceedance in 50 years (2475 years return period)

References

Anbazhagan P, Vinod JS, Sitharam TG (2008) Probabilistic seismic hazard analysis for Bangalore. Nat Hazards 48:45–166

Chandra U (1977) Seismotectonics of Himalaya. Cur Sci 62(1–2):40–71

Cornell CA (1968) Engineering seismic risk analysis. Bull Seismol Soc Am 58:1583–1606

Gardner JK, Knopoff L (1974) Is the sequence of earthquakes in Southern California with aftershocks removed Poissonian. Bull Seismol Soc Am 64(5)

Gutenberg B, Richter CF (1944) Frequency of earthquake in California. Bull Seismol Soc Am 34:185–188

Hwang H, Huo JR (1997) Attenuation relations of ground motion for rock and soil sites in eastern United States. Soil Dyn Earthq Eng 16:363–372

IS 1893 (2016) Indian Standard criteria for earthquake resistant design of structures. Part 1. General provisions and buildings. Bureau of Indian Standards, New Delhi

Iyengar RN, Ghosh S (2004) Microzonation of earthquake hazard in Greater Delhi area. Curr Sci 87(9) (General Article)

Jaiswal K, Sinha R (2006) Probabilistic modelling of earthquake—hazard in stable continental shield of The Indian Peninsula. ISET J Earthq Technol 470, 43(3):49–64

Kaila KL, Gaur VK, Narain H (1972) Quantitative seismicity maps of India. Bull Seismol Soc Am 62:1119–1131

Lai CG, Menon A, Corigliano M, Ornthamrrath T, Sanchez HL, Dodagoudar GR (2009) Probabilistic seismic hazard assessment and stochastic site response analysis at the archaeological site of Kancheepuram in southern India. Research report EUCENTRE 2009/01. IUSS Press, Pavia, p 250. ISBN 978-88-6198-037-2

Malik JN, Sohoni S, Karanth V, Merh S (1999) Modern and Historic Seismicity of Kachchh Peninsula. Western India. J Geol Soc India 54:545–550

McGuire RK, Arabasz WJ (1990) An introduction to probabilistic seismic hazard analysis. Geotech Environ Geophys. In: Ward SH (ed) Society of Exploration Geophysicist 1:333–353

Mehta P, Thaker TP, Raghavendra HB (2017a) Review of seismic hazard approaches for Indian subcontinents. In: International conference on research and innovations in science, engineering & technology, Birla Vishvakarma Mahavidyalaya

Mehta P, Thaker TP, Raghavendra HB (2017b) Development of seismotectonic model for central Gujarat region. In: Indian geotechnical conference, GeoNEst, IIT Guwahati

Mehta P, Thaker TP, Raghavendra HB (2018) Deterministic seismic hazard analysis of central Gujarat region. In: Proceedings of GeoShanghai 2018 international conference: advances in soil dynamics and foundation engineering, GeoShanghai international conference, Shanghai

Menon A, Ornthammarath T, Corigliano M, Lai CG (2010) Probabilistic seismic hazard microzonation of Tamil Nadu in Southern India. Bull Seismol Soc Am 100(3):1320–1341

Merh SS (1995) Geology of Gujarat. GeolSocInd, 222 p

National Disaster Management Authority (2010) Development of probabilistic seismic hazard map of India. Technical report Published by Govt. of India, Working committee of experts (WCE), NDMA

Oldham T (1883) A catalogue of indian earthquakes from the earliest times to the end of 1869 A.D. Mem Geol Surv India XIX (3)

Raghukanth STG, Iyengar RN (2006) Seismic hazard estimation for Mumbai city. Curr Sci 91(11) (Research Article)

Raghukanth STG, Iyengar RN (2007) Estimation of seismic spectral acceleration in Peninsular India. J Earth Syst Sci 116(3):199–214

Ram A, Rathor HS (1970) On frequency magnitude and energy of significant Indian earthquakes. Pure Appl Geophys 79:26–32

Rao RB, Rao SP (1984) Historical seismicity of Peninsular India. Bull Seismol Soc Am 74:2519–2533

Schweig E, Gomberg J, Petersen M, Ellis M, Bodin P, Mayrose L, Rastogi BK (2003) The M_W 7.7 Bhuj earthquake: global lessons for earthquake hazard in intra-plate regions. J Geol Soc India 61(3):277–282

Seeber L, Armbruster JG, Jacob KH (1999) Probabilistic assessment of seismic hazard for Maharashtra. Govt. of Maharashtra, Unpublished report

SEISAT (2000) Seismotectonic atlas of India and its environs. Geological Survey of India

Shukla J, Choudhury D (2012) Estimation of seismic ground motions using deterministic approach for major cities of Gujarat. Nat Hazards Earth Syst Sci 12:2019–2037

Sinha-Roy S, Malhotra G, Guha DS (1995) A transect across Rajasthan Precambrian terrain in relation to geology. Tectonics and crustal evolution of South Central Rajasthan, In: Sinha-Roy S, Gupta KR (eds) Continental crust of Northwestern and Central India, vol 31. Geological Soc. India Mem., pp 63–90

Stepp (1973) Analysis of completeness of the earthquake sample in the Puget Sound area. In: Harding ST (ed) Seismic zoning. NOAA Tech. report, ERL 267-ESL30, Boulder, Colorado

Thaker TP, Rathod GW, Rao KS, Gupta KK (2012) Use of seismotectonic information for the seismic hazard analysis for Surat city, Gujarat, India: deterministic and probabilistic approach. Pure Appl Geophys 169:37–54

Tinti S, Mulargia F (1985) Completeness analysis of a seismic catalog. Ann Geophys 3:407–414

Toro GR (2002) Modification of the Toro et al. (1997), attenuation equations for large magnitudes and short distances. Risk Engineering, Boulder

Tripathi JN (2005) Probabilistic assessment of earthquake recurrence in the January 26, 2001 earthquake region of Gujarat, India. J Seismol 10:119–130

Uhrhammer R (1986) Characteristics of Northern and Southern California seismicity. Earthq Notes 57

Vipin KS, Anbazhagan P, Sitharam TG (2009) Estimation of peak ground acceleration and spectral acceleration for South India with local site effects: probabilistic approach. Nat Hazards Earth Syst Sci

Wiemer S, Wyass M (2000) Minimum magnitude of complete reporting in earthquake catalogs: examples from Alaska, the Western United States, Japan. Bull Seismol Soc Am 90:859–869

Response of Monopile Supported Offshore Wind Turbine in Liquefied Soil

Sangeet Kumar Patra and Sumanta Haldar

Abstract Offshore wind turbines (OWT) are the potential source of renewable energy. Monopile is a common choice as foundation for OWT, since this type of foundation has proved to be economical at shallow water depth and is designed for 25–30 years. These structures are designed as soft–stiff approach, where the fundamental frequency of soil–monopile–tower system is placed between the rotor frequency (1P) and blade passing frequency (3P for 3 bladed turbines). Design guidelines suggest that the fundamental frequency of the OWT system shall be at least 10% away from the 1P and 3P frequencies. Previous studies mostly focus on the behaviour of these structures under long-term cyclic loads. Limited studies are available on the behaviour of OWT structures in liquefied soil when subjected to earthquake loading. Soil surrounding the monopile may liquefy during earthquake which affects the dynamic response of OWT structures. In this study, dynamic behaviour of OWT structure in a liquefied soil is examined using numerical model in open-source code, OpenSees with aid of OpenSeePL. In this study, monopile is embedded in saturated silty sand with 60 m depth soil profile is considered. Spectrum consistent strong motion accelerogram is generated for various earthquake moment magnitudes. The dynamic analysis is carried out in time domain, and response of the OWT system is examined for various diameter and length of monopile, and magnitude of earthquake. Finally, design implications are suggested.

Keywords Offshore wind turbine · Liquefaction · Monopile · Soil–structure interaction

S. K. Patra · S. Haldar (✉)
Department of Civil Engineering, School of Infrastructure,
Indian Institute of Technology Bhubaneswar, Jatni, Bhubaneswar, Odisha 752050, India
e-mail: sumanta@iitbbs.ac.in

S. K. Patra
e-mail: skp16@iitbbs.ac.in

M. Latha Gali and P. Raghuveer Rao (eds.), *Geohazards*, Lecture Notes in Civil Engineering 86, https://doi.org/10.1007/978-981-15-6233-4_26

1 Introduction

Wind energy has shown unprecedented growth to the production of renewable energy. Offshore wind turbines provide an increasing proportion of wind energy generation capacity because offshore sites are characterized by stronger and more stable wind conditions (Lombardi et al. 2013). Monopile is a common choice as foundation for OWT due to its simpler shape, easy to construct and economical (Cui and Bhattacharya 2016). Monopile is a long slender steel member, typically 3–6 m outer diameter, 22–40 m long, and is installed at a water depth of 10–25 m. Due to growing energy demand, OWT structures are constructed in seismic areas which are subjected to seismic loading during its operational period. Many countries such as USA, China, India, and South East Asia are in high seismic zones, where magnitude M9-class earthquakes may occur (Risi et al. 2018). In response to growing energy demand from low-carbon energy source, the government of India has recently proposed to double its renewable energy capacity (https://www.renewableenergyworld.com). This would include exploiting India's 7500 km-long coastline for the production of low-carbon energy from offshore wind power (Indian Wind Energy Outlook 2012).

Major challenges in the design of OWT are to satisfy the serviceability limit state (SLS) under long-term cyclic loads (DNV GL-ST-0126 2016). Modern variable speed wind turbines operate at rotor speed of 10–20 rpm, i.e. excitation frequency interval is about 0.1–0.3 Hz (Lombardi et al. 2013). This rotor frequency is referred as 1P frequency. LeBlanc (2009) reported a typical frequency range of waves is in the range of 0.05–0.5 Hz. Blade passing frequency (i.e. 2P frequency for two-bladed and 3P frequency for three-bladed) induces dynamic loading on the structure due to tower shadowing effect (Zaaijer 2006). Offshore wind turbines are designed as soft–stiff approach, where the fundamental frequency of soil–monopile–tower system is placed between the rotor frequency (1P) and blade passing frequency (3P for three-bladed turbines) (Cui and Bhattacharya 2016; Bhattacharya et al. 2013). The resonance condition is avoided for an OWT structure keeping the fundamental frequency of the system $\pm 10\%$ away from the rotor frequency (1P) and blade passing frequency (3P) (DNV GL-ST-0126 2016). Past studies by Lombardi (2010), Bhattacharya et al. (2011) and Cox and Jones (2011) showed that the fundamental frequency of a wind turbine system changes with cycles of loading. Abhinav and Saha (2017), Bhattacharya et al. (2011) reported that change in the fundamental frequency of OWT structure is strongly dependent on the shear strain level in the soil surrounding the pile.

Indian offshore sites show the variation of different soils, namely silty sand, soft clay and dense sand at various depths (Gulhati 1989). OWT embedded in silty sand may liquefy during strong motion which affects the dynamic response of OWT structures. The fundamental frequency of the OWT system may approach to 1P frequency, and resonance may occur due to the softening of soil during liquefaction. Lombardi and Bhattacharya (2016) studied the seismic performance of pile-supported models under transient to full liquefaction condition using experimental and numerical analyses. They reported underestimation of response of the structure from numerical analysis. Risi et al. (2018) studied the response of offshore wind turbine during earthquake using finite element model (*p–y* spring approach) and observed that the deformation (displacement and rotation) increases in case of loose soil. Finn and Martin (1980) suggested modification of API (2011)-based *p–y* curves to account for the resistance of pile to soil liquefaction. Amini and Qi (2000) carried stress-controlled undrained tests on homogeneous and stratified silty sand (with 10–50% of silt) to check the liquefaction behaviour. No significant difference has been reported between homogeneous and stratified samples; however, increase in silt content increases the resistance to liquefaction. Zheng et al. (2015) carried out scaled model test to investigate the structural response of OWT structure under the action of both seismic and wave loads. They reported that the dynamic amplification factor is larger for joint action of both earthquake and wave loading than earthquake alone. A few past studies on liquefied deposit have been conducted on offshore wind turbine resting on bucket foundations, suction caisson and monopile foundation. Wang et al. (2017) and Zhang et al. (2014) studied the behaviour of OWT during earthquake. However, no rational design strategy has been proposed. Objective of this study is to examine the response of monopile supported OWT on silty sand under seismic loading of different moment magnitudes. A parametric study is performed to examine the response of OWT structure due to the seismic loading in liquefied soil.

1.1 Methodology

The response of OWT structure in liquefied soil due to seismic load is obtained using three-dimensional (3D) finite element (FE) simulations. The simulations reported in this study are conducted using the open-source computational platform OpenSees (https://opensees.berkeley.edu; Mazzoni et al. 2006; McKenna et al. 2010) with the aid of a graphical user interface "OpenSeesPL" (Lu 2006). OpenSees is a software framework for developing applications to simulate the performance of structural and geotechnical systems subjected to earthquake loading using finite element methods (McKenna et al. 2010). Figure 1 shows the typical 3D FE model used in this study. The OWT system (monopile–transition piece–tower) is modelled by means of linear beam–column element with structural properties of a 5 MW OWT (Jonkman et al. 2009). The rotor nacelle assembly (RNA) is modelled as lumped mass at the tower top. The modelling procedure is explained in the following section.

1.2 Modelling of Soil

The soil domain is modelled by 20-node, hexahedral brick elements with solid–fluid fully coupled material (Lu 2006; Yang et al. 2008), where 20 nodes define the solid

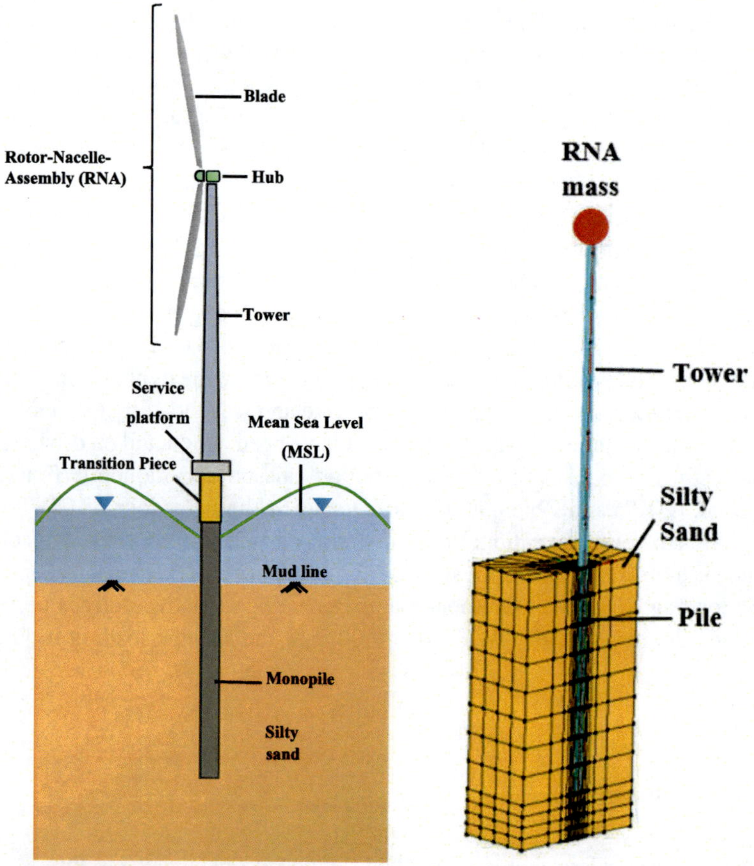

Fig. 1 Schematic diagram of the proposed numerical model

translational degrees of freedom, and eight corner nodes define the fluid pressure. Soil constitutive model is defined as PressureDependMultiYield material (Parra 1996; Elgamal et al. 2003; Yang et al. 2008), which is based on the multi-yield surface plasticity theory (Prevost 1978, 1985). It is an elastic–plastic material to simulate the response characteristics of pressure-sensitive soil materials under general loading conditions. The characteristics include dilatancy (shear-induced volume contraction or dilation) and non-flow liquefaction (cyclic mobility), that is typically exhibited in sands or silts during monotonic or cyclic loading. During the application of gravity load (and static loads), material behaviour is considered as linear elastic. In the subsequent dynamic loading phase, the stress–strain response is elastic–plastic. In this material model, shear stress and shear strain backbone curve is represented by the hyperbolic relationship. Drucker–Prager yield criterion is considered for analysis.

A 60-m-thick silty sand layer with saturated mass density 1700 kg/m^3 with permeability coefficient, $k = 6.6 \times 10^{-5}$ m/s used in this study. The bedrock depth is assumed to be at 60 m below the marine bed. The modelling parameters include

Table 1 Soil parameters

Parameters	Silty sand
Mass density (kg/m^3)	1700
Low-strain shear modulus G_r (Pa) at 80 kPa mean effective confinement	5.5×10^7
Friction angle ϕ (degrees)	29
Phase transformation angle ϕ_{PT} (degrees)	29
Contraction parameter c_1[a]	0.21
Liquefaction parameter l_1 (kPa)[a]	10
Liquefaction parameter l_2[a]	0.02
Liquefaction parameter l_3[a]	1

[a]More details are available elsewhere (Yang et al. 2008)

various dynamic soil properties, such as low-strain shear modulus, friction angle, parameters to control the dilatancy effects (e.g. phase transformation angle, contraction, and dilation) and liquefaction parameters. The relevant parameters are listed in Table 1. The FE matrix equation is integrated in time using trapezoidal rule with the second-order backward difference formula (TRBDF2) integrator (Bathe 2007) in OpenSeesPL. For each time step, the solution is obtained using the modified Newton–Raphson approach with Krylov subspace acceleration (Carlson and Miller 1998; Mazzoni et al. 2006). On this basis, the initial tangent stiffness of the system (after application of gravity) is used for all steps and iterations to achieve the prescribed tolerance (normalized energy increment less than 1.0×10^{-6}). A relatively low level of Rayleigh damping where the stiffness-proportional coefficient (0.00091) is based on the initial stiffness matrix is employed to enhance the numerical system stability (Yang and Elgamal 2002). The mass-proportional damping coefficient 0.215 is used for all the cases in this study. Figure 1 shows the schematic diagram of the proposed numerical model, where monopile and tower are replaced by a single solid column. RNA mass is replaced by a point mass at the top. Rigid box boundary condition is used for the soil domain. In this boundary condition, lateral boundaries are fixed in both horizontal directions and free in vertical direction.

1.3 Modelling of Pile and Tower

The OWT system (monopile–transition piece–tower) is modelled by means of linear beam–column element with structural properties similar to National Renewable Energy Laboratory (NREL) 5 MW reference OWT (Jonkman et al. 2009). The properties of NREL 5 MW OWT are listed in Table 2. The monopile and tower are hollow cylindrical section for the actual OWT. The present numerical model is developed by considering an equivalent solid cylindrical section to reduce the computational time. An equivalent diameter for monopile–transition piece and tower is computed by equating their moment of inertia. The equivalent density of the transformed section is also estimated to preserve the mass of the system. The length and elastic modulus are kept the same as that of original OWT. Tower is considered as uniform cross section and its diameter assumed to be equal to the diameter of monopile for simplicity. Table 3 shows the equivalent diameter used for the different outer diameter of monopile keeping constant value of thickness (t_p) equals to 0.09 m. The mass and modulus of elasticity of the tower kept the same as prototype OWT. The mass of the rotor nacelle assembly (RNA) at the top is replaced with a point mass. A convergence study is carried out to determine the mesh size. Figure 2 shows the mesh convergence study. It shows the maximum displacement of OWT structure with total number of mesh elements per unit volume. The convergence occurred at 3103 elements per unit volume of the model hence adopted in this study.

Table 2 Properties of 5 MW offshore wind turbine

Property	Value
Rotor diameter (m)	126
Hub height (above MSL) (m)	80
Cut in wind speed (m/s)	3
Cut out wind speed (m/s)	25
Rated wind speed (m/s)	11.4
Tower Young's modulus (Pa)	210×10^9
Rotor mass (kg)	111×10^3
Tower mass (kg)	347,460
Nacelle mass (kg)	240×10^3
Density of tower (kg/m^3)	7850
Yield strength of steel (N/m^2)	355×10^6

Table 3 Equivalent diameter of monopile

Outer diameter, D_0 (m)	Equivalent diameter of solid section (m)
5.0	3.04
5.5	3.27
6.0	3.5

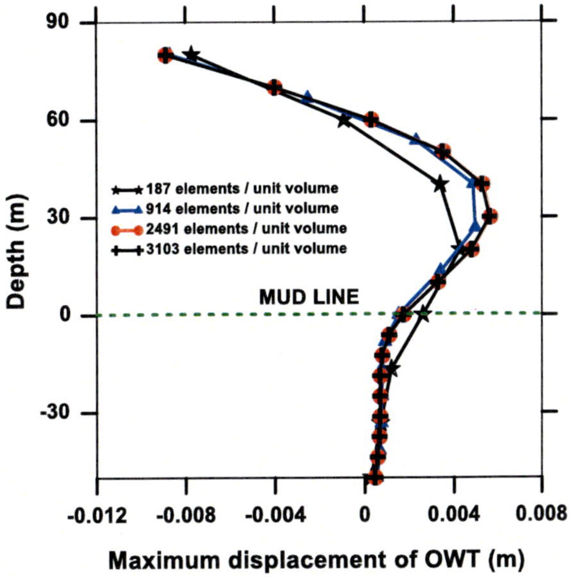

Fig. 2 Maximum displacement responses of OWT structure with depth for different mesh sizes

1.4 Seismic Loading

The artificial seismic accelerogram is generated matched to a specific response spectrum using SeismoArtif (2018). In this study, IS 1893: Part-1 (2016) response spectra is considered for the rock. According to the code, the design horizontal acceleration spectrum value shall be taken as half the value of response spectra that of the ground when the depth below ground is equal or more than 30 m. In this study, bedrock depth is considered as 60 m below the marine bed where the horizontal accelerogram is applied. Hence, the artificial accelerogram is generated considering the half of the design horizontal acceleration spectrum given in IS 1893: Part-1 (2016) as input of SeismoArtif (2018), as shown in Fig. 3. Where S_a/g is the average response acceleration spectrum coefficient, T is period of the structure in second. Bedrock is considered as generic rock with shear wave velocity ($V_s = 620$ m/s). Inter-plate regime with epicentral distance (distance between main event and station) of 10 km is considered. Damping is taken as 5%. Artificial seismographs of magnitudes 5.0, 6.0, 6.5 and 7.0 are generated which are shown in Fig. 4. The generated horizontal acceleration is applied at the base of soil along X-direction, and response of the structure is monitored. The peak ground acceleration (PGA) and duration of earthquake for different magnitudes are shown in Fig. 4.

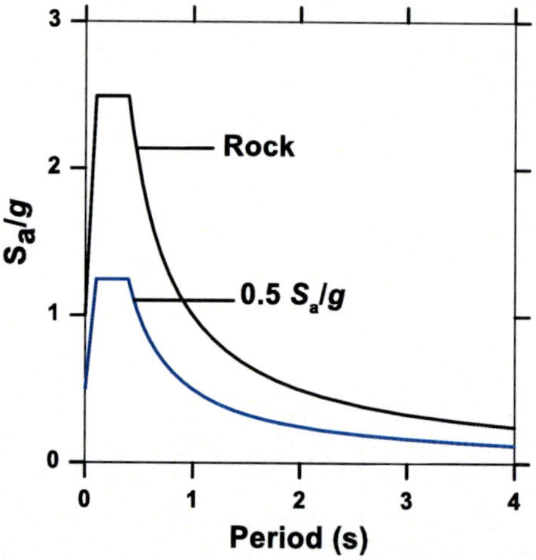

Fig. 3 Response spectra for rock for 5% damping

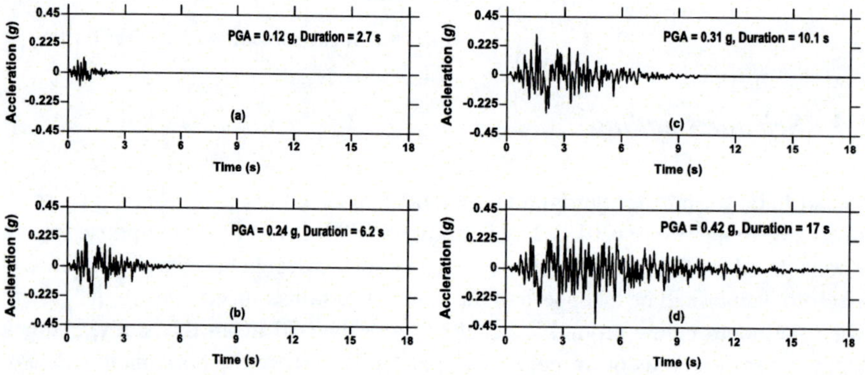

Fig. 4 Artificial acceleration time history for **a** earthquake magnitude 5.0 **b**, 6.0, **c** 6.5, and **d** 7.0

1.5 Parametric Study

A parametric study is carried out to examine the effect of pile geometry (embedded length and outer diameter) and earthquake magnitude on the response, namely depth of liquefaction of soil, maximum bending moment of pile, maximum rotation of pile at mud line level and the fundamental frequency of OWT structure. Different parameters considered are listed in Table 4. The analysis is carried out considering three different monopile diameters (D_o), namely 5, 5.5 and 6 m, and three different embedded lengths (L_p), namely 30, 40 and 50 m. The wall thickness of tower and monopile is considered as 0.09 m for all cases. The different values of moment magnitudes of seismic event, namely $M = 5.0$, 6.0, 6.5, and 7.0, are also considered.

Table 4 Range of selected parameters for analysis

Group	Parameter	value
Monopile geometry	Embedded length, L_p (m)	30, 40, 50
	Outer diameter, D_o (m)	5, 5.5, 6
	Wall thickness, t_p (m)	0.09
Loading	Seismic moment magnitude, M	5.0, 6.0, 6.5, 7.0

2 Results and Discussion

2.1 Depth of Liquefaction

Liquefaction is a state in saturated cohesionless soil wherein the effective shear strength is reduced to a negligible value. In this condition, the soil tends to behave like a fluid mass. The excess pore pressure ratio ru is a factor that is used to determine the onset of liquefaction. According to Kramer (1996), the excess pore pressure ratio is defined as:

$$r_u = \frac{\Delta p_w}{\sigma'_{3,c}} \tag{1}$$

where $\sigma'_{3,c}$ is the initial effective confining pressure from a triaxial compression test and Δp_w is the excess pore water pressure. Beaty and Byrne (2011) suggested a range for r_u as:

$$r_u = 0,\ \text{denotes no liquefaction and} \tag{2a}$$

$$r_u \geq 0,\ \text{denotes onset of liquefaction} \tag{2b}$$

Here the zone below mud line having excess pore pressure ≥0.7 considered as zone of liquefaction. Initially, an analysis is carried out to determine the variation of the depth of liquefaction for soil only. In order to examine the effect of monopile on the depth of liquefaction in soil, depth of liquefaction is also evaluated in the presence of monopile. The variation of depth of liquefaction in soil below the mud line level and depth of liquefaction in soil in the presence of monopile are shown in Fig. 5a, b. From both the figures, it is concluded that the depth of liquefaction is strongly dependent on the presence of monopile as well as seismic moment magnitude. Figure 5a, b shows for $M = 5$, and it can be seen that the depth of liquefaction is 6 m in case of soil, whereas the depth of liquefaction is 1.5 m for the soil in the presence of monopile. The presence of monopile decreases the depth of liquefaction due to increasing the

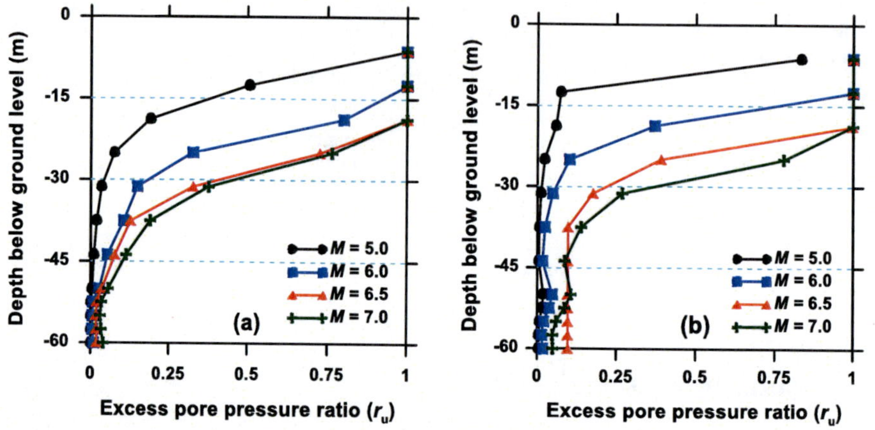

Fig. 5 Excess pore pressure ratio with respect to the depth at different seismic moment magnitudes: **a** soil only and **b** soil and pile, for $L_p = 50$ m, $D_o = 5$ m and $t_p = 0.09$ m

overall stiffness of soil–pile system as it increases the confinement zone of soil inside and around the monopile. It is also observed that the depth of liquefaction increases if moment magnitude of seismic event increases for both the cases (i.e. soil and soil–pile system). As the moment magnitude increases both pick ground acceleration (PGA) and duration of earthquake increases, which increases the depth of liquefaction.

2.2 *Effect of Length of Monopile*

A parametric study has been carried out to examine the effect of depth of embedment of monopile (L_p) on the response of the OWT structure. This study is done by keeping the outer diameter ($D_o = 5$ m) and thickness ($t_p = 0.09$ m) of monopile constant and by varying the embedment depth of the monopile ($L_p = 30$, 40 and 50 m). Figure 6 shows the non-dimensional liquefaction depth (h_L/h_s) as a function of non-dimensional embedment length (L_p/D_o) at different seismic moment magnitudes, where h_L is the depth of liquefaction of soil below mud line and h_s is the depth at which bedrock is present. In this study, $h_s = 60$ m is used. Figure 6 shows the depth of liquefaction decreases as embedment length of monopile increases. An increase in length of monopile increases the overall stiffness of soil–pile system and provides an additional strength to the soil and resistance to liquefaction. The plot of non-dimensional maximum bending moment (M_{MAX}/M_R) and non-dimensional embedment length (L_p/D_0) at different seismic moment magnitudes is shown in Fig. 7. M_R is the maximum resisting moment of monopile which is estimated as:

$$M_R = f_y Z \quad (3)$$

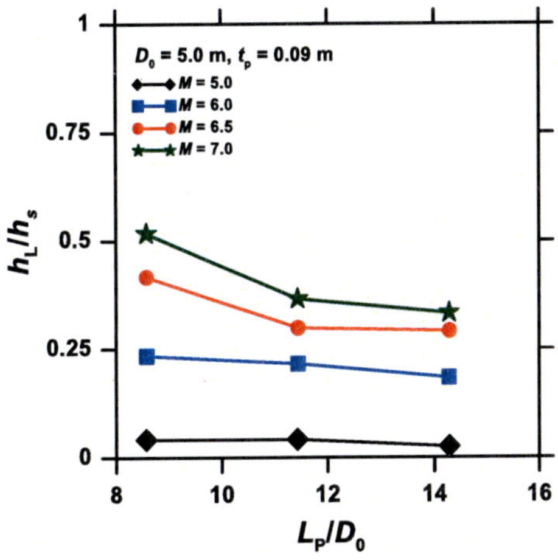

Fig. 6 Non-dimensional depth of liquefaction with depth for $D_o = 50$ m and $t_p = 0.09$ m at different embedded length of monopile (L_p)

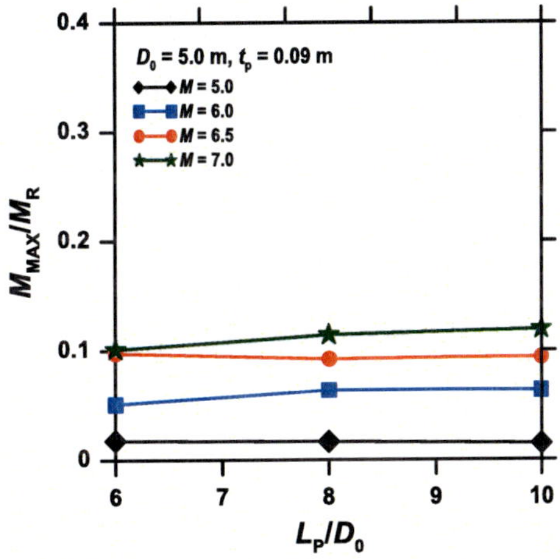

Fig. 7 Non-dimensional maximum bending moment of with depth for $D_o = 50$ m and $t_p = 0.09$ m at different embedded length of monopile (L_p)

where f_y is the yield stress of the steel ($f_y = 355$ MPa) and Z is the section modulus of the monopile section.

It is observed that the effect of length of monopile has marginal influence on the maximum bending moment. However, maximum bending moment is largely affected by the seismic moment magnitude. It can be seen from Fig. 7, that if magnitude increases the maximum bending moment also increases as the depth of liquefaction increases. This is due to the fact that the PGA and duration of earthquake increase if moment magnitude increases, which results in increase in maximum bending in monopile.

Similarly, non-dimensional maximum rotations at mud line level ($\theta_{max}/\theta_{all}$) is plotted with non-dimensional embedment length (L_p/D_o) at different seismic moment magnitudes in Fig. 8. Allowable rotation (θ_{all}) at mud line level is 0.5° according to DNV GL-ST-0126 (2016). θ_{max} is the maximum rotation at mud line level of

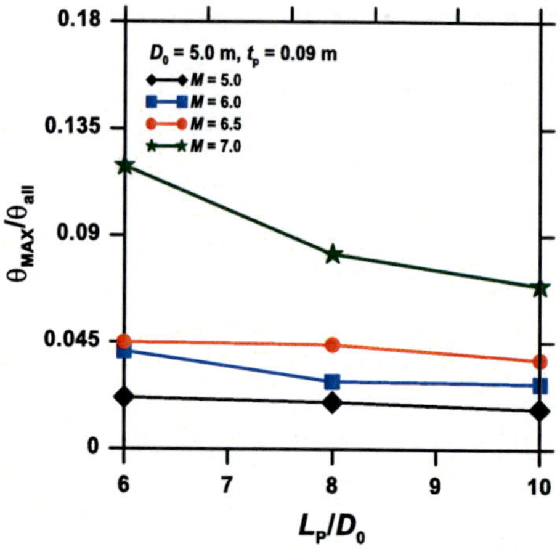

Fig. 8 Non-dimensional maximum rotation at mud line level for $D_o = 5.0$ m and $t_p = 0.09$ m at different embedded length of monopile (L_p)

monopile. In all the cases, maximum rotation at mud line level decreases with embedment length. As length increases, the densified zone around monopile increases. As depth of liquefaction also decreases as length increases (i.e. lesser length of pile remain unsupported) resistance to rotation increases. Further, maximum rotation at mud line level increases with increase in earthquake magnitude for a particular length of monopile. It is due to increase in PGA and duration of earthquake.

2.3 *Effect of Diameter of Monopile*

A parametric study is carried out to examine the effect of monopile diameter on the response of OWT structure due to different seismic moment magnitudes. For this purpose, embedment depth ($L_p = 30$ m) and thickness ($t_p = 0.09$ m) of monopile are kept constant and outer diameter is varied ($D_o = 5.0$, 5.5 and 6.0 m). Results are presented in Fig. 9. Figure 9 shows the non-dimensional liquefaction depth (h_L/h_s) as a function of non-dimensional monopile outer diameter (D_o/t_w) at different seismic magnitudes. It is observed that with increase in diameter, depth of liquefaction increases marginally. The plot of non-dimensional maximum bending moment (M_{MAX}/M_R) and non-dimensional outer diameter (D_o/t_w) of monopile at different earthquake magnitudes is shown in Fig. 10. It is observed that the maximum bending moment developed on the monopile increases with diameter. Similarly, non-dimensional maximum rotations at mud line level with non-dimensional outer diameter (D_o/t_w) of monopile at different seismic moment magnitudes are shown in Fig. 11. In all the cases, maximum rotation at mud line level increases with outer diameter. An increase in diameter causes the increase in density of monopile and hence, inertial force increases. This inertial force generated on OWT structure is transmitted to the soil inside and surrounding the monopile. It increases the depth

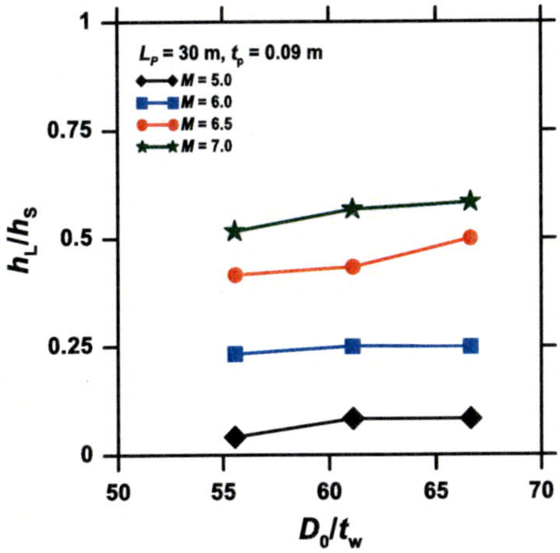

Fig. 9 Non-dimensional depth of liquefaction with depth for $L_p = 50$ m and $t_p = 0.09$ m at different outer diameter of monopile (D_o)

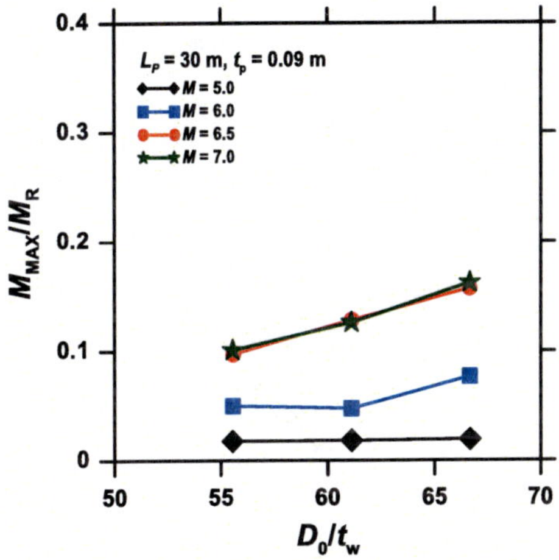

Fig. 10 Non-dimensional maximum bending moment for $L_p = 50$ m and $t_p = 0.09$ m at different outer diameter of monopile (D_o)

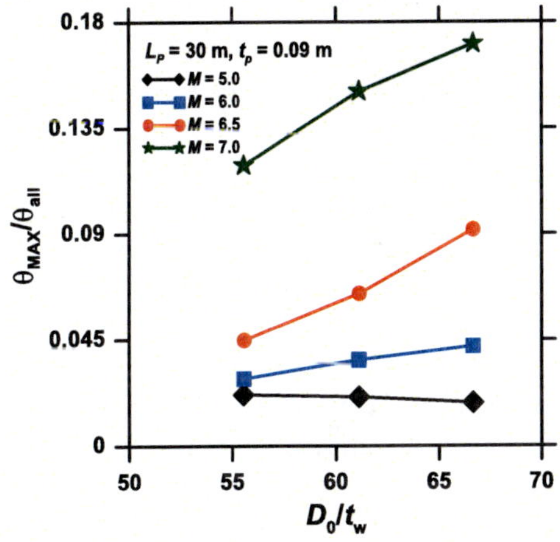

Fig. 11 Non-dimensional maximum rotation at mud line level for $L_p = 50$ m and $t_p = 0.09$ m at different outer diameter of monopile (D_o)

Table 5 Variation of natural frequency with embedment depth before liquefaction and after liquefaction for different earthquake magnitude

Embedment length, L_p (m)	Seismic moment magnitude, M	Natural frequency after liquefaction, f_{n1} (Hz)	Natural frequency before liquefaction, f_n (Hz)
30	5.0	0.384	0.394
	6.0	0.373	
	6.5	0.372	
	7.5	0.365	

of liquefaction along the depth of monopile resulting in maximum bending moment and maximum rotation.

2.4 *Effect on Fundamental Frequency of OWT*

The fundamental frequency of OWT structure strongly depends on the soil stiffness surrounding the monopile. During liquefaction certain depth of soil surrounding, the monopile loses its stiffness. It causes change in the fundamental frequency of OWT structure. Table 5 shows the fundamental frequencies of OWT structure for $L_p = 30$ m and $D_o = 5$ m before and after liquefaction for different seismic moment magnitudes. The fundamental frequency is found to be less after liquefaction in all the cases. A reduction of 4–6% is noticed due to increase in depth of liquefaction. The fundamental frequency decreases as seismic moment magnitude increases due to increase in depth of liquefaction.

3 Conclusions

A parametric study is carried out to examine the response of OWT structure during liquefaction. For this purpose, different magnitudes of earthquake are chosen. The analysis is carried out using OpenSees with the aid of OpenSeesPL, a user interface that simplifies the pre- and post-processing phases. For this representative study, a 60 m saturated cohesionless soil profile (silty sand) with monopile and tower was subjected to ground motion generated. From the study, it is concluded that embedment length of monopile has great influence to resist bending moment and the diameter has a marginal impact on resisting maximum bending moment and rotation. The fundamental frequency of OWT structure decreases with increasing seismic moment magnitude and moves towards 1P frequency after liquefaction.

References

Abhinav KA, Saha N (2017) Dynamic analysis of monopile supported offshore wind turbines. Proc Inst Civ Eng-Geotech Eng 170(5):428–444

Amini F, Qi GZ (2000) Liquefaction testing of stratified silty sands. J Geotech Geoenviron Eng 126(3):208–217

API-RP-2GEO (2011) Geotechnical and foundation design considerations. American Petroleum Institute, Washington, DC, USA

Bathe KJ (2007) Conserving energy and momentum in nonlinear dynamics: a simple implicit time integration scheme. Comput Struct 85:437–445

Beaty MH, Byrne PM (2011) UBCSAND constitutive model version 904aR. Itasca UDM Web Site, p 69

Bhattacharya S, Lombardi D, Wood DM (2011) Similitude relationships for physical modelling of monopile-supported offshore wind turbines. Int J Phys Model Geotech 11:58–68

Bhattacharya S, Nikitas N, Garnsey J, Alexander NA, Cox J, Lombardi D, Wood DM, Nash DF (2013) Observed dynamic soil–structure interaction in scale testing of offshore wind turbine foundations. Soil Dyn Earthq Eng 54:47–60

Carlson N, Miller K (1998) Design and application of a gradient-weighted moving finite element code I: in one dimension. SIAM J Sci Comput 19(3):728–765

Cox J, Jones C (2011) Long term performance of suction caisson supported offshore wind turbines. Struct Eng 89(19):12–13

Cui L, Bhattacharya S (2016) Soil–monopile interactions for offshore wind turbines. Proc Inst Civ Eng-Eng Comput Mech 169(4):171–182

DNVGL-ST-0126 (2016) Design of offshore wind turbine structures. DET NORSKE VERITAS

Elgamal A, Yang Z, Parra E, Ragheb A (2003) Modeling of cyclic mobility in saturated cohesionless soils. Int J Plast 19(6):883–905

Finn WL, Martin GR (1980) Offshore pile foundations in sand under earthquake loading. Appl Ocean Res 2(2):81–84

Gulhati SK (1989) Geotechnical aspects of the Indian offshore environment

GWEC (2012) Indian wind energy outlook. Global Wind Energy Council, Brussels, Belgium

Jonkman J, Butterfield S, Musial W, Scott G (2009) Definition of a 5-MW reference wind turbine for offshore system development No. NREL/TP-500-38060. National Renewable Energy Lab (NREL), Golden, CO, United States

Kramer SL (1996) Geotechnical earthquake engineering. In: Prentice-Hall international series in civil engineering and engineering mechanics. Prentice-Hall, New Jersey

LeBlanc C (2009) Design of offshore wind turbine support structures. Doctor of Philosophy, Technical University of Denmark

Lombardi D (2010) Dynamics of offshore wind turbines. M.Sc. thesis. U.K: University of Bristol

Lombardi D, Bhattacharya S (2016) Evaluation of seismic performance of pile-supported models in liquefiable soils. Earthq Eng Struct Dyn 49, 45(6):1019–1038

Lombardi D, Bhattacharya S, Wood DM (2013) Dynamic soil–structure interaction of monopile supported wind turbines in cohesive soil. Soil Dyn Earthq Eng 49:165–180

Lu J (2006) Parallel finite element modeling of earthquake site response and liquefaction. Ph.D. thesis, University of California, San Diego, La Jolla, CA

Mazzoni S, McKenna F, Fenves GL (2006) Open system for earthquake engineering simulation (OpenSees) user manual. Pacific Earthquake Engineering Research Center, University of California, Berkeley. https://opensees.berkeley.edu/.

McKenna F, Scott M, Fenves G (2010) Nonlinear finite-element analysis software architecture using object composition. J Comput Civ Eng 24(1):95–107

Parra E (1996) Numerical modeling of liquefaction and lateral ground deformation including cyclic mobility and dilation response in soil systems. Ph.D. thesis, Rensselaer Polytechnic Institute, Troy, NY

Prevost JH (1978) Plasticity theory for soil stress-strain behavior. ASCE J Eng Mech Div 104(5):1177–1194
Prevost JH (1985) A simple plasticity theory for frictional cohesionless soils. Int J Soil Dyn Earthq Eng 4(1):9–17
Risi DR, Bhattacharya S, Goda K (2018) Seismic performance assessment of monopile-supported offshore wind turbines using unscaled natural earthquake records. Soil Dyn Earthq Eng 109:154–172
Wang X, Yang X, Zeng X (2017) Seismic centrifuge modelling of suction bucket foundation for offshore wind turbine. Renew Energy 114:1013–1022
Yang Z, Elgamal A (2002) Influence of permeability on liquefaction-induced shear deformation. J Eng Mech 128(7):720–729
Yang Z, Lu J, Elgamal A (2008) OpenSees soil models and solid-fluid fully coupled elements user manual. University of California, San Diego
Zaaijer MB (2006) Foundation modelling to assess dynamic behaviour of offshore wind turbines. Appl Ocean Res 28(1):45–57
Zhang P, Xiong K, Ding H, Le C (2014) Anti-liquefaction characteristics of composite bucket foundations for offshore wind turbines. J Renew Sustain Energy 6(5):053102
Zheng XY, Li H, Rong W, Li W (2015) Joint earthquake and wave action on the monopile wind turbine foundation: an experimental study. Mar Struct 44:125–141

Reliability Analysis of Single Pile in Lateral Spreading Ground: A Three-Dimensional Investigation

J. S. Rajeswari and Rajib Sarkar

Abstract Estimation of design forces in pile foundation is prevalently deterministic for all the practical purposes and thereby ignores the intrinsic uncertainties associated with the geotechnical parameters. Three-dimensional numerical investigations have been carried out using open-source software OpenSees considering a single free head pile embedded in three-layer soil profile with liquefiable sandy layer sandwiched between two non-liquefiable clayey soil layers. In the present study, the influence of uncertainties in geotechnical parameters on the response of pile foundation in case of lateral spreading ground conditions has been considered and response surface models were generated using *response surface method* (RSM). Reliability analyses have been performed using the *first-order reliability method* (FORM) for pile head deflection and bending moment. Finally, the performance of the pile foundation in spreading ground condition has been presented in terms of reliability index values. The study indicated that the performance of pile in laterally spreading soil is highly sensitive to the thickness of liquefiable soil and overlying non-liquefiable crust.

Keywords Reliability · Pile · Lateral spreading

1 Introduction

Liquefaction of soil in sloping ground disrupts the stability of slope causing the soil to move downwards when static shear stress induced in soil during earthquake exceeds residual shear strength of soil in liquefied condition. Liquefaction induced lateral spreading during earthquakes has demonstrated disastrous effects on pile foundation located in water font areas. This effect is more damaging when the liquefied layer supports non-liquefiable crust on the top. Records of past earthquakes of 1964 Alaska, 1995 Kobe, 1964 Niigata, and 1987 superstition hills and recent earthquakes of 2010

J. S. Rajeswari · R. Sarkar (✉)
IIT (ISM) Dhanbad, Dhanbad 826004, India
e-mail: rajib@iitism.ac.in

J. S. Rajeswari
e-mail: rajeswarijs1993@gmail.com

M. Latha Gali and P. Raghuveer Rao (eds.), *Geohazards*, Lecture Notes in Civil Engineering 86, https://doi.org/10.1007/978-981-15-6233-4_27

Haiti, 2010 New Zealand and 2011 Tohoku have manifested severe cases of pile damages due to lateral spreading of liquefied soil (Finn and Fujita 2002; Ishihara and Cubrinovski 2004; Madabhushi et al. 2010; Motamed et al. 2013; Forcellini et al. 2013). Studies on coastal lowlands of California recorded by Youd and Perkins (1987) indicated that lateral spreading is most likely to occur in gentle slopes of 0.5° to 6°. Most of the pile damages are observed at the pile head for fixed head piles and at the interface of liquefied and non-liquefied soil for both free head and fixed head piles. Kinematic loading due to lateral spreading is solely responsible for the large forces and moments at the soil interface and hence makes the pile vulnerable at the zone of interface.

The conventional method of pile design neglects the uncertainties associated with the soil parameters. This may lead to unsafe design of pile foundation in laterally spreading soil since soil is inherently variable in nature. The traditional method of considering uncertainty is to adopt factor of safety approach which on other hand fails to truly appraise the variations and uncertainties associated with geotechnical parameters. Hence more efficient probabilistic technique such as reliability analysis becomes imperative to overcome this limitation and evaluate the performance of structures. In order to carry out reliability analysis, a linear or nonlinear functional relationship between input and output parameters is required. This relationship is either obtained from readily available analytical methods or by carrying out series of numerical analyses using designed methods such as response surface methodology. Babu and Srivastava (2007) demonstrated how reliability analysis can be utilized for obtaining allowable range of pressure acting on shallow foundation resting on cohesive frictional soil. Tandjiria et al. (2000) carried out reliability analysis in conjunction with response surface method and obtained maximum bending moment and pile head displacement for laterally loaded piles. Wong (1985) studied slope reliability using response surface method and found that it gives satisfactory results for preliminary assessment. Jiang et al. (2016) conducted reliability analysis of piles resting on slopes subjected to lateral loads and observed that failure probability is dependent on coefficient of variation of various input parameters. Lacasse and Nadim (1996) discussed the importance of probabilistic approach in pile axial capacity calculation. Kozubal et al. (2013) evaluated reliability of laterally loaded pile under two different subsurface conditions. Farag (2014) proposed an improved response surface scheme to evaluate reliability of piles exposed to lateral spreading.

The present study focuses on the performance of single piles embedded in three-layer soil profiles with liquefiable sand sandwiched between two non-liquefiable clayey layers subjected to lateral spreading. Reliability analysis is carried out using advanced first-order second-moment or Hasofer–Lind method using relationship obtained from response surface methodology by carrying out three-dimensional fully coupled numerical analyses. All the analyses have been performed using open-source software package OpenSees with advanced soil constitutive models. The highlights of the present study can be summarized as follows:

- Consideration of variable function for modeling changes in permeability of sand layer during the process of liquefaction.
- Study of effect of various input parameters on response of pile.
- Comparing the performance of single pile embedded in liquefiable and non-liquefiable layers of different thickness.

2 Basics of Reliability Analysis

Reliability analysis determines the capability of a system to carry the loads acting on it. Reliability is commonly expressed in terms of reliability index (β) which is related to the probability of failure as

$$\beta = 1 - p_{\mathrm{f}} \tag{1}$$

If the loads acting on the system and resistance are represented by Q and R, respectively, then the limit state condition is given by

$$Z = R - Q \tag{2}$$

And reliability index (β) can be determined as

$$\beta = \mu_Z / \sigma_Z \tag{3}$$

where μ_Z and σ_Z are mean and standard deviation of Z and denote the distance of the mean of failure condition from its critical value in units of standard deviation.

2.1 *Hasofer–Lind Reliability Index*

The advanced first-order second-moment (FOSM) method also known as Hasofer–Lind method calculates reliability index by reducing the variables to a standard coordinate system (Hasofer and Lind 1974). This reliability index gives better idea about stability of the system than factor of safety. Figure 1 shows reduced variables in transformed system.

In reduced coordinate system, Hasofer–Lind index (β_{HL}) is defined as the minimum distance of limit state surface from origin of axes and this minimum distance point on limit state surface is called design point. The matrix formulation for Hasofer–Lind index is given by

$$\boldsymbol{\beta}_{\mathbf{HL}} = \min_{x \in F} \sqrt{(\mathbf{x} - \mathbf{m})^{\mathrm{T}} \mathbf{C}^{-1} (\mathbf{x} - \mathbf{m})} \tag{4}$$

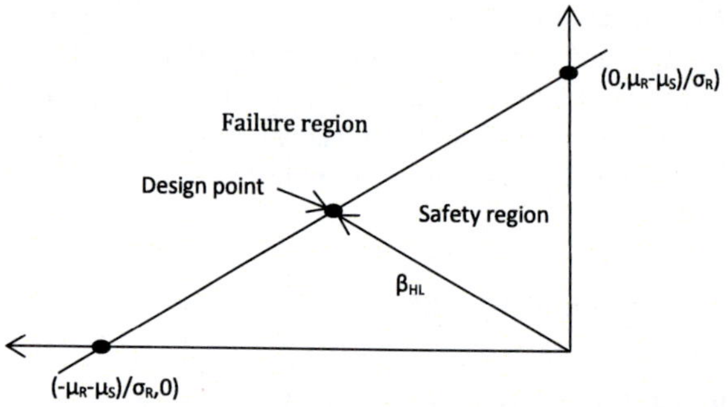

Fig. 1 Hasofer–Lind reliability index in reduced coordinate system

where *x*, *m*, and *C* represent vector of random variables, mean values of random variables, and covariance matrix, respectively, and *F* is the failure region.

2.2 *Ellipsoid Approach*

Low and Tang (1997) reported an alternative perspective of Hasofer–Lind reliability index in which variables can be represented by one standard deviation dispersion ellipsoid in original space of random variables. The amount of tilt of ellipsoid depends on the correlation between the variables. If the variables are uncorrelated, then the ellipsoid is parallel to coordinate axes. Thus the ellipse which is tangent to failure surface is β times the size of $1 - \sigma$ ellipse. Figure 2 shows critical ellipse and reliability index in normal variable space for correlated variables. This approach can be implemented easily to calculate reliability index using spreadsheet.

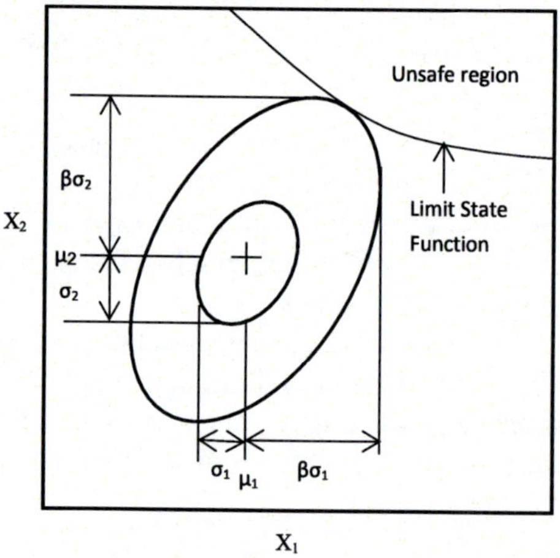

Fig. 2 Critical ellipse and reliability index in normal variable space

3 Response Surface Methodology

Response surface methodology is an assemblage of statistical and mathematical tools for developing an appropriate approximate relationship between input variables and response (Myers et al. 1995). The general 2^k factorial design is used for fitting linear or nonlinear response surface models, where k is the number of independent variables considered for the study. In RSM, the natural variables are converted to coded variables with mean zero and the same standard deviation. The first-order linear regression model in coded variables is given by

$$y = \beta_0 + \beta_0 x_1 + \beta_0 x_2 + \cdots + \beta_k x_k + \varepsilon \tag{5}$$

where y is the response, x_1, x_2, …, x_k are random variables, β_0, β_1, …, β_k are unknown variables, and ε is the source of errors.

3.1 *Model Adequacy*

For the fitted model to be adequate, the following two criteria need to be met:

(a) The normal probability plot must be approximately along a straight line.
(b) The coefficient of multiple determination, R^2 and adjusted R^2 values must be close. It is the measure of the reduction in variability of y. The large difference between the two implies that non-significant terms are present in the model. These non-dimensional coefficients can be calculated using Eqs. 6 and 7.

$$R^2 = 1 - (\mathrm{SS_E}/\mathrm{SS_T}) \tag{6}$$

$$R^2_{\mathrm{adj}} = 1 - \{\mathrm{SS_E}/n - p\}/\{\mathrm{SS_T}/n - 1\} \tag{7}$$

where n is the number of observations and p is the number of regression coefficients.

4 Problem Definition

A single free head pile of 0.5 m diameter and 10 m length embedded in three-layer soil profile as shown in Fig. 3 is considered. The soil profile consists of liquefiable sand layer sandwiched between two non-liquefiable clayey layers.

The pile properties are given in Table 1. The water table is assumed to be located at the ground surface. The size of the model is maintained to be 20 m × 12 m × 15 m throughout the study. The model is inclined at an angle α to the horizontal for

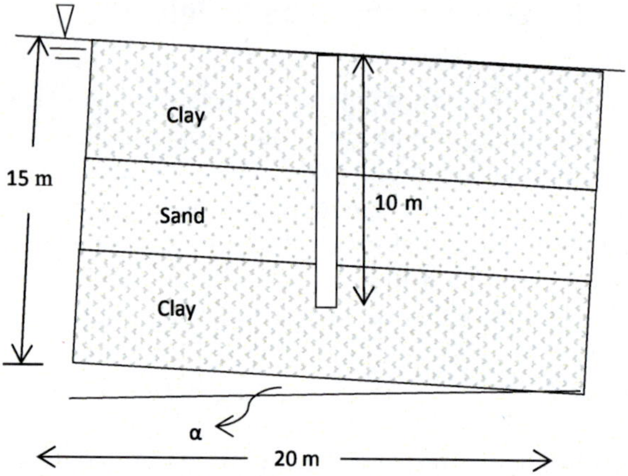

Fig. 3 Schematic diagram of single pile in layered soil profile

Table 1 Pile properties

Parameter	Values
Grade of concrete (MPa)	30
Modulus of elasticity (MPa)	2.74×10^7
Poisson's ratio	0.2
Moment of Inertia (m^4)	1.53×10^{-3}

simulating slightly sloping, almost level ground condition. The model is subjected to the 2004 Parkfield earthquake time history (PGA 0.36 g) in the direction of slope.

4.1 Input Parameters

As per Cubrinovski and Ishihara (2004), important parameters influencing pile response in laterally spreading soil are magnitude of ground displacement, ultimate pressure from crust layer, and stiffness degradation of liquefied soil. Taking this into consideration, three cases—A, B, and C—with different thickness of liquefiable and non-liquefiable layers as presented in Table 2 are considered for the present study. The crust thickness and liquefiable sand thickness are selected as per suggestion given in Brandenberg et al. (2011). The soil properties for upper (clay) and middle layer (Nevada sand) are considered as variable parameters and are presented in Table

Table 2 Cases considered for study

Parameter	Layer 1 (upper) (m)	Layer 2 (middle) (m)
Case A	1.5	4
Case B	1.5	1
Case C	4.5	4

Table 3 Soil properties

Property	Mean	Standard deviation	Lower limit	Upper limit
Density of clay (ton/m^3)	1.5	0.18	1.2	1.8
Relative density of sand	40	6	30	50
Density of clay (ton/m^3)	1.8 (fixed)	–	–	–

3, and bottom layer is considered to be stiff clayey soil for all the cases. Slope of the ground (α) is another key parameter affecting ground displacement and is taken as random variable with mean 2° and standard deviation 0.24°. Since all the random variables are assumed to be normally distributed, the lower and upper limit values are calculated considering 95% confidence interval: $x_{\min} = \mu - 1.65\sigma$ and $x_{\max} = \mu + 1.65\sigma$.

5 Numerical Modeling

The numerical modeling and analysis are performed using OpenSees. Since the model is symmetrical in nature, only one half of the model is considered in order to reduce computational time and effort. Eight-noded brickUP elements available in OpenSees are used to simulate saturated soil. The pile is modeled using displacement-based beam-column elements and is connected to the surrounding soil elements using rigid connection elements. The presence of groundwater table is properly simulated by allowing free drainage at the top. The model has meshed in such a way that soil elements near the pile are finer due to large stress gradient in the pile zone. The maximum element size is also maintained to be less than one-tenth to one-eighth of wavelength associated with highest frequency of input wave as suggested by Kuhlemeyer and Lysmer (1973). Figure 4 shows finite element model for case A.

5.1 Numerical Formulation

The *u–p* formulation developed by Zienkiewicz and Shiomi (1984) as implemented in OpenSees for earthquake loading conditions has been considered in the present study. The *u–p* formulation for finite element model can be written as follows:

$$M\ddot{U} + \int_V B^{\mathrm{T}}\sigma' \mathrm{d}V - QP - f^{(s)} = 0 \tag{8}$$

$$Q^{\mathrm{T}}\dot{U} + HP + S\dot{P} - f^{(p)} = 0 \tag{9}$$

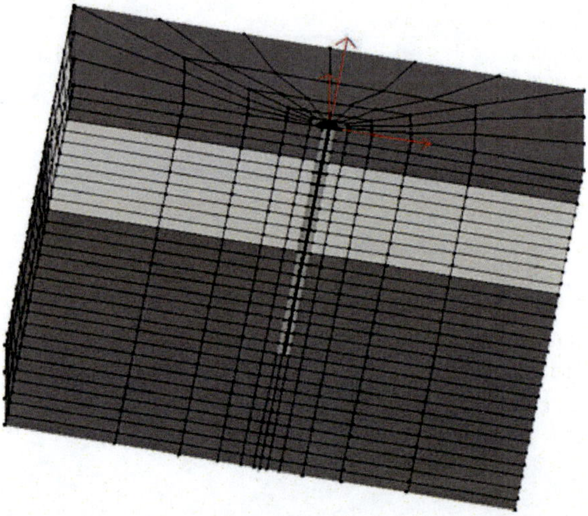

Fig. 4 Finite element mode for case A

where M is the mass matrix, U is the solid displacement vector, B is the strain–displacement matrix, σ' is the effective stress tensor, Q is the discrete gradient operator coupling the motion and flow equations, P is the pore pressure vector, and S and H are compressibility and permeability matrix, respectively. $f^{(s)}$ and $f^{(p)}$ are vectors which include the effects of body forces, external loads, and fluid fluxes.

5.2 *Constitutive Model for Soil*

An elastic–plastic material model (PressureDependentMultiYield—PMDY02) available in OpenSees has been employed for simulating liquefied sand behavior. This model was first developed by Iwan (1967) and Mroz (1967) but was later modified for earthquake engineering problems by Elgamal et al. (2002) and (2003). The yield surface is based on multi-surface concept and is of Drucker–Prager formulation. It forms a conical shape in principal space with its apex at the origin as shown in Fig. 5. The yield function as per Yang et al. (2003) may be given as

$$f = \frac{3}{2}(s - \alpha) : (s - p\alpha) - m^2 p^2 = 0 \tag{10}$$

where $s = \sigma' - p'\delta$ = deviatoric stress tensor; p' = mean effective stress; $p_0{}'$ = small positive constant such that yield surface size remains constant at $p' = 0$; α = second-order kinematic deviatoric tensor defining yield stress center; m = yield surface size.

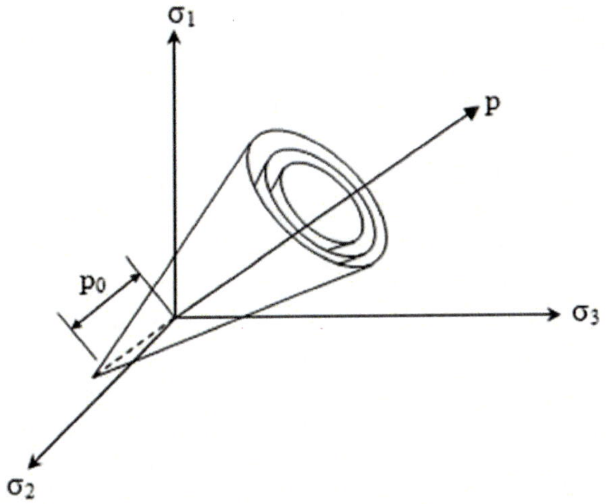

Fig. 5 Yield surface in principal stress space (Prevost 1985)

Clay is modeled using PressureIndependentMultiYield model available in OpenSees. This material shows plastic behavior only in deviatoric stress–strain response while volumetric stress–strain response is linear elastic. The yield surface is based on multi-surface concept and is of von Mises type.

5.3 *Variability in Permeability of Sand*

At the time of liquefaction, soil loses contact with each other and therefore water has more paths to flow causing permeability coefficient to increase up to 10 times the initial permeability coefficient of soil. Considering constant permeability or constant increased permeability does not capture soil and pile response adequately (Bardet et al. 1993; Balakrishnan 2000; Arulanandan and Sybico 1993; Jafarzadeh and Yanagisawa 1995). Therefore, the present study adopts variable permeability function suggested by Shahir et al. (2014) as given in Eq. 11.

$$\begin{array}{lll} \frac{k_b}{k_i} = 1 + 9r_u^2 \quad \text{for } r_u < 1 & \text{Buildup phase} \\ \frac{k_l}{k_i} = 1 \quad \text{for } r_u = 1 & \text{Liquefaction phase} \\ \frac{k_d}{k_i} = 1 + 9r_u^{10} \quad \text{for } r_u < 1 & \text{Dissipation phase} \end{array} \tag{11}$$

where k_i is the initial permeability coefficient of sand and r_u is the excess pore pressure ratio.

5.4 Stages of Loading

The model is developed in three stages with proper loading and boundary conditions at each stage.

- Stage I: During this stage, the base of the model is completely fixed in all the three directions and boundaries perpendicular to the direction of excitation is tied together to simulate free field condition while the boundaries parallel to the excitation direction is restricted to move in the direction perpendicular to its plane. Then soil and pore water is subjected to gravity load first in elastic state and then in plastic state to develop initial state of stress.
- Stage II: In this stage, pile is installed and pile and pile mass are distributed along the pile nodes. The soil elements in the space occupied by the pile are removed and replaced by elements of smaller stiffness and lower permeability to prevent the water from flowing into the pile zone during self-weight application. The permeability of the soil is kept very high (1 m/s) in all directions during self-weight stages to facilitate the development of hydrostatic condition in soil.
- Stage III: Once the soil and pile are under proper initial conditions, earthquake time history is applied to the base of the model as input motion. The permeability of soil is updated in each step based on variable permeability function proposed by Shahir et al. (2014) as explained earlier.

6 Factorial Design for Response Surface Model

A 2^3 factorial design has been adopted since only three random variables have been considered in the study. The maximum pile bending moment and pile head displacement are obtained by performing numerical analysis for eight combinations of input variables for each case. The results obtained for single replicate 2^3 factorial design is presented in Tables 4, 5 and 6 for cases A, B, and C, respectively.

All the three parameters: density of clay, relative density of sand and slope were found to have significant influence on both maximum bending moment and pile head displacement. Significance of parameters was tested taking *p*-value significance level as 5%. Figures 6 and 7 show the effect of three parameters on maximum bending moment and maximum pile head displacement, respectively, for case C.

From Fig. 6, it can be inferred that maximum bending moment increases with increase in clay density, decrease in relative density of sand, and increase in slope of ground. The input variables have a similar effect on pile head displacement Fig. 7. Steeper the slope of the line, higher is the effect of parameter on the pile response.

Table 4 Maximum bending moment and pile head displacement for Case A

Run	Clay density, ρ	Relative density, φ	Slope angle, α	Max BM (kN m)	Max Pile head displacement (mm)
1	–	–	–	978.8	150
2	+	–	–	858.1	89.4
3	–	–	+	1277.2	209
4	+	–	+	1162.6	134
5	+	+	+	859.4	104
6	–	+	–	721.6	121
7	–	+	+	989.5	171
8	+	+	–	669.3	76.7

Table 5 Maximum bending moment and pile head displacement for Case B

Run	Clay density, ρ	Relative density, φ	Slope angle, α	Max BM (kN m)	Max Pile head displacement (mm)
1	–	–	–	712.2	39.7
2	+	–	–	928.1	55.9
3	–	–	+	820.1	49.5
4	+	–	+	1010.3	64.4
5	+	+	+	818.4	51.5
6	–	+	–	590.6	37.7
7	–	+	+	669.5	45.7
8	+	+	–	756.9	45.0

Table 6 Maximum bending moment and pile head displacement for Case C

Run	Clay density, ρ	Relative density, φ	Slope angle, α	Max BM	Max Pile head displacement
1	–	–	–	1096.9	188
2	+	–	–	1378.9	147
3	–	–	+	1480.9	305
4	+	–	+	1901.1	223
5	+	+	+	1368.7	123
6	–	+	–	878.8	154
7	–	+	+	1193.9	241
8	+	+	–	990.6	90

Fig. 6 Effect of input parameters **a** clay density **b** sand relative density **c** slope on maximum pile bending moment

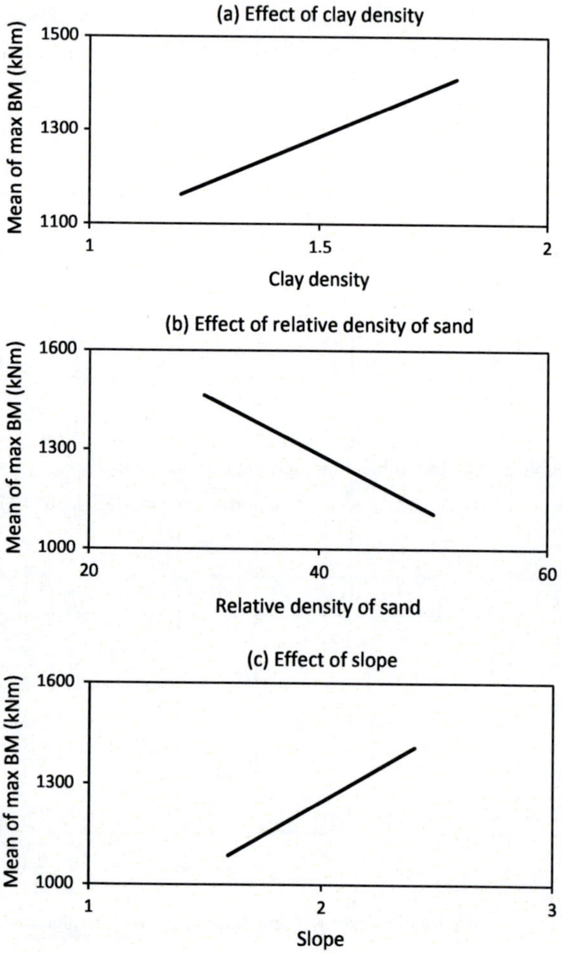

Using regression analysis, first-order linear response surface equation is obtained for maximum pile bending moment and pile head displacement for cases A, B, and C as presented in Eqs. 12–17.

Case A:

$$\text{Max. BM} = 1055.9 - 174.0\rho - 12.96\varphi + 331.6\alpha \quad (12)$$

$$\text{Max. PD} = 0.228 - 0.103\rho - 0.00137\varphi + 0.0564\alpha \quad (13)$$

Case B:

$$\text{Max. BM} = 448.5 + 300.5\rho - 7.942\varphi + 103.3\alpha \quad (14)$$

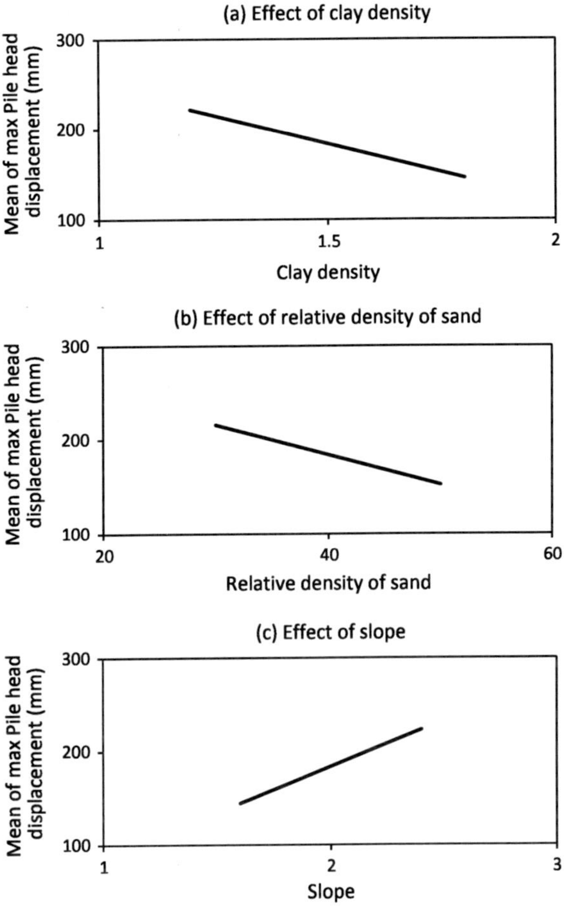

Fig. 7 Effect of input parameters **a** clay density **b** sand relative density **c** slope on maximum pile head displacement

$$\text{Max. PD} = 0.0154 + 0.0184\rho - 0.00037\varphi + 0.0103\alpha \tag{15}$$

Case C:

$$\text{Max. BM} = 381 + 412\rho - 17.82\varphi + 499.8\alpha \tag{16}$$

$$\text{Max. PD} = 0.2853 - 0.1218\rho - 0.00302\varphi + 0.1017\alpha \tag{17}$$

The R^2 and adj R^2 values are found to be close to each other for all fitted models, and normal probability plot is also approximately along a straight line. Figure 8 shows normal probability plot for maximum bending moment model for case A.

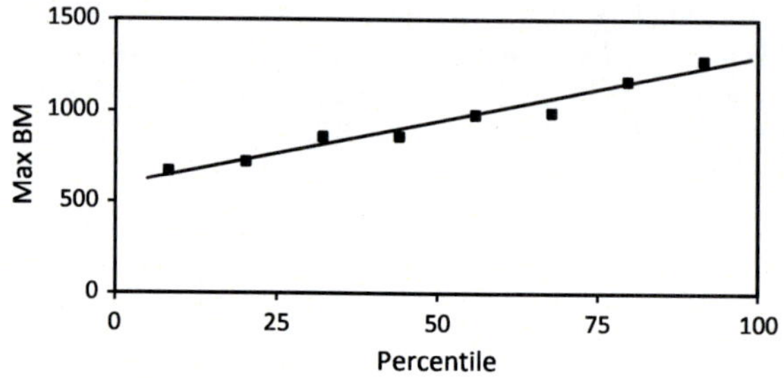

Fig. 8 Normal probability plot for maximum bending moment model for case A

7 Limit State Function for Single Pile

As per ultimate limit state condition, the pile is safe if maximum bending moment induced in pile at any section is less than plastic yield moment capacity of pile. Hence, ultimate limit state function for reliability analysis can be expressed as

$$G(x) = M_y - M_{max} \tag{18}$$

where M_{max} is the maximum bending moment recorded at any pile section and M_y is the plastic yield moment capacity of M30 concrete pile (281.2 MPa) for the present study.

As per serviceability condition, the maximum pile head deflection must be less than 12 mm. The limit state function for deflection can be written as

$$G(x) = d - D_{max} \tag{19}$$

where D_{max} is the maximum pile head deflection obtained from numerical analysis for each case and d is equal to 12 mm.

8 Reliability Analysis

The performance of the pile under laterally spreading condition is studied by carrying out reliability analysis via spreadsheet using the method suggested by Low and Tang (2007). The solver option available in Microsoft Excel is used for calculating reliability index for each case. The reliability index values for different cases are summarized in Table 7.

The pile response is sensitive to the thickness of non-liquefiable crust and liquefiable layer (Janalizadeh and Zahmatkesh 2015) From Table 7, it can be inferred that single pile in Case B (1.5-m-thick non-liquefiable clay and 1-m-thick liquefiable sand) has highest reliability index values while that in Case C (4.5-m-thick

Table 7 Reliability index for different cases

Case	Max BM	Max Pile head deflection
A	1.2	1.03
B	1.4	1.59
C	1.17	0.93

non-liquefiable clay and 4-m-thick liquefiable sand) has lowest. This indicates that reliability index values decrease with increase in thickness of liquefiable sand and non-liquefiable crust at the top.

9 Conclusion

In the present study, the performance of single pile under laterally spreading condition is evaluated by collaborating Hasofer–Lind method with response surface methodology (RSM). Three different cases are analyzed by varying thickness of liquefiable sand and non-liquefiable crust to study its effect on pile response. The results indicate that more the thickness of liquefiable sand and non-liquefiable surface crust, greater is the maximum bending moment and pile head displacement.

The advantage of using the reliability index is clearly evident from the present study. The RSM and FORM methods clearly help in taking all the uncertainties into consideration without increasing the complexity of the problem unlike Monte Carlo simulation which is computationally more demanding.

References

Arulanandan K, Sybico J (1993) Post-liquefaction settlement of sands. In: Proceedings of the Wroth memorial symposium, St. Catherine's College, Oxford

Babu GS, Srivastava A (2007) Reliability analysis of allowable pressure on shallow foundation using response surface method. Comput Geotech 34(3):187–194

Balakrishnan A (2000) Liquefaction remediation at a bridge site. University of California, Davis

Bardet JP, Huang Q, Chi SW (1993) Numerical prediction for model no. 1. Verification of numerical procedures for the analysis of soil liquefaction problems A.A. Balkema, Rotterdam, pp 67–86

Brandenberg SJ, Zhang J, Kashighandi P, Huo Y, Zhao M (2011) Demand fragility surfaces for bridges in liquefied and laterally spreading ground. Pacific Earthquake Engineering Research Center, p 162

Cubrinovski M, Ishihara K (2004) Simplified method for analysis of piles undergoing lateral spreading in liquefied soils. Soils Found 44(5):119–133

Elgamal A, Yang Z, Parra E (2002) Computational modeling of cyclic mobility and post-liquefaction site response. Soil Dyn Earthq Eng 22(4):259–271

Elgamal A, Yang Z, Parra E, Ragheb A (2003) Modeling of cyclic mobility in saturated cohesionless soils. Int J Plast 19(6):883–905

Farag R (2014) Probabilistic pseudostatic analysis of pile in laterally spreading ground: two layer soil profile. Ain Shams Eng J 5(2):343–354

Finn WDL, Fujita N (2002) Piles in liquefiable soils: seismic analysis and design issues. Soil Dyn Earthq Eng 22(9–12):731–742
Forcellini D, Della Bartola F, Tarantino AM (2013) Liquefaction-induced lateral deformations computational assessment during Tohoku Earthquake. ISRN Civ Eng
Hasofer AM, Lind NC (1974) Exact and invariant second-moment code format. J Eng Mech Div 100(1):111–121
Ishihara K, Cubrinovski M (2004) Case studies of pile foundations undergoing lateral spreading in liquefied deposits. In: 5th International conference on case histories in geotechnical engineering, New York
Iwan WD (1967) On a class of models for the yielding behavior of continuous and composite systems. J Appl Mech 34(3):612–617
Jafarzadeh F, Yanagisawa E (1995) Settlement of sand models under unidirectional shaking. In: 1st International conference on earthquake geotechnical engineering, vol 2, pp 693–698
Janalizadeh A, Zahmatkesh A (2015) Lateral response of pile foundations in liquefiable soils. J Rock Mech Geotech Eng 7(5):532–539
Jiang C, Li TB, Zhou KP, Chen Z, Chen L, Zhou ZL, Liu L, Sha C (2016) Reliability analysis of piles constructed on slopes under laterally loading. Trans Nonferrous Met Soc China 26(7):1955–1964
Kozubal J, Puła W, Wyjadłowski M, Bauer J (2013) Influence of varying soil properties on evaluation of pile reliability under lateral loads. J Civ Eng Manage 19(2):272–284
Kuhlemeyer RL, Lysmer J (1973) Finite element method accuracy for wave propagation problems. J Soil Mech Found Div 99
Lacasse S, Nadim F (1996) Model uncertainty in pile axial capacity calculations. In: Offshore technology conference
Low BK, Tang WH (1997) Efficient reliability evaluation using spreadsheet. J Eng Mech 123(7):749–752
Low BK, Tang WH (2007) Efficient spreadsheet algorithm for first-order reliability method. J Eng Mech 133(12):1378–1387
Madabhushi G, Knappett J, Haigh S (2010) Design of pile foundations in liquefiable soils. Imperial College Press, London
Motamed R, Towhata I, Honda T, Tabata K, Abe A (2013) Pile group response to liquefaction-induced lateral spreading: e-defense large shake table test. Soil Dyn Earthq Eng 51:35–46
Mroz Z (1967) On the description of anisotropic work hardening. J Mech Phys Solids 15(3):163–175
Myers RH, Montgomery DC, Anderson-Cook CM (1995) Response surface methodology: process and product optimization using designed experiments. Wiley series in probability and statistics. Applied Probability and Statistics
Prevost JH (1985) A simple plasticity theory for frictional cohesionless soils. Int J Soil Dyn Earthq Eng 4(1):9–17
Shahir H, Mohammadi-Haji B, Ghassemi A (2014) Employing a variable permeability model in numerical simulation of saturated sand behavior under earthquake loading. Comput Geotech 55:211–223
Tandjiria V, Teh CI, Low BK (2000) Reliability analysis of laterally loaded piles using response surface methods. Struct Saf 22(4):335–355
Wong FS (1985) Slope reliability and response surface method. J Geotech Eng 111(1):32–53
Yang Z, Elgamal A, Parra E (2003) Computational model for cyclic mobility and associated shear deformation. J Geotech Geoenviron Eng 129(12):1119–1127
Youd TL, Perkins DM (1987) Mapping of liquefaction severity index: American Society of Civil Engineers. J Geotech Eng 113(11):1374–1392
Zienkiewicz OC, Shiomi T (1984) Dynamic behavior of saturated porous media; The generalized Biot formulation and its numerical solution. Int J Numer Anal Meth Geomech 8:71–96

Influence of Soil–Structure Interaction in Elevated Water Tank

Sutanuka Nath and Atanu Kumar Dutta

Abstract This paper presents the effect of soil–structure interaction (SSI) on the response of elevated water tank in different types of soil medium. Two intze-type elevated water tanks of the same capacity and the same dynamic characteristics, one with frame staging and the other with shaft staging, based on three different soil types are chosen to check the influence of SSI due to three representative earthquakes. Modeling of elevated water tank–soil foundation system is done using ANSYS Workbench®. Soil mass is considered large enough to avoid reflection of Rayleigh waves. While the bottom of the soil mass is considered fixed, vertical rollers are provided in the vertical faces of the soil mass to account for semi-infinite nature of the soil. Time history analysis is performed on the multiple systems. It is found that the gravity load design of Krishna Raju (Krishna Raju, Advanced reinforced concrete design (IS: 456-2000) (English), CBS Publisher, 2015) is not valid for seismicity-prone region.

Keywords Soil–structure interaction · Elevated water tank · Time history analysis

1 Introduction

In the past decades, there are many severe cases of failure of elevated water tank during earthquakes. Failures of elevated water tank during an earthquake depend on various factors such as geometry of tank, material used, supporting system, foundation failure fluid sloshing and characteristics of earthquake. In the design of fluid-filled elevated tank, flexibility of soil mass needs to be considered to get a realistic soil–structure interaction. Sloshing produces hydrodynamic pressure on the wall, which may cause permanent deformation of the elevated water tank including serious failure like roof failure, damage of walls, etc. There are many cases of elevated tank structures getting collapsed due to sloshing effect during an earthquake.

S. Nath (✉) · A. K. Dutta
Department of Civil Engineering, Jorhat Engineering College, Jorhat, Assam 785007, India
e-mail: sutanuka888@gmail.com

M. Latha Gali and P. Raghuveer Rao (eds.), *Geohazards*, Lecture Notes in Civil Engineering 86, https://doi.org/10.1007/978-981-15-6233-4_28

This study attempts to address this soil–fluid–structure interaction on two intze-type elevated water tanks with different staging but with the same dynamic characteristics, due to different earthquakes and different soil conditions.

2 Literature Review

Consideration of soil–structure interaction in design of elevated water tank is seldom reported in the literature. In most of the cases, the base of foundation tank is considered to be fixed (Dutta et al. 2001). Different studies with varied approaches have been done to study this soil–structure interaction. FEMA 368/369 (2000) seismic codes suggest the use of springs (Fig. 1) with equivalent translational spring constant k_y(Eq. 1)and equivalent translational spring constant k_θ(Eq. 2) below the foundation to represent soil–structure interaction.

$$k_y = \left[\frac{8\alpha_y}{2-\upsilon}\right]Gr \tag{1}$$

$$k_\theta = \left[\frac{8\alpha_\theta}{3(1-\upsilon)}\right]G \tag{2}$$

In Eqs. 1 and 2, r is the radius of the foundation, G is the shear modulus of the half space, υ is the Poisson's ratio for the soil and α_y and α_y are the dimensionless coefficients depending on the period of the excitation, the dimension of the foundation and the properties of the supporting medium.

Livaoglu and Dogangün (2005) investigated fluid–structure and structure foundation and soil interaction in the frequency domain. Frequency-dependent cone

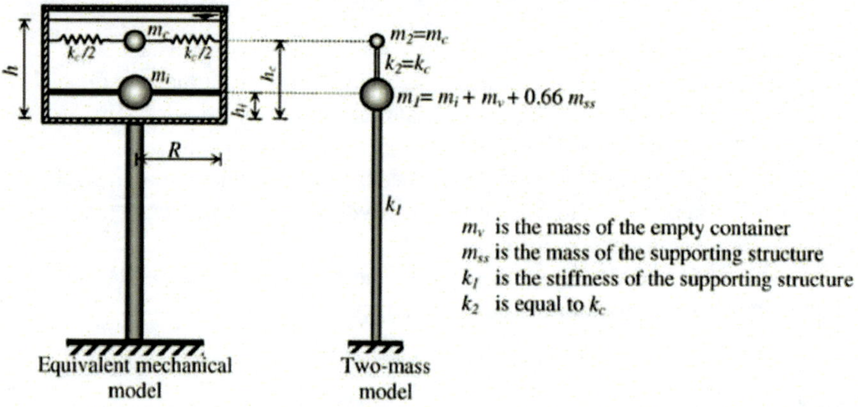

Fig. 1 Spring mass model consisting fluid–soil–structure interaction after Doğangün and Livaoğlu (2004)

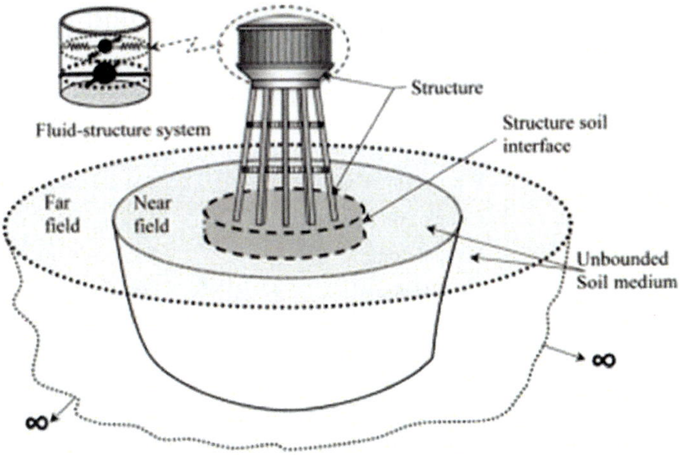

Fig. 2 Problem investigated for dynamic fluid–structure–soil interaction from Livaoglu and Dogangün (2005)

method is considered for soil–structure interaction and spring mass model, as given in Eurocode 8, which is considered for fluid–structure interaction. The seismic analysis is carried out using a computer program on six different types of soils ranging from stiff soil to soft clay. In this work, SSI is a bounded structure with adjacent irregular near field interacts with unbounded far field (infinite/semi-infinite) soil as shown in Fig. 2. Authors observe that sloshing response is less in stiff soils. Roof displacement and base shear force responses are also not affected by the embedment in relatively stiff soils, but in relatively soft soils.

It has been observed in the literature that though soil has been modeled as a material in few cases, representation of water a fluid element (not as equivalent mass and spring) is scarce.

3 Methodology

In this work, the soil and water and the structure are represented by respective finite elements in ANSYS workbench®. Two types of tank structures based on staging systems, namely frame and shaft staging, are considered. The frame-staged model is picked up from a textbook with Krishna Raju (2015), and the shaft-staged model is modeled with trial and error to represent the same dynamic characteristics as of the frame-staged model. As another parametric variation, properties of soil are also varied to see the interaction effect. The capacity of elevated water tank is 1 million liters. In the textbook solution, the tank is supported on 16 m high frame staging system with 8 columns. The diameter of cylindrical section of tank is 12 m. The dimensions of various components are given in Table1.

Table 1 Sizes of various components of intze-type tank after Krishna Raju (2015)

Components	Dimensions
Top dome	120 mm thick
Top ring beam	300 mm × 300 mm
Cylindrical wall	200 mm thick
Bottom ring beam	300 mm × 300 mm
Circular ring beam	1200 mm × 600 mm
Bottom dome	300 mm thick
Conical dome	600 mm thick
Braces	500 mm × 500 mm
Radius of cylindrical section	6000 mm

Table 2 Material properties after Kalani Sarokolayi et al. (2008)

Material	Specific mass Kg/m^3)	Modulus of elasticity (N/m^2)	Poisson's ratio
Concrete	2400	2.23×10^{10}	0.27
Water	1000	Bulk modulus = 2.2×10^9	–

Physical and mechanical properties of the materials of the structure are listed in below Table 2.

The shaft model is designed with trial and error in ANSYS workbench® so that the capacity of tank is same, and at the same time, it maintains the dynamic characteristics of the frame-staged system. The shaft-staged water tank is also considered with same diameter of frame-staged elevated water tank and thickness of shaft being 150 mm. Height of staging is also kept the same with that of the frame staging system. M20 concrete grade is considered for modeling the concrete structure. Physical and mechanical properties of the materials of the structure are listed in Table 2.

Safe bearing capacity of soil is kept as 250 kN/m^2 as considered in Krishna Raju (2015). The material damping (ξ) of soil is assumed to be 5% in this case.

3.1 Soil Properties

Three different soils are considered with the properties shown in Table 3. Soil types have been chosen in terms of decreasing order of stiffness.

Table 3 Material properties of soil mass after soil properties are taken from Kumar and patel (2016)

Soil type	Young's modulus (kN/m^2)	Density (kN/m^3)	Poisson's ratio
S1	60,000	19.90	0.33
S2	45,000	18.00	0.30
S3	35,000	17.10	0.28

3.2 Soil Modeling

In this work, the soil mass is modeled using wave propagation theory in semi-infinite soil. During seismic ground motion, Rayleigh wave propagates to a distance from the source, and then, its energy gets dissipated. The previous studies in the domain indicate that a zone extending to a distance of $10L_R$ (where L_R is the Rayleigh wavelength) from the vibration source is sufficient for wave barrier analysis (Ahmad et al. 1996).

Rayleigh wavelength depends on the elastic properties of soil and dominant frequency of the exiting earthquake. Here, dominant frequency of earthquake time history data is considered to calculate the Rayleigh wavelength. Earthquake data considered are N-S component El-Centro (1940) and Kobe (1995) earthquake. Time history data is converted using fast Fourier transformation (FFT) of MATLAB®.

The dominant frequencies are found to be 1.47 Hz, 1.41 Hz for the El Centro time history and, Kobe time history, respectively. Table 4 shows the calculation of different ground motion parameters for S2 soil type. Calculation is done using the dominant frequency of El-Centro as it is not varying much with that of Kobe.

Table 5 shows the ground motion parameters of other two types of soil (S1 and S3).

Rayleigh wavelength is taken 19.56 for modeling of the soil mass. Considering width of the soil mass is $7L_R$ and depth of the soil mass is $3L_R$. Length/width of the soil mass $= 19.56 \times 7 = 136$ m and depth of the soil mass $= 19.56 \times 3 = 58.68$ m. The dimension of the soil mass is adopted as 136 m $\times$ 58 m.

Table 4 Ground motion parameters for half-space soil

Parameter	Expression	Value density
Shear modulus	$G = E/2(1+\nu)$	17307.69 kN/m^2
Shear wave velocity	$V_s = \sqrt{G/\rho}$	31 m/s
Rayleigh wave velocity	$V_R = \left(\frac{0.87+1.12\nu}{1+\nu}\right)V_s$	28.75 m/s
Rayleigh wavelength	$L_R = V_R/f$	19.56 m

Table 5 Ground motion parameters for half-space soil

Soil Type	Shear modulus	Shear wave velocity	Rayleigh wave velocity	Rayleigh wavelength
S1	22556.39	33.66	31.372	21.34
S3	13671.875	28.27	25.158	17.11

3.3 Boundary Conditions for Soil–Structure Interaction (SSI)

The radiation and reflection effect of surface waves can be terminated by applying absorbing boundary conditions. Viscous boundary condition is frequently used as an absorbing boundary. This boundary condition uses one-dimensional beam theory, and this theory has been commonly used with the finite element methods. Massless foundation approach considers roller support on the vertical faces of soil mass in case of static analysis.

However, soil need not be bunded by absorbing boundary condition if sufficiently large dimension of soil body is used [Livaoğlu and Dogangun (2005)]. In case of large soil mass, radiation effects of propagating waves are eliminated. In this study, soil mass is considered large enough to avoid reflection. Roller supports are provided in the vertical faces of the soil to provide lateral stiffness to the soil mass simulating the semi-infinite nature of the soil.

3.4 Description of Modeling in ANSYS Workbench®

Design modeler is used for building the geometry. Material and geometric properties are assigned. Appropriate mesh is generated. Depending upon the meshing method, sizing different solid elements are assigned by default. Boundary condition is assigned. Contact pair between different types of solid material is program controlled. SOLID187 is used in modeling the concrete elevated water tank, staging and soil mass using coarse mesh ascribing different material properties. SOLID187 is a tetrahedral element. While modeling this elevated water tank, TARGE170 is taken as target element and CONTA174 is considered as contact element. These surface-to-surface contact elements allow modeling of fluid pressure penetration loads. These interfaces are capable of performing fluid–structure and soil–structure interaction. The fluid element is capable of exchanging pressure and deformation properties with solid concrete elevated tank structure. As visualization of fluid–structure interaction is effective with sloshing effect of water during free mode vibration, water inside the tank is modeled with FLUID80 element.

3D soil body is modeled using ANSYS Workbench®. Figures 3 and 4 show the soil mass with frame staging and shaft staging systems in tank full condition.

Fig. 3 3D model of soil mass with frame staging elevated water tank

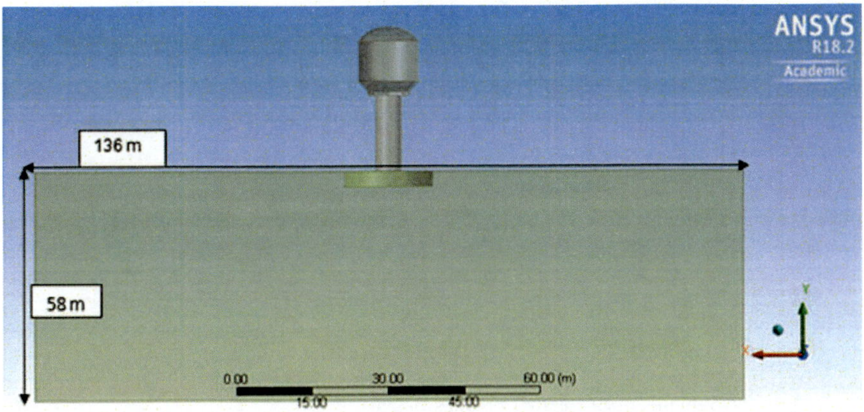

Fig. 4 3D model of soil mass with shaft staging elevated water tank

3.5 Time History Data

Time history analysis is computed for two representative earthquakes.

Time history analysis is computed for two representative earthquakes these are El-Centro (May 18, 1940), Kobe (January 17, 1995). These ground motions are selected on the basis of their maximum peak ground acceleration. Ground acceleration data for north–south component of El-Centro is obtained from https://www.vibrationdata.com/elcentro.htm. Source of Kobe earthquake data is Pacific Earthquake Engineering Research–Next Generation Attenuation (PEER-NGA) strong motion database records available online at https://peer.berkeley.edu/nga. Table 6 shows the two characteristics of the ground motion records.

Table 6 Ground motion characteristics

Different ground motions	Station	Magnitude	Time steps (s)	Total time period (s)	PGA (g)
El-Centro (1940)	Imperial valley	$Mw = 6.9$	0.02	31.14	0.319
Kobe (1995)	Takarazuka, Japan	$Mw = 7.2$	0.02	40	0.8

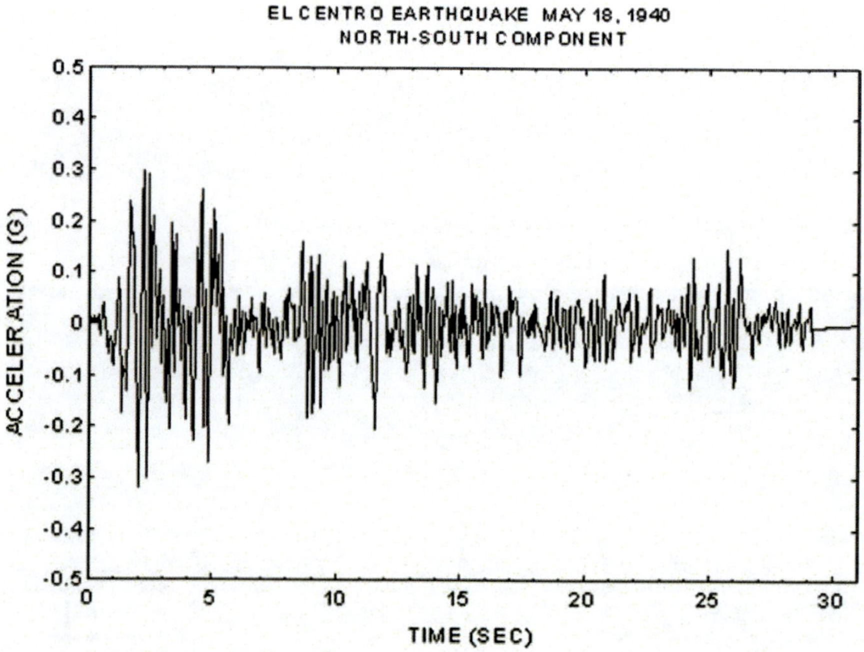

Fig. 5 Accelerogram plot of El-Centro (1940) earthquake

The input acceleration-time plot of El-Centro earthquake, Kobe earthquake is shown in Figs. 5 and 6.

4 Results and Discussions

Modal analysis data is shown in Table 7. Mode shapes are similar for both the systems.

Time history analysis is performed through "TRANSIENT STRUCTRE" tool in ANSYS Workbench® considering three representative earthquakes. The acceleration is applied in horizontal X. The maximum principal stresses in concrete due to time history analysis are computed.

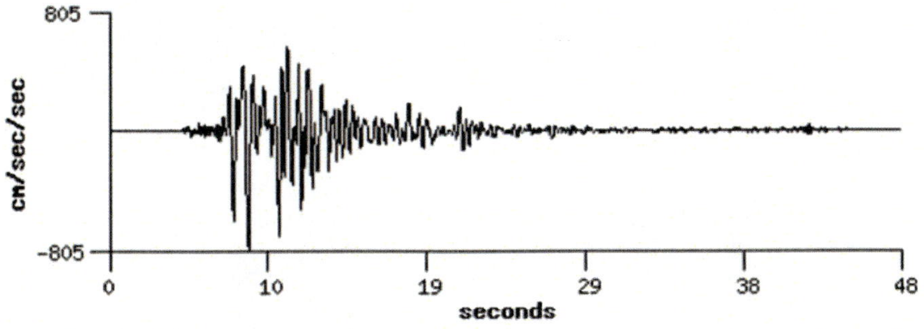

Fig. 6 Accelerogram plot of Kobe (1995) earthquake

Table 7 Comparison between the frequencies of frame-staged and shaft-staged water tanks

Mode shapes	Frequency in frame staging system (Hz)	Frequency in shaft staging system (Hz)
1	0.52052	0.52084
2	0.52852	0.52691
3	0.55834	0.60216
4	0.66586	0.68841
5	0.67851	0.72501
6	0.68233	0.73676

Comparison between the maximum principal stresses in both the concrete structures on different soil types due to El-Centro (1940) and Kobe (1995) earthquakes is shown in Table 8 and Table 9.

Considering maximum limit of stress in M20 grade of concrete under compression as $f_{ck}/1.5 = 13.33\ \text{N/mm}^2$, it is observed that maximum principal stress value on both the concrete structures (staging) exceeds permissible limit for all the earthquakes. Figure 7 shows the comparison of maximum principal stress in both the staging systems due to two different earthquakes.

Table 8 Maximum principal stresses in concrete for El-Centro (1940) earthquake

Soil types	Frame staging (N/mm^2)	Shaft staging (N/mm^2)
S1	2.8255e2	4.3077e1
S2	2.8556e2	4.3629e1
S3	6.9287e1	2.4806e1

Table 9 Maximum principal stresses in concrete for Kobe (1995) earthquake

Soil types	Frame staging (N/mm^2)	Shaft staging (N/mm^2)
S1	7.502e2	3.166e1
S2	5.243e2	9.8001e1
S3	7.435e1	2.6357e1

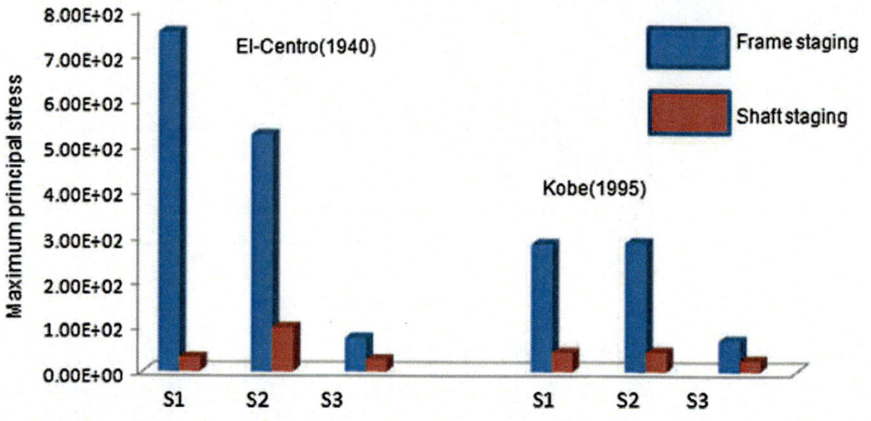

Fig. 7 Maximum principal stress on both the staging system

The results show that stress values are more in frame staging system compared to shaft staging on different soil types. Stress values are more in stiffer soil due to El-Centro (1940) ground motion. It is clear that structural configuration as per Krishna Raju (2015) was arrived at for gravity load analysis only and hence cannot be used for earthquake-prone region.

Maximum principal stresses are measured from color contours. Comparison between the maximum principal stresses on different types of soil mass due to El-Centro ground motions is shown in Table 10.

Maximum principal stress values in both the staging systems due to Kobe (1995) earthquakes are shown in Table 11 for different soil types.

Considering safe bearing capacity of soil as 250 kN/m^2, it can be observed that maximum principal stress on the soil mass exceeds this permissible limit. Figure 8 shows the comparison of maximum principal stress on different types of soil mass of both the staging systems due to two different earthquakes.

Table 10 Maximum principal stress in soil due to El-Centro (1940) earthquake

Soil types	Frame staging (N/mm^2)	Shaft staging (N/mm^2)
S1	2.2238e4	1.2219e3
S2	2.223e4	1.2087e3
S3	5.5511e3	6.7056e2

Table 11 Maximum principal stress in soil due to Kobe (1995) earthquake

Soil types	Frame staging (N/mm^2)	Shaft staging (N/mm^2)
S1	5.9188e4	3.9866e3
S2	4.1142e4	3.3198e3
S3	5.0447e3	2.7059e3

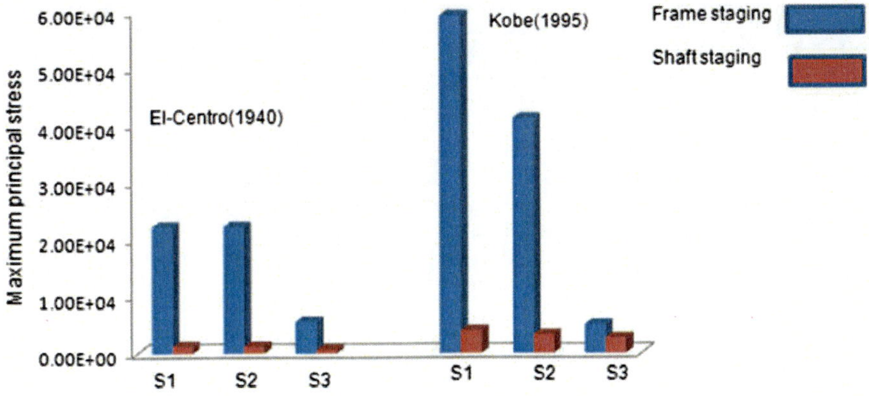

Fig. 8 Maximum principal stress on different soil types in both the staging systems

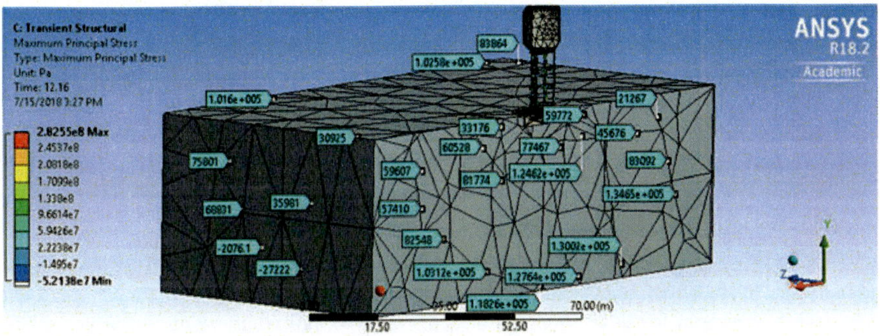

Fig. 9 Sectional view of frame staging system on S1 soil type

Maximum permissible stress on soil is very high compared to safe bearing capacity. Figure 9 shows the sectional view of frame staging with various stress points due to El-Centro (1940) earthquake.

Figure 10 shows the cross-sectional view of shaft staging with different stress points due to El-Centro (1940) earthquake.

5 Conclusions

The paper concludes that both frame staging and shaft staging systems are not safe for seismicity as per time history analysis with two representative earthquakes. It is clear that structural configuration as per Krishna Raju (2015) was arrived at using gravity load analysis only and hence cannot be used for earthquake-prone region. The maximum principal stress on concrete structure due to El-Centro (1940) earthquake

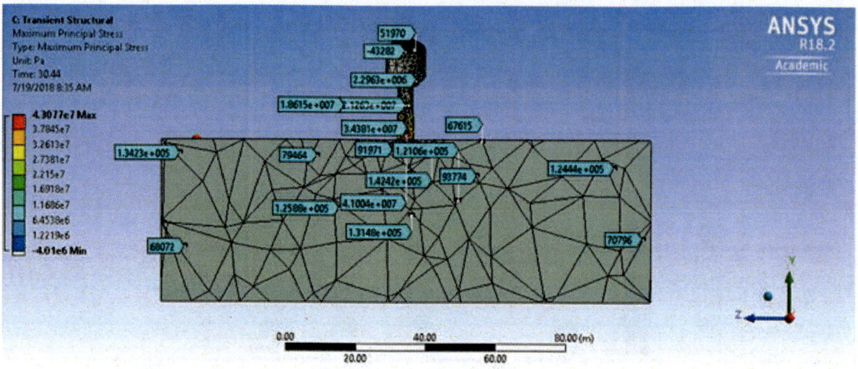

Fig. 10 Sectional view of shaft staging system on S1 soil type

for both the staging systems is more comparing to Kobe (1995) earthquake. The maximum principal stress on soil exceeds safe bearing capacity.

References

ANSYS Theory Manuel (1994). In: Kohnke P (ed), 12th edn. SAS IP, Inc., pp 1266

Ahmad S, Al-Hussaini TM, Fishman KL (1996) Investigation of active isolation of machine foundations by open trench. J Geotech Eng 117(4):616–636

Doğangün A, Livaoğlu R (2004) A simple seismic analysis procedure for fluid-elevated tank-foundation/soil systems. Sixth international conference on advances in civil engineering (ACE 2004), İstanbul, Turkey, vol 1, pp 570–580

Dutta SC, Jain SK, Murthy CVR (2000) Assessing the seismic torsional vulnerability of elevated tanks with RC frame-type staging. Soil Dynam Earthquake Eng 19:183–197

Housner GW (1963) Dynamic analysis of fluids in containers subjected to acceleration. Nuclear reactors and earthquakes, Report No. TID 7024, U.S. Atomic Energy Commission, Washington, DC

Krishna Raju N (2015) Advanced reinforced concrete design (IS: 456-2000) (English) 3rd edn. CBS Publisher

Kalani Sarokolayi L, Navayineya B, Hosainalibegi M, Vaseghi Amiri J (2008) Dynamic analysis of water tanks with interaction between fluid and structure. In: The 14th world conference on earthquake engineering

Kumar D, Patel B (2016) Response of overhead water tank staging considering fluid-structure-soil interaction. Int J Adv Eng Res Dev 3(4)

Livaoğlu R, Doğangün A (2005) Seismic evaluation of fluid-elevated tank-foundation/soil systems in frequency domain. Struct Eng Mech 21(1):101–119

Seismic Ground Response Analysis for Soil Site in Johor, Malaysia

Anurag Sahare and Deepankar Choudhury

Abstract Seismic ground response analysis (GRA) has been carried out for a soil site located in Johor, Malaysia, in order to know the degradation of soil layers overlying a bed rock subjected to a ground motion. In this paper, an effort has been made to critically study the results obtained by 1-D equivalent linear analysis and nonlinear analysis using DEEPSOIL v6.1, SHAKE 2000 and D-MOD 2000 for different modulus reduction and damping curves with two different soil models. The results are observed to be nearly similar for a lesser depth soil model unlike larger depth soil model, for which the results are slightly different. The results from DEEPSOIL v6.1, SHAKE 2000, and D-MOD 2000 are found to be different for a larger depth soil model. Nonlinear analysis is preferred as it takes into account the actual nonlinear characteristics of soil.

Keywords 1-D ground response analysis · PGA · Spectral acceleration · Response spectra

1 Introduction

Seismic GRA has been performed in order to estimate the response of local soil layers to earthquake loading in terms of surface accelerations, maximum horizontal accelerations along depth, response spectrum, etc. For this, ground response analysis makes use of wave propagation theory in order to estimate the local soil effects on the propagating shear wave originated from the bed rock during an earthquake. Seismic waves after originating from the source propagates in all directions. However, it is assumed that the waves travel only in vertical direction from bed rock to the surface for 1-D seismic GRA, which has been used for the present study. During the propagation of seismic waves, the wave characteristics such as amplitude and frequency content get drastically modified depending on the local soil properties, and if the structure is not designed for the appropriate ground motion, it may result

A. Sahare (✉) · D. Choudhury
Department of Civil Engineering, Indian Institute of Technology Bombay, Powai, Mumbai 400076, India
e-mail: sahare.anurag@gmail.com

M. Latha Gali and P. Raghuveer Rao (eds.), *Geohazards*, Lecture Notes in Civil Engineering 86, https://doi.org/10.1007/978-981-15-6233-4_29

into large destruction leading to severe consequences. The prerequisites for carrying out site specific ground response study are soil model with all soil layers including dynamic properties of the soil as well. The soil properties like unit weight of soil, depth of soil layer shall be provided as input in the soil model. The soil model can be defined based on the bore-log profile of the site representing various soil layers with thickness.

The dynamic properties of the soil can be evaluated based on either laboratory tests (cyclic triaxial test, resonant column test, etc.), field measurements or by using existing correlations with field tests, e.g., standard penetration test, cone penetration test, Becker penetration test, etc. Mostly, it is not feasible to carry out laboratory tests, and hence commonly, the dynamic properties of soil are evaluated based on certain correlation which has been developed for a particular soil and area of concern. The most commonly used correlations are with SPT N-value. Choosing the appropriate correlation for the site is the most important task as it should completely be dependent on the local soil properties. Soil properties drastically vary from location to location. Modulus reduction and damping curves can be obtained by cyclic triaxial testing for the corresponding soil type. In lack of such tests, modulus reduction and damping curves put forward by the researchers for different soil types may be used. Several ground response studies have been carried out in the past e.g. Choudhury and Savoikar (2009) carried out equivalent-linear GRA of MSW landfills, Phanikanth et al. (2011) carried out 1-D equivalent linear GRA for typical sites in Mumbai subjected to four strong earthquakes, Shylamoni et al. (2014) carried out 1-D equivalent linear GRA for a nuclear plant in Japan, and the results were observed to be in a very close approximation to the recorded values of most of the aftershocks, Naik and Choudhury (2014) made a comparative study of seismic ground response using DEEPSOIL, SHAKE and D-MOD for a soil site in Goa, Desai and Choudhury (2015) carried out non linear GRA for port sites in Mumbai and Mahmood et al. (2016) carried out 1-D equivalent linear GRA for a collapsed tower in Islamabad and concluded the evaluated spectral acceleration values were observed to be higher than the values recommended by the building code of Pakistan.

2 Site Location

The site chosen for the present study is located in Johor, Malaysia. It is an Oil Refinery and Petrochemicals Integrated Development (RAPID) Project.

The soil seems to be predominantly silty clayey type of soil with some of the soil layers as sandy soil. Two soil models of significantly different depth are considered in the present research work in order to know the effect of the number of soil layers to ground response. The bore-log data was obtained from one of the reputed geotechnical firm which represents SPT N-value along depth as well as the basic geotechnical properties which were obtained from the laboratory tests. No laboratory or field data information about the dynamic properties of soil was available for the site, and hence, SPT N-value versus shear wave velocity correlation is used in order to estimate dynamic soil properties as proposed by Ohta and Goto (1978), the correlation which has been developed for all types of soil which is represented in Eq. 1.

$$V_s = 85.35N^{0.348} \tag{1}$$

where V_s is shear wave velocity (in m/s) and N is the recorded SPT N-value.

3 Methodology

Ground response analysis can be carried out for one-dimensional, two-dimensional, or three-dimensional depending upon the dimensionality of the soil model which accounts for the variability in the soil properties along depth. In the present work, one-dimensional GRA has been carried out due to its simplicity. Following are the assumptions used in 1-D GRA.

- The soil layer is assumed to be horizontal which extends to infinity.
- Ground surface is horizontal.
- The seismic waves propagate vertically from the bed rock to the ground surface.

3.1 Methods of Analysis

Seismic GRA can be carried out using equivalent linear, linear, and nonlinear approaches depending on the stress strain relationship. The major difference between these three methods is the selection of shear modulus. G_{sec} is used for equivalent linear GRA, whereas G_{max} is used for linear GRA and G_{tan} is used for nonlinear GRA. For 1-D seismic GRA, always $G_{max} > G_{sec} > G_{tan}$.

3.1.1 Equivalent Linear Ground Response Analysis

The real nonlinear response of soil can be approximated by equivalent linear GRA using frequency domain analysis by transfer functions. It follows an iterative procedure in order to estimate shear modulus and damping corresponding to effective shear strain in each soil layer. The effective shear strain is generally considered as 65% of the maximum shear strain for each layer in the soil model. Following is the iterative procedure adopted in 1-D equivalent linear GRA.

- Assume some initial reference value of strain (low strain level) and find out the corresponding values of shear modulus (G_{sec}) and damping ratio from modulus reduction and damping curves.

- Take new strain level and determine the shear modulus and damping values for the updated strain.
- Repeat step 2 until the values of shear modulus and damping converges and determine resulting effective shear strain.
- The values of shear modulus and damping with respect to final effective shear strain are the final values for that particular soil layer.

3.1.2 Nonlinear Ground Response Analysis

In order to overcome several limitations of equivalent linear GRA, like not able to model strain-dependent modulus degradation and not able to account for the excess pore pressure ratio which can be used to check the liquefaction susceptibility of soil, nonlinear analysis may be used which would represent the actual nonlinear behavior of soil under cyclic loading. In time-domain analysis, the maximum frequency $f_{\max}$ (in Hz) which can propagate through the soil layer of thickness H (in m) is evaluated as shown in Eq. 2. For a layer with lesser thickness, maximum frequency increases.

$$f_{\max} = \frac{V_s}{4H} \tag{2}$$

3.2 *Methods Used for the Present Study*

There are various geotechnical software available for 1-D GRA like EERA, NERA, DEEPSOIL, SHAKE 2000, D-MOD 2000. In the present study, 1-D equivalent linear GRA has been carried out using DEEPSOIL v6.1 (Hashash 2012) and SHAKE 2000 and 1-D nonlinear GRA has been carried out using DEEPSOIL v6.1 (Hashash 2012) and D-MOD 2000 as shown in Table 1. This study presents comparative analysis of the different software's used in the present work as well as the comparative assessment between different methods of GRA, viz. equivalent linear and nonlinear.

Table 1 Software used for the present study

Software	DEEPSOIL	SHAKE 2000	D-MOD 2000
Analysis	1-D	1-D	1-D
Type	Equivalent linear and nonlinear	Equivalent linear	Nonlinear
Domain	Frequency domain–equivalent linear	Frequency domain	Time domain

Table 2 Soil Model for BH 153-D used in present study

Depth (m)	Soil Profile	Soil Layer	SPT 'N' Value	V_s (m/s)
1.0		Clay	6	159.2
2.0		Clay	9	183.3
3.0		Clay	16	224.0
4.5		Sand	39	305.4
6.0		Sand	55	343.3

3.3 *Soil Models Used in the Present Analysis*

This study has been carried out for two boreholes BH 559 (48 m) having water table at 6 m from the ground surface and BH 153-D (6 m) having water table at the ground surface as shown in Tables 2 and 3.

3.4 *Input Motion Considered for the Present Analysis*

1989 Loma Prieta (Anderson Dam Downstream, 360), motion is considered for the present seismic GRA. It is directly applied at the bed rock, i.e., synthetic analysis has been carried out in order to access influence of local soil layers to the input motion. Acceleration time history of 1989 Loma Prieta earthquake is shown in Fig. 1. The earthquake parameters are shown in Table 4.

4 Results and Discussions

In the present study, 1-D GRA has been carried out using three software's for two soil models subjected to 1989 Loma Prieta (Anderson Dam Downstream, 360). The results are showed in terms of maximum horizontal acceleration (MHA) along the depth, spectral acceleration versus period for the surface layer and Fourier amplitude ratio versus frequency for the surface layer. In the present work, an effort has been made to study the response of soil under the earthquake loading for different modulus reduction and damping curves, for different method of GRA, for different software and for two soil models having different depth.

The following abbreviations are used in the figures to indicate the input soil model properties and software employed in the present study.

Table 3 Soil Model for BH 559 used in present study

Depth (m)	Soil Profile	Soil Layer	SPT 'N' Value	V_s (m/s)
1.0		Silt	10	190.2
2.0			10	190.2
3.0		Clay	11	196.6
4.5			10	190.2
6.0			12	202.7
7.5			9	183.3
9.0			10	190.2
10.5			12	202.7
12.0		Sandy Silt	12	202.7
13.5			10	190.2
15.0			11	196.6
16.5			17	228.8
18.0			20	242.1
19.5			15	219.0
21.0			18	233.4
24.0			50	333.0
27.0			66	368.1
30.0			66	368.1
33.0			>100	-
36.0			>100	-
39.0			>100	-
42.0			>100	-
45.0			>100	-
48.0			>100	-

(i) P—Soil profile for BH 559 (ii) Q—Soil profile for BH 153-D (iii) DS—DEEPSOIL v6.1 (iv) SK—SHAKE 2000 (v) DM—D-MOD 2000 (vi) VD—Vucetic and Dobry (1991) (vii) DR—Darendeli (2001).

4.1 *Influence of the Modulus Reduction and Damping Curves*

Various material curves have been defined in DEEPSOIL, SHAKE, and D-MOD which depends on soil properties like unit weight, plasticity index, effective vertical

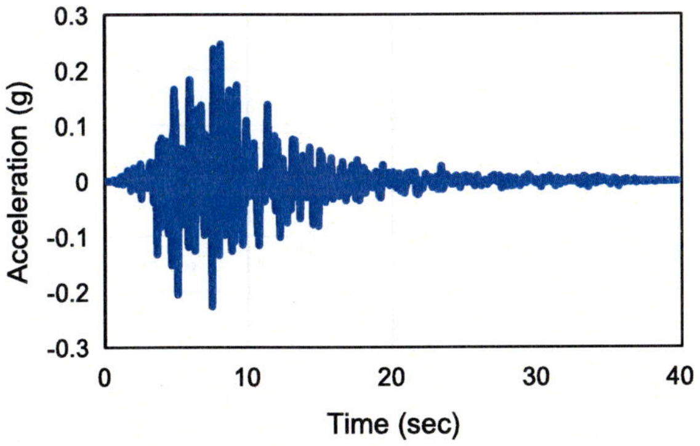

Fig. 1 Acceleration versus time history for 1989 Loma Prieta (Anderson Dam) Earthquake

Table 4 Earthquake parameters of selected motion

S. No.	Earthquake parameters	1989 Loma Prieta
1	Date of occurrence	17/10/1989
2	Recording station	Anderson Dam Downstream, 360
3	Moment Magnitude (M_w)	6.9
4	Maximum Horizontal acceleration (g) applied at the bedrock	0.25 g
5	Predominant spectral period	0.63 s
6	Bracketed duration	11.4 s
7	Significant duration	10.5 s
8	Mean period	0.48 s

stress, overconsolidation ratio. In order to determine the variations observed in soil response to earthquake motion with different modulus reduction and damping curves, a study has been carried out for BH 153-D and BH 559 using modulus reduction and damping curves presented by Vucetic and Dobry (1991) and Darendeli (2001) which are shown in Fig. 2. The soil properties obtained from the geotechnical investigation report are used as an input for these selected curves. Equivalent linear GRA using DEEPSOIL and SHAKE is carried out for this study.

4.1.1 Comparison of Maximum Horizontal Acceleration (MHA)

From Fig. 3a, b, it is observed that BH 559 results into much higher maximum horizontal accelerations along the depth as compared to BH 153-D for DEEPSOIL and SHAKE. However, the results are nearly similar for BH 153-D and BH 559 as per curves proposed by Darendeli (2001) when analyzed by SHAKE. For both

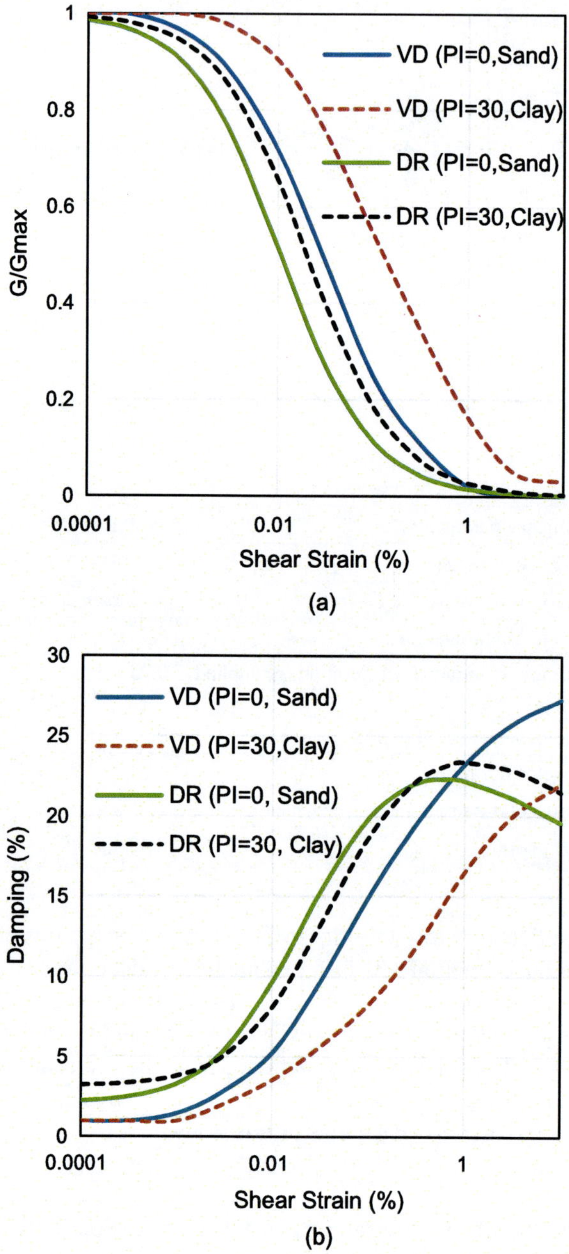

Fig. 2 **a** Modulus reduction and **b** damping curves used in the present study

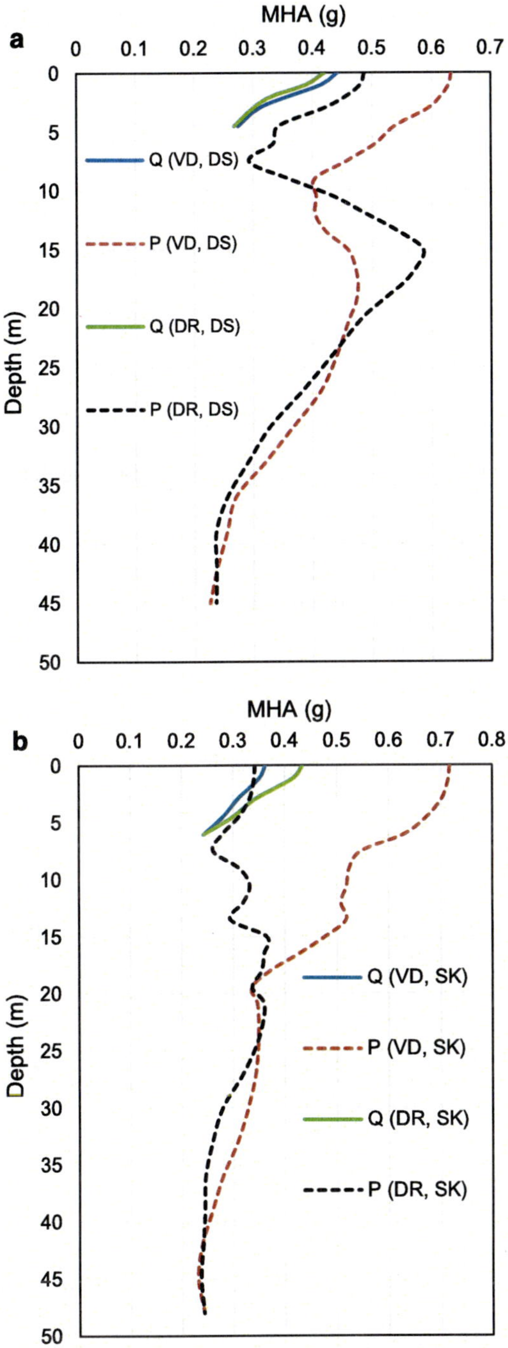

Fig. 3 **a** Maximum horizontal acceleration along depth for DEEPSOIL. **b** Maximum horizontal acceleration along depth for SHAKE 2000

the cases, it is observed that when BH 559 is subjected to 1989 Loma Preita Earthquake, MHA values obtained by Vucetic and Dobry (1991) curves exceed the values obtained by using Darendeli (2001) curves by 28.57 and 102.857% as per DEEPSOIL and SHAKE, respectively, at the surface. The values of MHA along the depth are approximately similar for BH 153-D when analyzed using both the curves as per DEEPSOIL and SHAKE. Hence, it can be said that similar results are obtained for a lesser depth soil model as compared to drastically different results obtained for a larger depth soil model using the curves proposed by Vucetic and Dobry (1991) and Darendeli (2001).

4.1.2 Comparison of Spectral Acceleration (g) Versus Period (Surface Layer)

The results are shown in Fig. 4a, b. Spectral accelerations calculated using the curves proposed by Vucetic and Dobry (1991) and Darendeli (2001) for BH 153-D is found to be nearly similar for DEEPSOIL and SHAKE 2000. For BH 559, Vucetic and Dobry (1991) predict larger spectral accelerations as compared to Darendeli (2001). The values of spectral accelerations are found to be higher for BH 559 when Vucetic and Dobry (1991) curves are used as compared to BH 153-D. However, the results are nearly similar when Darendeli (2001) curves are used.

4.1.3 Comparison of Amplification Ratio Versus Frequency (Surface Layer)

The typical results are shown in Fig. 5a, b. The amplification ratio obtained for BH 153-D is much higher as compared to BH 559 for both DEEPSOIL and SHAKE 2000. It is observed that Vucetic and Dobry (1991) curves give higher values of amplification ratio as compared to Darendeli (2001) curves.

4.2 Influence of Method of Analysis and Software Type

In the present research work, an effort has been made to compare the response of soil under an earthquake loading for different methods as well as for the different software's. Three software's, namely DEEPSOIL v6.1, SHAKE 2000, and D-MOD 2000, are used for the present research work. Equivalent linear GRA has been performed using DEEPSOIL v6.1 and SHAKE 2000 while nonlinear GRA is performed using DEEPSOIL v6.1 and D-MOD 2000.

For the present study, modulus reduction and damping curves proposed by Seed and Idriss (1970) are used for sandy soils and curves proposed by Vucetic and Dobry (1991) are used for silty and clayey soils which takes into account the plasticity

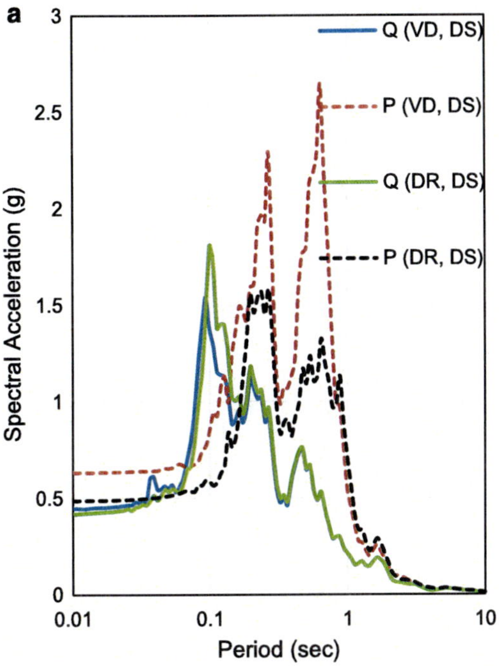

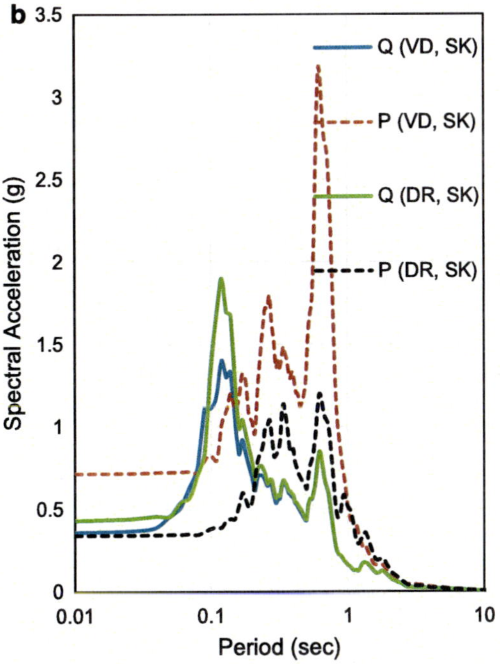

Fig. 4 **a** Spectral acceleration versus period for DEEPSOIL. **b** Spectral acceleration versus period for SHAKE 2000

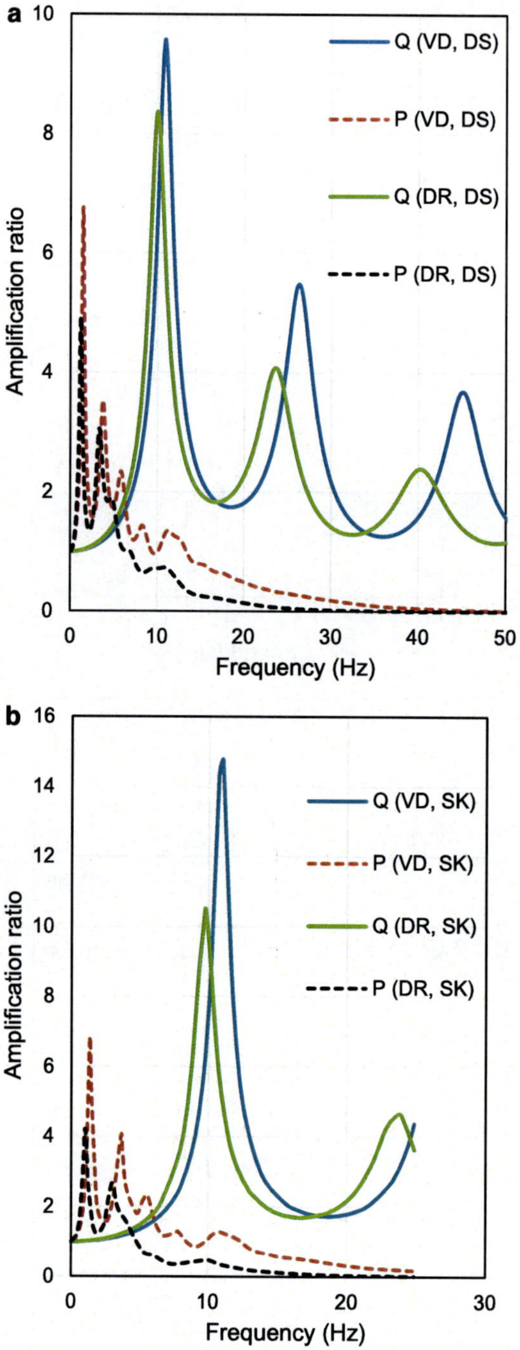

Fig. 5 **a** Amplification ratio versus frequency for DEEPSOIL. **b** Amplification ratio versus frequency for SHAKE 2000

properties of fine grained soil. Two soil models, i.e., BH 153-D (6 m depth) and BH 162 (45 m depth), are used for the present analysis.

4.2.1 Seismic Ground Response Analysis for BH 162

The variation of maximum horizontal acceleration along the depth of soil model is shown in Fig. 6. The PGA obtained using equivalent linear analysis in SHAKE is similar to that obtained using equivalent linear method in DEEPSOIL. From the figure, it can be seen that the difference is insignificant for equivalent linear and nonlinear analysis using DEEPSOIL. The PGA is obtained as 0.368 g and 0.342 g, respectively, for equivalent linear and nonlinear analysis for a soil model subjected to 1989 Loma Prieta earthquake using DEEPSOIL.

Amplification ratio is found out to be greater than 1 for all cases except for nonlinear analysis using D-MOD. The amplification ratio is obtained as 1.49 and 1.39 for equivalent linear and nonlinear analysis using DEEPSOIL. For SHAKE, the amplification ratio is obtained as 1.77, and for D-MOD, de-amplification is occurring with a value of 0.77. Hence, nonlinear analysis using D-MOD predicts very less

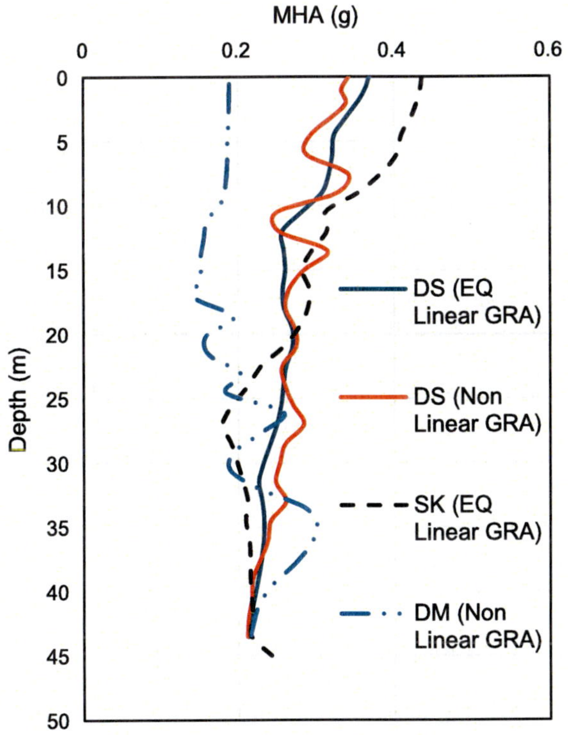

Fig. 6 Maximum Horizontal acceleration along the depth for BH 162

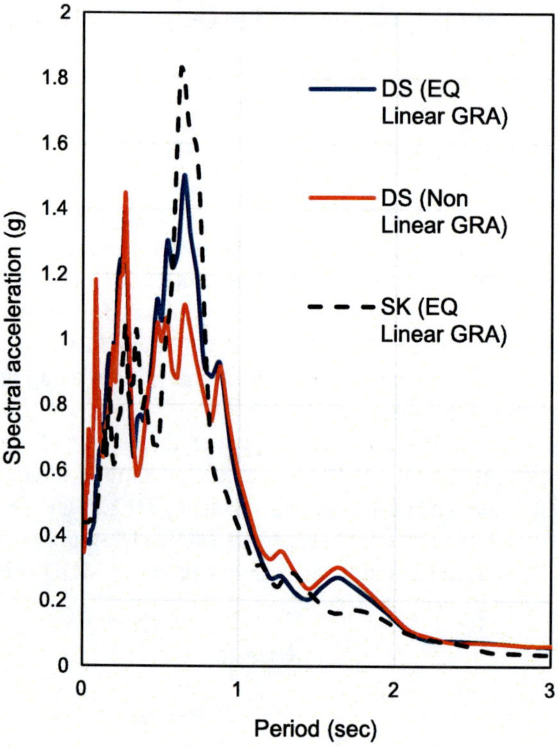

Fig. 7 Spectral acceleration versus time period for surface layer for BH 162

PGA value as compared to nonlinear analysis using DEEPSOIL, although significant duration remains nearly similar for all the cases. Hence, it can be concluded that the results might vary based on the software and method of analysis.

Maximum spectral acceleration versus period for the surface layer is as shown in Fig. 7. Maximum spectral acceleration of 1.42 g is obtained at a period of 0.27 s for equivalent linear analysis and for nonlinear analysis; 1.44 g is obtained at 0.27 s using DEEPSOIL. Hence, it can be seen that the results are very much similar for DEEPSOIL using equivalent linear and nonlinear GRA. Predominant period seems to be nearly similar for all the methods of analysis. When analyzed using SHAKE, maximum spectral acceleration of 1.83 g is obtained at 0.62 s. Hence, the result seems to be very similar for DEEPSOIL (EQ Linear analysis) and SHAKE.

4.2.2 Seismic Ground Response Analysis for BH 153-D

The variation of maximum horizontal acceleration along the depth of soil model is shown in Fig. 8. It is observed that nearly similar PGA values are obtained for equivalent linear and nonlinear analysis using DEEPSOIL. The trend of variation of

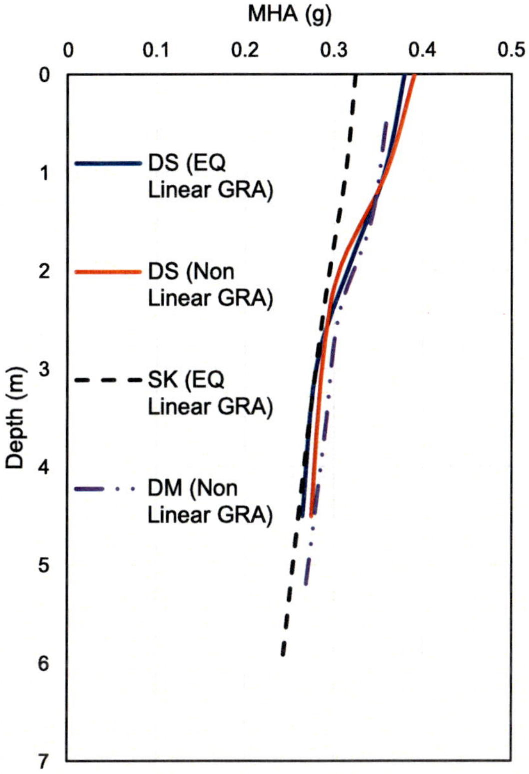

Fig. 8 Maximum horizontal acceleration (MHA) along the depth for BH 153-D

MHA along depth is also similar for all the different approaches. The PGA is obtained as 0.380 g and 0.391 g using equivalent linear and nonlinear analysis by DEEPSOIL. For SHAKE, PGA is calculated as 0.324 g. Hence, it can be concluded that both SHAKE and DEEPSOIL analyses provide similar PGA values for equivalent linear analysis and the variation of MHA along the depth of soil model is also similar.

Amplification ratio is obtained as 1.54 and 1.58 for equivalent linear and nonlinear ground response analysis (GRA) using DEEPSOIL. For SHAKE and D-MOD, amplification ratio is obtained as 1.31 and 1.42, respectively. Hence, it can be concluded that nearly similar amplification ratio is observed for a smaller depth soil model, i.e., BH 153-D using all the three approaches for a soil model subjected to an input motion and for three software's as well.

The variation of spectral acceleration versus time period for the surface layer is found to be exactly similar for equivalent linear and nonlinear ground response analysis (GRA) by DEEPSOIL, SHAKE as well as for D-MOD as shown in Fig. 9. The value of maximum spectral acceleration is obtained to be 1.13 g at 0.19 s for all the three methods of analysis. From the above study, it is concluded that the results obtained by equivalent linear and nonlinear GRA using three software's are nearly

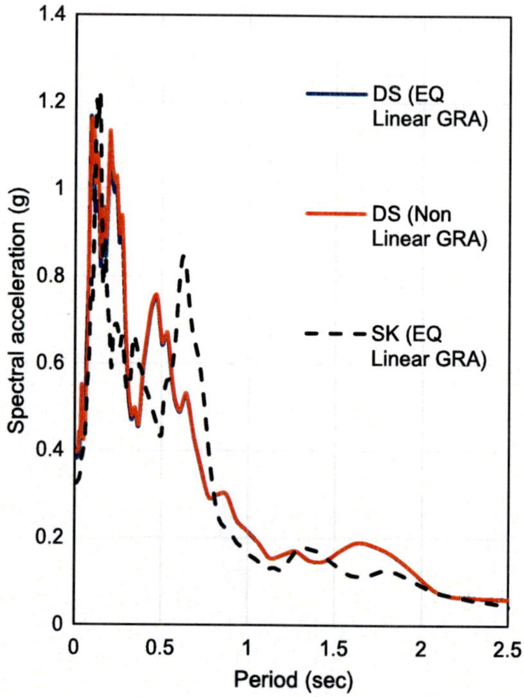

Fig. 9 Spectral acceleration versus time period for surface layer for BH 153-D

similar for a lesser depth soil model. However, the results are slightly different for a larger depth soil model. It is found that the results obtained using DEEPSOIL and SHAKE are nearly similar. However, the results obtained by DEEPSOIL and D-MOD are different for a larger depth soil model. Hence, it can be said that ground response depends largely on the depth of soil model, method of analysis as well as on the software used.

5 Conclusions

The following conclusions are obtained from the present study

- It is found that soil layers present in the site above the bedrock have a great effect in altering the ground response resulting either in amplification or de-amplification depending on the input motion.
- The variation observed in the results along the depth is similar for Vucetic and Dobry (1991) and Darendeli (2001) curves for a smaller depth soil model which changes significantly for a larger depth soil model.

- The maximum horizontal acceleration (MHA) and spectral acceleration at the ground surface are observed to be significantly more for a larger depth soil model as compared to a smaller depth soil model.
- It is observed that PGA is found to be slightly different for equivalent linear and nonlinear analysis using SHAKE and D-MOD for a larger depth soil model. However, the results are nearly similar for DEEPSOIL.
- For a small depth soil model, maximum horizontal acceleration (MHA) along the depth and spectral acceleration at the ground surface are found to be similar for three software's used as well as for the three methods of GRA used.
- The results from equivalent linear analysis in DEEPSOIL and SHAKE 2000 are found to be different for a larger depth soil model. However, the results are nearly similar for a lesser depth soil model.
- The results of (MHA) along the depth for nonlinear GRA appear to be different for DEEPSOIL and D-MOD for a larger depth soil model, i.e., BH 162 with DEEPSOIL predicting higher values. However, the results are exactly similar for a lesser depth soil model, i.e., BH 153-D.
- It is concluded that nonlinear analysis provides most accurate results since it considers the actual nonlinear behavior of soil.
- After successful validation with the field data results, suitability of the software can be decided.
- It is concluded that soil response to an earthquake loading depends highly on depth of soil model, number of soil layers, modulus reduction and damping curves, method of analysis as well as on the type of software used. Hence, it is very important to use the appropriate input data.

References

Choudhury D, Savoikar P (2009) Equivalent-linear seismic analyses of MSW landfills using DEEPSOIL. Eng Geol 107:98–108

Darendeli MB (2001) Development of a new family of normalized modulus reduction and material damping curves. Ph.D. dissertation, University of Texas Austin

Desai SS, Choudhury D (2015) Non-linear site-specific seismic ground response analysis for port sites in Mumbai, India. Japanese Geotechnical Society Special Publication, Japan. In: The 15th Asian regional conference on soil mechanics and geotechnical engineering, Japan, vol 2, no 19, pp 733–736

Hashash Y (2012) DEEPSOIL v6. 1–Tutorial and user manual. 2002–2012. University of Illinois at Urbana-Champaign, Illinois, Urbana

Mahmood K, Rehman Z, Farooq K, Memon AS (2016) One dimensional equivalent linear ground response analysis—a case study of collapsed Margalla Tower in Islamabad during 2005 Muzaffarabad earthquake. J Appl Phys 130:110–117

Naik NP, Choudhury D (2014) Comparative study of seismic ground responses using DEEPSOIL, SHAKE and D-MOD for soils of Goa, India. In: Geo-congress 2014: Geotechnical Special Publication No. GSP 234, ASCE, Reston, VA, pp 1101–1110

Ohta Y, Goto N (1978) Empirical shear wave velocity equations in terms of characteristic soil indexes. Earthq Eng Struct Dyn 6 (2):167–187

Phanikanth VS, Choudhury D, Reddy GR (2011) Equivalent-linear seismic ground response analysis of some typical sites in Mumbai. Geotech Geol Eng Springer 29:1109–1126

Seed HB, Idriss IM (1970) “Soil moduli and damping factors for dynamic response analyses, Technical Report” EERRC-70-10. University of California, Berkeley

Shylamoni P, Choudhury D, Ghosh S, Ghosh AK, Basu PC (2014) Seismic ground response analysis of KK-NPP site in the event of NCO earthquake using DEEPSOIL. In: Geo-congress 2014 Geotechnical Special Publication No. GSP 234, ASCE, pp 840–849

Vucetic M, Dobry R (1991) Effect of soil plasticity on cyclic response. J Geotech Eng ASCE 117(1):89–107

Design of Pile Foundation System for Wharf Structure in Liquefiable Soils

Putti Swathi Priyadarsini and Satyam Neelima

Abstract Liquefaction of soils during an earthquake motion results in complete loss of shear strength and stiffness leading to huge lateral ground movements and further failure of structures. Therefore, it is recommended to include liquefaction hazard assessment of soils in design procedure to avoid future damage of structures. There exists some concurrence between liquefaction assessment of free field sites and liquefaction of soils beneath foundations which is still a controversial issue. In the present study, a pile supported wharf has been considered from Vishakhapatnam port (Andhra Pradesh). Liquefaction hazard assessment has been carried out, and depending upon the factor of safety against liquefaction; the pile foundation system has been redesigned against liquefaction. The results from the present study indicate an optimized pile configuration compared to conventional design procedure. Factor of safeties achieved are high enough to avoid liquefaction-induced failures.

Keywords Pile foundation · Wharf · Liquefaction · Inertial force

1 Introduction

Most of the foundation system failures in past earthquakes are due to liquefaction of soils, as represented by many case histories in the 1964 Niigata and Alaska earthquakes, 1995 Kobe earthquake and 2004 Great Indian Ocean earthquake. Such failures can generally be attributed to either a loss of vertical or lateral support of the pile, or displacements post liquefaction or lateral spread. During the Great Hanshin earthquake (1995, $M_w = 6.9$), liquefaction occurred extensively around the coast area, and many structures founded on piles were damaged due to liquefaction. In India, pile-supported wharf structures were observed to be damaged in Kandla port and

P. Swathi Priyadarsini (✉)
International Institute of Information Technology-Hyderabad, Hyderabad, Telangana, India
e-mail: pspdarshini@gmail.com

S. Neelima
Discipline of Civil Engineering, Indian Institute of Technology-Indore, Indore, Madhya Pradesh, India

M. Latha Gali and P. Raghuveer Rao (eds.), *Geohazards*, Lecture Notes in Civil Engineering 86, https://doi.org/10.1007/978-981-15-6233-4_30

Gandhi nagar port during Gujarat earthquake (2001, $M_w = 7.7$) and the Great Indian Ocean earthquake (2004, $M_w = 9.3$), respectively. Huge research has been carried out analyzing the case histories, and it was observed that not only the inertial forces of the super-structures but also the deformations of the liquefied ground contributed to the damages. It has been recognized that adopting a design procedure for pile foundations considering the liquefaction-induced ground deformation is necessary.

Therefore, in the present study, an existing wharf structure has been considered from Vishakhapatnam port and analyzed for liquefaction hazard. The results obtained from liquefaction hazard assessment under free field condition and existing structure condition recommended redesign of the pile foundation system of the wharf to withstand against liquefaction-induced bearing failure and settlement failures.

1.1 Motivation and Objective

The purpose of this research is to assess the dynamic site characteristics in Vishakhapatnam, Andhra Pradesh (Zone-II, IS:1893–2016) (India) and to design a pile supported wharf structures based on the site-specific design parameters obtained from dynamic site characterization. A wharf is a berth-like structure which connects container vessels and ships to the ground making it a prominent structure in ports. Structural configurations of a wharf are not common unlike other structures, and any failure leads to immediate halt in port activities. Damage to the wharf can also induce damage to the adjacent structures such as tanks, cranes and other infrastructure.

In India, there is no specific codal provision for seismic design of wharf structures supported on piles. The conventional design procedure adopted does not provide any specific guidelines to design the structure against seismically induced damages. This research gap motivated to carry out design of pile supported wharf structure against liquefaction.

1.2 Liquefaction Assessment for the Existing Wharf

Liquefaction is a crucial challenge where the soils are exposed to high strain levels under dynamic loading due to seismic events (Satyam and Towhata 2016). Prominent structural and lifeline damages in urban areas have been observed from the historic earthquakes across the world. Demarcating liquefaction susceptible zones is therefore crucial in risk reduction through appropriate mitigation measures. Liquefaction in saturated soils revealed ground failures, sand boils and further subsidence of structures resulting in tilts, cracks or even collapse.

Liquefaction in foundation soils is influenced by the static as well as cyclic stress in the soil deposits. Initially, conventionally used demand term, cyclic stress ratio is corrected for K (correction for over burden pressure). The corrected CSR value, $CSR_{SSEI,rep}$ varying along the thickness of the soil sediment has been defined. This cyclic stress ratio accounting soil structure interaction is potentially used in liquefaction assessment. A semi-empirical simplified procedure has been formulated for the estimation of $CSR_{SSEI,rep}$ and is concluded to be reliable for liquefaction assessment in foundation soils. Thus, a comparison of the result from the liquefaction analysis from free field and for existing overburden has been done. The methodology adopted for liquefaction hazard assessment under existing structure condition has been shown in Fig. 1.

Thus, for the dynamic analysis, CSR formula has been modified which account for the stress due to the existing structure as well as the soil. The induced shear waves during a seismic event travel in vertical direction to the foundation system and are further transmitted to the super structure. These components of shear stress can be either in-phase or out-of phase. The following expression can be used to account the contribution by these two related and interdependent shear stress sources.

$$CSR_{SSEI} = (f(\sigma) * f(S_a/PGA) * f(H/B) * \tau_b + \tau_s)/(\sigma_{SSI} * K\sigma * K\alpha) \quad (1)$$

where

τ_b = Base shear

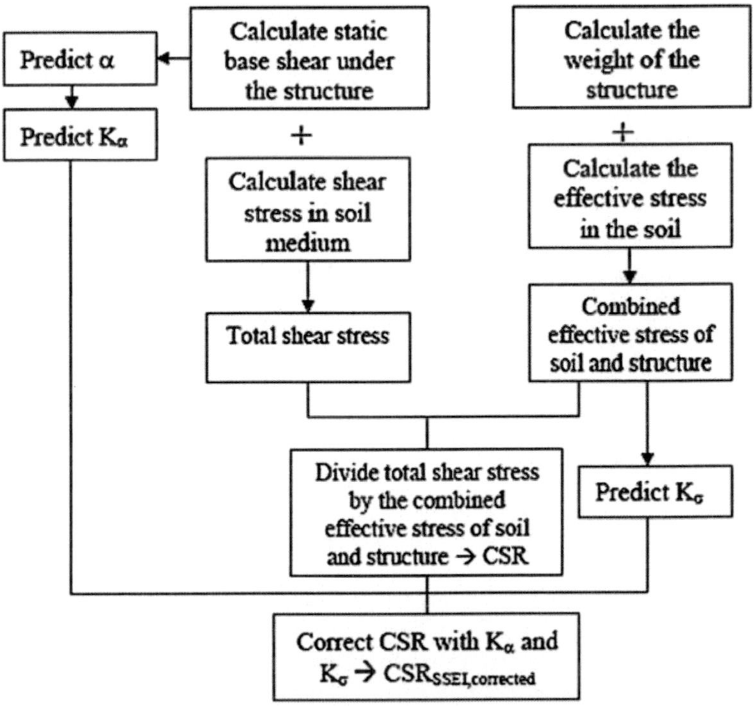

Fig. 1 Flowchart showing procedure for liquefaction hazard assessment for soil beneath an existing structure

τ_s = Shear stress
σ_{SSi} = Effective vertical stress induced by the soil and the structure
$f(\sigma) * f(S_a/\text{PGA}) * f(H/B)$ = SSEI weighting functions
$K\sigma$ = Overburden correction
Ks = Initial shear stress correction

These weight functions are evaluated using statistical assessments and through maximum likelihood assessments and presented below

$$f(\sigma) = 14.79 * \exp(-0.003(\sigma)) \quad (2)$$

$$f(S_a/\text{PGA}) = 0.74 * \exp(-0.683 * (S_a/\text{PGA})) \quad (3)$$

$$f(H/B) = \exp(-0.612 * (H/B)) \quad (4)$$

Base shear immediately beneath the depth of the foundation is expressed using:

$$\tau_b = (0.8S_a * \sigma_b(B * L))/((B + 1.6z)(L + 1.6z)) \quad (5)$$

Cyclic-induced stress due to soil column can be estimated by Seed and Idriss (1971); the simplified formula is represented below:

$$\tau_s = 0.65 * (a_{\max}/g) * \sigma_v * r_d \quad (6)$$

Effective stress due to weight of the soil as well as structure (σ_{SSI}). To calculate the seismically induced stress under the foundation due to structure and soil.

$$\sigma_{\text{SSI}}(z) = \sigma'_{\text{soil}}(z) + (\sigma_{\text{str}} * (B * L)/((B + 0.9z)(L + 0.9z))) \quad (7)$$

$$\sigma = V_{s,\text{final}} * T/H_{\text{effective}} \quad (8)$$

where

σ = Relative stiffness
T = Time period
H_{eff} = Effective height of the building
V_s = Final shear velocity

The soil profile beneath the existing wharf structure considered in the study is as shown in Table 1.

Liquefaction hazard assessment has been carried out for the given soil profile under free field and existing structure condition using the methodology provided by Idriss and Boulanger (2006). From the results (factor of safety Fig. 2), it is evident that the first three layers of the soil are liquefiable under both the conditions (Table 2) which may contribute to the damage of the structure due to seismic loading.

Table 1 Properties of the soil beneath the structure

Depth	*N*	Density (kN/m^3)
0	14	18.3447
3.1	7	16.8732
6.7	20	17.658
9.5	37	18.2466
12.4	20	18.2466
16.5	16	17.658
19.5	75	21.5

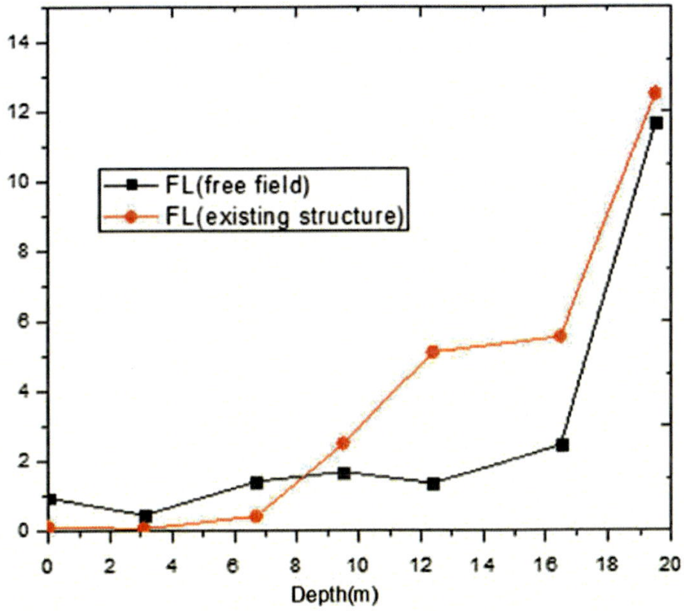

Fig. 2 Plot showing variation of FOS along with depth

Table 2 Liquefaction assessment results for free field and existing structure condition

Depth	F_L(Free field)	Remark	F_L(existing structure condition)	Remark
0	0.945	Liquefiable	0.128	Liquefiable
3.1	0.447	Liquefiable	0.063	Liquefiable
6.7	1.406	Certainly liquefiable	0.417	Liquefiable
9.5	1.673	Non liquefiable	2.508	Non liquefiable
12.4	1.345	Non liquefiable	5.111	Non liquefiable
16.5	2.45	Non liquefiable	5.550	Non liquefiable
19.5	11.65	Non liquefiable	12.000	Non liquefiable

Therefore, it is concluded that there is a necessity to redesign the existing structure against liquefaction.

1.3 Design of Foundation System Against Liquefaction

The most common deep foundation systems that are subjected to failures during a seismic event are piles. Seismic events can impose typically challenging set of loads on pile foundation systems, prominently when the soil sediments in the vicinity of the foundation system tend to liquefy. Lateral spreading in the sloped soil can impose huge lateral loads additionally on the piles. Research since past has considered few case histories related to the performance of the pile foundation systems in the soils that are liquefiable. Based on the failures and performances, possible set of failure mechanisms have been defined for both single piles and pile groups founded on level ground with liquefiable soils and laterally spreading sloping ground from past observations.

In the present study, an initial (assumed) pile foundation system with five rows of 750 mm diameter bored in situ piles are considered along the length of the proposed wharf. Each row has eight piles with spacing of 4 m. The soil profile beneath the existing structure has been assessed for liquefaction hazard, and it is observed that first three layers of the soil profile are liquefiable. The total height of the soil column considered for liquefaction assessment and further design is 19.5 m. Hard rock has been encountered at 19 m depth. The design earthquake (Bhuj Earthquake) has magnitude 7.7 (scaled) with a peak ground acceleration of 0.12 g.

The foundation must be safe against bearing or yielding failure, and to fulfill the serviceability requirements, displacements in foundation should not exceed the following values:

Displacement in horizontal direction: 100 mm
Displacement in vertical direction: 75 mm

The sizes of members considered are as follows:

- Diameter of piles: 0.75 m
- Length of the pile: 28 m
- Main beam: 1600 × 1800 mm
- Longitudinal beam: 1600 × 1800 mm
- Thickness of deck slab: 400 mm

Initially, the conventional design of the foundation system subjected to axial loads has been considered and the components of base capacity and shaft friction have been estimated using the approach proposed by Berezantzev.

$$Q_b = A_b \sigma_b \left(N_q - 1\right) \tag{9}$$

A_b—Base area of the pile

σ_b—Effective overburden pressure at pile tip level
N_q—Bearing capacity factor (Berezantzev et al. 1961).

Shaft resistance

$$Q_s = \pi D \int_0^L \tau_s \tag{10}$$

$$\tau_s = K_s \sigma'_v \tan\delta_{cv} \tag{11}$$

K_s = earth pressure coefficient
σ'_v = effective vertical stress
δ_{cv} = friction angle between pile material and soil

$$\text{Pile capacity} \quad Q_u = Q_b + Q_s \tag{12}$$

The pile capacity is calculated to be 3420 kN. All the design loads such as crane loadings, berthing loads, mooring loads and others are calculated using IS 4651 Part-3 (1974). From the design loads and the axial pile capacity, numbers of piles assumed at the initial stage (40 piles) of the design are sufficient to support whole length of the wharf structure.

1.3.1 Estimation of Inertial and Kinematic Loads

Pile foundation system suffers huge damage during and immediately after the seismic event. In this study, the action of the seismic loads on the foundation system is considered in detail. They are generally classified as kinematic and inertial loads depending on their action on the soil or the foundation system. After the evaluation of axial static load capacity of the pile foundation system, it is further analyzed for kinematic and inertial loads. Thereafter should be checked for the safety against liquefaction-induced bearing and settlement failures. For this purpose, a methodology proposed by (Madabushi et al. 2010) has been adopted to design the pile foundation system in liquefiable soils. The flowchart showing the design procedure has been presented below in Fig. 3.

- Effective length of the pile is estimated using

$$L_{ad} = 2D\left(\frac{E_p}{E_{sD}}\right)^{0.22} \tag{13}$$

E_p = Young's modulus of pile
E_{sD} = Young's modulus of pile at depth D

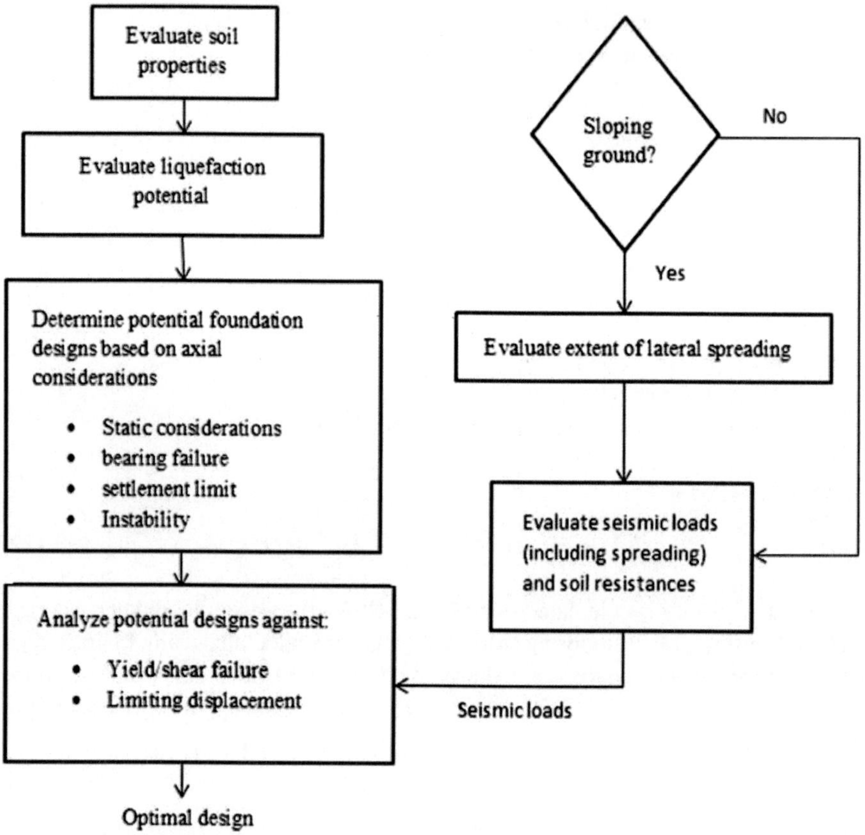

Fig. 3 Summary of procedure for design of pile foundation system in liquefiable soils

- Flexibility of the pile is calculated according to (IS2911 Part 1 Section 1)

$$T = \sqrt[5]{\frac{EI}{\eta_h}} \tag{14}$$

- Estimation of inertial loading on the pile

$$K_h = 0.79DE_{sD}\left(\frac{E_p}{E_{sD}}\right)^{0.28} \tag{15}$$

$$K_M = 0.15D^3E_{sD}\left(\frac{E_p}{E_{sD}}\right)^{0.77} \tag{16}$$

$$K_{hM} = -0.24D^2E_{sD}\left(\frac{E_p}{E_{sD}}\right)^{0.53} \tag{17}$$

Table 3 Estimated kinematic and inertial loads and other design parameters

Parameter	Estimated value
Small-strain shear modulus of silty sand (G_o)	3.8×10^4 kN/m^2
Peak cyclic shear stress (τ_{max})	8.2 kN/m^2
Natural frequency	18.5 Hz
Effective length (L_{ad})	21.0856 m
Horizontal stiffness (K_h)	545.8242 kN/m
Rotational stiffness (K_m)	711.7249 kNm/rad
Coupling stiffness (k_{hm})	−445.7825 kNm/rad
Equivalent horizontal stiffness (K_{h_eq})	545.8242 kN/m
Natural frequency of pile group	0.0831 Hz
Natural period of pile group	12.0273 s
Horizontal inertia load on pile group (H)	12,480 kN
Moment load on pile group (M)	124,800 kN
Interaction factor	1.0000
Displacement at the top of pile group	92.0001 mm
Factor of safety to avoid liquefaction-induced bearing failure	1.56
Factor of safety to avoid liquefaction-induced settlement	2.5

$$e = \frac{K_{hM}}{K_h} \tag{18}$$

$$K_{h\_eq} = \frac{K_h K_M - K_{hM}^2}{K_M - eK_{hM}} \tag{19}$$

- Estimation of kinematic interaction

$$I_u = aF^4 + bF^3 + cF^2 + 1.0 \tag{20}$$

$$F = \left(\frac{f_p}{f_n}\right)\left(\frac{E_p}{E_{sD}}\right)^{0.16}\left(\frac{L}{D}\right)^{-0.35} \tag{21}$$

where

T = Stiffness factor
K_h = Horizontal stiffness
K_M = Rotational stiffness
K_{hM} = Coupling stiffness
e = Eccentricity

K_{h_eq} = Equivalent horizontal stiffness
I_u = Interaction factor

- FOS (factor of safety during static condition) must be sufficiently high for liquefaction-induced bearing failure to be avoided. This is achieved if

$$\text{FOS} \geq \frac{1}{\alpha_{\text{ult}}\left(1 - r_{u,\text{base}}\right)^{\frac{3-\sin\varphi}{3(1+\sin\varphi)}} - \alpha_{\text{ult}} + 1} \quad (22)$$

- FOS must be sufficiently high to avoid liquefaction-induced settlement. This is achieved if

$$\text{FOS} \geq 1 + 5.5\left(r_{u,\text{base}}\right)^{3.5} \quad (23)$$

where

$$r_{\text{u,base}} = \frac{\text{depth of full liquefaction}}{\text{length of the pile}} \quad (24)$$

$$\alpha_{\text{ult}} = \frac{Q_b}{Q_u} \quad (25)$$

2 Conclusions

From the past earthquakes, it is observed that most of the failures in pile-supported wharf are due to liquefaction, slope stability problem in the embankment and failure of connection between the piles and the deck slab. Therefore, in the present study, liquefaction hazard assessment has been carried out for the soil, prior to the design of wharf structure. From the liquefaction analysis, it is evident that the first three layers of the soil strata are prone to liquefaction (factor of safety < 1.0) under both free field and existing structure condition. Therefore, the foundation system has to be redesigned against liquefaction-induced bearing and settlement failures. An attempt has been made to design the pile foundation system against liquefaction using the method proposed by (Madabhushi et al. 2010) for the estimated loads.

From the results, it is evident that the proposed pile foundation is safe against bearing and settlement failures as the FOS = 3 in axially loaded static condition is greater than factor of safety against bearing failure (F.S = 1.56) and factor of safety against settlement failure (F.S = 2.5). The horizontal displacement at the top of pile group is estimated as 92.0001 mm (Table 3) which is less than 100 mm which satisfies the limit state of serviceability criteria. Therefore, the proposed design is acceptable and is considered as an optimal solution.

References

Berezantzev VG, Khristoforov VS, Golubkov VN (1961) Load bearing capacity and deformation of piled foundations. Proceedings of the 5th International Conference on Soil Mechanics and Foundation Engineering, 2, pp 11–15

Development of West Quay North (WQ-7 & WQ-8) Breth in the Inner Harbor of Vishakhapatanm Port, Design Report of Sub-Structure, DBM/VPT/PT/02, REV-2

Idriss IM, Boulanger RW (2006) Semi-empirical procedures for evaluating liquefaction potential during earthquakes. Soil Dynam Earthquake Eng 26(2–4):115–130

IS:4651 (Part-3) (1974) Code of practice for planning and design of ports and harbours, loading

Madabhushi G, Knappett J, Haigh S (2010) Design of pile foundations in liquefiable soils. Imperial College Press

Satyam ND, Towhata I (2016) Site-specific ground response analysis and liquefaction assessment of Vijayawada city (India). Nat Hazards 81(2):705–724

Standard I (1893) Criteria for earthquake resistant design of structures. Bureau of Indian Standards, Part 1

Ground Response Analysis of a Nuclear Power Plant Site in Southern India: A Nonlinear Approach

U. Veena, Naveen James, and T. G. Sitharam

Abstract One-dimensional nonlinear ground response analysis is performed for a nuclear power plant site situated in east coast of south India. Since the strong motion records available for the study area are scarce, a synthetic ground motion time history developed for the same site by (James et al., Nat Hazards 71:419–462, 2014) and its scaled-up scaled-down versions are used as the input motion at bedrock level. A few more synthetic ground motion histories are also developed and applied as input motion in order to analyse the deviation in ground response with variation in frequency content and duration of bedrock motion. Martin–Finn–Seed pore pressure model is used to compute development of excess pore pressure during loading. The peak horizontal acceleration (PHA) profile, amplification spectra and response spectra at surface level for 5% damping are plotted for each location.

Keywords GRA · Nonlinear analysis · Artificial ground motion

1 Introduction

The damage caused to a structure due to a strong ground motion depends on response of underlying ground to the propagation of seismic waves. Hence, ground response analysis (GRA), the prediction of soil behaviour at different depths in a site under seismic loading, is an essential task to be performed in order to ensure the seismic safety of structures. In addition, proper ground response analysis is indispensable to assess the liquefaction hazard and to take necessary liquefaction mitigation measures. Generally, the ground response analysis is performed in a one-dimensional manner assuming horizontal soil layers subjected to vertically propagating SH waves.

U. Veena (✉) · N. James
Indian Institute of Technology Ropar, Punjab, India
e-mail: veenau1010@gmail.com

T. G. Sitharam
Indian Institute of Technology Guwahati, Assam, India

M. Latha Gali and P. Raghuveer Rao (eds.), *Geohazards*, Lecture Notes in Civil Engineering 86, https://doi.org/10.1007/978-981-15-6233-4_31

The most popular approach is frequency-domain equivalent linear method, first proposed by Idriss and Seed (1967). In this approach, soil stiffness and damping ratio are modified after each iteration according to the computed effective strain value. The main reason behind the wide acceptance of this approach is the simplicity of parameter selection and well-documented literature. The major shortcoming of this approach is that the dynamic properties are kept constant throughout the duration of loading in each iteration. Under small strain, this kind of idealization of soil behaviour works well as the degree of nonlinearity is very less. But it becomes highly unrealistic when the dynamic properties change dramatically by the generation of excess pore pressure during shaking. Equivalent linear approach results in underestimation of ground response under large strain (Kaklamanos et al. 2013; Kim et al. 2013; Bolisetti et al. 2014) which is crucial for structures like nuclear power plants. To capture this complex nonlinear behaviour, nonlinear ground response analysis, which involves numerical integration of equation of motion in time domain, is to be used. Lack of sufficient guidelines for parameter selection limits the applicability of nonlinear approach in ground response analysis for ordinary projects. But for safety-related structures like nuclear power plants, it is preferred to follow nonlinear approach to characterize accurate soil behaviour required for a conservative seismic design.

The Kalpakkam nuclear power plant site is situated at the Coromandel Coast with Bay of Bengal on the east side. Being in Peninsular India, which is no longer being considered as a low seismic zone according to a number of researchers (Chandra 1977; Gangrade and Arora 2000; Ramalingeswara Rao 2000; Ramasamy 2006), seismic characterization of this nuclear power plant site is of paramount importance. A comprehensive microzonation work has been done for this site by James et al. (2014), in which they followed the equivalent linear approach for performing GRA. In the present study, extensive GRA is performed for Kalpakkam nuclear power plant site by following nonlinear method.

In nonlinear ground response analysis, any linear or nonlinear stress–strain model can be used. Suitable soil properties to be used for each time step are chosen by referring to the stress–strain relation followed. Stress reversals are given special attention if nonlinear inelastic models are used. The hysteresis material models utilized in nonlinear GRA consist of (i) a backbone curve, to explain stress–strain relation during initial loading and (ii) hysteresis rules, to explain the unloading/reloading behaviour. In simple form, the cyclic nonlinear models can be expressed as $\tau = F_{bb}(\gamma)$, where τ is the shear stress, γ is the shear strain and F_{bb} denotes the nonlinear function. The shape of the backbone curve is governed by low-strain stiffness and high-strain shear strength of soil. A number of backbone curve models are available in the literature such as Ramberg–Osgood model (Ramberg and Osgood 1943), hyperbolic model (Kondner and Zelasko 1963), modified hyperbolic or modified KZ (MKZ) model (Matasovic and Vucetic 1993). Hysteresis rules like Masing rule (Masing 1926), Cundall–Pyke model (Pyke 1979), etc., are employed to explain unloading/reloading behaviour. To incorporate the effects of development of excess pore pressure, any of the pore water pressure generation models developed by Martin et al. (1975), Ishihara and Towhata (1980), Finn and Bhatia (1981), Matasovic (1993), etc., can be used.

2 Nonlinear Ground Response Analysis Using Finite Element Method

In the present study, the nonlinear GRA of the nuclear power plant site is performed using the geotechnical finite element software QUAKE/W, which is a component of software package GeoStudio 2016. The program uses hyperbolic model to analyse the nonlinear behaviour of soil along with Masing rules to describe stress reversal conditions. The soil properties required to define the backbone curve are the small strain modulus $G_{\max}$, which denotes the initial slope, and the shear strength parameters c' and ϕ', which denote the asymptote, $\tau_{\max}$. Equation 1 expresses the hyperbolic backbone function (Kramer 1996).

$$\tau = F_{\rm bb}(\gamma) = \frac{G_{\max}\gamma}{1 + (G_{\max}/\tau_{\max)|\gamma|}} \tag{1}$$

MFS pore pressure model (Martin et al. 1975) is employed to incorporate the generation of excess pore water pressure in soil. This model estimates the pore pressure generated during undrained loading by relating it to incremental volumetric strain ($\Delta\varepsilon_{vd}$) that would have occurred for the same stress increment under drained loading condition as shown by Eq. 2.

$$\Delta u = \bar{E}_r \Delta\varepsilon_{vd} \tag{2}$$

The rebound modulus, $\bar{E}_r$, is determined by carrying out a rebound test in a consolidometer. Due to the unavailability of actual experimental data from the site, rebound modulus curve and volumetric strain curve corresponding to sand of relative density 45%, which were taken from Martin et al. (1975), were used to define the pore pressure generation in sand layers.

QUAKE/W code follows Gauss–Legendre numerical integration to form element characteristic matrix and Wilson θ method for time-domain integration of motion equation. The finite element equations are solved by using the Cholesky factorization technique which is closely related to the Gauss elimination method.

2.1 *Soil Profile and Properties*

Seven different locations from the site are selected for performing the analyses. Major geological formations observed in the site are loose to medium dense silty sand, very soft high plastic clay, residual soil composed of completely weathered rock and strong granitic bedrock. Table 1 shows soil types present in each borehole.

The input data required for the analysis are collected from IGCAR, Kalpakkam. Shear strength parameters (c and ϕ) for each soil layer are obtained from the bore

Table 1 Depth (in metres) of different soil types overlying bedrock in the selected boreholes

Soil layer	BH1	BH2	BH3	BH4	BH5	BH6	BH7
Sand	–	0–6.2	0–3.5	0–2.5	0–9.2	0–2.3	–
Clayey sand	0–3.9	–	–	–	–	–	0–2.8
Soft clay	3.9–6	6.2–13.5	3.5–5	2.5–6.5	–	2.3–6.1	–
Residual soil	–	13.5–18.6	5–8	6.5–9.2	9.2–18	–	–
Weathered rock	6–11.7	25–43.1	8–34.5	9.2–29.6	18–32.1	8.6–15.9	2.8–9.9
Rock	–	–	–	–	32.1–41.9	15.9–21	–

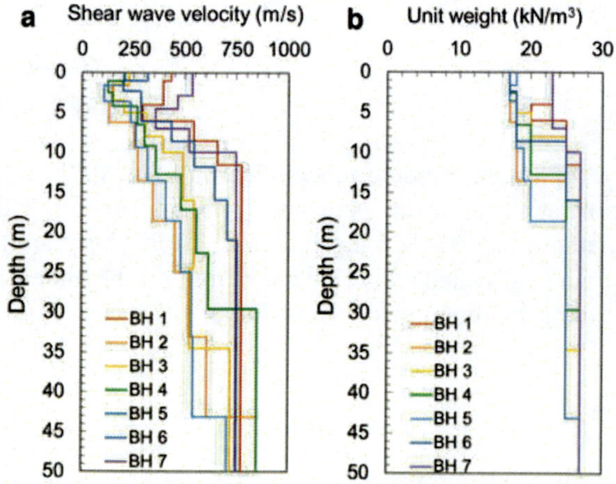

Fig. 1 (**a**) Shear wave velocity profile and (**b**) density profile at the selected locations

log data. Other soil parameters such as density (γ), Poison's ratio (ν) and shear wave velocity (V_s) are obtained from cross-hole test reports which correspond to the selected locations, and low-strain shear moduli ($G_{\max}$) are calculated from shear wave velocity (V_s) and mass density (ρ) using Eq. 3.

$$G_{\max} = \rho V_s^2 \tag{3}$$

Depth of bedrock was found to vary from 10 to 43 m according to the profile of selected locations. Water table was located at an average depth of 2 m. Figure 1 shows the shear wave velocity profiles and the density profiles at the selected locations.

2.2 *One-Dimensional Model and Boundary Conditions*

To perform one-dimensional GRA, the soil profile above the bedrock is modelled as a soil column. In the initial static stage, in which in situ stresses are calculated, the base of soil column is kept fixed in all directions and only vertical displacement is allowed on the sides of soil column. In nonlinear dynamic stage, the motion is restrained in vertical direction and is permitted in horizontal direction. At the base, the reflecting boundary condition, i.e. the rigid boundary condition, is chosen as recommended by Kwok et al. (2007) and input acceleration time history is directly applied.

2.3 *Input Strong Motion History*

From the Deterministic Seismic Hazard Analysis (DSHA) carried out for Kalpakkam nuclear power plant site by James et al. (2014), the maximum credible earthquake (MCE) was found to be an event of 5.7 Mw at a linear source 20 km away. Since the earthquake strong motion histories are not available for the site, strong motion history generated artificially corresponding to MCE obtained from DSHA by James et al. (2014) based on methodology proposed by Feng-xin and Yu-shan (2006) has been used for the present studies. Figure 2 shows the synthetic ground motion history used.

In addition to this, a scaled-up motion and a scaled-down motion, which were obtained by multiplying the original acceleration time history by the factors 4 and 0.25, were also used as input motions in order to assess the effect of peak bedrock acceleration on the response of overlying soil layers. A few more acceleration time histories were generated by using a MATLAB program based on the stochastic point simulation technique developed by Boore (2003). These artificial earthquake time histories were different from each other in terms of frequency content and duration while PGA remains almost same. Analyses were performed by using these earthquake time histories as input motion with the intention of examining the dependency of ground response on predominant period and duration. The details of all the input motions used for the analyses are given in Table 2.

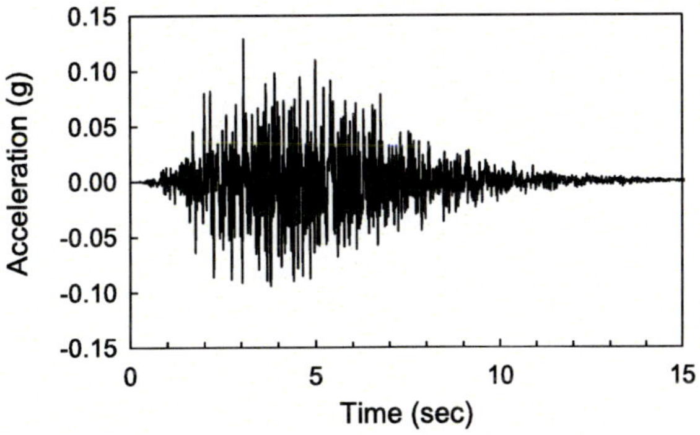

Fig. 2 Artificial ground motion history (from James et al. 2014)

Table 2 Ground motion parameters of input motions

Strong motion	PGA (g)	Predominant period (sec)	Bracketed duration (sec)[1]
Original strong motion (James et al. 2014)	0.129	0.04	5.01
Scaled-up motion	0.516	0.04	9.94
Scaled-down motion	0.032	0.04	0
SGM[2] 1	0.137	0.02	10.02
SGM 2	0.135	0.08	10.16
SGM 3	0.132	0.16	9.67
SGM 4	0.132	0.04	1.79
SGM 5	0.132	0.04	6.4
SGM 6	0.131	0.04	12.21

[1]Time between the first and last exceedances of threshold frequency of 0.05 g
[2]Strong ground motion

2.4 *Equivalent Linear Ground Response Analysis*

For a comparative study, equivalent linear ground response analyses were conducted for all the selected soil profiles with each of earthquake motion histories mentioned above. The one-dimensional GRA program SHAKE 2000, which is an upgraded version of the SHAKE program developed by Schnabel et al. (1972), was used for this study. The modulus reduction curves and damping curves were chosen for each soil type by referring to James et al. (2014) as given in Table 3.

Table 3 Dynamic properties assigned to each soil type (from James et al. 2014)

Soil type	G/Gmax	Damping curve
Sand	Sand average—Seed and Idriss (1970)	Sand average—Seed and Idriss (1970)
Clay	Clay upper range—Seed and Idriss (1970)	Clay upper—Seed and Sun (1985)
Filled-up soil	Soil with OCR 1–15—(Vucetic and Dorby 1991)	Soil with OCR 1–15—(Vucetic and Dorby 1991)
Residual soil	Gravel (Mean)—(Rollins et al. 1998)	Gravel (Mean)—(Rollins et al. 1998)
Weathered rock	Rockfill—Gazetas and Dakoulas (1992)	Rockfill—Gazetas and Dakoulas (1992)
Hard rock	Rock—Schnabel (1973)	Rock—Schnabel (1973)

3 Results and Discussions

3.1 Results of Nonlinear Analysis

Nonlinear one-dimensional ground response analyses have been performed for each of the seven locations selected with input strong motion history generated by James et al. (2014). The results of analyses are presented in terms of peak horizontal acceleration at various depths, ratio of surface motion amplitude to the bedrock motion amplitude and response spectra of a single-degree-of-freedom (SDOF) system of 5% damping. Variation of peak horizontal acceleration with depth is illustrated in Fig. 3. Figure 4 shows the amplification spectrum smoothed by using Hanning window, and Fig. 5 shows the surface response spectra for 5% damped system at the seven locations.

In PHA profiles (Fig. 3), it can be seen that even though some localized deamplification effects are there in a few soil layers, the site characteristics contribute

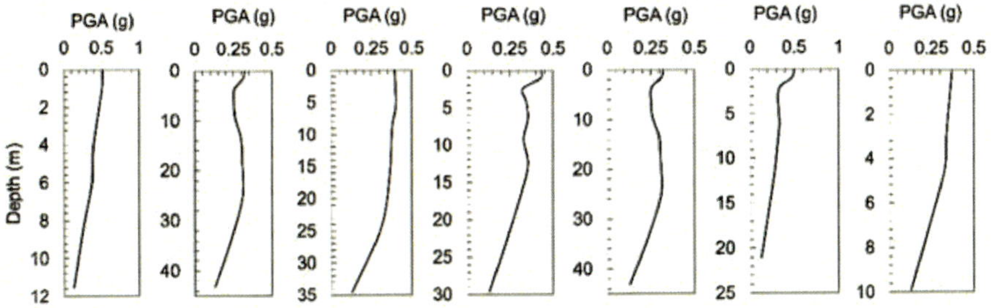

Fig. 3 PHA profiles of the selected boreholes

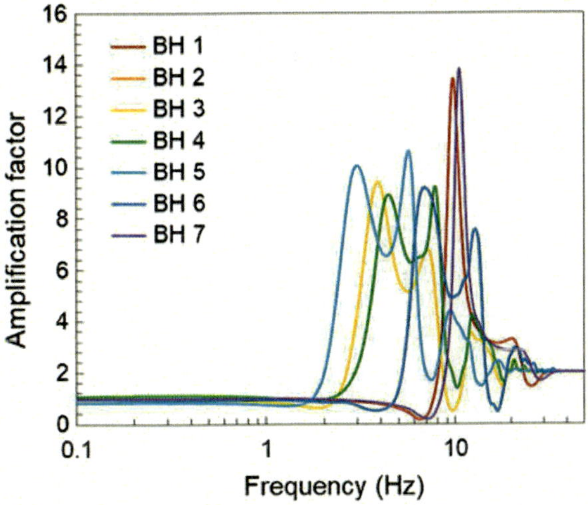

Fig. 4 Amplification spectra at the selected locations

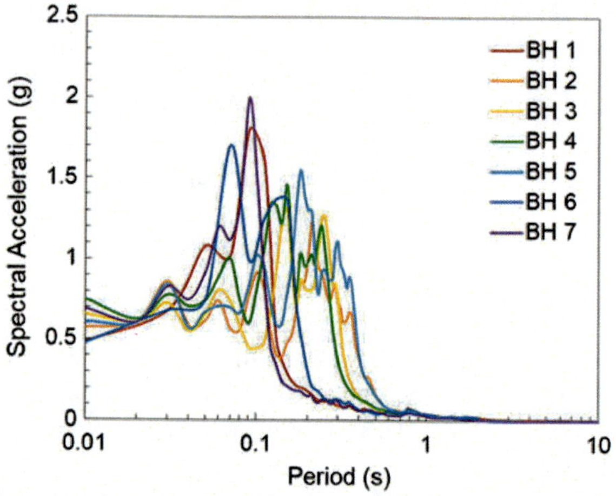

Fig. 5 Response spectra of 5% damped SDOF system at the surface level of selected locations

to overall amplification of ground motion. The amplification spectra (Fig. 4) show that there is no amplification which takes place for longer period components of the motion (frequency <1.5 Hz). Maximum amplification factor (nearly 13.5) was observed at first and seventh borehole locations. BH 1 is one of the boreholes very near to the nuclear reactor. Hence, the huge amplification observed at BH1 should be given sufficient consideration during the seismic safety assessment of reactor.

At BH 1 and BH 7, the maximum spectral accelerations of 1.8 and 2 g are experienced by 5% damped SDOF system with natural period of 0.09 s. The results of nonlinear ground response analysis of the nuclear power plant site are consolidated in Table 4.

Table 4 Results of one-dimensional nonlinear ground response analysis

Location	PGA (g)	Predominant period of output motion (s)	Amplification factor[1]	Maximum spectral acceleration (g)	Period of maximum spectral acceleration (s)
BH 1	0.4617	0.10	3.99	1.80	0.09
BH 2	0.3216	0.14	2.49	1.23	0.21
BH 3	0.3945	0.24	3.06	1.31	0.15
BH 4	0.4308	0.14	3.34	1.46	0.15
BH 5	0.3622	0.18	2.81	1.55	0.18
BH 6	0.4996	0.14	3.87	1.95	0.07
BH 7	0.3703	0.10	2.87	2.00	0.09

[1]Ratio of PHA at surface level to PHA at bedrock level

3.2 Variation in Response with Peak Bedrock Acceleration (PBRA)

The effect of bedrock acceleration on response of overlying soil was examined by performing nonlinear GRA of selected locations with a scaled-up (multiplication factor 4) and a scaled-down (multiplication factor 0.25) versions of original input motion. Equivalent linear GRA was also performed for all these cases using SHAKE 2000 for a comparative study. The results obtained for BH 1 and BH 4, which are nearest to the nuclear reactor, are presented here. The PHA profiles, amplification spectra and surface response spectra are displayed in Figs. 6, 7 and 8, respectively.

Peak horizontal accelerations predicted by equivalent linear method are smaller than those predicted by nonlinear analysis which agrees with the observation made by Joyner and Chen (1975). The amplification factor, i.e. ratio of PHA at surface level to PHA at bedrock level, was found to decrease with increase in peak bedrock acceleration (PBRA). This variation is illustrated in Fig. 9. The same trend was observed by Basu and Dey (2017).

It was observed that results of equivalent linear analysis match well with results of nonlinear analysis for weak motion (scaled-down motion). During strong motions with higher PGA values (scaled-up motion), soil behaves in more nonlinear

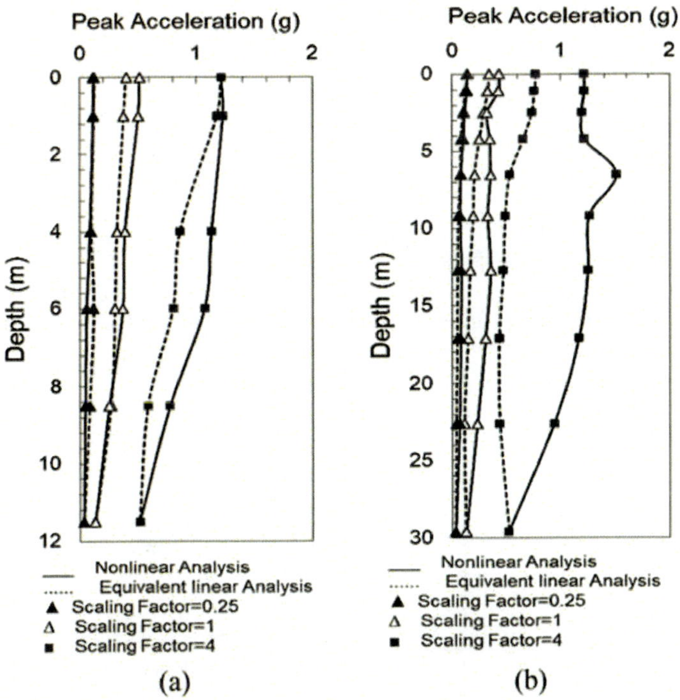

Fig. 6 PHA profiles of (**a**) BH 1 and (**b**) BH 4 with scaled-up and scaled-down input motions

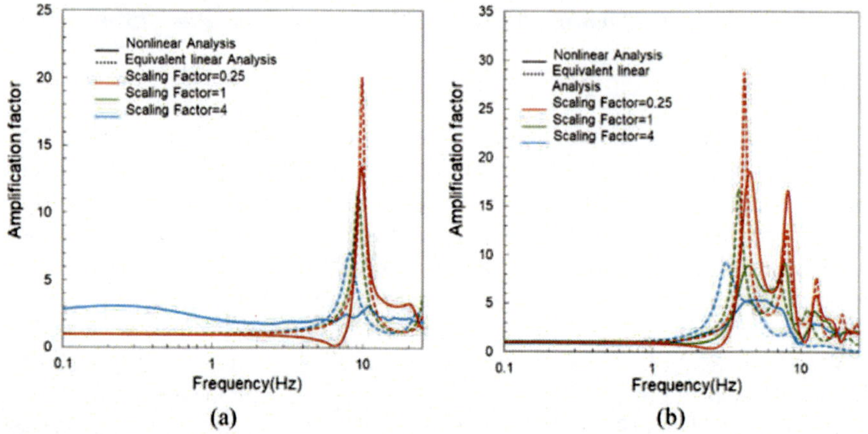

Fig. 7 Amplification spectra at (**a**) BH 1 and (**b**) BH 4 with scaled-up and scaled-down input motions

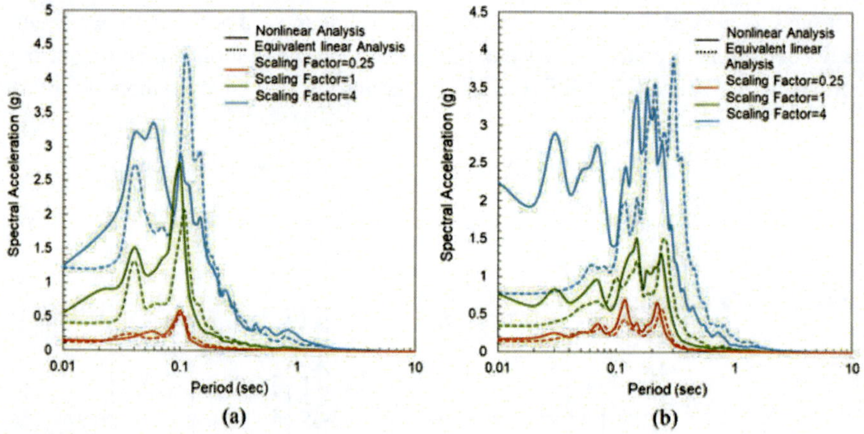

Fig. 8 Response spectra of 5% damped SDOF system at the surface level of (**a**) BH 1 and (**b**) BH 4 with scaled-up and scaled-down input motions

Fig. 9 Variation in amplification factor with PBRA

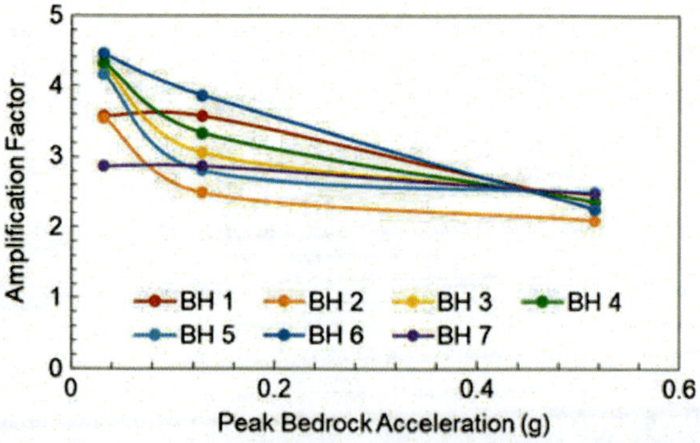

manner, and it leads to deviation of results of equivalent linear analysis from results of nonlinear analysis. The resonant frequencies (frequencies corresponding to peak amplifications) estimated by both equivalent linear and nonlinear methods match reasonably well. For long-period motions, both equivalent linear method and nonlinear method produce similar response same as observed by the previous researchers (Joyner and Chen 1975; Stewart and Kwok 2008; Kaklamanos et al. 2013; Kim et al. 2013; Bolisetti et al. 2014). Also, it can be noted that nonlinear method can reproduce the variation of response at short period range which cannot be obtained by equivalent linear analysis. Equivalent linear method results in almost constant response at periods less than around 0.02 s.

3.3 Variation in Response with Frequency Content and Duration of Input Motion

A few more analyses were performed with two sets of artificially generated ground motion histories. In the first set, all the time histories have almost same PGA value and comparable durations, but different predominant periods. The second set was characterized with same PGA and predominant period but different durations. The results obtained by nonlinear method are illustrated in Figs. 10, 11, 12, 13, 14, 15, 16 and 17 along with the results from equivalent linear method.

The results show random variation in peak horizontal accelerations, amplification factors and spectral accelerations with change in predominant frequency and duration. The response could not be properly related to predominant frequency or bracketed duration. This can be because of the reason that only PGA, predominant frequency and bracketed duration were considered in the study. Other strong motion parameters such as sustained maximum acceleration and bandwidth which also define the intensity of motion were not given any considerations. Therefore, more studies are required by including all the strong motion parameters to evaluate the effect of each parameter on ground response.

4 Conclusions

Nonlinear ground response analysis was performed for soil columns corresponding to seven different boreholes in a nuclear power plant site located in southern India. Due to the lack of recorded strong motion histories, artificial acceleration history generated by James et al. (2014) for the same site was used for the analysis. The results show amplification of ground motion at every location. A maximum amplification factor, i.e. ratio of PGA to PBRA, of 3.99 was observed at BH 1 which is around 500 m away from the location of nuclear reactor. In addition, nonlinear and equivalent linear analyses were performed with a scaled-up and a scaled-down input motions

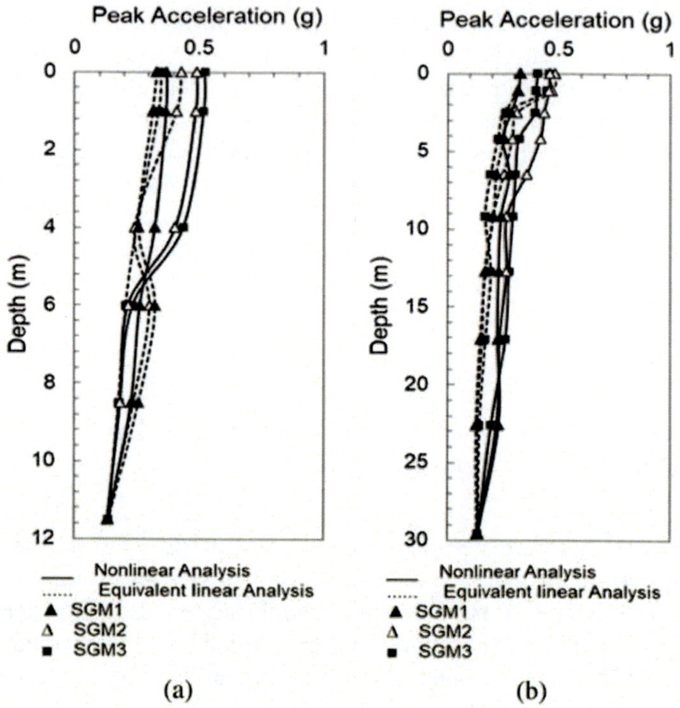

Fig. 10 PHA profiles of **(a)** BH 1 and **(b)** BH 4 with input motions having different predominant periods

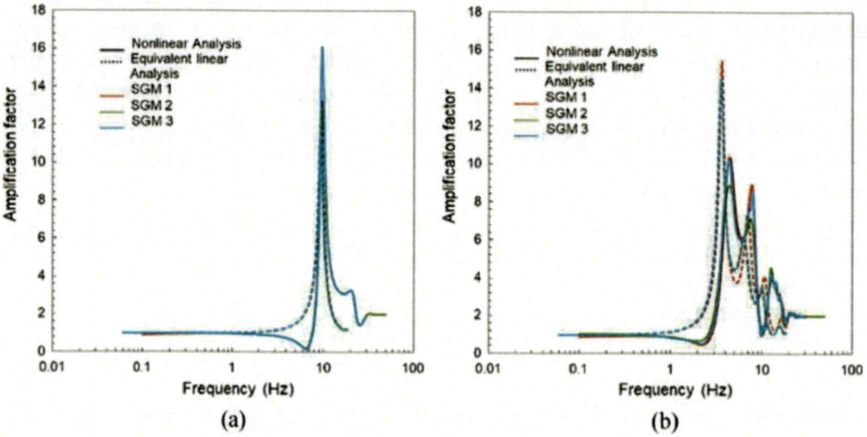

Fig. 11 Amplification spectra at **(a)** BH 1 and **(b)** BH 4 with input motions having different predominant periods

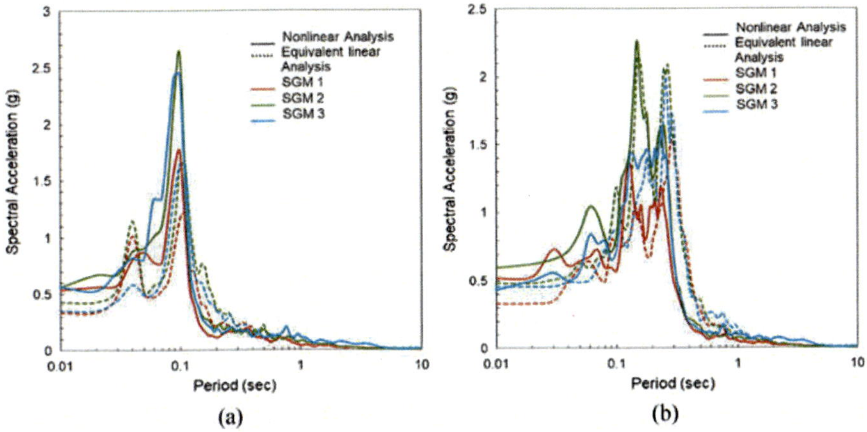

Fig. 12 Response spectra of 5% damped SDOF system at the surface level of **(a)** BH 1 and **(b)** BH 4 with input motions having different predominant periods

Fig. 13 Variation in amplification factor with predominant period

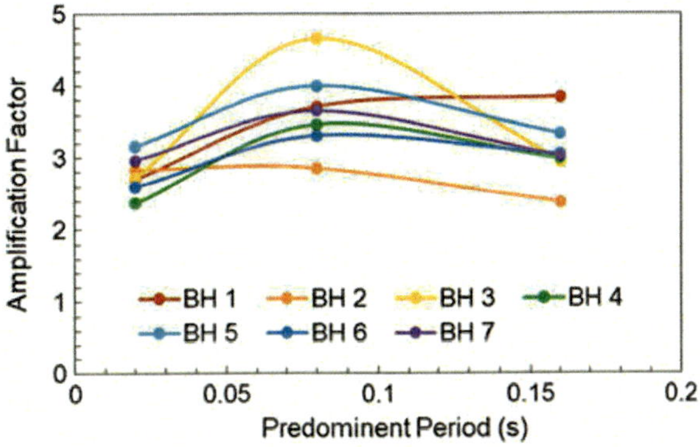

Fig. 14 PHA profiles of **(a)** BH 1 and **(b)** BH 4 with input motions having different bracketed durations

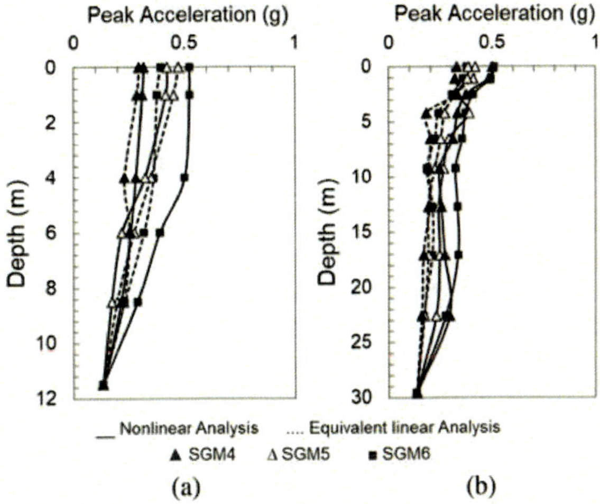

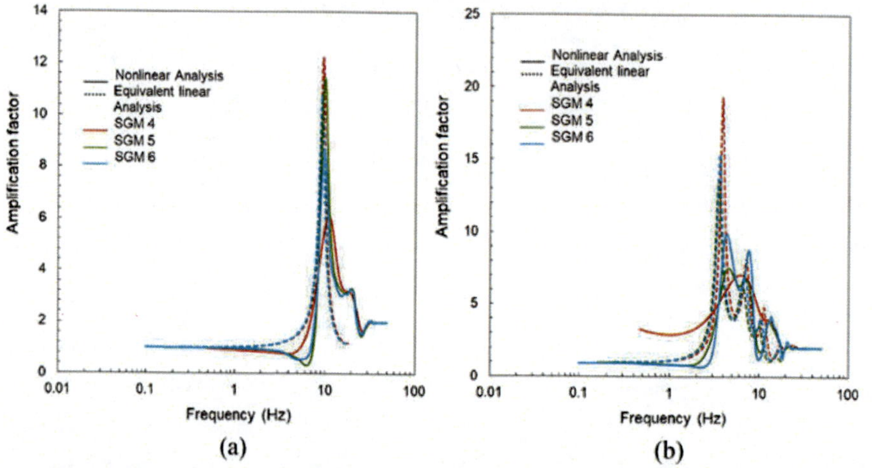

Fig. 15 Amplification spectra at (**a**) BH 1 and (**b**) BH 4 with input motions having different bracketed durations

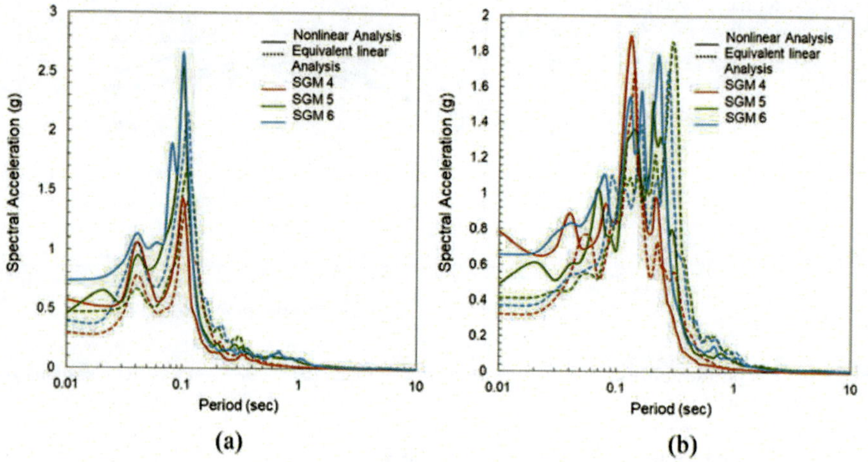

Fig. 16 Response spectra of 5% damped SDOF system at the surface level of (**a**) BH 1 and (**b**) BH 4 with input motions having different bracketed durations

which were obtained by multiplying original acceleration history by 4 and 0.25, respectively. Equivalent linear results were found to deviate from nonlinear results as the soil behaves more in nonlinear manner with the increase in PBRA. A reduction of amplification factor was detected with a rise in PBRA. It was noted that high frequency response can be reproduced only by nonlinear method. At longer periods, both the equivalent linear method and nonlinear method give comparable results. A few attempts were made to formulate a reliable relationship for ground response with frequency content and duration. But there were uncertainties in the results. Therefore,

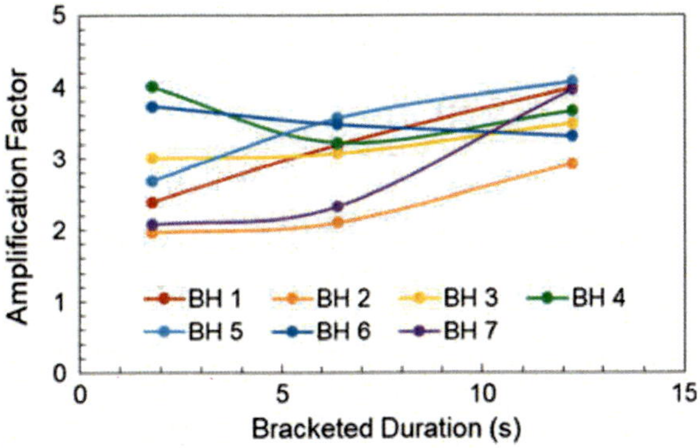

Fig. 17 Variation in amplification factor with bracketed duration

it was concluded that all the strong motion parameters should be properly monitored to estimate the influence of frequency content and duration.

References

Basu D, Dey A (2017) 1D Nonlinear ground response analysis of soils in IIT Guwahati and liquefaction potential identification. In: Proceedings of 16th world conference on earthquake, Santiago, Chile

Bolisetti C, Whittaker AS, Mason HB, Almufti I, Willford M (2014) Equivalent linear and nonlinear site response analysis for design and risk assessment of safety-related nuclear structures. Nucl Eng Des 275:107–121

Boore DM (2003) Simulation of ground motion using the stochastic method. Pure Appl Geophys 160(3–4):635–676

Chandra U (1977) Earthquakes of Peninsular India—a seismotectonic study. Bull Seismol Soc Am 67(5):1387–1413

Feng-xin Z, Yu-shan Z (2006) Artificial ground motion compatible with specified peak velocity and target spectrum. Acta Seismol Sinica 19(4):461–471

Finn WDL, Bhatia SK (1981) Prediction of seismic pore-water pressure. In: Proceedings of 10th international conference on soil mechanics and foundation engineering, Rotterdam, The Netherlands, pp 201–206

Gangrade BK, Arora SK (2000) Seismicity of the Indian Peninsular shield from regional earthquake data. Pure Appl Geophys 157(10):1683–1705

Gazetas G, Dakoulas P (1992) Seismic analysis and design of rockfill dams: state-of-the-art. Soil Dynam Earthquake Eng 11(1):27–61

Idriss IM, Seed HB (1967) Response of horizontal soil layers during earthquakes. Soil Mechanics and Bituminous Materials Research Laboratory. University of California, Berkeley, CA

Ishihara K, Towhata I (1980) One-dimensional soil response analysis during earthquakes based on effective stress method. J Faculty Eng (University of Tokyo) 4:655–700

James N, Sitharam TG, Padmanabhan G, Pillai CS (2014) Seismic microzonation of A nuclear power plant site with detailed geotechnical, geophysical and site effect studies. Nat Hazards 71(1):419–462

Joyner WB, Chen AT (1975) Calculation of nonlinear ground response in earthquakes. Bull Seismol Soc Am 65(5):1315–1336

Kaklamanos J, Bradley BA, Thompson EM, Baise LG (2013) Critical Parameters affecting bias and variability in site-response analyses using KiK-net downhole array data. Bull Seismol Soc Am 103(3):1733–1749
Kim B, Hashash YMA, Kottke AR, Assimaki D, Li W, Rathje EM, Campbell KW, Silva WJ, Stewart JP (2013) A predictive model for the relative differences between nonlinear and equivalent-linear site response analyses. In: Proceedings of the 22nd international conference in structural mechanics in reactor technology, San Francisco, CA
Kramer SL (1996) Geotechnical earthquake engineering, 1st edn. Prentice Hall, New Jersey
Kondner RL, Zelasko JS (1963) A Hyperbolic stress-strain formulation of sands. In: Proceedings of 2nd pan American conference on soil mechanics and foundation engineering, Brazil, pp 289–324
Kwok AO, Stewart JP, Hashash YM, Matasovic N, Pyke R, Wang Z, Yang Z (2007) Use of exact solutions of wave propagation problems to guide implementation of nonlinear seismic ground Response Analysis Procedures. J Geotech Geoenviron Eng 133(11):1385–1398
Martin GR, Finn WL, Seed HB (1975) Fundamentals of liquefaction under cyclic loading. J Geotech Geoenviron Eng 101(GT5):423–438
Masing G (1926) Eigenspannungen und Verfertigungbeim Messing. In: Proceedings of 2nd international congress on applied mechanics, Zurich
Matasovic N (1993) Seismic response of composite horizontally-layered soil deposits. Ph.D. dissertation, Department of Civil Engineering, University of California, Los Angeles, California
Matasovic N, Vucetic M (1993) Cyclic characterization of liquefiable sands. ASCE J Geotech Geoenviron Eng 119(11):1805–1822
Pyke RM (1979) Nonlinear soil models for irregular cyclic loadings. J Geotech Geoenviron Eng 105(GT6):715–726
Ramalingeswara Rao B (2000) Historical seismicity and deformation rates in the Indian Peninsular shield. J Seismolog 4(3):247–258
Ramasamy SM (2006) Remote sensing and active tectonics of south India. Int J Remote Sens 27(20):4397–4431
Ramberg W, Osgood W (1943) Description of stress–strain curves by three parameters. Technical Note No, 902, National Advisory Committee of Aeronautics, Washington, DC
Rollins KM, Evans MD, Diehl NB, Daily WD III (1998) Shear modulus and damping relationships for gravels. J Geotech Geoenviron Eng 124(5):396–405
Schnabel PB (1973) Effect of local geology and distance from source on earthquake ground motions. Ph.D. Thesis, University of California, Berkeley
Schnabel PB, Lysmer J, Seed HB (1972) Shake: a computer program for earthquake response analysis of horizontally layered sites. EERC Report 72–12, University of California, Berkeley
Seed HB, Idriss IM (1970) Soil moduli and damping factors for dynamic response analysis. Report no. EERC 70–10, University of California, Berkeley
Seed HB, Su JH (1989) Implication of site effects in the Mexico city earthquake of september 19 1985 for earthquake-resistance-design criteria in the San Francisco Bay Area of California. Report no. UCB/ EERC-89/03, University of California, Berkeley
Stewart JP, Kwok AO (2008) Nonlinear seismic ground response analysis: code usage protocols and verification against vertical array data. In: Proceedings of geotechnical earthquake engineering and soil dynamics IV, Sacramento, CA, pp 1–24
Vucetic M, Dobry R (1991) Effect of soil plasticity on cyclic response. J Geotech Eng 117(1):89–107

Assessment of Soil Liquefaction Resistance by Finite Element Approach

Anasua GuhaRay and Mohammad Zaid

Abstract Soil liquefaction is the source of several major damages during earthquakes on the material and human level. Liquefaction of soil occurs under dynamic or cyclic loading, due to which the soil loses its strength. A soil deposit subjected to seismic loading can be viewed as a binary system: it will either liquefy or not liquefy. Generalized linear models are versatile tools for predicting the response of a binary system and hence potentially applicable to liquefaction prediction. In the present study, a numerical procedure based on finite element technique is presented for evaluating the characteristics of soil liquefaction and the foundation response under seismic loading. The relative density of soil is an important parameter in prediction of liquefaction. However, this property varies from place to place with the density and nature of soil. Therefore, a double layered soil profile is used to model the change in relative density of soil with depth. A constitutive elasto-plastic model with dynamic loading is considered to study the phenomenon of liquefaction. The approach is presented through the simulation of five major earthquake data. The different parameters of simplified method as proposed by Seed and Idriss have been measured for the given field conditions, with varying magnitudes of earthquake and horizontal ground acceleration. The factor of safety has been calculated and is being used to denote the occurrence of liquefaction for all the cases of soil strata considered. Parametric studies are carried out considering different ranges of relative density of sand and earthquake magnitudes, and the corresponding factors of safety are proposed for the same.

Keywords Soil liquefaction · Finite element analysis · Relative density of soil

---

A. GuhaRay (✉) · M. Zaid
BITS Pilani Hyderabad Campus, Secunderabad, Telangana 500078, India
e-mail: guharay@hyderabad.bits-pilani.ac.in

M. Latha Gali and P. Raghuveer Rao (eds.), *Geohazards*, Lecture Notes in Civil Engineering 86, https://doi.org/10.1007/978-981-15-6233-4_32

1 Introduction

Soil liquefaction occurs in soil having very less force of cohesion between its particles. The weaker the force subjected on the soil particles due to overlying particles, the more easily will the pore pressure be able to separate the particles and force its way to the surface of the soil strata. Soil liquefaction is the phenomena of upward movement of water trapped in between the soil particles. This occurs when, the force of the water, i.e., pore water pressure becomes more than the downward force created due to the overlying soil particles. During the phenomena of soil liquefaction, the soil will lose its water content, due to which the cohesive force generated due the water in between the particles will reduce drastically. As a result, the soil will become powdery in nature and lose its shape.

The disadvantage of soil liquefaction is that the bearing capacity of the soil gets reduced by a big factor. The soil will not be able to support heavy loads. The soil will also not be able to maintain its shape if left standing. Due to reduction in cohesion value of the soil, the soil cannot be used for any land-filling process also. The soil will have to be reinforced with geosynthetics to increase the interlocking of the soil particles, which in turn will increase the bearing capacity of the soil. However, liquefaction can also occur even if the soil has been reinforced with geosynthetics, and hence, this method is also prone to failure, and its life and strength will depend on the geosynthetic. Hence, it is necessary to prevent the failure of soil and the occurrence of liquefaction in the first place.

Liquefaction of soil had been occurring for a long time, but serious approach toward it for its place of occurrence and the subsequent mitigation of the same was taken after the devastating earthquakes of 1964 in Niigata and Alaska. After seeing the serious consequences of liquefaction, such as the post-liquefaction strength and stress deformation, various efforts were made to calculate the occurrence of the same. The simplified methods as proposed by Seed and Idriss in 1971 were the first concrete efforts in this direction. The simplified methods were a deterministic method, which used the field conditions to predict if the liquefaction was to happen at the test site. According to this method, a factor of safety is calculated, if the factor of safety is greater than 1, then liquefaction will not occur, however if it is not, then it is assumed that liquefaction will occur for the given field conditions (Seed and Idriss 1971). Even though this method is a deterministic one, it is however not necessary that liquefaction will occur for the given field conditions.

There were a lot of changes made to the simplified procedure to adopt a large number of field conditions. The modified simplified method was put forward by Youd and Idriss in 2001. In this research work, they proposed a lot of changes such as the criteria based on SPT- and CPT-based methods (Youd et al. 2001). The review presented in the paper did not change the method from probabilistic to another, but it increased the way in which the factor of safety could be calculated. It also changed the soils which could be taken into consideration for liquefaction as well as assigning a particular type of test for one type of soil. Thus, this way it increased the efficiency of calculation. Factors such as magnitude scaling factor were included to compensate for the change in magnitude of earthquake, and thus, in this way, a lot of new site conditions could be incorporated into the study.

In 2005, Juang et al. proposed a framework of first-order reliability method, with the subject of model uncertainty being scrutinized in this framework. The liquefaction method developed by Andrus and Stokoe has been the prime focus of this research, with the model being based on simplified method of shear wave velocity being distinguished for unpredictability. The liquefaction resistance as a function of shear wave velocity has been defined by the simplified model as being represented by a boundary curve. The determination of two variables, viz. mean and coefficient of variation, leads to evaluation of a lognormal variable which in turn is the uncertainty of the simplified model developed. The two-statistic variables are determined from the process of trial and error with the case history as the database, with the help of Bayesian mapping function. The trial and error method being used in the above procedure has been developed by Juang et al. (2005) for his own liquefaction triggered evaluation. The uncertainty of the model in consideration has been distinguished and with the help of the above procedure and the liquefaction probability can also be calculated using the FORM analysis with the help of the known variables and parameters. After the research was carried out, it was shown that if the different types of uncertainty associated with model, like the parameter uncertainty and the model uncertainty, were calculated accurately, then the FORM analysis was able to give the liquefaction probability accurately for any given soil under a stipulated seismic loading (Juang et al. 2005).

Liao et al. (1988) proposed the development of statistical regression method for calculating the occurrence of liquefaction using the soil resistance and earthquake load as input parameters of the mode being developed. Use of soil gradation variables on the estimation of probability is limited due to inconsistent data, due to which quantitative trends cannot be determined. It is best to use the parameter of fines content as it is assumed to be adequately supported by current data (Liao et al. 1988). According to Liao et al. (1988), the regression model is bound by the limitations of the database. However, due to the inconsistency of regression model in some cases, it is better to apply the regression model in regional mapping of hazards rather be used on an empirical basis for different places.

Simplified method is based on limit states, but this limit states causes it to shift from a deterministic approach to a probabilistic one to complement the uncertainties which crop up due to assumptions in the limit state method. A new reliability method was proposed by Juang et al. (1999a, b). Before this approach toward reliability method, the reliability method was primarily used for only one structure which was unique in every aspect (Juang et al. 1999a, b) to analyze it for safety against liquefaction. Since there was no universal reliability method for calculating the failure for all the structures, hence the authors proposed that the current method should encompass 225 cases of liquefaction/non-liquefaction which will allow it to build a realistic relation between the frequency of occurrence of liquefaction and probability of its occurrence.

In 2001, Juang et al. proposed a method of calculating probability of liquefaction using shear wave velocity method. Youd et al. (2001) had put forward a proposal that to increase the efficiency of the probabilistic method in gravelly soil, shear wave velocity method would be the most appropriate for this kind of soil. Juang et al. instead of using the penetration test that is the SPT blow counts to calculate the cyclic ratios; he used the shear wave velocity (Vs) to calculate the CSR and CRR functions, and using that, he found the factor of safety.

In the present study, the simplified method has been used to calculate the factor of safety for a given set of soil samples. The method used in the study is as one proposed by Youd et al. (2001). The present study aims to determine the safe limit of liquefaction resistance or potential for a given set of soil conditions. The study incorporates a double-layered soil into consideration to simulate the conditions of landfills and during reclamation effort. The current study has tried to achieve the same through simulation of different soil conditions in the software, Plaxis 2D, a finite element software. A total of five different earthquakes from different time have been taken into consideration for study. The relative density of the soil has been varied to simulate the conditions of different places. The factor of safety calculated has been compared to assess the conditions of different places after a seismic load has been applied.

2 Numerical Modeling

Numerical model is a mathematical simulation which helps to find the behavior of real physical problems over time. In the current study, a numerical model has been simulated in Plaxis 2D for observing the behavior of the soil when it is subjected to different earthquakes. Plaxis 2D is a finite element method (FEM)-based software which is used for simulation of deformation in soil due to different causes as well as in rock.

In the present study, a numerical model consisting of a double-layered soil profile with a solid construction on top of it has been subjected to dynamic loading in the horizontal direction. The edge of the solid construction has been kept in the middle portion of the model to prevent any effect of side edges on the behavior of the solid construction. In the numerical study, the dynamic loading has been provided in the form of acceleration–time data. The data from five different earthquakes have been taken under consideration for the present study.

A two-layered solid structure has been constructed on the top of the soil layer. The structure also includes a basement layer beneath the ground. Its dimension is 10 m in width with 8 m in height. The length of the building has been assumed to be far greater than its width. The building has been modeled for plain strain conditions. The reason for choosing the plain strain condition is to observe the phenomena of liquefaction on both sides of the solid construction. The building has been placed symmetrically about the model with the axis of the model passing through the center of the building. The building has also been composed of two different materials so as to simulate the difference between the outer and inner portion of the building (Grote et al. 2001).

2.1 Soil

Two different layers of the soil have been considered in the present study to simulate the effect of change in relative density of the soil layers as relative density of the soil increase on going deeper into the earth surface. The aim of doing so was to find the effect of liquefaction on a double-layered soil with changing relative density. The behavior of the double-layered soil has been modeled using a simple elastic-plastic-dynamic Mohr–Coulomb in the current study so as to be in harmony with the design of the solid structure on the soil surface (Potts 1999). The soil has been modeled using the 15-node triangular elements to discretize the soil structure using more points of analysis. The material of the soil taken in the present study is sand, which is easily liquefiable theoretically. The two different layers of the soil have varying relative densities so as to ascertain the resistance to liquefaction. The relative density of the upper layer soil has been kept constant at 15%, while it has been changed for the lower layers from 15 to 35% with an increment of 10% per cycle of simulation. For poor graded sand, the young's modulus has been kept at 20 MPa (Obrzud et al. 2012). According to Krishnaswamy and Isaac (1994), the specific gravity of the sand has been kept at 2.65. Poisson's ratio has been kept constant at 0.3 (Krishnaswamy and Isaac 1994) as well as the internal angle of friction has been kept at 250 for all the cases of relative densities as well as for the different earthquakes for the two layers of the soil. The permeability of the two layers has been kept a constant.

2.2 Mesh

Meshing is the process of discretization of a substance into finer elements so that there is an ease of analyzing the substance. The method of discretization depends on the need of accuracy for a particular problem. Since the present study has been conducted to find the effect of liquefaction on the solid construction, therefore there is a need for greater for finer detail underneath the base of the building. Hence a method of fine global coarseness with global refinement has been used in the current method (Fig. 1).

3 Mathematical Formulation

The results have been divided into five parts according to the data of earthquake. The method of factor of safety as proposed by Seed and Idriss in 1971 and modified by Youd et al. (2001) has been used in the current study to illustrate the phenomena of liquefaction. According to Seed and Idriss,

$$\mathrm{FOS} = \frac{\mathrm{CRR}_{7.5}}{\mathrm{CSR}} \tag{1}$$

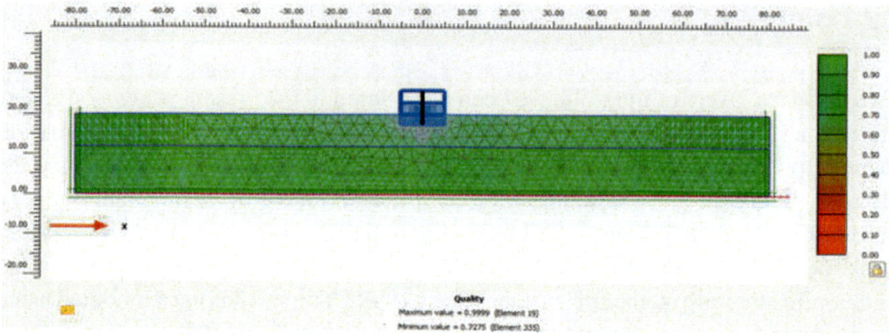

Fig. 1 Mesh of the model

where $CRR_{7.5}$ = Cyclic resistance ratio measured at earthquake magnitude of 7.5 when it is measured on a moment magnitude scale. CSR = Cyclic stress ratio.

CSR. CSR or cyclic stress ratio has been defined as the stress ratio required to generate the phenomena of liquefaction (Youd et al. 2001). For calculating the cyclic stress ratio, the formula given by Seed and Idriss (1979) has been used in the present study.

$$CSR = 0.65(a_{max}/g)(\sigma_{vo}/\sigma_{vo}^{'})r_d \tag{2}$$

where σ_{vo} = total vertical overburden pressure; $\sigma_{vo}{}'$ = effective overburden pressure; a_{max} = maximum acceleration which occurred for a particular earthquake in the absence of liquefaction (The graphs provided for acceleration–time data have been taken from Plaxis 2D); g = acceleration due to a gravity; r_d = depth-related stress reduction factor which varies with depth z. r_d has been calculated using the method provided by Youd et al. (2001).

$$r_d = \frac{1.0 - 0.4113z^{0.5} + 0.04052z + 0.001753z^{1.5}}{1.0 - 0.4177z^{0.5} + 0.05729z - 0.006205z^{1.5} + 0.001210z^2} \tag{3}$$

CRR. CRR or cyclic resistance ratio has been defined as the resistance offered by the soil to the phenomena of liquefaction. CRR is calculated at the earthquake magnitude of 7.5 measured on moment magnitude scale, and hence, CRR is often referred to as $CRR_{7.5}$ as the official notation for the term. The resistance ratio is generally measured through the method of field tests as the method of laboratory testing is not a viable one, as the soil structure is disturbed while transportation to laboratory. Out of the four methods of field estimation of CRR, standard penetration test (SPT), cone penetration test (CPT), shear wave velocity method (Vs) and Becker penetration test (BPT), SPT and CPT methods are given more priority due to their extensive database and past experience (Youd et al. 2001).

In the present study, the method of SPT blow counts has been used. The number of blow counts has been calculated using the relationship between blow counts and the relative density. As the relative density of the two layers of soil has been assumed in our current study, hence corresponding SPT blow counts or N can be found out using Meyerhof charts (Meyerhof 1956). After applying the correction to the value of N according to the field conditions, the value of $(N1)60$ is obtained, which is the corrected value of the blow count. But in the present study, since the actual field tests were not performed, and there are many unknown factors involved with the field conditions, and hence, the SPT blow count or N-value has been directly used. Meyerhof relation between relative density and SPT blow counts has been used to find the value of N (Meyerhof 1956).

To compensate the change in magnitude of the CRR due to the change in the magnitude of the earthquake, Seed and Idriss (1982) presented with scaling factors also know magnitude scaling factor (MSF). Two more correction factors known as overburden stress correction factor (Kσ) and sloping ground correction factor ($K\alpha$) were also added to complete CRR. In the present study, due to absence of a sloping ground in the numerical model considered, there is not a need for consideration of the sloping ground factor.

MSF. Magnitude scaling factor is used to compensate for the change in the magnitude of the earthquake. There are several methods for the calculation of the same, but in the current study, the theoretical method will be used to avoid any error in calculation due to field experiments.

$$\text{MSF} = 10^{2.24}/M_w^{2.56} \tag{4}$$

Kσ. The value of overburden stress factor has been calculated using the chart available in Youd et al. (2001). The relation between $K\sigma$ and $\sigma_v{}'$ has been used to establish the values of $K\sigma$, and the same has been used in estimation of CRR.

4 Application of Proposed Methodology to Different Earthquakes

A total of five different earthquakes were chosen for the current study. The acceleration–time data of the earthquakes were taken as one of the input parameters in Plaxis 2D, a finite element software. The earthquakes chosen were from different timelines, through which it has been tried to incorporate the conditions of different places and their soil geometry. Also, these five different earthquakes have occurred around the globe, and hence, to increase the versatility, it has been done so. The area around Southeast Asia is frequently struck by an earthquake, and hence, to incorporate more conditions of soil liquefaction, data from two different earthquakes in the region were used. El Centro has been selected because it lies below the sea level, and therefore, there should be a high chance of occurrence of liquefaction. Bhuj is one of most famous earthquakes in the Indian history, while the same can be said for Chile earthquake, hence their reason for selection.

4.1 Chi-Chi Earthquake

Chi earthquake took place in 1999, with a peak ground acceleration of 1.01 g (Lee et al. 2003). A magnitude of 7.3 was recorded on the Richter (*R*) scale. The acceleration–time graph for the earthquake is provided in Fig. 2.

The factor of safety calculated according to Seed and Idriss method is presented in Table 1.

From the calculation of CSR and CRR, the factor of safety was found which found to be less than 1 for all the cases of soil geometry in the above table. This implies that liquefaction will take place at the given relative density of the soil for this earthquake. As the relative density of the soil is increased, the FOS also increases. The factor of safety in the above case is very less because the earthquake is very strong here with a high peak ground acceleration (PGA), which is greater than g.

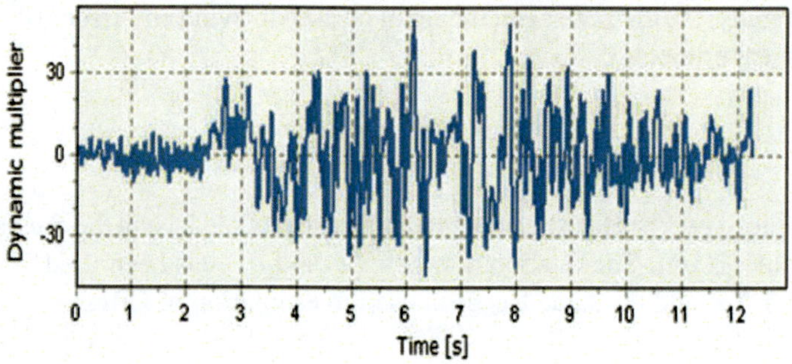

Fig. 2 Acceleration–time graph for Chi-Chi earthquake

Table 1 Factor of safety for Chi-Chi earthquake

FOS				
Depth (*z*) in meters				
RD (1st layer) (%)	RD (2nd layer) (%)	5.9	7.92	9.99
15	15	0.037758	0.03716	0.035282
	25	0.046144	0.045864	0.043996
	35	0.06654	0.066553	0.064338

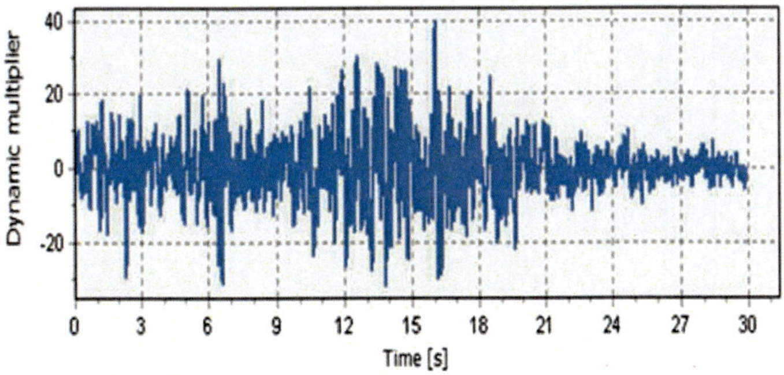

Fig. 3 Acceleration–time graph for Chile Earthquake

Table 2 Factor of safety for Chile earthquake

FOS				
Depth (*z*) in meters				
RD (1st layer)	RD (2nd layer)	5.9	7.92	9.99
15	15	0.071137	0.061021	0.057202
	25	0.074553	0.074954	0.070039
	35	0.107897	0.108236	0.101368

4.2 Chile Earthquake

Chile earthquake took place in 2010, having a peak ground acceleration (PGA) of 0.65 g (Liberatore et al. 2012). The earthquake recorded a magnitude of 8.8 on the moment magnitude scale (MMS). The data of acceleration–time are presented in Fig. 3.

Factor of safety calculated according to Seed and Idriss method is presented in Table 2.

When the soil layers have been subjected to the acceleration–time data of the Chile earthquake, the FOS for all the cases of soil geometry was observed to be less than 1. This implies that the soil was not strong enough to prevent the cyclic stress from liquefying the soil.

4.3 El Centro Earthquake

El Centro earthquake took place in 1940, with a peak ground acceleration of 0.35 g. A magnitude of 7.1 was recorded on Richter scale. The acceleration–time data have been provided in Fig. 4.

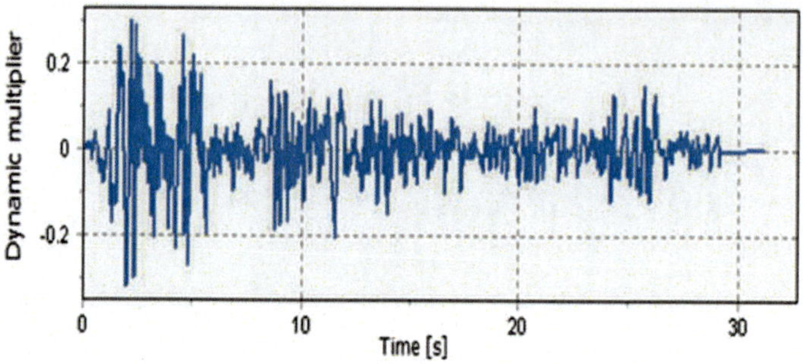

Fig. 4 Acceleration–time graph for El Centro earthquake

The factor of safety calculated has increased from the previous case of Chile earthquake. However, the value is still less than 1, and hence, liquefaction will occur in all the above cases of soil geometry. Here also, a decrease in factor of safety has been observed when the depth changes from layer 1 to layer 2 of the soil profile.

4.4 Taiwan Earthquake

Taiwan earthquake took place in 2016, registering a peak ground acceleration of 0.22 g. An earthquake magnitude of 6.4 was recorded on moment magnitude scale. The acceleration–time data for the earthquake are provided in Fig. 5.

The factor of safety calculated according to Seed and Idriss method is presented in Table 4. Though the FOS has increased in the present earthquake, in comparison with the past ones in the current study, however, the increased FOS is still less than 1, due to which the condition of liquefaction will occur according to the Seed and Idriss (1971) simplified method for all the cases of soil strata. A decrease in factor of safety value when going from layer 1 to layer 2 has also been observed here.

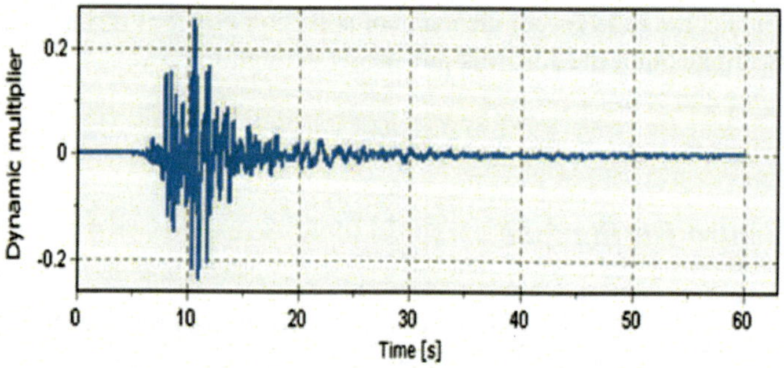

Fig. 5 Acceleration–time graph for Taiwan earthquake

Table 4 Factor of safety for Taiwan earthquake

FOS				
Depth (z) in meters				
RD (1st layer)	RD (2nd layer)	5.9	7.92	9.99
15	15	0.416536	0.403182	0.38888
	25	0.513218	0.496643	0.481393
	35	0.743618	0.719263	0.697825

4.5 Bhuj Earthquake

Bhuj earthquake took place in 2001 in Gujarat in India, with a magnitude of 7.7 being recorded on the moment magnitude scale. The peak ground acceleration of the earthquake was observed to be 0.11 g. The data for acceleration–time of the earthquake are presented in Fig. 6.

The factor of safety calculated according to Seed and Idriss method is presented in Table 5.

The decrease observed in factor of safety while going down from layer 1 to layer 2 has been consistent in this case also. This case has provided the highest factor of safety observed till now at 0.93. Hence, it can be assumed here that if the soil is

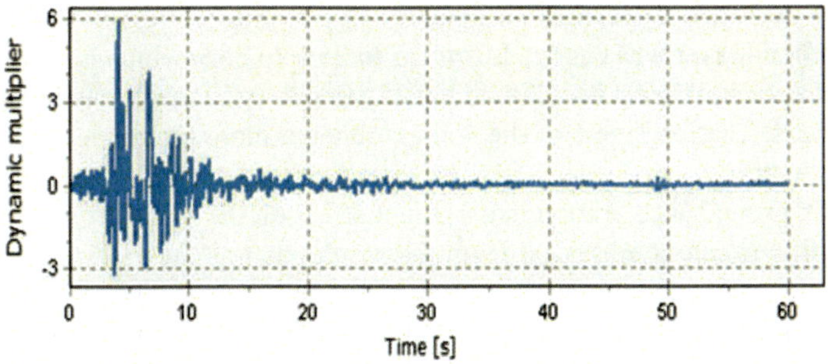

Fig. 6 Acceleration–time graph for Bhuj earthquake

Table 5 Factor of safety for Bhuj earthquake

FOS				
Depth (z) in meters				
RD (1st layer)	RD (2nd layer)	5.9	7.92	9.99
15	15	0.518752	0.501512	0.484333
	25	0.640276	0.599987	0.599485
	35	0.928347	0.89578	0.86941

densified more at this depth, then the factor of safety will increase, and it will become more than 1, thus preventing the phenomena of liquefaction.

Using the Seed and Idriss simplified method (Seed and Idriss 1971), the factor of safety was found for the different earthquake. For all the cases that were taken into consideration, CSR was always greater than $CRR_{7.5}$, which implies that the resistance of soil to liquefaction was less than the seismic loading applied on it. Hence, this causes the water trapped inside the soil layers to push up to the surface, thus causing the phenomena of liquefaction.

5 Summary and Conclusions

The primary aim of this study was to ascertain the change in liquefaction if a double-layered soil is taken into consideration. Often during landfills, the land is filled with another soil to bring it into level desired. This study aims to determine the changes in the behavior of the landfill around the boundary separation of the two different layers. A double-layered soil with varying relative density has subjected to seismic loading, and using the Seed and Idriss method, an attempt has been made to establish a relation between the relative density of the soil and the liquefaction potential of the soil. The observation of the numerical modeling is enumerated in the following set of points:

- The soil exhibited a decrease in liquefaction potential when the relative density of the second layer was changed from 15 to 25% to 35% which is expected.
- The soil also exhibited a decrease in liquefaction as one moves downward into the sand layers. The top layers of the soil exhibited a more pronounced phenomena of liquefaction.
- The peak ground acceleration had a direct effect on the liquefaction potential of soil profile. It can be observed from the results that as the PGA decreased, the liquefaction potential of the sand strata decreased.
- At the current relative density of the soil, all the earthquakes taken into consideration would cause a phenomenon of liquefaction. Hence, the relative density of the sand layer is needed to be increased.

Since in all the cases of soil geometry, the value obtained as the factor of safety is less than 1, hence, it is needed to increase the relative density of the soil layers, and if needed, also add geosynthetics to prevent the phenomena of liquefaction. Further research is needed on this topic to investigate the change in liquefaction potential of a double-layered soil. Further work will be done in future to ascertain the changes in liquefaction potential of the soil geometry.

References

Grote DL, Park SW, Zhou M (2001) Dynamic behavior of concrete at high strain rates and pressures: I Experimental characterization. Int J Impact Eng 25:869–886
Juang CH, Rosowsky DV, Tang WH (1999a) Reliability based method for assessing liquefaction potential of soils. J Geotech Geoenviron Eng 125(8):684–689
Juang CH, Yang H, Yuan H, Lee DH, Andrus RD (1999b) CPT-based liquefaction triggering analysis with an emphasis on 1999 Chi-Chi, Taiwan, earthquake
Juang CH, Chen CJ, Jiang T (2001) Probabilistic framework for liquefaction potential by shear wave velocity. J Geotech Geoenviron Eng 127(8):670–678
Juang CH, Yang SH, Yuan H (2005) Model uncertainty of shear wave velocity-based method for liquefaction potential evaluation. J Geotech Geoenviron Eng 131(10):1274–1282
Krishnaswamy NR, Isaac NT (1994) Liquefaction potential of reinforced sand. Geotext Geomembr 13:23–41
Liao SSC, Veneziano D, Whitman RV (1988) Regression model for evaluating liquefaction probability. J Geotech Eng 114(4):389–411
Liberatore L, Sorrentino L, Liberatore L, Decanini LD (2012) Failure of modern structures induced by the Emilia (Italy) 2012 Earthquakes. Eng Fail Anal
Lee D-H, Ku C-S, Yuan H (2003) A study of the liquefaction risk potential at Yuanlin, Taiwan. Eng Geol 71(1):97–117
Meyerhof GG (1956) Penetration test and bearing capacity of cohesionless soils. J Soil Mech Found Div ASCE 82(SM1):1–19
Obrzud RF, Truty A, Vulliet L (2012) Numerical modeling and neural networks to identify model parameters from piezocone tests: II. Multi-parameter identification from piezocone data. Int J Numer Anal Methods Geomech 36:743–779
Potts DM, Zdravkovic L (1999) Finite element analysis in geotechnical engineering: theory. Thomas Telford, London, UK
Seed HB, Idriss IM (1971) Simplified procedure for evaluating soil liquefaction potential. J Geotech Eng Div ASCE 97(9):1249–1273
Seed HB (1979) Soil liquefaction and cyclic mobility evaluation for level ground during earthquakes. J Geotech Eng Div ASCE 105(GT2) 201–255
Seed HB, Idriss IM (1982) Ground motions and soil liquefaction during earthquakes. Earthquake Engineering Research Institute Monograph, Oakland, Calif
Youd TL, Idriss IM, Andrus RD, Arango I, Castro G, Christian JT, Dobry R, Liam Finn WD, LF Jr Harder, Hynes ME, Ishihara K, Koester JP, Laio SSC, Marcuson III WF, Martin GR, Mitchell JK, Moriwaki Y, Power MS, Robertson PK, Seed RB, Stokoe II KH (2001) Liquefaction resistance of soils: summary report from the 1996 NCEER and 1998 NCEER/NSF workshops on evaluation of liquefaction resistance of soils. ASCE J Geotech Eng Geoen Eng 127(10):817–833

Liquefaction Susceptibility Mapping of Kollam Coastal Stretch, Kerala, Considering Geotechnical Parameters

S. K. Sithara, S. Surya, Sayana Parveen, Liz Maria Damiyan, Vinayak Mohan, A. Muhammed Siddik, and S. Adarsh

Abstract The assessment of liquefaction potential is one of the important scientific problems for the geotechnical investigators. Liquefaction of loose saturated cohesionless soil during earthquake has been a major cause of damage due to earthquake for the buildings, earth embankments and other civil engineering structures. The study area of the coastal stretch of Kollam in western Kerala, India, mainly consists of coastal alluvial deposits and marine sand. Samples were collected along the length and width of coastal stretch of Kollam and are tested to obtain water content, dry density, fines content and the gradation curves. Limiting curves-based gradation criteria proposed by Tsuchida (1970) and empirical relation for stress ratio (SR) values obtained by Chien et al. (2002) were used for the calculation of liquefaction potential at sample locations. Spatial analysis of this data is done using QGIS to delineate the region into most liquefiable, liquefiable, less likely to be liquefiable and not liquefiable zones. The susceptibility map developed based on SR criteria is found to be in agreement with the liquefaction potential map developed by overlay of the two criteria which infer the dominant influence of dry density of the deposits of Kollam coastal stretch. Further, a ground truth examination of the final susceptibility map revealed that the zones in which water table lies within 0–5 m from ground level are more vulnerable to liquefaction in the Kollam coastal stretch. The proposed liquefaction susceptibility map can be used as a firsthand info on liquefaction potential of region which can aid in site-specific studies for future development.

Keywords Liquefaction · Kollam · Coastal · Stress ratio · Mapping

1 Introduction

Liquefaction has drawn much attention from engineers because it can create great damage to manmade structures. It causes ground to subside and to spread laterally, thus inducing buildings to tilt, damaging airport runways and earth embankments,

S. K. Sithara (✉) · S. Surya · S. Parveen · L. M. Damiyan · V. Mohan · A. M. Siddik · S. Adarsh
Department of Civil Engineering, TKM College of Engineering, Kollam, Kerala, India
e-mail: sitharask18@gmail.com

M. Latha Gali and P. Raghuveer Rao (eds.), *Geohazards*, Lecture Notes in Civil Engineering 86, https://doi.org/10.1007/978-981-15-6233-4_33

and disrupting buried pipes and pile foundations. In the wake of the two devastating earthquakes of 1964, namely the 1964 Niigata and the 1964 Great Alaska earthquakes, engineers have carried out a great amount of research to study liquefaction and to predict its occurrence. There is a critical need to predict the occurrence and severity of soil liquefaction for engineering design, hazard mapping, urban planning and regulatory purposes.

The Kollam district classified under the moderate seismic zone; it means that the probability of earthquake in this zone is up to 6.9 magnitude (www.dmg.kerala.gov.in). The coastal areas of Kollam mainly consist of unconsolidated coastal alluvium and marine sand deposits of low bearing capacity. Kollam city being an important industrial area in Kerala, which is undergoing an increased growth in infrastructure development and urban settlement, the safety of structures must be ensured to prevent any disasters. To understand the regional liquefaction susceptibility of coastal stretch of Kollam, a study is carried out based on geotechnical characteristics.

Generally, two types of liquefaction hazard studies are carried out widely. The first type, based on historical liquefaction map, shows where liquefaction has occurred during the past earthquakes. The second type, a liquefaction hazard map, divides a region into areas having different degrees of liquefaction hazard (Green and Ziotopoulou 2001). Many researchers carried out studies in both. Youd and Hoose (1977) carried out a study based on historical earthquakes for the California earthquake. According to them, geological and hydrological factors mainly control the liquefaction susceptibility of soil. Many of the investigators have applied liquefaction hazard map concepts successfully.

Based on the observations made on the liquefaction failures due to Haicheng and Tangshan earthquake in China, Wang proposed Chinese Criteria in 1970. The key focus of these criteria is the percentage of "clay" present, the plasticity index and liquid limit of the soil. Based upon further field experiences and differences in testing methodologies, several modifications have been proposed to the Chinese Criteria (Seed and Idriss 1982; Finn et al. 1994; Seed et al. 2003). Based on the field performance data obtained from the Niigata earthquake of 1964, enhanced by the results of shaking table test on saturated sand deposits, ranges of grain size accumulation curves (limiting curves) for liquefiable soils were proposed (Tsuchida 1970; Iai et al. 1989).

The standard penetration test (SPT) is the most widely used test for liquefaction studies. The SPT N-value must be modified to include corrections for energy efficiency, overburden stress level, fines content, borehole diameter, barrel liner, rod lengths, aging and other factors. Recently, cone penetration test (CPT) has been used to develop CSR curves for the in situ evaluation of liquefiable soils. The CPT offers several advantages over the SPT including better standardization. The major earthquakes of Niigata in 1964 show that soil below 50% relative density would liquefy. Chien et al. (2002) performed a triaxial shear test on soil samples to understand the influence of fines content and relative density on liquefaction resistance.

Many of the investigators have used different concepts for developing liquefaction hazard map, viz., liquefaction potential index (LPI) map using standard penetration test data (Mhaske and Choudhury 2010), probabilistic liquefaction hazard (PLH) map and liquefaction potential map, based on geological and geomorphological conditions (Ganapathy and Rajawat 2012). A concept is usually selected based on availability of data and suitability of method.

Liquefaction susceptibility of a site depends on compositional and environmental factors of the soil as well as of the imposed loading characteristics. The relevant environmental and compositional factors are mineralogy, shapes and size distribution of particles, density, effective confining stress and saturation. A multi-tiered screening approach is commonly used in evaluating liquefaction triggering hazard. The evaluation method of each of these factors can range widely from project to project and simply could involve use of empirical correlations with in situ test indices [e.g., cone penetration test (CPT), tip resistance (qc) and sleeve friction (fs)] or could be much more involved, entailing detailed geologic studies and geotechnical sampling and laboratory testing (e.g., grain size distribution, water content and Atterberg limits) (Green and Ziotopoulou 2001). The screening criteria discussed herein are limited to compositional factors, mainly focusing on the geotechnical properties, of the soil.

2 Methodology

To identify the properties influencing liquefaction susceptibility, several investigators have tried to correlate liquefaction potential with geological, geotechnical and geomorphological criteria. The liquefaction susceptibility criteria proposed by Tsuchida (1970) based on the particle size analysis and the stress ratio (SR) criteria proposed by Chien et al. (2002) are used for the present study. The D_{60} values are obtained by sieve analysis, and the SR values are calculated from dry density and fines content. The obtained values are then interpolated using inverse distance method in QGIS to obtain the liquefaction potential map of Kollam coastal region. The resulted two thematic maps are integrated in QGIS by employing the UNION and overlay operations to develop the liquefaction susceptible map of the study area. The summary of the methodology used for delineating liquefaction susceptible region is shown in the flowchart in Fig. 1.

Of the many parameters influencing liquefaction, this study mainly focuses on the following geotechnical parameters to quantify liquefaction susceptibility:

Particle Size Distribution: The limiting curves proposed by Iai et al. (1989) are used for comparison. The soil samples whose gradation curves fall within the most liquefiable region in the limiting curve are assumed to have liquefaction susceptibility. To quantify this, the D_{60} values of the specimens are used. From the limiting curve, it is evident that the liquefiable soils have a D_{60} value ranging from 0.1 to 1.

$(SR)_{20}$: Chien et al. (2002) have given an empirical relation to calculate the stress ratio at 20 number of cycles in cyclic triaxial test. This value is used to express the liquefaction resistance of the soil. The soils with lower SR values are more susceptible to liquefaction.

Dry density (γ_d) is related with $(SR)_{20}$ under different fines contents (FC) as follows:

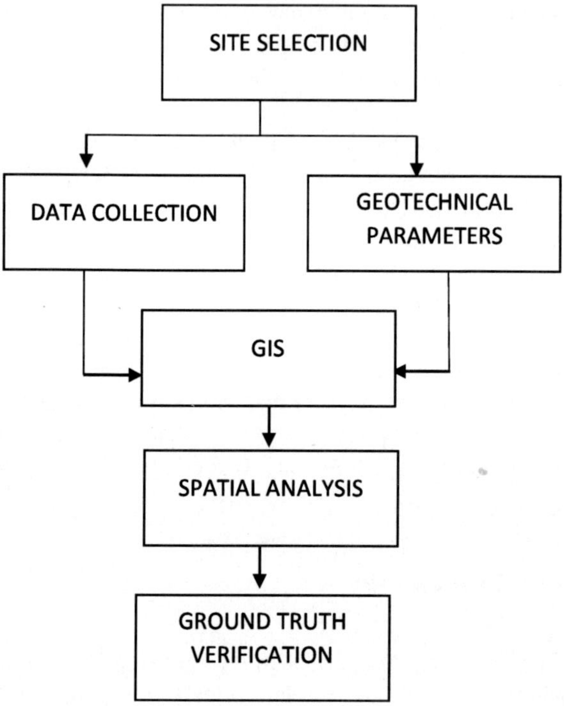

Fig. 1 Methodology of the study

$$(\mathrm{SR})_{20} = \boldsymbol{a}_1 \times (\boldsymbol{\gamma}_d)^{\boldsymbol{b}_1} \tag{1}$$

where a_1 and b_1 are function of fine content,

$a_1 = 0.059 - 0.0078 \times \mathrm{FC} + 0.0015 \times \mathrm{FC}^2$, and

$b_1 = 5.311 + 0.0247 \times \mathrm{FC} - 0.0714 \times \mathrm{FC}^2$

3 Study Area

Kollam district covers an area of 2491 km^2 in the southernmost part of Kerala. It is situated on the southwest coast of India between north latitudes 9° 10′ and 8° 45′ and east longitudes 76° 25′ and 77° 15′. The district is bounded by the Lakshadweep Sea on the west and Tamil Nadu state in the east. Along the northern boundary lie Alappuzha and Pathanamthitta district, while to the south lies Thiruvananthapuram district. It has a maximum length of 75 km in the E–W direction and maximum width of 45 km in the N–S direction.

The district can be broadly divided into three geological provinces—the westernmost Quaternary alluvial deposits followed by a narrow N–S zone of late Tertiary sediments and the easternmost Precambrian metamorphics. Geo-morphologically,

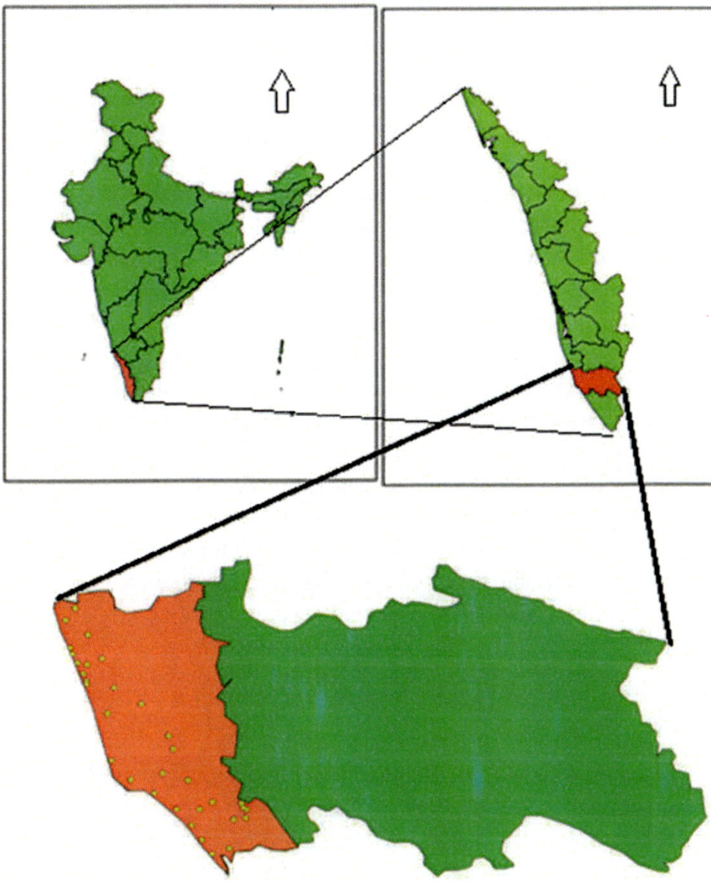

Fig. 2 Location map of study area and sampling points

the district can be divided into three units from west to east as the coastal plain, the midland and the high hill region. The coastal plain has a maximum width of 4 km in the north and gradually narrows down to less than 0.5 km toward south. It is nearly level to very gently sloping terrain depicting deposition of all forms like strandlines, flood plain and tidal flats. The coastal plain has several backwaters known as Kayals in Kerala, the prominent being Ashtamudi Kayal, Paravur Kayal, Panmana Kayal and the Sasthamcotta Kayal.

The samples were collected throughout the coastal stretch of Kollam covering a certain width along the coast. About 34 samples were collected. The sample locations were chiefly along the coastal stretch of Kollam and in the environs of the Kayamkulam Kayal, Ashtamudi Kayal and Paravur Kayal. Location map of study area and sampling points are provided in Fig. 2.

4 Results and Discussions

Grain size distribution can affect the dynamic loss of soil strength and liquefaction. Hence, the property of each soil sample was studied by performing sieve analysis and analyzing the parameters obtained from the test. The results obtained are compared with the limiting criteria for each parameter to define liquefaction potential of each location. These point data were used to interpolate the liquefaction potential of whole study area. The spatial distribution of local liquefaction potential formed the key information to arrive at a conclusion regarding the liquefaction susceptibility. The gradation curves for uniformly graded soil along with standard curve are presented in Fig. 3.

First, the spatial distribution of D_{60} values is prepared and presented in Fig. 4.

Next, SR values are determined, and the spatial distribution of SR values is provided in Fig. 5.

Finally, the liquefaction susceptibility map is prepared by the weighted overlay of integration of the two maps (Figs. 4 and 5). The liquefaction susceptibility map is given in Fig. 6. A careful perusal of Fig. 6 shows that the regions having D_{60} value less than 1 are more susceptible to liquefaction. Region with SR value less than 0.357 is susceptible to liquefaction. Also, an independent comparison of Fig. 6 with Figs. 4

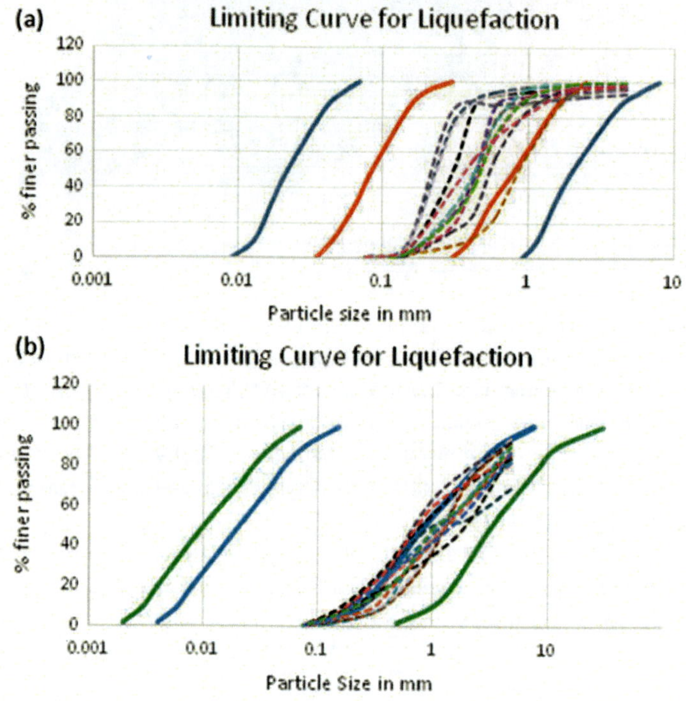

Fig. 3 Gradation curve comparison **a** for uniformly graded soils, **b** for well-graded soil

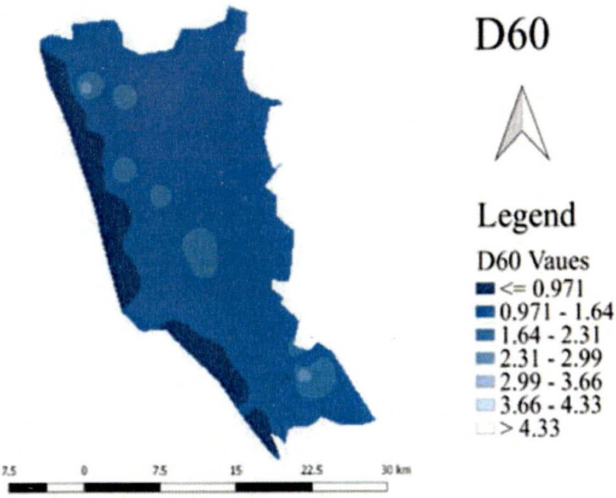

Fig. 4 Spatial distribution of D_{60} values

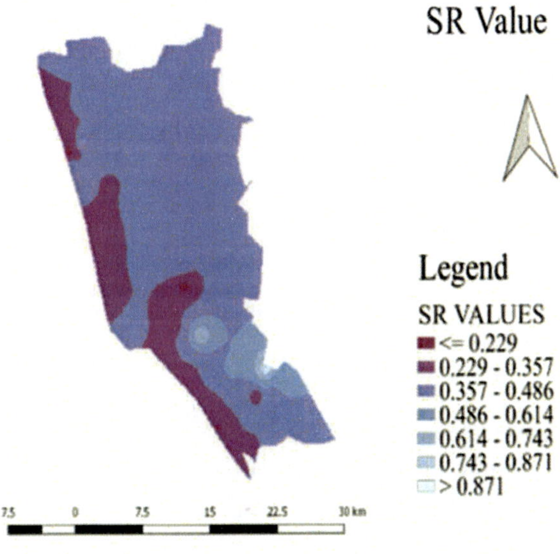

Fig. 5 Spatial distribution of SR values

and 5 shows the striking similarity of the final map with Fig. 5. This clearly infers the role of SR and hence influences of density of soil on liquefaction of the study area, which is the additional parameter considered in computation of SR.

Next, the liquefaction susceptibility map is overlayed with the water table data (Fig. 7) to check the water table data of the liquefaction susceptible areas.

From Fig. 7, it is noted that the depth to water table of susceptible region ranges mostly from 0 to 5 m. Hence, the liquefaction susceptibility of the area is also attributed to the effect of water table, as the soil layer gets saturated which eventually increases the liquefaction susceptibility of the study area.

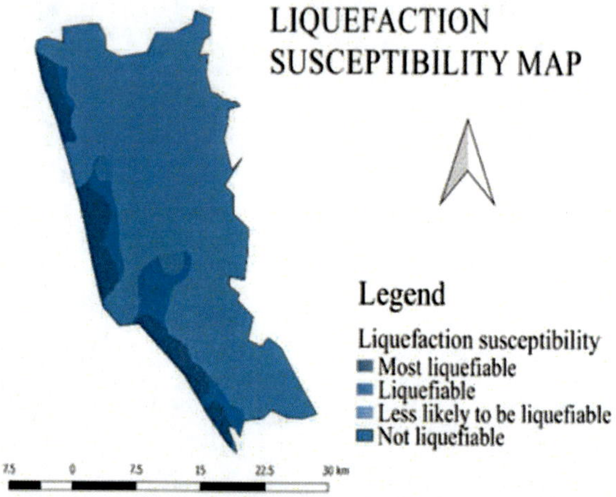

Fig. 6 Liquefaction susceptibility map of the study area

Predicting the liquefaction potential of a region is challenging, and the most studies used the test data of SPT test and cyclic triaxial test; this study followed a simplified procedure which checks the geotechnical parameters for the assessment of liquefaction susceptibility.

5 Conclusion

This study attempts to delineate the liquefaction susceptibility of Kollam coastal stretch through experimental methods which can constitute a preliminary information map on regional liquefaction potential. The cyclic stress ratio obtained indirectly through an empirical relation and the gradation curve quantified based on D_{60} is the parameter considered for the map preparation.

Important conclusions drawn from the study are presented below:

1. The coastal stretch extending to a range of width from 0.5 to 2.5 km is mostly susceptible to liquefaction.
2. The most susceptible areas consist of coastal alluvium and marine sand, which are mostly unconsolidated sediments.
3. The geology of susceptible areas consists of marine and fluvial deposits of Holocene age which conforms to the geologic criteria proposed by Youd and Hoose (1977).
4. Dry density has a greater influence in determining liquefaction susceptibility as observed from overlayed map. As dry density increases, the resistance to liquefaction increases.
5. The depth to water table of susceptible region ranges mostly from 0–5 m which increases the chance of soil layer to get saturated, and this increases the susceptibility.

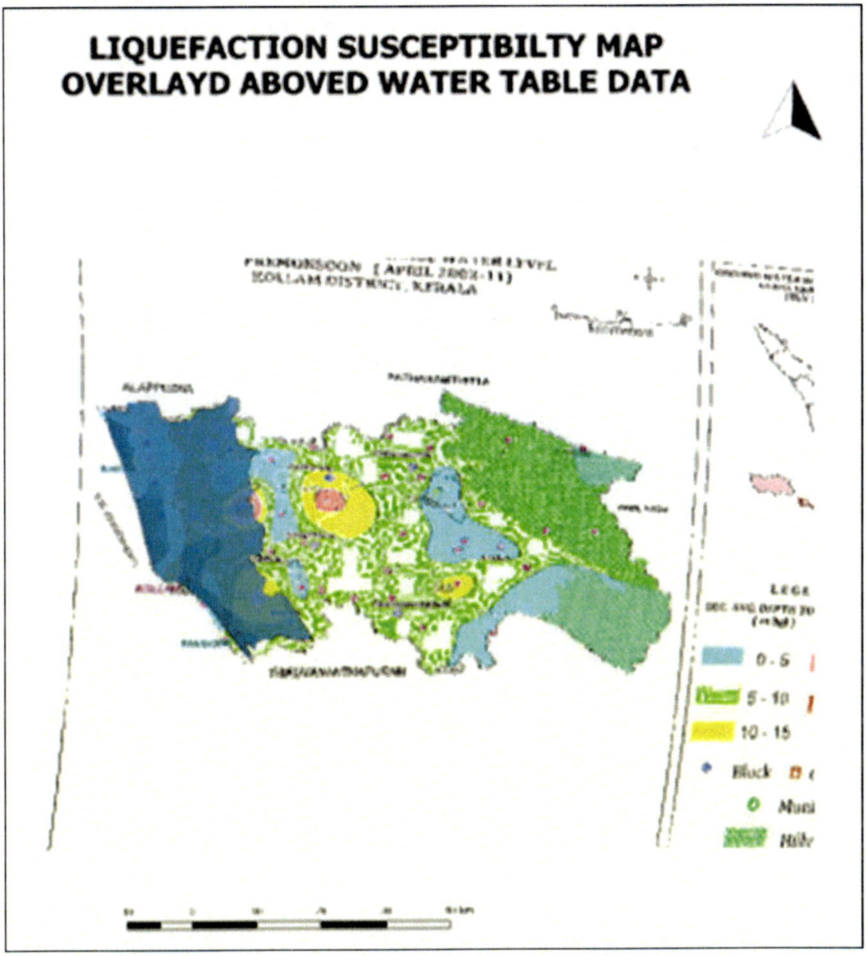

Fig. 7 Ground water table map overlaid with LSM of the area

References

Chien LK, Oh YN, Chang CH (2002) Effects of fineness content on liquefaction strength of reclaimed. Can Geotech J 43:903–914

District survey report of Minerals, KOLLAM district, by Department of Mining and geology (www.dmg.kerala.gov.in)

Finn WL, Ledbetter RH, Wu G (1994) Liquefaction in silty soils: design and analysis. Ground failure under seismic conditions, pp 51–76

Ganapathy PG, Rajawat AS (2012) Evaluation of liquefaction potential hazard of Chennai city, India: using geological and geomorphological characteristics. Nat Hazards 64:1717–1729

Green RA, Ziotopoulou K (2001) Overview of screening criteria for liquefaction triggering susceptibility, In: Proceedings of the tenth Pacific conference on earthquake engineering building an earthquake-Resilient Pacific 6–8 Nov 2001

Iai S, Tsuchida H, Koizumi K (1989) A liquefaction criterion based on field performances around seismograph stations. Soils Found 29(2):52–68

Mhaske SY, Choudhury D (2010) GIS-based soil liquefaction susceptibility map of Mumbai city for earthquake events. J Appl Geophys 70(216–225):217

Seed HB, Idriss IM (1982) Ground motions and soil liquefaction during earthquakes. Earthquake Engineering Research Institute, Berkeley

Seed RB, Cetin KO, Moss RES, Kammerer A, Wu J, Pestana J, Riemer M, Sancio RB, Bray JD, Kayen RE, Faris A (2003) Recent advances in soil liquefaction engineering: a unified and consistent framework. Keynote presentation, 26th annual ASCE Los Angeles Geotechnical Spring Seminar, Long Beach, CA

Tsuchida H (1970) Prediction and counter measure against the liquefaction in sand deposits. Abstract of the Seminar in the Port and Harbor Research Institute

Youd TL, Hoose SN (1977) Liquefaction susceptibility and geologic setting. In: Proceedings of the 6th world conference on earthquake engineering, vol 3, pp 2189–2194

Finite Element Analysis of Foundation on Layered and Homogeneous Soil Deposit Under Dynamic Loading

Abhay Kumar Verma and Supriya Mohanty

Abstract An earthquake force causes severe damages to civil engineering structures and their foundations. Geographical statistic shows that almost 54% area of India vulnerable to earthquake. Hence, it is very important to analyze dynamic response of the foundation. In the present study, an attempt has been made to study the behavior of foundation resting on layered and homogeneous soil deposit under dynamic loading condition. Two types of soil systems have been considered for the analysis; one is the layered soil system consisting of loose sand at the top, soft clay at the bottom and medium sand in between, and the other one is the homogenous soil system with soft clay. The effect of water table on soil foundation system of both layered and homogeneous soil systems has also been studied. The foundation considered for the analysis is shallow foundation (continuous footing). The dynamic loading considered for the analysis is of the recent Nepal earthquake (Mw-7.8). The modeling of the soil foundation system has been carried out using two-dimensional finite element software CyclicTP. The response of shallow foundation resting on layered and homogeneous soil deposit under dynamic loading condition is presented in the form of horizontal and vertical displacement; ground acceleration; excess pore pressure and excess pore pressure ratio; and shear stress versus shear strain at various locations. The results have been compared to understand the effect of layered and homogeneous soil on dynamic response of soil foundation.

Keywords Finite element analysis · CyclicTP · Dynamic response analysis · Earthquake

1 Introduction

Ground motions can be induced by various natural and human-made activities. These seismic motions give a very adverse effect on structures. These motions are also responsible for soil liquefaction and severe foundation damages. Reasons mentioned

A. K. Verma (✉) · S. Mohanty
Indian Institute of Technology (Banaras Hindu University), Varanasi, Uttar Pradesh 221005, India
e-mail: abhaykrverma.rs.civ17@itbhu.ac.in

M. Latha Gali and P. Raghuveer Rao (eds.), *Geohazards*, Lecture Notes in Civil Engineering 86, https://doi.org/10.1007/978-981-15-6233-4_34

above are made dynamic response study quite essential for different soil domains and for various seismic motions. Dynamic response analysis can give a better idea to engineers to design earthquake-resistant structures.

Many researchers have already made their attempt to analyze the dynamic response of the foundations placed on different types of soil. Reddy et al. (2016) investigated layered soil deposit under Bhuj and Nepal earthquake motions. Reddy and Mohanty (2017) studied the seismic behavior of black cotton soil reinforced with the granular column. Asgari et al. (2014) carried out a numerical simulation to find seismic responses of shallow foundation placed on silt and sand. Many researchers also did the experimental study to find the settlement and pattern of pore pressure generation on various seismic motions. Yoshimi and Tokimatsu (1977) conducted a shake table test to find settlement and pore pressure near building during earthquake motion. Ueng et al. (1997), Paul and Dey (2007) conducted a series of cyclic triaxial test to find dynamic responses of sands and partially saturated soil. Liu et al. (2017) studied different modes of pore pressure generation in the loose saturated soil during liquefaction. Mohanty and Patra (2014, 2016a, b) conducted extensive experimental and numerical research over Indian pond ash to find its dynamic behavior and liquefaction potential. Lui and Dobry (1997) studied about dynamic response of shallow foundation placed in sand. Militano and Rajapakse (1999) studied the dynamic response of a pile in a multilayered soil. Nath et al. (1987) carried out a nonlinear dynamic response study on doubly curved shallow shell on elastic foundation.

In the present study, an attempt has been made to study the behavior of foundation resting on layered and homogeneous soil deposit under dynamic loading condition. The peak responses and their comparisons have been made in the following sections.

2 Parameters Considered

Following data are used in the present analysis.

2.1 Foundation Details

For the present study, shallow footing is considered. Foundation is having a depth of 1 m below the ground surface, and 2 m height above the ground surface. The width of the foundation is considered as 2 m. The elastic modulus of concrete is 25×10^6 kPa. Poisson's ratio and density are taken as 0.3 and 2300 kg/m^3.

Table 1 Table captions should be placed above the tables

Case	Layer	Thickness	Type of soil (Vs in m/s)
I	1	5	Cohesive soft (100)
	2	5	Cohesive soft (100)
	3	5	Cohesive soft (100)
II	1	5	Loose sand (185)
	2	5	Medium sand (205)
	3	5	Cohesive soft (100)

2.2 Details of Soil Strata

The analysis is carried out for two different cases of soil domains. In Case-I, homogenous soil domain is analyzed. It is having width and depth 15 m each. A multilayered soil stratum having the same dimensions has been analyzed in Case-II. At 15 m depth, hard rock has been considered for both the cases. Dynamic analysis for each case is carried out at full saturation condition (0 m GWT), 1 m GWT, and 2 m GWT. Details of both the cases are given in Table 1.

2.3 Displayed Input Earthquake Motion

For dynamic analysis, an input motion of Nepal earthquake 2015 (Mw = 7.8) is considered. Peak acceleration of 0.1639 g is observed at 49.31 s for Nepal Earthquake. Figure 1 shows the input motion of Nepal Earthquake motion considered for the present study.

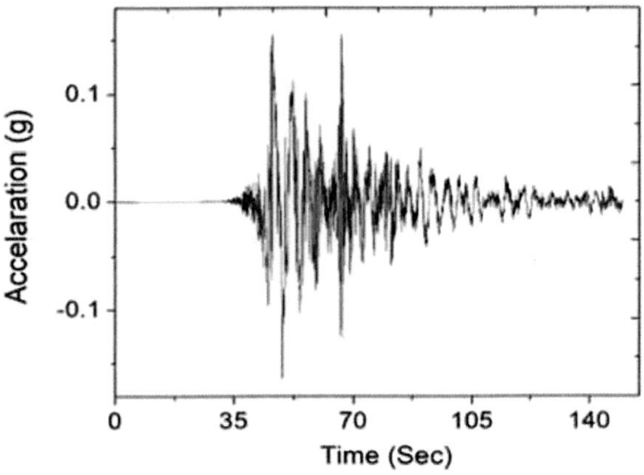

Fig. 1 Acceleration versus time plot for Nepal Earthquake (Mw-7.8)

3 Finite Element Analysis

The modeling of the soil domain and the shallow foundation is done using two-dimensional finite element method (FEM)-based software CyclicTP. The CyclicTP acts as dynamic site-response simulator. The program operates in the time domain, allowing for linear and nonlinear studies. Using CyclicTP, a nonlinear finite element analysis of soil foundation system has been carried out here. For analyzing different cases, the soil foundation system is divided into the finite number of elements. In the finite element modeling of the soil foundation domain, 4-4 nodded element is considered. The software is programmed to find approximate values of unknowns, by minimizing the errors at the discrete number of points over the domain. Typical diagram of FEM-based mesh generated of the soil foundation system (for Case-II) is shown in Fig. 2.

3.1 Convergence Study

Before starting the actual analysis, a convergence study has been carried out to find the optimum number of element for the soil foundation system. Soil foundation domain is analysis on denser element mesh until the peak responses start converging. This is done for every water table condition in both cases. Figure 3 shows a typical convergence graph for 0 m GWT for Case-I. The optimum number of element for

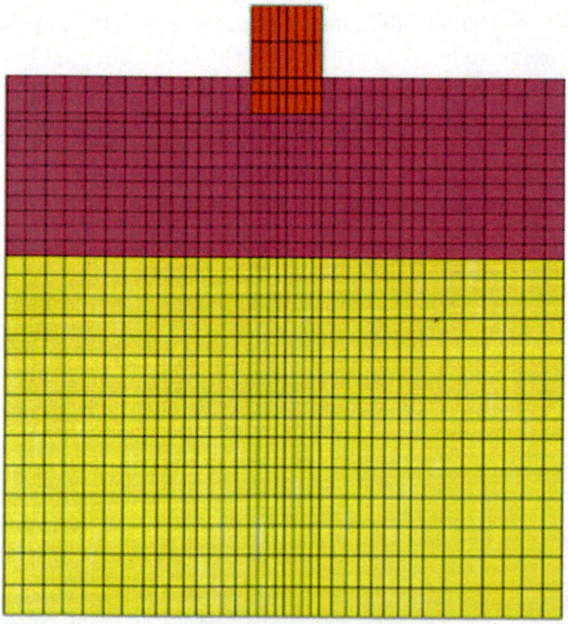

Fig. 2 Typical diagram of FEM-based mesh generated of the soil foundation system (for Case-II)

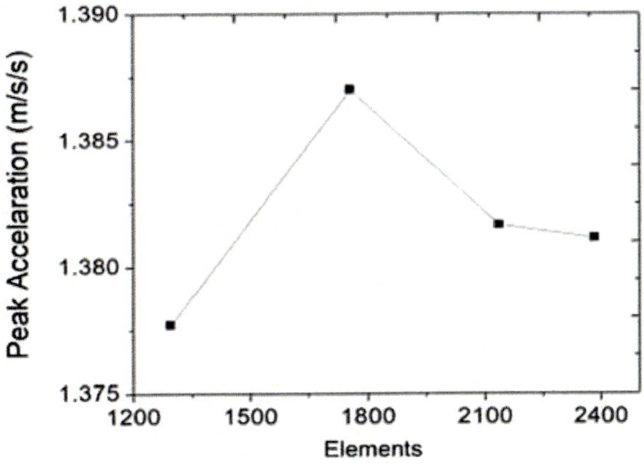

Fig. 3 A typical convergence graph for 0 m groundwater table in Case-I

Case-I is found to be 2384, 3156, and 2916 for 0 m, 1 m, and 2 m, respectively. Similarly, the optimum numbers of elements for Case-II is found to be 812, 1042, and 1042 for 0 m, 1 m, and 2 m, respectively.

4 Results and Discussion

The two-dimensional dynamic response results are represented in terms of the peak observed values of horizontal acceleration, lateral displacement, vertical displacement, excess pore pressure, and excess pore pressure ratio. In the following sections, the analyzed results are discussed.

4.1 Acceleration Response

At the different depths of the soil domain, the peak ground acceleration (PGA) is recorded. A typical variations of peak horizontal acceleration (or PGA) with depth for water table for Case-I and Case-II are shown in Figs. 5 and 6 According to the present analysis, it is found that the value of peak acceleration is more in the layered soil, i.e., in Case-II (ref. Table 2). It is also observed that the horizontal acceleration value is higher at top and bottom of soil stratum, and horizontal acceleration is less in comparison with the middle of soil domain in both cases. The maximum value of PGA is found in Case-II, i.e., 4.376 m/s^2 at 0 m GWT, 2 B m left from footing center ('B' denotes width of footing) (Fig. 4).

Fig. 4 Typical horizontal acceleration variation with time for soil foundation system (at: 1 m depth, 0 m GWT, Case-I)

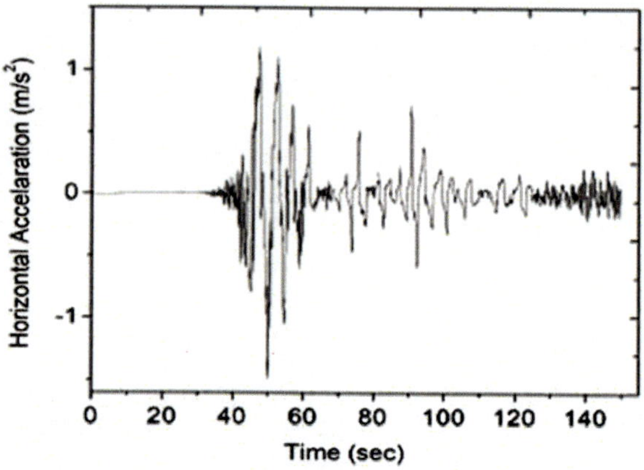

Fig. 5 Acceleration versus depth for soil foundation system (1 m GWT Case-I)

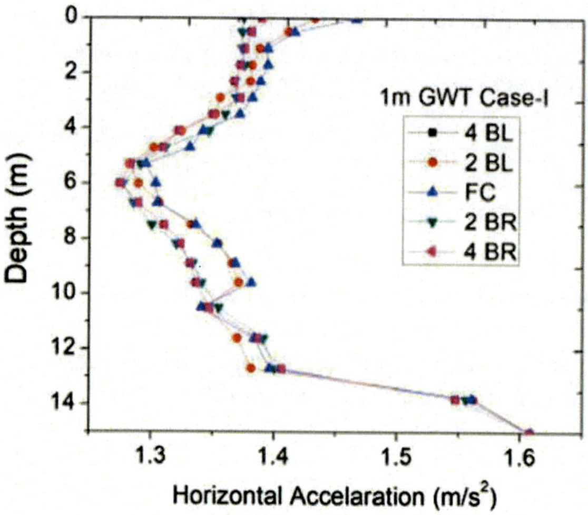

4.2 *Lateral Displacement Response*

Lateral displacement response has been recorded at different locations of the foundation soil system. In the complete analysis, maximum value of lateral displacement is found to be 0.370 m (ref. Table 3). This peak value is recorded in Case-II at full saturation condition (0 m GWT) under the Nepal earthquake motion. Figures 7 and 8 show typical variations of lateral displacements with depth at different locations from the footing center for Case-I and Case-II, respectively, are represented. Peak lateral displacements are found almost the same for all saturation conditions, in both cases (ref. Table 3).

It is observed that the value of lateral displacement is more at the surface, while it gradually decreases to a lower depth.

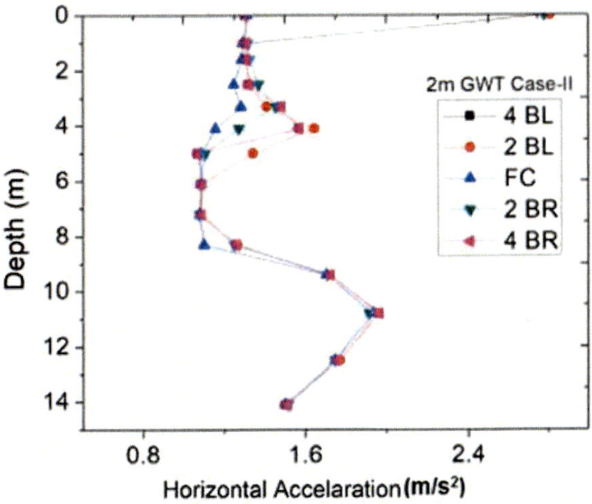

Fig. 6 Acceleration versus depth for soil foundation system (2 m GWT Case-II)

Table 2 Peak horizontal acceleration (PGA) values at different distances from footing center

Case	GWT depth	PGA (m/s^2)				
Distance from footing center →		4 B[1] (L)	2 B (L)	Footing center	2 B (R)	4 B (R)
I	0 m	1.631	1.608	1.608	2.528	1.631
	1 m	1.608	1.608	1.608	1.608	1.608
	2 m	1.608	1.608	1.608	1.608	1.608
II	0 m	1.469	4.376	1.683	3.895	1.848
	1 m	1.965	2.811	1.935	2.782	1.965
	2 m	2.338	2.011	2.148	2.012	2.338

[1]Width of footing

Table 3 Peak lateral displacement values at different distances from footing center

Case	GWT depth	Lateral displacement (m)				
Distance from center footing →		4 B (L)	2 B (L)	Footing center	2 B (R)	4 B (R)
I	0 m	0.117	0.117	0.117	0.116	0.117
	1 m	0.116	0.117	0.116	0.116	0.116
	2 m	0.117	0.117	0.117	0.116	0.116
II	0 m	0.370	0.370	0.348	0.323	0.334
	1 m	0.344	0.348	0.349	0.343	0.344
	2 m	0.298	0.302	0.300	0.301	0.298

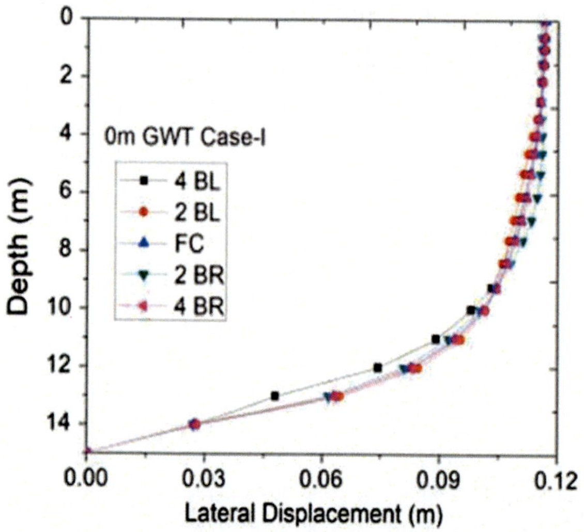

Fig. 7 Lateral displacement versus depth for soil foundation system (0 m GWT Case-I)

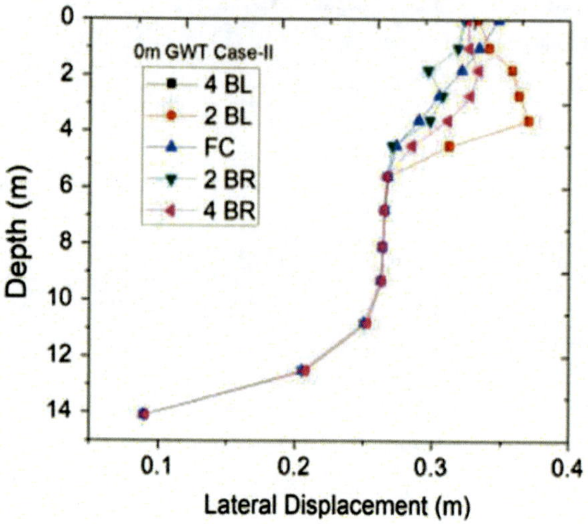

Fig. 8 Lateral displacement versus depth for soil foundation system (0 m GWT Case-II)

4.3 Vertical Displacement Response

Peak vertical displacement is also observed in Case-II, i.e., 0.6319 m (ref. Table 4). It is observed below the footing center at full saturation condition (0 m GWT) under the Nepal earthquake motion. Typical variations of vertical displacements for Case-I and Case-II with the depth are presented in Figs. 9 and 10. For both cases, values of peak vertical displacement are found to be more at full saturation condition than in partially saturated soil (ref. Table 4).

Table 4 Vertical displacement values at different locations from footing center

Case	GWT depth	Vertical displacement (m)				
Distance from footing center →		4 B (L)	2 B (L)	Footing center	2 B (R)	4 B (R)
I	0 m	0.006	0.001	0.006	0.001	0.006
	1 m	0.003	0.002	0.004	0.001	0.003
	2 m	0.002	0.002	0.004	0.001	0.003
II	0 m	0.208	0.208	0.632	0.216	0.351
	1 m	0.149	0.038	0.246	0.053	0.149
	2 m	0.046	0.010	0.083	0.014	0.046

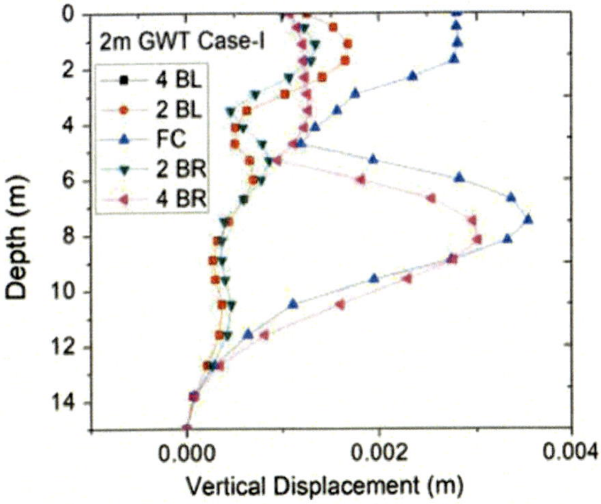

Fig. 9 Vertical displacement versus depth for soil foundation system (2 m GWT Case-I)

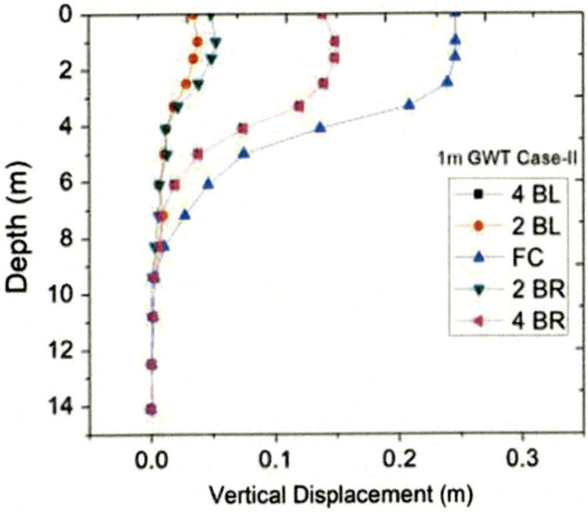

Fig. 10 Vertical displacement versus depth for soil foundation system (1 m GWT Case-II)

Fig. 11 A typical graph of shear stress versus shear strain obtained from the center of the footing at 10.994 m depth from the ground surface (Case-I, 0 m GWT)

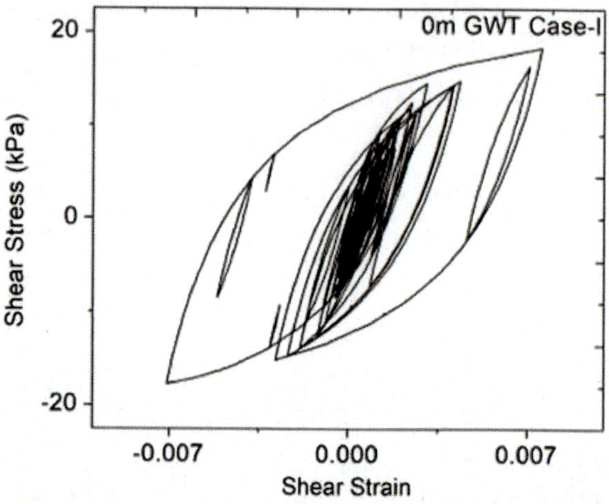

As per the observations recorded (Tables 3 and 4), there are fair chances of lateral spreading, as lateral displacement is higher than vertical displacement in both cases and at all water table conditions.

It is found that vertical displacement is higher near the corners of the foundation. In the partially saturated soil of Case-I, vertical displacement is more below the footing center.

In Case-II, vertical displacement is higher below the footing center for all saturation condition. Values of vertical displacement in both cases gradually decrease at a lower depth of soil domain.

4.4 Stress Versus Strain Response

Responses of shear stress versus shear strain are recorded at different depths and location from the center of footing. In Fig. 11, a typical graph of shear stress versus shear strain is shown. The typical graph shown in Fig. 5 recorded for Nepal Earthquake Motion at a depth of 10.994 m for 2 m GWT in Case-I.

4.5 Excess Pore Pressure and Excess Pore Pressure Ratio Response

Excess pore pressure ratio is calculated by dividing excess pore pressure at a location to initial effective vertical stress at that particular location. It is known that the soil is liquefiable at the point where excess pore pressure ratio reaches unity. In the present analysis, for finding the possibility of liquefaction excess pore pressure ratio is calculated at different depth and locations.

By investigating the results, it is found that in Case-I, the possibility of liquefaction occurs only in the condition of complete saturation (0 m GWT) (ref. Table 5); soil domain is safe against liquefaction if it is partially saturated. Results show that soil is prone to liquefaction at all location in Case-II (ref. Table 5), irrespective of the saturation condition adopted for the study.

Table 5 shows the values of the peak values of the excess pore pressure ratio at the different locations for both the cases. The typical graphs are shown in Figs. 12 and 13 and represent the values of excess pore pressure versus depth for 1 m GWT of Case-I and 0 m GWT for Case-II, respectively.

By these graphs, it can be seen that in Case-I, under high seismic motion the value of excess pore pressure reaches too high at just below the footing. The values are found higher again at a lower depth in Case-I.

Table 5 Variation of excess pore pressure ratio at different locations from footing center

Case	GWT depth	Excess pore pressure ratio				
Distance from footing center →		4 B (L)	2 B (L)	Footing center	2 B (R)	4 B (R)
I	0 m	3.597	0.488	11.315	1.817	2.485
	1 m	0.310	0.341	0.513	0.364	0.343
	2 m	0.330	0.330	0.401	0.347	0.202
II	0 m	3.000	3.000	2.898	2.562	2.332
	1 m	1.888	2.555	1.792	2.164	1.864
	2 m	1.746	1.592	1.717	1.499	1.836

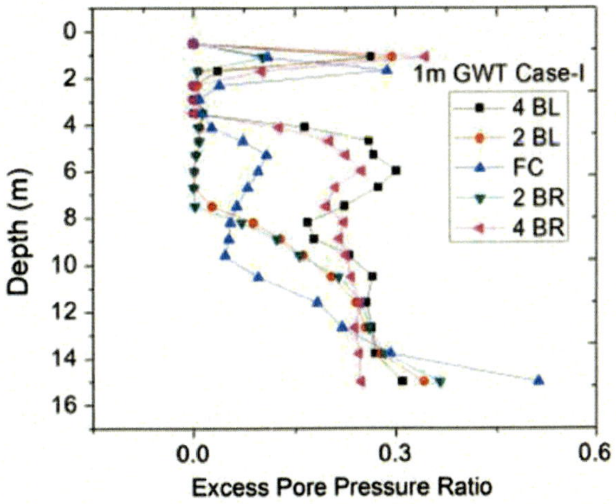

Fig. 12 Excess pore pressure versus depth for soil foundation system (1 m GWT Case-I)

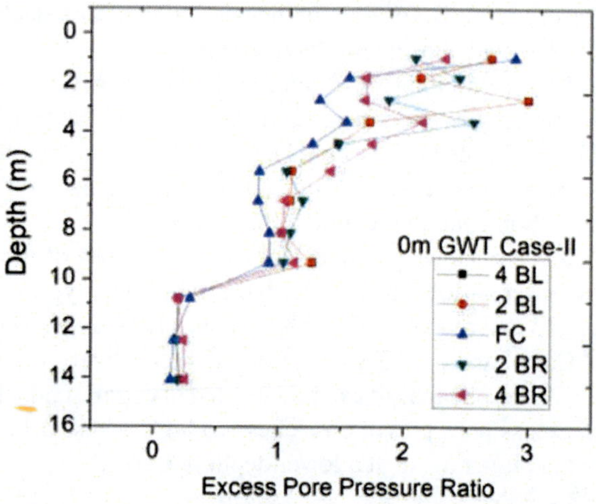

Fig. 13 Excess pore pressure versus depth for soil foundation system (0 m GWT Case-II)

In Fig. 13, it can be seen that values of excess pore pressure is more near the footing, while it gradually decreases at lower depths in Case-II.

5 Conclusions

The present paper discusses comparisons of dynamic responses for two different kinds of soil stratum under the excitation of Nepal earthquake motion. The comparisons have been made in terms of acceleration (PGA), peak lateral displacements, peak vertical displacement, excess pore pressure, and excess pore pressure ratio for both soil domains. The present analysis is concluded in the following points:

- For both the cases, PGA is recorded higher at the sides of footing, and it is then gradually decreasing in middle of soil domain and then again increases at the bottom.
- The maximum value of PGA for Case-I is found 2.527 m/s^2 on full saturation condition at 2 B m right of footing center. It is 4.376 m/s^2, for Case-II, in full saturation condition at 2 B m left of footing center.
- High lateral and vertical displacements are observed at the top and gradually decrease to the bottom of the soil domain.
- Highest values of lateral and vertical displacements are recorded in Case-II, which are 0.370 m and 0.6319 m, respectively, at 0 m GWT.
- Excess pore pressure values are found to be 30.94 kPa and 130.31 kPa, respectively, for Case-I and Case-II, respectively.
- Observation for excess pore pressure ratio shows that soil foundation system may liquefy if it reaches to the full saturation condition in Case-I, while there is a danger of liquefaction in Case-II, at all the considered water table conditions.

As the occurrence of an earthquake is unpredictable, the above data and conclusions can help engineers to get new insight for the seismic design of the structure in the layered soil conditions.

References

Asgari A, Golshani A, Bagheri M (2014) Numerical evaluation of the seismic response of shallow foundation on loose silt and silty sand. J Earth Syst Sci 123(2):365–379

Liu L, Dobry R (1997) Seismic response of shallow foundation on liquefiable sand. J Geotech Geoenviron Eng 123:557–567

Liu J, Guo T, Zhang G, Fu H (2017) Experimental study on pore pressure generation mode of saturated remolded loess during dynamic liquefaction. IOP Conf Ser: Earth Environ Sci 61:1–7

Militano G, Rajapakse RKND (1999) Dynamic response of a pile in a multi-layered soil to transient torsional and axial loading. Geotechnique 49(1):91–109

Mohanty S, Patra NR (2014) Cyclic behavior and liquefaction potential of Indian pond ash located in seismic zone III and IV. J Mater Civ Eng ASCE 26(7)

Mohanty S, Patra NR (2016a) Liquefaction and earthquake response analysis of Panipat pond ash embankment in India. J Earthquake Tsunami 10:1650009. https://doi.org/10.1142/S1793431116500093

Mohanty S, Patra NR (2016b) Dynamic response analysis of Talcher pond ash embankment in India. Soil Dyn Earthquake Eng 84:238–250

Nath Y, Mahrenholtz O, Varma KK (1987) Nonlinear dynamic response of a doubly curved shallow shell an elastic foundation. J Sound Vib 112(1):53–61

Paul S, Dey AK (2007) Cyclic triaxial testing of fully and partially saturated soil. In: Proceeding of 4th international conference on earthquake geotechnical engineering, Silchar, paper no 1194

Reddy CS, Mohanty S (2017) Parametric study on seismic behavior of black cotton soil reinforced with granular column. In: Proceeding of Indian geotechnical conference, 2017, pp 1-4

Reddy MVRK, Mohanty S, Ramancharla PK (2016) Comparative study of dynamic response analysis of shallow foundation on layered soils. Geotech Eng Lect Notes Civ Eng 15:92–99

Ueng TS, Wu MC, Lin CY, Yu RY (1997) Pore water pressure changes in sands under earthquake loading. In: 12th world conference on earthquake engineering, vol 1285, pp 1–7

Yoshimi Y, Tokimatsu K (1977) Settlement of buildings on the saturated sand during earthquakes. Soils Found 17(1):23–38

Site Classification of Strong Motion Stations of Uttarakhand, India, Based on Standard Spectral Ratio, and Horizontal-to-Vertical Spectral Ratio Methods

N. H. Harinarayan and Abhishek Kumar

Abstract Program for excellence in strong motion studies) (PESMOS) is an important accelerogram database in India consisting of earthquake records from 300 recording stations (RSs) located across the country. A major limitation of PESMOS database is the lack of accurate information on the seismic site class of RSs, which is essential for the utilization of ground motion records in seismic hazard and ground response studies. In the absence of in-situ field study data, site class of RSs can be determined indirectly analyzing recorded ground motions using empirical and semiempirical methods. In the present study, site class for 4 RSs located in the eastern region of Uttarakhand is established in accordance with the predominant frequency values obtained using standard spectral ratio method (SR), generalized inversion technique (GINV) as well as horizontal-to-vertical spectral ratio method (HVSR).

Keywords Predominant frequency · Seismic site class · PESMOS

1 Introduction

The phenomenon of soil deposits modifying the amplitude and the frequency content of the incoming seismic waves from bedrock is termed as local site effect (LSE) and is a vital component accountable for large damages during earthquakes. LSE is a main reason for large amplitudes observed at soil sites in contrast to sites on rock during an earthquake. Severity of LSE was observed in the damage patters of the 1985 Michoacan earthquake, where the incoming seismic waves from bedrock were amplified up to 5 times at surface causing grave damages in parts of Mexico City, located as far as 600 km from the epicenter (Zeevaert 1991). In an another instance, the 1989 Loma Prieta earthquake reported extensive damages in soft soil sites like San Francisco-Oakland region located 80 km from the epicenter (Seed et al. 1990).

N. H. Harinarayan (✉) · A. Kumar
Indian Institute of Technology Guwahati, Guwahati, Assam 781039, India
e-mail: nhharinarayan@gmail.com

M. Latha Gali and P. Raghuveer Rao (eds.), *Geohazards*, Lecture Notes in Civil Engineering 86, https://doi.org/10.1007/978-981-15-6233-4_35

Numerous examples illustrating the role of LSE in the damage patterns during earthquakes have been reported worldwide (e.g., Nihon 2011; Borcherdt 1970). Moreover, there are ample examples from the Indian subcontinents where severe damages were reported at sites on soft soil, located far from the epicenter. For example, the 2001 Bhuj earthquake (Mw = 7.7) reported severe damages in Ahmedabad, Rajkot, Anjar, and Gandhidham areas situated 350 km from the epicenter (Verma et al. 2013). The 1991 Uttarkashi earthquake (Mw = 6.8) (Kayal 1996) and the 1999 Chamoli earthquake (Mw = 6.5) (Mahajan and Virdi 2001) are other examples from the Indian subcontinents where severe damages were reported at sites located on soft soils.

The Himalayan arc within the Indian subcontinent stretching from the valley of Kashmir in the west to Arunachal Pradesh in the east has undergone six strong earthquakes ($M \geq 7.0$) in the last 120 years, namely the 1897 Shillong earthquake (Mw = 8.3), the 1905 Kangra earthquake (Mw = 7.9), the 1934 Bihar–Nepal earthquake (Mw = 8.0), the 1950 Assam earthquake (Mw = 8.6), the 2005 Kashmir earthquake (Mw = 7.6), the 2011 Sikkim earthquake (Mw = 6.9), and the 2015 Gorkha and Dholakha earthquake (Mw = 7.8 and Mw = 7.3). The seismicity of the Himalayan arc is because of the convergence of the Indian plate against the Eurasian plate (Srivastava et al. 2013), that had began about fifty million years ago. Currently, the convergence rate between the two plates is 5 cm per year. The collision caused uneven stress accumulation resulted in the creation of a series of thrust sheets (Mugnier et al. 2012) along the Himalayas. The Bureau of Indian Standards (BIS 2002) classified the regions in and around the Himalaya as seismic zones five or four, suggesting regions of high or very high seismicity. Hence, attempts to understand surface seismic hazard of this region are of great significance.

With the purpose of understanding the ongoing seismicity of various seismically active regions in India, several numbers of RSs were placed on or after 2004 under the project titled "National Strong Motion Instrumentation Network" by the government of India. Strong motion records from these RSs since 2004 are maintained by PESMOS and are available on the Web site www.pesmos.in. Currently, PESMOS database is a major source of strong motion records in the region. By the end of 2016, the PESMOS database comprises of three component accelerographs from 300 RSs in various regions in India. A major drawback of PESMOS database is the discrepancies in the site class given for the RSs. Site class given by PESMOS is based on the physical description of local geology and not based on in-situ field test results. The value of V_{s30} (average shear wave velocity for 30 m) is approximated for each RS based on the SEISAT (2000) and Geological Maps of India. Site class for the RS is assigned following the modified Borcherdt (1994) classification scheme, consisting of three site classes, namely site class A (V_{s30} > 700 m/s), site class B (375 m/s < V_{s30} > 700 m/s), and site class C (V_{s30} < 375 m/s) (Mittal et al. 2012). Here, the site class A refers to firm/hard rock site, the site class B refers to soft to firm rock, and the site class C refers to soil sites. It can be seen that the PESMOS followed site classification scheme is very broad. Geophysical field studies on some of the RSs by Pandey et al. confirmed mismatches in the site class given by PESMOS and filed study results. Hence, the strong motion data from these RSs cannot be utilized for seismic hazard assessment considering the site class given by PESMOS. Thus, accurate assessment of site class of these RSs is important in order to utilize these data in seismic hazard analysis. In the present study, site class following the NEHRP classification scheme (National Earthquake Hazards Reduction Program) of 4 RSs in Uttarakhand region is assigned using three different approaches, namely SR, GINV, and HVSR, as discussed in detail under subsequent headings.

2 Study Area

RSs situated in the Garhwal Himalaya area of the State of Uttarakhand, covering the region between 79 °E–80.5 °E longitude and 29.5 °N–30.4 °N latitude in the towns of Champawat, Darchula, Lansdome, and Chamoli is considered for the present work. The locations of the four RSs are shown in Fig. 1, and coordinates of the four RSs are listed in Table 1. Garhwal Himalaya is surrounded by the Main Central Thrust (MCT)

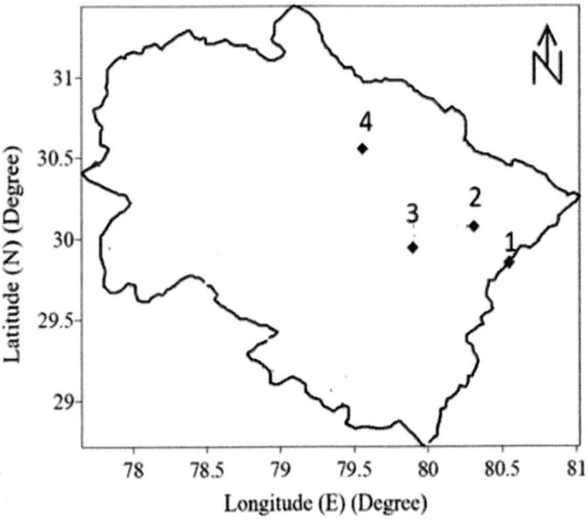

Fig. 1 Study area showing four recording stations in the state of Uttarakhand considered for the present work (1-Darchula, 2-Champawat, 3-Lansdome, and 4-Chamoli)

Table 1 Location details of 4 RSs

S. No.	Recording station	Station code	Lat. (°) (N)	Long. (°) (E)
(1)	(2)	(3)	(4)	(5)
1	Darchula	DAR	29.7	80.4
2	Champawat	CHM	30.3	80.1
3	Langsdome	LAN	29.9	79,8
4	Chamoli	CHP	30.5	79.6

Table 2 Details of earthquakes

S. No.	Date	Lat.	Long.	Mag.	Depth (km)
1	15-05-2009	30.5	79.3	4.1	15
2	11-01-2010	29.7	80	3.9	15
3	14-03-2010	31.7	76.1	4.6	29
4	01-05-2010	29.9	80.1	4.6	10

and the Main Boundary Thrust (MBT) (Valdiya 1980). Reports of past seismicity indicate that the study area has witnessed two moderate earthquakes recently, namely 1991 Uttarakashi earthquake (Mw = 6.8) and 1999 Chamoli earthquake (Mw = 6.6). Both these earthquakes have caused extensive damages to the traditional houses in the region (Mahajan and Virdi 2001) and caused several casualties (Verma et al. 2013). Before the 1991 Uttarakashi earthquake and the 1999 Chamoli earthquake, the study region has endured two other severe earthquakes in the past, namely the 1803 Kumaon–Nepal earthquake (M-7.5+) and the 1905 Kangra earthquake (Ms = 7.8). According to the 2011 Census, Uttarakhand has a population close to 1.01 crore and more than ninety percent houses in the region of Uttarakhand are constructed using materials like mud, brick, and stones which are weak and highly vulnerable during earthquakes. Hence, this region could incur heavy loss of life and property during an earthquake.

3 Data Set

For the present study, accelerograms from 4 RSs in the state of Uttarakhand are selected. The database used in the present work includes 12 accelerograms recorded during 4 EQs. The magnitudes of these earthquakes range from 3.5 to 5, and focal depths range from 2 to 21 km. The details of each EQ are summarized in Table 2.

4 Data Processing

All the strong motion records are subjected to baseline correction following Kumar et al. (2012). Also, a 5% cosine taper followed by a band-pass filter between the frequency range 0.15 and 15.0 Hz using a Butterworth filter is done. Finally, *S*-wave portions of the accelerogram are separated in accordance with Kato et al. for GINV analysis (shown in Fig. 2). Afterward, the Fourier amplitude spectra are calculated for the S-wave portion of accelerogram, for each EQ record. The Fourier amplitude spectra are smoothened using the Konno and Ohmachi algorithm similar to Harinarayan and Kumar.

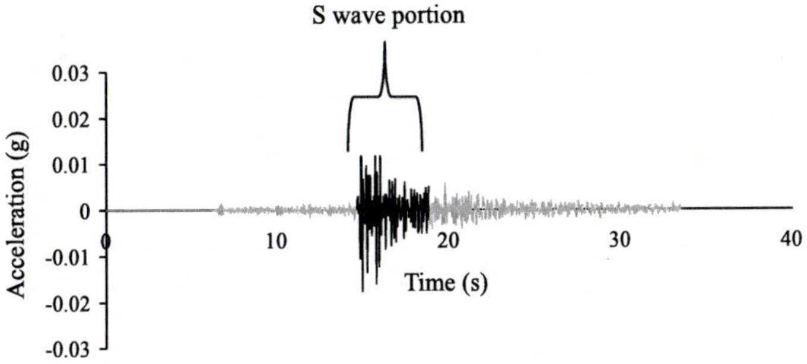

Fig. 2 Separation of S-wave portion from the accelerogram

5 Methodology Used and Analyses

A number of analytical studies for site classification of RSs using strong motion records have been reported in the past. Field and Jacob (1995) and Parolai et al. (2004), reported a comparison among various analytical approaches and concluded that SR, GINV and HVSR can give consistent value of peak frequency (f_{peak}) that can be used for site classification of RS. Discussion on SR, GINV, and HVSR can be found in the following subsections.

5.1 *Spectral Ratio Method*

The SR method was first developed by Borcherdt (1970) for estimating site effects (S_i^{SR}) by dividing the Fourier spectra of the site of interest (U_{iJ}) to the Fourier spectra of reference site (U_{iJ}) (rock site) with comparable epicentral distances as represented in Eq. (1).

$$\ln U(f)_{ij} = \ln G(f)_i + \ln P(f)_{ij} + \ln S(f)_{ij} \tag{1}$$

In Eq. (1), j is the earthquake event recorded at site i and the reference site. In the present analysis, Pithoragarh and Langsdome RSs are selected as a reference site based on Harinarayan and Kumar.

5.2 *Generalized Inversion Technique*

The GINV technique was first introduced by Andrews (1986) after modifying the SR method. The GINV methodology in accordance with Andrews (1986) is discussed below.

The spectral acceleration of the earthquake j recorded at the RSi, denoted by $U(f)_{ij}$, can be linearly expressed in the frequency domain as given in Eq. (2).

$$\ln U(f)_{ij} = \ln G(f)_i + \ln P(f)_{ij} + \ln S(f)_{ij} \tag{2}$$

In Eq. (2), $G(f)_i$ is the source effect of the event j, $P(f)_{ij}$ is the term representing the propagation path, and $S(f)_{ij}$ is the site effect. The path attenuation parameter is removed from the spectral content of the record in accordance with Andrews (1986) as;

$$\ln U(f)_{ij} - \ln P(f)_{ij} = \ln J(f)_{ij} \tag{3}$$

The term $P(f)_{ij}$ in Eq. (2) is computed using the equation given below;

$$P(f)_{ij} = \frac{1}{R_{ij}}\left[\mathrm{e}^{\frac{(-(\pi\cdot f\cdot R_{ij})}{(Q_s(f)\cdot\beta)}}\right] \tag{4}$$

In Eq. (4), R represents the hypocentral distance, f represents the frequency, $Q_s(f)$ represents the quality factor for S-wave, β represents the average shear wave velocity of the crustal medium for the region. In the present study, the values of β are taken as 3.15 km/s and $Q_s = 105f^{0.94}$ after Harinarayan and Kumar for the northwest Himalayan region. Further, $\ln J(f)_{ij}$ is substituted in Eq. (2) giving Eq. (5).

$$\ln J(f)_{ij} = \ln S(f)_i + \ln G(f)_j \tag{5}$$

An unconstrained degree of freedom exists in Eq. (5) resulting in a trade-off between $S(f)_i$ and $G(f)_i$ (Andrews 1986). To remove the trade-off between source and site term, site term for a reference site is constrained to one irrespective of frequency. A RS located on a rock site is considered as a reference site, similar to the SR method (Field and Jacob 1995).

Following the notations of Menke (1989), Eq. (5) in matrix form can be expressed as:

$$\begin{vmatrix} A \\ C \end{vmatrix} m = \begin{vmatrix} b \\ 0 \end{vmatrix} \tag{6}$$

In Eq. (6), m represents a model space matrix (consisting of only two nonzero elements in each row and column) related to $\ln S(f)_i$ and $\ln G(f)_j$. "A" term in Eq. (6) represents the matrix linear operator, and "b" term represents data vector containing the elements which are to $\ln J(f)_{ij}$. A row matrix C representing the constraint condition for the reference station is added such that $C \times \mathrm{m} = 0$. Equation (6) is solved using singular value decomposition method (Menke 1989). Pithoragarh RS is selected as a reference site based on the findings of Harinarayan and Kumar.

5.3 Horizontal to Vertical Spectral Ratio Method

Nakamura (1989) proposed a methodology for determining the site characteristics in terms of f_{peak} and A_{peak} using horizontal to vertical ratio of Fourier spectrum (HVFR) of ambient noise or microtremor records at the site of interest. Nakamura (1989) assumed that only the horizontal component retains the local site characteristics, whereas the source and the propagation terms are retained in both vertical and horizontal components of ground motion. Therefore, the ratio of horizontal component to vertical components for record highlights the local soil condition.

HVFR contains several spikes which makes identification of a clear and distinct peak difficult. Figure 3a shows a typical HVFR based on recorded ground motions at Champawat during January 2010 earthquake. It can be seen from Fig. 3a, that the HVFR curve has many spikes, with no clear peak. Identifying predominant frequency in this case is difficult. Zhao et al. (2006) used spectral acceleration for 5% damping instead of Fourier spectra and concluded that the resulting curve is smoother with less number of spikes and thus giving a clear and distinct peak. Figure 3b shows HVSR curve for 5% damping for the same Champawat RS it can be seen that compared to Fig. 3a, b is more smoother with clear and distinct peak.

In the present study, HVSR curve is generated considering response spectra with 5% as damping value and smoothen the response spectra of east–west, north–south, and vertical component using Konno-Ohmachi (1998) algorithm.

The HVSR curve for each RSs is computed by the following procedure given below.

1. Calculate response spectra after applying 5% damping for the three components of the accelerogram.
2. The response spectra of each component are smoothened as per Konno and Ohmachi (1998).
3. The geometric mean of the east–west and north–south components is calculated as shown below.

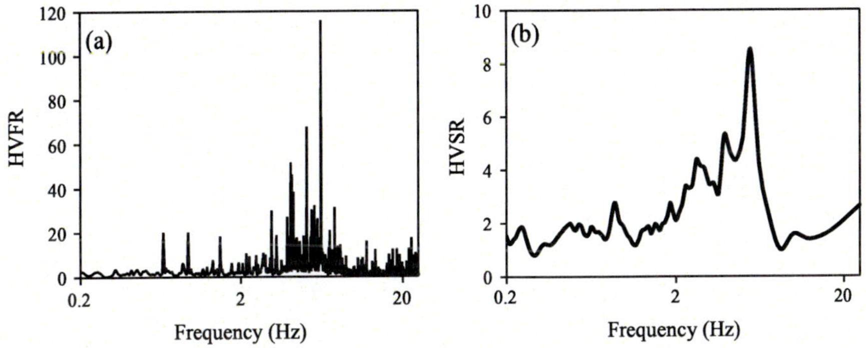

Fig. 3 Comparison between horizontal-to-vertical ratio of Fourier spectra (**a**) and horizontal-to-vertical response spectra ratio (**b**) for Champawat RS

$$H = (H_{EW} \times H_{NS})^{0.5} \tag{7}$$

4. H/V is calculated.

In Eq. 1, H_{EW} and H_{NS} denote the pseudo-response acceleration (PSA, 5% damped system) of the horizontal east–west and north–south components, respectively, and V denotes the PSA (5% damped system) of the corresponding vertical component.

The HVSR at each RS is then estimated as;

$$(\text{HVSR})_i = \frac{\sum_{j=1}^{Ni} \frac{H}{V}}{Ni} \tag{8}$$

In Eq. (8), Ni represents the number of events recorded at RS "i". Further, f_{peak} for a RS is visually estimated as the frequency corresponding to the highest amplitude denoted as A_{peak}. It has to be highlighted here that in the present study emphasis is given to the value of f_{peak} since the classification of RSs is done based on the value of f_{peak}.

6 Results and Interpretation

Comparison of site terms computed using HVSR (indicated by solid line), SR method (indicated by dashed line) as well as GINV (indicated by dotted line) for the recoding RS at Champawat, Darchula, Lansdome, and Chamoli are shown in Fig. 4. A general observation from Fig. 4 is that site response curves show similar variation trends with clear and distinct peaks based on HVSR, SR method, and GINV for all the 4 RSs. Moreover, the peak frequency (f_{peak}) obtained from the three methods is closely matching as shown in Fig. 4. Another observation from Fig. 4 is that HVSR method gives maximum amplification (A_{peak}). Moreover, the A_{peak} obtained using SR method is higher than those obtained using GINV method for all the four RSs. The value of f_{peak} obtained in the present study for the RSs considered is given in Table 3. The value of f_{peak} is in the range of 1.3 Hz (observed for Lansdown RS) and 5.5 Hz observed for Champawat RS). The value of A_{peak} obtained in the present study for the RSs considered is given in Table 4. Large discrepancies in the value of A_{peak} obtained using SR method, GINV, and HVSR are observed for all locations. It has to be mentioned here that the aim of the present study is to classify RSs based on f_{peak}. Thus, the variation in A_{peak} obtained using the three methods will not alter the findings from the present work.

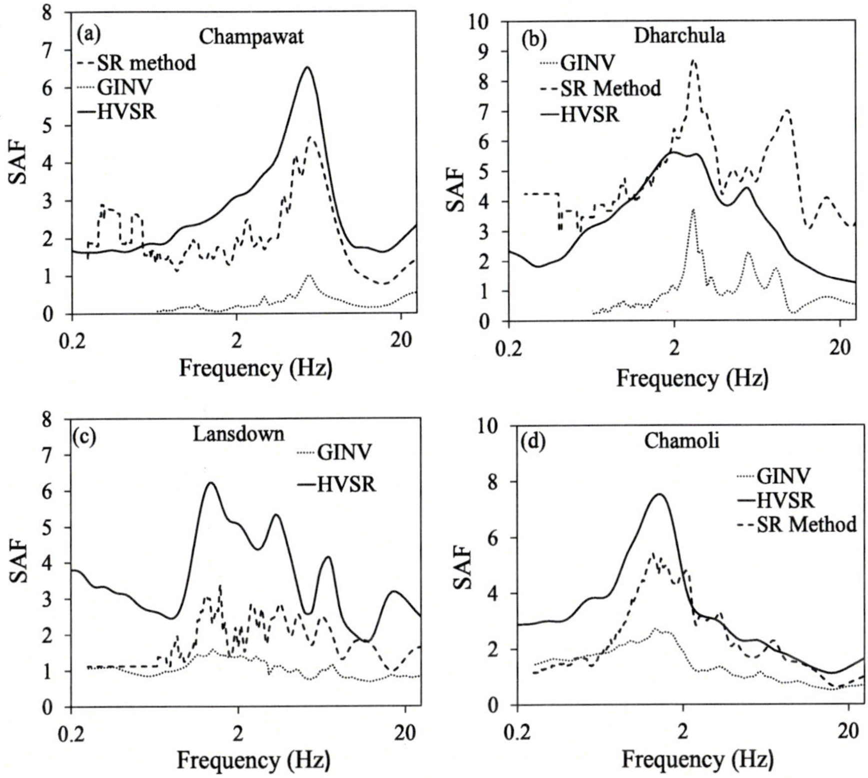

Fig. 4 a–d Site amplification factor (SAF) curves based on GINV, HVSR, and SR Method for: RSs **a** Champawat; **b** Dharchula; **c** Lansdown; **d** Chamoli

Table 3 Summary of f_{peak} obtained using HVSR, SR method, and GINV

S. No.	Station code	HVSR	SR method	GINV
(1)	(2)	(3)	(4)	(5)
1	DAR	5.5	5.4	5.5
2	CHM	2.6	2.5	2.6
3	LAN	1.3	1.3	1.4
4	CHP	1.4	1.4	1.3

Table 4 Site classification scheme for strong motion RSs as per PESMOS (modified after Mittal et al. 2012)

Site class	General description	Shear wave velocity (m/s)
A	Firm/hard rocks (fresh and compact metamorphic)	700–1620
B	Soft to firm rocks (sedimentary rocks)	375–700
C	Soils (alluvium, slope wash material, Eolian)	200–375

6.1 Grouping of RS Based on f_{peak} Value

The site classification of RSs is essential for a proper utilization of accelerograms from these RSs in seismic hazard analysis. As discussed earlier, the site classification scheme used by PESMOS is not in accordance with any in-situ filed investigation exercise. The site classification scheme adopted by PESMOS (described in Table 4) gives a broader sense of the site class in contrast with the worldwide followed NEHRP classification scheme described in Table 5. Based on the average value of f_{peak} obtained using the three methods, corresponding V_{S30} values are computed and given in Table 5 for all the 4 RSs using Eq. (9) given by Kramer (1996), for a single layer model over half space;

$$V_Z = f_{\text{peak}} 4H \tag{9}$$

In Eq. (9), H is the soil depth (taken as 30 m) and V_z is shear wave velocity at depth z. The value of V_{s30} for each of the RS is considered in the present study (given in Table 6, Column 2) is used to classify the RSs based on the classification scheme used by PESMOS (Table 4, Column 4) and is compared with the site class given in PESMOS (Table 6, Column 3). PESMOS classified all the four RSs as site class A indicating rock site. Based on the present analysis, Darchula, Lansdown, and Chamoli RSs have V_{s30} values of 660, 156, and 168 m/s. Hence, the following PESMOS classification scheme Darchula belongs to site class B, whereas Lansdown

Table 5 NEHRP site classification provisions

Site class	General description	Shear wave velocity (m/s)
A	Hard rock	>1500
B	Rock	760–1500
C	Very dense soil and soft rock	360–760
D	Stiff soil	180–360
E	Soft soil	<180

Table 6 Summary of shear wave velocity and site class as per PESMOS and NEHRP classification

Recording station	V_{s30} (m/s)	Site class: Given in PESMOS	Site class: Actual site class (PESMOS scheme)	Site class: NEHRP
-1	-2	-3	-4	-5
DAR	660	A	B	C
CHM	312	A	C	D
LAN	156	A	C	E
CHP	168	A	C	E

and Chamoli RSs belong to site class C. Site classification based on NEHRP classification scheme is also determined for the 4 RSs and is given in Column 5, Table 5. Following NEHRP classification scheme, Lansdown and Chamoli are classified as site class E, Champawat is classified as site class D, and Darchula is classified as site class C.

7 Conclusion

In the present study, site class of 4 RSs situated in the Garhwal, Uttarakhand region, is estimated from recorded accelerograms at these RSs using indirect analytical methods such as SR, GINV, and HVSR methods. The SAF curves obtained using SR, HVSR, and GINV show similarity in terms of the general shape with clear and distinct peaks. The value of f_{peak} obtained for each RS using the three methods is closely matching. Based on the value of f_{peak} estimated in the present study, V_{s30} for the 4 RSs are calculated and site classification based on PESMOS and NEHRP classification scheme is attempted. The site class obtained in the present study shows a mismatch with the site class given by PESMOS. Hence, utilizing records from the 4 RS considered in the present study based on the site class given by PESMOS would give erroneous results. From the present study, it is clear that in the absence of geotechnical and geophysical tests data, indirect analytical methods like SR, GINV, and HVSR can be used to classify the RSs.

References

Andrews DJ (1986) Objective determination of source parameters and similarity of earthquakes of different size. Earthquake Source Mech 259–267

BIS IS (2002) IS 1893 (Part 1): general provisions and buildings: criteria for earthquake resistant design of structures. Bureau of Indian Standards, New Delhi, India

Borcherdt RD (1994) Estimates of site-dependent response spectra for design (methodology and justification). Earthquake Spectra 10:617–653

Field EH, Jacob KHA (1995) Comparison and test of various site-response estimation techniques, including three that are not reference-site dependent. Bull Seism Soc Am 85(4):1127–1143

Kayal JR (1996) Precursor seismicity, foreshocks and aftershocks of the Uttarkashi earthquake of October 20, 1991 at Garhwal Himalaya. Tectonophysics 263:339–345

Konno K, Ohmachi T (1998) Ground motion characteristics estimated from spectral ratio between horizontal and vertical components of microtremor. Bull Seism Soc Am 88(1):228–241

Kramer SL (1996) Geotechnical earthquake engineering. Engineering 6:653

Kumar A, Mittal H, Sachdeva R (2012) Indian strong motion instrumentation network. Seismol Res Lett 83(1):59–66

Mahajan AK, Virdi NS (2001) Macroseismic field generated by 29 March, 1999 Chamoli earthquake and its seismotectonics. J Asian Earth Sci 19:507–516

Menke W (1989) International geophysics series. Geophysical data analysis: discrete inverse theory, p 45

Mittal H, Kumar A, Ramhmachhuani R (2012) Indian national strong motion instrumentation network and site characterization of its stations. Int J Geosci 06:1151

Mugnier JL, Gajurel A, Huyghe P, Jayangondaperumal R, Jouanne F, Upreti B (2012) Structural interpretation of the great earthquakes of the last millennium in the central Himalaya. Earth Sci Rev 127:30–47

Nakamura Y (1989) A method for dynamic characteristics estimation of subsurface using microtremor on the ground surface. Railw Tech Res Inst Quart Rep 30(1)

SEISAT (2000) Seismotectonic Atlas of India and its Environs. Geological Survey of India

Parolai S, Bindi D, Baumbach M, Grosser H, Milkereit C, Karakisa S, Zünbül S (2004) Comparison of different site response estimation techniques using aftershocks of the 1999 Izmit earthquake. Bull Seismol Soc Am 94(3):1096–1108

Seed RB, Dickenson SE, Rimer MF, Bray JD, Sitar N, Mitchell JK, Idriss IM, Kayen RE, Kropp A, Harder LF, Power MS (1990) Preliminary report on the principal geotechnical aspects of the October 17, 1989 Loma Prieta Earthquake. UCB/EERC-90/05

Srivastava HN, Bansal BK, Verma M (2013) Largest earthquake in Himalaya: an appraisal. J Geol Soc India 82:15–22

Valdiya KS (1980) Geology of Kumaun lesser Himalaya. Wadia Institute of Himalayan Geology

Verma M, Singh RJ, Bansal BK (2013) Soft sediments and damage pattern: a few case studies from large Indian earthquakes vis-a-vis seismic risk evaluation. Nat Hazards 74:1829–1851

Zeevaert L (1991) Seismosoil dynamics of foundations in Mexico City earthquake, September 19, 1985. J Geotech Eng 117(3):376–428

Zhao JX, Irikura K, Zhang J, Fukushima Y, Somerville PG, Asano A, Ohno Y, Oouchi T, Takahashi T, Ogawa H (2006) An empirical site-classification method for strong-motion stations in Japan using H/V response spectral ratio. Bull Seismol Soc Am 96:914–925

Seismic Site Classification and Site Period Determination of NIT Silchar Using MASW

Arindam Saha, Kallol Saha, and Ashim Kanti Dey

Abstract Silchar is situated in the north-eastern region of India which comes under seismic zone-V. This region of India is bounded by several faults like Dauki fault, Indo-Myanmar subduction zone and Sitakunda–Teknaf fault. Most of the buildings in Silchar including some buildings at NIT Silchar developed cracks during Imphal Earthquake of magnitude Mw = 6.7 on 4th January 2016. Hence, seismic site classification becomes very crucial for the entire north-east region. In the present study, NIT Silchar campus has been taken as a study area. Multichannel analysis of surface waves (MASW) technique, one of the recent developments in geophysical methods, was carried out at nine sites within the campus and the average shear wave velocity (Vs30) upto 30 m depth was obtained. Three confirmatory boreholes were dug for validation. The study showed that NIT Silchar falls under seismic site class D category according to National Earthquake Hazards Reduction Program (NEHRP) Code. The fundamental site period obtained from MASW reveals that, the soil conditions at NIT Silchar pose a potential threat to four- and five-storey buildings due to resonance during earthquakes.

Keywords Multichannel analysis of surface waves (MASW) · Average shear wave velocity upto 30 m depth (Vs30) · Fundamental site period · Seismic site classification

1 Introduction

Subsurface soil conditions play a major role in the damage potential of earthquakes and the seismic soil amplification of a site which is a critical factor affecting the level of ground shaking. Earthquake-like Bhuj earthquake (2001) in India, Guerrero earthquake (1985) in Mexico City, Loma Prieta earthquake (1989) in San Francisco caused notable destruction even at a far location from epicentre due to site-specific amplification of ground motion. This shows the importance of seismic site characterization

A. Saha · K. Saha (✉) · A. K. Dey
National Institute of Technology Silchar, Silchar, Assam 788010, India
e-mail: kallol001.saha@gmail.com

M. Latha Gali and P. Raghuveer Rao (eds.), *Geohazards*, Lecture Notes in Civil Engineering 86, https://doi.org/10.1007/978-981-15-6233-4_36

of a site. Many researchers have performed seismic site characterization of different cities in India, site characterization of Delhi (Satyam and Rao 2007), mapping of shear wave velocity for Bangalore City (Anbazhagan and Sitharam 2008), site classification and site response of Chennai (Maheswari et al. 2010) and dynamic site characterization of Agartala City (Sil and Sitharam 2014). Earthquakes occurred in Assam during the past century like Cachar Earthquake (10 January 1869) having magnitude (Mw) of 7.5, Rangjoli Earthquake (1897) having Mw of 8.0, Imphal Earthquake (1984) having Mw of 6.0 caused major destruction. The Silchar City, which is approximately 28 and 22 km away from the epicentre of Cachar and Imphal earthquake, respectively, received considerable damages during these earthquakes along with other locations in the state.

It is known that slippage of faults beneath the surface initiates' earthquake shaking. The shaking may be amplified or reduced depending upon the soil type present at any site along with numerous factors like displacement, velocity, duration, acceleration and inertial forces. Usually, soft soil layers extending upto significant depth may amplify the stress amplitude remarkably during earthquakes. The shear wave velocity (V_S) is considered a parameter of paramount importance for site classification. The V_S profile not only represents underground stiffness change with respect to depth, but also the centre of attention for many geotechnical studies.

The primary objective of this paper is to obtain average shear wave velocity–depth profiles throughout the NIT Silchar campus. The seismic site classification has been done according to the National Earthquake Hazards Reduction Program (NEHRP) code. Determination of fundamental site period (T_f) of all sites from the average velocity of the shear wave (V_S30) till 30 m depth from ground level using Multichannel Analysis of Surface Waves (MASW) technique was obtained, and necessary recommendations about the construction of buildings have been presented in this paper.

2 Seismicity of Assam

Plate tectonics of NE India is very complex due to the convergence of Indian plate towards north-south, Eurasian plate and Indian plate collision along the Himalayan bend, continent–continent subduction along the Burmese arc which are responsible for convergence towards east-west and folding in the Indo-Burma ranges.

Recent seismicity map of NE India utilizing earthquake catalogue data from the International Seismological Centre (ISC) for 44 years (1965–2008) has shown numerous shallow and deep earthquakes having different magnitude at the Indo-Burma subduction zone. Earthquakes having lesser intensity occurred near Misumi hills on Assam.

Syntaxis Zone. The intra-plate zone between Shillong Plateau-Assam valleys also witnessed high-intensity earthquakes. Kayal (1996) suggested five major seismo-tectonic zones of NE India depending on the geological setting and distribution of epicentre of earthquake: (i) Northeast Himalayan collision zone, (ii) Indo-Burma

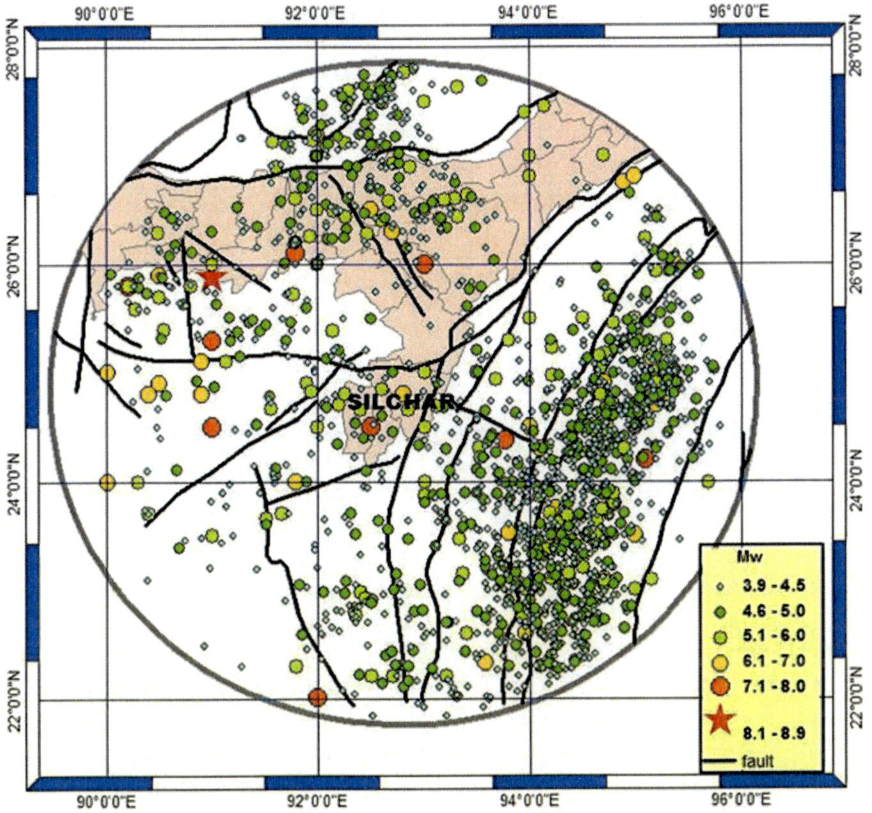

Fig. 1 Seismicity map of Silchar from the year 1762–2008, obtained from Technical Report and Atlas on remote sensing and GIS-based inputs for hazard risk vulnerability assessment of Guwahati City, Silchar, Dibrugarh Towns and Dhemaji District, Assam

subduction zone, (iii) Eastern Himalaya syntaxis zone, (iv) Shillong Plateau–Mikir massif–Assam valley and (v) Bengal basin Tripura fold belt intra-plate zones.

The seismo-tectonic map of Silchar is shown in Fig. 1. The figure shows more than 200 earthquakes having a magnitude greater than 5 occurred around the city since 1762. Among them, Cachar Earthquake (1869) of $M = 7.5$, Srimangal Earthquake (1918) of $M = 7.9$, and Imphal Earthquake (1984) of $M = 5.8$ shook the city greatly.

3 Seismic Site Classification

Seismic site classification is the basic requirement to understand the average dynamic behaviour of the soil profiles during an earthquake. Performing seismic site classification helps us to predict the vulnerability of any structures standing on the soil surface during an earthquake. A simple way to characterize the seismic behaviour

Table 1 Site classification according to V_S30 as per NEHRP design manual

NEHRP site class	V_S30	General description
A	>1500 m/s	Hard rock
B	760–1500 m/s	Firm and hard rock
C	360–760 m/s	Dense soil, soft rock
D	180–360 m/s	Stiff soil
E	< 180 m/s	Soft clay
F		Soils vulnerable to potential failure, highly organic clays, very high plasticity clays, very thick soft/medium stiff clays

of the site is by estimating the shear wave velocity of the deposit. The shear wave of the near-surface (30 m), also denoted by average shear wave velocity (V_S30), is used for site classification as recommended by NEHRP code and international building code (IBC 2000). The site classification system as per the NEHRP design manual is presented in Table 1.

NIT Silchar campus having an area of 2.52 km^2 is situated at a latitude 24° 45′ 22″ N and longitude 92° 47′ 17″ E. The campus is located almost 8 km away from the Silchar city, Assam, India. For seismic site classification of NIT Silchar campus, the whole area has been divided into 9 grids, each of them having dimension around 0.5 km × 0.5 km, as shown in Fig. 2. MASW test was conducted on every 9 grids to determine the shear wave velocity profile. The location of the nine MASW test is also shown in Fig. 2. Three numbers of confirmatory boreholes of 30 m depth were conducted within NIT Silchar near site 3, site 4 and site 8. During the digging of the borehole, it was observed that there was a variation of the type

Fig. 2 Division of NIT Silchar into 9 grids and location of MASW test sites

of soil and the thickness of each layer in the different location within NIT Silchar. Standard penetration test (SPT) was also carried out in those locations. SPT values were taken at a 1.5 m interval from GL. Undisturbed samples were collected from regular depth. Basic geotechnical tests like grain size analysis, density, plastic limit and liquid limit are performed.

4 Methods and Methodology

4.1 Multichannel Analysis of Surface Waves (MASW)

In a vertically non-homogeneous media, surface wave techniques are based on the dispersive nature of Rayleigh waves. The dispersion curve (phase velocity versus frequency) is found out from the wavefield data produced by a dynamic (active or passive) source. Wavefield data is recorded by geophones on the surface. A complex inversion process is used in layered media, which leads to an estimation of the shear wave velocity profile. During the 1980s, a group of scientists at Texas University developed Spectral Analysis of Surface Waves (SASW) test, where a two-station set-up was used for shear wave velocity measurements. During practice, some shortcomings of SASW were noticed, for which this method was not widely accepted around the globe. MASW technique overcomes most of these disadvantages (Park et al. 1999). MASW method is a developed version of the SASW method. Starting from two-station set-ups, MASW method improved to multiple-station set-ups and had more sensitive data acquisition systems. It reduced the measurement time and eased fieldwork greatly. The way dispersion analysis which was being previously performed changed in such a way that the whole recorded seismic wavefield was transformed without any preconditioning into a different space where dispersive patterns are imaged through energy accumulations. As a consequence, the multimodal nature of surface wave dispersion could be recognized much more effective than the previous two-receiver approach (Park and Ryden 2007). Because of the ability with this imaging technique, which averages out such adverse influences as near- and far-field effects, body-wave inclusions and noise surface waves—data acquisition procedures became much simpler and less sensitive to field parameters. As it became possible to extract multimodal dispersion curves from the field data, multimodal inversion techniques were subsequently used. Utilizing the 2D V_S mapping approach, many MASW case studies were undertaken in many different parts of the world during the past several years.

The overall procedure of MASW technique to generate 2D V_S profile consists of 3 steps—data acquisition, dispersion analysis and inversion. Data acquisition refers to acquiring multichannel field records from the shot. Dispersion analysis refers to the extraction of dispersion curves (phase velocity versus frequency of Rayleigh waves) from multichannel field records. The dispersion curve has been generated in this project using common midpoint (CMP) cross-correlation method (Hayashi

Fig. 3 4.5 Hz geophones being inserted into the ground surface for receiving seismic signals

2008). Finally, inversion refers to the back calculation of 1D shear wave velocity (V_S) profile, which gives theoretical dispersion curves closest to the extracted curves (one 1D shear wave velocity profile from each curve).

24 numbers of vertical geophones having a natural frequency of 4.5 Hz were arranged in a linear array. The geophones were inserted into the ground to receive the wavefields generated by an active source of 8 kg sledgehammer as shown in Fig. 3. The spacing between two consecutive geophones was kept at 3 m. The connection of two geophone cables (each having 12 numbers of geophones) with data logger and laptop is shown in Fig. 4. The measure of shear wave velocity using experimental setup is shown in Fig. 5.

The data was collected using data logger provided by "PASI", and acquisition software "GEA24" was used. The data was acquired with sampling time as 500 μs

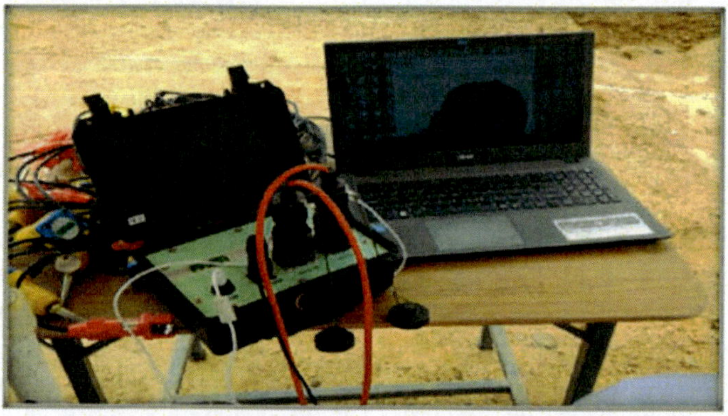

Fig. 4 Connection of 2 geophone cables with data logger and laptop

Fig. 5 Determination of in-situ shear wave velocity from MASW test

which correspond to frequency 2000 and acquisition time of 3000 ms was maintained. 10 number of stacking was taken for each data acquisition as higher stacking increases the data quality. A pre-trigger time of 30 ms was kept as a precaution measure for the time gap between the actual time of creation of seismic source by hitting the steel plate with the sledgehammer and activation of receiver geophone to record the data.

The analysis of field data was performed by SeisImager software. SeisImager software consists of three individual software—Pickwin, Wave Eq and Geoplot software. Several tests set-up details (i.e. geophone spacing, source interval, receiver interval, location of first source point and location of the first receiver) were fed into Pickwin software. Using Pickwin software, multichannel records (shot gathers) for the different shot position can be viewed. Multichannel records (shot gathers) at site 3 for source position taken at the middle of the stretch between 12 and 13th geophone is shown in Fig. 6. Any manual error during the field test can be identified. Wave Eq software generates a dispersion curve (phase velocity vs. frequency) after analysing all these inputs. The generation of a dispersion curve is the most critical step in all MASW methods. A typical dispersion curve generated for site 3 is shown in Fig. 7. All the low-quality data on dispersion curve were discarded. Signal/Noise (*S*/*N*) ratio was used as the parameter to judge the quality of the field data. A data on dispersion curve having *S*/*N* ratio more than 0.75 is considered to be sufficiently good quality. There are other methods, which can be used to generate proper dispersion curve, i.e. setting up the minimum and maximum frequency, deleting higher mode data, deleting low-quality data, etc. After modifying the dispersion curve, "initial model" of V_S profile has been generated. This model provides one-dimensional shear wave velocity profiles along the whole spread length of 69 m. Then the inversion process has been used over that "initial model" to find actual 2D shear wave velocity profile.

It seems reasonable that the centre of the receiver spread be the most appropriate point because the analysed shear wave velocity profile represents an average property within the spread length. Two-dimensional shear wave velocity profile has been presented in Fig. 8 for the location site 3.

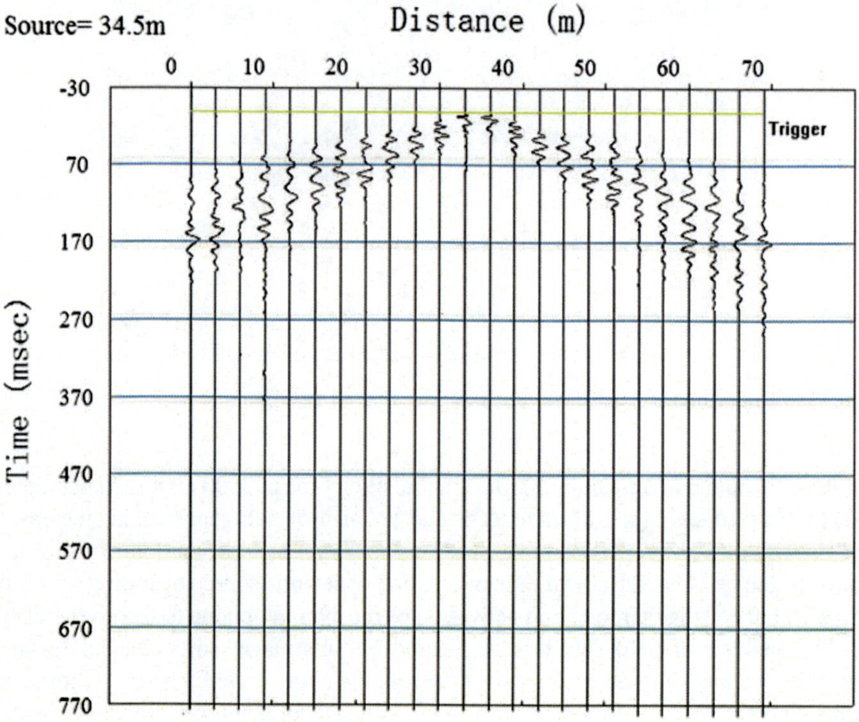

Fig. 6 Multichannel records (Shot Gathers) at site 3, shot position taken at the middle of the stretch between 12 and 13th geophone

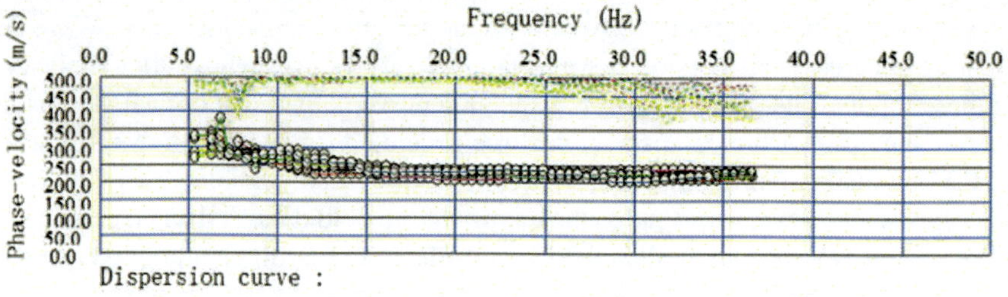

Fig. 7 Dispersion curve obtained using field data generated by WAVE EQ software

4.2 *Average Shear Wave Velocity (V_S30) Distribution*

For investigating a site, minimum 30 m depth is considered for an average depth of investigation and exact site classification (Hollender et al. 2018). A vast majority of the site-impact research in earthquake ground movements depends mostly on the properties of the upper 30 m soil. The site classes are characterized considering the velocity of shear wave upto a depth of 30 m, meant by V_S30. If no estimations of

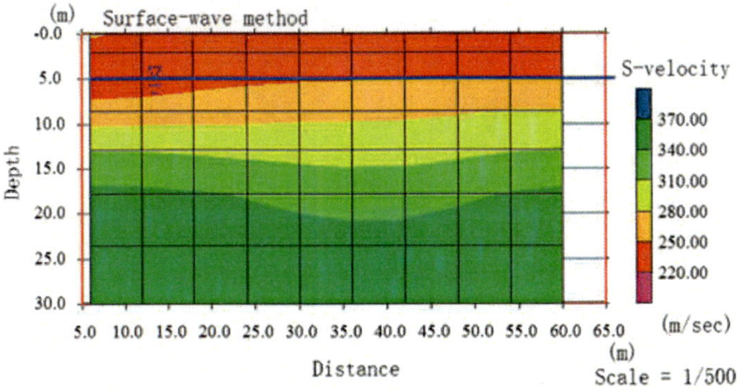

Fig. 8 Shear wave velocity profile at site 3

V_S until 30 m are possible to conduct, standard penetration test (N) and undrained shear strength (S_u) are considered for site classification (IBC 2009). The zones for site classification are separated into grids by a network framework for evaluating the impacts of site conditions by allocating indicative soil profiles at the central position of every grid. The average shear wave velocity (V_S30) was computed using Eq. 1.

$$V_{S30} = \frac{\sum_{i=1}^{n} d_i}{\sum_{i=1}^{n} \frac{d_i}{V_{Si}}} \quad (1)$$

d_i Thickness of ith soil layer from GL.
V_{Si} Shear wave velocity of the ith soil layer from ground level.

4.3 *Fundamental Site Period Determination*

The natural period of any building refers to the time (in seconds) taken by the building to vibrate forward and backwards naturally. Each site containing particular soil strata has a fundamental period of vibration. Hard rock has a lesser time period than soft sediments due to high shear wave velocity. Resonance will occur if the fundamental site period of ground motion matches with the natural period of a building. Resonance refers to the tendency of any system to oscillate with greater amplitude at some specific frequencies. It occurs when a vibrating system is given an extra push at its natural period. Vibrations of the system increase reasonably in response to even rather small pushes and, in fact, its accelerations may increase as much as four or five times.

Kramer (1996) suggested the following expression of fundamental site period (T_f), showing that thickness and shear wave velocity of the soil layer govern this factor:

$$T_f = \frac{4 * H}{V_{S30}} \quad (2)$$

where

H the total thickness of soil layer considered i.e. 30 m.
V_S30 the average shear wave velocity of the overburden soil.

5 Results and Discussions

The seismic site classification has been done based on the average shear wave velocity (V_S30) of the top 30-m depth from ground level. Vs30 of each of 9 sites has been tabulated below. The seismic site classification was done based upon NEHRP method. The fundamental site period (T_f) corresponding to the first mode of vibration of the soil deposit was calculated for each site. Shear wave profile of each site represents the V_S30 of that corresponding grid. The boreholes mentioned were conducted at the middle of the MASW test profiles.

5.1 Shear Wave Velocity and Geotechnical Properties

The shear wave velocity profile at each location is given in Table 3. The shear wave velocity, borelog data and geotechnical properties at site 3 are discussed in this section. The shear wave velocity profile obtained from MASW test at site 3 is shown in Fig. 8. A 30 m depth borehole was constructed at the middle of the profile and at a distance of 30 m from the initial point of the spread length of MASW profile. 5 numbers of prominent layers were found during the digging of the borehole. The borelog data of site 3 is shown in Fig. 9. SPT data were taken at a regular interval of 1.5 m and conducted upto 19.5 m depth. During the borehole construction, it was observed that the depth of the groundwater table lies at a depth of 5 m from ground level. Basic geotechnical properties of soil-like cohesion (c), angle of internal friction (φ), density (ρ) and IS classification (IS: 2720 Part 4 1985-RA-2010) of soil are given in Table 2.

The top 5 m depth comprises of stiff clayey sand with traces of gravel and has an average shear wave velocity 230 m/s. The second layer comprises of medium to dense silty sand of 4 m thickness and having shear wave average velocity of 260 m/s. The thickness of the third layer is 5 m and comprises of very stiff silty clay having an average shear wave velocity of 310 m/s. The fourth layer comprises of dense fine silty sand of 5 m thickness having shear wave velocity 330 m/s. The fifth layer extending upto 30 m depth and comprises of dense coarse silty sand having a shear wave velocity of 360 m/s. The soil profile obtained using MASW shear wave velocity matched well with the borehole soil profiles.

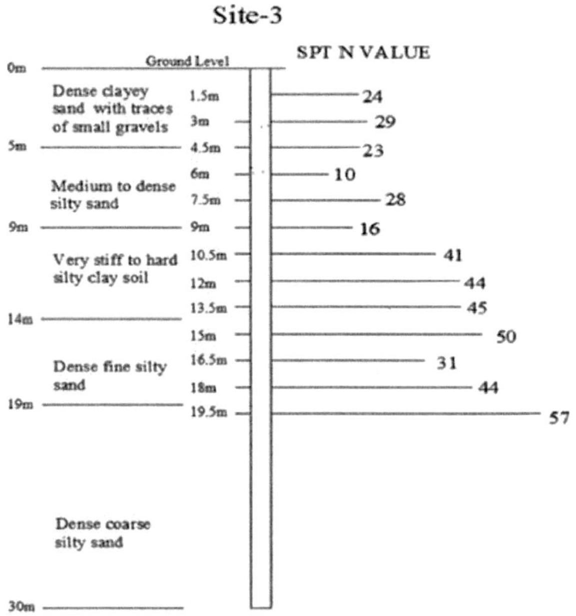

Fig. 9 Typical borelog data at site

Table 2 Geotechnical properties of soil at site 3

Layer No.	Soil type	Thickness (m)	*C* and φ (kg/cm^2 and °)	Density (kN/m^3)	IS classification
1	Reddish-brown-coloured clayey sand with gravels	5	0.15 and 30.3°	19.3	SC
2	Reddish-brown-coloured silty sand	4	0.365 and 27.6°	19.1	SM
3	Ash-coloured silty clay soil	5	1.21 and 7°	19.5	CL
4	Reddish-coloured fine sitly sand	5	0.224 and 29.6°	20.0	SM
5	Yellowish-coloured coarse silty sand	11	0.265 and 32.3°	20.2	SM

The soil at site 3 predominantly consists of sand with a high percentage of silt and clay. At site-4 and site-8, clayey and silty soil is predominant upto 30 m depth. There is a large variation in the type of soil at NIT Silchar. In most of the site, alternate layer of sand and clay deposit exist. A high percentage of silt fraction is present in the soil deposit. A gravelly soil layer of around 0.5 m thickness is present almost everywhere within NIT Silchar and is located at a depth of 2–3 m. The groundwater table is located at a depth of 4–5 m from the ground surface.

Table 3 Determination of site classification using NEHRP (BSSC 2009)

Sl. No.	Site No.	Thickness (m)	Shear wave velocity (m/s)	V_S30 (m/s)	Site class
1	Site 1	2	250	313	D
		2	375		
		2	275		
		2	210		
		3	260		
		8	340		
		11	360		
2	Site 2	1	220	285	D
		5	260		
		8	280		
		11	300		
		5	310		
3	Site 3	5	240	306	D
		4	260		
		5	310		
		5	330		
		11	360		
4	Site 4	5	220	307	D
		4	260		
		5	350		
		7.5	330		
		8.5	380		
5	Site 5	8	220	280	D
		2	260		
		2	300		
		7.5	350		
		10.5	300		
6	Site 6	2	250	257	D
		6	200		
		5	260		
		17	285		
7	Site 7	2	240	275	D
		3	210		
		5	240		
		3	275		
		17	310		

(continued)

Table 3 (continued)

Sl. No.	Site No.	Thickness (m)	Shear wave velocity (m/s)	V_S30 (m/s)	Site class
8	Site 8	2	245	280	D
		3.5	195		
		2	227		
		3	262		
		5	295		
		14.5	330		
9	Site 9	2	250	256	D
		6	200		
		2	240		
		3	262		
		17	287		

5.2 *Seismic Site Classification of NIT Silchar*

The average shear wave velocity is determined for all the 9 zones at NIT Silchar. The results are given in Table 2. The seismic site classification of NIT Silchar has been done according to NEHRP categorization mentioned in BSSC, 2009. The seismic site classification zones for any site are directly related to geological settings and soil thickness of that site. NIT Silchar campus comes with no exception. The stiffness of a particular soil layer is directly proportional to shear wave velocity (V_S) of that layer. As the shear wave velocity as seen in Table 2 increases with depth, the stiffness of soil also increases the same way. The average shear wave velocity (Vs30) of NIT Silchar campus comes in the range of 250–320 m/s. Thus, all the zones come under site class D according to NEHRP (BSSC 2009). The site class D refers to the presence of stiff soil. The SPT data also denotes the presence of stiff soil in NIT Silchar. Classification done by average shear wave velocity (V_S30) also matches with SPT data. As shear wave velocity indicates the stiffness of soil, therefore it can be predicted that soil in site 1 ($V_S30 = 318$ m/s), site 3 ($V_S30 = 306$ m/s) and site 4 ($V_S30 = 307$ m/s) is stiffer compared to other six sites. On the other hand, soil in site 6 and site 9 which has almost average shear wave velocity of 256 m/s are less stiff as compared to other seven sites.

5.3 *Fundamental Site Period Determination at NIT Silchar*

The fundamental site periods of soil deposit at NIT Silchar were determined using Eq. 2. The period ranges from 0.38 to 0.47 s. The fundamental site period of nine sites is given in Table 4.

Table 4 Determination of fundamental site period

Sl. No.	Site	V_{S30}(m/s)	Fundamental site period T_f (s)
1	Site 1	313	0.38
2	Site 2	285	0.42
3	Site 3	306	0.39
4	Site 4	307	0.39
5	Site 5	280	0.43
6	Site 6	257	0.47
7	Site 7	275	0.44
8	Site 8	280	0.43
9	Site 9	256	0.47

For steel and concrete moment resisting framed buildings (having less or equal to 12 storeys), the natural period of the structure can be calculated by the following equation (Di Julio 2001):

$$T_N = 0.1 * N \tag{3}$$

where

N Number of storeys present in the building.

Natural frequency of a building can be a deciding factor in determining seismic vulnerability of buildings having different storeys. All objects have their own natural frequency. When earthquake occurs, a building will sway back and forth at its natural period. If the frequency of ground matches with the natural frequency of the building, resonance will occur. Generally, hard ground has short period of vibration due to more V_S and soft ground has large period of vibration due to lesser V_S. During resonance, the building will undergo the largest oscillations possible and suffer the greatest damage. High-frequency waves tend to have higher amplitudes of acceleration but smaller amplitudes of displacement, compared to those of low-frequency waves. For this reason, high rise buildings are suggested to be built on hard bedrock having high frequency and low rise buildings on soft soil.

It is obvious from Table 4, that buildings having a natural period of 0.38–0.47 are most vulnerable at NIT Silchar. So, from Eqs. 2 and 3, it is obvious that any four-storey building can be in trouble during earthquakes. The natural time period of five-storey building; i.e. 0.5 s is not far away from 0.47 s. So, five-storey building should also be avoided in the campus. Hence, the soil conditions at NIT Silchar pose a potential threat during an earthquake scenario mostly to buildings of 4 and 5 storeys.

As minimum fundamental site period is 0.37, the construction of many buildings having 1, 2 and 3 storeys, i.e. academic building, administrative building (old), guesthouse, Kendriya Vidyalaya, Estate Engineering Department, students activity centre (SAC) building, fair price shop (FPS), State Bank of India (NITS branch), health centre, professor's quarter, etc., is justified.

5.4 Cracks in Building

Cracks developed during the earthquake in four-storey buildings also indicate that these buildings are the most vulnerable during resonance in an earthquake. Cracking of concrete will occur whenever the stress exceeded the tensile strength of concrete. Cracking is generally undesirable in concrete as it affects the durability of structures in aggressive weather conditions and leads to corrosion of the embedded steel. The IS 456 recommends a maximum limit of 0.3 mm width crack on the assessed surface width for concrete structure subjected to "mild" exposure condition. Considering all buildings of NIT Silchar as falling within "mild" exposure condition, only cracks having more than 0.3 mm width were considered. During the project, we carefully observed the condition of twenty buildings in NIT Silchar, among them, six are four-storey buildings. We observed a total of 24 cracks in beams and columns, ranging from 0.3 to 0.6 mm in width. Around 70% of these cracks were found in 4-storey reinforced concrete structures which were much more than those observed in buildings having less than 4 storeys.

6 Conclusion

We have classified NIT Silchar campus based on the direct measurement of average shear wave velocity values till 30 m depth (V_S30) suggested by NEHRP. A series of MASW has been carried out at 9 locations within the campus. The following are the conclusions from this study:

1. As per the NEHRP classification system, the NIT Silchar campus comes under site class "D". Site class "D" corresponds to V_S30 lying within a range of 180–360 m/s. Class D refers to the presence of stiff soil at NIT Silchar.
2. The fundamental site period (T_f) obtained from MASW varies from 0.38 to 0.47 s. So, any building having a natural frequency within 0.38–0.47 s will resonate with the site during the earthquake.
3. Four- and five-storey steel and concrete structures may face severe damages during earthquakes. It is recommended to construct building either upto 3 storey or above 5 storey within NIT Silchar campus.
4. During the project, the condition of 20 buildings in NIT Silchar was observed. Among them, six buildings have 4 storied. It is observed that a total 24 cracks in beam and column, out of which four-storied building has more cracks than those observed in 1-, 2- and 3-storey buildings.

References

Anbazhagan P, Sitharam TG (2008) Mapping of average shear wave velocity for Bangalore region: a case study. J Environ Eng Geophys 13(2):69–84

BSSC (2009) NEHRP recommended provisions for the development of seismic regulations for new buildings and other structures, 2009 edn. Part 1: Provisions, Report No. FEMAP-750. Building Seismic Safety Council for the Federal Emergency Management Agency, Washington, D.C., USA

Di Julio RM (2001) Linear static seismic lateral force procedures. In: Naiem F (ed) The seismic design handbook, 2nd edn. Kluwer Academic Publishers, Boston, pp 247–274

Hayashi K (2008) Development of the surface-wave methods and its application to site investigations

Hollender F, Cornou C, Dechamp A, Oghalaei K, Renalier F, Maufroy E, Burnouf C, Thomassin S, Wathelet M, Bard P-Y, Boutin V, Desbordes C, Douste-Bacqué I, Foundotos L, Guyonnet-Benaize C, Perron V, Régnier J, Roullé A, Langlais M, Sicilia D (2018) Characterization of site conditions (soil class, VS30, velocity profiles) for 33 stations from the French permanent accelerometric network (RAP) using surface-wave methods. Bull Earthq Eng 16(6):2337–2365. https://doi.org/10.1007/s10518-017-0135-5

IBC (2000) International Building Code-2000, 5th edn. International Code Council Inc., Falls Church

IBC (2009) International Building Code. International Codes Council

Kayal JR (1996) Earthquake source process in Northeast India: a review. Himalayan Geol 17:53–69

Kramer SL (1996) Geotechnical earthquake engineering. Prentice Hall, Upper Saddle River, NJ

Maheswari RU, Boominathan A, Dodagoudar GR (2010) Seismic site classification and site period mapping of Chennai city using geophysical and geotechnical data. J Appl Geophys 152–168

Park C, Ryden N (2007) Historical overview of the surface wave method. In: Symposium on the application of geophysics to engineering and environmental problems, pp 897–909

Park CB, Miller RD, Xia J (1999) Multichannel analysis of surface waves. Geophysics 64:800–808

Satyam N, Rao KS (2007) Seismic site characterization in Delhi region using multi channel analysis of shear wave velocity (MASW) testing. Int J Geotech Eng 13:167–183

Sil A, Sitharam TG (2014) Dynamic site characterization and correlation of shear wave velocity with standard penetration test ‘N’ values for the city of Agartala, Tripura State, India. Pure Appl Geophys 1859–1876

Technical Report and Atlas on remote sensing and GIS based inputs for hazard risk vulnerability assessment of Guwahati city, Silchar, Dibrugarh towns and Dhemaji district, Assam, 2014, vol II, Silchar town, India (2014)

Comparative Study of 1D, 2D and 3D Ground Response Analysis of Pond Ash from Odisha Under Different Earthquake Motions

M. V. Ravi Kishore Reddy, Supriya Mohanty, and Rehana Shaik

Abstract It was attempted in the current study to study the one-dimensional (1D), two-dimensional (2D) and three-dimensional (3D) ground response of pond ash collected from Odisha under different earthquake motions. 1D ground response of pond ash deposit was studied by employing equivalent linear and nonlinear analysis methods, viz. SHAKE2000 and Cyclic1D software, respectively. 2D and 3D seismic response analysis of pond ash was investigated by employing Plaxis2D and Plaxis3D, respectively. Response of the pond ash deposit when subjected to Northeast India earthquake (Mw-7.5) and Nepal earthquake (Mw-7.8) was studied here in this study. Results of the response analyses were presented in terms of acceleration, displacement, excess pore pressure and excess pore pressure ratio. It was attempted to perform a comparative study on variation of stated parameters over 1D, 2D and 3D ground response analysis performed in this study. Variation in peak ground acceleration of 1D over 2D and 3D seismic response analysis of pond ash under Northeast India earthquake was 1.21 times and 1.46 times, respectively; whereas under Nepal earthquake was 0.72 times and 0.92 times, respectively. Maximum peak ground displacement was noticed in case of 3D ground response analysis. Pond ash was not susceptible to liquefaction when got excited under both the earthquakes.

Keywords 1D, 2D, 3D ground response analysis · Pond ash · NE India earthquake · Nepal earthquake · Liquefaction

M. V. R. K. Reddy · R. Shaik
Earthquake Engineering Research Centre, International Institute of Information Technology Hyderabad, Hyderabad, Telangana 500032, India
e-mail: meegada.reddy@research.iiit.ac.in

R. Shaik
e-mail: rehana.s@iiit.ac.in

S. Mohanty (✉)
Department of Civil Engineering, Indian Institute of Technology (Banaras Hindu University) Varanasi, Varanasi, Uttar Pradesh 221005, India
e-mail: supriya.civ@iitbhu.ac.in

M. Latha Gali and P. Raghuveer Rao (eds.), *Geohazards*, Lecture Notes in Civil Engineering 86, https://doi.org/10.1007/978-981-15-6233-4_37

1 Introduction

In India, coal ash has been generated in large quantity due to being dependent on coal-based thermal power plants for decades to generate electricity. Ash generation is more because of the use of low grade Indian coal that produces 30–45% of ash content when compared with imported coal which produces 10–15% ash. Utilization of generated coal ash is about 63.28% in India for the year 2016–2017 (Central Electricity Authority, India 2017). Out of various modes of utilization of coal ash, raise of ash dyke/pond (accounts for 7% of total utilized coal ash) to deposit coal ash is prevalent in most of the states of India. Structure constructed for wet disposal of two kinds of coal ash (based on collection at electrostatic precipitator (i.e. fly ash) and at boiler (i.e. bottom ash)) is ash pond. Ash stored in ash pond is referred as pond ash. Seismic response of typical pond ash deposit is prime area of interest in this study. Contamination of elements of life like land, water and air by the disposal of coal ash was experienced; structural failures like excessive settlement, slope stability failure and liquefaction (Mohanty and Patra 2015a) were observed in the past. Especially, liquefaction is a major problem which may be experienced by saturated loose deposited ash when subjected to loads that cause sudden change in stress condition (usually dynamic loads like earthquake and storm wave).

Geotechnical properties of various kinds of coal ash materials and the properties which were similar or lie in the range to that of few kinds of soil were reported by many researchers (Bera et al. 2007; Das and Yudhbir 2005; Kaniraj and Gayathri 2004; Mohapatra and Kanungo 1997; Pandian 2004). Employability of coal ash instead of soil has been successful for various applications such as embankment material (Mohanty and Patra 2014), backfill material, filter material (Mohanty and Patra 2015b), highways (Kim et al. 2005) and stowing of underground mines (Mishra and Das 2010). To improve the stiffness and liquefaction resistance of coal ash embankment, coal ash reinforced with geosynthetics or geosynthetic fibres/mesh elements which were randomly distributed was proved to be effective (Boominathan and Hari 2002; Vijayasri et al. 2016). Multidiscipline utilization of coal ash (Blissett and Rowson 2012) needs to be encouraged to conserve natural resources like sand/soil.

1.1 Literature Review

Ground response analysis of soil/ash deposits can be performed either by model/shake table tests or numerical analysis (Singh et al. 2008; Mohanty and Reddy 2016; Reddy et al. 2018). Singh et al. (Singh et al. 2008) performed small-scale shake table (1.05 m × 0.60 m × 0.60 m) tests under one-dimensional horizontal harmonic excitation with amplitude ranging from 0.1 to 0.5 g for evaluation of liquefaction potential of pond ash. It was concluded that no case of liquefaction potential was observed both from model tests and methods based on field tests. Ground response analysis of soil/ash deposits was performed in the past to predict site amplification (Raju et al. 2004). Liquefaction potential of pond ash is governed by several factors like location of water table (Mohanty and Patra 2016a); earthquake magnitude (Mohanty and Patra 2016b); grain size or morphology of pond ash (Singh et al. 2008); density of pond ash, confining pressure, location of sample collected (Jakka et al. 2010); seismicity of the region, etc.

It was observed from past studies that Talcher pond ash site was prone to liquefaction when subjected to earthquakes of magnitude greater than 7.5 (Mohanty and Patra 2016b). Also, it was susceptible to liquefaction when fully saturated and no liquefaction was noticed under the conditions of existing conditions of water table (Mohanty and Patra 2016a; Khanna and Mohanty 2017). No studies have been reported on 3D analysis of Talcher pond ash/site, and hence, it was attempted in the current study to examine variation of response of pond ash against different earthquakes.

Necessity of seismic response analysis of pond ash before its usage is important and of good scope for researchers all the time because of the variation of properties of ash produced at times due to operational, handling and environmental changes of coal ash. Hence, in this study, 1D, 2D and 3D seismic response analysis of Talcher pond ash was performed by employing some numerical modelling media like Cyclic1D, SHAKE2000, PLAXIS 2D and PLAXIS 3D.

1.2 Materials Considered in This Study

Pond ash from ash pond of Talcher Thermal Power Station (coal-based power plant), Angul district, Odisha was collected. Ash pond site falls under the category of seismic zone III (moderate earthquake zone, $a_{max} = 0.16$ g). Pond ash sample was collected at 0.2 m depth from surface near outflow point of ash pond and lies in the range of silt size particles. Percentage of sand and silt range particles in Talcher pond ash sample was 48.8%, 51.2%, respectively. According to Jakka et al. (2010) collected pond ash sample can be referred as fine ash (Table 1).

Table 1 Properties of Talcher pond ash

S. No.	Property	Value
1	Specific gravity	1.85
2	Unit weight	1.065 g/cc
3	Cohesion	22.45 kPa
4	Angle of internal friction	36.3°
5	Plastic limit	Non-plastic
6	Void ratio	0.746
7	Compression index, Cc	0.02450
8	Reloading index, Cs	2.87E−3

Specific gravity of Talcher pond ash is in the range of Indian coal ash samples (Sridharan 2001). Short brief of numerical analysis employed in this study and response in terms of various parameters such as acceleration (g), displacement (m), excess pore pressure (kPa) and excess pore pressure ratio have been presented in succeeding sections.

2 Numerical Methodology

Advantage of numerical analysis over model testing is its adaptability for non-periodic motions which is difficult in model/shake table tests. Different numerical modelling media, viz. Cyclic1D, SHAKE2000, PLAXIS 2D and PLAXIS 3D were employed for study on equivalent linear and nonlinear response of pond ash over 1D, 2D and 3D domain. Ground response analysis was performed in time domain, i.e. time history analysis which means analysis of dynamic response of pond ash deposit at each instant of time.

Cyclic1D and SHAKE2000 were employed to simulated 1D dynamic response of pond ash. 1D response in terms of acceleration, displacement in lateral direction was determined. It was assumed that boundary in horizontal direction is extended infinite; number of elements in Cyclic1D were limited to 400 in the current study. PLAXIS 2D and 3D was employed for 2D and 3D seismic response analysis of pond ash. It facilitates in presenting detailed computational results. In PLAXIS, numerical modelling is done in five modes, i.e. soil, structure, mesh, flow conditions and staged construction mode. Displacement multiplier was opted in this analysis to simulate earthquake effect, and hence prescribed displacements were defined at the bottom of the soil domain. Size of pond ash domain considered for 2D and 3D analysis was 80 m × 20 m and 80 m × 3 m × 20 m, respectively.

Pond ash deposit was discretized into number of elements for 1D, 2D and 3D analysis such that it satisfies criteria according to Eq. 1 (Kuhlemeyer and Lysmer 1973; Lysmer et al. 1975).

$$\lambda/8 < l_e < \lambda/5 \tag{1}$$

where l_e is maximum dimension of any element, λ is least wavelength for the analysis $= V_s/f$, where V_s is least shear wave velocity considered for the study (m/s), f is frequency (Hz).

Two input motions were considered for 1D, 2D and 3D seismic response analysis of pond ash, viz. Nepal earthquake (Mw: 7.8) and Northeast India earthquake (Mw: 7.5) (NE India earthquake). Peak ground acceleration of Nepal earthquake motion was 0.1609 g and that of Northeast India earthquake was 0.174 g. Typical view of generated mesh, workspace of Cyclic1D, SHAKE2000 and PLAXIS 2D and 3D was presented in Figs. 1, 2, 3 and 4. Typical acceleration time history of both the earthquakes was presented in Fig. 5.

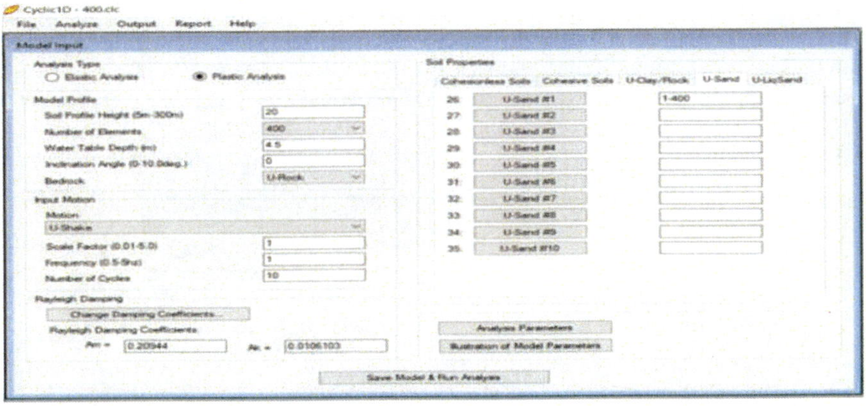

Fig. 1 Typical workspace of Cyclic1D

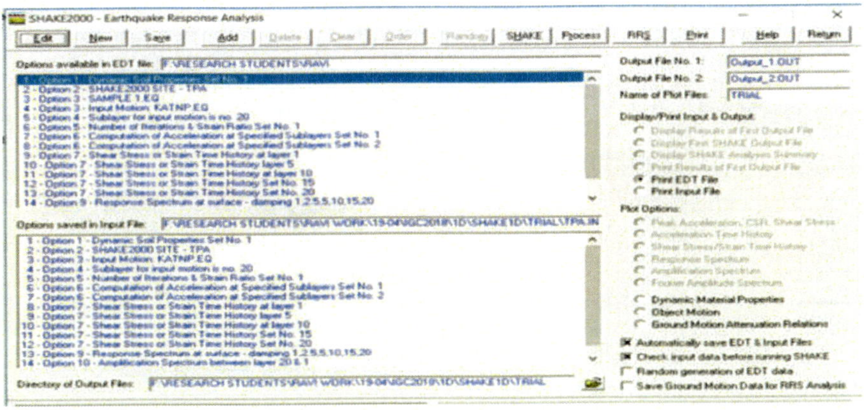

Fig. 2 Typical workspace of SHAKE2000

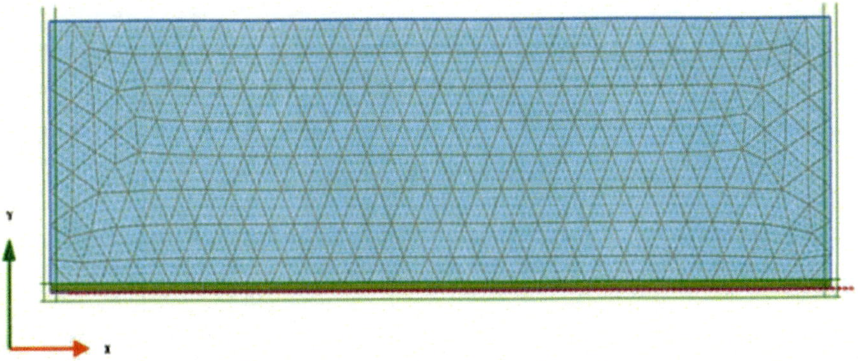

Fig. 3 Typical view of 2D generated mesh of pond ash domain (80 m × 20 m) using PLAXIS 2D

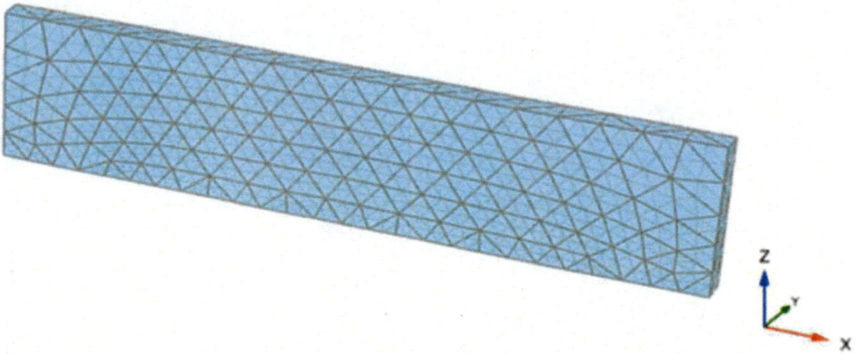

Fig. 4 Typical view of 3D generated mesh of pond ash domain (80 m × 3 m × 20 m) using PLAXIS 3D

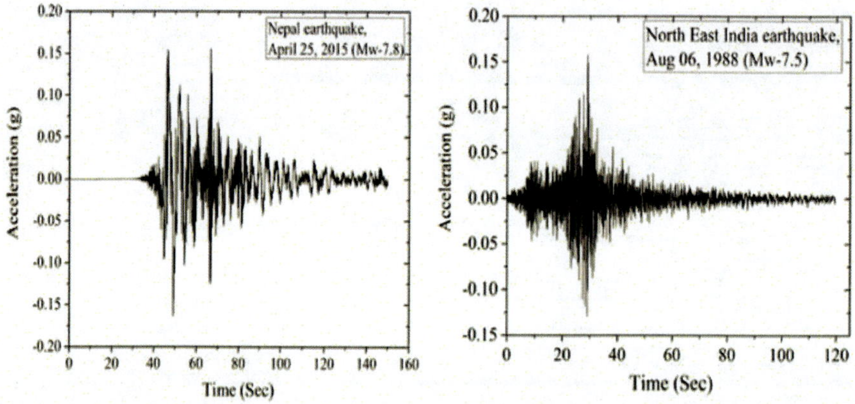

Fig. 5 Acceleration time history of Nepal earthquake (Mw: 7.8) and NE India earthquake (Mw: 7.5)

3 Results and Discussion

Seismic response analysis of pond ash when excited under Nepal and NE India earthquake was evaluated in terms of acceleration (g), displacement (m), excess pore pressure (kPa) and excess pore pressure ratio. Comparative study on results obtained from 1D, 2D and 3D analysis was presented in the following sections.

3.1 Evaluation of Acceleration Response

Peak ground acceleration values in case of 1D analysis (SHAKE2000 and Cyclic1D) were observed to be 0.18 and 0.17 g for Nepal earthquake and 0.19 and 0.17 g for NE

Table 2 Variation of acceleration (g) against depth for 1D (SHAKE2000 and Cyclic1D) seismic response analysis of Talcher pond ash

Depth (m)	Nepal Eq (M_w: 7.8)		NE India Eq (M_w: 7.5)	
	SHAKE2000	Cyclic1D	SHAKE2000	Cyclic1D
1	0.180	0.166	0.193	0.166
2	0.180	0.166	0.192	0.165
3	0.179	0.166	0.190	0.162
4	0.178	0.166	0.185	0.158
5	0.178	0.166	0.178	0.153
6	0.177	0.165	0.171	0.150
7	0.176	0.165	0.162	0.150
8	0.175	0.164	0.159	0.144
9	0.173	0.163	0.158	0.140
10	0.171	0.162	0.159	0.137
11	0.168	0.161	0.163	0.133
12	0.166	0.160	0.163	0.130
13	0.164	0.159	0.157	0.126
14	0.162	0.157	0.144	0.122
15	0.160	0.156	0.149	0.117
16	0.158	0.155	0.152	0.117
17	0.156	0.153	0.151	0.120
18	0.153	0.151	0.162	0.121
19	0.151	0.149	0.196	0.123
20	0.148	0.147	0.218	0.128

India earthquake as base excitation. Observed values of peak ground acceleration for 2D and 3D analysis were 0.231 and 0.18 g for Nepal input motion, whereas 0.137 and 0.114 g were observed for NE India input motion for 2D and 3D analysis, respectively. Variation of acceleration values of 1D, 2D and 3D seismic response analysis were presented in Tables 2 and 3. Typical acceleration time history at ground surface for Nepal input motion using Cyclic1D and for NE India input motion using SHAKE2000 were presented in Figs. 6 and 7. Similarly, typical acceleration time history at ground surface for Nepal input motion using PLAXIS 2D and for Northeast India input motion using PLAXIS 3D were presented in Figs. 8 and 9, respectively.

3.2 *Evaluation of Displacement Response*

Peak ground displacement values in case of 1D analysis (Cyclic1D) were observed to be 0.0097 m and 0.0076 m for Nepal and NE India earthquakes, respectively,

Table 3 Variation of acceleration (g) against depth for 2D and 3D (PLAXIS 2D and 3D) seismic response analysis of Talcher pond ash

Depth (m)	2D analysis		Depth (m)	3D analysis	
	Nepal Eq (M_w: 7.8)	NE India Eq (M_w: 7.5)		Nepal Eq (M_w: 7.8)	NE India Eq (M_w: 7.5)
0	0.231	0.137	0	0.18	0.114
3.75	0.24	0.133	3.44	0.18	0.12
6.75	0.241	0.117	6.89	0.182	0.125
10	0.252	0.123	10.32	0.168	0.116
13.75	0.333	0.135	13.68	0.158	0.088
17.5	0.388	0.158	16.96	0.164	0.082
20	0.378	0.158	20	0.163	0.093

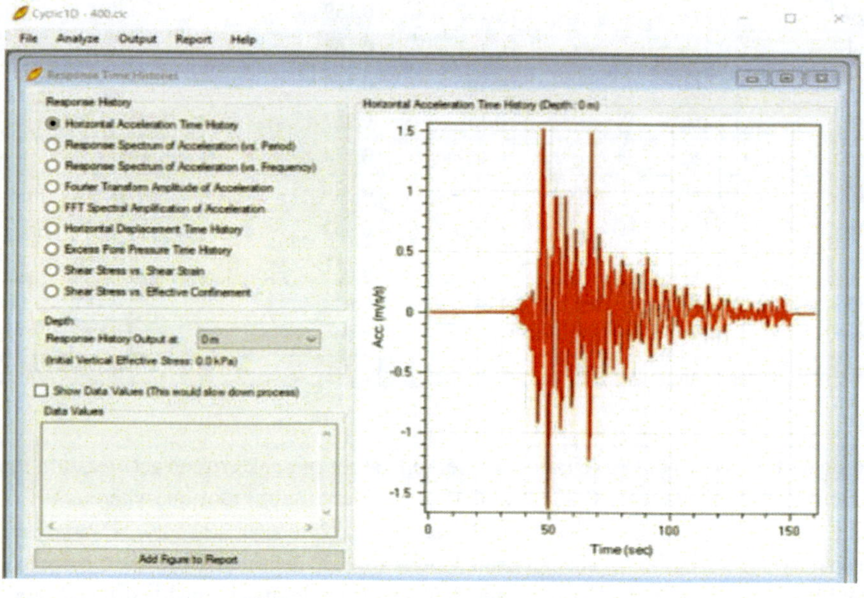

Fig. 6 Typical view of acceleration time history at ground surface under excitation of Nepal earthquake input motion (Cyclic1D)

as base excitation. Observed values of peak ground displacement for 2D and 3D (PLAXIS 2D and 3D) analysis were 0.228 m and 0.801 m in case of Nepal input motion, whereas a displacement of 2.424 m and 5.522 m was observed in case of NE India input motion for 2D and 3D analysis, respectively. Variation of displacement values of 1D, 2D, 3D ground response analysis were presented in Tables 4 and 5. Typical view of shadings of variation of total displacements was presented in Figs. 10 and 11.

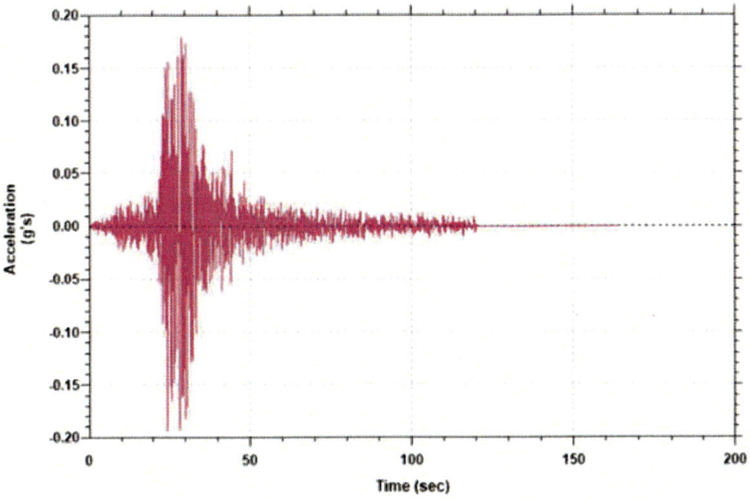

Fig. 7 Typical view of acceleration time history at ground surface under excitation of NE India earthquake (SHAKE2000)

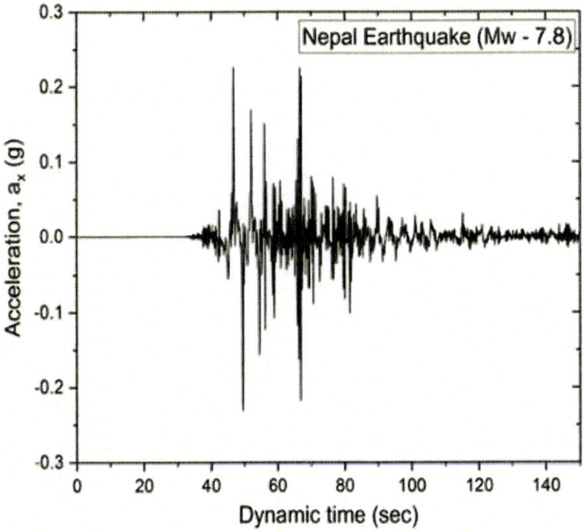

Fig. 8 Typical view of acceleration time history at ground surface for Nepal input motion (2D analysis using PLAXIS 2D)

3.3 Excess Pore Pressure Variation

Maximum excess pore pressure values for Nepal and Northeast input motion in 1D analysis were observed to be 38.72 kPa and 33.08 kPa, respectively. In case of 2D ground response analysis, maximum excess pore pressure values were observed as

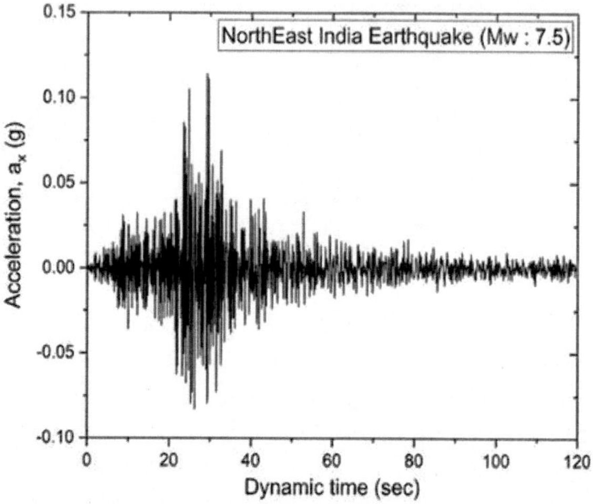

Fig. 9 Typical view of acceleration time history at ground surface for NE India input motion (3D analysis using PLAXIS 3D)

Table 4 Variation of displacement (m) against depth for 1D (Cyclic1D) seismic response analysis of Talcher pond ash

Depth (m)	1D analysis	
	Nepal Eq (M_w: 7.8)	NE India Eq (M_w: 7.5)
1	0.0097	0.0077
2	0.0097	0.0076
3	0.0096	0.0076
4	0.0095	0.0075
5	0.0094	0.0073
6	0.0092	0.0071
7	0.009	0.007
8	0.0087	0.0066
9	0.0084	0.0063
10	0.0081	0.0060
11	0.0076	0.0056
12	0.0072	0.0052
13	0.0066	0.0047
14	0.0060	0.0042
15	0.0054	0.0037
16	0.0047	0.0031
17	0.0039	0.0025
18	0.0030	0.0019
19	0.0021	0.0013
20	0.0011	0.0006

Table 5 Variation of displacement (m) against depth for 2D and 3D (PLAXIS 2D and 3D) seismic response analysis of Talcher pond ash

Depth (m)	2D analysis		Depth (m)	3D analysis	
	Nepal Eq (M_w: 7.8)	NE India Eq (M_w: 7.5)		Nepal Eq (M_w: 7.8)	NE India Eq (M_w: 7.5)
0	0.228	2.424	0	0.801	5.522
3.75	0.179	2.443	3.44	0.799	5.576
6.75	0.133	2.453	6.89	0.793	5.632
10	0.087	2.438	10.32	0.78	5.689
13.75	0.037	2.481	13.68	0.755	5.745
17.5	0.039	2.489	16.96	0.731	5.8
20	0.073	2.492	20	0.731	5.85

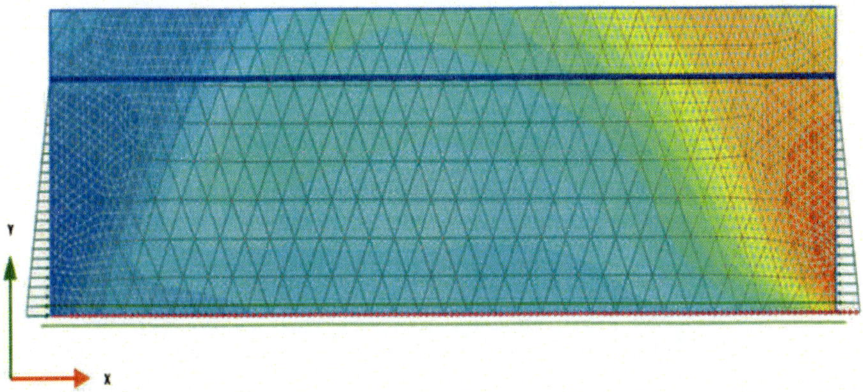

Fig. 10 Typical view of shadings of total displacements for Nepal earthquake motion (2D analysis)

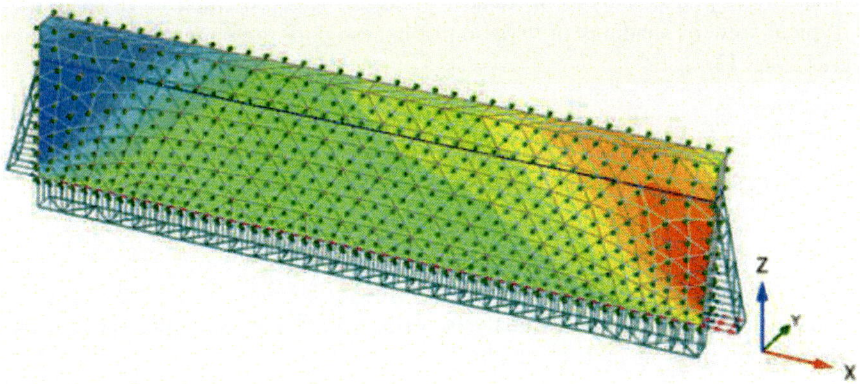

Fig. 11 Typical view of shadings of total displacements for Nepal earthquake motion (3D analysis)

Table 6 Variation of excess pore pressure (kPa) against depth for 1D (Cyclic1D) ground response analysis of Talcher pond ash

Depth (m)	1D analysis	
	Nepal Eq (M_w: 7.8)	NE India Eq (M_w: 7.5)
5	3.05	3.57
6	8.0	8.35
7	12.30	11.95
8	16.04	15.23
9	17.75	18.12
10	22.41	20.80
11	25.20	23.08
12	27.70	25.12
13	29.94	26.86
14	32.00	28.4
15	33.76	29.78
16	35.34	30.95
17	36.73	31.86
18	37.77	32.53
19	38.46	32.94
20	**38.72**	**33.08**

Bold indicates peak values of excess pore pressure and excess pore pressure ratio

86.20 kPa and 32.30 kPa for Nepal and Northeast input motion, respectively. In case of 3D ground response analysis, excess pore pressure values were observed as 99.94 kPa and 27.10 kPa for Nepal input motion and Northeast input motion, respectively. Variation of excess pore pressure (kPa) against depth for 1D, 2D and 3D seismic response analysis of Talcher pond ash was presented in Tables 6 and 7. Typical view of shadings of variation of excess pore pressure was presented in Figs. 12 and 13.

3.4 *Excess Pore Pressure Ratio Variation*

Based on the observed values of excess pore pressure ratio, it was inferred that pond ash was susceptible to no liquefaction as r_u lies far behind one in all the cases of 1D, 2D and 3D ground response analysis. Variation of excess pore pressure against depth was presented in Tables 8 and 9.

Table 7 Variation of excess pore pressure (kPa) against depth for 2D and 3D (PLAXIS 2D and 3D) seismic response analysis of Talcher pond ash

Depth (m)	2D analysis		Depth (m)	3D analysis	
	Nepal Eq (M_w: 7.8)	NE India Eq (M_w: 7.5)		Nepal Eq (M_w: 7.8)	NE India Eq (M_w: 7.5)
6.25	33.85	13.78	6.418	56.86	9.99
8.75	44.25	22.60	9.844	51.84	17.57
11.25	57.45	32.06	13.22	80.36	17.17
13.75	72.75	**35.06**	16.05	82.87	19.33
16.25	77.01	33.42	19.76	**99.94**	**27.10**
18.75	**86.20**	32.30			

Bold indicates peak values of excess pore pressure and excess pore pressure ratio

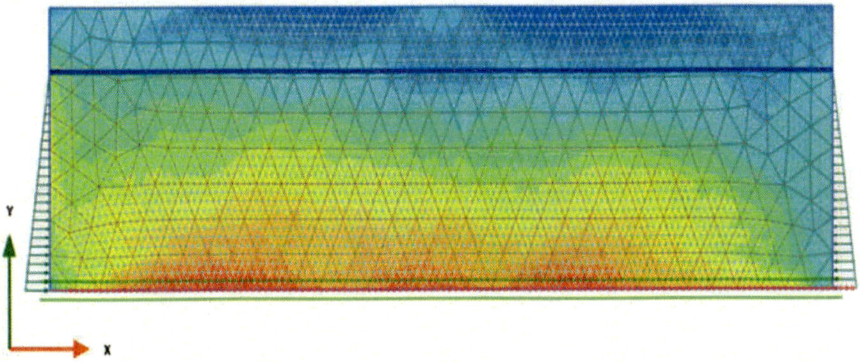

Fig. 12 Typical view of shadings of variation of excess pore pressure for Nepal earthquake motion (2D analysis)

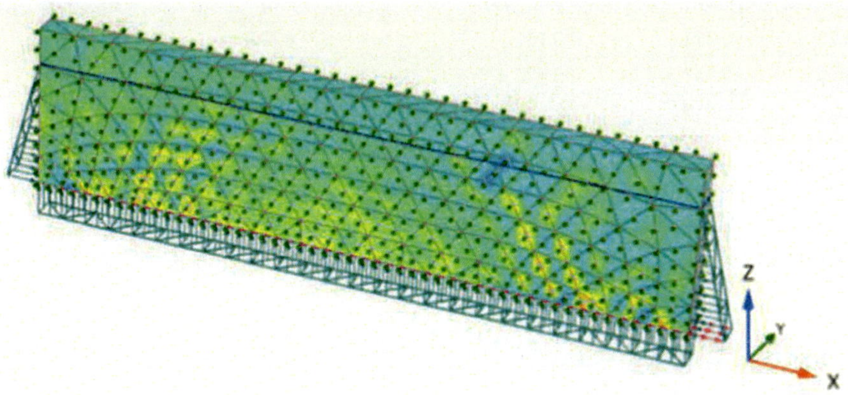

Fig. 13 Typical view of shadings of variation of excess pore pressure for Nepal earthquake motion (3D analysis)

Table 8 Variation of excess pore pressure ratio against depth for 1D (Cyclic1D) seismic response analysis of Talcher pond ash

Depth (m)	1D analysis	
	Nepal Eq (M_w: 7.8)	NE India Eq (M_w: 7.5)
5	0.044	0.052
6	0.109	0.114
7	0.157	0.153
8	0.193	0.184
9	0.221	0.207
10	0.242	0.225
11	0.259	0.237
12	0.271	0.246
13	0.280	0.254
14	0.286	0.254
15	0.290	**0.255**
16	**0.291**	0.255
17	0.291	0.252
18	0.288	0.248
19	0.283	0.242
20	0.275	0.235

Bold indicates peak values of excess pore pressure and excess pore pressure ratio

Table 9 Variation of excess pore pressure ratio against depth for 2D and 3D (PLAXIS 2D and 3D) seismic response analysis of Talcher pond ash

Depth (m)	2D analysis		Depth (m)	3D analysis	
	Nepal Eq (M_w: 7.8)	NE India Eq (M_w: 7.5)		Nepal Eq (M_w: 7.8)	NE India Eq (M_w: 7.5)
6.25	0.371	0.151	6.148	**0.608**	0.106
8.75	0.347	0.177	9.844	0.361	**0.122**
11.25	0.350	**0.195**	13.22	0.417	0.089
13.75	**0.363**	0.175	16.052	0.354	0.082
16.25	0.325	0.141	19.76	0.347	0.094
18.75	0.315	0.118			

Bold indicates peak values of excess pore pressure and excess pore pressure ratio

4 Conclusions

1D, 2D and 3D seismic response analysis of Talcher pond ash was performed using SHAKE2000, Cyclic1D, PLAXIS 2D and 3D, respectively. Depth of pond ash domain considered was 20 m. Location of groundwater table was at 4.5 m from ground surface. Maximum value of peak ground acceleration for 2D analysis was observed when Nepal input motion served as a base excitation, i.e. 0.231 g. But, less value of PGA was observed when subjected to NE India earthquake for 3D analysis, i.e. 0.114 g. In both 2D and 3D analysis, maximum value of displacement was observed when NE India earthquake acted as a base excitation. It may be due to severe response of pond ash when subjected to Northeast India earthquake, which has higher amplitudes on average when compared with Nepal earthquake excitation. Maximum peak ground displacement was observed to be 5.522 m when subjected to NE India earthquake for 3D analysis. Maximum excess pore pressure value was observed as 99.94 kPa in case of Nepal earthquake as base excitation in 3D analysis. R_u was observed to be maximum for Nepal earthquake input motion in 3D analysis, i.e. 0.608 and was observed to be less than one for 1D, 2D and 3D analysis for both the cases of input motions. It can be concluded that pond ash is prone to no liquefaction as r_u falls behind one for both the cases of input motions.

Based on the observation of varied results, it can be stated that extensive field, laboratory and numerical studies are required for characterization of pond ash; ground engineering is necessary for employing it as a building material.

Acknowledgements Authors are grateful to Department of Science and Technology, India for providing financial support and assistance throughout the study. In addition, authors are grateful to General Manager, Talcher Thermal Power Station for providing permission to collect pond ash sample for this study.

References

Bera AK, Ghosh A, Ghosh A (2007) Compaction characteristics of pond ash. J Mater Civ Eng 19(4):349–357. https://doi.org/10.1061/(ASCE)0899-1561(2007)19:4(349)

Blissett RS, Rowson NA (2012) A review of the multi-component utilization of coal fly ash. Fuel 97:1–23. https://doi.org/10.1016/j.fuel.2012.03.024

Boominathan A, Hari S (2002) Liquefaction strength of fly ash reinforced with randomly distributed fibers. Soil Dyn Earthq Eng 22:1027–1033. https://doi.org/10.1016/S0267-7261(02)00127-6

Central Electricity Authority, India (2017). https://www.cea.nic.in/reports/others/thermal/tcd/fly ash_201617.pdf

Das SK, Yudhbir (2005) Geotechnical characterization of some Indian fly ashes. J Mater Civ Eng 17(5):544–552. https://doi.org/10.1061/(ASCE)0899-1561(2005)17:5(544)

Jakka RS, Ramana GV, Datta M (2010) Shear behavior of loose and compacted pond ash. Geotech Geol Eng 28:763–778. https://doi.org/10.1007/s10706-010-9337-1

Kaniraj SR, Gayathri V (2004) Permeability and consolidation characteristics of compacted fly ash. J Energy Eng 130(1):18–43. https://doi.org/10.1061/(ASCE)0733-9453(2004)130:1(18)

Khanna A, Mohanty S (2017) 2D ground response analysis of pond ash deposits. In: Geotechnical frontiers, vol 276. ASCE, Orlando, Florida, pp 387–396. https://doi.org/10.1061/978078448043 4.042

Kim B, Prezzi M, Salgado R (2005) Geotechnical properties of fly and bottom ash mixtures for use in highway embankments. J Geotech Geoenviron Eng 131(7):914–924. https://doi.org/10.1061/(ASCE)1090-0241(2005)131:7(914)

Kuhlemeyer RL, Lysmer J (1973) Finite element method accuracy for wave propagation problems. J Soil Dyn Div 99:421–427

Lysmer J, Udaka T, Tsai CF, Seed HB (1975) FLUSH: a computer program for approximate 3-D analysis of soil-structure interaction problems. Report EERC 75-30, Berkeley, p 83

Mishra DP, Das SK (2010) A study of physico-chemical and mineralogical properties of Talcher coal fly ash for stowing in underground coal mines. Mater Charact 61:1252–1259. https://doi.org/10.1016/j.matchar.2010.08.008

Mohanty S, Patra NR (2014) Cyclic behavior and liquefaction potential of Indian pond ash located in seismic zones III and IV. J Mater Civ Eng 06014012:1–5. https://doi.org/10.1061/(ASCE)MT.1943-5533.0000964

Mohanty S, Patra NR (2015a) Liquefaction and 1D ground response analysis of Talcher pond ash. In: 50th Indian geotechnical conference, India, pp 1–7

Mohanty S, Patra NR (2015b) Geotechnical characterization of Panki and Panipat pond ash in India. Int J Geo-Eng 6(13):1–18. https://doi.org/10.1186/s40703-015-0013-4

Mohanty S, Patra NR (2016a) Dynamic response analysis of Talcher pond ash embankment in India. Soil Dyn Earthq Eng 84:238–250. https://doi.org/10.1016/j.soildyn.2016.01.021

Mohanty S, Patra NR (2016b) Liquefaction and earthquake response analysis of Panipat pond ash embankment in India. J Earth Tsunami 10(2):1–23. https://doi.org/10.1142/S1793431116500093

Mohanty S, Reddy MVRK (2016) Seismic analysis of shallow foundation on layered soil deposits. In: International conference on advances in civil engineering and sustainable construction, Chennai, pp 532–537

Mohapatra R, Kanungo SB (1997) Physio-chemical characteristics of fly ash samples from thermal power plants of Orissa. Indian J Eng Mater Sci 4:271–281

Pandian NS (2004) Fly ash characterization with reference to geotechnical applications. J Indian Inst Sci 84:189–216

Raju LG, Ramana GV, Rao CH Sitharam TG (2004) Site specific ground response analysis. Geotech Earthq Hazard 87(10):1354–1362

Reddy MVRK, Mohanty S, Ramancharla PK (2018) Comparative study of dynamic response analysis of shallow foundation on layered soils. In: Adimoolam B, Banerjee S (eds) Soil dynamics and earthquake geotechnical engineering. Lecture notes in civil engineering, vol 15. Springer, Singapore, pp 91–99. https://doi.org/10.1007/978-981-13-0562-7_11

Singh HP, Maheshwari BK, Saran S, Paul DK (2008) Evaluation of liquefaction potential of pond ash. In: The 14th world conference on earthquake engineering, China, pp 1–8

Sridharan A (2001) Physical, chemical and engineering properties of Indian coal ashes. National seminar on utilization of fly ash in water resources sector. Central Soil and Material Research Station, New Delhi, pp 15–28

Vijayasri T, Raychowdhury P, Patra NR (2016) Seismic response analysis of Renusagar pond ash embankment in Northern India. Int J Geomech 17(6):1–12. https://doi.org/10.1061/(ASCE)GM.1943-5622.0000828

Effect of Soil Grain Size on Liquefaction Strength of Sandy Soil

Pradipta Chakrabortty, Angshuman Das, and Anil

Abstract This study is focused on estimating the difference in behavior between three soils: natural, medium and fine sand. Strain-controlled cyclic triaxial tests were conducted on all three types of soils. Samples were prepared using air pluviation method, and soil relative density was controlled between 10 and 20% (loose soil). The medium sand was considered as base case with a frequency of 0.5 Hz, axial strain of 0.5 mm and effective confining pressure of 100 kPa. The axial strains (0.1–1 mm) was also varied to find out the effect of various strain levels on excess pore water pressure generation and degradation of soil strength during cyclic loading. From the results, it can be concluded that the excess pore water pressure generation was much faster in fine sand than that in medium sand. This ultimately causes degradation of liquefaction strength at much faster rate in fine sand compared to that in medium sand.

Keywords Strain-controlled test · Cyclic triaxial · Grain size distribution parameters · Dynamic behavior · Mean grain size

1 Introduction

During earthquake natural soil is subjected to cyclic shear stresses of different frequencies and different amplitudes that cause the settlement and generation of excess pore water pressure (EPWP) in soil. Due to generation of EPWP, soil losses its strength partially or completely. This phenomenon is known as soil liquefaction (Castro and Poulos 1977). Liquefaction strength of sandy soil has been studied extensively in the past using cyclic triaxial tests. The liquefaction potential of cohesionless soil depends on its void ratio, relative density, cyclic stress amplitude, mean effective stress, initial stress of the soil and grain size distribution (Seed and Lee

P. Chakrabortty (✉) · A. Das · Anil
Indian Institute of Technology Patna, Bihta, Patna 801103, India
e-mail: pradipt@iitp.ac.in

A. Das
e-mail: dasangshuman6@gmail.com

M. Latha Gali and P. Raghuveer Rao (eds.), *Geohazards*, Lecture Notes in Civil Engineering 86, https://doi.org/10.1007/978-981-15-6233-4_38

1966). Various investigators have studied the effect of particle shape and size on liquefaction strength (Lee and Fitton 1969; Wong et al. 1975; Ishihara et al. 1978). However, the effect of grain size on liquefaction strength is still a part of discussion (Hakam 2016).

Excess pore water pressure dissipates quickly in coarse sand compared to fine sand. So, the medium and coarse sand are more difficult to be liquefied than the silty fine sand. The liquefaction test studies for sand with different particle diameter show that average particle diameter of 0.07 mm is the likeliest to have liquefaction. The bigger the D_{50} the greater is the liquefaction strength. The laboratory test and field investigation shows that the average particle diameter range of liquefied soil is 0.01–2 mm (Ye 2017). To reinforce above finding, the present study has been designed. In this present study, the effect of soil mean diameter (D_{50}) and loading strain amplitude on liquefaction strength has been studied and presented for locally available sandy soil.

Earlier study confirms that cyclic shear strain is the fundamental parameter governing pore pressure buildup, because strain-controlled tests eliminate the influence of specimen fabric and sample disturbance (Vucetic and Dobry 1988). Therefore, a series of strain-controlled cyclic triaxial tests were performed to determine the effect of soil grain size on liquefaction strength. The cohessionless soil was collected from the IIT Patna, Bihta campus. Collected natural sand has been sieved into fine (particle size between 0.425 and 0.075 mm) and medium sand (particle size between 2 and 0.425 mm) by mechanical sieving. Air pluviation technique was used for sample preparation as this method produced reconstituted sand specimens with controlled initial void ratio (Vaid and Negussey 1984). Also, sample prepared using air pluviation strain softens to a lesser extent when compared to moist tamped specimens (Vaid and Sivathayalan 2000). Specimens were saturated first and consolidated at different effective confining pressures of 100, 150 and 200 kPa. The range of consolidation pressures considered in the present study corresponds to the pressures approximately at a depth of 5–20 m below the ground surface. Axial deformation was also varied (0.1, 0.5 and 1 mm) to find out the effect of various strain levels on excess pore water pressure generation and degradation of soil strength during cyclic loading.

2 Soil Properties

All tests were conducted in soil dynamics laboratory at IIT Patna. The sieve analysis was performed on the dry soil sample as per IS code, and the soil was classified as poorly graded sand. Other two type of soil (medium and fine sand) are separated by dry sieving technique. Classification of soil into poorly graded or well graded is achieved based on the values of coefficient of uniformity (C_u) and coefficient of curvature (C_c). The values of C_u and C_c obtained for all three sands are tabulated in Table 1. Following the procedure recommended by IS:2720—Part 4 (BIS 1985), the particle-size distribution curve for natural, medium and fine sands has been obtained

Table 1 Gradation of sandy soils used in the tests

		Fine sand	Medium sand
D_{50} (mm)	0.34	0.27	0.75
C_u	2.31	1.74	1.56
C_c	0.83	0.89	1.02
Specific gravity	2.72	2.58	2.64
Maximum density (g/cc)	1.69	1.58	1.76
Minimum density (g/cc)	1.43	1.33	1.42
Gradation	Poorly graded	Poorly graded	Poorly graded

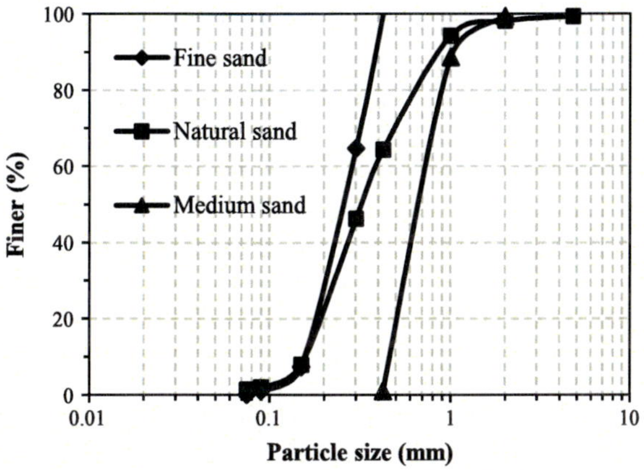

Fig. 1 Particle size distribution curves of all soils used in the analyses

and is shown in Fig. 1. From the figure, it can be seen that major percentage of soil in natural soil is fine-grained sand. The maximum and minimum dry density values were calculated based on the guidelines given in IS: 2720—Part 14 (BIS 1983). Specific gravity for each soil samples was calculated according to the recommended given in IS: 2720—Part-3 (BIS 1980) and shown in Table 1.

3 Strain-Controlled Cyclic Triaxial Test

The seismic loading induced in the soil during earthquake can be represented by a series of uniform strain cycle and liquefaction potential of a soil under seismic loading can be determined by applying such series of strain cycles using strain-controlled triaxial test apparatus. Previously, Silver and Seed (1971) experimentally showed that cyclic shear strain rather than cyclic shear stress controls the behavior of dense sand. Further, Martin et al. (1975) and Dobry et al. (1982) strongly suggest that the shear strain controls both the liquefaction and densification of soil during cyclic loading. More research using strain-controlled test (Sitharam et al. 2004; Wichtmann and Triantafyllidis 2010) shows this test can well predict the soil behavior under dynamic load. In 2015, Du and Chian (2015) have performed direct comparison between stress- and strain-controlled test. The results indicate of a similar liquefaction potential assessment from both types of tests, but the results show that the initiation of liquefaction can clearly identified in strain-controlled triaxial test compare to stress-controlled test. In this present study, displacement-controlled cyclic triaxial test on different soil samples was performed for various conditions.

3.1 Cyclic Triaxial Test Setup

In IIT Patna, automated computerized cyclic triaxial test equipment is available as shown in Fig. 2. It has six sensors, among them three sensors measure the pressure (confining pressure (CP), back pressure (BP) and pore water pressure (PP)), one sensor is there for volume change measurement, and other two sensors are for load and displacement measurement. Mold with diameter of 38, 50, 75 and 150 mm is available for sample preparation. In the present work, mold with 38 mm diameter and 76 mm height was used, and consolidated undrained triaxial tests were performed as per ASTM designation (ASTM D-5311-92) (American Society for Testing and Materials 1996).

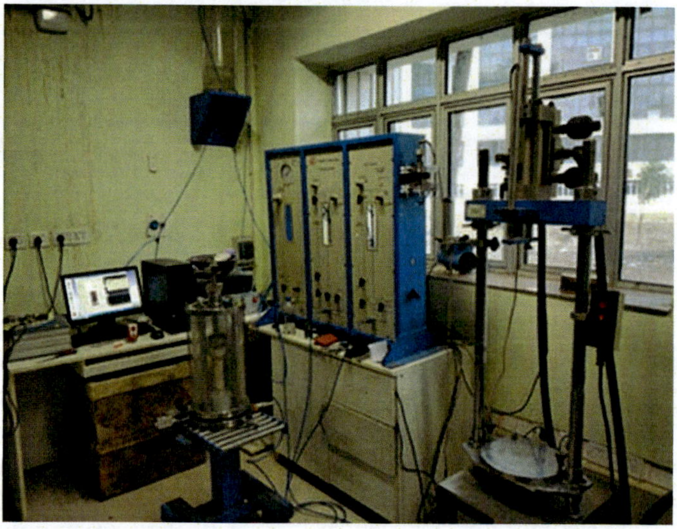

Fig. 2 Computer controlled cyclic triaxial test apparatus at IIT Patna used in this study

3.2 Sample Preparation and Saturation

First the mold was fixed with the base using removable clamp. The assembly was connected with vacuum during placing the membrane to prevent necking during sample preparation (Wijewickreme and Sanin 2006). Dry pluviation method was used for sample preparation (as shown in Fig. 3) because it is considered as better than other sample preparation techniques (Juneja and Raghunandan 2010). A relation between height of fall and relative density was estimated initially to achieve a certain relative density of soil samples. All soil samples studied here having a target relative density of 10–20%. From a pre-decided height, the sand particles were allowed to fall freely to achieve the target relative density. From the weight of particle poured into the mold and the volume of sample, the achieved density was calculated. The sample was saturated using the de-aired water through the back pressure line. After

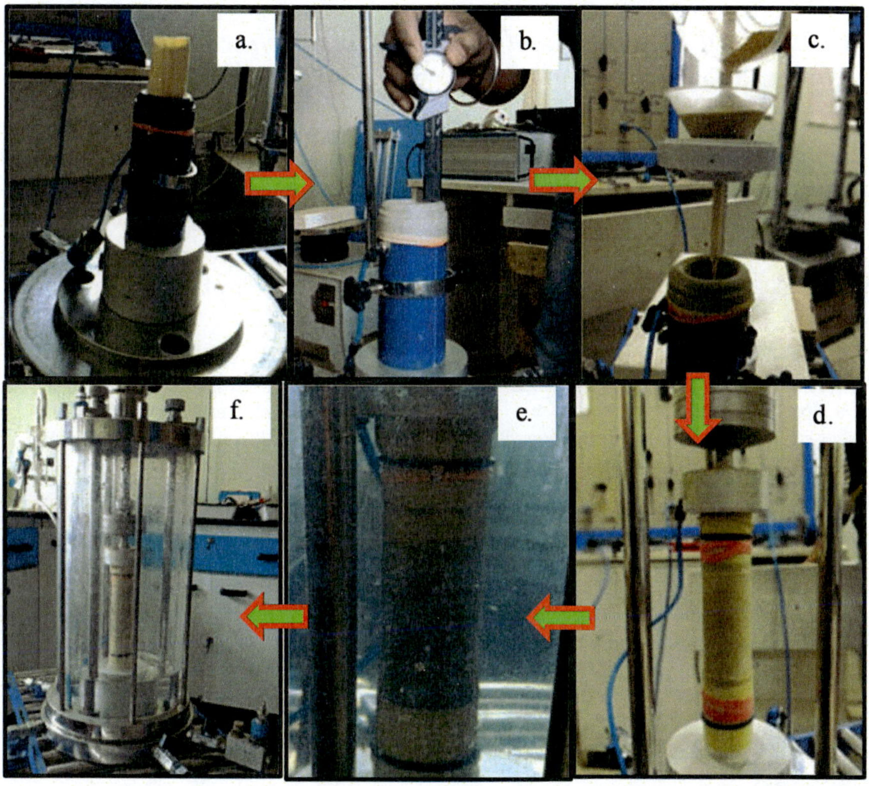

Fig. 3 Various steps of sample preparation **a** fixing of membrane, **b** measuring the internal dimensions, **c** pouring of sample, **d** attaching the load cell, **e** filling the cell with water, **f** sample under saturation

the saturation is complete or the B-value become equal to or greater than 0.9, the specimens were then isotopically consolidated under desired effective stress. All the samples were tested after consolidation.

4 Results and Discussion

A series of cyclic triaxial tests were performed to estimate the effect of soil grain size on liquefaction strength. Cyclic frequency for all tests was fixed at 0.5 Hz. Results of all tests were analyzed, and relation has been studied between soil type and the cyclic behavior of the soil. Next subsection discusses the effect of loading amplitude on dynamic response.

4.1 Effect of Loading Amplitude on Cyclic Behavior of IIT Patna Soil

Strain-controlled consolidate undrained cyclic triaxial tests were performed on fine, medium and natural sand with different axial deformation amplitudes of 0.1, 0.5 and 1.0 mm. The axial strain and shear strain corresponding to these are shown in Table 2. Figure 4 shows the degradation of strength (in terms of effective stress) due to development of excess pore water pressure (EPWP) for the displacement amplitude

Table 2 Details of the cyclic triaxial tests performed in this study

Test No.	Soil type	Axial deformation (mm)	Axial strain (%)	Shear strain (%)	Confining pressure (kPa)
1	Fine	±0.1	0.13	0.20	100
2	Fine	±0.5	0.66	0.99	100
3	Fine	±1.0	1.32	1.97	100
4	Medium	±0.1	0.13	0.20	100
5	Medium	±0.5	0.66	0.99	100
6	Medium	±1.0	1.32	1.97	100
7	Medium	±0.1	0.13	0.20	150
8	Medium	±0.5	0.66	0.99	150
9	Medium	±1.0	1.32	1.97	150
10	Medium	±0.5	0.66	0.99	200
11	Natural soil	±0.1	0.13	0.20	100
12	Natural soil	±0.5	0.66	0.99	100

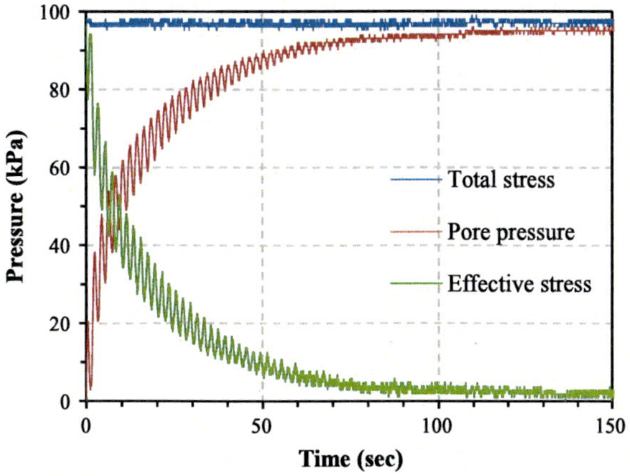

Fig. 4 Variation of EPWP, total stress and effective stress with time for 0.1 mm axial deformation in fine sand at 100 kPa effective confining stress

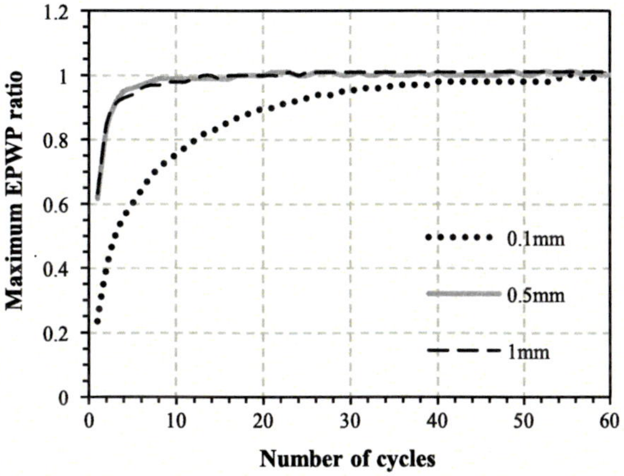

Fig. 5 Variation of EPWP ratio with number of cycles at different amplitude on fine sand at 100 kPa effective confining stress

of 0.1 mm in fine sand. Figure 5 shows the variation of maximum EPWP ratio with number of cycle for all three loading amplitudes. The EPWP ratio is defined here as the ratio between generated maximum excess pore water pressure and initial effective confining pressure. If EPWP ratio reaches unity (i.e. $r_u = 1$), soil behaves like liquid. However, the liquefaction may initiate when EPWP ratio reaches 0.6 depending on various conditions (Singh et al. 1980; Finn 1981). Therefore, the EPWP ratio (r_u) of 0.90 or 0.95 can be considered as liquefaction criteria (Jiaer et al. 2004). In this study, r_u of 0.9 is considered as the lower bound for EPWP ratio for the soil to be considered liquefied. It can be observed from this study that if soils are subjected to low strain amplitude, generation of pore water pressure takes place very slowly. More numbers of cycles are required for reaching EPWP ratio 0.9 than soil subjected to high strain amplitude. Fine sand subjected to ±0.1 mm axial deformation liquefy around 21 cycles of loading while soil subjected to 0.5 mm amplitude and 1.0 mm amplitude liquefy around 4 and 3 number of loading cycles, respectively.

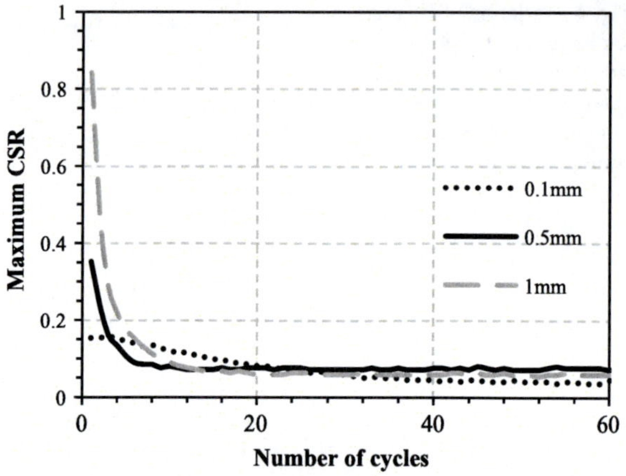

Fig. 6 Variation of maximum CSR with number of cycles at different amplitude on fine sand at 100 kPa effective confining stress

Figure 6 shows the degradation of strength with number of loading cycles for various tests performed in fine sand at 100 kPa confining pressure. The strength is expressed in terms of maximum cyclic stress ratio (CSR). The CSR is defined as the ratio of cyclic deviator stress and initial mean effective stress. For 1.32% axial strain (1 mm), maximum CSR become less than 0.1 within 9 cycles. Peak CSR is higher for higher axial strain inputs. However, rate of degradation in strength is much less for lower axial strain. As soil liquefied CSR became almost constant.

Figure 7 shows the degradation of strength (in terms of effective stress) due to development of excess pore water pressure (EPWP) for the deformation amplitude of 0.1 mm in medium sand. Figure 8 shows the variation of maximum excess pore water pressure generation with number of loading cycles in medium sand at 100 kPa confining pressure. Medium sand subjected to 0.1 mm amplitude liquefies around 200 cycles of loading, while soil subjected to 0.5 mm amplitude and 1.0 mm amplitude liquefies around 20 and 5 number of loading cycles, respectively.

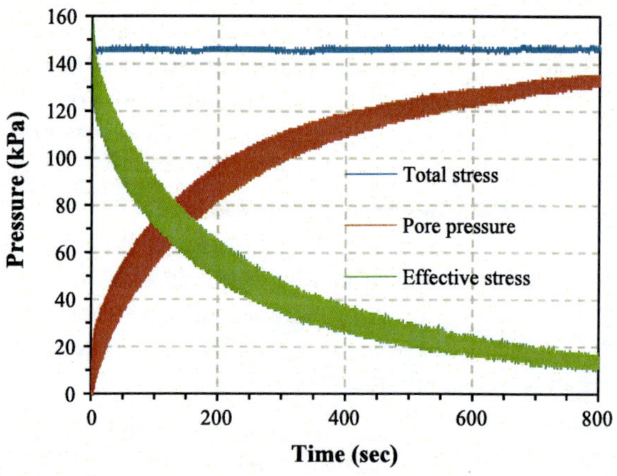

Fig. 7 Variation of EPWP, total stress and effective stress with time for 0.1 mm axial deformation in medium sand at 150 kPa effective confining stress

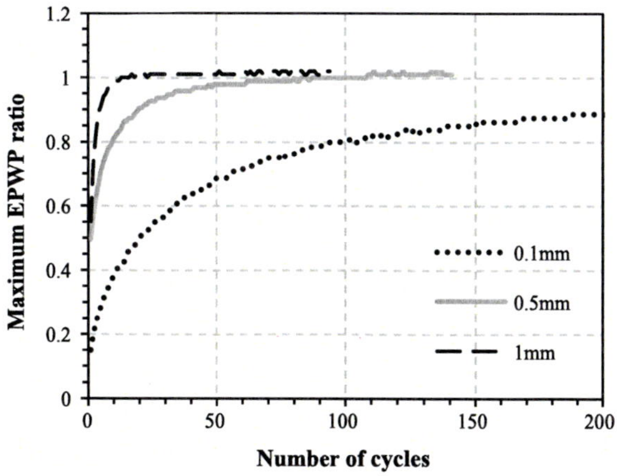

Fig. 8 Variation of EPWP ratio with cycles at different amplitude on medium sand at 100 kPa effective confining stress

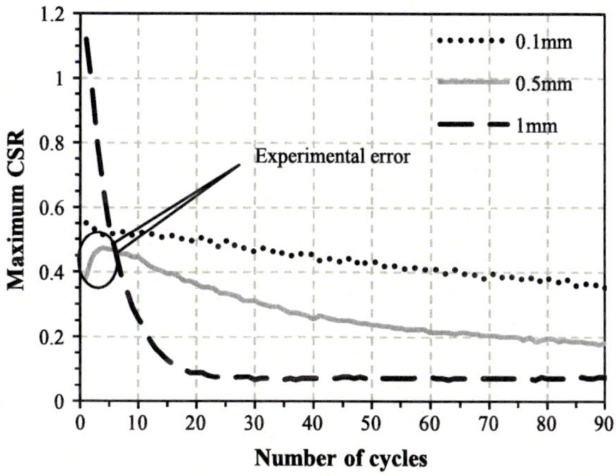

Fig. 9 Variation of maximum CSR with cycles at different amplitudes on medium sand at 100 kPa effective confining stress

Figure 9 shows the degradation of strength with number of loading cycles in medium sand at 100 kPa confining pressure. For 1.32% axial strain, maximum CSR become less than 0.2 within around 20 cycles. Similar trend for rate of degradation in lower strain amplitude was observed compared to that observed in fine sand. From the results, it can be concluded that soil is liquefied faster at higher strain compared to that in lower input axial strain.

4.2 *Effect of Soil Type on Cyclic Behavior of IIT Patna Soil*

Three different types of sands were used here to show the effect of soil grain size on cyclic behavior of these soils collected from IIT Patna campus. The comparison of stress path is shown in Fig. 10 for all the sand studied here. Figure 11 shows the

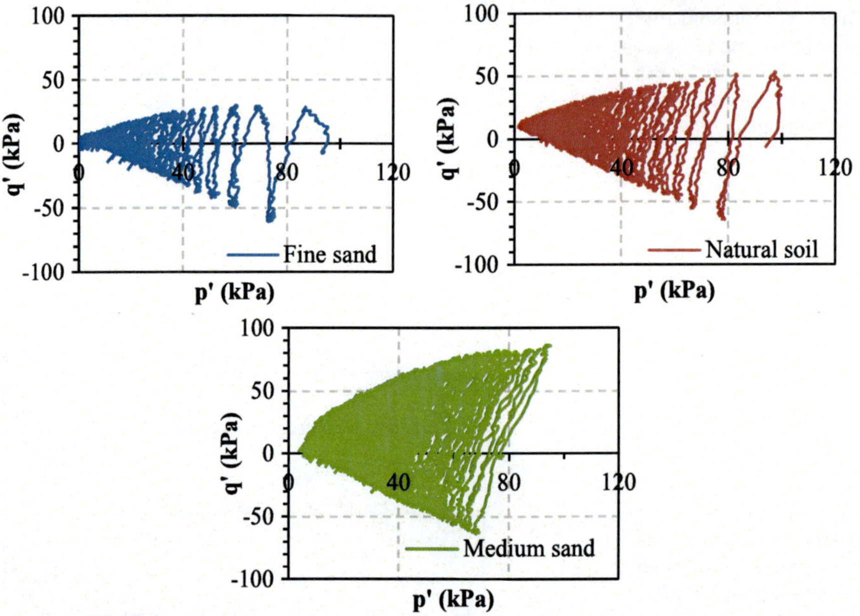

Fig. 10 Comparison of effective stress path between all studied soils for 0.1 mm axial deformation at 100 kPa effective confining stress

Fig. 11 Variation of maximum EPWP ratio with cycles in different soil with 0.1 mm amplitude and 100 kPa effective confining stress

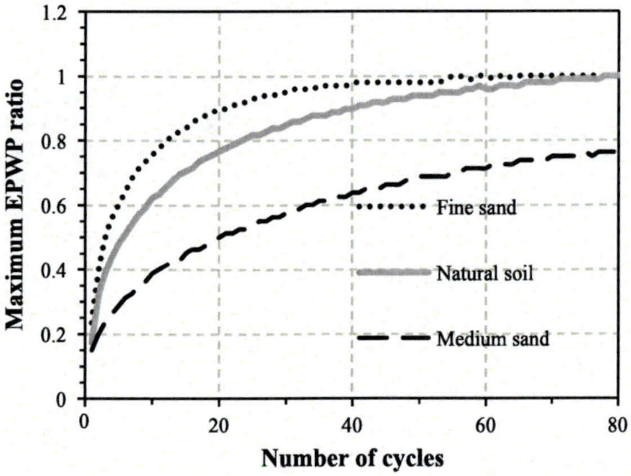

variation of pore water pressure generation in different soils with effective confining stress of 100 kPa and axial deformation amplitude of 0.1 mm. Fine sand liquefies in around 20 cycles, natural soil liquefies around 40 cycles, and medium sand liquefies around 200 cycles.

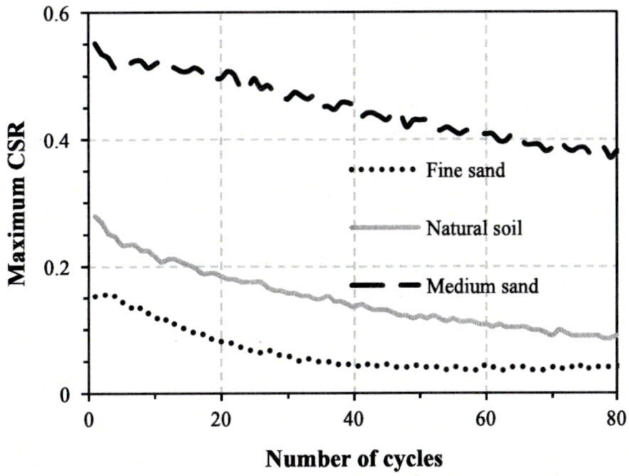

Fig. 12 Variation of maximum CSR with cycles in different soil with 0.1 mm amplitude and 100 kPa effective confining stress

The strength degradation with number of strain cycle is shown in Fig. 12 for all three type of soil. It can be observed that in medium sand more maximum CSR (1.1) is developing than fine sand (0.303). Since the natural sand is the mixture of fine and medium sand, the developed maximum CSR is also between that in fine and medium sand. The natural sand is having about 64% fine sand and 36% medium sand. It is interesting to observe that the initial maximum CSR value (0.56) for natural soil is very close to weightage average (between fine and medium soil) of the same (0.59).

Figure 13 is showing the variation of maximum excess pore water pressure generation in different soils with number of loading cycle for an effective confining stress of 100 kPa and amplitude of 0.5 mm. Fine sand and natural sand liquefied within 3 cycles and medium sand liquefied around 20 cycles.

The natural soil is having (Fig. 14) more initial maximum CSR than fine and medium sand. But because of faster generation of EPWP, strength degraded quickly in fine and natural sand. Since EPWP generation in medium sand is much slower, strength degradation was also slower.

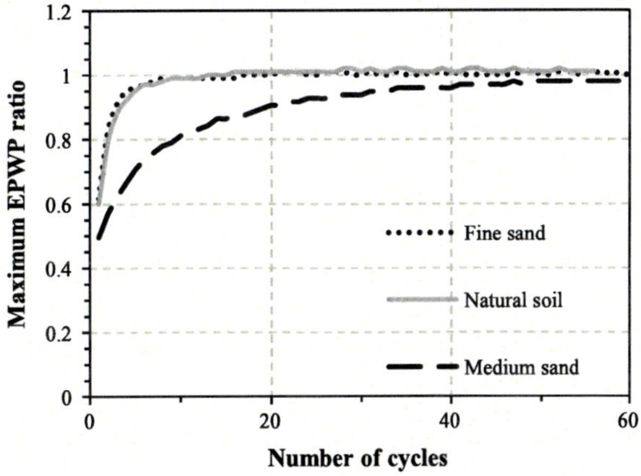

Fig. 13 Variation of maximum EPWP ratio with cycles in different soil with 0.5 mm amplitude and 100 kPa effective confining stress

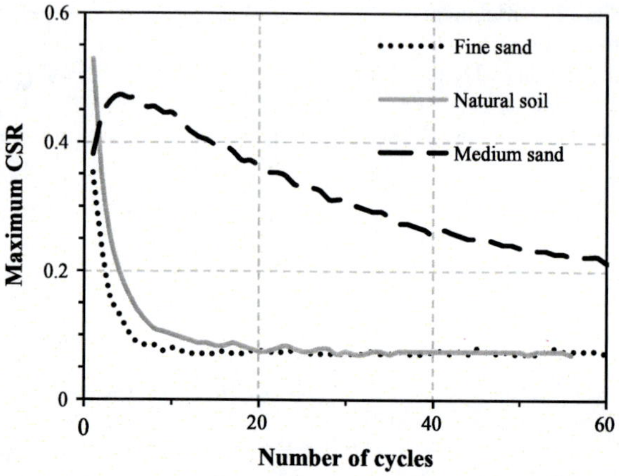

Fig. 14 Variation of maximum CSR with cycles in different soil with 0.5 mm amplitude and 100 kPa effective confining stress

The result shows that fine sand is liquefying fastest followed by natural soil and medium sand. So, it can be concluded that fine sand is more susceptible to liquefaction than natural soil existing at IIT Patna campus. The medium sand considered in this study is least susceptible to liquefaction. The liquefaction potential of sand increases with the decreasing in mean grain size (D_{50}). The results are compared with that documented by Lee and Fitton (1969). Figure 15 shows the comparison of deviatoric

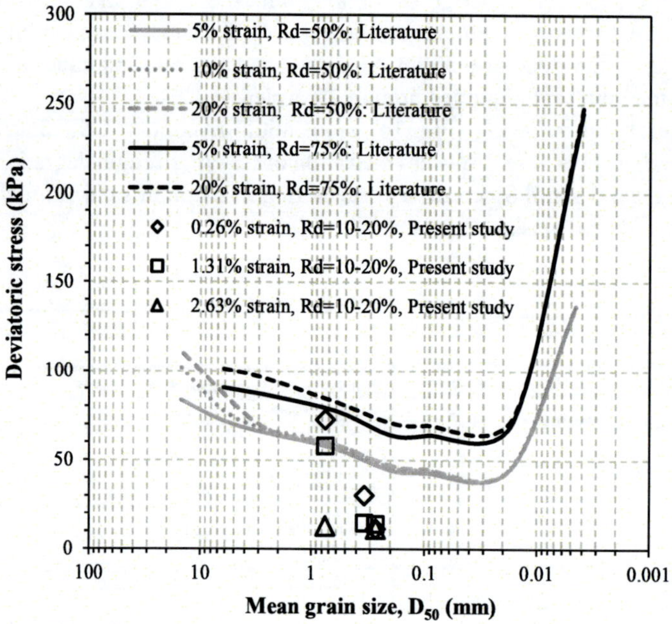

Fig. 15 Comparison of deviatoric loading strength ($N = 30$) of different types of sandy soil with respect to literature (modified from Lee and Fitton (1969))

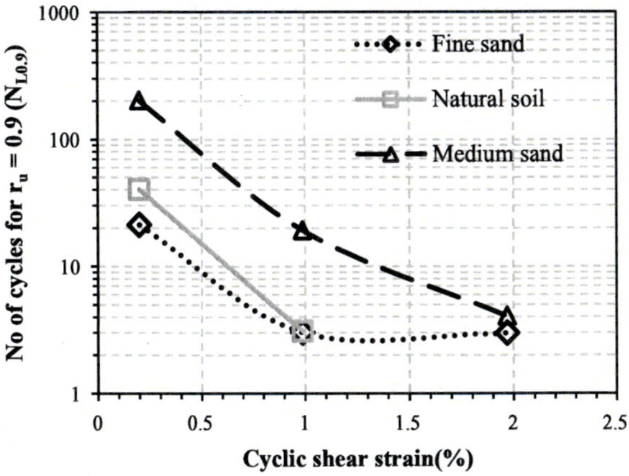

Fig. 16 Variation of number of cycles for initial liquefaction with cyclic shear strain

loading strength of different types of sandy soil studied here with respect to D_{50} values. The present study was performed at low relative density (Rd) (10–20%) which was different from that reported in the literature. But, the reduction in strength was observed with the reduction in mean grain size which was reported earlier for other relative densities also.

Figure 16 shows the variation in number of loading cycles required for initial liquefaction ($N_{L0.9}$) with cyclic shear strain. The initial liquefaction is considered here when the EPWP ratio reaches 90% of initial effective confining pressure. Figure shows the decrease in liquefaction strength of soil samples with the decrease in D_{50} value. It was also observed that as the input cyclic shear strain increased, the liquefaction strength of the soil is also reduced. This demonstrated that soil will liquefy early for large magnitude earthquakes. The number of cycles for EPWP ratio reaches 90% of initial effective confining pressure for all the studied samples which are shown in Table 3.

Figure 17 shows the liquefaction strength in terms of maximum CSR for all the tests performed at 100 kPa confining pressure. The result also shows higher resistance against liquefaction for medium grain sand. Figure 18 shows the maximum CSR value for various cyclic shear strain values. The trend clearly shows that maximum CSR value is increasing with the increase in cyclic shear strain. The maximum CSR measured for medium sand at 0.99% cyclic shear strain shows lower strength compared to the trend. This could be because of experimental error due to sample preparation. However, the trend line for medium sand shows increase in maximum CSR value with the increase in cyclic shear strain.

Table 3 Number of loading cycles required for initial liquefaction for all the cyclic triaxial tests performed in this study

Test No.	Parameters			N_L
	Sand type	Confining stress (kPa)	Amplitude (mm)	
1	Fine	100	0.1	21
2	Fine	100	0.5	3
3	Fine	100	1.0	3
4	Medium	100	0.1	201
5	Medium	100	0.5	19
6	Medium	100	1.0	4
7	Medium	150	0.1	334
8	Medium	150	0.5	20
9	Medium	150	1.0	22
10	Medium	200	0.5	11
11	Natural	100	0.1	40

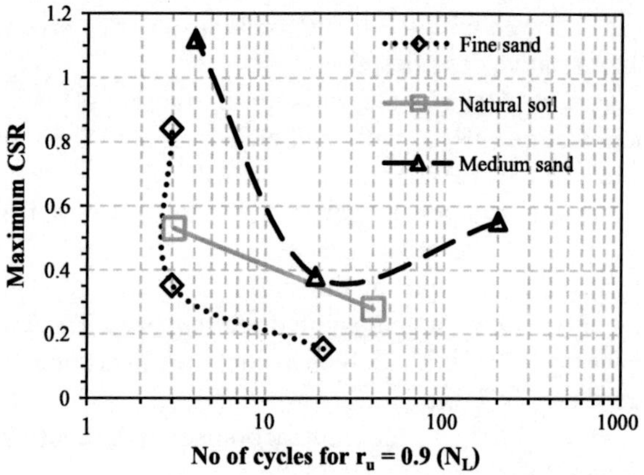

Fig. 17 Variation of maximum cyclic stress ratio with number of cycles for initial liquefaction

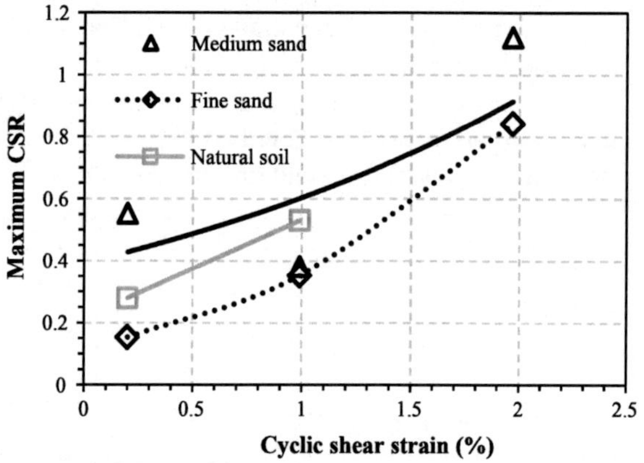

Fig. 18 Variation of maximum cyclic stress ratio with cyclic shear strain

5 Conclusions

A series of strain-controlled undrained cyclic triaxial tests were performed in this study. Results showed that fine sand is more susceptible to liquefaction than that of medium sand. As it can be seen in Table 3, number of cycles needed for liquefaction was 21 for fine sand and 201 for medium sand at same amplitude of 0.1 mm and effective confining stress of 100 kPa. From the study, the following conclusions are made:

(1) Fine sand is more susceptible to liquefaction than medium sand as it liquefies with less number of cycles.
(2) Cyclic shear strain amplitude is a factor that affects the cyclic behavior of soil as with increase in amplitude of dynamic loading causes fast liquefaction in each soil.
(3) Furthermore, if soil is subjected to more effective confining stress, the greater number of cycles is required to trigger the liquefaction of the soil.
(4) Additional number of cycles tends to reduce the capacity of the soil; after soil liquefaction, soil capacity became almost negligible.

References

American Society for Testing and Materials (ASTM) (1996) Standard test method for load controlled cyclic triaxial strength of soil. ASTM D 5311-92 (Re-approved 1996), ASTM, West Conshohocken. https://doi.org/10.1520/D5311-92R96

BIS 1980. IS: 2720 (Part 3)-1980. Methods of test for soils. Part 3: Determination of specific gravity, Section 2: Fine, medium and coarse grained soils. Bureau of Indian Standards, India (Reaffirmed 2002). https://archive.org/details/gov.in.is.2720.3.2.1980/page/n3

BIS 1984. IS: 2720 (Part 14)-1983. Methods of test for soils. Part 14: Determination of density index (relative density) of cohesionless soils. Bureau of Indian Standards, India (Reaffirmed 2006). https://archive.org/details/gov.in.is.2720.14.1983/page/n3

BIS 1986. IS: 2720 (Part 4)-1985. Methods of test for soils. Part 4: Grain size analysis. Bureau of Indian Standards, India (Reaffirmed 2006), pp 1–38. https://archive.org/details/gov.in.is.2720.4.1985/page/n3

Castro G, Poulos SJ (1977) Factors affecting liquefaction and cyclic mobility. J Geotech Geoenviron Eng 103(6):501–516

Dobry R, Ladd RS, Yokel FY et al (1982) Prediction of pore water pressure buildup and liquefaction of sands during earthquakes by the cyclic strain method. National Bureau of Standards Gaithersburg, MD. https://nvlpubs.nist.gov/nistpubs/Legacy/BSS/nbsbuildingscience138.pdf

Du S, Chian SC (2015) Cyclic response of liquefiable sand under stress-controlled and strain-controlled triaxial testing. In: Proceedings of the XVI ECSMGE, geotechnical engineering for infrastructure and development, 13–17 Sept, Edinburgh, Scotland. https://www.icevirtuallibrary.com/doi/abs/10.1680/ecsmge.60678.vol4.324

Finn WD (1981) Liquefaction potential: developments since 1976. In: Proceeding of international conferences on recent advances in geotechnical earthquake engineering and soil dynamics, St. Louis, Missouri, pp 655–681. https://scholarsmine.mst.edu/cgi/viewcontent.cgi?article=2508&context=icrageesd

Hakam A (2016) Laboratory liquefaction test of sand based on grain size and relative density. J Eng Technol Sci 48:334–344. https://doi.org/10.5614/j.eng.technol.sci.2016.48.3.7

Ishihara K, Sodekawa M, Tanaka Y (1978) Effects of overconsolidation on liquefaction characteristics of sands containing fines. In: Silver ML, Tiedemann D (eds) Dynamic geotechnical testing. ASTM International, West Conshohocken, PA, pp 246–264. https://doi.org/10.1520/STP35680S

Jiaer WU, Kammerer AM, Riemer MF et al (2004) Laboratory study of liquefaction triggering criteria. In: 13th World conference on earthquake engineering, Vancouver, BC, Canada. https://www.researchgate.net/profile/Juan_Pestana2/publication/265063346_LABORATORY_STUDY_OF_LIQUEFACTION_TRIGGERING_CRITERIA/links/546fb6590cf24af340c0941f.pdf

Juneja A, Raghunandan ME (2010) Effect of sample preparation on strength of sands. In: Indian geotechnical conference, Bombay. pp 327–330. https://gndec.ac.in/~igs/ldh/conf/2010/articles/081.pdf

Lee K, Fitton J (1969) Factors affecting the cyclic loading strength of soil. In: Vibration effects of earthquakes on soils and foundations. ASTM International, West Conshohocken, PA, pp 71–95. https://doi.org/10.1520/STP33637S

Martin GR, Finn WDL, Seed HB (1975) Fundamentals of liquefaction under cyclic loading. J Geotech Geoenviron Eng 101(5):423–438

Seed B, Lee KL (1966) Liquefaction of saturated sands during cyclic loading. J Soil Mech Found Div 92(SM6):105–134

Silver ML, Seed HB (1971) Volume changes in sands during cyclic loading. J Soil Mech Found Div 97(9):1171–1182

Singh S, Donovan NC, Park T (1980) A re-examination of the effects of prior loading on liquefaction of sands. In: Proceedings of the 7th world conference on earthquake engineering, pp 321–325. https://www.iitk.ac.in/nicee/wcee/article/7_vol3_321.pdf

Sitharam TG, Govindaraju L, Ravishankar BV, Murthy BRS (2004) Characteristics of liquefied silty sands from meizoseismal region of Shillong plateau, Assam and Bhuj in India. In: Proceedings of 13th world conference on earthquake engineering, Vancouver, Canada. https://www.iitk.ac.in/nicee/wcee/article/13_2375.pdf

Vaid Y, Negussey D (1984) A critical assessment of membrane penetration in the triaxial test. Geotech Test J 7:70–76. https://doi.org/10.1520/GTJ10595J

Vaid YP, Sivathayalan S (2000) Fundamental factors affecting liquefaction susceptibility of sands. Can Geotech J 37(3):592–606. https://doi.org/10.1139/t00-040

Vucetic M, Dobry R (1988) Cyclic triaxial strain-controlled testing of liquefiable sands. In: Advanced triaxial testing of soil and rock. ASTM International. https://doi.org/10.1520/STP29093S

Wichtmann T, Triantafyllidis T (2010) On the influence of the grain size distribution curve on dynamic properties of quartz sand. In: International conferences on recent advances in geotechnical earthquake engineering and soil dynamics, Missouri University of Science and Technology, San Diego, California. https://scholarsmine.mst.edu/cgi/viewcontent.cgi?article=2710&context=icrageesd

Wijewickreme D, Sanin MV (2006) New sample holder for the preparation of undisturbed fine-grained soil specimens for laboratory element testing. Geotech Test J 29:242–249. https://doi.org/10.1520/GTJ12699

Wong RT, Seed HB, Chan CK (1975) Cyclic loading liquefaction of gravelly soils. J Geotech Geoenviron Eng 101(GT6):561–583

Ye Y (2017) Marine geo-hazards in China. Elsevier, Amsterdam

Seismic Vulnerability Assessment of Pile Foundation in Liquefied Soil Incorporating Ground Motion Uncertainty

Partha Bhowmik and Rajib Saha

Abstract Seismic design of pile foundation has given a special attention by the earthquake professionals in last decade after experiencing many failures due to moderate to severe earthquake. Dynamic soil–structure interaction (DSSI) was reported as an important phenomenon in seismic design, and however, used to explain failure mechanism of piles during past earthquakes. It is observed from past case studies that pile failures are mainly reported in liquefied ground. Hence, vulnerability assessment of pile foundation in liquefiable deposit considering different failure mechanism is important for ensuring seismic safety of pile foundation. Present study is an attempt in this direction. Vulnerability of pile foundation will be assessed by calculating probability of exceedance of serviceability limit state criteria incorporating ground motion uncertainty. The uncertainty factors considered in generating ground motions are consists of source and attenuation of earthquake, local geology and site condition and variability associated with seismic force determination. Monte Carlo simulation (MCS) technique is used for probabilistic analysis. Seismic response is obtained for soil-pile foundation-structure system using 3D finite element software, namely OPENSeeSPL (Edu version 2.7.2). The case study of Showa Bridge during 1964 Niigata earthquake is considered in present study to perform the reliability analysis. Finally, the vulnerability of pile foundation is presented in the form of fragility curves. The limited study indicates that probability of failure reaches 100% considering serviceability condition when PGA of ground motion exceeds 1.0 g. This study will help to revamp the present design guidelines of pile foundation.

Keywords Kinematic interaction · Inertial interaction · Liquefaction · Fragility curves · OPENSeeSPL

P. Bhowmik (✉) · R. Saha
Department of Civil Engineering, National Institute of Technology Agartala, Barjala, Jirania, Agartala, Tripura 799046, India
e-mail: parthabhowmik14@gmail.com

R. Saha
e-mail: rajib.iitbbsr@gmail.com

M. Latha Gali and P. Raghuveer Rao (eds.), *Geohazards*, Lecture Notes in Civil Engineering 86, https://doi.org/10.1007/978-981-15-6233-4_39

1 Introduction

Failure of pile foundations during several earthquakes were evidenced which drew a global attention to the earthquake professionals. Pile failure is mainly observed in soft/loose liquefiable deposits. Many cases of bridge failure were reported in liquefied soil throughout the globe due to significant displacement of pile resulting to failure of deck, flexural failure of pile due to either inertial load or lateral spreading of soil mass, subsidence of pile and damage of piers due to inelastic deformation, etc. Meymand (1998) presented a comprehensive discussion of several case study failures of pile foundation supported structures during different earthquakes. A number of hypotheses were emerged while defining the reason of such failure. Bhattacharya et al. (2008) presented bending-buckling theory of failure mechanism of pile for analyzing the case study failure of Showa Bridge during 1964 Niigata earthquake. Later, the same group of researcher (Bhattacharya et al. 2014) has concluded that bending-buckling theory sole cannot explain the failure mechanism of pile of Showa Bridge and suggested resonating response of the whole structure would be governing factor to expedite bending–buckling interaction which lead to failure of the pile before commencement of lateral spreading event. On the other hand, many researchers reported that bridge failure has occurred mainly due to significant bending moment developed at the pile head due to lateral spreading caused during liquefaction occurrence (e.g., Hamada and O'Rourke 1992; Tokimatsu et al. 2005). Furthermore, the differential settlement occurred to the building structure supported on pile due to flow liquefaction is also observed in many earthquake events (1964 Alaska earthquake, 1990 Luzon earthquake, 1991 Costa Rica earthquake). From this view point, it has been realized that vulnerability assessment of pile foundation mainly in liquefiable deposit is indispensable. However, the complex dynamic interaction between superstructure-pile foundation-liquefied soil deposits is found to be an important factor which governs the seismic response of whole structure. On the contrary, kinematic soil movement in liquefiable deposit is also another important factor which may also govern the failure mechanism. Hence, these issues are to be addressed properly while calculating vulnerability of structure or foundation.

In fact, vulnerability assessment of structure needs to address uncertainty parameters associated with the system. Material and load characteristics are the key sources of uncertainty. Present study is focused on to study the effect of earthquake ground motion uncertainty on seismic response of pile foundation embedded in liquefiable deposit along with seismic vulnerability of pile foundation supported structure. Assessment of sustainable performance during seismic event an existing pile foundation requires quantification of risk considering variation of seismic loading. It is reported that input motion characteristics have a significant effect on vulnerability curves (Booth 2007). Choi et al. (2004) evaluated seismic vulnerability of six classes of typical bridges supported on pile foundation in California considering both material and load uncertainty. The study indicates that vulnerability of bridges subjected to either ground vibration or lateral spreading show significant correlation with the structural characterization. Seismic fragility curves were used in many studies to represent the probability of failure of structure and foundation as a function of ground motion intensity measures. Colangelo (2008) has discussed different fragility analysis methods are described and their advantages and disadvantages are discussed: (i) the safety factor method, in which the fragility curve is estimated based on safety margins with respect on an existing deterministic design; the numerical simulation method, in which the parameters of the fragility curve are obtained by (ii) regression analysis or (iii) maximum likelihood estimation from a set of nonlinear time history analysis at different seismic levels; (iv) the incremental dynamic analysis method where a set of accelerograms is scaled until failure.

Present study attempts to calculate vulnerability of pile foundation in liquefiable deposit incorporating the effect of variability of ground motion considering soil-pile foundation-structure interaction in the form of fragility curves. The case study of Showa Bridge pile foundation details is used to perform the reliability analysis. Probabilistic analysis is performed based on randomly generated IS code spectrum consistent synthetic ground motions. Monte Carlo simulation (MCS) is used to perform the analysis. Fragility curves are generated based on the 'failure' and 'success' information gathered from comparisons using maximum likelihood method. Serviceability limit state is considered in present study. Hence, the present study will help to give some useful inputs for designing pile foundation in liquefiable deposit.

2 Structural Modeling

OpenseesPL Edu version 2.7.2 (2018) software is used in present study to model the pile foundation system following (Bhattacharya et al. 2008) as presented in Fig. 1a. At first, the model pile foundation system is embedded in sandy soil. The consistency of soil is medium dense up to a depth of 10 m. However, the consistency changes to dense condition with a depth below 10 m. The pile is modeled using 3D deformable linear elastic beam column element, whereas a nonlinear elastic model is used to define the soil parameters in present study. The length of the pile is 25 m and the diameter is 0.6 m. The 25 m long pile passes through a four-phase system of air, water, liquefied soil, and non-liquefied soil surrounding it as shown in Fig. 1b. Meshing is done using a 4 nodded linear tetrahedron element. Total numbers of mesh elements generated are 144 in case of soil and 53 in case of pile. Both the side and bottom boundaries of the soil are constrained against lateral and vertical direction and pile head is considered to be free. The present analysis also assumes that the pile is stable under vertical settlement, hence, the support condition is considered as a hinged support at the tip of the pile. The axial load acting on the pile is calculated as 740 kN. Table 1 presents the detailed parameters of the pile and soil.

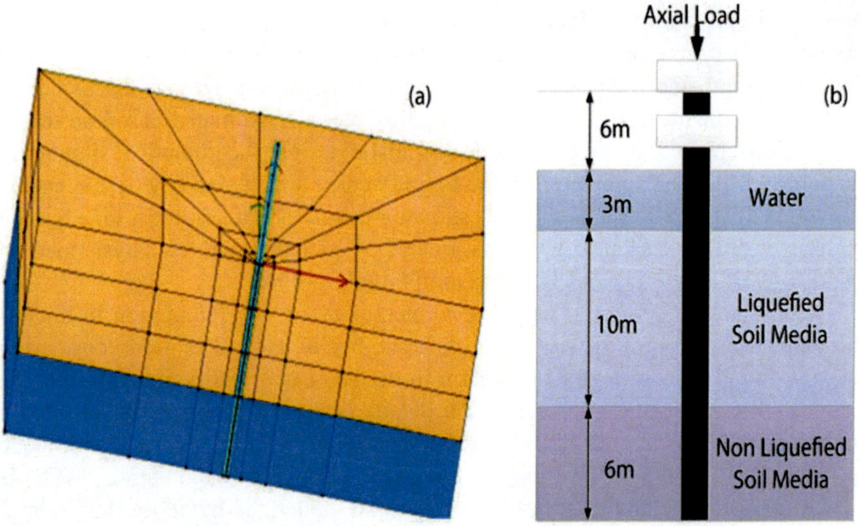

Fig. 1 **a** Numerical model of soil-piled structure system modeled in OpenseesPL 2.7.2, and **b** displacement-based model of Showa Bridge pile

Table 1 Properties of pile and soil used in study (Bhattacharya et al. 2014)

Pile data	Values
Pile diameter (m)	0.6
Pile length (m)	25
Young's modulus (kN/m^2)	2×10^7
Section modulus, Z (m^3)	0.0212
Flexural strength of pile, F_y (kN/m^2)	3050
Flexural moment of pile, M_y (kN-m)	64.660
Soil data	Values
Soil condition up to 10 m	Medium dense sand
Soil condition below 10 m	Dense sand
Liquefiable depth (m)	10
Density (kN/m^3)	19
Frictional angle	35°

3 Structural Modeling

3.1 Uncertainity Modeling of Ground Motion

The spectral ordinates considered in generation of ground motion are considered as variable parameter in the present study. Sigbjörnsson (2004) mentioned that the

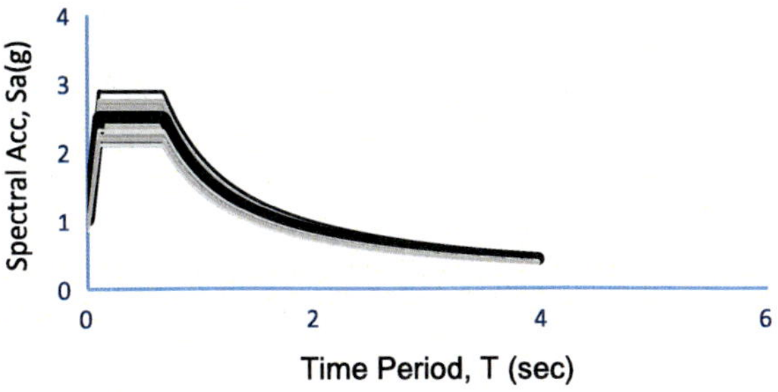

Fig. 2 Random generated acceleration spectra of IS 1893 Part 1 2016

variability associated with elastic response spectrum ordinates can be divided into three main classes as seismic source and attenuation variability (σ_{SE}), variability due to local geology and site condition (σ_{GS}) and variability associated with seismic force determination (σ_{RS}). Bea (1999) suggested the value of σ_{SE} in respect to peak ground acceleration divided by gravitational constant for different seismo-tectonic characteristics of a location. The value of σ_{SE} in this study is taken as 0.001 g. The variability due to σ_{GS} is taken as 0.004 g considering category of soil as class A (IS 1893 Part 1 2016). The variability due to σ_{RS} is taken as 0.003 g considering variability in modeling uncertainty. These three categories of variability are combined to determine the resultant variability (σ_R) of the response spectrum ordinate, which is utilized to determine the variability in the ground motion. The resultant variability (σ_R) is presented as follows,

$$\sigma_R = \sqrt{\sigma_{SE}^2 + \sigma_{GS}^2 + \sigma_{RS}^2} \tag{1}$$

In this study, we generally consider the ordinates of ground response as a random variable which is considered as log-normally distributed for the analysis. The assumption of log-normal distribution is valid because of non-negativity of the response spectrum ordinate and it has simple relation to normal distribution. In this study, we considered IS spectra (soft soil) as mean spectra considering 5% damping and COV of 10%. Figure 2 presents the mean response spectrum and randomly created spectrum ordinates for soil class A (IS 1893 Part 1 2016) using Monte Carlo simulation.

3.2 *Generation of Artificial Random Ground Motion*

The spectral Seismoartif (2012) is used for generation of artificial synthetic ground motions. This software is capable of generating artificial earthquake accelerograms matched to a specific target response spectrum using different calculation methods and varied assumptions. This software can thus be used to generate suites of accelerograms for nonlinear dynamic analysis of new or existing structures. In this study, total ten numbers of response spectra are randomly generated using this software and for each spectra eight numbers of ground motion are created. Total eighty numbers of

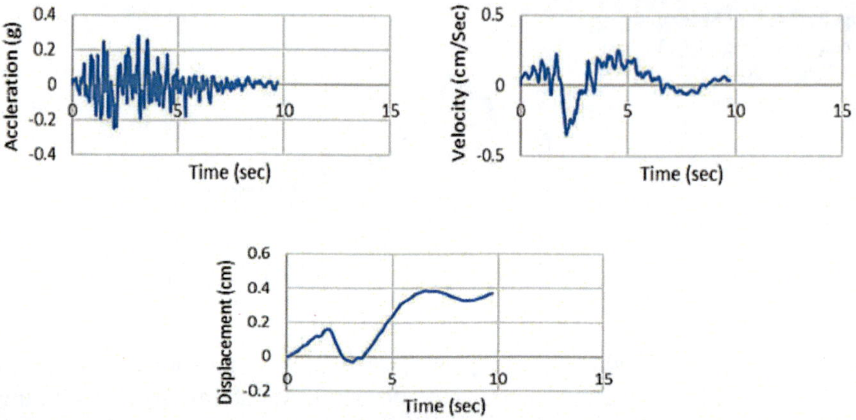

Fig. 3 Representative artificially generated synthetic ground motion

ground motions are considered in our study. Generated ground motions are basically spectra consistent ground motions. One randomly generated ground motion from spectral ordinates for zero period 0.651 g spectral acceleration is shown in Fig. 3.

3.3 Limit State Definition

Two limit states are defined based on global behavior as well as element deformation, namely collapse state and serviceability state. In this study, we have taken two limit states (i) collapse state and (ii) serviceability limit state. The allowable lateral displacement is set to a value of 30 mm for serviceability limit state and the collapse state is defined by maximum resisting moment of 64.660 kN-m as per Das et al. (2016). In this study, the fragility curve is developed only for serviceability limit state as it is an important criterion for vulnerability analysis of pile supported structure (Table 2).

Table 2 Limit states

Limit states	Criteria	Values
Collapse state	Max. resisting moment M_y, kN-m	64.660
Serviceable state	Max. pile head displacement, mm	30

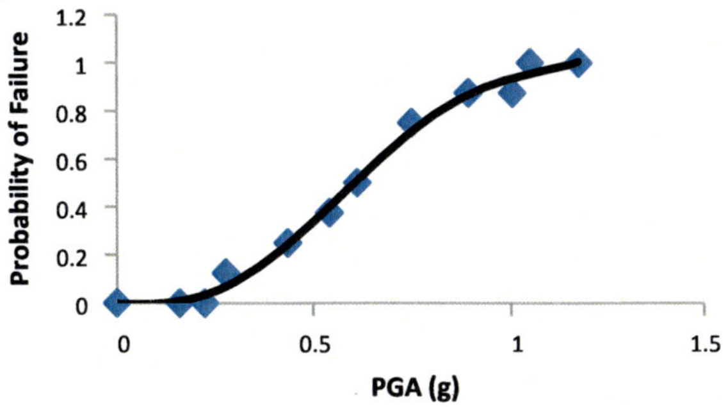

Fig. 4 Vulnerability curve for pile foundation in liquefied soil

4 Results and Discussion

The numerical model is developed and the following (Haldar and Babu 2009) is used for eighty artificial ground motion analysis. Figure 4 presents the fragility curve under serviceability limit state condition for different intensity measure level. It can be inferred from the figure that the failure starts at 0.279 g PGA and 100% failure occurs when PGA exceeds 1 g.

5 Conclusion

Present study attempts to assess the vulnerability of pile foundation embedded in liquefied deposit considering serviceability limit state in the form of fragility curve. Case study failure of Showa Bridge supported on pile foundation during 1964 Niigata earthquake is considered for the analysis. Ground motion variability is considered in present study to perform probabilistic analysis on 3D finite element model of soil-pile foundation-structure system. The outcome of present study indicates that the probability of failure of pile is found to be 100% at PGA of 1.0 g. Hence, IS code proposed design guideline need to revised based on the outcome of present study.

References

Bea RG (1999) Reliability based earthquake design guidelines for marine structures. J Waterway Port Coast Ocean Eng ASCE 125(5):219–231

Bhattacharya S, Dash SR, Adhikari S (2008) On the mechanics of failure of pile-supported structures in liquefiable deposits during earthquakes. Curr Sci 94(5):605–611

Bhattacharya S, Tokimatsu K, Goda K, Sarkar R, Shadlou M, Rouholamin M (2014) Collapse of Showa Bridge during 1964 Niigata earthquake: a quantitative reappraisal on the failure mechanisms. Soil Dyn Earthq Eng 65:55–71

Booth E (2007) The estimation of peak ground-motion parameters from spectral ordinates. J Earthq Eng 11:13–32

Choi E, DesRoches R, Nielson B (2004) Seismic fragility of typical bridges in moderate seismic zones. Eng Struct 26:187–199

Colangelo F (2008) On the computation of seismic fragility curves. In: The 14th world conference on earthquake engineering, Beijing, China

Das B, Saha R, Haldar S (2016) Effect of in-situ variability of soil on seismic design of piled raft supported structure incorporating dynamic soil-structure-interaction. Soil Dyn Earthq Eng 84:251–268

Haldar S, Babu GLS (2009) Probabilistic seismic design of pile foundations in non-liquefiable soil by response spectrum approach. J Earthq Eng 13:737–757

Hamada M, O'Rourke TD (1992) Case studies of liquefaction and lifeline performance during past earthquakes. Japanese case studies. Technical report NCEER-92-0001

IS: 1893 (2016) Part I. Bureau of Indian Standards. Indian Standard criteria for earthquake resistant design of structures. BIS, New Delhi, India

Meymand PJ (1998) Shaking table scale model tests of nonlinear soil-pile-superstructure interaction in soft clay. Doctoral dissertation, University of California, Berkeley

OPENSeeSPL (Edu Version 2.7.2) (2018) Pacific earthquake engineering research (PEER). https://cyclic.ucsd.edu/openseespl/

SeismoSoft (2012) SeismoArtif, version 1.0. 0 2012: SeismoArtif's help system©2002–2012. Seismosoft Ltd.

Sigbjörnsson R (2004) Uncertainty analysis of strong ground motion. In: 13th World conference on earthquake engineering, Vancouver, B.C., Canada, Paper No. 1536

Tokimatsu K, Suzuki H, Sato M (2005) Effects of inertial and kinematic interaction on seismic behavior of pile with embedded foundation. Soil Dyn Earthq Eng 25(7–10):753–762

1 g Shake Table Study on Seismic Behaviour of Model Structure Supported by Pile Foundation in Liquefiable Soil

Archana Kunwar, Swagata DebRoy, and Rajib Saha

Abstract Failure of pile foundation in liquefiable soil was evidenced in many past earthquakes. In fact, the complex interaction between soil, pile foundation and structure in liquefiable soil during seismic loading commonly known as seismic soil-pile foundation-structure interaction (SSPSI) is an important phenomenon which influences the dynamic behaviour of the whole structural system. Previously, SSPSI is generally disregarded in seismic design due to complexity in modelling and beneficial attributes of soil structure interaction as highlighted in previous codes. However, research indicates that argument still prevails with failure mechanism of pile foundation in liquefiable ground. In this context, present study is an attempt to assess the influence of dynamic interaction between soils, pile foundation and structure system during pre- and post-liquefaction stages by carrying out 1 g shake table experiments. Four physical models, consisting of 1×2 and 2×2, 3×3 and 4×4 pile groups supporting SDOF structure having fundamental period of 0.2 s and 2.0 s, respectively, are tested. Dynamic strain at pile head and pore water pressure is measured during pre- and post-liquefaction stages. Results indicate that pile foundation attracts higher forces and displacement during onset of liquefaction even though damping increases in the system. Finally, this study gives insight into the problem which may help in modifying existing seismic design guidelines for pile foundation supported structure in liquefiable soil.

Keywords Liquefaction · Shake table · SDOF structure · Dynamic strain · Pore water pressure

A. Kunwar · S. DebRoy (✉) · R. Saha
Department of Civil Engineering, National Institute of Technology Agartala, Barjala, Jirania, Agartala, Tripura 799046, India
e-mail: swagatadevroy@gmail.com

R. Saha
e-mail: rajib.iitbbsr@gmail.com

M. Latha Gali and P. Raghuveer Rao (eds.), *Geohazards*, Lecture Notes in Civil Engineering 86, https://doi.org/10.1007/978-981-15-6233-4_40

1 Introduction

The seismic design of pile foundation in liquefiable soil is a challenge for the engineers. In fact, foundations undergo severe distress during an earthquake. Liquefaction of cohesionless soils is one of the potential research areas of recent times because the awareness towards the liquefaction-related damages during seismic events and the need for mitigation of liquefaction hazards has increased worldwide. Design of foundations in earthquake prone areas needs special considerations. Shallow foundations may experience a reduction in bearing capacity and increase in settlement and tilt due to seismic loading. The reduction in bearing capacity depends on the nature and type of soil and ground acceleration parameters. In the case of piles, the soil-pile behaviour under earthquake loading is generally nonlinear. Hushmand et al. (1988) carried out centrifuge tests on saturated sand deposits to understand the time histories of accelerations and pore pressures during cyclic shaking in centrifuge models. Hamada (2000) investigated a case study on damage to foundation piles caused by the 1964 Niigata earthquake and the process of the damage to the piles by taking liquefaction-induced permanent ground displacement into consideration. He concluded that permanent displacement of non-liquefied soil overlaying the liquefied soil was a governing factor of the damage to the piles. Boulanger et al. (2003) mentioned key needs for advancing the design of pile foundations in soil profiles which are susceptible to liquefaction and lateral spreading during earthquakes. A series of large-scale dynamic centrifuge model tests were performed to study the behaviour of single piles and pile groups in a soil profile comprised of a non-liquefied crust spreading laterally over a loose-saturated sand layer. Further, Bhattacharya et al. (2008) analysed failure case studies of pile foundation during liquefaction and presented new hypothesis of bending-buckling mechanism which may be the probable cause of failure of pile in liquefied deposit. Use of shaking table tests to understand liquefaction of soils has been conducted by several researchers (Lombardi and Bhattacharya 2013; Sasaki et al. 1992). A recent shake table study by (Lombardi and Bhattacharya 2013) has investigated the pre- and post-seismic behaviours of pile foundation in liquefiable soil. The study presented some important observations on changes of fundamental period and damping forces on pile foundation with transition of prestage to poststage of liquefaction. However, present study is primarily motivated from the work of Lombardi and Bhattacharya (2013) and investigates whole system, i.e., soil-pile foundation–structure interaction on the seismic response of pile foundation and superstructure during before, onset and after liquefaction. Superstructure model in the form of a single degree of freedom (SDOF) oscillator is attached on the top of pile group foundation to incorporate the whole system interaction behaviour. Present study is limited to only investigating the effect of inertial interaction under harmonic loading. Finally, present study presents the result in the form of acceleration response at superstructure and foundation under time and frequency domains, dynamic strain at pile head with time and pore water pressure during different stages of liquefaction.

2 Methodology and Experimentation

2.1 Materials

2.1.1 Sand Bed Preparation

The sand bed consists of local river sand of Tripura. The index properties of the sand used for the test are given in Table 1. Tests were conducted on dry sand and saturated sand. Saturated sand bed was prepared by adding 27 kg of water to 90 kg of dry sand. Saturation is confirmed by physically observing submergence condition. The saturated sand bed was prepared for a total height of 340 mm in the container.

Tests were conducted on a rectangular laminar shear box (Fig. 1) which was made of aluminium alloy. A light weight laminar shear box is a flexible container in which the shear stiffness of the wall is proportional to the soil inside. The inner dimension of laminar box is 750 mm length, 400 mm width and 360 mm height. The container consists of 12 rigid rectangular; laminar frames were supported individually by bearings connected to an external frame. Weight of each lamina is 0.5 kg and total weight of the container was measured to be 11 kg. The container is then covered from inside with plastic to prevent any leakage. At the base of the container, wooden plank was placed to fix the pile bottom.

Table 1 Properties of sand

Description of properties	Value
Classification as per IS code	SM
Specific gravity	2.56
Maximum dry unit (kN/m^3)	16.23
Minimum dry unit weight (kN/m^3)	13.73
Maximum void ratio (e_{max})	0.83
Minimum void ratio (e_{min})	0.54
Coefficient of uniformity (C_u)	2.15
Coefficient of curvature (C_c)	1.33
Dry unit weight after placing in container (kN/m^3)	14.01
Saturated unit weight (kN/m^3)	18.47
Saturated water content (%)	30
% of sand	95
% of silt	05
% of clay	00

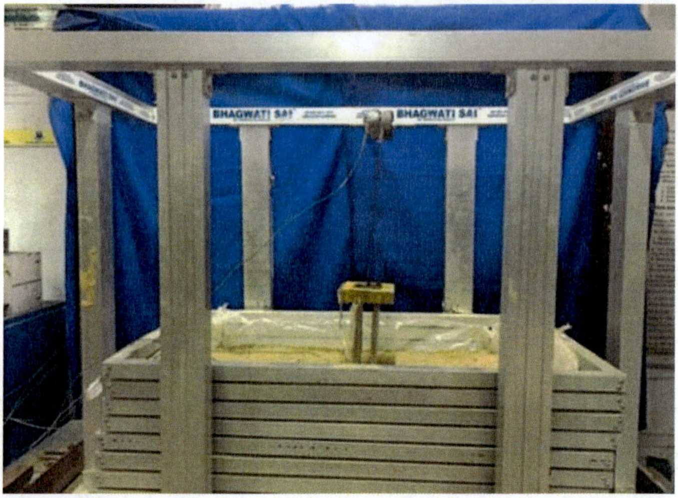

Fig. 1 Laminar shear box

2.1.2 Structural Model

Lumped mass SDOF oscillator is fabricated in order to represent superstructure in simplified manner. Measurement scale made of steel alloy is used to represent the column member and lumped mass is simulated by attaching circular mild steel rings of known weights. Figure 2 presents lumped mass column model used to represent superstructure in present study. Two different fundamental periods of structure, such as, 0.2 and 0.4 s are considered herein by adjusting the mass attached to column. Lateral stiffness of steel scale is calculated which is kept constant. Forced vibration test on SDOF superstructure model in fixed base condition is performed to verify

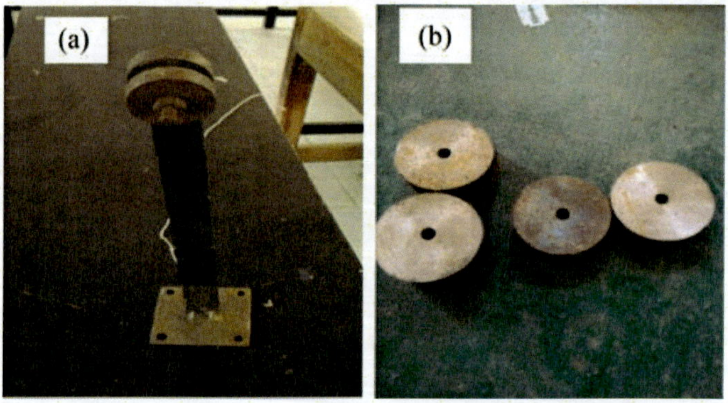

Fig. 2 **a** Lumped mass structure **b** different masses used

Table 2 Superstructure properties

Properties	Value
Length of steel column (*L*) (mm)	300
Width of steel column (*W*) (mm)	20
Thickness of steel scale (*T*) (mm)	2
Young's modulus (N/mm^2)	5.64×10^4
Moment of inertia (m^4)	1.33×10^{-11}

Table 3 Dimension and properties of pile

Properties	Value
Length (mm)	300
Inner diameter (mm)	10
Outer diameter (mm)	12
Mass (gm)	120
Young's modulus (kN/m^2)	2.1×10^8
Moment of inertia (m^4)	527×10^{-12}
Section modulus (m^3)	8.78×10^{-8}

the designed lateral period of structures. Dimension and properties of superstructure model are presented in Table 2.

Pile cap and hollow pile groups are made of MS plate and stainless steel alloy, respectively. Model dimensions are determined considering prototype pile groups after applying geometric scaling of 1:50 suggested elsewhere (Meymand 1998). Tables 3 and 4 present properties of pile material and prototype with model dimensions, respectively. Figure 3 presents model pile cap and hollow piles used in present study.

Table 4 Model and prototype size of different pile groups

Pile cap of different pile groups	Model weight of pile cap (gm)	Model dimensions ($L_m \times B_m \times T_m$) (mm)	Prototype dimensions ($L_p \times B_p \times T_p$) (m)
Single pile	250	50 × 50 × 24	2.5 × 2.5 × 1.2
1 × 2 pile	568	100 × 50 × 24	5 × 2.5 × 1.2
2 × 2 pile	1212	100 × 100 × 24	5 × 5 × 1.2
3 × 3 pile	1350	100 × 100 × 24	5 × 5 × 1.2
4 × 4 pile	1508	132 × 132 × 24	6.6 × 6.6 × 1.2

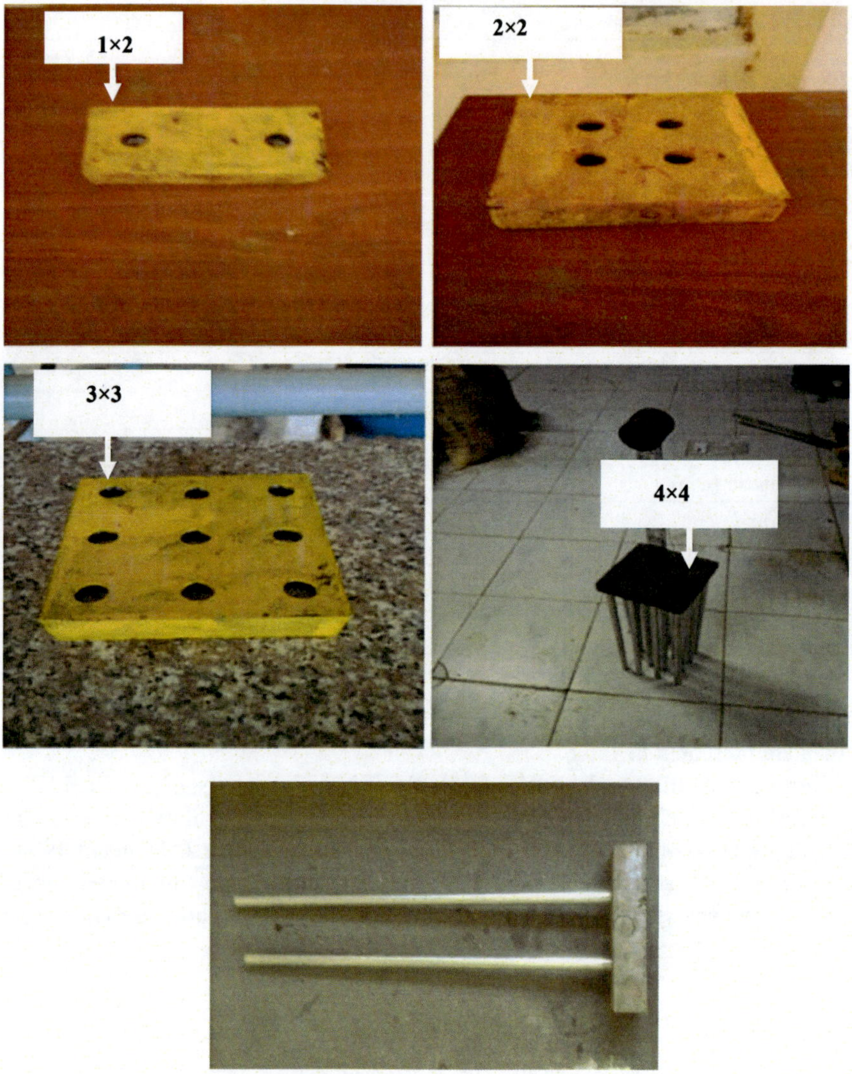

Fig. 3 Model pile caps and hollow pile

2.2 *Experimental Set-Up*

The experimental investigation is carried out in unidirectional 1 g shake table set-up available at structural engineering laboratory of NIT, Agartala. Seismic response of model pile foundation-supported SDOF structural models embedded in dry and saturated sandy bed are recorded with application of input swept sine motion. The acceleration responses at different locations of pile foundation-supported structure are recorded in dry and saturated sand, respectively, by accelerometers. Details of

Table 5 Properties of shake table

Component	Description of property	Value
Shake table	Size of table	1.0 m × 1.0 m
	Type of actuator	Electro-dynamic
	Range of forcing frequency	0.5–30 Hz
	Force capacity	9800 N
	Maximum displacement of table	±25 mm

Table 6 Properties of accelerometers

Component	Description of property	Value
Accelerometer	A1	10 g, 98.85 V/g
	A2	10 g, 97.98 V/g
	A3	10 g, 94.46 V/g

shake table and accelerometers are provided in Tables 5 and 6, respectively. Water pressure transducers were used to record the dynamic pore water pressure in soil during vibration. The transducer was inserted into the soil at a height of 100 mm below ground level and the readings were recorded in MICRON data acquisition system. Strain gauge (Fig. 4) was connected at the pile head to know the amount of strain developed on pile with time during seismic motion. The specifications of the strain gauge are given in Table 6. Schematic diagram of test set-up in shake table is presented in Fig. 5. Test matrix is presented in Table 7. Analysis of experimental

Fig. 4 Strain gauge attached to pile

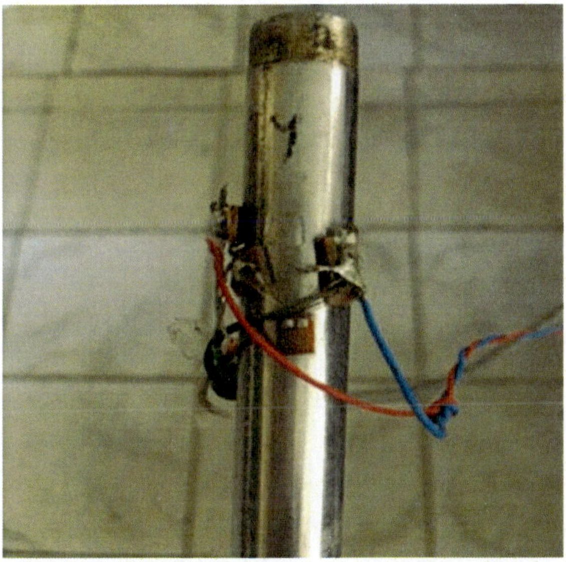

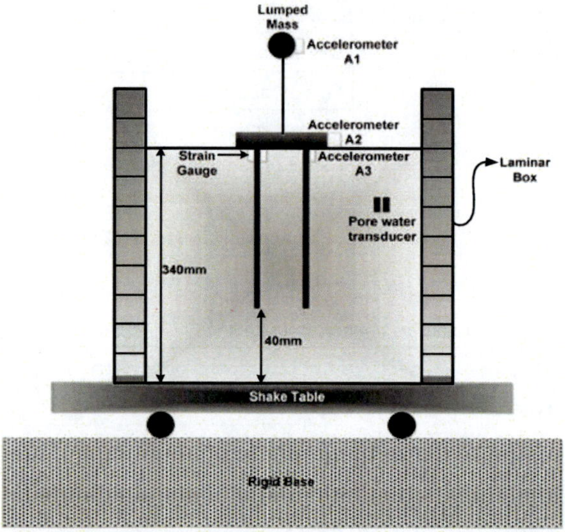

Fig. 5 Schematic diagram of the experimental set-up

observations also analysed in time and frequency domain is presented herein. The response of strain and pore water pressure was also recorded during pre-onset and post-liquefaction stage.

3 Results and Discussion

3.1 Fundamental Period of Structure

The fundamental natural period of the two different structural systems obtained from forced vibration test is presented in Table 7. The acceleration response of superstructure mass level in time and frequency domains (fast Fourier transformation, FFT) is presented in Figs. 6 and 7, respectively.

Table 7 Natural time period of model structure

Structure ID	Mass of structure (gm)	Obtained natural period (s)
m_1	200	0.2
m_2	600	0.4

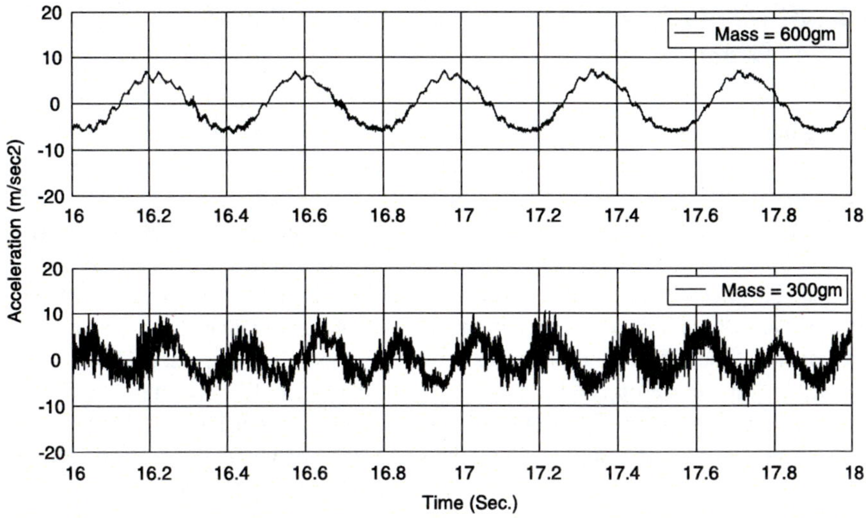

Fig. 6 Number of oscillation of SDOF structures for 1 s during forced vibration

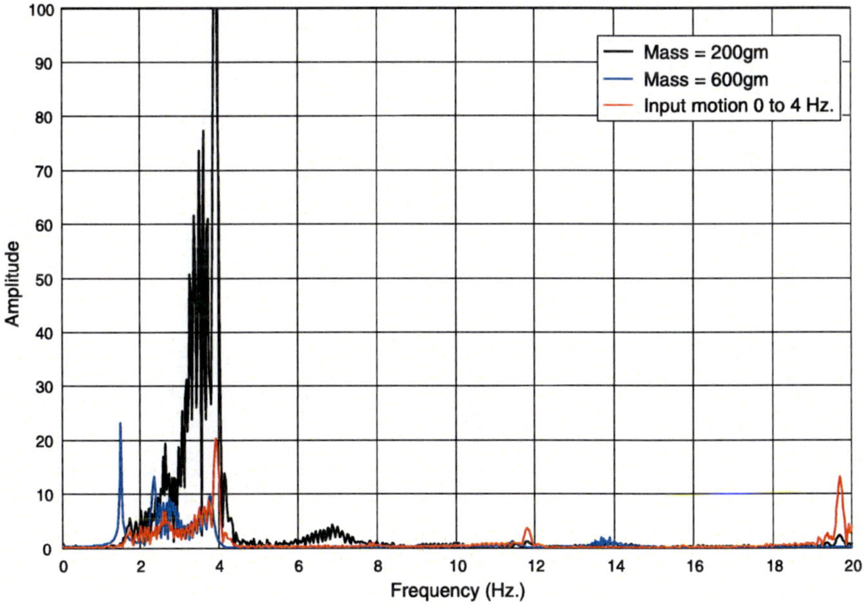

Fig. 7 FFT responses of two different SDOF structures under 0–4.0 Hz loading

3.2 *Effect of Liquefaction on Fundamental Period of Structure*

During liquefaction, fundamental period of the structure gets increased due to depth of fixity changes with time. In general, the depth of fixity gradually increases with the progress of liquefaction from top layer to bottom. The period lengthening is calculated from FFT response of the structure. FFT is performed on recorded time–acceleration responses for the structural systems embedded in loose dry and saturated fine sand bed. In case of saturated fine sand, several peaks in FFT response of structure are observed which indicate change in fundamental natural frequency of the structure with elapsed time, i.e., with the progress of liquefaction. Table 8 presents lengthened period for four pile groups consisting of 1 × 2 and 2 × 2, 3 × 3 and 4 × 4 pile groups supporting 0.2 and 0.4 s SDOF structures, respectively, are tested in both dry and saturated sand. For example, Fig. 8 presents FFT response of 2 × 2 pile group-

Table 8 Observed variation of fundamental period due to liquefaction

Mass of SDOF (gm)	T_{fixed} (s)	Time period (dry loose soil) (s)	Time period (liquefied soil) (s)
200	0.20	0.38 (1 × 2)	0.30^{min} to 0.60^{max}
600	0.40	0.48 (2 × 2)	0.40^{min} to 0.60^{max}
		0.33 (3 × 3)	0.42^{min} to 0.60^{max}
		0.42 (4 × 4)	0.48^{min} to 0.60^{max}

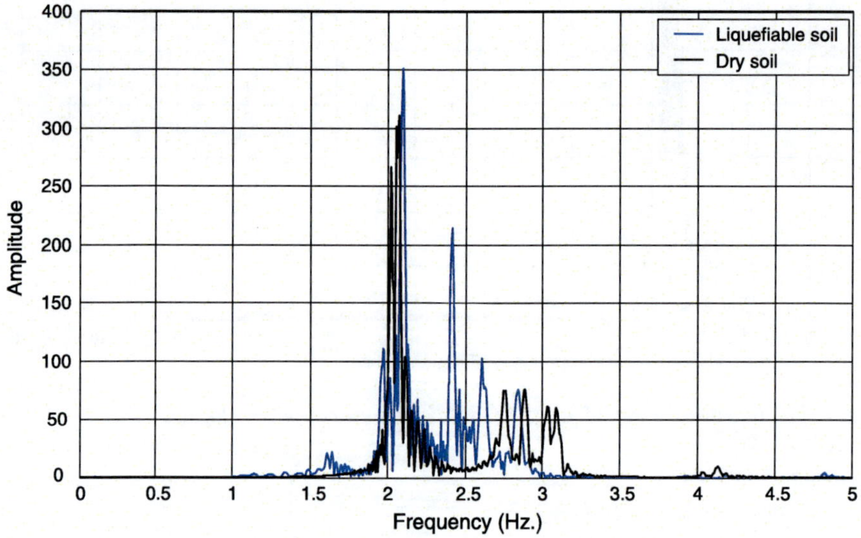

Fig. 8 Response of SDOF structure (0.4 s) in frequency domain supported on 2 × 2 pile group in dry and liquefiable soil under 0–4 Hz frequency range

supported SDOF structure having period of 0.40 s in dry and saturated soil. Some interesting observations are made in saturated fine sand deposit considering before and during liquefaction stages. It is observed that fundamental period changes in an interesting manner, i.e., it varies 0.40–0.60 s with the starting and during liquefaction stages which indicates the structure is behaving as fixed base condition in the initial stage of loading and then gradually effect of liquefaction has resulted lengthening of period. The reason may be due to densification of loose sand in initial stage of dynamic loading resulting the structure to behave as fixed base condition and gradually lengthening observed upto maximum 50% due to increase in pore water and initiation of liquefaction.

3.3 Effect of Liquefaction on Dynamic Response of Foundation

Physical models 1 × 2, 2 × 2, 3 × 3 and 4 × 4 were tested in dry sand and fully saturated sand. Figure 9 presents the time–acceleration, r_u and dynamic strain response of 1 × 2 pile group-supported SDOF structural system having fixed base period of 0.20 s. Figures 10, 11 and 12 presents the time–acceleration, r_u and dynamic strain response of 2 × 2, 3 × 3 and 4 × 4 pile group-supported SDOF structural system

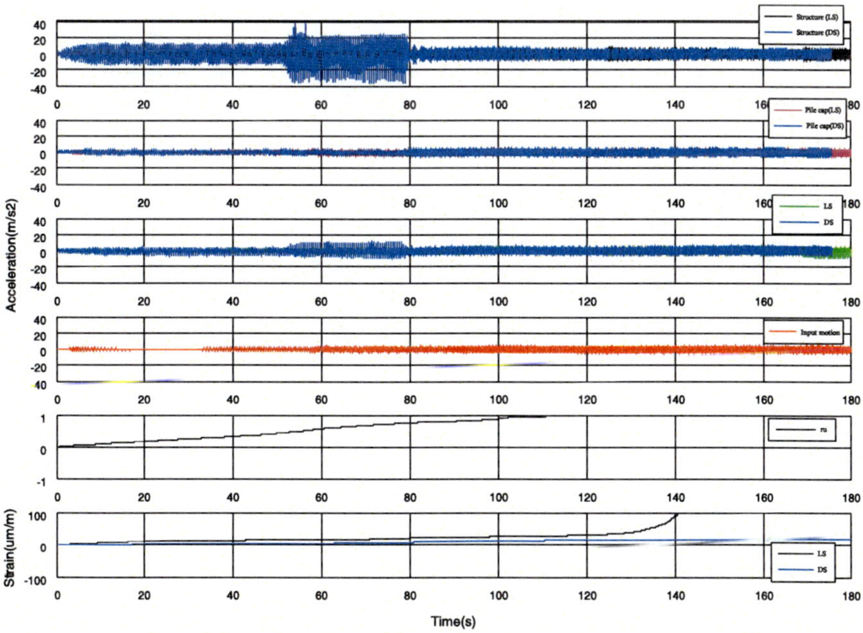

Fig. 9 Time–acceleration, r_u and dynamic strain response of SDOF (0.2 s) supported on 1 × 2 pile group in dry and liquefiable soil under 0–4 Hz frequency range

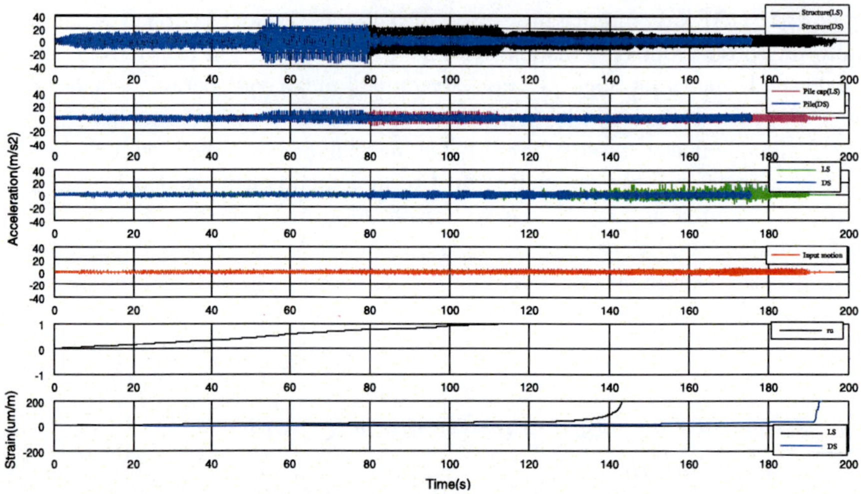

Fig. 10 Time–acceleration, r_u and dynamic strain response of SDOF structure (0.4 s) supported on 2 × 2 pile group in dry and liquefiable soil under 0–4 Hz frequency range

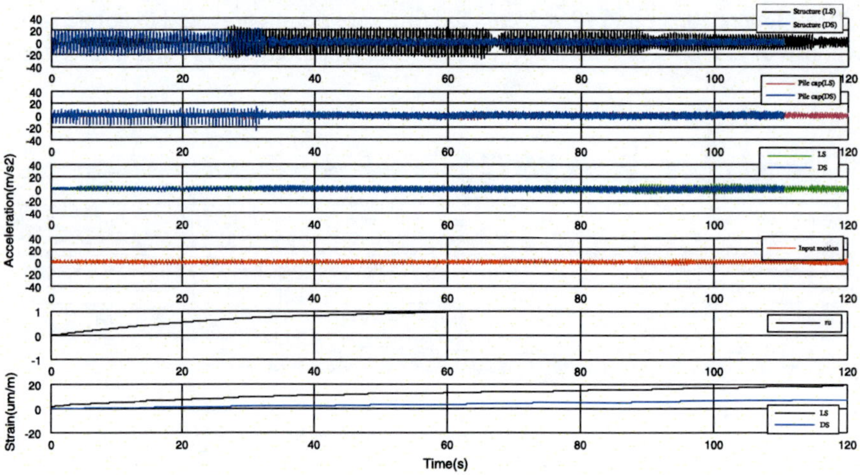

Fig. 11 Time–acceleration, r_u and dynamic strain response of SDOF (0.4 s) supported on 3 × 3 pile group in dry and liquefiable soil under 0–4 Hz frequency range

having fixed base period of 0.40 s, respectively. The acceleration response of dry sand and fully saturated sand is compared along with the increase in pore water pressure ratio which increases during liquefaction. Also the strain on pile foundation is compared for dry sand and liquefied sand. It was observed that before liquefaction in saturated sand, superstructure and pile foundation exhibit lesser acceleration as compared to dry soil response as well as liquefaction condition which infers that pile foundation attracts lesser shear force as compared to a non-liquefiable and dry

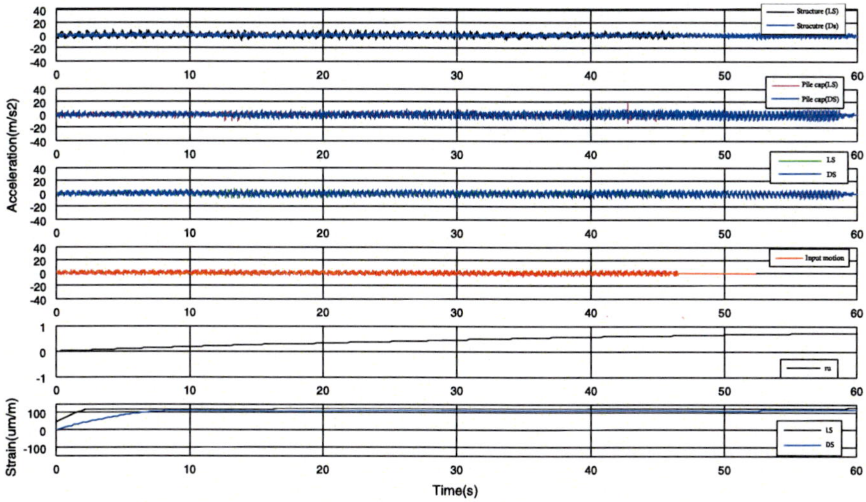

Fig. 12 Time–acceleration, r_u and dynamic strain response of SDOF (0.4 s) supported on 4 × 4 pile group in dry and liquefiable soil under 0–4 Hz frequency range

sandy soil. However, the pile head and superstructure acceleration get significantly increased due to initiation of liquefaction after around 80–100 s and further get subdued with the onset of liquefaction. The reason behind such increase in acceleration may be due to resonance with the input motion as it is already observed that fundamental period for the said structure may get lengthened to 0.60 s maximum range as observed in previous section. The dynamic strain subsequently with the progress of liquefaction gets significantly amplified indicating higher displacement of pile foundation which leads to failure of the foundation as seen in past earthquakes deformation.

References

Bhattacharya S, Dash S, Mitra N, Adhikari S, Blakeborough A (2008) Investigation of bending–buckling interaction of piles in liquefiable soils. In: Proceedings of world conference on earthquake engineering, Beijing, China

Boulanger RW, Kutter BL, Brandenberg SJ, Singh P, Chang D (2003) Pile foundations in liquefied and laterally spreading ground during earthquakes: centrifuge experiments & analyses. University of California, California, UCD/CGM-03/01

Hamada M (2000) Performances of foundations against liquefaction-induced permanent ground displacements. In: Proceedings of world conference on earthquake engineering, Auckland, New Zealand

Hushmand B, Crouse CB, Marti G, Scott RF (1988) Site response and liquefaction studies involving the centrifuge. Developments in Geotechnical Engineering, Elsevier

Lombardi D, Bhattacharya S (2013) Modal analysis of pile-supported structures during seismic liquefaction. Earthq Eng Struct Dyn

Meymand PJ (1998) Shaking table scale model tests of nonlinear soil-pile-superstructure interaction in soft clay, vol 4, pp 757–760

Sasaki Y, Towhata I, Tokida K-I, Yamada K, Matsumoto H, Tamari Y, Saya S (1992) Mechanism of permanent displacement of ground caused by seismic liquefaction. Soils Found 32(3):79–96

Studying and Comparing the Declustered EQ Catalogue Obtained from Different Methods for Guwahati Region NE India

Niranjan Borah and Abhishek Kumar

Abstract Declustering is a major step in any seismic hazard assessment approach and the choice of it controls the seismic activity estimation of the region. In this method, three different declustering methods are used for declustering the EQ catalogue collected around 500 km of Guwahati city. Completeness tests are performed on each of the declustered catalogue to find out the events which are complete with respect to magnitude and time. Further, complete catalogues are used to find out the Gutenberg–Richter (G–R) parameters from three different catalogues for the same seismotectonic province. Comparison is done on the basis of numbers of EQs in the declustered catalogues and on the basis of '*a*' and '*b*' parameters of the Gutenberg–Richter (G–R) recurrence law. Both '*a*' and '*b*' parameters are found varying based on the catalogues developed using each of the three methods clearly indicating that the choice of method may assign low to higher seismicity to the same region and same EQ catalogue.

Keywords Declustering · EQ catalogue · Seismicity · NE India · Seismicity parameters

1 Introduction

One of the most important information required for seismic study of an area is the earthquake (EQ) events data (EQ catalogue). EQ catalogues consist of two types of EQ events, independent EQs and dependent EQs. In seismic study, the phenomenon of EQ is generally modelled based on Poisson's distribution. This means that EQ occurs randomly with no memory of time, size, and location. Thus, all the dependent events which form seismicity cluster need to be separated from the main events. This process is called EQ declustering (van Stiphout et al. 2016). Proper-declustered EQ

N. Borah (✉) · A. Kumar
Indian Institute of Technology Guwahati, Guwahati, Assam 781039, India
e-mail: niran521152@gmail.com

A. Kumar
e-mail: abhiak@iitg.ac.in

M. Latha Gali and P. Raghuveer Rao (eds.), *Geohazards*, Lecture Notes in Civil Engineering 86, https://doi.org/10.1007/978-981-15-6233-4_41

catalogue is the key to the determination of seismic activity of the area/region and subsequently the seismic hazard. Use of different methods for the declustering of EQ events are in practice, irrespective of choice to be made/reason for selecting any specific method. As a result, each method yields different EQ catalogues and hence different seismic activity. This work presents a more insight into the effect of the choice of declustering method on seismic activity parameters for Guwahati region.

Knopoff (1964) first introduced a declustering algorithm by excluding the aftershocks from the analysis and found that the resultant catalogue showed Poissonian behaviour (van Stiphout et al. 2016). After this, Gardner and Knopoff (1974) found out a procedure to remove aftershock from the catalogue by using space and time distances between events. Gardner and Knopoff (1974) considered this space and time distances as a function of the magnitude of the main events and did not consider secondary and higher-order aftershock for the analysis. The method was applied to EQ catalogue collected for Southern California region. This method is well known as window method of declustering. Later, many researchers used this method for declustering the catalogues of different regions. Uhrhammer (1976) modified the length of the space–time windows. This modified version is also used by many researchers for declustering purpose (Pandey et al. 2017; Sitharam and Sil 2014; Kolathayar et al. 2012; Wiemer and Schorlemmer 2007). Reasenberg (1985) introduced another method of declustering by linking the EQs on the basis of hypocentral distances and time length between two events. Event having the highest magnitude was considered as main event. Reasenberg (1985) used this method for declustering EQ catalogue for Central California. Reasenberg (1985) method is also very popular among seismological community (van Stiphout et al. 2016).

In this study, each of the above three different methods are used for declustering EQ catalogue for Guwahati region and try to find the characteristics of these declustered catalogues. Guwahati along with other North Eastern (NE) part of India is situated in zone V which is the most seismically active region in India. Different researchers (Bahuguna and Sil 2018; Baro et al. 2018; Baro and Kumar 2017) have performed seismic study for Guwahati and its nearby areas. Higher seismicity of the area was also highlighted in their studies.

2 Methodology

In this work, three methods are used for declustering the EQ catalogue of Guwahati region of NE India. These include Gardner and Knopoff (1974) method, a modified version of this method by Uhrhammer (1976) and Reasenberg (1985) for declustering EQ events catalogue. Here, the first two methods are the window method of declustering and the third method is the linked window method. In these window-based methods, dependent events are identified in a space–time window around a large event (Gardner and Knopoff 1974; Knopoff and Gardner 1972). The large events are considered as independent events in the methods and except these, other events are removed. The window size is considered depending on the size of the main event;

larger the main event, bigger is the space–time window size. Approximate windows size as per Gardner and Knopoff (1974) is given in Eq. (1) (van Stiphout et al. 2016).

$$d = 10^{0.1238M+0.983} \text{ and } t = \begin{cases} 10^{0.032M+2.7389}, & \text{if } M \geq 6.5 \\ 10^{0.5409M-0.547}, & \text{else} \end{cases} \quad (1)$$

where

d Spatial window size in km
t time window size in days
M Magnitude of the EQ.

NDMA-2010 (2010) used a modified version of this Gardner and Knopoff (1974) method [by Uhrhammer (1976)] for declustering EQ events catalogue of entire India up to 2010. Uhrhammer (1976) modified the Gardner and Knopoff (1974) method and proposed the approximate spatial [d (km)] and temporal [t (days)] windows sizes. Governing correlations for this method are given in Eq. (2), where M is the magnitude of EQs.

$$d = \mathrm{e}^{-1.024+0.804M} \text{ and } t = \mathrm{e}^{-2.87+1.235M} \quad (2)$$

One other method widely used for declustering is the linked method which is based on space–time distance between two events (Reasenberg 1985; Frohlich and Davis 1990; Davis and Frohlich 1991). Reasenberg (1985) was the first one to introduce a method where events are linked to a cluster on the basis of spatial and temporal interaction zones. EQs occur within the interaction zone of a prior EQ is considered as an aftershock of that EQ as per Reasenberg (1985). Thus, these two events are considered in a cluster. When a new event is associated with an event which is previously associated in a cluster, then the new event becomes the member of the cluster. If two events from two different clusters fall within the interaction zone, then the two clusters merged to form a cluster. The spatial windows in this method are spherical as Reasenberg (1985) considered hypocentre of the events. The size of this window depends on the source dimension (r (M)) of the EQs, which is a function of M. In Reasenberg (1985) method, the spatial window of the ith event is considered spherical having a radius of $Qr(M_i)+r(M^*_{j,i})$. Where Q is a constant and $M^*_{j,i}$ is the magnitude of the largest event in the cluster. The temporal window (τ) of the method depends on the $M_{\max}$ and $M_{\min}$ of the catalogue and the time of the occurrence of the events.

$$\tau = \frac{-\ln(1-P)}{10^{\frac{2(\Delta M-1)}{3}}} \quad (3)$$

where $\Delta M = M_{\max} - M_{\min}$ and P is the probability of occurrence of a dependent event. τ value is in between 1 and 10 days (Reasenberg 1985).

3 Study Area and Its Seismicity

Guwahati is situated at a centre coordinate of 26.14° N and 91.74° E in NE India near the bank of the river Brahmaputra. It has the capital of Assam and is the gateway of NE India and is a very important city for Assam as well as for the whole NE India. Situated in zone V of seismic zonation of India, the area has suffered damage from many past EQs. The 1897 Assam EQ occurred very near to the Guwahati which had a magnitude of 8.1 Mw. 1869 Cachar EQ with magnitude 7.5 Mw caused visible damage in the area. These two EQs triggered many landslides in the nearby area and damaged many roads (Baro and Kumar 2015; Raghukanth et al. 2011; Kayal 2008; Oldham 1897). Many masonry buildings collapsed due to these EQs. Occurrence of frequent EQs and past great and major EQs makes the region as one of the most seismically active regions. Seismologists believe that there is an unbroken segment between the rupture zones of the events of 1897 and 1950. This is called as 'Assam Gap' and has the potential to cause EQ capable of causing severe damage in the region (Raghukanth and Dash 2010; Khattri and Wyss 1978).

4 Earthquake Data Catalogue

EQ data for this study is collected from various sources like NDMA-2010 (2010), USGS (2018), IMD (2018), ISC (2018), etc. A total of 6202 events are found to occur between 825 AD to April 2018 within a radius of 500 km around the Guwahati city (Centre coordinate 26.14° N and 91.74° E). The events data collected from these sources contains magnitudes in different scales. These scales are Richter magnitude (M_L), body wave magnitude (M_b), surface wave magnitude (M_s), and moment magnitude (M_w). In order to achieve the uniformity in scale, it is required to convert all the events magnitudes into one magnitude scale. For this study, the collected catalogue has been converted to Mw scale by using the following correlation developed by Sitharam and Sil (2014) for NE India.

$$\mathrm{Mw} = 0.862M_b + 1.034 \tag{4}$$

$$\mathrm{Mw} = 0.673M_L + 1.730 \tag{5}$$

$$\mathrm{Mw} = 0.625M_s + 2.350 \tag{6}$$

Above-collected EQ catalogue might have some repeated events, as data are collected from numerous different sources. These repeated events are identified on the basis of their same occurrence times, same locations, and on the basis of the close magnitude values. After removal of these repeated events, the numbers of events in the catalogue are reduced to 3379. Out of which 99 events have magnitude less than

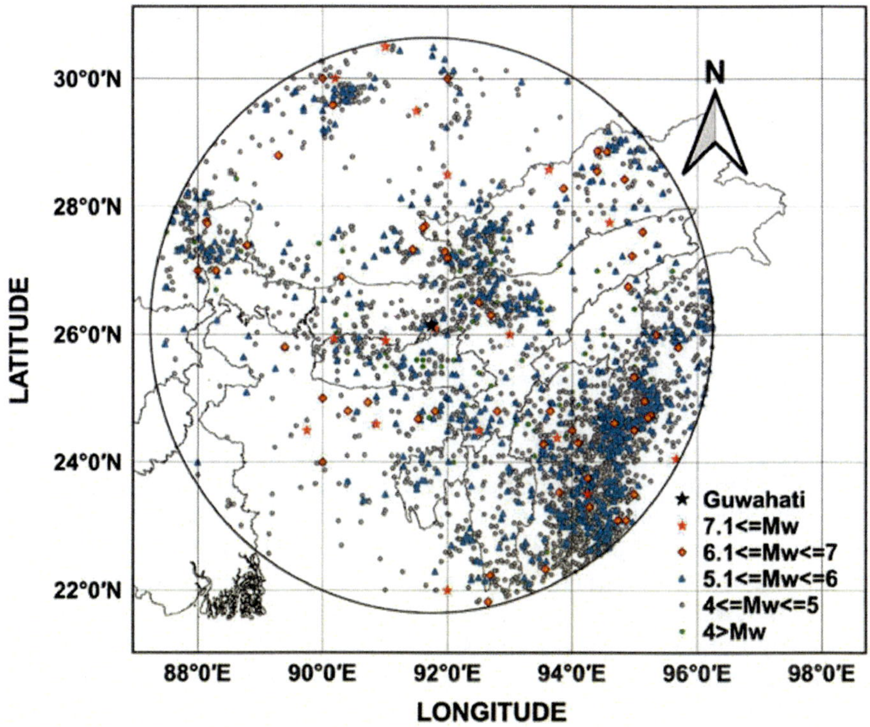

Fig. 1 Past EQs within 500 km radius of Guwahati city centre (3379 EQ events)

4 Mw. The numbers of events having magnitude (Mw) between 4.1–5, 5.1–6, and 6.1–7 are 2498, 707, and 58, respectively. Seventeen numbers of events have magnitude greater than 7. The highest magnitude event in the catalogue has a magnitude of 8.1.

Figure 1 shows the past seismic activity of the area around 500 km of central coordinate of Guwahati city. Total 3379 EQ events are shown in Fig. 1. From the figure, it can be seen that the seismic activity of the south-east part of the area is more than the other part. South west part of the area is less active region.

5 Declustering the Catalogue

Collected EQ catalogue contains both main events and dependent events. Poisson's distribution is used to model the phenomenon of EQ consider for that EQ occurs randomly with no memory of time, size, and location. So, to model the phenomenon of EQ, only main events are considered. For this, dependent events need to be separated from the catalogue. This is done using different declustering methods. The declustering methods mentioned above are applied on this EQ catalogue by using the

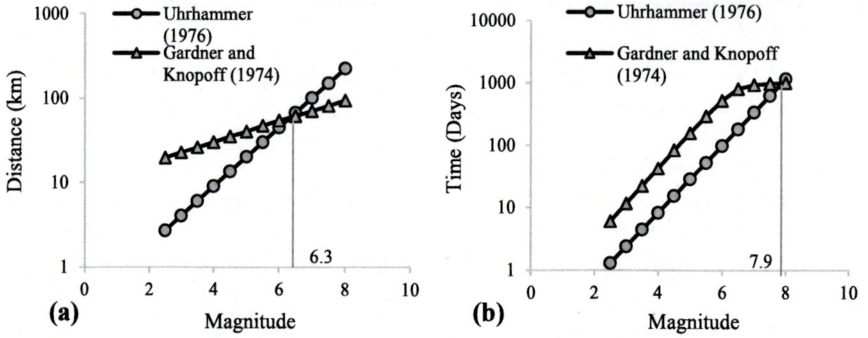

Fig. 2 Comparison between Gardner and Knopoff (1974) and Uhrhammer (1976) **a** space window and **b** time window

MATLAB codes received from Dr. Jiancang Zhuang, co-author of van Stiphout et al. (2016) (based on personal communication). In case of Gardner and Knopoff (1974) method, the numbers of events remain in the catalogue is 1314. When the modified window size by Uhrhammer (1976) is used, 2664 numbers of independent events are found in the catalogue as the main events. So, it is seen that the numbers of events obtained by declustering using original Gardner and Knopoff (1974) windows is less than Uhrhammer (1976) windows. If the space–time windows of the two methods are compared (Fig. 2), it is seen that the time window in Gardner and Knopoff (1974) is larger than Uhrhammer (1976) time window up to a magnitude of 7.9. Similarly, for space window, up to a magnitude of 6.3, the space window of Gardner and Knopoff (1974) is larger than Uhrhammer (1976). This is the reason for which the numbers of main events obtained after applying Gardner and Knopoff (1974) is less than the numbers of main events obtained after applying Uhrhammer (1976) window.

Similarly, a total 3067 numbers of main events are obtained when Reasenberg (1985) method is used for declustering. As the length of windows is small and the whole declustering process is based on every event, the numbers of main events obtained in this process are more than the other two methods. Besides these, the Reasenberg (1985) method consider the hypocentre distances of the events and for this reason also the numbers of main events obtain in this method is more than the other two methods as hypocentral distance between two events is more than the epicentral distance between them.

6 Completeness of the Earthquake Catalogues

As collection of data is based on instrumental and historical data, it is possible that not all the past events may present in the catalogue. If the whole catalogue is directly used for hazard analysis, then it may give underestimated result. To get proper result, only the complete part of the catalogue needs to be used. The catalogue may not contain complete events for the whole range of magnitude and may not complete for all the years. So, it is required to check which magnitude ranges are complete in the catalogue and for how many years and use only these complete part of the catalogue for further analyses. In this report, completeness test is carried out for all the catalogues obtained by using all the declustering methods. Completeness with respect to magnitude and time is compared among each of the above three declustered catalogues obtained based on each methods. To check the completeness with respect to magnitude, a minimum magnitude of the catalogue is estimated and the catalogue is considered as complete for the EQs having magnitude higher than this minimum magnitude. This minimum magnitude is called the magnitude of completeness (M_c) (Rydelek and Sacks 1989). In this report, maximum curvature (MAXC) method by Wiemer and Wyss (2000) is used to check the completeness of the catalogue with respect to magnitude. For this, cumulative numbers of EQ events are plotted against each magnitude and thus, a curve is generated between them. As per Wiemer and Wyss (2000), the magnitude corresponds to the maximum curvature point is the M_c value.

Figure 3 shows the magnitude of completeness for all three declustered catalogues. Figure 3a shows that the maximum curvature for the declustered catalogue obtained by using Gardner and Knopoff (1974) window method is occurred at a magnitude of 4, which is the magnitude of completeness for this catalogue. Similarly, Fig. 3b, c show the magnitude of completeness, $M_c = 4$ for the declustered catalogues obtained by using Uhrhammer (1976) window method and Reasenberg (1985) method, respectively. In all the three cases, the magnitude of completeness is found to be 4.

To find out completeness with respect to time, Stepp (1972) proposed a method which found the completeness of a certain magnitude for a definite length of period. To do this, firstly, the magnitudes are grouped into certain magnitude ranges and test is performed on each group. In each group, as per Stepp (1972), the catalogue is divided into time bins. These bins consist of EQs in that time interval from the last year of data present in the catalogue. Stepp (1972) method considers that the Poisson distribution model is followed by EQ events and for this, if $x_1, x_2, x_3, \ldots, x_n$ are the number of EQ events for a definite unit time interval, then an unbiased mean rate (λ) for each of the time interval can be calculated by following formula.

$$\lambda = \frac{1}{n}\sum_{i=1}^{n} x_i \tag{7}$$

And for this dataset, the variance will be

$$\sigma_\lambda^2 = \frac{\lambda}{n} \tag{8}$$

where n is the unit time interval. If one year is taken as the time interval, then n will be equal to T, where T is the sample length. Then, the standard deviation (σ_λ) will become:

$$\sigma_\lambda = \sqrt{\frac{\lambda}{T}} \tag{9}$$

From the study, Stepp (1972) found that, for constant λ in each magnitude class, σ_λ behaves as $\frac{1}{\sqrt{T}}$. For the completeness test, the standard deviation (σ_λ) versus time interval was plotted for all the group of magnitudes and σ_λ was plotted against T in the same graphs. It was concluded that the plotted points follow a straight-line slope and as long as it is parallel to the $1/\sqrt{T}$ line, for that T, the data was concluded to be complete. Stepp (1972) method is very widely used to find out the completeness of an EQ catalogue with respect to time (also used by NDMA-2010 (2010)).

For the present study, the EQ events from each of the catalogues are grouped into four magnitude classes: (1)$4 \le \text{Mw} \le 5$, (2)$5.1 \le \text{Mw} \le 6$, (3)$6.1 \le \text{Mw} \le 7$ and (4)$7.1 \le \text{Mw}$. Then for each catalogue, Stepp (1972) method is used for all the magnitude classes EQ events. In Fig. 4, σ_λ versus T for different magnitude classes for different declustered catalogues are plotted. These plots of points are

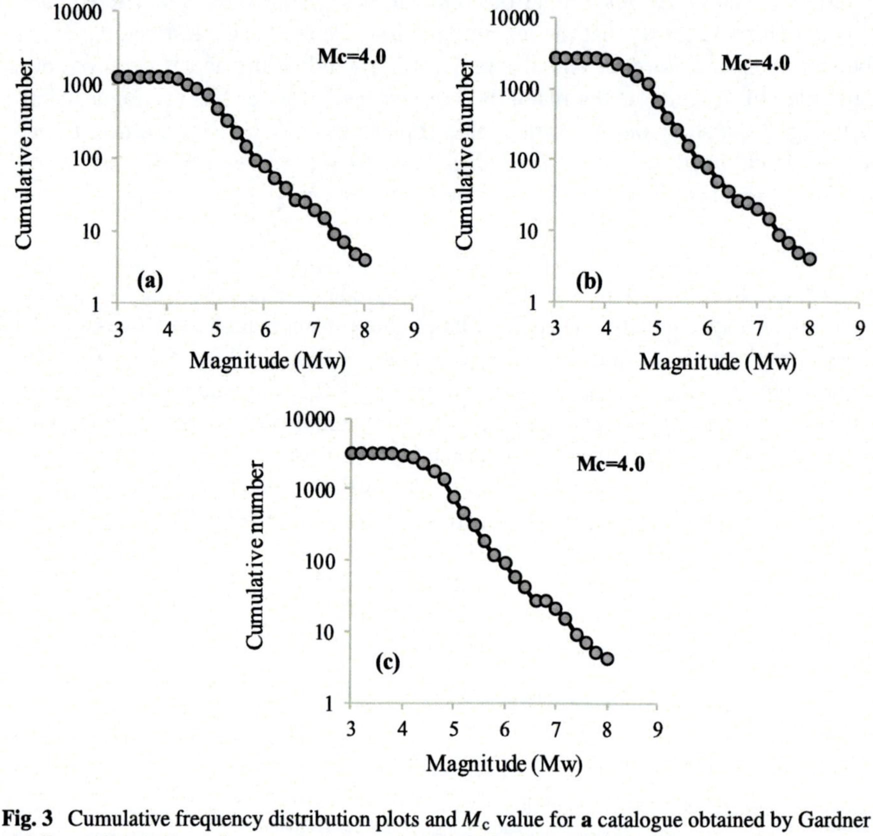

Fig. 3 Cumulative frequency distribution plots and M_c value for **a** catalogue obtained by Gardner and Knopoff (1974) window method, **b** catalogue obtained by Uhrhammer et al. (1996) window method, and **c** catalogue obtained by Reasenberg (1985) method

then compared with the $1/\sqrt{T}$ line; as long as the slope of these plotted points are parallel to the $1/\sqrt{T}$ line, the data is complete. Figure 4a–c show the completeness of catalogue with respect to time as per Gardner and Knopoff (1974) window method, Uhrhammer (1976) window method, and Reasenberg (1985) method, respectively. Table 1 summarizes the year of completeness for different magnitude classes for each of the three catalogues.

It is seen from Table 1 that the year of completeness for the magnitude class $4 \leq$ Mw ≤ 5, the year of completeness for declustered catalogue obtained by Reasenberg (1985) method is 30 years, and for both the other cases, it is 40 years. For magnitude class $6.1 \leq$ Mw ≤ 7 for Gardner and Knopoff (1974) the year of completeness is 100 and for the other two methods it is 110 years. Other than these two cases, for all other cases, the year of completeness is same for the three catalogues.

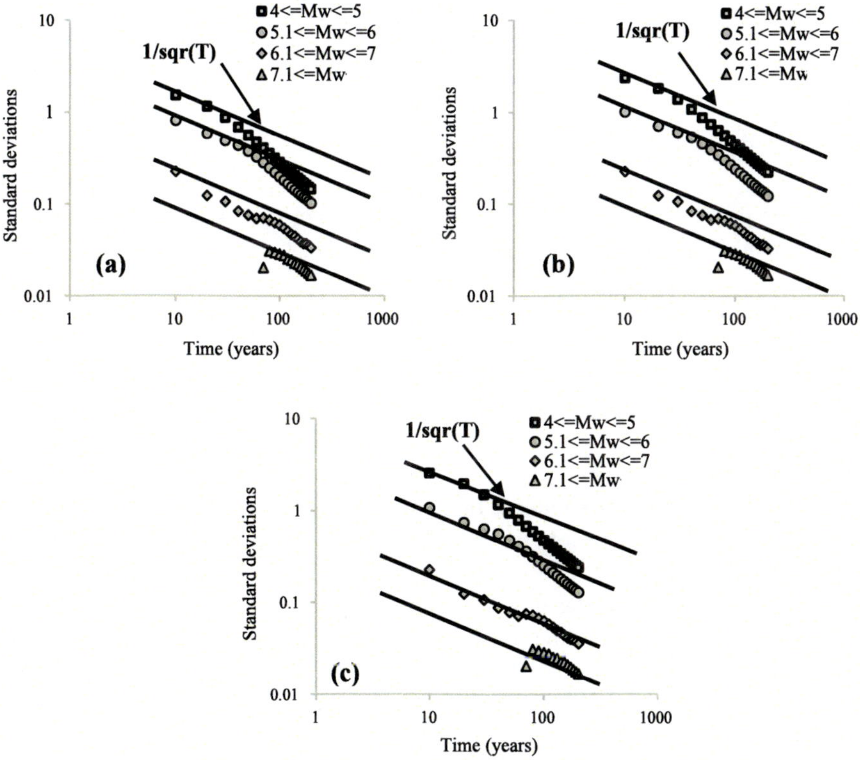

Fig. 4 Standard deviation versus time interval plots for different magnitude classes for **a** data obtained by using Gardner and Knopoff (1974) window, **b** data obtained by using Uhrhammer (1976) window, and **c** data obtained by using Reasenberg (1985) method

Table 1 Year of completeness for different magnitude classes for all the catalogues

Methods	Magnitude classes	Years of completeness
Gardner and Knopoff (1974)	$4 \leq Mw \leq 5$	40
	$5.1 \leq Mw \leq 6$	60
	$6.1 \leq Mw \leq 7$	100
	$7.1 \leq Mw$	160
Uhrhammer (1976)	$4 \leq Mw \leq 5$	40
	$5.1 \leq Mw \leq 6$	60
	$6.1 \leq Mw \leq 7$	110
	$7.1 \leq Mw$	160
Reasenberg (1985)	$4 \leq Mw \leq 5$	30
	$5.1 \leq Mw \leq 6$	60
	$6.1 \leq Mw \leq 7$	110
	$7.1 \leq Mw$	160

7 Seismicity Parameters

Seismic activity of an area is known from the values of seismic hazard parameters. According to Gutenberg–Richter (G–R) recurrence relation law,

$$\log N(M) = a - bM \tag{10}$$

In the above Eq. (10), the '*a*' and '*b*' are the seismicity parameters.

Here, *N* (*M*) is the total number of EQs which are greater than or equal to the magnitude *M*. Seismic activity of an area is represented by the parameter '*a*'. '*a*' value represents the logarithm of the numbers of EQ occurring in an area per year. The value '*b*' indicates the relative likelihood of larger and smaller EQs. As '*b*' value increases, the number of smaller EQ increases compared to larger EQs. The value of '*a*' and '*b*' can be found out from regression analysis of the annual rate of exceedance versus magnitude plot. Figure 5 shows the annual rate of exceedance versus magnitude plot for all the three declustered catalogues.

From Fig. 5, it can be seen that the value of 'a' is more in the declustered catalogue obtained by Reasenberg (1985) method ($a = 5.824$) than Gardner and Knopoff (1974) method ($a = 5.2607$) and Uhrhammer (1976) method ($a = 5.6418$). As it is mentioned earlier that the number of main events in case of Reasenberg (1985) method of declustering is more than the other two methods. This is the reason for finding high seismicity value (i.e. '*a*') for the declustered catalogue obtained from Reasenberg (1985) method. The '*b*' parameter for declustered EQ catalogue obtained by Gardner and Knopoff (1974), Uhrhammer (1976), and Reasenberg (1985) methods are 0.9099, 0.9339, and 0.9592 respectively. In case of catalogue declustered by Reasenberg (1985) method, more numbers of smaller magnitude EQ events are present and this is the reason that the '*b*' value is more than the other two methods. Due to relatively

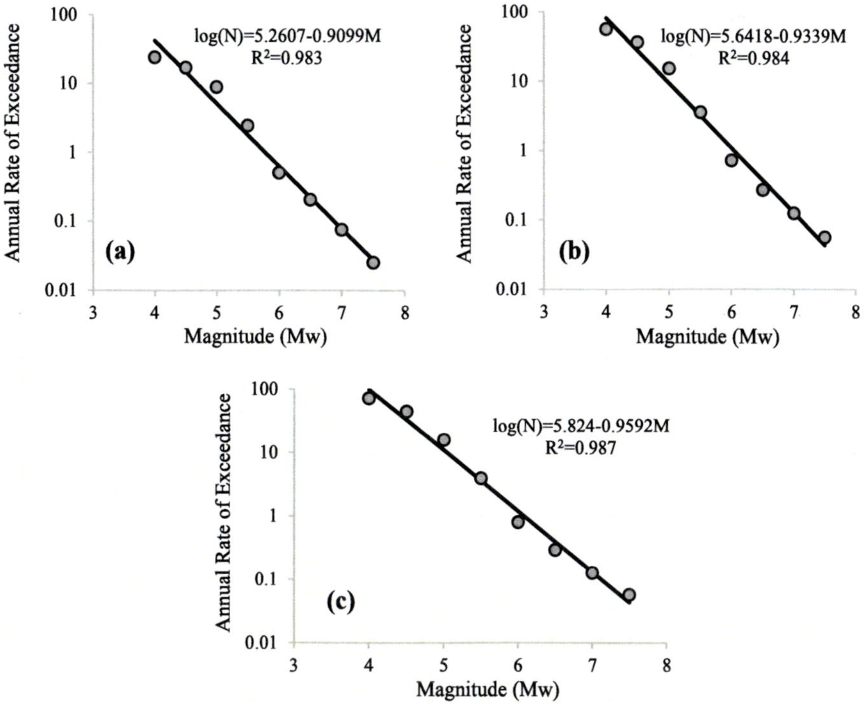

Fig. 5 Gutenberg Richter relation for declustered EQ catalogue based on; **a** Gardner and Knopoff (1974) window; **b** Uhrhammer (1976) window; **c** Reasenberg (1985) method

high window sizes, Gardner and Knopoff (1974) methods eliminate more numbers of smaller magnitude EQ events. For this reason, lowest '*b*' value is obtained by Gardner and Knopoff (1974) amongst the three methods than that obtained by using the other two methods. As per NDMA-2010 (2010), the '*b*' value for Shillong Plateau and Assam valley, Naga Thrust, Bengal Basin, Indo-Burmese Arc and Eastern Himalaya are found as 0.73, 0.67, 0.74, 0.80, and 0.71, respectively. Sil and Sitharam (2013) found that the '*b*' value for Shillong Plateau and Assam valley, Naga Thrust, Bengal Basin, Indo-Burmese range and Eastern Himalaya are 0.61, 0.54, 0.61, 0.79, and 0.86, respectively. As per Baro and Kumar (2017), the values of '*b*' are 0.91, 0.80, 0.89, and 0.94 for Shillong Plateau and Assam valley, Bengal Basin, Eastern Himalaya, and Indo-Burmese range, respectively. In current study, '*b*' values are found out considering the whole study area as one. It can be seen from the results that the values of '*b*' parameters for all the three declustered EQ catalogues are changing with respect to each other.

The '*a*', '*b*', and R^2 values obtained in the present study are shown in Table 2. On the basis of the R^2 value, it can be seen that data in case of declustered catalogue by Reasenberg (1985) method gives the closest result to the fitted regression line. But other two methods are also giving good R^2 values and those are very near to that of the declustered catalogue obtained by Reasenberg (1985). So, from these results, it

Table 2 Seismicity parameters obtained from all the three catalogues (rounded values)

Parameters	Gardner and Knopoff (1974)	Uhrhammer (1976)	Reasenberg (1985)
a	5.26	5.64	5.82
b	0.91	0.93	0.96
R^2	0.983	0.984	0.987

can be said that all of these methods are good for applying in EQ catalogue for NE India region. For NE India region, most EQ events data are available; Reasenberg (1985) method can be used for the area and this method is giving the best R^2 values in recurrence relation out of all three methods. But this method should be avoided for the areas where EQ events data is not properly available. Uhrhammer (1976) method is very good for the region as it gives second best R^2 value and it depends only on the size of the main events. Gardner and Knopoff (1974) method gives less numbers of declustered EQ events. It may possible that some main events are removed by the method.

8 Conclusions

Choice of declustering method controls the seismic activity parameters and consequently, the seismic hazard values. Window methods eliminate more numbers of EQ from the catalogue as compared to the linked window method by Reasenberg (1985). This results in higher '*a*' parameters in Reasenberg (1985) declustered catalogue than the other two methods. In window methods, considering Gardner and Knopoff (1974) window eliminate more dependent events than considering modified window by Uhrhammer (1976). In Gardner and Knopoff (1974) and Uhrhammer (1976) method, the window size is independent of the dependent events magnitude. Thus, the declustering is based only on the sizes of the main events. In Reasenberg (1985), size of all events take part in the declustering process. If some events are missing from the catalogue, then it will not affect the declustering process in case Gardner and Knopoff (1974) method is used. However, it will affect in the declustering process when Reasenberg (1985) method is used. Thus, it is observed that for same region and same set of initial EQ data, three declustering methods show a change in seismic activity parameters from present work.

References

Bahuguna A, Sil A (2018) Comprehensive seismicity, seismic sources and seismic hazard assessment of Assam, North East comprehensive seismicity, seismic sources and seismic. J Earthq Eng 0:1–44. https://doi.org/10.1080/13632469.2018.1453405

Baro O, Kumar A (2015) A review on the tectonic setting and seismic activity of the Shillong Plateau in the light of past studies. Disaster Adv 8:34–45

Baro O, Kumar A (2017) Seismic source characterization for the Shillong Plateau in Northeast India. J Seismol 21:1229–1249. https://doi.org/10.1007/s10950-017-9664-2

Baro O, Kumar A, Ismail ZA (2018) Seismic hazard of the Shillong Plateau. Geomatics Nat Hazards Risk

Davis SD, Frohlich C (1991) Single-link cluster analysis, synthetic earthquake catalogues, and aftershock identification. Geophys J Int 104:289–306. https://doi.org/10.1111/j.1365-246X.1991.tb02512.x

Frohlich C, Davis SD (1990) Single-link cluster analysis as a method to evaluate spatial and temporal properties of earthquake catalogues. Geophys J Int 100:19–32

Gardner JK, Knopoff L (1974) Is the sequence of earthquakes in Southern California, with aftershocks removed, Poissonian? Bull Seismol Soc Am 64:1363–1367. https://doi.org/10.1785/0120160029

Indian Meteorological Department IMD. https://www.imd.gov.in/pages/earthquake_prelim.php. Last accessed 2018/05/01

ISC event catalogue search. https://www.isc.ac.uk/iscbulletin/search/catalogue/. Last accessed 2018/05/01

Kayal JR (2008) Microearthquake seismology and seismotectonics of South Asia. McGraw Hill Publication, India

Khattri K, Wyss M (1978) Precursory variation of seismicity rate in the Assam area, India. Geology 6:685–688. https://doi.org/10.1130/0091-7613(1978)6<685:PVOSRI>2.0.CO;2

Knopoff L (1964) The statistics of earthquakes in Southern California. Bull Seismol Soc Am 54:1871–1873

Knopoff L, Gardner JK (1972) Higher seismic activity during local night on the raw worldwide earthquake catalogue. Geophys J R Astron Soc 28:311–313. https://doi.org/10.1111/j.1365-246X.1972.tb06133.x

Kolathayar S, Sitharam TG, Vipin KS (2012) Spatial variation of seismicity parameters across India and adjoining areas. Nat Hazards 60:1365–1379. https://doi.org/10.1007/s11069-011-9898-1

NDMA (2010) Development of probabilistic seismic hazard map of India technical report. National Disaster Management Authority, 126

Oldham RD (1899) Report on the great earthquake of 12 June 1897. Mem. Geol. Soc. India, 379 p

Pandey AK, Chingtham P, Roy PNS (2017) Homogeneous earthquake catalogue for Northeast region of India using robust statistical approaches, 5705. https://doi.org/10.1080/19475705.2017.1345794

Raghukanth STG, Dash SK (2010) Deterministic seismic scenarios for North East India. J Seismol 14:143–167. https://doi.org/10.1007/s10950-009-9158-y

Raghukanth S, Dixit J, Dash S (2011) Ground motion for scenario earthquakes at Guwahati city. Acta Geodaet Geophys Hung 46:326–346. https://doi.org/10.1556/AGeod.46.2011.3.5

Reasenberg P (1985) Second-order moment of Central California seismicity, 1969–1982. J Geophys Res 90:5479–5495

Rydelek PA, Sacks IS (1989) Testing the completeness of earthquake catalogues and the hypothesis of self-similarity. Nature 337:251–253. https://doi.org/10.1038/337249a0

Sil A, Sitharam TG (2013) Site response evaluation of Agartala city using geophysical and geotechnical data. Int J Geotech Earthq Eng 4:53–73. https://doi.org/10.4018/ijgee.2013070104

Sitharam TG, Sil A (2014) Comprehensive seismic hazard assessment of Tripura and Mizoram states. J Earth Syst Sci 123:837–857. https://doi.org/10.1007/s12040-014-0438-8

Stepp J (1972) Analysis of completeness of the earthquake sample in the Puget sound area and its effect on statistical estimates of earthquake hazard. In: International conference on microzonation, Seattle, USA, pp 897–910

Uhrhammer RA (1976) Characteristics of Northern and Central California seismicity. Earthquake Notes

Uhrhammer RA, Loper SJ, Romanowicz B (1996) Determination of local magnitude using BDNS broadband records. Bull Seismol Soc Am 86:1314–1330

USGS (2018) Search earthquake catalog. https://earthquake.usgs.gov/earthquakes/search/. Last accessed 2018/05/01

van Stiphout T, Zhuang J, Marsan D (2016) Seismicity declustering, community online resource for statistical seismicity analysis. https://doi.org/10.5078/corssa-52382934

Wiemer S, Schorlemmer D (2007) ALM: an asperity-based likelihood model for California. Seismol Res Lett 78:134–140

Wiemer S, Wyss M (2000) Minimum magnitude of completeness in earthquake catalogs: examples from Alaska, the Western United States, and Japan. Bull Seismol Soc Am 90:859–869

3D Finite Element Analysis of Seismic Behavior of Soil-Piled Raft-Structural System

Chaidul Haque Chaudhuri and Rajib Saha

Abstract Piled raft foundations are mainly used in case of high rise buildings and important structures to control total and differential settlement and to enhance the bearing capacity. Behavior of structural system placed on piled raft foundation during earthquake is considered fairly complex due to dynamic interaction between soil, pile, raft, and structure. Traditionally, piled raft foundation and structure under seismic load are designed without considering flexible base condition, although soil flexibility may have significant effect on the response of soil–pile raft–structure system. In this connection, present study is an attempt to numerically investigate the seismic behavior of piled raft supported structural system embedded in homogenous soft clay using 3D finite element model in ABAQUS/CAE v6.8. Two representative fixed base periods of structure, such as, 0.3 and 2.0 s which represents low, and high rise structure respectively are considered to be supported on piled raft foundation. Nonlinearity of soil is modeled by Mohr–Coulomb soil plasticity model. Elastic responses are obtained for both soil-piled raft–structure system and fixed base system with recorded motion of 1995 Kobe earthquake. Present study calculated the maximum shear forces at the column ($V_{B,col}$) and pile head ($V_{B,pile}$) and compared with those for the same structure in the fixed base condition ($V_{B,fixed}$). Results indicated that shear force considerably increases in pile head for both low and high rise structures as compared to fixed base shear. Further, increase at pile head observed to be significant when pile length asymmetry introduced considering optimum pile group design philosophy. However, increase in shear at column is marginal in long period structure.

Keywords Piled raft foundation · Seismic behavior · FE analysis

C. H. Chaudhuri · R. Saha (✉)
National Institute of Technology Agartala, Barjala, Jirania, Agartala, Tripura 799046, India
e-mail: rajib.iitbbsr@gmail.com

C. H. Chaudhuri
e-mail: chaidul.kdl@gmail.com

M. Latha Gali and P. Raghuveer Rao (eds.), *Geohazards*, Lecture Notes in Civil Engineering 86, https://doi.org/10.1007/978-981-15-6233-4_42

1 Introduction

Piled raft foundations are basically used in soft to loose soil to support multi-storied buildings and important structures. Due to complexity and perception of beneficial attributes as suggested in previous codes, the influence of soil–structure interaction (SSI) is generally neglected during seismic design of structures based on piled raft foundation. But such perception was not always correct especially in case of heavy structure placed on pile foundation in soft/liquefiable soil. Many case studies highlighted the detrimental effect of SSI on seismic response of structure supported by pile foundation (Meymand 1998). The influence of soil–pile foundation–structure interaction (SPSI) on seismic response of structure resting in soft clay was investigated in several studies (Gazetas 1984; Eslami et al. 2011; Boulanger et al. 1999). Most of the studies emphasized on calibration of seismic pile–soil interaction modeling and assessment of efficacy of such models in order to predict the dynamic response of pile foundation and structure. On the other hand, considerable progress on design of piled raft foundation under static loading was reported in literature (Shukla et al. 2013; Radhika et al. 2015; Nandwani et al. 2015; Tom and Sindhu 2016). Past works revealed optimum piled raft foundation design with different configurations of pile and raft. However, limited studies were reported for seismic design of piled raft foundation system, while, more attention was rendered to develop optimum design for piled raft system under static gravity loading. In fact, framing of seismic design guideline of piled raft foundation considering the dynamic interaction between raft, soil, piles, and superstructure of different lateral period is equally important for ensuring resilient seismic performance of structure based on piled raft foundation system. From this point of view, present study is a humble approach to assess seismic behavior of soil-piled raft–structural system incorporating parametric variation of system parameters. Seismic behavior assessment of optimum designed piled raft system is also performed in current study to investigate the safety margin from the point of view of seismic loading.

3D finite element-based software Abaqus CAE v6.8 (2008) is used to model soil–pile raft supported structural system and dynamic analysis is performed with input of 1995 Kobe earthquake motion. Soil is assumed as nonlinear material and structural members are considered to be elastic. Seismic responses are recorded at superstructure column and pile head and presented in terms of normalized shear at the column and pile head, i.e., ($V_{B,col}/V_{B,fixed}$) and ($V_{B,pile}/V_{B,fixed}$). Finally, the study gives some valuable inputs from the point of view of seismic design of piled raft foundation supported structure and indicates the need of detailed study encompassing different influential parameters of piled raft system.

2 Idealization and Modeling Details

A single degree lumped mass supported by single column is used in the present study to simulate the superstructure. The column having 0.3 and 2.0 s of fundamental period in fixed base condition which represents approximately three storied and twenty storied structures respectively. Fundamental periods are maintained by altering mass and sectional properties of column. Height of column is considered as 3.0 m. Both pile and column are modeled using one-dimensional linear Timoshenko beam–column element having six degrees of freedom (three translational and three rotational) at each node of the elements. Further, raft and soil are modeled using eight-node hexahedral solid elements having three translational degrees of freedom at each node of raft and soil. Piles are embedded in raft and soil using embedded region constraint and superstructure column is connected to raft using coupling constraint. A 3D replica of piled raft foundation is showcased in Fig. 1a.

Nonlinear behavior of soil is modeled by Mohr–Coulomb soil plasticity model. Soil stiffness is assumed to be constant throughout the depth of soil. For computational efficiency, near field soil domain is meshed into 0.5 m × 0.5 m and far field is meshed into 2.5 m × 2.5 m. Vertical meshing is carried out in 2.5 m interval as shown in Fig. 1b. The soil domain is considered 45 m × 45 m × 25 m size of cube, whereas square raft of 10 m × 10 m is considered to be placed at the centroid of soil domain. Raft soil interface is defined using surface-to-surface interaction method assuming soil surrounding raft surface as master surface and that of raft surrounding soil surface as slave surface. Contact was applied both normal and tangential directions, whereas in the normal direction, hard contact was used and for tangential direction, frictional behavior is defined by introducing a frictional coefficient of 0.6

(a) (b)

Fig. 1 **a** Piled raft and **b** soil domain modeled in Abaqus CAE v6.8

Fig. 2 3D soil-piled raft–structure model with infinite boundary modeled in Abaqus CAE v6.8

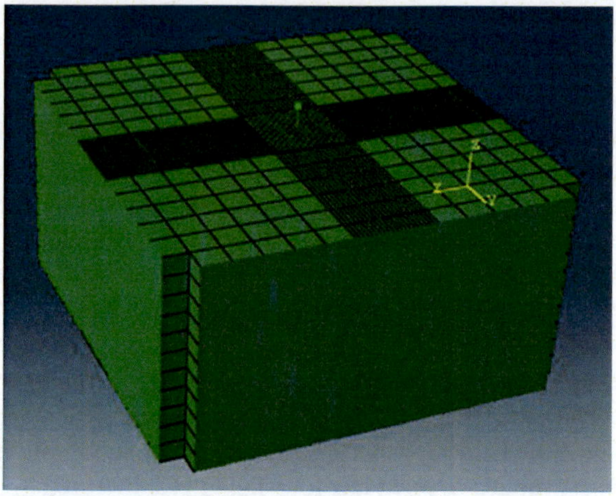

based on the recommendations of the American Petroleum Institute (1991). Further, for pile–soil interface, embedded constraint is used for simulating piles embedded in soil.

3 Material Damping

Rayleigh coefficients are used to define this kind of damping. Firstly, fundamental frequency of soil-piled raft-supported structural system is evaluated using modal analysis in Abaqus CAE v6.8 and mass proportional damping is then calculated considering 5% critical damping coefficient for the whole system. Further, to simulate radiation damping, i.e., to avoid reflection of energy wave back into the model, infinite boundary condition is used. Figure 2 represents a complete 3D soil-piled raft-structural system.

4 Ground Motion Used

Present study uses a recorded motion during 1995 Kobe earthquake with PGA of 0.30 g. The motion is applied as acceleration with respect to time as presented in Fig. 3.

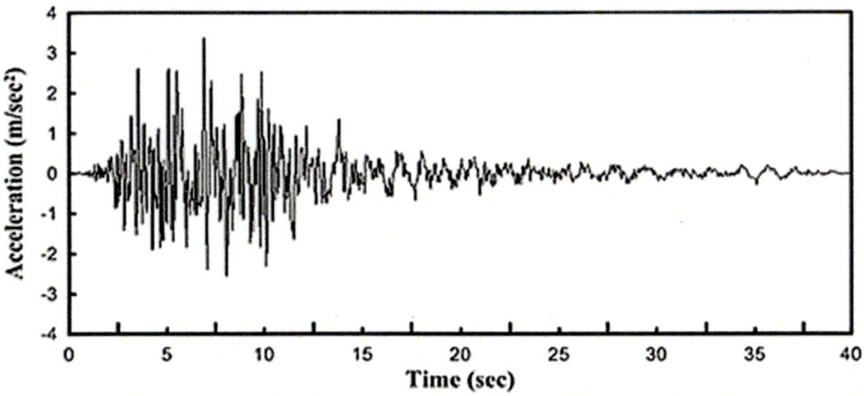

Fig. 3 1995 Kobe earthquake motion having 0.3 g PGA

5 Details of Case Studies

The effect of soil-piled raft–structure interaction on seismic responses of structure and piled raft foundation is studied in the present analysis. Two representative fixed base period of structures, namely 0.3 and 2.0 s are considered in this study that resemble approximately short (3 storied) and long period (20 storied) of structure, respectively. Subsoil condition is considered as very soft clay. Various soil parameters and additional details related to structural model considered for the present analysis are given in Table 1.

Load in vertical direction is considered to be 8.0 kN/m^2 per story incorporating both dead and live load acting on superstructure floor area of 8.5 m × 8.5 m for design of piled raft foundation. Lumped mass of superstructure is calculated based on above assumption and stiffness of column is adjusted to represent 0.3 s and 2.0 s structure, respectively. Piled raft supported structural system is modeled considering steel material. Young's modulus of steel is assumed as $E_{steel} = 180 \times 10^6$ kN/m^2 and that of mass density $\rho = 8000$ kg/m^3. Circular pile with 0.5 m diameter is assumed herein. Raft is considered as fully embedded in soil and the total depth of soil is considered as 25 m. The input motion is applied at the base of soil domain. A factor of safety of 5.0 is maintained to design the piled raft foundation and following literature as suggested elsewhere (Poulos 2001; Poulos and Davis 1980) is used during design. Further, load settlement analysis was done in Abaqus CAE v6.8 for both pile group configurations, i.e., for 0.3 and 2.0 s fixed base period of structure and it was found that total settlement of piled raft foundation does not exceed the permissible limit. Table 2 presents designed pile group configuration. Further, asymmetric pile group configurations are generated for long period of structure ($T_{fixed} = 2.0$ s) to simulate optimum piled raft foundation by considering varying pile length (15 nos. of pile @ 5 m and 10 nos. @ 20 m) for assessing the effect of pile length asymmetry on dynamic response of structure.

Table 1 Soil parameters and additional details related to structural models

Soil data	Value
Soil condition	Very soft clay
Young's modulus, E_s (MPa)	1.5(correlation available in Bowles 1997)
Unit weight, γ_{sat} (kN/m^3)	13.5
Cohesion, C_u (kN/m^2)	9.8
Additional details related to structural models	
Raft size	10 m × 10 m
Superstructure bay size	8.5 m × 8.5 m
Spacing by diameter ratio of piles (s/d)	3
L/d ratio of pile	40
Length (L) of pile (m)	20
Diameter (d) of pile (m)	0.5
Raft thickness (m)	1.0
Relative stiffness of pile ($k_p = E_p/E_s$)	12×10^4

Table 2 Designed pile group configuration

Period of structure at fixed base condition (T_{fixed}) (s)	Pile behavior	Design pile group
0.3	Flexible	NPR (adopted 1 × 2)
2.0	Flexible	5 × 5

NPR No piles required

6 Analysis Methodology

Frequency step is used to find out the fundamental natural frequency of soil-piled raft supported structural system. In frequency step, Lanczos solver is used for Eigen value extraction process. For dynamic analysis, dynamic implicit step is used in the present study.

7 Results and Discussion

7.1 *Seismic Response of Soil-Piled Raft–Structure System (Symmetric Pile Group)*

Present study computed the elastic responses in terms of shear force at column and pile head for $T_{fixed} = 0.3$ and 2.0 s structures supported by 1 × 2 and 5 × 5 pile group

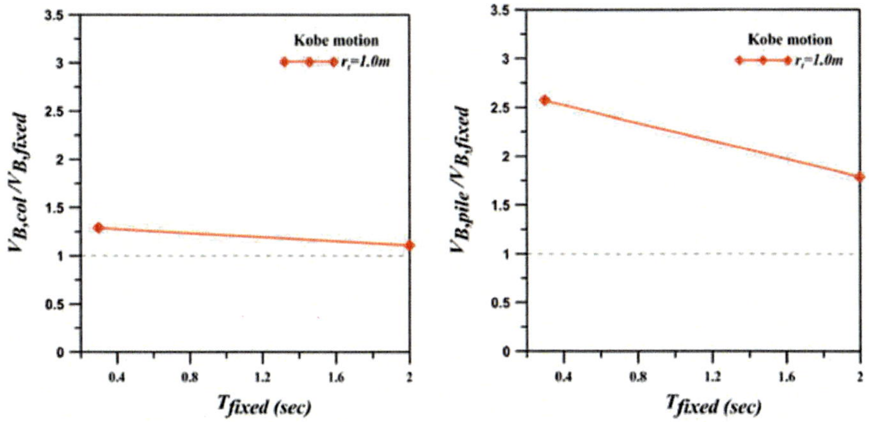

(a) (b)

Fig. 4 Normalized base shear at **a** column and **b** pile head with s/d = 3.0 and raft thickness = 1.0 m under 1995 Kobe motion

respectively incorporating the effect of SSI and normalized with respect to shear obtained in fixed base condition i.e., in terms of ($V_{B,col}/V_{B,fixed}$) and ($V_{B,pile}/V_{B,fixed}$) under 1995 Kobe earthquake motion. This section presents the results for identical pile lengths in group. Distribution of normalized shear force at column and pile head with respect to period of structure (T_{fixed}) is presented in Fig. 4a, b. Figure 4a indicates that shear force at column head increases compared to fixed base shear but this increasing value of shear is marginal at higher period of structure (T_{fixed} = 2.0 s). Further, Fig. 4b indicates shear force considerably increases in pile head for both low and high rise structures as compared to fixed base shear and this increasing value of pile head shear tends to decrease with increasing the period of structure.

It is observed from Fig. 4a, b that the values of normalized shear are greater than unity irrespective of the period of structure. This indicates, if seismic forces are calculated considering traditional method, i.e., fixed base condition, there is always an underestimation of calculation. Hence, it is essential to incorporate the influence of dynamic soil-piled raft–structure interaction during earthquake design of structure based on piled raft foundation.

7.2 *Effect of Asymmetric (Non-identical) Pile Group Configuration*

The effect of asymmetry in pile group configuration on normalized shear force at column and pile head is presented in Table 3 only for long period of structure (T_{fixed} = 2.0 s) under 1995 Kobe motion.

Table 3 Normalized base shear at column and pile head for symmetric and asymmetric pile length configuration

Pile length configuration (long piles length = 20 m and short piles length = 5 m)	Normalized column head shear ($V_{B,col}/V_{B,fixed}$)	Normalized pile head shear ($V_{B,pile}/V_{B,fixed}$)
Symmetric configuration 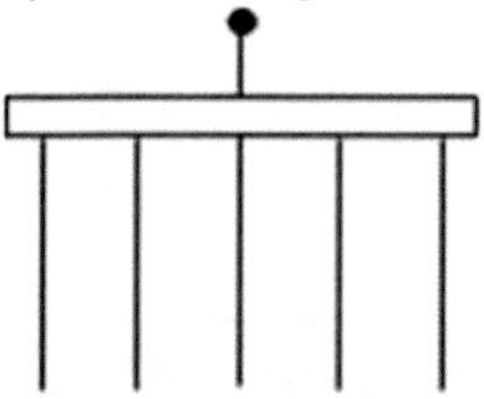	1.11	1.79
Asymmetric configuration 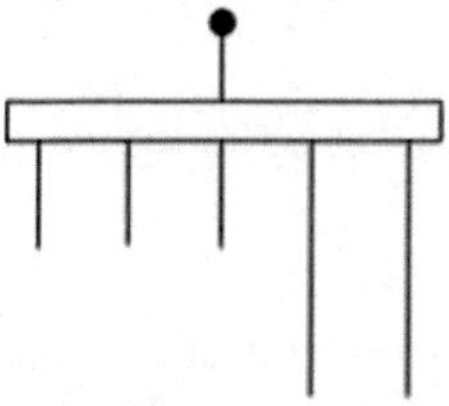	0.92	3.21

It is observed that the influence of asymmetry on normalized pile head shear is significantly high compared to symmetric pile group configuration. For instance, it is observed from Table 3 that normalized pile head shear increased by 79.33% for structure having $T_{fixed} = 2.0$ s with change in pile configuration from symmetric to asymmetric under 1995 Kobe motion. On the other hand, the influence of asymmetric pile length configuration on normalized column head shear is marginal as compared to symmetric counterpart. For instance, it is observed that normalized column head shear is decreased by 17.11% with change in pile configuration from symmetric to asymmetric.

8 Summary and Conclusions

Summarily, the present study highlighted the influence of soil-piled raft–structure interaction on seismic response of structure supported by piled raft foundation in homogenous soft clay using 3D finite element model in ABAQUS/CAE v6.8. In this study, elastic response in terms of shear force is calculated at column and piles head considering 1995 Kobe motion and normalized with shear force obtained at column under fixed base condition. Results indicate that pile head shear considerably increases for both low and high rise structures as compared to fixed base shear. However, column head shear is marginal in high rise buildings. Further, it is perceived that asymmetric pile length distribution in pile length in group may lead to shear force at pile head compared to symmetric pile group. This issue needs to be studied further which seems to be an important issue in the design of piled raft foundation. Hence, the present study provides valuable inputs for further details study in this direction.

References

Abaqus CAE v6.8 (2008) User's manual. Abaqus Inc.

American Petroleum Institute (1991) Recommended practice for planning, designing and constructing fixed offshore platforms. Washington, D.C API (RP 2A)

Boulanger RW, Curras CJ, Kutter BL, Wilson WD, Abghari AA (1999) Seismic soil-pile-structure interaction experiments and analyses. J Geotech Geoenviron Eng ASCE 125(9):750–759

Eslami MM, Aminikhah A, Ahmadi MM (2011) A comparative study on pie group and piled raft foundations (PRF) behavior under seismic loading. CMCE 2(2):185–199

Gazetas G (1984) Seismic response of end-bearing single piles. Soil Dyn Earthq Eng 3(2):82–93

Meymand PJ (1998) Shaking table scale model tests of nonlinear soil-pile-superstructure interaction in soft clay. PhD thesis, University of California, Berkeley

Nandwani N, Salunke PJ, Gore NG (2015) Comparative study of piled raft foundation. IJESRT. ISSN: 2277-9655

Poulos HG (2001) Piled raft foundations: design and applications. Geotechnique 51(2):95–113

Poulos HG, Davis EH (1980) Pile foundation analysis and design. The University of Sydney

Radhika R, Jeyapriya SP, Soundrapandiyan P (2015) Parametric study and numerical analysis of piled raft foundation on soft clay. Int J Res Emerg Sci Technol 2(4)

Shukla JS, Desai KA, Solanki HC (2013) Behavioural study of piled raft foundation in layered deposits of soft soils. J Eng Technol 3(1)

Tom A, Sindhu AR (2016) 1G Model study on the behaviour of piled raft foundation. IJSTE 3(2)

Seismic Requalification of Pile-Supported Structure: Pseudo-Static Approach

Pradeep Kumar Dammala, Manoj Kumar Manne, and A. Murali Krishna

Abstract Pseudo-static approaches often become handy due to their simplicity compared to the complex coupled continuum dynamic analysis and yield satisfactory results. A pseudo-static approach is employed herein to evaluate the response of a pile-supported structure located in a highly active seismic zone in northeast India. Pseudo-static approach essentially involves two main steps: estimation of maximum ground deformations and surface accelerations from a free field ground response study, combined with static analysis considering the surface acceleration and superstructure mass. In the present study, dynamic behavior of underlying soil strata was investigated using two high-quality element testing techniques—resonant column and dynamic simple shear tests. The dynamic soil properties were established and site-specific ground response study was conducted for the chosen site, using the artificially generated ground motions for the region. Equivalent static load was estimated from the surface acceleration obtained from free field ground response study. A three-dimensional finite element model has been developed and validated against the field lateral load test results. An increased magnitude of displacement and bending moment was observed with increased ground motion intensity. The peak bending moments induced for the scenario ground motions is higher than the plastic moment capacity of the pile section, displaying the probable failure of considered pile-supported structure.

Keywords Pile foundations · Bending moments · Ground response analysis

P. K. Dammala (✉) · M. K. Manne · A. Murali Krishna
Indian Institute of Technology Guwahati, Guwahati 781039, India
e-mail: dammala@iitg.ac.in

M. K. Manne
e-mail: manojcivil122@gmail.com

A. Murali Krishna
e-mail: amurali@iitg.ac.in

M. Latha Gali and P. Raghuveer Rao (eds.), *Geohazards*, Lecture Notes in Civil Engineering 86, https://doi.org/10.1007/978-981-15-6233-4_43

1 Introduction

The seismic performance of a structure supported by pile foundations is governed by the dynamic soil–structure interaction (SSI) mechanism between the pile, surrounding soil and the superstructure. Such interaction can be modeled by different approaches: pseudo-static approach and fully coupled dynamic continuum approach. Although the coupled continuum approaches yield accurate results, the computational expertise and complexity involved make it often difficult for the design engineers to adopt such methods. Simplified pseudo-static approaches are more attractive due to their easiness in handling and were also calibrated to provide satisfactory results (Liyanapathirana and Poulos 2005). Furthermore, pseudo-static approaches were observed to provide experimental consistent peak lateral displacements and peak bending moments (Tabesh and Poulos 2001).

Requalification studies are identified as strategic approaches to assess the stability of existing structure (Krishna et al. 2014). In case of pile foundations in seismic prone regions, stability concerning issues are bending, buckling, bending–buckling interaction and dynamic time period (Mohanty et al. 2017). All the aforementioned failure mechanisms are related to lateral load and moment carrying capacity (plastic moment capacity, M_p) of the pile section. Therefore, effective seismic requalification of existing structure or safe design of any new pile-supported structure depends on the capacity of the pile section in addition to the surrounding soil's shear strength. Since the pseudo-static approach is expected to provide satisfactory response of pile foundation subjected to seismic loads, the same approach was adopted herein to assess the safety of an existing pile-supported structure located in a highly active seismic region.

In the present study, a structure supported by pile foundations, embedded in multi-layered soil deposits is selected, to assess its seismic safety. Pseudo-static analysis is performed by two steps incorporating the soil's nonlinearity: free field nonlinear time domain ground response analysis (GRA) followed by a 3D finite element analysis with static lateral force acting at the pile head. The induced inertial bending moments are compared to the moment carrying capacity of the pile section.

2 Pseudo-Static Approach

Pseudo-static approach is a simplified equivalent dynamic method consisting of two or more steps. Abghari and Chai (1995) initiated the method with two simple steps: free field GRA to estimate the maximum ground deformations at each depth (node) followed by a static analysis of soil–pile–superstructure interaction by applying the estimated ground deformations along the depth. The ground deformations were simultaneously applied with the static lateral load at the pile head, which is a multiplication of spectral acceleration and pile cap mass. Reduction factors of 25 and 50% for lateral loads in order to estimate displacements and bending moments were proposed to better compare the pseudo-static results with the coupled dynamic finite element analysis. Similarly, Tabesh and Poulos (2001) carried out an elastic free field GRA followed by static displacement-based analysis with full consideration given to the inertial force. Liyanapathirana and Poulos (2005) came up with a modified pseudo-static approach for pile foundations in liquefiable soils. This method is similar to the Abghari and Chai (1995) except that the strength of the liquefiable soil zones is reduced by the prevailing minimum effective vertical stress and surface acceleration was used instead of spectral acceleration. More recently, Choudhury et al. (2016) adopted the pseudo-static approach to analyze the pile foundations in liquefiable soil zones and witnessed agreeable results with the fully coupled continuum dynamic analysis. Also, Kumar and Choudhury (2016) utilized pseudo-static approach for the combined pile raft foundation (CPRF) analysis and observed centrifuge test consistent results.

3 Case Study—Pile-Supported Structure

3.1 Location and Seismic History

A pile-supported structure located in north Guwahati, Assam, has been considered for the present study. Assam is one of the seven north eastern states of India and is categorized as the highly seismic active regions of the country (Zone V) as per IS:1893 (2002). Figure 1a presents the chosen site location, superimposed on the seismic zonation map of India. Twenty large ($M_w \geq 7.0$) and two great ($M_w \geq 8.0$) great earthquakes have been witnessed by this region in the last century Fig. 1b. The intense seismicity of the region is attributed due to the proximity of Himalayan thrust in the north and Indo-Burmese Arc in the east along with the adjoining geological fault (Kayal 2010). Figure 1c presents the aerial view of the site (under construction) with coordinates and Fig. 1d shows the elevated view of the pile-supported structure under construction.

3.2 Soil Characterization

Three boreholes were drilled and soil samples were obtained to characterize the behavior of underlying soil strata. A representative soil stratigraphy was obtained after analysis of the bore log data. Table 1 presents the soil stratigraphy of the site. The surficial strata (1.5 m) consist of loose organic fill followed by shallow layers of silty clay with low plasticity for a depth of 5.5 m. The silty clay layer was overlain by a silty sandy clay for a depth of 6 m, which was followed by a deep (17 m) sandy stratum of medium density. A layer of silty clay was again explored for a depth of 3 m, followed by a deep highly dense sandy stratum until the explored depth

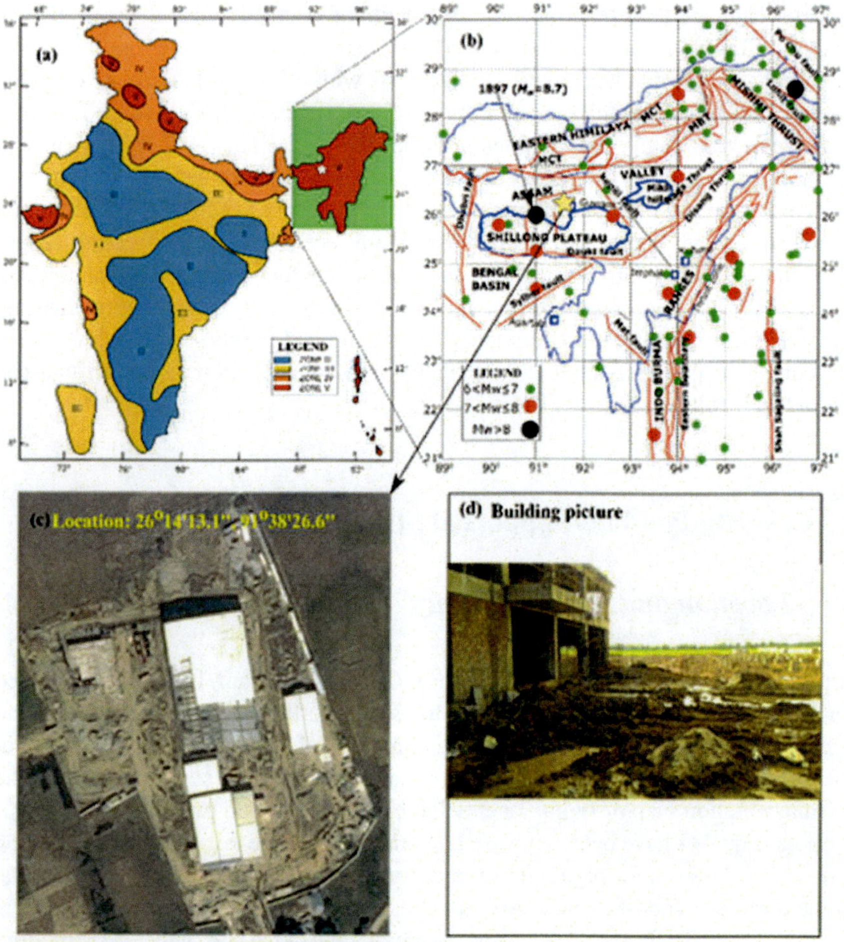

Fig. 1 **a** Location of the site in the seismic zonation map of India (IS:1893 2002). **b** Seismo-tectonic setup of the region (modified after Raghu Kanth et al. (2008)). **c** Aerial view of the site (*Source* Google Maps). **d** View of the structure under construction

Table 1 Representative soil stratigraphy details

Depth (m)	Soil type (USCS)	Unit weight (kN/m^3)	Cohesion (kPa)	Friction angle (°)	SPT N_{avg}	E_s (MPa)
0–1.5	OL	16	–	–	5	3.85
1.5–7	CL	19.5	33	2	8	30
7–13	CL/SM	19.6	43	2	12	45
13–30	SP	21	–	35	25	22.5
30–33	CL	22	53	2	47	53
33–40	SP	22	–	38	20	46.5

(40 m). Ground water table is available at a depth of 1.5 m from the surface level. Geotechnical properties of different layers are listed in Table 1.

The dynamic behavior (stress–strain response) of the soil is linear elastic in the low strain range (<0.001%) and is traditionally represented using maximum shear modulus (G_{max}). Upon continuous dynamic loading and with increasing induced shear strains, nonlinearity of the soil is expected which is represented using modulus degradation (G/G_{max}) and damping (D) variation. These are traditionally termed as dynamic soil properties (Kramer 1996). These dynamic soil properties are a prerequisite for advanced/comprehensive seismic analysis of structures. It has been emphasized the importance of site-specific properties to obtain accurate estimation of seismic demanding forces (Dammala et al. 2017; Kumar et al. 2018).

As the dynamic soil behavior is fully strain dependant, resonant column (RC) and dynamic simple shear (DSS) available at the SAGE laboratory, University of Surrey, UK, have been utilized to fully comprehend and establish the required dynamic soil properties. RC apparatus can provide accurate estimation of low strain (<0.1%) properties while DSS apparatus can furnish the required properties up to a strain level of 10%. Figures 2 and 3 present the RC and DSS apparatus, respectively, along with schematic representations of typical loading conditions prevail during the testing. The description of working principle, sample preparation and testing methodology of RC and DSS apparatus is omitted here to keep the brevity of the article and the same can be found in corresponding literature (ASTM D3999 2003; Dammala et al. 2016; Nikitas et al. 2016).

Two soil samples (silty clay at shallow depth and sandy soil at deep stratum) were collected and used for the testing. These two samples would represent mostly the entire soil stratum behavior (Table 1). Tests at different confining pressures (50–300 kPa) representing different embedment depths and shear strain levels (0.001–5%)

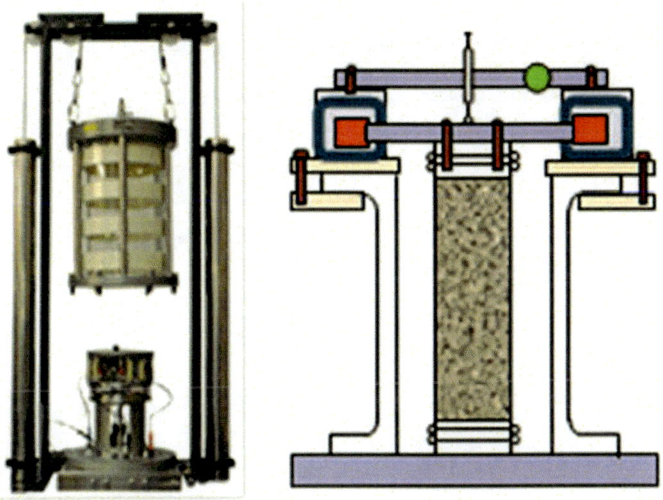

Fig. 2 Resonant column apparatus with schematic view of sample fitted with instrumentation

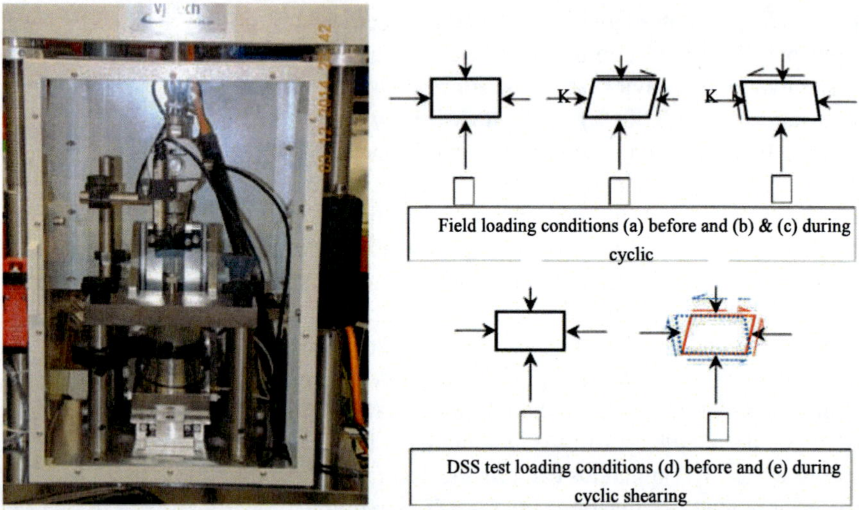

Fig. 3 Dynamic simple shear apparatus with prevailing loading conditions on the sample during testing

were conducted. Figure 4 shows the modulus reduction and damping curves obtained from the element tests for both the soils.

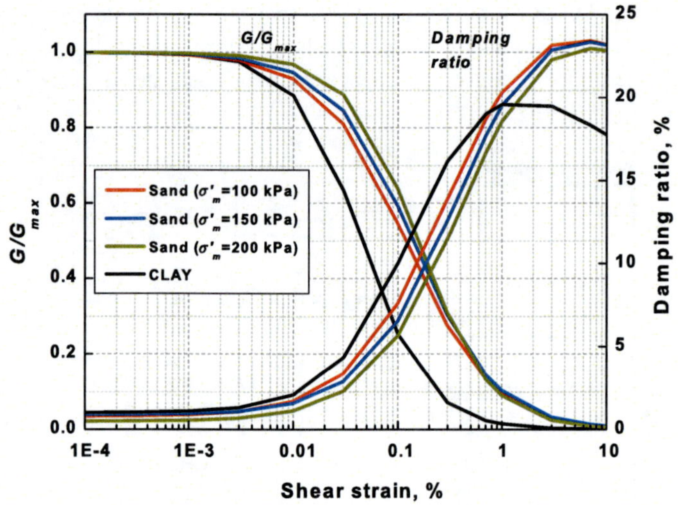

Fig. 4 Established dynamic soil properties for sand and clay

3.3 Ground Response Analysis (GRA)

Pseudo-static approach requires the acceleration at the surface to evaluate the equivalent static lateral force acting on the pile head. Surficial ground motion can be conveniently obtained by performing one-dimensional GRA studies using any commercially available programs such as DEEPSOIL, CYCLIC 1D, EERA, etc. Present study utilizes DEEPSOIL V6.1 (Hashash et al. 2016).

The pre-requisites for performing ground response study are the underlying soil properties-dynamic soil properties, bedrock if present at the bottom of the profile, and the input ground motion characteristics. A nonlinear confining pressure dependant GRA has been performed for the chosen soil profile. The required pressure-dependant soil properties at the confining pressures of interest (100, 150 and 200 kPa) are evaluated by performing a regression analysis on the experimental data. Darendeli (2001) single parameter model was used to fit the experimental data for the sandy soil and the corresponding strain-dependent properties at different effective confining pressures (σ_m') were established. The experimental values are directly used for silty clay. Figure 4 demonstrates the modulus and damping curves used for the nonlinear analysis. Elastic stratum of appropriate stiffness has been chosen as the bedrock. Synthetic ground motion developed by Raghu Kanth et al. (2008) for Guwahati region for 1897 Shillong earthquake ($M_w = 8.1$) has been considered. The required input ground motion for the study. The closest seismic fault to Guwahati being the Oldham fault, and hence, the chosen ground motion can be justifiable. The considered motion Fig. 5 has a peak ground acceleration (PGA) of 0.160 g, however, Guwahati categorized as Zone V (IS:1893 2002) and is expected to have a maximum PGA of 0.36 g. Therefore, the 0.160 g motion from Raghu Kanth et al. (2008) study has been scaled to different intensities (0.08, 0.24 and 0.36 g). The four ground motions with varying intensities were applied at the elastic stratum—base of the model for GRA.

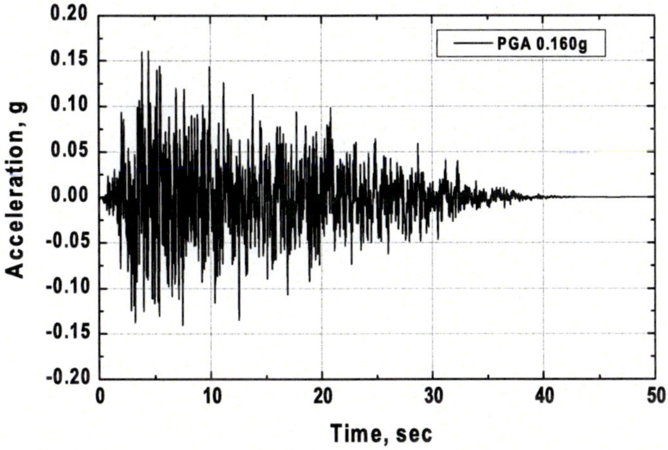

Fig. 5 Input ground motion considered from the study of Raghu Kanth et al. (2008)

Fig. 6 PGA profiles for different input motions of varying PGA

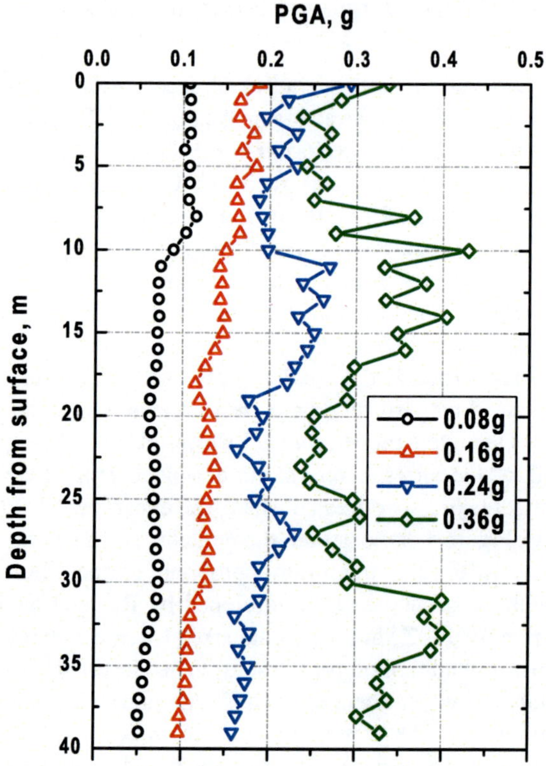

Figure 6 presents the acceleration response along the depth for different ground motions of varying intensity. It is interesting to note the amplification for 0.08, 0.16 and 0.24 g motions toward the surface while 0.36 g motion exhibited de-amplification (0.337 g) reaching the surface. The amplification/attenuation of the seismic waves depends upon the damping properties of the stratum. As the former ground motions (0.08, 0.16, and 0.24 g) are relatively less intensive compared to the latter and the shear strains induced in the soil are smaller, the damping behavior shown in Fig. 4 shows low magnitude for low shear strain resulting in amplification of the seismic waves. In contrast, intensive ground motion (0.36 g) induces high shear strain resulting in high damping and subsequent attenuation of the seismic waves toward the surface. Similar de-amplification of intense seismic motions was also observed (Kumar et al. 2015; Basu et al. 2017).

3.4 *Pseudo-Static Analysis: 3D FEM*

The pseudo-static analysis part of the work was performed using geotechnical finite element modeling (FEM) program-PLAXIS 3D. The building (currently under construction) is supported by pile foundation with a pile cap. Bored concrete piles of 26 m in length and 0.60 m in diameter were embedded in to the stratum with pile base ending in sandy stratum (at 26 m from surface). Analysis of single pile can be considered sufficient as the pile group with a fixed pile head represents a fixity condition as rigid behavior (Bhattacharya 2003).

Pile is modeled as an embedded beam with linear elastic behavior while the surrounding soil's behavior was simulated using Mohr–Coulomb shear strength criteria with appropriate engineering properties (Table 1). The soil–structure interaction (SSI) has been modeled using the interface properties available in the FE program. An interface stiffness value (R_{int}) of 0.67 has been set to account for any gapping/slip that may have occurred during the loading phases.

Geometrical extent of the model has been decided based on the literature suggested ranges. A lateral extent of $\pm 20D$ (12 m) in both the lateral directions (X and Y) was chosen while an extent of $1.7L$ along the depth (Z direction) was considered. These boundary extents would satisfy no-stress effect due to the finite boundary conditions prevailing (Kim and Jeong 2011; Kumar and Choudhury 2016; Murphy et al. 2018). Ground water table is present at a depth of 1.5 m from the surface and the same has been modeled in the FEM. Vertical roller boundary conditions were applied at the sides of the model restricting the lateral deformations and pinned boundaries (full fixities) applied at the base of the model. A very fine mesh was used for the enhanced accuracy of the results. The calculations have been performed in multiple stages—(1) only soil with equilibrium conditions achieved (2) pile construction (3) loading phases (Fig. 7).

3.5 FE Model Validation

In order to validate the FE model developed, field lateral load test results presented by Kim et al. (2009) have been considered. A drilled shaft of 1 m in diameter was embedded in multilayered soil stratum to a depth of 42 m from ground surface. Two lateral loads of different magnitudes were applied at the pile head—425 and 765 kN. Lateral displacements and bending moments along the depth of the shaft were monitored using strain gauges and LVDTs (Kim et al. 2009). Lateral extent of $\pm 20D$ was chosen on both the lateral directions while $1.7L$ was chosen in the vertical direction. Kim and Jeong (2011) also validated their numerical model (PLAXIS 3D Foundation) based on the Kim et al. (2009) field load test data. The appropriate stiffness properties of the layers and shaft can be found in Kim et al. (2009) and Kim and Jeong (2011).

Figures 8 and 9 present the comparison of lateral displacement and bending moment variation of the shaft along the depth from field load tests, numerical approach by Kim and Jeong (2011) and the present study (PLAXIS 3D). It can be observed that the present FEM could satisfactorily simulate the bending response of the shaft in terms of both lateral displacement and bending moments along the depth. Therefore, the present FEM can be considered validated.

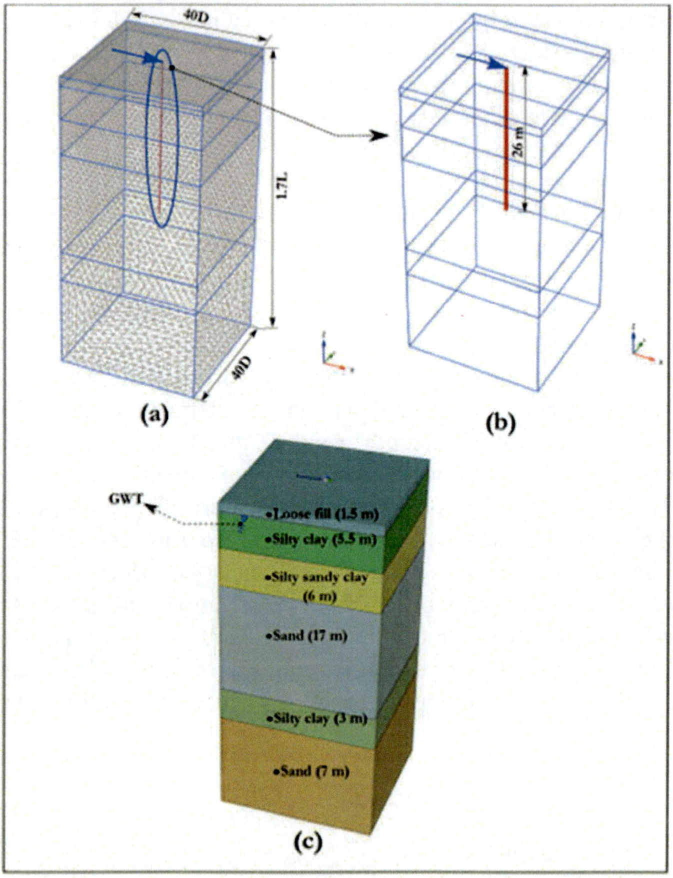

Fig. 7 3D FE model geometry

4 Results and Discussions

After successful validation of the FE model, results of field lateral load tests performed at the site are considered for efficient comparison/calibration of further FE results. Figure 10 provides a comparison of field lateral load test results with the load deformation response obtained from 3D FE model. A very close match of FEM results can be observed justifying the selection of input parameters for the pile, soil and the interaction in-between.

In order to proceed for the pseudo-static analysis on the soil–pile system, the equivalent static lateral loads need to be estimated from the expected superstructure mass (design axial load) and surface acceleration (from free field GRA). Similarly, equivalent pseudo-static analysis for CPRF was conducted by Kumar and Choudhury (2016). However, in the analysis by Kumar and Choudhury (2016), surface accelerations were not considered. The design axial load on each pile is 600 kN and

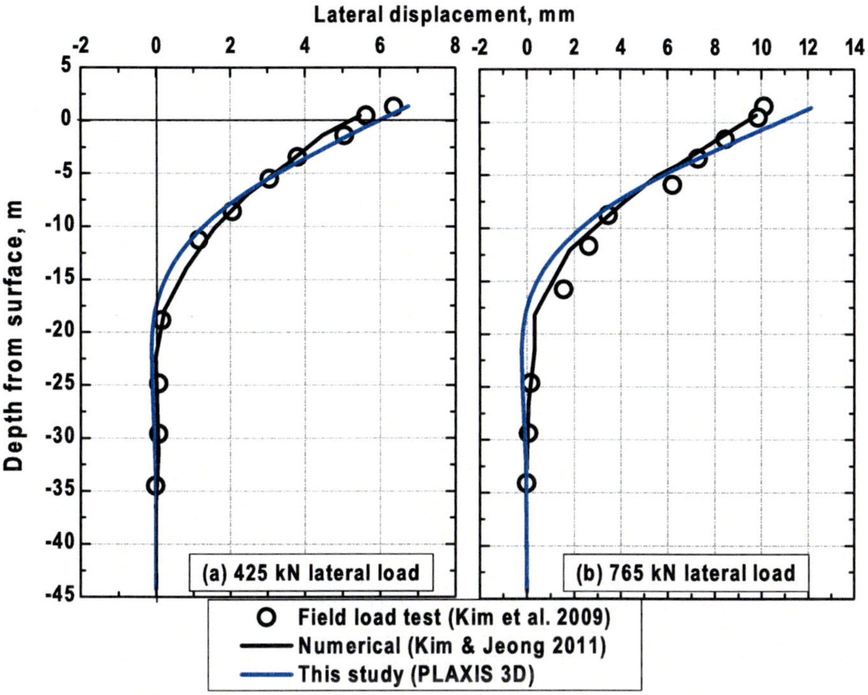

Fig. 8 Lateral displacement variation along the depth

the equivalent static lateral loads thus calculated for each ground motion are listed in Table 2. Static FE analysis was performed for each lateral load on the calibrated model and bending responses of the pile (lateral displacement and bending moment) along the depth were monitored.

Figures 11 and 12 present the variation of lateral displacement and bending moment along the depth for different earthquake scenario. It can be noted that the increase in earthquake intensity increased the lateral displacement. The observed 5 mm displacement for 0.08 g motion increased to 20 mm for 0.36 g, displaying a linear incremental displacement pattern. A lateral displacement of 2–4% of the pile diameter can be considered allowable for serviceability limit state (Shirato et al. 2009), which yields 12–24 mm for the present study. Keeping this in view, the present pile section can be considered safe for excessive lateral displacements.

Considering the bending moment (BM) profile, the peak bending moments induced are 88 kN m, 161 kN m, 265 kN m and 321 kN m for 0.08 g, 0.16 g, 0.240 g and 0.360 g input motions, respectively. The location of peak BM occurs at shallow depth (2–2.5 m from surface) and deepens as the intensity of earthquake increases which is consistent with the literature. Any structure (be it a column or a beam or a pile) is said to be failed/reached ultimate state, by comparing the induced moments (demand) to the plastic moment capacity (M_p) of the section (resistance). The plastic moment can be estimated using the cross section properties and the

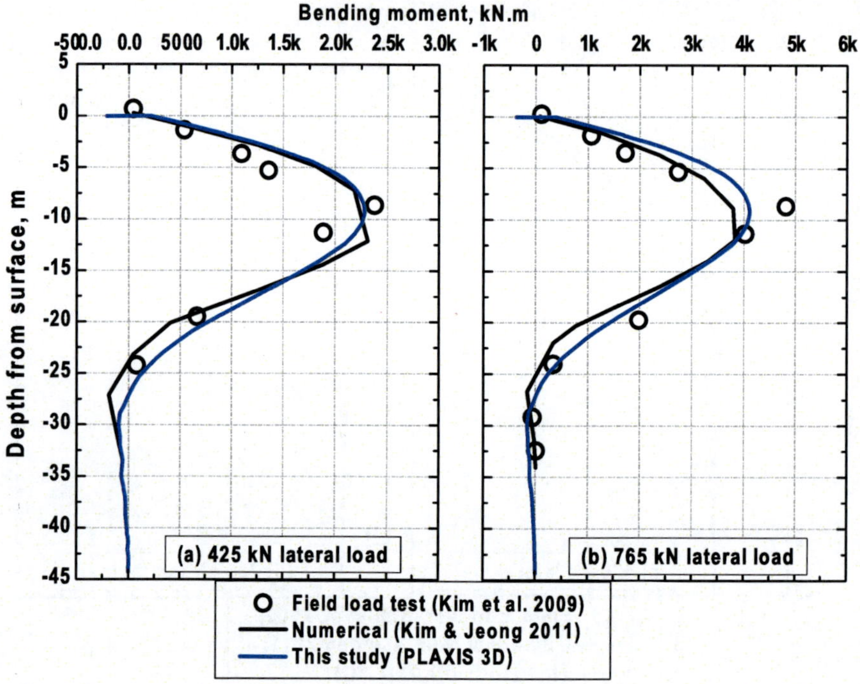

Fig. 9 Bending moment variation along the depth

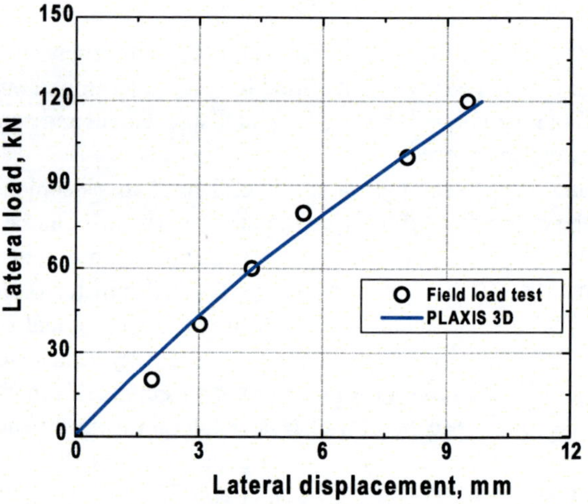

Fig. 10 Lateral load deformation response obtained from field load tests and FE model

Table 2 Pseudo-static loads applied to the pile head

Ground motion (g)	Surface PGA (g)	Pseudo-static force (kN)
0.08	0.107	65
0.16	0.188	113
0.24	0.293	176
0.36	0.337	203

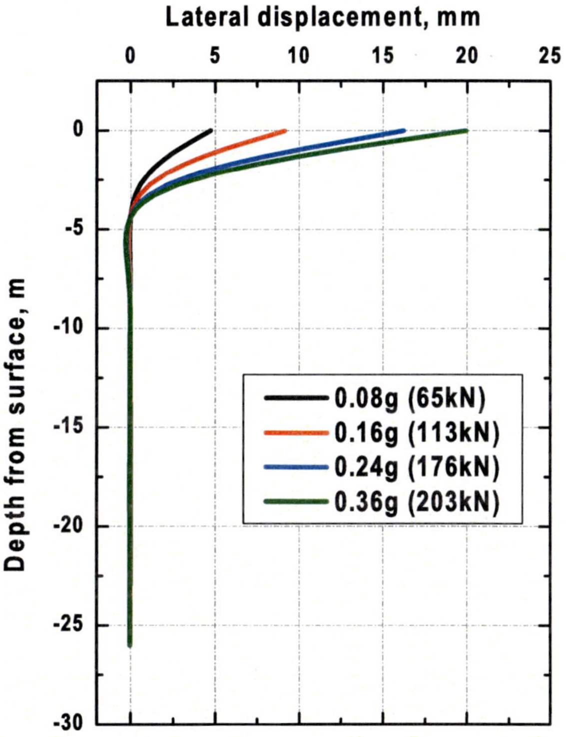

Fig. 11 Pile lateral displacement profile for various earthquake excitation conditions

yield strength of the material. The elastic and plastic moments of the present pile section thus obtained are 159 kN m and 270 kN m, respectively. Comparing M_p to the induced moments, the 0.240 and 0.36 g induced moments are higher than the capacity and structure may be considered unsafe for such intense events. Figure 13 presents the variation of obtained peak BM for different input motions. The current pile-supported structure can be considered unsafe in the considered seismic scenario conditions based on the pseudo-static analysis performed.

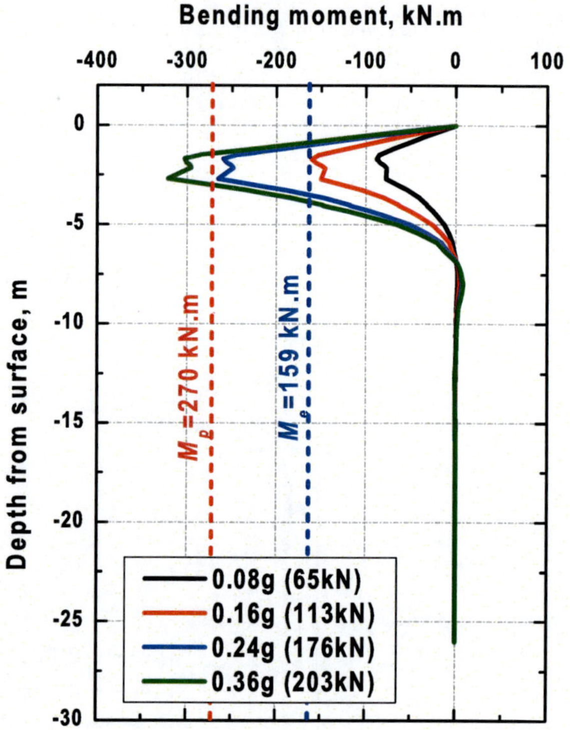

Fig. 12 Pile bending moment profile for various earthquake excitation conditions

5 Summary and Conclusions

A pseudo-static approach for seismic requalification study of pile-supported structures is presented. The chosen site is located in a highly active seismic region in India and has experienced many significant seismic events in the past. The behavior of underlying soil for the chosen site is dynamically characterized using advanced element testing techniques—resonant column and dynamic simple shear apparatus. Free field ground response study was conducted for input motions of varying intensity to assess the effect of low to high intensity ground motions on the pile behavior. Amplification of seismic waves toward surface was observed for low-intensity seismic motions while attenuation has been observed for high motions due to the high damping existing at high shear strains. The achieved peak ground accelerations were used to convert the design axial load on pile to equivalent static lateral load.

Three-dimensional finite element modelling technique (PLAXIS 3D) has been employed to perform the pseudo-static analysis. The developed model has been satisfactorily validated and calibrated against the field lateral load test results. Static lateral loads have been applied at the pile head and displacements along with bending moments along the pile were monitored. Increased displacements and bending

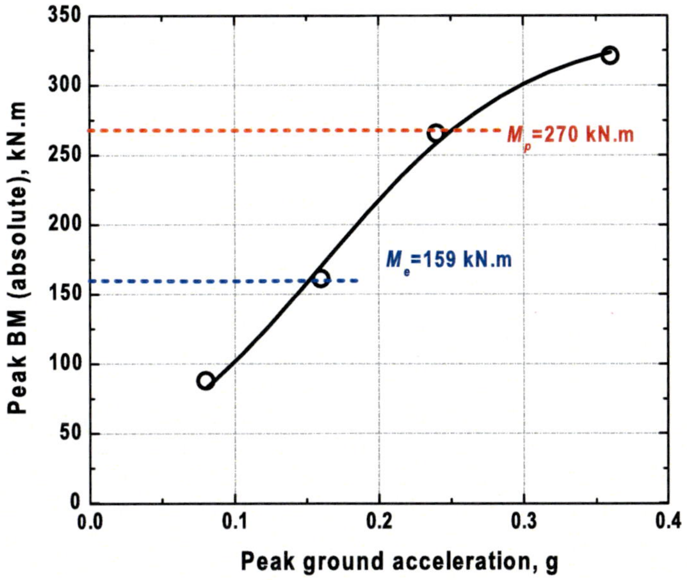

Fig. 13 Variation of peak BM with PGA of input ground motions

moments were observed with increased ground motion acceleration. The considered pile section was found to be unsafe for the intense seismic scenario based on the pseudo-static approach adopted.

Acknowledgements Prof. Subhamoy Bhattacharya and Mr. Georgious Nikitas of University of Surrey, UK, have cooperated in planning and performing the laboratory tests. Their help in this regard is fully acknowledged.

References

Abghari A, Chai J (1995) Modeling of soil-pile-superstructure interaction for bridge foundations. In: Performance of deep foundations under seismic loading, pp 45–59

ASTM D3999 (2003) Standard test methods for the determination of the modulus and damping properties of soils using the cyclic triaxial apparatus. Am Soc Test Mater 91:1–16

Basu D, Dey A, Kumar SS (2017) One-dimensional effective stress non-Masing nonlinear ground response analysis of IIT Guwahati. Int J Geotech Earthq Eng 8:1–27

Bhattacharya S (2003) Pile instability during earthquake. University of Cambridge, UK

Choudhury D, Ashuthosh K, Chatterjee K (2016) Coupled behavior of pile foundations in liquefied and non-liquefied soils during earthquakes including case study. In: International workshop on geotechnics for resilient infrastructure the second Japan-India workshop in geotechnical engineering, pp 11–15

Dammala PK, Adapa MK, Bhattacharya S, Aingaran S (2016) Cyclic response of cohesionless soil using cyclic simple shear testing. In: 6th International conference on recent advances in geotechnical earthquake engineering. ICRAGEE, New Delhi, pp 1–10

Dammala PK, Krishna AM, Bhattacharya S, Nikitas S, Rouholamin S (2017) Dynamic soil properties for seismic ground response studies in northeastern India. Soil Dyn Earthq Eng 100:357–370
Darendeli MB (2001) Development of a new family of normalized modulus reduction and material damping curves. University of Texas, Austin
Hashash YMA, Musgrouve MI, Harmon JA (2016) DEEPSOIL 6.1, User manual. University of Illinois Urbana
IS:1893 (2002) Criteria for earthquake resistant design of structures. Indian Stand, New Delhi, pp 1–44
Kayal JR (2010) Himalayan tectonic model and the great earthquakes: an appraisal. Geomatics Nat Hazards Risk 1:51–67
Kim Y, Jeong S (2011) Analysis of soil resistance on laterally loaded piles based on 3D soil-pile interaction. Comput Geotech 38:248–257
Kim Y, Jeong S, Won J (2009) Effect of lateral rigidity of offshore piles using proposed p-y curves in marine clay. Mar Georesour Geotechnol 27:53–77
Kramer SL (1996) Geotechnical earthquake engineering, 1st edn. Prentice-Hall, New Jersey
Krishna AM, Bhattacharya S, Choudhury D (2014) Seismic requalification of geotechnical structures 44:113–118
Kumar A, Choudhury D (2016) Effect of earthquake on combined pile–raft foundation. Int J Geomech 16:1–16
Kumar A, Harinarayan NH, Baro O (2015) High amplification factor for low amplitude ground motion: assessment for Delhi. Disaster Adv 8:1–11
Kumar SS, Adapa MK, Dey A (2018) Importance of site-specific dynamic soil properties for seismic ground response studies. Int J Geotech Earthq Eng
Liyanapathirana DS, Poulos HG (2005) Pseudostatic approach for seismic analysis of piles in liquefying soil. J Geotech Geoenviron Eng 131:1480–1487
Mohanty P, Dutta SC, Bhattacharya S (2017) Proposed mechanism for mid-span failure of pile supported river bridges during seismic liquefaction. Soil Dyn Earthq Eng 102:41–45
Murphy G, Igoe D, Doherty P, Gavin KG (2018) 3D FEM approach for laterally loaded monopile design. Comput Geotech 100:76–83
Nikitas G, Vimalan NJ, Bhattacharya S (2016) An innovative cyclic loading device to study long term performance of offshore wind turbines. Soil Dyn Earthq Eng 82:154–160
Raghu Kanth STG, Sreelatha S, Dash SK (2008) Ground motion estimation at Guwahati city for an Mw 8.1 earthquake in the Shillong plateau. Tectonophysics 448:98–114
Shirato M, Kohno T, Nakatani S (2009) Geotechnical criteria for serviceability limit state of horizontally loaded deep foundations. In: Proceedings of the second international symposium on geotechnical risk and safety, pp 119–126
Tabesh A, Poulos HG (2001) Pseudostatic approach for seismic analysis of single piles. Geotech Geoenviron Eng 127(9):757–765

Earthquake Early Warning System

Aarti Taneja, Aniket Desai, and Ravi S. Jakka

Abstract Earthquake early warning is considered as one of the real-time earthquake damage mitigation measures which detects seismic waves, analyzes and transmits information of the impending ground shaking at the potential user site. Earthquake warning is issued in order to reduce the damage caused by earthquake such as trains speed can be reduced, and people can move themselves to safer place. The basic requirement of an EEW system is the development of real-time algorithm for fast calculation of earthquake parameters. Earthquake warning is possible by detecting primary waves that travel more quickly than the secondary and Rayleigh waves. The reliable issuance of the warning depends on the accuracy of the calculated parameters. The parameters are estimated on the basis of initial portion of P-wave. Body waves are of two types, namely compressional waves and shear waves. Compressional waves or the P-waves are fast in velocity as compared to shear waves and are recorded first in the seismograph. On the other hand, shear wave or secondary waves are slower than P-wave, and most of the damage is caused due to S-wave. An EEW system warns about an area of forthcoming strong shaking, normally with a few tens of seconds of warning time before the arrival of destructive S-wave. Even a few sec warning will be useful for some emergency measures such as shut off gas pipelines to minimize fire hazards, to slow down rapid transit vehicles and safeguarding of computer facilities to avoid loss of data bases. This can be done by calculating EEW parameters such as τ_c and peak displacement from the initial 3 s of the P-waveform. With the help of these parameters, magnitude can be estimated, and warning can be provided.

Keywords Earthquake early warning system · Seismic waves

A. Taneja (✉) · A. Desai · R. S. Jakka
Indian Institute of Technology Roorkee, Roorkee, Uttrakhand 247667, India
e-mail: aarti.taneja9@gmail.com

M. Latha Gali and P. Raghuveer Rao (eds.), *Geohazards*, Lecture Notes in Civil Engineering 86, https://doi.org/10.1007/978-981-15-6233-4_44

1 Introduction

Earthquake early warning system is provided by calculating the parameters. There are basically two types of early warning system regional and onsite early warning system. Regional early warning system is network-based, and warning is provided when parameters exceed the limiting value, then only estimation of location and magnitude is done. While in onsite early warning system, warning is provided from a single station and will be near epicentral distance.

Earthquake early warning is calculated using initial 3 s of the P-wave as P-wave travels faster than S-waves, and most damage is caused due to shear waves. Shear waves are the secondary waves, and they move perpendicular to the direction of motion of wave. Some of the important studies on earthquake earlywarning systems were reported by Allen and Kanamori (2003), Allen et al. (2009), Bhardwaj R (2013), Dunn et al. (2016), Hsiao et al. (2011), Kanamori (2005), Lin et al. (2012), Nakamura (1988) and Nakamura et al. (2011)

2 Methodology

Earthquake early warning system is used to provide alert from few sec to couple of sec so that risk due to earthquake can be reduced. Parameters that are used for early warning system are peak displacement and average period of the P-wave. Both are calculated using initial 3 s of the P-wave. As P-wave travels faster than S-wave, mostly damage is caused due to shear wave. So, analysis is done the basis of primary waves. Firstly, the earthquake data is downloaded from different sites, and Butterworth filter of frequency 0.1 Hz is applied. Later on, only 3 s data is taken as this is only responsible for providing alert. Five different soil profiles assume such that their average shear velocity comes under different site classes. There are five types of site classes from site class A to site class E. Site class A corresponds to hard rock, and site class E corresponds to soft soil based on shear wave velocity. To this profile, earthquake of different magnitude and epicentral distances is applied is applied, and parameters are calculated. Epicentral distances are taken such that they are less than 100 km.

Peak displacement (P_d) is calculated from 3 s of the P-wave. This 3 s starts where amplification starts. Peak displacement is the maximum amplitude of the displacement in the first 3 s of the P-wave. It is the maximum amplitude of the displacement, and on the basis of that, a relation is estimated between magnitude and peak displacement. With the help of calculated magnitude, early warning is provided.

The second parameter that is estimated is the average period of P-wave, i.e., (τ). It is based on velocity and displacement. It is estimated using the relation:

$$T = \frac{2\pi}{\sqrt{\gamma}} \tag{1}$$

$$\gamma = \frac{\int_0^{t_0} |\dot{u}|^2}{\int_0^{t_0} |u|^2} \tag{2}$$

where $\ddot{u}$, $\dot{u}$ and u are the acceleration, velocity and displacement of the earthquake history.

After calculating average period of the P-wave, a relation is developed between magnitude and the early warning parameter. With the help of calculated magnitude, early warning can be provided.

With the help of early warning system, losses due to earthquake can be reduced to a large extent, and lives of many people can be saved as they can take several precaution to save themselves. Although warning that is provided is only of few couple of sec, then also, it is very helpful in reducing risk.

3 Results

Early warning parameters are calculated using initial portion of the P-wave. For both the parameters, i.e., peak displacement and average period of the P-wave, a relation is developed between magnitude and the early warning parameters. The relation is shown below:

$$M = (-0.430 P_{\mathrm{d}}^2) + (1.488 P_{\mathrm{d}}) + (5.962) \tag{3}$$

where M is the magnitude of the earthquake and the P_{d} is the peak displacement of the earthquake.

$$M = (-0.746 \log \tau) + 6.33 \tag{4}$$

where τ is the average period of the P-wave and is estimated using Eq. 1.

With the help of these equation, alert can be issued.

References

Allen RM, Kanamori H (2003) The potential for earthquake early warning in southern California. Science 300(5620):786–789

Allen RM, Brown H, Hellweg M, Khainovski O, Lombard P, Neuhauser D (2009) Real-time earthquake detection and hazard assessment by ElarmS across California. Geophys Res Lett 36(5)

Bhardwaj R (2013) Algorithm for earthquake early warning system. Ph.D. thesis, IIT Roorkee

Dunn PT, Ahn AY, Bostrom A, Vidale JE (2016) Perceptions of earthquake early warnings on the US West Coast. Int J Disaster Risk Reduction 20:112–122

Hsiao NC, Wu YM, Zhao L, Chen DY, Huang WT, Kuo KH, Shin TC, Leu PL (2011) A new prototype system for earthquake early warning in Taiwan. Soil Dyn Earthq Eng 31(2):201–208

Kanamori H (2005) Real-time seismology and earthquake damage mitigation. Annu Rev Earth Planet Sci 33:195–214

Lin CCJ, Lin PY, Chang TM, Lin TK, Weng YT, Chang KC, Tsai KC (2012) Development of on-site earthquake early warning system for Taiwan. In: Earthquake research and analysis-new frontiers in seismology. InTech, UK

Nakamura Y (1988) On the urgent earthquake detection and alarm system (UrEDAS). In: Proceedings of the 9th world conference on earthquake engineering, vol 7, Tokyo-Kyoto, Japan, pp 673–678

Nakamura Y, Saita J, Sato T (2011) On an earthquake early warning system (EEW) and its applications. Soil Dyn Earthq Eng 31(2):127–136

Dynamic Characteristics of Subsoil Deposit of Tripura by Nakamura Method

Rajat Debnath and Rajib Saha

Abstract Determination of local site effect by calculating fundamental period and amplification of motion using Nakamura (Railway Tech Res Inst Q Rep 30:1–9, 1989) proposed *H*/*V* spectral ratio approach was recognized to be a simplified, reliable and well-accepted method. Both strong motion recordings and ambient noise recorded from micro-tremor can be used to construct the *H*/*V* spectral ratio. It is a well-known fact that the affect of body wave in an earthquake signal is relatively higher compared to Rayleigh wave and *H*/*V* spectral ratio approach is a useful method to differentiate the influence of both the waves. Tripura is one of the Northeastern states of India surrounded three sides by Bangladesh and rest by Assam which was categorized as the highest seismically vulnerable zone (BIS 2016). In fact, the geological history of Agartala, the capital city of Tripura, and other river valleys of Tripura are relatively young of Holocene age group. Significant depth of sedimentation with soft alluvial and fluvial type at top layers is also reported. Hence, the possibility of local site effect may lead to significant effects. From the above viewpoint, present study is an attempt to investigate dynamic characteristics and amplification ratio after analyzing strong motion records following Nakamura (Railway Tech Res Inst Q Rep 30:1–9, 1989) proposed *H*/*V* spectral ratio approach. The acceleration history record of January 3, 2017, earthquake epicentered at North part of Tripura is used to construct the *H*/*V* spectral ratio. Results indicate that peak amplitude of *H*/*V* spectra is recorded at around 1.65–1.75 Hz for both the sites, which comprises SH waves of higher energy content and further troughs are noted at higher frequencies confirming moderate effect of Rayleigh waves. This limited study highlights amplification potential in order of 1.65–1.75 Hz indicating influence of local site effect at various locations of Tripura and need of microzonation map of important towns.

Keywords *H*/*V* spectral ratio · Nakamura · Amplitude

R. Debnath (✉) · R. Saha
Civil Engineering Department, National Institute of Technology Agartala, Barjala, Jirania, Tripura 799046, India
e-mail: rd2100@gmail.com

R. Saha
e-mail: rajib.iitbbsr@gmail.com

M. Latha Gali and P. Raghuveer Rao (eds.), *Geohazards*, Lecture Notes in Civil Engineering 86, https://doi.org/10.1007/978-981-15-6233-4_45

1 Introduction

Tripura, one of the Northeastern states of India, is located with latitude 22° 90′ N–24° 52″ N and longitude 91° 15′ E–92° 40′ E with fast-growing infrastructural facilities and population due to its global importance by acting as business corridor for Southeast Asian countries through Bangladesh. Figure 1 presents the geographical position map of Tripura. As per census 2011 report, population of Tripura is 36 lakhs (approx.) and is located at the most highly seismically active zones of the country (zone-V, as per IS: 1893-2016, Indian Seismic Code) and sixth in the globe (Sil and Sitharam 2013). Tripura fold zone has experienced few mega earthquakes (EQ) of (Magnitude, $M_w = 8.1$, originated at dauki fault in the year 1897 and $M_w = 8.7$, the Great Assam Earthquake in the year 1950) and number of low to moderate range earthquakes (e.g., $M_w = 5.8$, 1984 Cachar EQ, $M_w = 6.9$, 2011 Nepal Sikkim EQ, $M_w = 7.6$, 1918 Srimangal EQ). Past earthquakes (1897 Shillong EQ and 1918 Srimangal EQ) caused some damages of structures and cases of liquefaction were also witnessed (Mukhopadhyay and Dasgupta 1998; Kayal 1991,

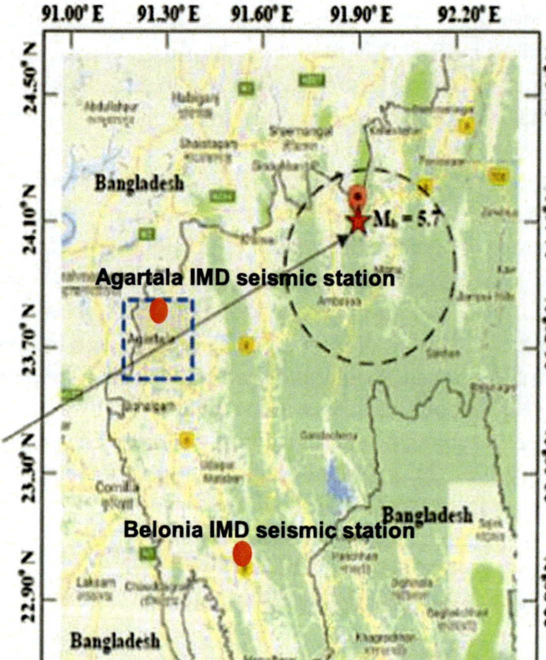

Fig. 1 Location of Tripura and red dots indicating locations of two seismic stations of IMD

1996; Satyabala 2003). A recent moderate range ($M_w = 5.7$) earthquake (January 3, 2017) epicentered at Dhalai district of Tripura has caused significant damages of mainly non-engineered structures, minor damages in masonry and RCC structures, few cases of slope failures and liquefaction near to epicenter. However, the rapid infrastructural growth of the state increases the seismic risk and necessitates proper evaluation of seismic multi-hazard. From this viewpoint, it is primarily essential to understand the influence of local site effect in order to prepare a safe earthquake disaster mitigation model. It is well-known fact that the affect of body wave in an earthquake signal is relatively higher compared to Rayleigh wave and H/V spectral ratio approach is a useful method to differentiate the influence of both the waves. Nakamura (1989) proposed H/V ratio is a popular technique which has been used extensively in India and other parts of world for understanding the dynamic behavior of subsoil deposit (e.g., Nakamura et al. 1989; Mohanty et al. 2007; Chowdhuri et al. 2011). Determination of H/V ratio gives an idea of fundamental natural frequency of ground as well as amplification potential of ground motion. These two parameters are important for future planning and designing of infrastructure. Further, frequency-dependent amplification ratio will also help to generate site-specific ground motions considering scenario earthquakes took place near to the area. However, the ratio of spectra for horizontal to vertical component of motion (i.e., represented as H/V) is developed from strong ground motion records of different earthquakes and recording micro-tremor by using broadband seismometer.

Present study attempts to determine H/V ratio as a function of frequency of motion using past strong ground motion records recorded at two different seismic observatory stations of Indian Meteorological Department (IMD) located at Tripura. The first station was located at Agartala (West Tripura) and second one at Belonia (South Tripura) which are shown in Fig. 1. Fundamental natural frequency (i.e., period) and amplification ratio can be interpreted from the developed H/V response. Dynamic properties evaluated based on recorded ground motions at two different locations may be considered as approximately representative for subsoil deposits of Tripura. In addition, subsoil borelog profiles at these two stations are collected and analyzed to verify the results obtained from Nakamura method. However, a detailed investigation is required for microzonation of important areas of Tripura. In this regard, present study may be considered as initial step for further detailed study.

2 Seismological Overview of Tripura

The entire Northeast is one of the most highly seismically active belts in the world. The Bureau of Indian Standards (BIS) marked as zone-V (maximum severe zone) having a zone factor of 0.36 shown in Fig. 2. The state is bounded by Sylhet depression (Surma basin) in the North, Chittagong hill ranges in the South, Bengal basin to the West and Tripura fold basin to the East. Tripura exhibits a wide array of sedimentary rocks having marine mixed fluvia-based origin ranging age from uppermost Oligocene (38 million years from present time) to recent period. Tectonically, this

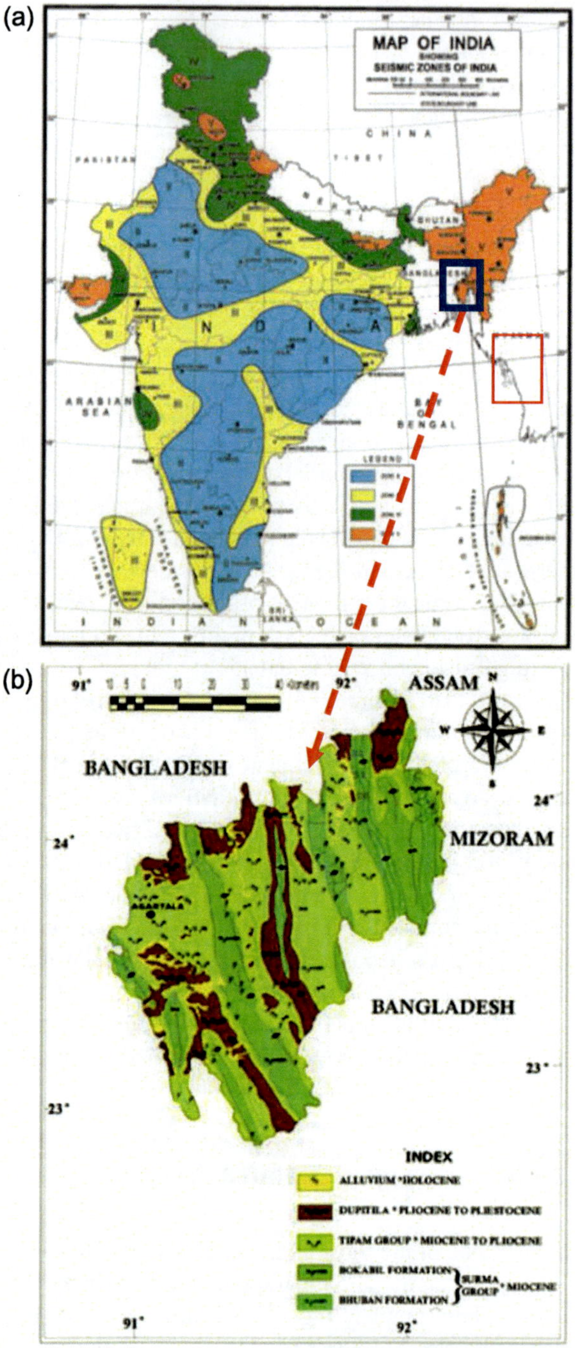

Fig. 2. **a** Seismic zonation map of India, as per IS: 1893 (Part1) (2016). **b** Geological map of Tripura, GSI (2011)

area comprises sub parallel arcuate, elongated doubly plunging folds in North–South direction. Agartala, the capital city of Tripura, formed by sediments of Quaternary age at upper layer and resting over Pleistocene age (Chowdhuri et al. 2011).

3 Collection of Data

3.1 Details of Ground Motion

Time history recordings of ground motion of January 3, 2017, Ambassa earthquake (epicenter at Tripura) recorded at two different stations of Tripura established by IMD are used in present study. The stations are located at Agartala (West district) and Belonia (South district) having coordinates of 23° 53′ N–91° 15′ E and 23° 15′ N–91°2 6′ E, respectively. Agartala and Belonia stations are situated at a height of 16 and 20 m from mean sea level, respectively. The aerial distance between epicenter of the earthquake and two different stations are calculated as 104.95 km (Agartala) and 135.30 km (Belonia), respectively, obtained from their respective coordinates. Ground motion data were recorded by V-Sat-Based Digital Seismic Telemetric System. The ground motion data are supplied by IMD Delhi in the form of .DAT format which was further processed into ASCII format. Finally, the data are processed using MATLAB 2014a. Peak Ground Acceleration (PGA) values are presented in Table 1, and the spectra developed for Agartala and Belonia sites in horizontal and vertical direction are shown in Figs. 3 and 4.

3.2 Geotechnical Data of Two Stations

Geology of a particular site plays a vital role in propagation of ground motion, and the propagation of motion waves varies considerably depending upon the geology of a place. However, the characteristics of geological layer up to the depth of 20–30 m is very important for evaluating the subsoil dynamic characteristics of a particular location (Borcherdt 1994). The subsoil profile along with depth wise shear wave velocity for both the seismic stations are presented in Fig. 5. The borehole depths are continued up to 37.5 and 20 m at Agartala and Belonia stations. Shear wave velocity is calculated using SPT-based empirical relationship as proposed by Anbazhagan et al. (2013).

Table 1 PGA values for ground motions at two sites

Site	PGA (Horizontal direction)	PGA (Vertical direction)
Agartala	0.0008g	0.0001g
Belonia	0.00041g	0.0003g

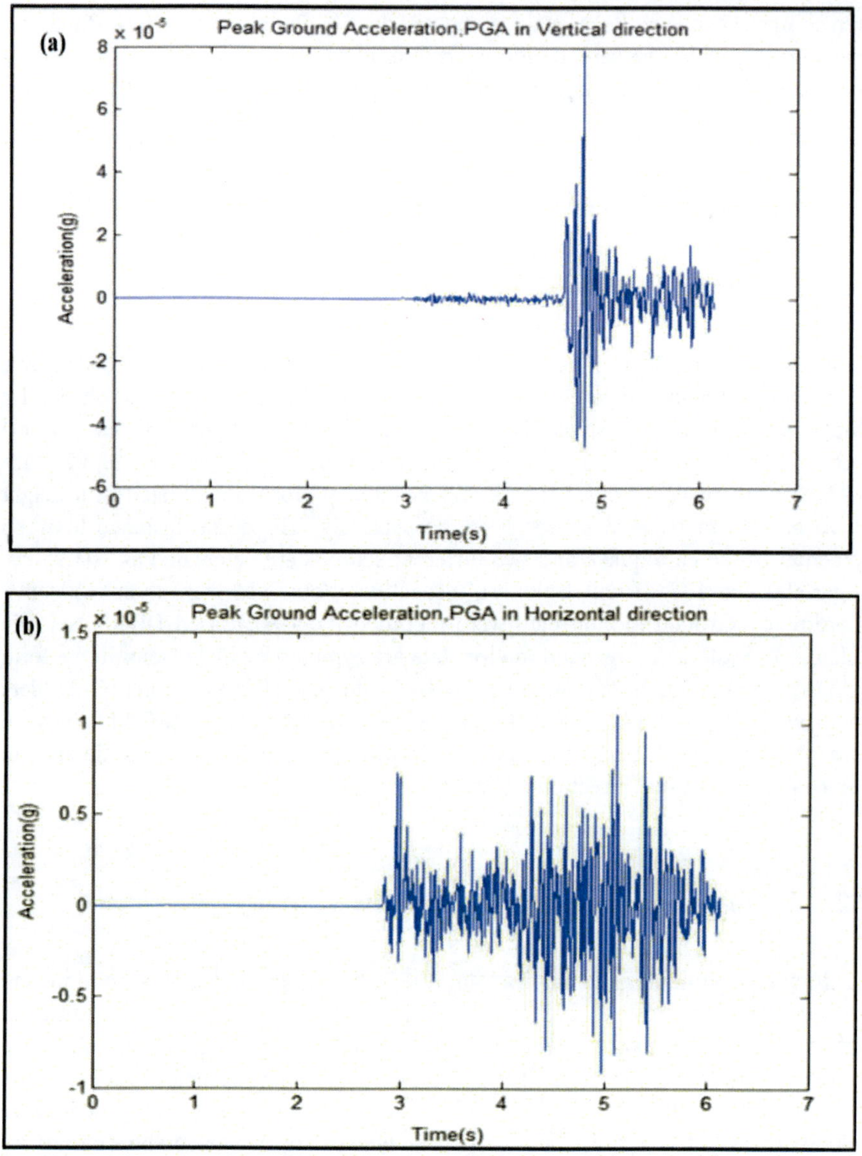

Fig. 3 Time–acceleration motion for Agartala IMD station. **a** Horizontal direction, **b** vertical direction

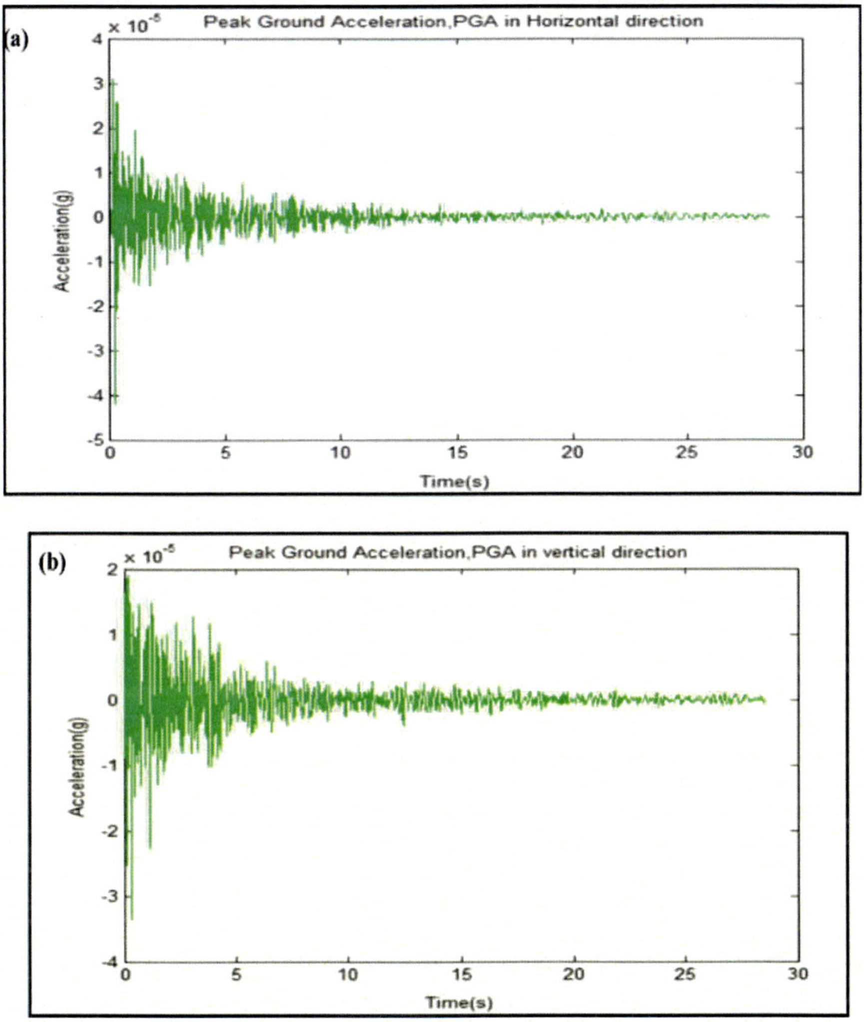

Fig. 4 Time–acceleration motion for Belonia IMD station. **a** Horizontal direction, **b** Vertical direction

4 Results and Discussion

4.1 Estimation of Fundamental Period

Figures 6 and 7 present the variation of *H*/*V* ratio with respect to frequency of recorded ground motion in log scale for Agartala and Belonia stations, respectively. It is considered that the corresponding frequency of first peak of *H*/*V* ratio wave will

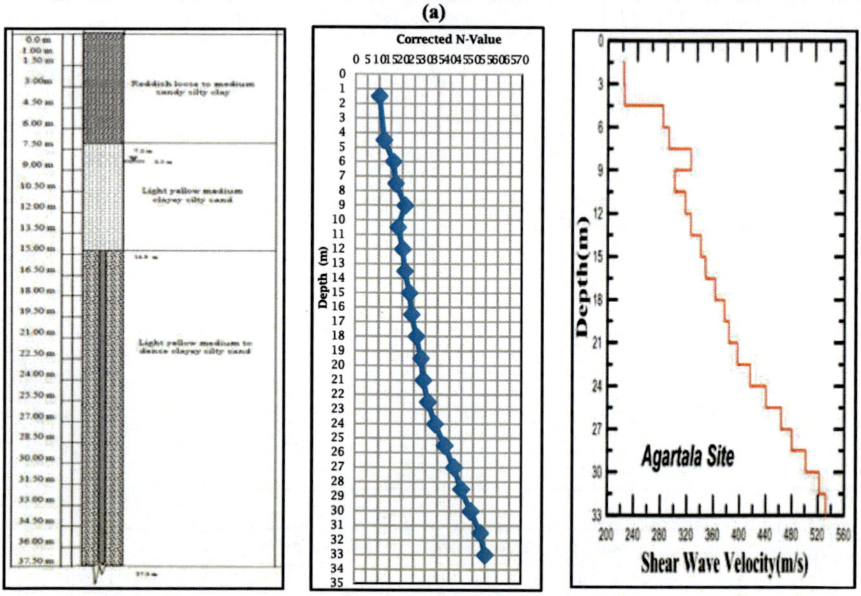

Fig. 5 Subsoil profile along with N value and V_S. **a** Agartala site. **b** Belonia site

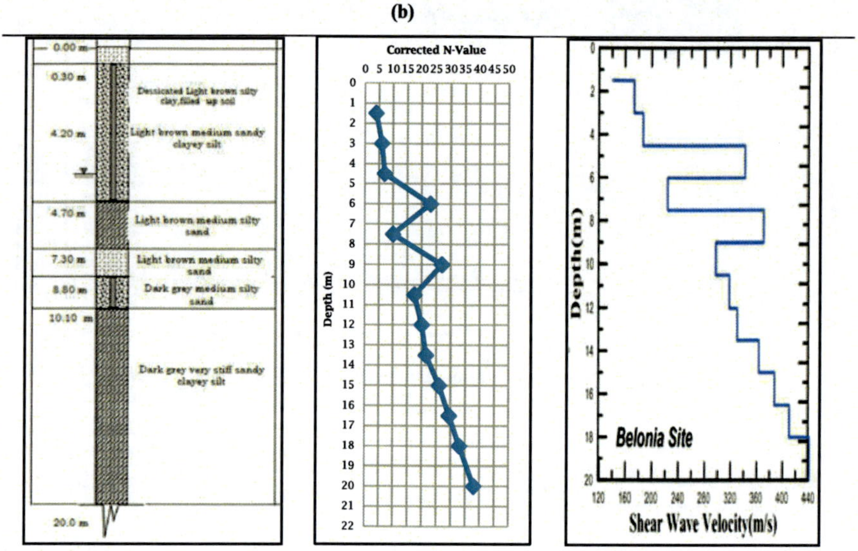

Fig. 5 (continued)

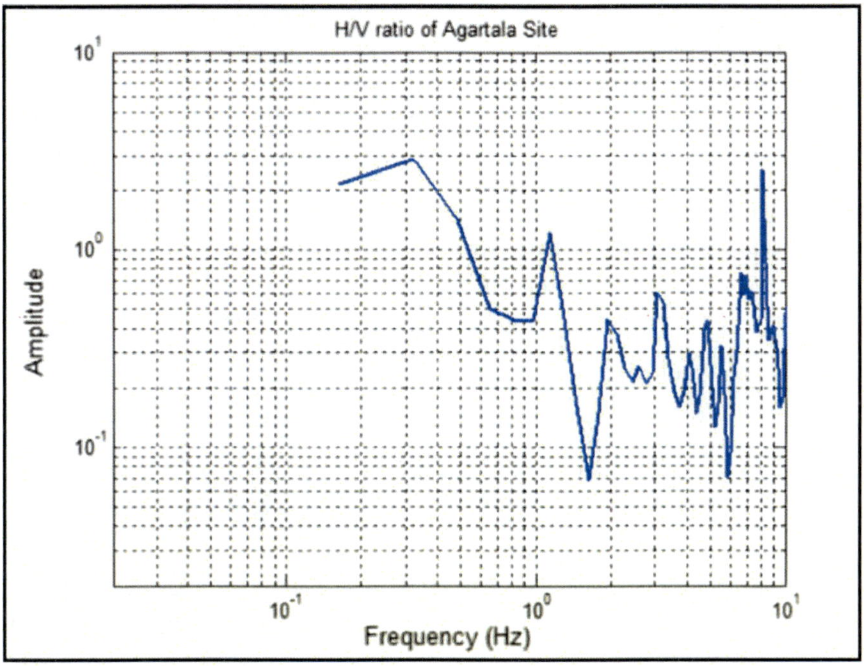

Fig. 6 *H*/*V* ratio of Agartala site

represent the fundamental natural frequency of subsoil deposit as proposed by Nakamura (1989). The fundamental period of ground for Agartala and Belonia stations is presented in Table 1. Further, the fundamental period of ground is calculated from the well-known expression presented below

$$T_{\text{ground}} = \frac{4H}{V_s} \tag{1}$$

where
H = height of the soil layer (m).
V_s = Shear wave velocity of soil (m/s).

Shear wave velocity (V_s) may be expressed in present case as V_{s30} which represents weighted average shear wave velocity of layered subsoil profile exhibiting different shear wave velocity at different layers up to 30 m depth below ground level. The V_{s30} as it is calculated from the expression as proposed by National Earthquake Hazards Reduction Program (NEHRP 2000).

$$V_{s30} = \frac{\sum_{i=1}^{n} d_i}{\sum_{i=1}^{n} \frac{d_i}{V_{si}}} \tag{2}$$

where,
d_i = depth of ith layer.
V_{si} = shear wave velocity of ith layer.

V_{s30} for Agartala and Belonia soil profile are calculated as 408.64 and 432.00 m/s, respectively. Table 2 presents fundamental period obtained from Nakamura approach and theoretical expression. The percentage variation between two approaches is found to be within 10–28%. Hence, it can be concluded that Nakamura approach can reasonably predict the dynamic characteristics of the site.

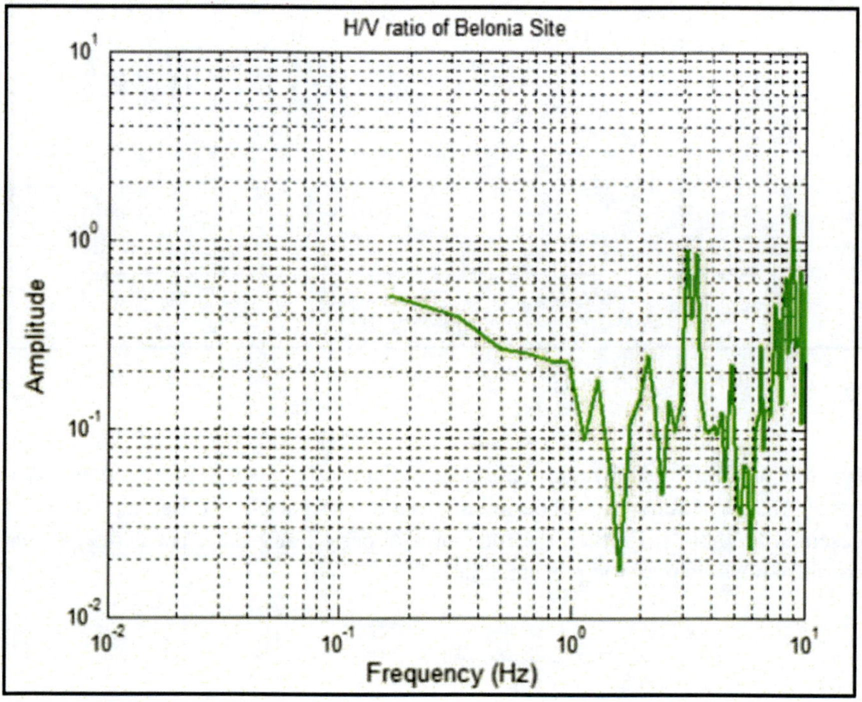

Fig. 7 *H*/*V* ratio of Belonia site

Table 2 PGA values for ground motions at two sites

Site	Nakamura approach (s)	Theoretical approach (s)
Agartala	0.25	0.32
Belonia	0.20	0.18

Table 3 Amplification factor of two sites

Site	Nakamura approach	Theoretical approach
Agartala	2.50	1.33
Belonia	1.80	1.29

4.2 Amplification Factor

It is observed from Figs. 6 and 7 that the peak amplitude of the H/V ratio for Agartala and Belonia stations is in the order of 2.5 and 1.8, respectively, which is also presented in Table 3. Further, the amplification ratio is also calculated using the following empirical expression proposed by Mukherjee and Gupta (2002).

$$A_f = \left(\frac{V_o}{V_r}\right)^{-0.47} \tag{3}$$

where
A_f = Amplification factor.
V_r = Shear wave velocity of Rock, 750 m/s;
V_o = Shear wave velocity of soil which is considered herein as V_{s30}.

Table 3 presents amplification factor obtained considering Nakamura method and empirical solution (Mukherjee and Gupta 2002). The variation in results ranges from 28 to 32% which may be reasonably acceptable. Results indicate that amplification factor may go up to 2.3 which is significant and site-specific spectra must be developed for seismic design of structures.

5 Summary and Conclusions

Present work is a humble attempt to predict the fundamental period and amplification factor following Nakamura proposed spectral H/V approach for two different locations of Tripura using ground motion records collected from two earthquake observatory stations. However, the results are verified comparing with well-known solutions. Most of the part of Tripura is mainly sediments of Quaternary age resting over Pleistocene age. The dynamic characteristics obtained may be considered as representative for Tripura. However, the study lands up with the following broad conclusions.

1. The fundamental period of Agartala and Belonia site was found to be 0.25 and 0.20 s using Nakamura approach, respectively. The range of fundamental periods for two locations exhibits stiffer range which may cause severe effect on low period of structures.

2. Amplification factor of Agartala and Belonia site was found to be 1.75 and 1.65, respectively, which are also considerable. Hence, site-specific response spectra to be developed for seismic design of structure.

Summarily, this limited study highlights the urgent requirement of detailed microzonation study for important locations of Tripura considering seismic multi-hazard. Present study gives inputs for revamping seismic design of future structures and retrofitting of existing weak structures.

References

Anbazhagan P, Kumar A, Sitharam TG (2013) Seismic site classification and correlation between standard penetration test N value and shear wave velocity for Lucknow city in Indo-Gangetic basin. Pure Appl Geophys 170:299–318

BIS: 1893-(Part 1) (2016) Criteria for earthquake resistant design of structures. Bureau of Indian Standards

Building Seismic Safety Council (BSSC) (2001) NEHRP recommended provisions for seismic regulations for new buildings and other structures, 2000 edn. Part 1: provisions. Rep. No. FEMA 368. FEMA, Washington, DC

Borcherdt RD (1994) Estimates of site-dependent response spectra for design. Earthq Spectra 10:617–653

Chowdhuri SN, Singh OP, Majumdar RK (2011) Site response studies in Agartala urban agglomeration. Nat Hazards 59:329–345

GSI (Geological Survey of India) (2011) Miscellaneous Publication, Geology, and Mineral Resources of Manipur, Mizoram, Nagaland, and Tripura vol 30(1). Part II, pp 72–83

Kayal JR, De R (1991) Microseismicity and tectonics in North East India. Bull Seismol Soc Am 81:131–138

Kayal JR (1996) Precursor seismicity, fore shocks and aftershocks of the Uttarkashi earthquake of October 20, 1991 at Garhwal Himalaya. Technophysics 263:339–345

Mohanty WK, William MY, Nath SK, Pal I (2007) First order seismic microzonation of Delhi, India using geographic information system. Nat Hazards 40:245–260

Mukhopadhyay M, Dasgupta S (1998) Deep structure and tectonics of the Burmese arc: constrains from earthquake and gravity data. Technophysics 149:299–322

Mukherjee S, Gupta VK (2002) Wavelet based generation of spectrum-compatible time histories. Soil Dyn Earthq Eng 22:799–804

Nakamura Y (1989) A method for dynamic characteristic estimation of sub surface using microtremor on the ground surface. Railway Tech Res Inst Q Rep 30:1–9

Satyabala SP (2003) Oblique plate convergence in the Indo-Burma (Myanmar) subduction region. Pure Appl Geophys 160:1611–1650

Sil A, Sitharam TG (2013) Probabilistic seismic hazard analysis of Tripura and Mizoram states. Nat Hazards 68:1089–1108

Equivalent 1D Ground Response Analysis (GRA) of Black Cotton Soil for Three Different Sites Near Indore City

Deepshikha Shukla and Chandresh H. Solanki

Abstract Seismic hazards have proved to be the most devastating having the potential for causing the greatest damages. The natures of forces and the energy that is released during an earthquake event are random in nature and unpredictable. To assess the hazards caused due to an earthquake, the quantification of the hazards becomes very important. As the soil profile of the site plays an important role in contributing to the hazards, ground response analysis is carried out for detailed analysis. In the current study, 1D ground response analysis has been carried out at three different locations near Indore, namely Reti Mandi (Indore), Manchaman, and Mahakal Temple (Ujjain) which have been subjected to two different earthquake motions namely Bhuj and Kobe's Earthquake. Linear analysis using DEEPSOIL to estimate the free-field ground response has been carried out. The input parameters considered are input ground motion, shear wave velocity, damping ratio 5%, and dynamic soil properties. Standard penetration tests have been carried out at all the three sites at Indore. It is observed that the soil layers under study result in similar PGA after analyzing from both earthquake motions.

Keywords Peak ground acceleration · Ground response analysis · 1D linear response

1 Introduction

Ground response analysis attributes to the shaking of the round at a particular site due to the effect of the earthquake event occurring at that site (Kramer 1996). The duration and hazards of the earthquake event depends on the location and ground characteristics of the site. To assess the seismic hazards, it is important to determine the ground shaking for that particular location. The intensity of the ground motion

D. Shukla (✉) · C. H. Solanki
Applied Mechanics Department, SVNIT Surat, Surat, Gujarat 395007, India
e-mail: deepshikhashukla@yahoo.com

C. H. Solanki
e-mail: chandresh1968@yahoo.co.in

M. Latha Gali and P. Raghuveer Rao (eds.), *Geohazards*, Lecture Notes in Civil Engineering 86, https://doi.org/10.1007/978-981-15-6233-4_46

can be measures in terms of peak ground acceleration (PGA) for any site (Kumar and Krishna 2013). The values of PGA will depend upon the geological characteristics of the ground location and the input ground motion data (Kumar and Krishna 2012). In the present study, ground response analysis has been performed at three different sites of Indore and Ujjain City (near Indore) which are chosen to carry out the study namely Reti Mandi in Indore, Manchaman and Mahakal Temple (a religious place in Ujjain) situated in the central region of Madhya Pradesh near the banks of river Shipra. Seismic hazards analysis for the entire region covering an area of more than 400 km is being carried out, and ground response analysis is just a part of the complete seismicity study (Shukla et al. 2018). This study will help us in minimizing the damage that will be caused due to the seismic hazards. Standard penetration tests (SPT) are performed to study the variation of penetration blows (N) and the shear wave velocity along the depth of the soil (Choudhury et al. 2011). 1D Ground response analysis has been performed using DEEPSOIL software (Hashash et al. 2016) from the borelog samples that have been collected from all the three sites. The input ground motion is taken as Bhuj earthquake and Kobe's earthquake, and the results are compared to check for the variation. One-dimensional equivalent linear ground response analyses have been performed for Mumbai site using input motion and PGA range of 0.102–0.834 g and (Phanikanth et al. 2011). Seismic ground response analysis of Dehradun city has been carried out and reported that the range of spectral acceleration was 0.06–0.37 g at frequency range 1–10 Hz (Ranjan 2005).

1.1 Study Region

Indore region is located in the central part of Malwa Nimar region of Madhya Pradesh with 22.7196° N, 75.8577° E and is coming up as the biggest commercial educational and industrial hub of central region. Many industries are operating in and around the Indore region. Recently, Indore is also considered for being transforming to a Smart City, and due to this reason, a lot of infrastructure development is likely to happen in and around this region. Indore region is specifically selected for the study because of the geographical location of the area and its importance. Considering the criticality of the area, it becomes utmost importance to perform the seismic analysis of the area to ascertain the seismicity of the region.

1.2 Seismicity of the Study Area

In order to study the seismicity of the study area, it is essential to have the knowledge of the past earthquake data, geological profile of the area, and presence of seismically active faults. Study of seismic hazards is complete by analyzing the data mentioned in the above. The study area for seismic analysis covers an area of 400 km with Indore

Table 1 Major earthquake events and sources near Indore region

Fault name	Distance (km)	Mw	PGA (g)	Fault length	No. of events
Narmada-Son	155.55	5.5	0.0408	320	5
Gavilgadh fault	161.11	7	0.1521	180	6
Nathwara fault	177.78	6	0.0622	80	0
Dudhpur fault	176.75	5.5	0.0387	30	0
Thargon fault	150	5	0.0249	30	0
Tapti North fault	144.44	4.5	0.0147	420	1
Barwani–Sukta fault	100	6.5	0.1313	220	6
Udaipur–Sardarpur lineament	70.55	5	0.0401	310	1
Dhar lineament	50	5	0.0526	290	1
Rakhabdev lineament	122.22	5.2	0.0341	530	2
Kishangarh–Chipri lineament	183.33	4.8	0.0185	320	1
Jaisalmer–Barwani lineament	122.22	4.7	0.0201	510	1
Chambal Jamnagar lineament	150	6	0.0669	380	0

city as center and falls in Seismic Zone–III as per the seismic zoning map of India published by Geographical Survey of India (GSI). The region has experienced mild tremors in the past years. Many seismically active faults and lineaments namely Barwani–Sukta Fault, Gavilgadh Fault, Son Narmada Tapti Fault, and Burhanpur Fault are present in the proximity of the region, and a lot of seismotectonic activities occur in this area. Other minor and major faults and lineaments can be seen using the Seismotectonic Atlas of India (2000) published by Geological Survey of India (GSI). The earthquake data has been collected by various sources.

The earthquake catalog contains a lot of data pertaining to the magnitude, date of occurrence, location, and depth which are few important information. At times, few data or information might be missing. In such cases, process of completeness of the data is to be performed for further analysis and computation of hazards. A total number of 225 earthquake events have been recorded after the removal of the foreshocks and the aftershocks. The data considered for the hazards assessment are of magnitude above 3.0 (Table 1).

1.3 Ground Response Analysis

Ground response analysis is used to predict the site response to determine the forces that are induced due to the earthquake motion. There are three types of analysis, 1D, 2D, and 3D. In the current study, 1D ground response analysis is performed on the borelog samples at three different sites. For this, field tests and laboratory tests are conducted and carried out for the site samples. Peak ground acceleration (PGA) is computed to analyze the 1D response for the site locations.

1.4 Site Selection

To carry out the ground response analysis, three sites have been chosen namely Reti Mandi in Indore, Manchaman and Mahakal Temple (a religious place in Ujjain). The details of the site will be discussed in the subsequent paragraphs.

1. Reti Mandi—This site is located at Indore and a commercial place for purchase of sand.
2. Manchaman—Manchaman is a very famous religious temple located in Keshav Nagar in Ujjain, and a lot of devotes come to this place to offer prayers to Lord Ganesh.
3. Mahakal Temple—Mahakal Temple is a famous temple of Lord Shiva and is also one of the twelve Jyotirlinga in India. This temple is situated near Rudra Sagar Lake.

1.5 Input Motion

Two input motions namely Bhuj and Kobe's earthquake motions have been selected for the study of ground response analysis at three different sites. Details about the other input motions are discussed below:

Bhuj earthquake also known as Gujarat earthquake occurred on Jan 26, 2001, at 08:46 AM IST and lasted for about two minutes. The epicenter of this earthquake was recorded at about 9 km south-southwest of the village of Chobari in Bhachau Taluka of Kutch District of Gujarat, India.
Kobe earthquake also known as great Hanshin earthquake occurred on Jan 17, 1995, at 05:46:53 JST (January 16 at 20:46:53 UTC) in the southern part of Hyōgo Prefecture, Japan. The magnitude recorded for this earthquake was 6.9 on the moment magnitude scale.

1.6 Methodology Adopted

Geological site conditions for any location play a very important and a major role in estimating the seismic hazards of any place due to nay major earthquake event. Borelog data samples for three different sites have been collected and analyzed to assess the hazards. The following steps have been followed in computing the ground response analysis.

Step 1: SPT tests are conducted at three different sites in Indore and Ujjain region using the characteristics of the borelog such as soil profile, bore locations, and depth of the borelog.

Step 2: The complete data of the SPT tests conducted are collected, and the undisturbed sample is further taken to laboratories for testing of soil over various parameters.

1.7 DEEPSOIL Software

DEEPSOIL software is used to study the ground response analysis for the sites. There are various methods and software such as SHAKE 2000, but this software is an open-source software and can be used to carry out the study (Desai et al. 2014). One of the most important parameter is shear wave velocity (V_s) which can be computed using the relationship between the shear wave velocity and SPT (N) values for that site.

For evaluating shear wave velocity (V_s) for a particular site, the N values are used through available correlations (Raghukant et al. 2008). The relation between shear wave velocity and SPT (N) adopted for the computation of shear wave velocity is $V_s = 97 * N^{0.314}$ which is applicable nearly for all soils is used and V_s is in m/s.

1.8 Borehole Data Collected from Site

The borehole profile as collected from three different sites is tabulated below. The values of the shear wave velocity computed from the above-mentioned relationship are taken as the input parameter for analyzing in the DEEPSOIL software (Abrahamson et al. 1997). Other required parameters are also listed in table (Tables 2, 3 and 4).

2 Results

To study the ground response analysis, DEEPSOIL software has been used for various site locations. To begin, Bhuj and Kobe earthquake motions are taken as input motion. The various graphs between the different parameters plotted using this software are discussed below. The graph between time period and pseudo spectral velocity, time period and spectral acceleration, time period and peak ground acceleration, shear strain and shear stress ratio are shown. The analysis is performed for all the boreholes for each site, but the result of only one borehole is discussed. Soft bedrock is considered at the bottom, and the damping ratio of 5% is considered (Figs. 1, 2, 3, 4, 5, 6, 7, 8, 9, 10, 11, 12, 13, 14, 15, 16, 17, 18, 19, 20, 21 and 22).

Table 2 Borehole data for Reti Mandi site at Indore

BH	Soil type	Depth	F.D	*N*	SWV (m/s)
BH 1	Yellow sandy soil	0–2.0	1.68	16	231.67
		2–3.5	1.68	16	231.67
		3.5–5.0	1.68	19	244.51
		5.0–7.0	1.69	17	236.12
		7.0–8.5	1.69	18	240.40
		8.5–10	1.7	17	236.12
BH 2	Yellow sandy soil	0–1.5	1.66	17	236.12
		1.5–3.0	1.66	18	240.40
		3.0–4.5	1.66	19	244.51
		4.5–6.0	1.67	19	244.51
		6.0–7.5	1.67	26	269.82
BH 3	Yellow soil	0–2.0	1.69	16	231.67
		2–3.5	1.67	16	231.67
		3.5–5.0	1.68	19	244.51
		5.0–7.0	1.68	17	236.12
		7.0–8.5	1.7	18	240.40
		8.5–10	1.69	17	236.12
BH 4	Yellow soil	0.0–1.5	1.74	20	248.48
		1.5–3.0	1.75	24	263.12
		3.0–4.5	1.76	29	279.23
	Weathered rock	4.6–5.5	2.78	25	266.52
		5.5–6.8	2.82	28	276.17
		6.8–8.6	2.84	27	273.04

3 Conclusions

Three different sites have been selected for carrying out 1D ground response analysis using borelog data near Indore and Ujjain region using two different strong motions namely Bhuj and Kobe's earthquake motion. After analyzing the borelog data using the DEEPSOIL software, it has been observed that the ground response at these particular sites is greatly influenced by the geological site conditions and also by the input ground motion. The analysis using both the earthquake input motion revealed that the soil is stiffer, and the response of the soil in terms of peak ground acceleration (PGA) is almost similar for both the input motions. The graphs have been plotted for all parameters effective vertical stress, relative displacement of soil, PGA, and maximum strain. The values of all parameters for each site, each borehole, and layer are obtained, but it is difficult to discuss the complete analysis in this paper.

Table 3 Borehole data for Manchaman site at Ujjain

BH	Soil type	Depth	F.D	*N*	SWV (m/s)
BH 1	Sandy soil	0.0–4.0		21	252.32
		4.0–6.0	1.66	23	259.63
		6.0–8.0	1.66	23	259.63
		8.0–10	1.67	21	252.32
	Rock	10–12.5	2.75	0	0.00
	Silty sand	12.5–15	1.69	21	252.32
BH 2	Silty clay	0.0–1.8	1.65	8	186.36
		1.8–3.0	1.65	6	170.26
		3.0–4.5	1.65	9	193.38
		4.5–5.0	1.65	8	186.36
BH 3	Silty clay	0.0–5.0	-	21	252.32
		5.0–7.0	1.66	22	256.03
		7.0–9.0	1.67	22	256.03
		9–10.5	1.67	23	259.63
		10.5–12	1.69	23	259.63
BH 4	Silty clay	0.0–4.0	–	27	273.04
		4.0–5.5	1.67	30	282.22
		5.5–7.0	1.67	30	282.22
		7.0–8.5	1.69	30	282.22
		8.5–10	1.70	29	279.23
		10–11.8	1.71	28	276.17
	Rock	11.8–14		0	0.00

Table 4 Borehole data for Mahakal Temple site at Ujjain

BH	Soil type	Depth	F.D	*N*	SWV (m/s)
BH 1	Gravel	0.0–1.5	1.69	10	199.88
	Sand	1.5–3.0	1.70	12	211.66
	Silty clay	3.0–4.5	1.70	13	217.05
		4.5–5.0	1.70	12	437.29
BH 2	Gravel	0.0–1.5	1.69	10	199.88
	Sand	1.5–3.0	1.70	13	217.05
	Silty clay	3.0–4.5	1.70	14	222.15
		4.5–5.0	1.70	13	217.05
BH 3	Gravel	0.0–1.5	1.67	11	205.95
	Sand	1.5–3.0	1.67	11	205.95
	Silty clay	3.0–4.5	1.68	13	217.05
		4.5–5.0	1.68	12	211.66

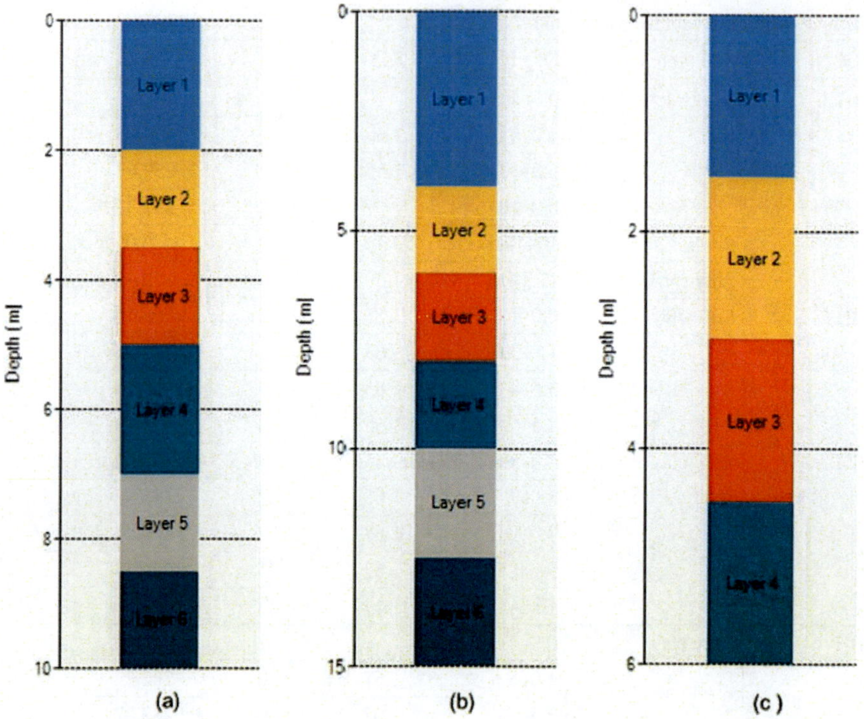

Fig. 1 Borelog profiles for **a** Reti Mandi, Indore; **b** Manchaman, Ujjain; **c** Mahakal, Ujjain

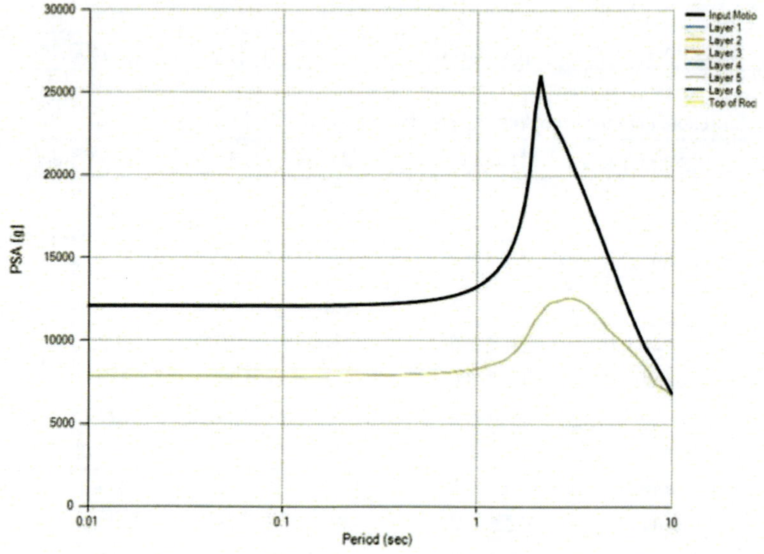

Fig. 2 Pseudo spectral acceleration graph for Reti Mandi borehole for Bhuj earthquake input motion

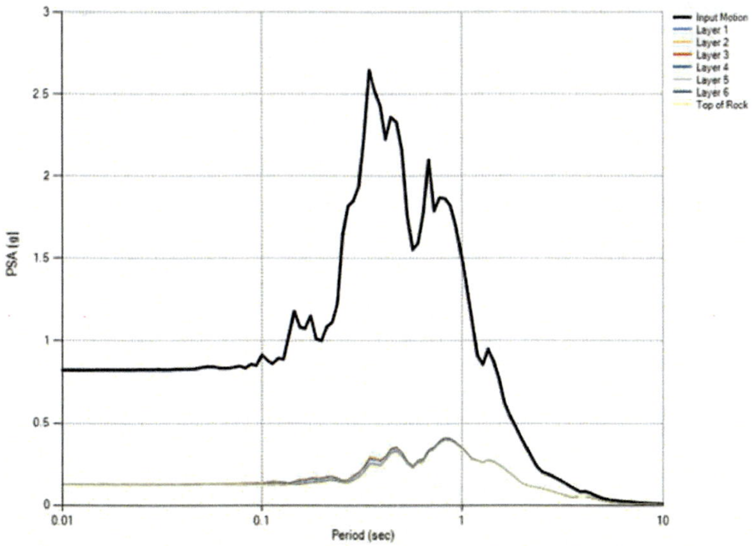

Fig. 3 Pseudo spectral acceleration graph for Reti Mandi borehole for Kobe earthquake input motion

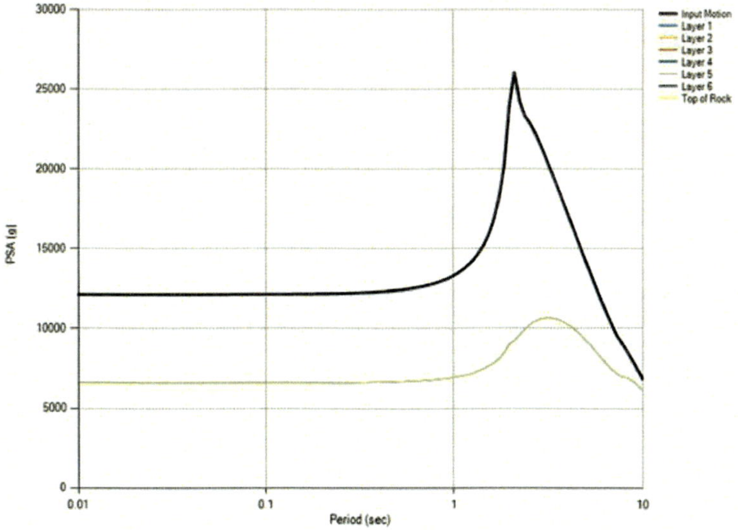

Fig. 4 Pseudo spectral acceleration graph for Manchaman borehole for Bhuj earthquake input motion

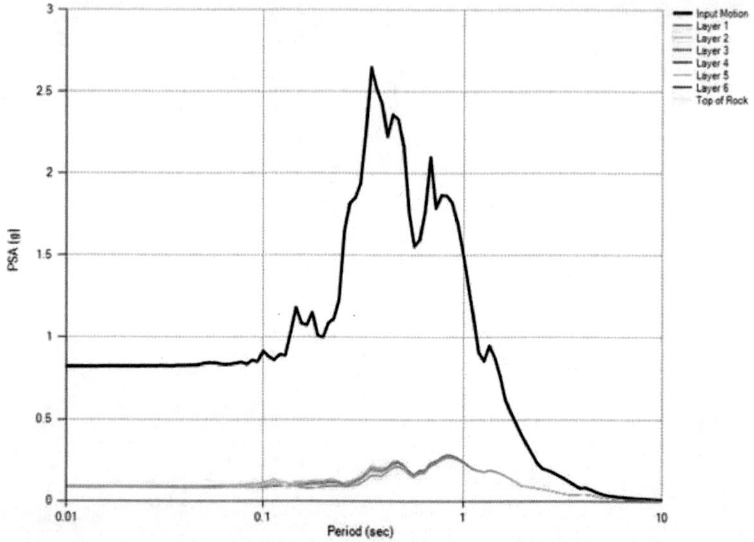

Fig. 5 Pseudo spectral acceleration graph for Manchaman borehole for Kobe earthquake input motion

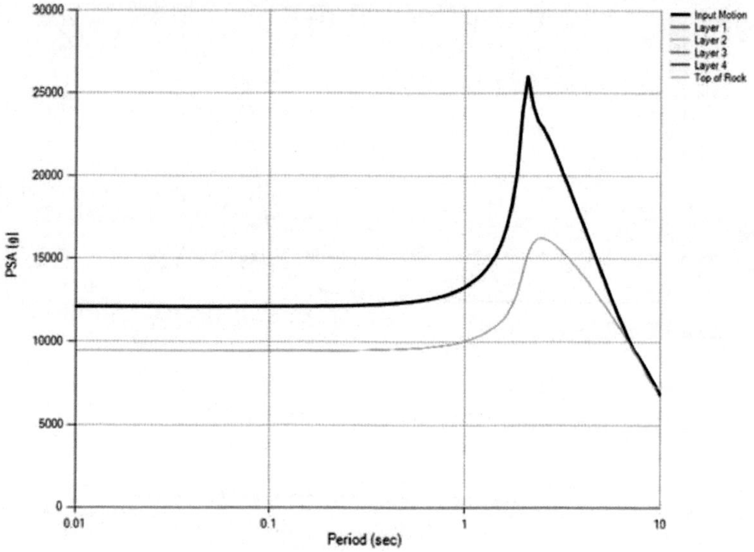

Fig. 6 Pseudo spectral acceleration graph for Mahakal borehole for Bhuj earthquake input motion

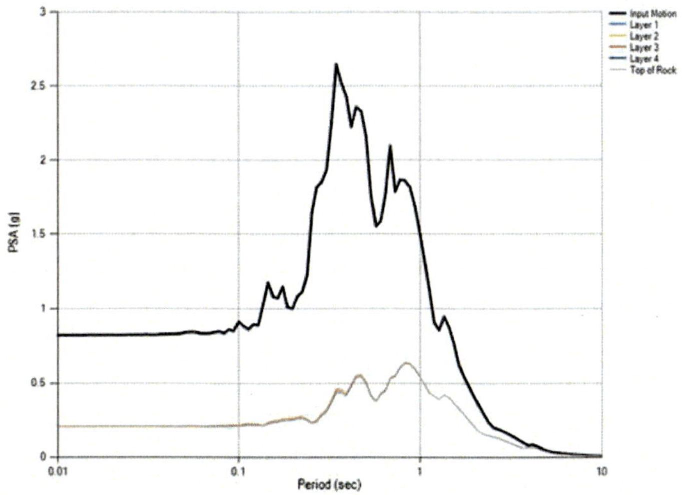

Fig. 7 Pseudo spectral acceleration graph for Mahakal borehole for Kobe earthquake input motion

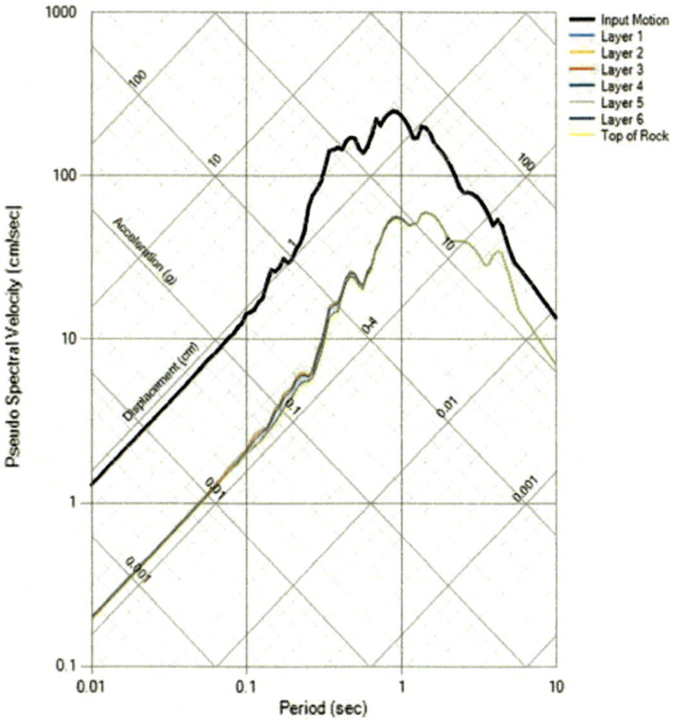

Fig. 8 Tripartite graph for Reti Mandi Kobe input motion

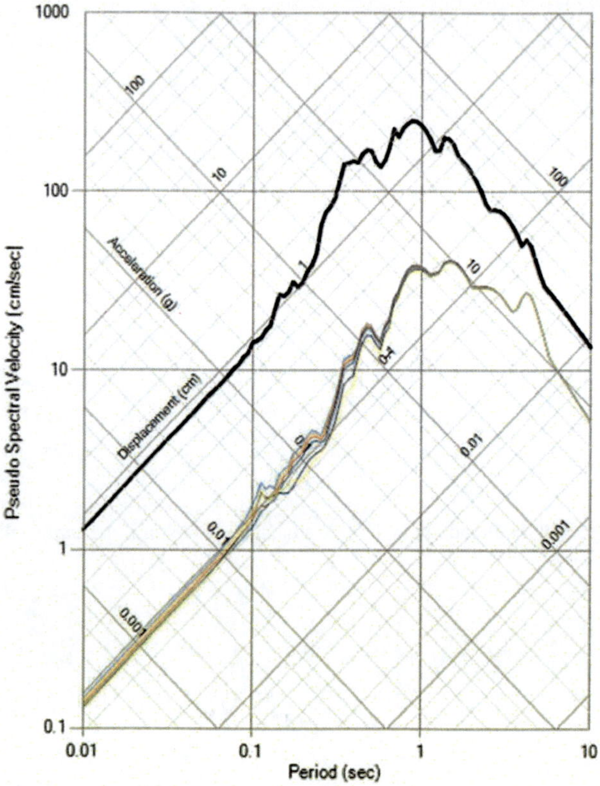

Fig. 9 Tripartite graph for Manchaman Kobe input motion

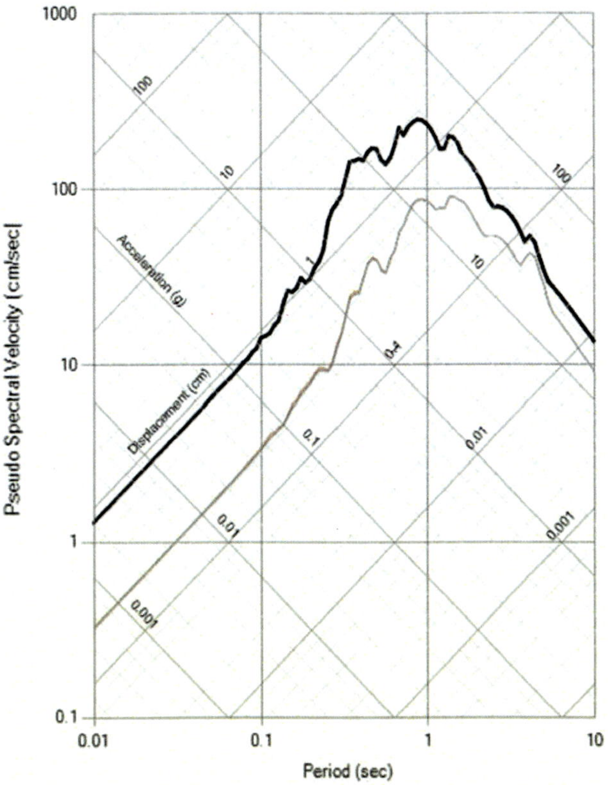

Fig. 10 Tripartite graph for Mahakal Kobe input motion

Fig. 11 Reti Mandi site, Indore

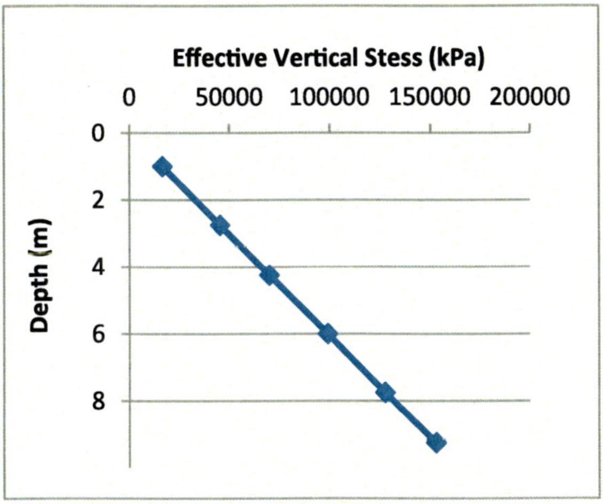

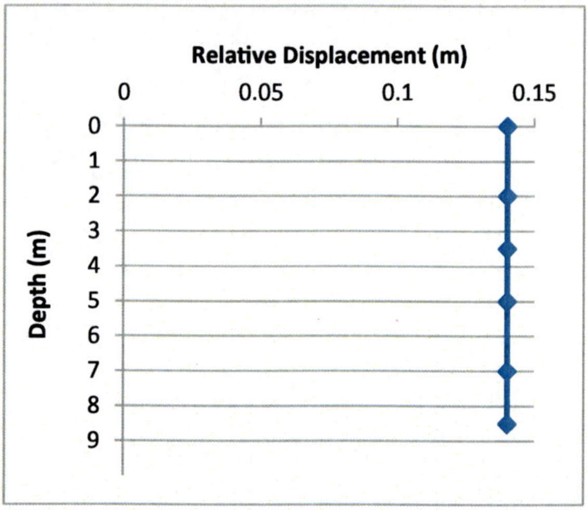

Fig. 12 Reti Mandi site, Indore

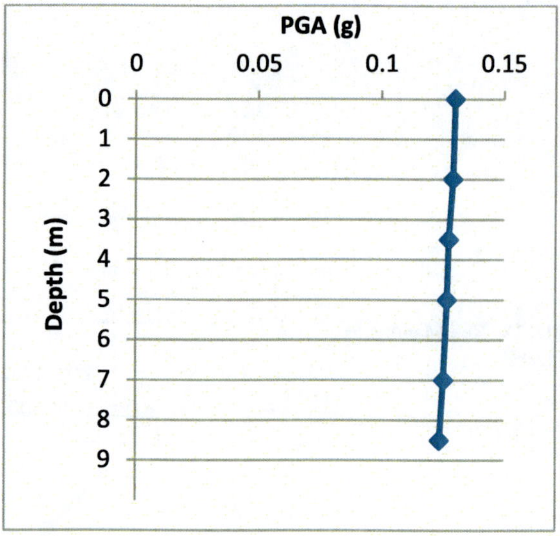

Fig. 13 Reti Mandi site, Indore

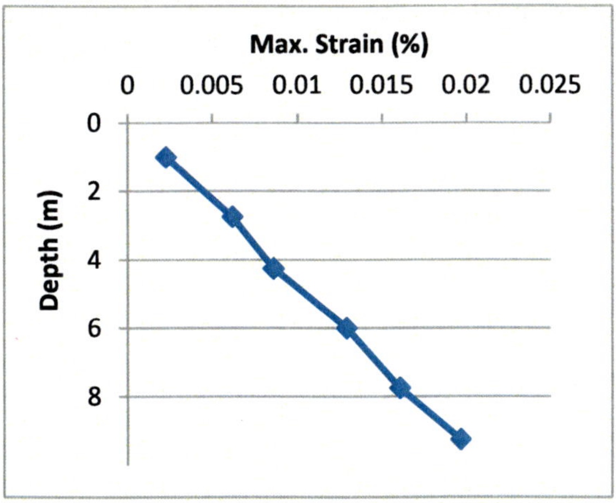

Fig. 14 Reti Mandi site, Indore

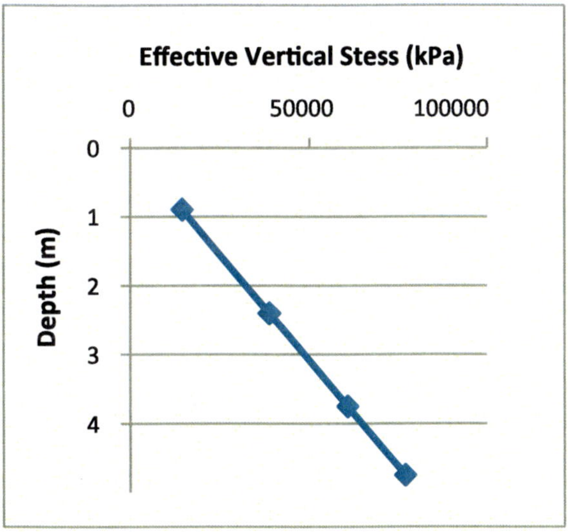

Fig. 15 Manchaman site, Ujjain

Fig. 16 Manchaman site, Ujjain

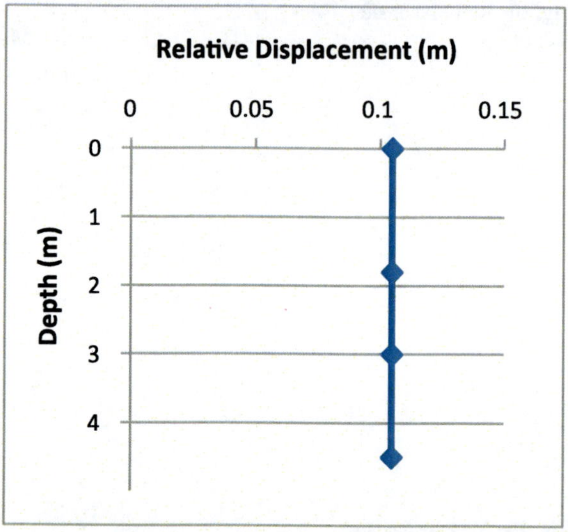

Fig. 17 Manchaman site, Ujjain

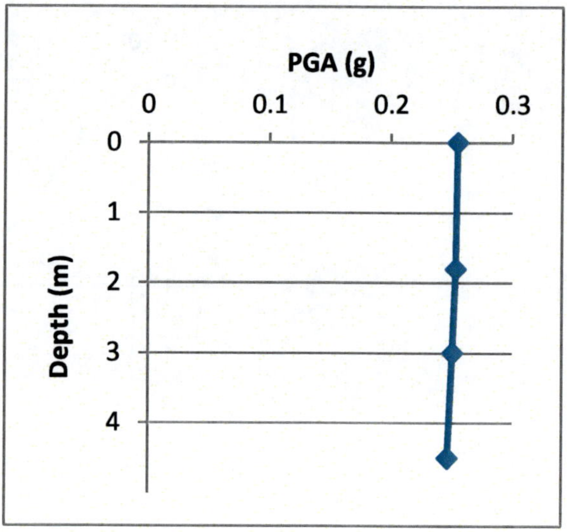

Fig. 18 Manchaman site, Ujjain

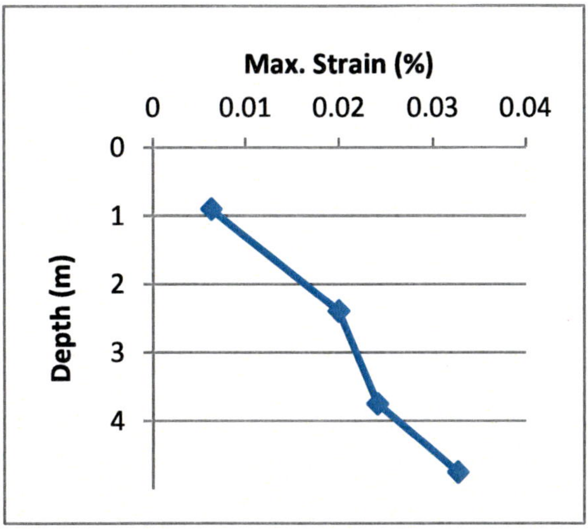

Fig. 19 Mahakal site, Ujjain

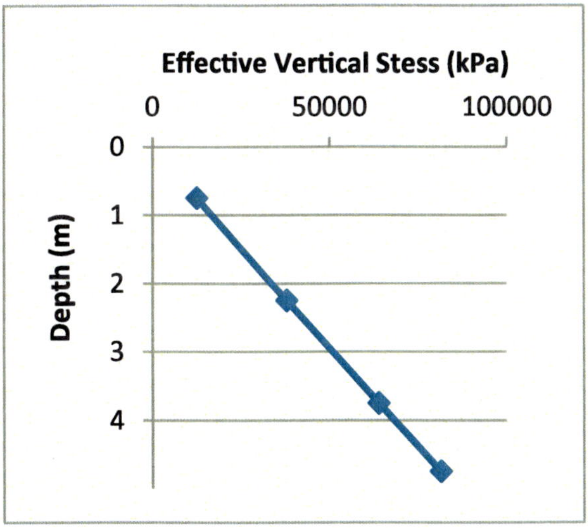

Fig. 20 Mahakal site, Ujjain

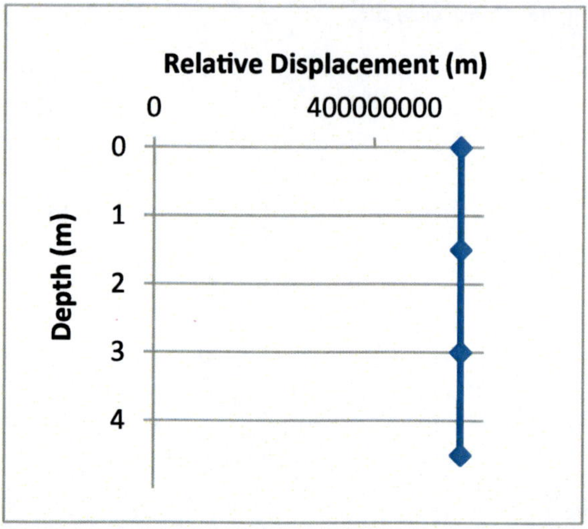

Fig. 21 Mahakal site, Ujjain

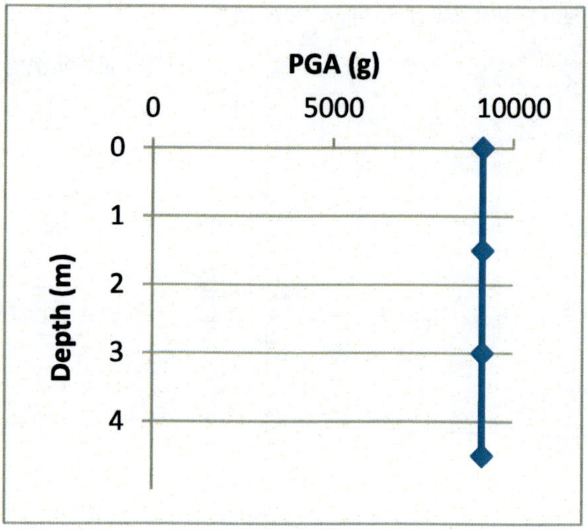

Fig. 22 Mahakal site, Ujjain

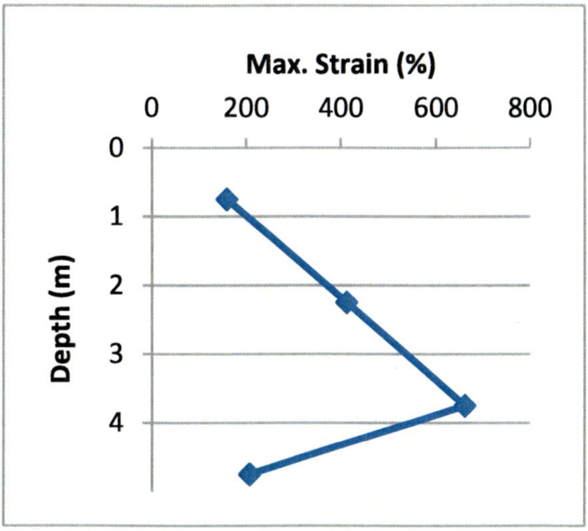

References

Abrahamson NA, Silva WJ (1997) Empirical response spectral attenuation relations for shallow crustal earthquakes. Seismol Res Lett 68(1):94–127

Choudhury D, Shukla J (2011) Probability of occurrence and study of earthquake recurrence models for Gujarat state in India. Disaster Adv 4(2):47–59

Desai S, Choudhury D (2014) Deterministic seismic hazards analysis for greater Mumbai, India. Geotechnical special publication no GSP 234. ASCE, pp 389–398. ISBN 978-0-7844-1329-6. https://doi.org/10.1061/978084413272.038

Hashash YMA, Musgrove MI, Harmon JA, Groholski DR, Phillips CA, Park D (2016) DEEPSOIL 6.1, User manual

Kramer SL (1996) Geotechnical earthquake engineering. Prentice Hall, New Jersey, p 653

Kumar SS, Krishna AM (2012) Site-specific seismic ground response to different earthquake motions. In: Proceedings of Indian geotechnical conference, Dec 13–15, 2012, Delhi

Kumar SS, Krishna AM (2013) Seismic ground response analysis of some typical sites of Guwahati city. Int J Geotech Earthq Eng 4(1):83–101

Phanikanth VS, Choudhury D, Rami Reddy G (2011) Equivalent-linear seismic ground response analysis of some typical sites in Mumbai. J Geotech Geol Eng. https://doi.org/10.1007/s10706-011-9443-8

Raghukanth STG, Sreelatha S, Dash SK (2008) Ground motion estimation at Guwahati city for an Mw 8.1 earthquake in the Shillong plateau. Tectonophysics 448:98–114

Ranjan R (2005) Seismic response analysis of Dehradun City, India. M.Sc. thesis, International Institute for GeoInformation Science and Earth Observations–Enschede, Netherlands, p 92

SEISAT (2000) Seismotectonic atlas of India. Geological Survey of India

Shukla D, Solanki CH, Desai M (2018) Gutenberg-Richter relationship for Indore and surrounding areas. In: Proceedings of 6th international conference on advancements in engineering & technology (ICAET 2018), Sangrur, Punjab

Pattern Recognition to Identify Susceptible Areas in Northwestern Himalaya

Swati Singh Rajput, Ravi S. Jakka, and Amita Sinvhal

Abstract Regions in northern India, especially within the Himalayan arc, have experienced frequent disastrous earthquakes. Major seismic activity in India is concentrated along the geologically young and seismo-tectonically active Himalayan arc. An area between the latitude 25° N to 35° N and longitude 72° E to 90° E was considered for the study, which falls between the great Kangra earthquake of 1905 and the great Bihar–Nepal earthquake of 1934. The main objective of the study is to identify the areas of high seismic susceptibility using pattern recognition (PR) exercise. Areas which have experienced high seismicity and have complex tectonics are more prone to much frequent seismic activity in future and are defined as seismically susceptible areas. The pattern recognition technique started with the identification, selection, and extraction of features from the seismicity and tectonic data. Various features were identified from a circle of radius 25 km around each epicenter, known as the central earthquake. These features were then subjected to discriminant analysis, which constituted the training exercise of the PR technique. The discriminant functions obtained from this training exercise were then applied for the decision-making exercise to identify the susceptible areas. This resulted in the identification of susceptible area within the study area in the form of clusters. Various clusters were identified along the Himalayan arc, which are capable of producing damaging earthquake of significant magnitude. A dense cluster was observed between the Main Boundary Thrust (MBT) and the Main Central Thrust (MCT), and Kishtwar Fault in west and Sundernagar Fault in east. The great Kangra earthquake is part of this dense cluster. A great amount of seismicity was also observed around MCT, east of Sundernagar Fault, in Uttarakhand and western Nepal. Epicenters of Uttarkashi ($M_w = 6.8$) and Chamoli ($M_w = 6.7$) earthquakes are within this cluster. Other dense clusters were also observed which trends transverse to the Himalayan arc and follows the Kaurik Fault system and in the vicinity of Lake Lighten Fault, in the Kashmir Tibet region.

S. S. Rajput (✉) · R. S. Jakka · A. Sinvhal
Department of Earthquake Engineering, Indian Institute of Technology Roorkee, Roorkee, Uttarakhand 247667, India
e-mail: swatisinghr.173@gmail.com

M. Latha Gali and P. Raghuveer Rao (eds.), *Geohazards*, Lecture Notes in Civil Engineering 86, https://doi.org/10.1007/978-981-15-6233-4_47

Keywords Himachal Pradesh (HP) · Uttarakhand (UK) · Main boundary thrust (MBT) · Main central thrust (MCT)

1 Introduction

Regions in northern India, especially within the Himalayan arc, have experienced frequent disastrous earthquakes. Western Himalayas comprises of the states of Himachal Pradesh (HP), Jammu and Kashmir (J&K), Uttarakhand (UK), Punjab, Haryana, union territory of Chandigarh, and Uttar Pradesh. In western Himalaya, different areas exhibit different patterns of seismicity, i.e., some parts of the Himalayan arc show dense seismicity, and other parts exhibit sparse seismicity. Also, this region is infested with complex tectonics. Hence, states within the Himalayan arc are of utmost seismo-tectonic importance.

The tragic event of flash floods which occurred in June 2013 in Uttarakhand shows that great amount of destruction can occur in the Himalayan ranges with only monsoonal rainfall. This opens the question of what can happen if a large magnitude earthquake visits Himalayan states which most of the time is accompanied by rockfall, avalanches, road blocks, delayed flash floods due to damming of rivers, liquefaction, and landslides. This calls for the identification of those regions in western Himalayan which are capable of producing destructive earthquake so that remedial counter measures can be planned and put in place before such a situation arises. The main objective of this paper is to make an attempt to apply pattern recognition (PR) technique for identifying small seismo-tectonically susceptible areas within Himalayan arc, so that disaster mitigation and management methods can be prioritized within such areas.

2 Study Area

The Himalayan arc is a part of Alpine–Himalayan seismic belt, which has experienced four great earthquakes within a short temporal range between the years 1897 and 1950, i.e., within 53 years. One of these great earthquakes is the great Kangra earthquake of April 4, 1905. The epicenter of this great earthquake was in the Himachal Pradesh (HP) and in the vicinity of the Main Boundary Thrust (MBT), in western Himalaya. Besides great earthquakes, the region has also experienced several moderate-to-large-sized earthquakes. However, it is pertinent to note that after 1950, no great earthquake has occurred within the Himalayan arc. Phenomenal damage to the built environment and ground was observed in the meizo-seismal and adjoining areas of this great earthquake. Since the seismicity in the region is highly

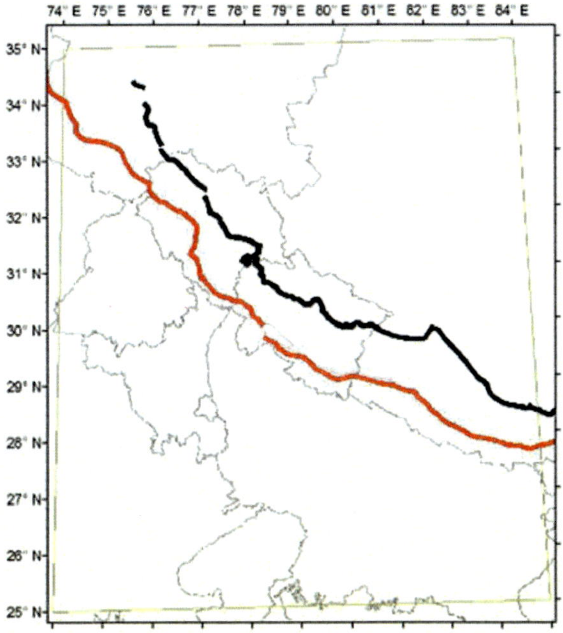

Fig. 1 Study area between latitude 25° N to 35° N and longitude 74° E to 84° E. Main Boundary Thrust (MBT) and Main Central Thrust (MCT) are shown in red and black, respectively

variable, and the region is also going through a phase of techno-economic development, it is pertinent to identify seismically susceptible areas in view of the available seismicity and tectonic database. For this purpose, a 10° × 10° area was selected which falls between latitude 25° N to 35° N and longitude 74° E to 84° E as shown in Fig. 1.

3 Seismicity Data

To identify the seismo-tectonically susceptible area, a comprehensive earthquake catalog was prepared using data from various sources, viz India Meteorological Department (IMD), United States Geological Survey (USGS), and International Seismological Center (ISC). Subsequently catalogs from above-stated sources were thoroughly studied and analyzed. The new workable catalog includes information on source parameters such as: location of epicenter, time of occurrence, depth, and magnitude. It was compiled for the study area in which data was merged for chronological continuity, homogenized to a required magnitude scale, and subsequently de-clustered to retain main events. The database was pared down to 1320 main events. It ranged in time from 1552–2012 AD and had moment magnitude, M_w for all events, which ranged from 3.5–8. This is henceforth referred to as the MHD earthquake catalog for the study area, and salient features of this catalog are given in Table 1.

Table 1 Salient features of MHD catalog for area between latitude 25° N to 35° N and longitude 74° E to 84° E

Catalogue	Time period	Magnitude		No. of events
		Scale	Range	
IMD	1552–1720	M_w	≥3.5	5
USGS (India)	1720–1963	M_w	≥3.5	45
ISC	1964–2010	M_w	≥3.5	1255
USGS (NEIC)	2011–2012	M_w	≥3.5	15
				1320

A diffuse pattern of seismicity was observed over a very large area. Several clusters of epicenters were aligned in an almost northwest-southeast trending belt, parallel to the Himalayan trend, between the epicenters of the two great earthquakes, and extending on either side. This NW–SE trending seismic belt overlaps the Himalayan arc and passes through Uttarakhand. Several clusters of seismicity were observed along the Himalayan arc, viz in the western syntaxes, on the border between Kashmir and Himachal Pradesh, and on the border between Himachal Pradesh and Uttarakhand, which continues eastward into Nepal. Another prominent cluster was observed in the Kashmir–Tibet region. Besides the great Kangra earthquake of 1905, this region has witnessed several other damaging earthquakes, some of which are: Dharamshala earthquake of April 26, 1986 ($M_w = 5.8$), Chamba earthquake of June 22, 1945 ($M_w = 6.5$), Uttarkashi earthquake of October 20, 1991 ($M_w = 6.8$), and Chamoli earthquake of March 29, 1999 ($M_w = 6.7$), and historical earthquakes of February 28, 1905 ($M_w = 7.0$) and July 10, 1947 ($M_w = 6.5$). Since epicentral data for the study area is available for a short span of time as compared to the average return period between large earthquakes, seismicity alone was not enough to identify vulnerable seismic zones. Inclusion of tectonics helped in circumventing this problem. A region where a seismicity cluster is concentrated can be considered to indicate a preliminary vulnerable seismic zone.

4 Tectonic Data

Tectonics of the area was considered according to Narula et al. (2000). 292 tectonic elements were identified and consisted of several types of faults, thrusts, lineaments, and suture zones. These comprised named and unnamed faults, neo-tectonic faults, gravity faults, faults involving basement and cover, named and unnamed thrusts, named and unnamed major lineaments, named and unnamed minor lineaments, and suture zones. The two main thrusts namely, the Main Boundary Thrust (MBT) and the Main Central Thrust (MCT), separate out three distinct geological settings in this area. The greater Himalayas are between Main Central Thrust and Indus Suture Zone; the lesser Himalayas lie between MCT and MBT; and the outer Himalaya lie between the MBT and the MFT. The Main Frontal Thrust, also called as the Frontal Foothill Thrust, is south of MBT and is a neo-tectonic thrust. The MBT, MCT, and ISZ have a NW–SE trend, and these are almost parallel to each other and to the trend of the Himalayan arc in the study area. The ISZ, MCT, MBT, and FFT are evident throughout the Himalayan arc and have many surface manifestations in the region under observation. Besides these mega Himalayan thrusts, several other prominent faults and thrusts exist in the study area. These include the Alaknanda Fault, Bang-gong Nu Jiang Suture, Beng Co Fault, Bharat Mount Abu Lineament, Chahapoli Fault, Chambal Jamnagar Lineament, Delwara Lineament, Drang Thrust, Great Boundary Fault, Indus Suture, Jwalamukhi Thrust, Kantli Fault, Karakoram Fault, Kaurik Fault, Kishtwar Fault, Lake Lighten Fault, Lucknow Fault, Mahendragarh–Dehradun Fault, Main Mantle Thrust, Martoli Thrust, Mastgarh Anticline, Mendha Stepped Graben, Moradabad Fault, North Almora Thrust, Ramgarh Thrust, Ropor Fault, Shyok Suture, South Almora Thrust, Sundernagar Fault, Tonk Lineament, Tso Morari Gravity Fault, and Vaikrita Thrust.

5 Seismo-tectonics of the Area

The seismo-tectonic map is an overlay of the merged, homogenized, and de-clustered (MHD) earthquake data on to the tectonic map of the region as shown in Fig. 2. Four prominent clusters of seismicity demarcated by prominent tectonics were observed.

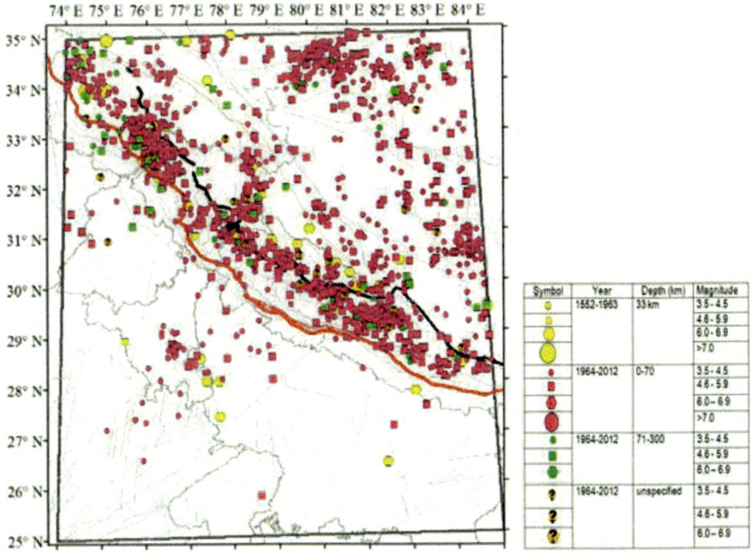

Fig. 2 Seismo-tectonic map of western Himalaya, defined by coordinates: 25°–35° N and: 74°–84° E

These are as follows: (1) the region between the MBT, MCT, and Kishtwar fault in west and Sundernagar fault in east. The great Kangra earthquake is part of this dense cluster; (2) a dense cluster was observed around the MCT east of Sundernagar fault, in Uttarakhand and western Nepal. Epicenters of Uttarkashi (M_w= 6.8) and Chamoli (M_w = 6.7) earthquakes are within this cluster; (3) Another dense cluster was observed which trends transverse to the Himalayan arc and follows the Kaurik fault system; (4) in the vicinity of Lake Lighten Fault, in northern part of the study area, in the Kashmir Tibet region. Several regions of relative scattered seismicity were also identified; these were (1) along the MBT, west of Kishtwar fault, (2) a large area south of MBT, in the Indo-Gangetic plains; and (3) along the Great Karakoram Range.

6 Pattern Recognition

Past studies have shown that hazard in a region is assessed either by deterministic seismic hazard analysis (DSHA) or probabilistic seismic hazard analysis (PSHA). In these two methods, peak ground accelerations (PGA) are computed by using various attenuation relations. DSHA and PSHA use the same seismicity and tectonic data set but yield very different numerical results for PGA, which makes it very difficult to decide which to use. Moreover, the results are totally dependent on empirical attenuation relationships, and each relation has a different set of assumptions. Moreover, data gathered from strong motion arrays, if operating in the area, is usually not part of the seismic hazard assessment. Moreover, the potentially damaging natural phenomena associated with earthquakes, which cause physical hazard such as ground shaking, fault rupture, soil liquefaction, landslides, fissures, and tsunami, are not mentioned. As these phenomena could result in adverse consequences to society such as destruction of the built environment or loss of life, they need to be considered to mitigate future disasters. In such an undecided scenario, it may be possible that pattern recognition (PR) may offer some alternative in decision making, in identifying seismo-tectonically susceptible areas. This paper is an attempt to try this aspect.

Pattern recognition (PR) is the science that concerns either the description and/or classification (recognition) of a set of events. Three important steps identified by Walt and Barnad (2007) are: the study of how machines can observe the environment, how it learns to distinguish patterns of interest from their background, and make sound and reasonable decisions about categories of patterns. The main objective of pattern recognition technique is to recognize objects or patterns in different classes or categories either on the basis of prior knowledge or on the basis of statistical data extracted from patterns. For more than 40 years, pattern recognition (PR) has been used in the recognition of earthquake prone areas of seismically active regions by identifying intersection of lineaments, termed as nodes. The pattern recognition technique has been applied to diverse geographical locations specifically, the Tien Shan and Pamir regions (Gelfand et al. 1972), Anatolia (Turkey), and adjacent regions in central Asia (Gelfand et al. 1973a, b); Balkans, Asia Minor, and Transcaucasia (Gelfand 1974a, b); California and Nevada (Gelfand et al. 1976), Kumaon Himalaya (Varunoday et al. 1979); Italy and Sicily (Caputo et al. 1980; Gorshkov et al. 2002), Andean South America (Gvishiani and Soloviev 1984), Kamchatka (Gvishiani et al. 1984), Pyrenees (Gvishiani et al. 1987), Caucasus (Gvishiani et al. 1988; Soloviev et al. 2013, 2014), Tehri (Sinvhal et al. 1990, 1991); Himalayan arc (Bhatia et al. 1992), Carpatho–Balkan orogenic belt, i.e., Hungary, Poland and Romania region (Gorshkov et al. 2000), Alps and Dinarides (Gorshkov et al. 2004), Alborz (Gorshkov et al. 2009), Iberia (Gorshkov et al. 2010), Mountainous Crimea (Gorshkov et al. 2017), Altai–Sayan–Baikal Region (Gorshkov et al. 2018), Italy (Peresan et al. 2011, 2015); North Vietnam (Tuyen et al. 2012), Kopet Dagh, mountain range on the frontier between Turkmenistan and Iran (Novikova and Gorshkov 2013), Alpine-Himalayan belt (Novikova and Gorshkov 2016), western Himalaya (Mridula et al. 2013, 2014a, b, 2015, 2016a, b; Rajput et al. 2014, 2016, 2017; Sinvhal et al. 1984; Sinvhal and Sinvhal 1992), Caucasus (Soloviev and Gorshkov 2017; Soloviev et al. 2013, 2014, 2016), Iran (Talebi et al. 2017) and Himalayan frontal arc (Vorobieva et al. 2017).

7 Identification of Seismo-tectonically Susceptible Area Using Pattern Recognition

Seismo-tectonically susceptible areas were identified in the northwestern Himalayan region using pattern recognition (PR) technique. Several features were studied, then identified, and later extracted from the available seismicity and tectonic data using ArcGIS 10.2.2 and MATLAB software packages.

7.1 Identification and Extraction of Features

For identification of features, a circle of radius 25 km was drawn around each epicenter. The selected epicenter is referred to as the central earthquake henceforth. All features were identified and quantified within this radius, using software package ArcGIS 10.2.2. For this study, initially eight features were identified from seismicity data which are as follows: (1) magnitude of central earthquake (M_w), (2) cumulative number of epicenters within the circle, (3) number of epicenters within the circle, magnitude range (M_w) 3.5–3.9, (4) number of epicenters within the circle, magnitude range (M_w) 4.0–4.9, (5) number of epicenters within the circle, magnitude ranges (M_w) 5.0–5.9, (6) number of epicenters within the circle, magnitude ranges (M_w) 6.0–6.9, (7) number of epicenters within the circle, magnitude ranges (M_w) 7.0–7.9, and (8) number of epicenters within the circle, magnitude ranges (M_w) 8.0 and above. Sixteen features were identified from tectonic data which are given as follows: (1)

number of named thrusts, (2) number of named faults, (3) number of named suture zones, (4) number of other named features, (5) cumulative number of named tectonic features, (6) number of unnamed thrusts, (7) number of unnamed faults, (8) number of other unnamed features, (9) cumulative number of unnamed tectonic features, (10) number of extremities of tectonic features within the circle, (11) number of intersections with MBT, (12) number of intersections with MCT, (13) number of other intersections, (14) cumulative number of intersections within the circle, (15) number of intersections of river and any tectonic feature, and (16) length of major river. For the current study, two seismicity and two tectonic features were retained from the data, viz number of epicenters within the circle, number of tectonic features within the circle, number of earthquakes with magnitude greater than 6 within the circle, and number of tectonic intersections within the circle. These four features were extracted for 1320 epicenters in the earthquake catalogue, each from a circle of radius 25 km drawn around each epicenter. These features for the Kangra earthquakes (M_w= 8.0) are shown in Fig. 3.

After features were extracted, a new data matrix was formed with 1320 rows and 14 columns. These 1320 rows contained information about the 1320 earthquake events against each column. The first column contained serial number of the event; second to sixth columns comprised details of origin time of the event, in the sequence of year, month, day, hours, and minutes; columns seven to nine have information about location of the epicenter, i.e., longitude, latitude, and depth; the tenth column has the moment magnitude of the earthquake (M_w); eleventh to fourteenth column comprised details about the four features extracted for each epicenter, i.e., earthquake around epicenter, number of tectonics features, number of earthquake with magnitude greater

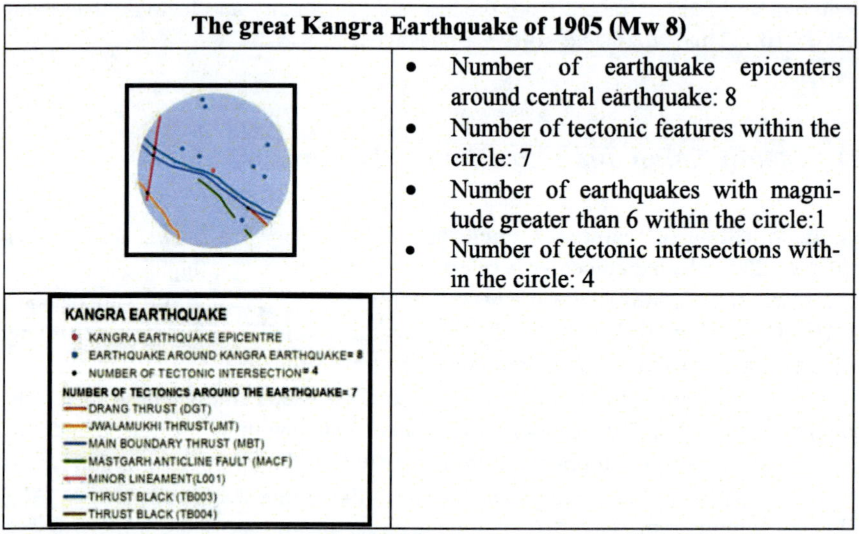

Fig. 3 Four features extracted for the great Kangra earthquake of 1905

than six, and number of tectonic intersections. This data matrix of 1320 rows and 14 columns was then used for pattern recognition, as per Sinvhal and Sinvhal (1992).

7.2 Classification Criteria

Several classification criteria were tried, and among these, four were based on seismicity data and two on tectonic data. While carrying out of the many iterations of discriminant analysis, the best combination of classification criteria and number of features was retained. The best combination can be determined by the best separation between the two classes achieved after discriminant analysis. The separation between both classes is shown by R0 in histograms. Classification of epicenters basis of magnitude higher than 6.0 gives the best combination. In this classification, epicenters were divided into two classes—Class A and Class B. Class A contained epicenters with magnitude ≥6.0 which consisted of 58 epicenters, and Class B contained the remaining epicenters which consisted of 1262 epicenters. This classification was retained for further applications for identifying seismo-tectonically susceptible areas.

8 Discussion and Results

A linear combination of extracted features was used to maximize the difference between classes A and B. Programming in MATLAB was used to develop a discriminant function that transformed the original set of four features of an earthquake into a single seismic score. Seismic scores were calculated for each earthquake, and percentage contribution of each feature was computed. Among the four features extracted, number of earthquakes with magnitude ≥6.0 in a circle of 25 km radius contributed highest toward PR, with 63.36%, which is followed by number of earthquakes in the circle, with 38.16%, then tectonic features, with 3.32%, while number of tectonic intersections made a negative contribution of −4.85%.

Seismic score for each event is shown in Fig. 4, with Class A and Class B shown by different colors. As a discernible separation was observed between the two classes,

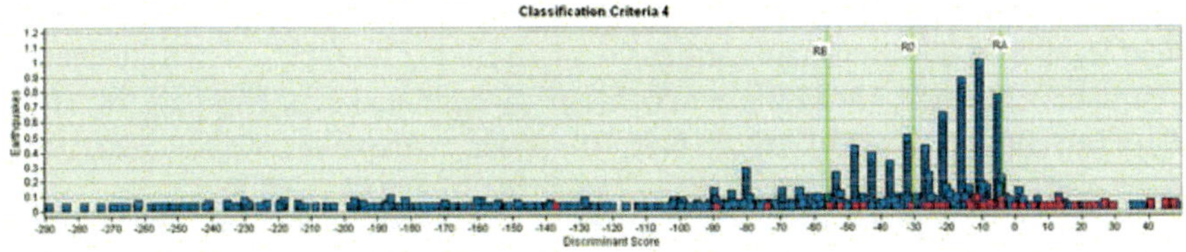

Fig. 4 Discriminant scores. R0 = −30.322, RA = −4.225, RB = −56.418 shown by vertical lines. Class A depicts epicenters having magnitude greater than equal to 6.0, represented by pink squares, while Class B depicts epicenters with magnitude less than 6.0, represented by blue squares

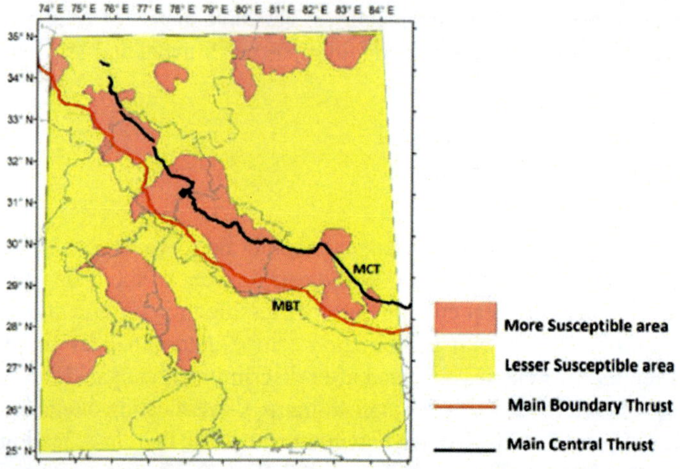

Fig. 5 Two types of susceptible areas in western Himalaya: More susceptible area; and lesser susceptible area. Main Boundary Thrust (MBT) and Main Central Thrust (MCT) are shown in red and black, respectively

which is a desirable aspect, the seismo-tectonic map was revisited, and envelopes were drawn around the clusters of same class, using nearest neighbor interpolation tool given in ArcGIS-10.2.2, which approximates the value of a function for a non-given point when given the value of that function in points around (neighboring) that point.

The nearest neighbor algorithm selects the value of the nearest point and does not consider the values of neighboring points at all, yielding a piecewise-constant interpolant. Two areas were identified from these clusters, and these were named according to their Class as A′ or B′. The study area was divided into two types based on their seismic scores, viz area A′ and area B′. Figure 5 shows these identified areas. This map is hereafter referred to as the susceptibility map of the study area. Red represents area A′ and named as more susceptible area, while yellow represents area B′ and named as lesser susceptible area. It can be observed from the susceptibility map that a large part of Himachal Pradesh and Uttarakhand are more prone to originate a destructive earthquake. This area also runs parallel to the trends of Main Boundary Thrust (MBT) and Main Central Thrust (MCT). Mainly destructive earthquakes have originated in the more susceptible area like the great Kangra earthquake of 1905 (M_w= 8.0), Dharamshala earthquake of April 26, 1986 ($M_w = 5.8$), Chamba earthquake of June 22, 1945 ($M_w = 6.5$), historical earthquakes of February 28, 1905 (M_w= 7.0) and July 10, 1947 ($M_w = 6.5$), Uttarkashi earthquake of October 20, 1991 ($M_w = 6.8$), and Chamoli earthquake of March 29, 1999 ($M_w = 6.7$).

9 Conclusions

Seismicity and tectonic data of western Himalayan when subjected to pattern recognition yielded susceptible areas. Several seismo-tectonically susceptible areas of radius 25 km around central earthquakes of magnitude 6 and above were identified in this study. The technique holds promise, and needs to be continued and further tested with more trials, additional features, and classification criteria, and interpretation in terms of geology and topography, preferably on 1: 250,000 scale, and for other Himalayan regions.

Acknowledgements The authors would like to thank the Indian Meteorological Department, New Delhi, for the epicentral data provided. Author (Swati Singh Rajput) is grateful to the Ministry of Human Resource Development (MHRD) and Department of Earthquake Engineering for the financial and academic support, respectively, provided for the study.

References

Bhatia SC, Chetty TRK, Filimonov MB, Gorshkov AI, Rantsman EY, Rao MN (1992) Identification of potential areas for the occurrence of strong earthquakes in Himalayan arc region. Proc Indian Acad Sci (Earth Planet Sci) 101:369–385

Caputo M, Keilis-Borok VI, Oficerova Ranzman E, Rotwain I, Solovjeff A (1980) Pattern recognition of earthquake prone areas in Italy. Phys Earth Planet Inter 305–320

Gelfand IM, Guberman SI, Izvekova ML, Keilis-Borok VI, Ranzman EJ (1972) Criteria of high seismicity, determined by pattern recognition. Tectonophysics 13(1):415–422

Gelfand IM, Guberman SA, Izvekova ML, Keilis-Borok VI, Ranzman E (1973a) Recognition of places where strong earthquake may occur. I. Pamir and Tien Shan. Computational seismology, vol 6 (in Russian)

Gelfand IM, Guberman SA, Kaletskaja MS, Keilis-Borok VI, Ranzman E, Zhidkov MP (1973b) On transfer of criterion of high seismicity from Central Asia to Anatolia and adjacent regions (in Russian)

Gelfand IM, Guberman SA, Kaletskaja MS, Keilis-Borok VI, Ranzman E, Rotwain IM, Zhidkov MP (1974a) Recognition of places where strong earthquakes may occur. II: four regions of Asia Minor and S-E Europe (in Russian)

Gelfand IM, Guberman SA, Keilis-Borok VI, Ranzman E, Rotwain IM, Zhidkov MP (1974b) Recognition of places where strong earthquakes may occur. III: thc case when the boundaries of disjunctive knots are unknown (in Russian)

Gelfand IM, Guberman SA, Keilis-Borok VI, Knopoff L, Press F, Rantsman E, Rotwain I, Sadovsky AM (1976) Pattern recognition applied to earthquake epicenters in California. Phys Earth Planet Inter 11(3):227–283

Gorshkov AI, Kuznetsov I, Panza G, Soloviev AA (2000) Identification of future earthquake sources in the Carpatho-Balkan Orogenic Belt using morphostructural criteria. Pure Appl Geophys 157:79–95

Gorshkov AI, Panza GF, Soloviev AA, Aoudia A (2002) Morphostructural zonation and preliminary recognition of seismogenic nodes around the Adria margin in peninsular Italy and Sicily. J Seismol Earthq Eng 4:1–24

Gorshkov AI, Panza GF, Soloviev AA, Aoudia A (2004) Identification of seismogenic nodes in the alps and dinarides. Boll Soc Geol Ital 123:3–18

Gorshkov AI, Mokhtari M, Piotrovskaya EP (2009) The Alborz region: identification of seismogenic nodes with morphostructural zoning and pattern recognition. J Seismol Earthq Eng 11:1–16

Gorshkov AI, Soloviev AA, Jimenez MJ, Fernandez G, Panza GF (2010) Recognition of earthquake prone areas ($M \geq 5.0$) in the Iberian Peninsula, Rend. Fis Acc Lincei 21(2):131–162

Gorshkov AI, Soloviev AA, Zharkikh YI (2017) A morphostructural zoning of the mountainous crimea and the possible locations of future earthquakes. J Volcanol Seismol 11(6):407–412

Gorshkov AI, Soloviev AA, Zharkikh YuI (2018) Recognition of strong earthquake prone areas in the Altai–Sayan–Baikal region. Dokl Earth Sci 479(1):412–414

Gvishiani AD, Soloviev AA (1984) Recognition of places on the Pacific coast of the South America where strong earthquakes may occur. Earthq Prediction Res 2:237–243

Gvishiani AD, Zhidkov MP, Soloviev AA (1984) On application of the criteria of high seismicity of Andean mountain belt to Kamchatka. Izv Akad Nauk SSSR Fiz Zemli 1:20–33

Gvishiani A, Gorshkov A, Kossobokov V, Cisternas A, Philip H, Weber C (1987) Identification of seismically dangerous zones in the Pyrenees. Ann Geophys B 5(6):681–690

Gvishiani AD, Gorshkov AI, Ranzman EY, Cisternas A, Soloviev AA (1988) Forecasting the earthquake locations in the regions of moderate seismic activity. Nauka, Moscow, p 187

ISC. https://www.isc.ac.uk/iscbulletin/search/bulletin

Mridula, Sinvhal A, Wason HR (2013) A review on pattern recognition techniques for seismic hazard analysis. In: Proceedings of international conference on emerging trends in engineering and technology, pp 854–858

Mridula, Sinvhal A, Wason HR (2014a) Probabilistic seismic hazard assessment in the vicinity of MBT and MCT in western Himalaya. Res Inventy Int J Eng Sci 4(11):21–34

Mridula, Sinvhal A, Wason HR (2014b) Seismic hazard zonation of Himachal Pradesh, Northwest Himalaya. In: National workshop on status of natural hazards in Himachal Pradesh (NHHP-14), 6–8 Nov, 2014 conducted by Central University of Himachal Pradesh, Shahpur, Dharamsala (Abstract)

Mridula, Rajput SS, Sinvhal A, Wason HR (2015) Seismic hazard and vulnerability assessment of Himachal Pradesh and contiguous regions. In: EMI 2015 conference. Stanford University, June 16–19, 2015 (Abstract)

Mridula, Sinvhal A, Wason HR (2016a) Identification of seismically susceptible areas in the western Himalaya using pattern recognition. J Earth Syst Sci 125(4):855–871

Mridula, Sinvhal A, Wason HR, Rajput SS (2016b) Segmentation of main boundary thrust and main central thrust in Western Himalaya for assessment of seismic hazard. Nat Hazards 84(1):383–403

Narula PL, Acharya SK, Banerjee J (2000) Seismotectonic atlas of India and its environs. Geological Survey of India, New Delhi

Novikova O, Gorshkov A (2013) Recognition of earthquake prone areas ($M \geq 6.0$) in the Kopet Dagh region using the GIS technology. J Seismol Earthq Eng 15(3):92–99, 31

Novikova O, Gorshkov A (2016) Seismogenic nodes defined by pattern recognition in the central part of the Alpine Himalayan belt. Bull Geol Soc Greece 50:1426–1432

Peresan A, Zuccolo E, Vaccari F, Gorshkov A, Panza GF (2011) Neo-deterministic seismic hazard and pattern recognition techniques: time-dependent scenarios for North-Eastern Italy. Pure Appl Geophys 168:583–607

Peresan A, Gorshkov AI, Soloviev A, Panza GF (2015) The contribution of pattern recognition of seismic and morphostructural data to seismic hazard assessment. Boll Geofis Teorica Appl 56:295–328

Rajput SS, Mridula, Sinvhal A (2014) Vulnerability assessment of area and structures. In: 15th symposium on earthquake engineering, proceedings volume, pp 1099–1108

Rajput SS, Mridula, Sinvhal A, Wason HR, Dixit P (2016) Seismic hazard and risk assessment in Kangra seismogenic source zone. In: 6th international conference on recent advances in geotechnical earthquake engineering & soil dynamics, Noida, India, Aug 1–6, 2016

Rajput SS, Mridula, Sinvhal A, Jakka RS (2017) Seismic hazard and risk assessment of Himachal Pradesh and its contiguous area. In: 14th annual meeting Asia Oceania Geosciences Society in SUNTEC Singapore from 6–11 Aug 2017 (Abstract)

Sinvhal A, Sinvhal H (1992) Seismic modelling and pattern recognition in oil exploration. Kluwer Academic Publisher, Dordrecht, p 178

Sinvhal A, Khattri KN, Sinvhal H, Awasthi AK (1984) Seismic indicators of stratigraphy. Geophysics 48:1196–1212

Sinvhal A, Joshi G, Sinvhal H, Singh VN (1990) A pattern recognition technique for microzonation. In: Proceedings of the ninth symposium on earthquake engineering, pp 24–30

Sinvhal A, Sinvhal H, Joshi G (1991) A valid pattern of microzonation. In: Proceedings of the fourth international conference on seismic zonation, pp 641–648

Soloviev AA, Gorshkov AI (2017) Modelling the dynamics of the block structure and seismicity of the Caucasus. Phys Solid Earth 53(3):321–331

Soloviev AA, Novikova OV, Gorshkov AI, Piotrovskaya EP (2013) Recognition of potential sources of strong earthquakes in the caucasus region using GIS technologies. Dokl Earth Sci 450(2):658–660. https://doi.org/10.1134/S1028334X13060159

Soloviev AA, Gvishiani AD, Gorshkov AI, Dobrovolsky MN, Novikova OV (2014) Recognition of earthquake-prone areas: methodology and analysis of the results. Izv Phys Solid Earth 50(2):151–168. https://doi.org/10.1134/S1069351314020116

Soloviev AA, Gorshkov AI, Soloviev AA (2016) Application of the data on the lithospheric magnetic anomalies in the problem of recognizing the earthquake prone areas. Phys Solid Earth 52(6):803–809

Talebi M, Zare M, Peresan A, Ansari A (2017) Long-term probabilistic forecast for $M \geq 5.0$ earthquakes in Iran. Pure Appl Geophys 174:1561–1580

Tuyen NH, Gorshkov AI, Lu NT (2012) Recognition of earthquake-prone areas ($M \geq 5.0$) applied for North Vietnam and adjacency. J Sci Earth 34(3):251–265

USGS. https://earthquake.usgs.gov/earthquakes/eqarchives

Varunoday, Gaur VK, Wason HR (1979) Spatial prediction of earthquakes in the Kumaon Himalaya by pattern recognition. Mausam 30:253–264

Vorobieva I, Mandal P, Gorshkov AI (2017) Block-and-fault dynamics modelling of the Himalayan frontal arc: implications for seismic cycle, slip deficit, and great earthquakes. J Asian Earth Sci 148:131–141

Walt CM, Barnard E (2007) Data characteristics that determine classifier performance. SAIEE Afr Res J 98:87–93

Modelling of Response of Tunnels Excavated in Squeezing Ground Condition

Dipaloke Majumder, M. N. Viladkar, and Mahendra Singh

Abstract As a part of development of infrastructure in hill regions, many hydropower projects, all weather proof road and railway projects have been undertaken by the Government of India. The geology of hilly regions is very complex and fragile in nature. The rock masses are highly fractured and weathered. As a result, construction of large underground structures and their stability becomes a major issue in hilly regions. Therefore, it is necessary to understand the rock mass-tunnel support interaction in detail before any underground operation is undertaken in hilly regions. In this study, a case history of a Head Race Tunnel of Maneri Bhali Hydro-Electric Project Phase-I has been considered for investigation. The interaction phenomenon has been studied by using a novel finite element-based framework where elastic-perfectly plastic behaviour of poor rock mass is modelled by implementing actual Hoek–Brown yield criterion in association with non-associated plastic flow rule. Additionally, the elasto-plasticity in the behaviour of different tunnel support components has also been included in this framework. Under the scope of present study, tunnel convergence, plastic straining and the ground reaction curve were obtained for the considered tunnel section. The tunnel convergence obtained through present study is in good agreement with the field measured values of tunnel convergence. Hence, it can be said that the proposed framework is a convenient tool for detail analysis of the rock mass-tunnel support interaction and for design of supports of underground structures in poor-quality rock masses.

Keywords Squeezing tunnel · Finite element analysis · Hoek–brown criterion

1 Introduction

In last five decades, with development of construction technologies, utilization of underground space is growing in the form of roadway and railway tunnels, transportation and storage of petroleum products, storage for strategic purposes and

D. Majumder · M. N. Viladkar (✉) · M. Singh
Department of Civil Engineering, Indian Institute of Technology Roorkee, Roorkee 247667, India
e-mail: mnviladkar50@gmail.com

M. Latha Gali and P. Raghuveer Rao (eds.), *Geohazards*, Lecture Notes in Civil Engineering 86, https://doi.org/10.1007/978-981-15-6233-4_48

nuclear wastes disposal. In India, majority of these underground operations have been performed in mountains of the Himalayan region. In this region, rock mass is highly fragile in nature due to existence of joints, faults, folds, shear zones and other discontinuities. In addition, magnitude of in situ stresses within the rock mass is very high due to large overburden depth and tectonic activities. Consequently, the rock mass behaves as an elasto-plastic material upon excavation in certain regions. On the other hand, excavation techniques, design and installation of support systems and cost of the overall project are governed by the behaviour of rock mass. Therefore, once the ground conditions for tunnelling are predicted through an appropriate ground condition prediction technique (Hoek and Marinos 2000; Majumder et al. 2017; Singh et al. 2007), the detailed investigation becomes compulsory at problematic sections along the proposed underground excavation.

In this study, a case history of a Head Race Tunnel (HRT) of Maneri Bhali Hydro-Electric Project (HEP) stage-I (Goel et al. 1995) was analysed using a novel framework. The framework was developed based on finite element method (FEM) formulation of Hoek–Brown (HB) yield criterion (Hoek et al. 2002). The behaviour of rock mass was idealized as linear elastic-perfectly plastic (EPP). In this framework, a return mapping technique (Clausen 2007), i.e. predictor–corrector method, was adopted for obtaining updated stresses at corner edges and at apex point of the yield surface. Finally, a detailed investigation of tunnel convergence, plastic straining (equivalent) and ground response curve (GRC) of the tunnel was performed. The tunnel convergence thus obtained has been compared with the field measurements.

2 Literature Review

Nowadays, various empirical, analytical and numerical techniques and/or combination of these are quite popular in practice for analysis of the underground structures. Particularly, the numerical techniques, like finite difference method (FDM), finite element method (FEM) and the distinct element method (DEM), have achieved popularity with advancement of computational technologies during last four decades. Depending upon stress–strain response of rock mass, all the studies can be broadly categorized into three groups, namely elastic-brittle plastic (EB), elastic-strain softening (ES) and elastic-perfectly plastic (EPP). Hoek and Brown (1997) suggested that the elastic-brittle plastic (EB) behaviour can be observed in case of good to very good-quality rock mass (GSI $\geq$ 75), whereas the elastic-strain softening (ES) behaviour is witnessed for medium quality rock mass (75 $\geq$ GSI $\geq$ 25). The poorest quality rock mass (GSI $\leq$ 25) shows elastic-perfectly plastic (EPP) behaviour which is the prime focus of the present study. However, due to extreme in situ stress conditions, the EPP behaviour can also be noticed up to GSI equal to 40–45 in certain Himalayan regions (Goel et al. 1995).

Generally, certain constitutive laws are considered at the time of analytical or numerical modelling of the EPP rock mass which will replicate the yielding behaviour of that rock mass appropriately, such as Mohr–Coulomb criterion, Drucker–Prager criterion and Hoek–Brown criterion. Among these constitutive laws, the Hoek–Brown yield criterion (Hoek et al. 2002) has attracted attention of research workers due to its nonlinear yet simple nature. A number of studies are available in literature which utilized this criterion for the elasto-plastic analysis of tunnel cavity in rock mass. For example, Pan and Hudson (1988) and Pan et al. (1989) used this criterion for elastic-viscoplastic analysis of coal mine tunnels in both plane strain and 3D condition. The yield surface of HB criterion looks like a cone whose normal section in octahedral plane takes the shape of an irregular hexagon with discontinuous edges. Therefore, the problem of singularity exists at the apex and along the edges of HB yield surface. Few research workers (Merifield et al. 2006; Wan 1992) have tried to resolve this issue by smoothening or rounding-off the yield surfaces along these discontinuous locations during FE applications. A few commercial software packages, like PLAXIS and Phase^2, also incorporate the HB criterion into their material behaviour archive by considering the rounding of apex and edges of yield surfaces (Basarir et al. 2005; Gurocak et al. 2007; Özsan and Başarir 2003). Some research workers have even approximated the HB criterion by switching Drucker–Prager (DP) yield criterion (Drucker and Prager 1952) along the edges and at apex stress points of the HB yield surface (Choi and Deb 2005; Crisfield 1996; Owen and Hinton 1980). Although these methods of approximating the edges and apex of yield surfaces are widely implemented in practice, results obtained by these approximate yield criteria may not converge to actual solutions (Clausen 2007). Hence, research workers are now leaning towards some advanced numerical procedures for FE application of elasto-plastic HB criterion where no smoothening or replacing of the yield criterion at corner edges is required. Kumar and Mohapatra (2017) have suggested lower-bound FE limit analysis of the HB criterion in plain strain by applying semi-definite program (SDP) with conic optimization. Karaoulanis and Chatzigogos (2010) and Karaoulanis (2013) have tried to solve these issues using spectral representation of stress and strain for infinitesimal deformation-plasticity coupled with return mapping scheme in principal stress directions. But, large displacements and/or finite strains have not been considered in this study. Hence, it is not suitable for application in most practical problems. Herein, the studies of Clausen and Damkilde (2008) and Clausen (2007) can be referred where the return mapping algorithm has been used for elasto-plastic FE applications of the HB criterion assuming finite strains. However, illustration of this approach with actual tunnel in poor rock mass is not available in literature, according to this authors' knowledge.

From the review of literature, it is quite clear that almost none of the studies have considered elasto-plastic analysis of realistic tunnel problem with FE formulation of actual HB criterion. Few methods suffer from the *curse of singularity*, whereas the others may not be applicable for problems with highly nonlinear material. A few studies have also tried to avoid these problems by smoothening or rounding or switching the corner edges of HB yield surface by DP criterion. However, these approximations may not lead to actual solutions. Therefore, to address the issues discussed above, a unique framework has been proposed for linear elastic-perfectly plastic analysis of realistic tunnel problems using FE formulation of actual HB criterion in the following sections of this paper.

3 Head Race Tunnel of Maneri Bhali Hydro-electric Project Stage-I

3.1 Topography and Geometry

In Maneri Bhali Hydro-Electric Project (HEP) Phase-I on the Bhagirathi River in Uttarakhand, India, a 8.56-km-long Head Race Tunnel (HRT) of circular cross section was constructed (Fig. 1). A water head of 180 m has been used to generate 80 MW power. The finished diameter of the HRT is 4.75 m. The tunnel spreads from Maneri at upstream to a place near Uttarkashi. The excavation of this HRT has been done by four headings, viz. one from downstream end, Tiloth near Uttarkashi, two from intermediate adits at Heena and fourth one from upstream end at Maneri. A schematic diagram of longitudinal cross section of this HRT is displayed in Fig. 2.

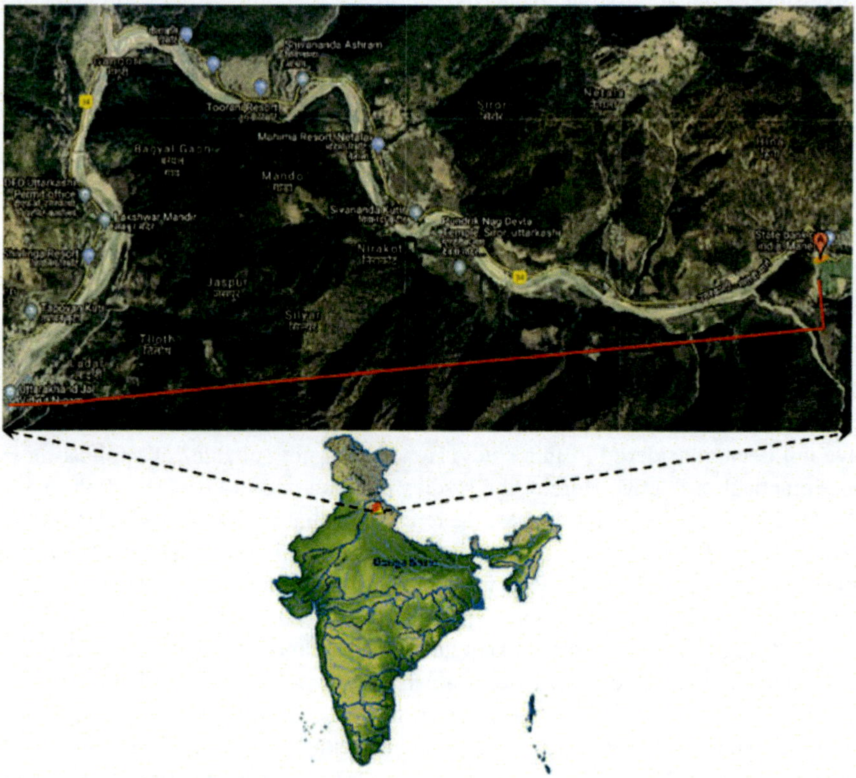

Fig. 1 Location of the Maneri Bhali hydroelectric project and layout plan of tunnel alignment (Gupta and Shankar 2012)

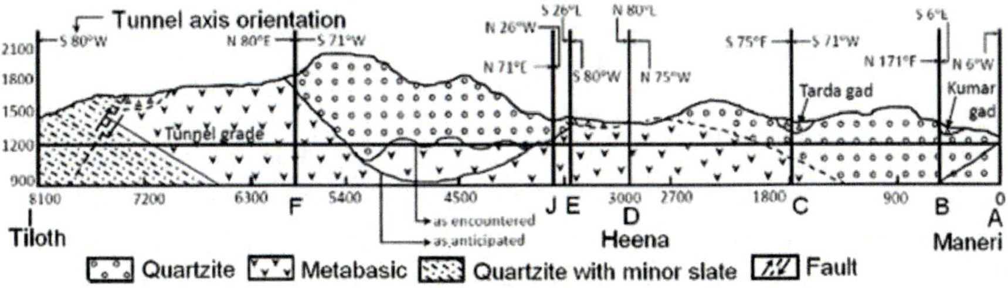

Fig. 2 Longitudinal geological section along the HRT of Maneri Bhali HEP (Goel et al. 1995)

3.2 *Geology Along the Power Tunnel*

The HRT crosses through rock masses of quartzite, quartzite interbedded with thin layers of phyllites, slate, chlorite schists, metabasics and basic intrusive from the Garhwal Himalayas (Jain et al. 1976). The rock mass beddings are disturbed by the Maneri central thrust towards north-east and east of the project region at a distance of 4.5 km east from Maneri. The rock formations of south and south-west region of the project belong to the Chandpur group. There is another thrust, namely the North Almora thrust at a distance of 12 km from south-west of Uttarkashi which separates the rock formations of Garhwal and Chandpur groups.

At the time of geological mapping between Heena in north and Tiloth in south, the rock formations were found to be metabasic quartzite and quartzite coupled with minor slate bands of the Garhwal group (Jain et al. 1976). In this region, heavy tectonic activities have created intense folds and faults which result in closely spaced joints, brecciation and shearing joints even in quartzite layers. The geological features along longitudinal cross section of the HRT are shown in Fig. 2.

3.3 *Tunnelling Problems: Squeezing Phenomena*

During tunnel excavation, field engineers have confronted with various problems, like water in-rush, cavity formation and support failure due to heavy squeezing conditions. In this article, the authors have focused their attention on the problems of squeezing in the HRT.

Various geometrical and material properties of two tunnel sections at chainage (ch.) 5350 m and ch. 4140 m are presented in Table 1, where severe squeezing was observed. The geological strength index (GSI) which was observed at ch. 5350 m ranged between 39.57 and 40.92, whereas at ch. 4140 m, its value is 37.25. This coupled with high in situ stresses and high overburden gave rise to severe squeezing of rock mass. Consequently, tunnelling excavations were in great trouble between ch. 5550 and 5250 m. In this stretch, the tunnel passes through partially wet and thinly foliated meta-basics. Besides, the height of overburden varies from 700 to 900 m along the tunnel axis.

Table 1 Various properties from Maneri Bhali HEP–I head race tunnel

Properties of rock mass	Symbol	Set-1	Set-2
Chainage (m)	–	5350	4140
Rock type	–	Thinly laminated foliated and wet meta basics	Highly fractured quartzites
Radius of tunnel opening (m)	a	2.4	2.4
Height of overburden (m)	H	800	350
Vertical in-situ stress (MPa)	σ_V	19.62	8.58
Horizontal in situ stress (MPa)	σ_H	13.73	5.15
UCS of intact rock (MPa)	σ_{ci}	87.45	97.5
Modulus of deformation (GPa)	E_d	9.88–10.72	8.458
Poisson's ratio	ν	0.27	0.27
Geological strength index	GSI	39.57–40.92	46.1
Hoek–Brown parameter	m_i	7	20
Disturbance factor	D	0.5	0.5

At the time of tunnel excavation and supporting, no squeezing was observed in this tunnel stretch. The opening was supported with ISMB 150 mm × 150 mm steel sets at 80–100 cm c/c spacing (Fig. 3). The gap between outer flange of the steel set and the excavated rock surface was filled up with plain cement concrete (PCC). After a period of 5–6 months, the squeezing of rock mass came into picture as the back-fill concrete started cracking along with buckling of the steel sets due to squeezing pressure.

To restrict the buckling, initially invert struts were attached to each steel set along with laggings of ISMB 150 mm × 75 mm. Additionally, PCC was filled up to the inner flanges of steel sets. These remedial measures restricted the deformations temporarily. However after three years, during construction of final lining, it was observed that the steel sets have deformed so much that their replacement became essential. Even, bottom of the tunnel was deformed up to 80 cm. The twisted steel sets and the back-fill PCC were removed, and the tunnel stretch was again supported by ISMB 150 mm × 150 mm sets spaced at 75 mm c/c to obtain required finished diameter. This renovation was done over most of the affected length. However, there was a heavy rock fall from the tunnel roof while removing the twisted steel sets between ch. 5509 and 5517 m and a cavity of 430 m^3 was formed. This 8-m-long stretch was regained by using forepoles and grouting above the forepoles. Then, this portion was re-excavated and supported with ISMB 150 mm × 150 mm sets spaced at 60 mm c/c.

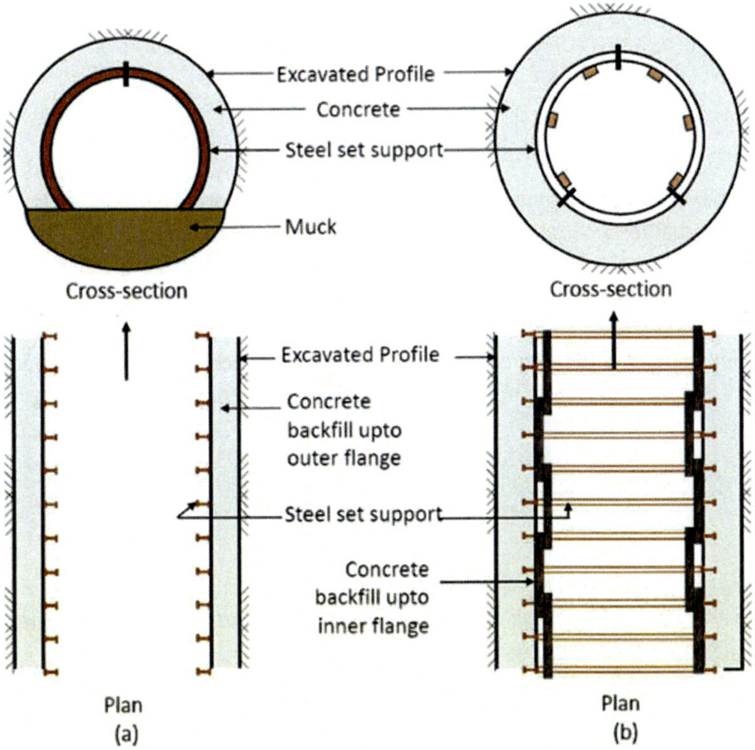

Fig. 3 Support systems used in squeezing ground condition (from Goel et al. 1995)

4 Finite Element Implementation of Hoek–Brown Criterion

4.1 Hoek–Brown Yield Criterion

For the constitutive modelling of nonlinear, heavily broken, poor-quality rock mass, Hoek et al. (2002) presented a failure criterion in terms of major (σ_1) and minor (σ_3) principal stresses as given by Eq. (1).

$$\sigma_1 = \sigma_3 + \sigma_{ci}\left(m_b\frac{\sigma_3}{\sigma_{ci}} + s\right)^a \quad (1)$$

where σ_1 and σ_3 are the effective principal stresses. The Hoek–Brown parameters, m_b, s and a are obtained from Eqs. (2), (3) and (4).

$$m_b = m_i \exp\left(\frac{\text{GSI} - 100}{28 - 14D}\right) \quad (2)$$

$$s = \exp\left(\frac{\text{GSI} - 100}{9 - 3D}\right) \tag{3}$$

$$a = 0.5 + \frac{1}{6}\left\{\exp\left(-\frac{\text{GSI}}{15}\right) - \exp\left(-\frac{20}{3}\right)\right\} \tag{4}$$

where GSI denotes geological strength index. m_i is a Hoek–Brown parameter. D denotes the disturbance factor whose value lies between 0 and 1.

In Eq. (1), compression is assumed to be positive which is a general practice in the geotechnical engineering field. On the other hand, in most of the FE codes, tension is assumed as positive and compressive is assumed as negative. Hence in rest of this paper, the FE sign convention is followed. Therefore, the functional form of Hoek–Brown criterion which can be used in the FE formulations is presented in Eq. (5).

$$f = \sigma_1 - \sigma_3 - \sigma_{ci}\left(s - m_b\frac{\sigma_1}{\sigma_{ci}}\right)^a = 0 \tag{5}$$

where $\sigma_1 \geq \sigma_2 \geq \sigma_3$ are the effective principal stresses. Here, σ_3 denotes the largest compressive stress. The Hoek–Brown criterion in three-dimensional principal stress space is represented in Fig. 4.

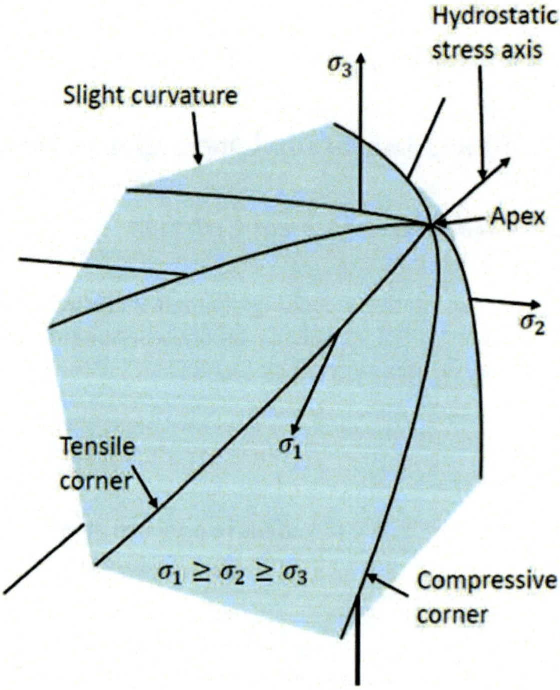

Fig. 4 Hoek–Brown criterion in principal stress space. The curved surface between compressive corner and tensile corner denotes primary yield surface that obeys $\sigma_1 \geq \sigma_2 \geq \sigma_3$

4.2 Elasto-plasticity and Return Mapping

The elasto-plastic formulation of Hoek–Brown criterion is performed in two parts. At first, linear elastic analysis of rock mass is carried out by following Hooke's law of elasticity. Then, yielding ($f > 0$) of the rock mass is checked at each Gauss point using Eq. (5). Once the rock mass has reached its yield point, i.e. $f \geq 0$, computation of updated stresses is carried out by general principles of plasticity. Herein, a non-associated plastic flow for rock mass is considered using a plastic potential function (g), similar to the yield criterion (f), as presented in Eq. (6) (Clausen 2007).

$$g = \sigma_1 - \sigma_3 - \sigma_{ci}\left(s_g - m_g \frac{\sigma_1}{\sigma_{ci}}\right)^{a_g} = 0 \tag{6}$$

After yielding, calculation of updated stresses is done by using a return mapping algorithm, i.e. by using a predictor–corrector scheme. For more details, interested readers may refer to studies of Clausen and Damkilde (2008) and Clausen (2007).

5 Finite Element Modelling

5.1 UMAT Coding

In this study, the linear elastic-perfectly plastic material model of rock mass using the actual Hoek–Brown yield criterion (Hoek et al. 2002), discussed in Sects. 4.1 and 4.2, is developed in the form of a user material (UMAT) subroutine in FORTRAN language. The UMAT subroutine is incorporated in another main FE code. Herein, commercial software ABAQUS CAE (Dassault Systems Simulia corporation 2009) is used as a main FE code. A schematic diagram of flowchart of the UMAT along with the main FE code is illustrated in Fig. 5.

5.2 Modelling of Tunnel System

The framework described in previous sub-section has been used for FE analysis of the tunnel case history (Goel et al. 1995) mentioned in Sect. 3. For this purpose, a two-dimensional plane strain half-symmetric FE model of tunnel was prepared in ABAQUS standard. A schematic diagram of this model is shown in Fig. 6.

The FE tunnel assembly contains three parts, viz. rock mass (surrounding tunnel cavity), steel sets (major support to tunnel cavity) and PCC (fill up material between rock mass and steel sets). Each part of this assembly was modelled independently.

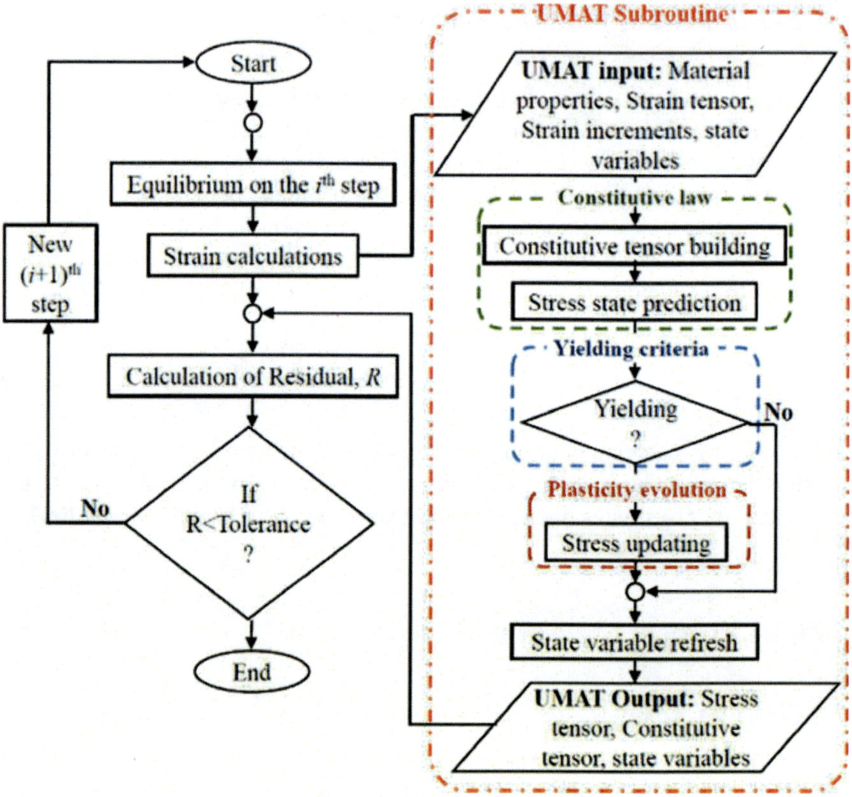

Fig. 5 A schematic diagram of flowchart of the UMAT along with main FE code

The rock mass was modelled as a solid, isotropic, homogeneous and continuum material. Eight node bi-quadratic plain strain quadrilateral element with reduced integration (CPE8R) was used to mesh this part. Finer meshing was done along tunnel circumference to monitor the plasticity behaviour of rock mass, and the mesh progressively became coarser away from the tunnel cavity to match with the remaining parts of this region which were meshed with coarser size elements. The element sizes were varied from 0.14 m at inner periphery to 2.85 m at outer boundary of this domain as shown in Fig. 7.

The steel set support was modelled as wire frame structure. It was meshed using three-node quadratic beam element (B22). Simple elastic–plastic behaviour of steel set was assumed. The elastic behaviour was defined by Young's modulus and Poisson's ratio of steel, whereas the plastic behaviour was modelled by yield stress-plastic strain curve of steel.

The last part of this assembly is the PCC section. It is modelled as a solid, isotropic, homogeneous continuum with CPE8R elements. The material response of PCC was modelled with concrete damaged plasticity model, available in ABAQUS material library (Dassault Systems Simulia corporation 2009).

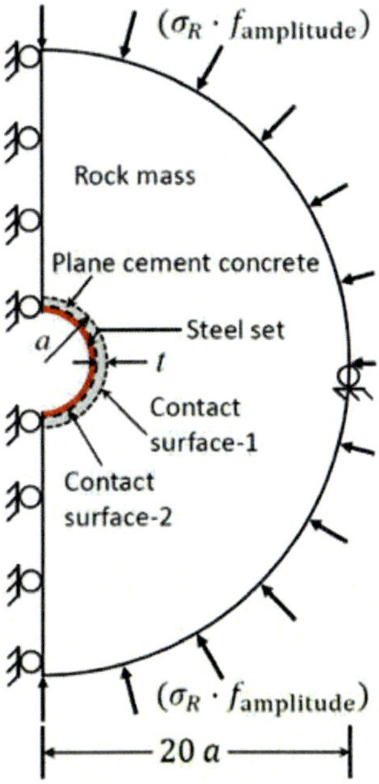

Fig. 6 FE model of tunnel cross section with steel set and PCC supports assuming vertical symmetry. Here, a = radius of excavated tunnel opening, t = thickness of PCC and resultant pressure, $\sigma_R = \sqrt{\sigma_V^2 + \sigma_H^2}$

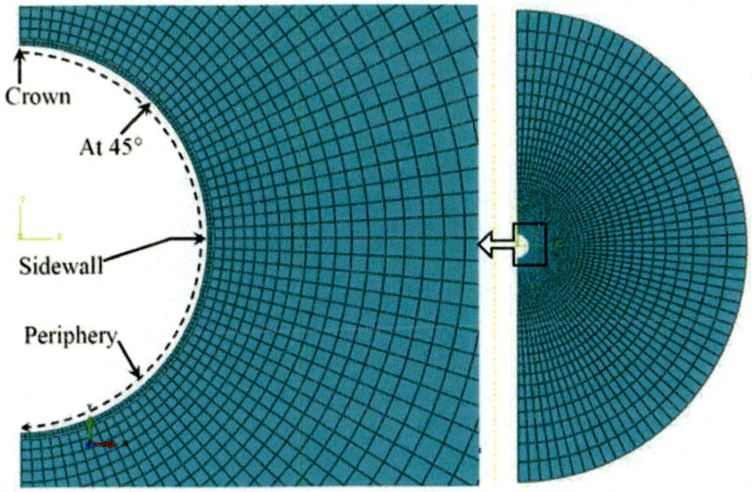

Fig. 7 Meshing pattern of FE model of rock mass around the tunnel cavity

After modelling, all the parts were combined in 'Assembly' module of this FE code. In this assembly, two contact surfaces come into play, viz. (i) contact surface-1: at the interface of PCC and rock mass and (ii) contact surface-2: at the interface of steel set and PCC. A master–slave algorithm was used to define these surfaces where stiffer and softer surfaces were considered as master and slave surfaces, respectively. Along these surfaces, presence of frictional behaviour was assumed during load transfer from rock mass to steel sets via PCC. Herein, coefficient of friction was assumed to be 0.65 and 0.45 along surface-1 and surface-2, respectively.

The boundaries of rock mass domain were placed away from the tunnel cavity so that stress–strain distributions remain undisturbed around the cavity. At left boundary of this FE model, condition of vertical symmetry was imposed to reduce the computational effort. The right most node of the rock mass was restricted to move along vertical direction for maintaining stability of the whole FE model.

The non-uniform in situ stress was applied in the form of pressure along circular outer periphery of the rock mass domain. Amplitude of the applied pressure is controlled with an analytical function $\left(f_{\text{amplitude}}\right)$ to consider the influence of vertical stress (σ_V) and horizontal stress (σ_H) as

$$f_{\text{amplitude}} = \frac{\left(\sigma_V\left|\frac{y}{20a}\right| + \sigma_H\left|\frac{x}{20a}\right|\right)}{\sqrt{\sigma_V^2 + \sigma_H^2}} \tag{7}$$

where x and y denote Cartesian coordinates of rock mass boundary and a is the tunnel radius. The details of in situ stress loading and the boundary conditions are shown in Fig. 6.

6 Results and Discussion

In this section, tunnel convergence characteristics and plastic straining of the tunnel section at ch. 5350 m of Maneri stage-I HEP are discussed. Different input parameters, necessary for this analysis, are given in Table 1. In addition, how the tunnel convergence and extent of plastic zone change from unsupported tunnel section to supported tunnel section has also been discussed. Attempt has also been made to obtain the ground response curve (GRC). In this context, the results corresponding to unsupported tunnel section were obtained for 76.32% load application. The analysis after this limit was terminated due to excessive distortion of elements along tunnel opening periphery which indicates collapse of the tunnel cavity.

6.1 Convergence of Tunnel Cavity

Radial displacement (U) contours of unsupported tunnel section, PCC supported tunnel section and steel set with PCC supported tunnel section under the action of non-uniform in situ stresses were determined using proposed framework. For these support conditions, radial displacement contours are presented in Fig. 8. It is clear from Fig. 8a that for unsupported tunnel section, maximum convergence is of the order of 2.193 m which practically indicates collapse of rock mass around the tunnel periphery.

Based on these contours, the distribution of tunnel convergence ($(U/a) \times 100\%$) has been plotted in Fig. 9 for three positions, namely crown, sidewall and around the periphery of tunnel cavity as indicated in Fig. 7. It can be noticed from Fig. 8a that maximum convergence develops at around ±45°–80° with horizontal axis, for unsupported tunnel section and similar behaviour was also found in Fig. 9 where the maximum convergence (75.62%) occurs at ±80° with horizontal axis. Due to presence of unequal in situ stresses in horizontal and vertical directions, the tunnel cavity is exposed to maximum resultant stress (σ_R) in the regions between ±45° and ±80°. As a result, the rock mass developed larger convergence in those regions. However, degree of tunnel convergence reduces largely (13.11%) as seen in Fig. 9, when PCC support was installed. The tunnel convergence reduced further (9.39%) when the steel sets were installed additionally (Fig. 9c).

The convergence obtained in the present analysis at chainage 5350 m has been compared with the corresponding field-measured values (Goel et al. 1995) as mentioned in Table 2. It has been found that the computed convergence lies within ±5% of the field measured tunnel convergence.

6.2 Extent of Plastic Zone around Tunnel Cavity

To predict the extent of region of broken rock mass around tunnel cavity (i.e. plastic zone), measurement of plastic strain (ε^p) is essential. In ABAQUS, ε^p of a yielding material cannot be measured by default features, while UMAT model is used in analysis. Hence, ε^p was measured by calculating incremental (equivalent) plastic strain ($\Delta\varepsilon^p$) at each iteration step within the UMAT subroutine through a state-dependent variable (SDV) at each gauss point of the FE model. The plots of ε^p (or SDV2) contour are presented in Fig. 10a–c for the unsupported tunnel, PCC-supported tunnel and PCC with steel set-supported tunnel, respectively. The distribution of ε^p at crown and sidewall with normalized radial distance (r/a) and at tunnel periphery are presented in Fig. 11a–c. In unsupported tunnel, occurrence of severe plasticity along tunnel periphery can be noticed in Fig. 10a. However, the amount of plasticity has reduced considerably after the installation of the PCC and steel set supports as shown in Fig. 10b, c. In Fig. 11c, it has been observed that the level of ε^p is maximum (413%) at ±45° with horizontal axis at the periphery of unsupported tunnel cavity. But after

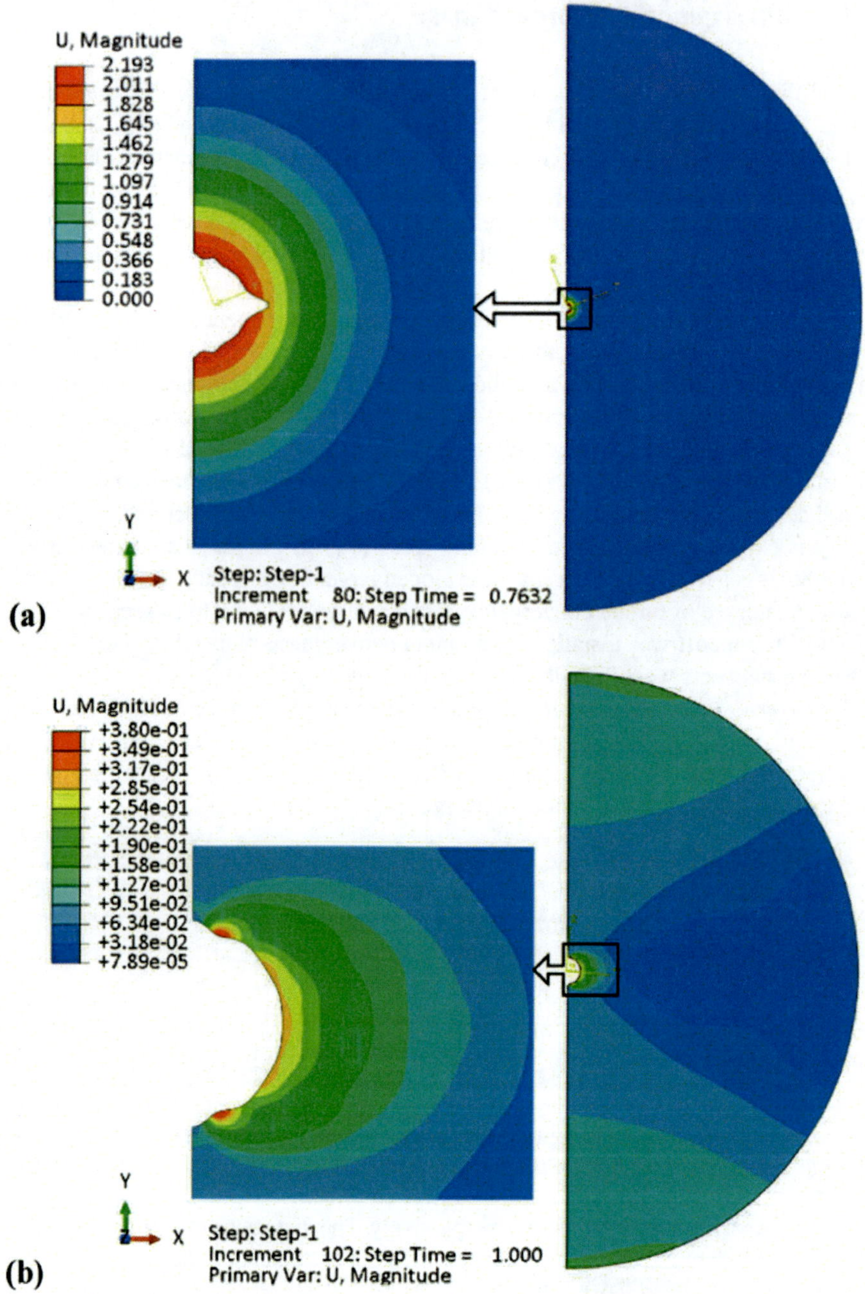

Fig. 8 Radial displacement contours (m) for tunnel cross section at chainage 5350 m for **a** unsupported section, **b** PCC supported section and **c** steel set with PCC supported section

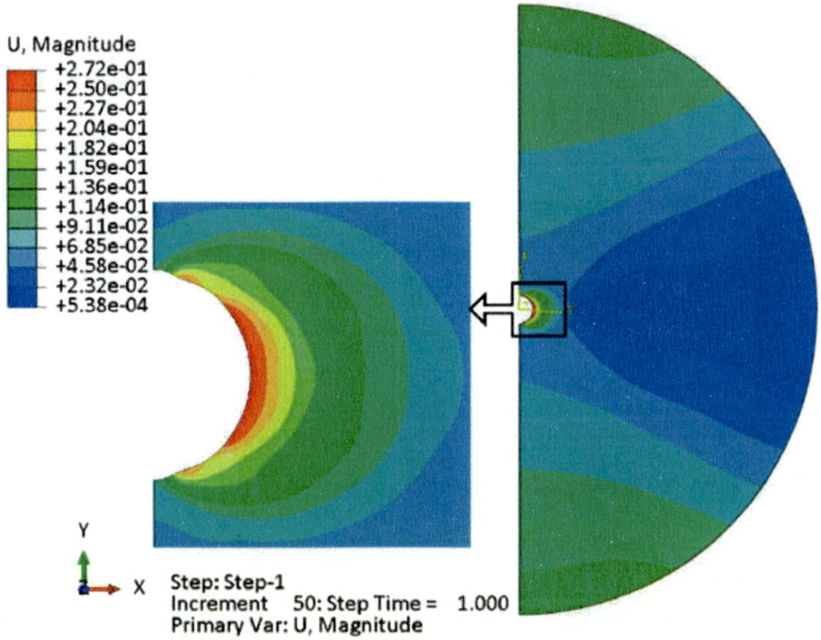

(c)

Fig. 8 (continued)

the installation of PCC and steel set supports, the points of maximum plasticity (21.24%) were relocated at ±70° (approx.) with horizontal axis as the mobilization of displacements was maximum at those points. One can always observe that the extent of plastic zone has reduced by five times the tunnel radius at both crown and sidewall points.

6.3 *Ground Response Curve (GRC)*

To obtain the ground reaction curve (GRC), the tunnel section has been subjected to a uniform external pressure (22.9 MPa) which is basically the resultant of vertical stress ($\sigma_V = 19.62$ MPa) and horizontal stress ($\sigma_H = 11.772$ MPa). The tunnel cavity has been loaded with an opposite internal pressure initially to maintain displacement along periphery equal to zero. Then, the simulations were carried out by reducing the internal pressure gradually up to zero and the corresponding displacements were measured at each step. For the present elasto-plastic analysis, ground response curve (GRC) of unsupported tunnel section at chainage 5350 m of Maneri stage-I HEP has been plotted in Fig. 12. In this figure, two curves have been displayed for maximum and minimum values of GSI at that tunnel section. In both the cases, a linear elastic part when the rock mass remains within yielding limit, followed by a plastic part when the rock mass has yielded, are easily recognizable.

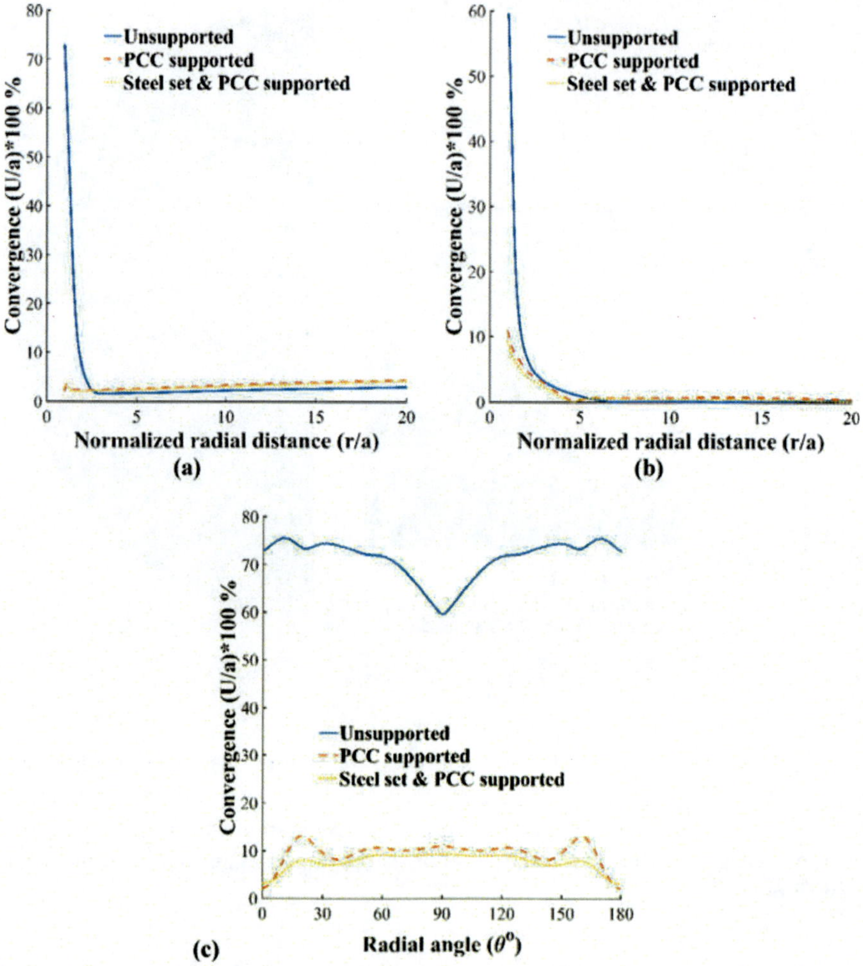

Fig. 9 Distribution of tunnel convergence (%) at **a** crown, **b** side wall and **c** periphery of tunnel cavity. Radius of tunnel cavity, $a = 2.9$ m

Table 2 Comparison of computed results with field-measured data

	Obtained from field measurement (Goel et al. 1995)	Computed from present study	
		GSI = 39.57	GSI = 40.92
Tunnel convergence (%) at ch. 5350 m	8.9	9.39	8.6

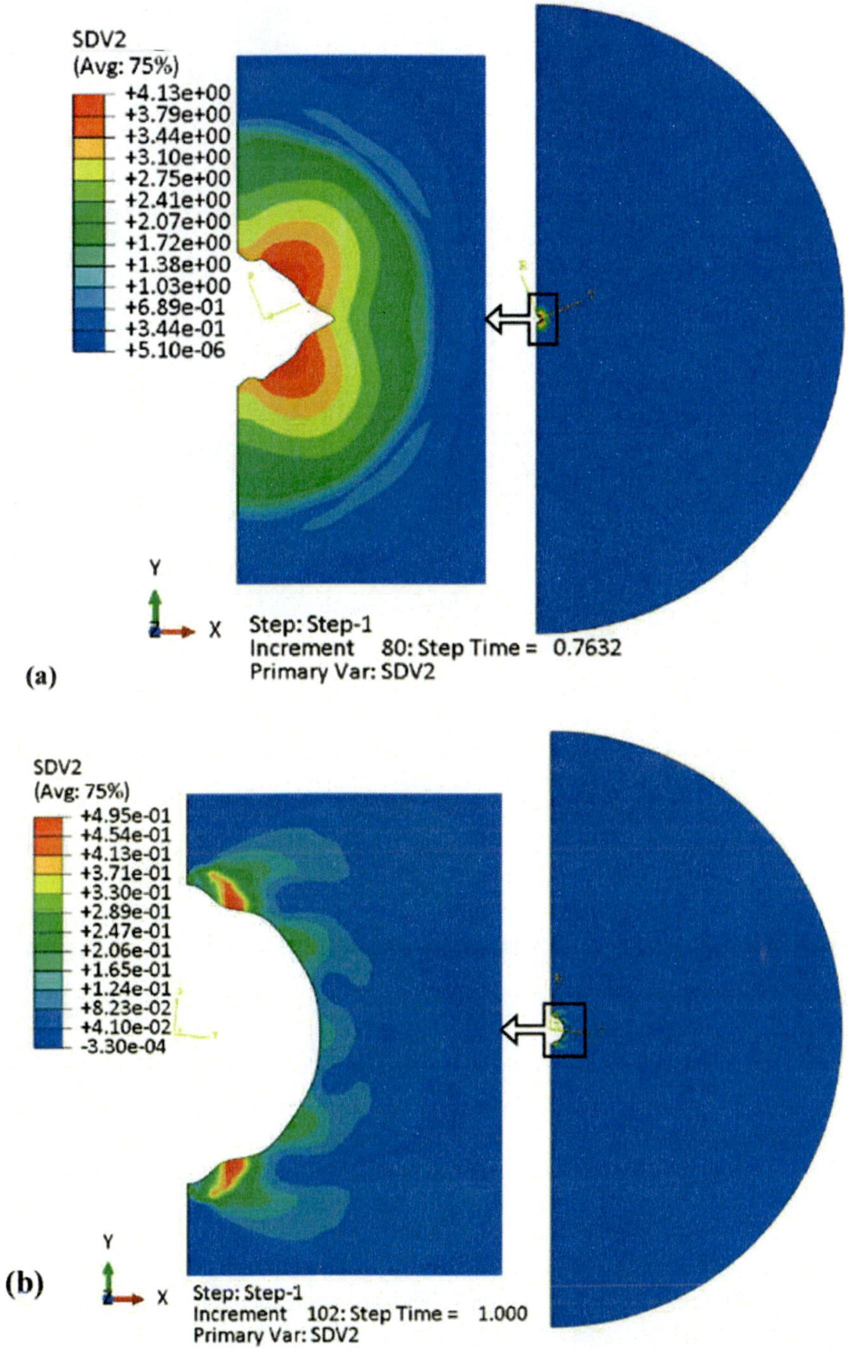

Fig. 10 Contours of plastic strain (equivalent) at tunnel cross section at chainage 5350 m for **a** unsupported section, **b** PCC-supported section and **c** steel set with PCC-supported section

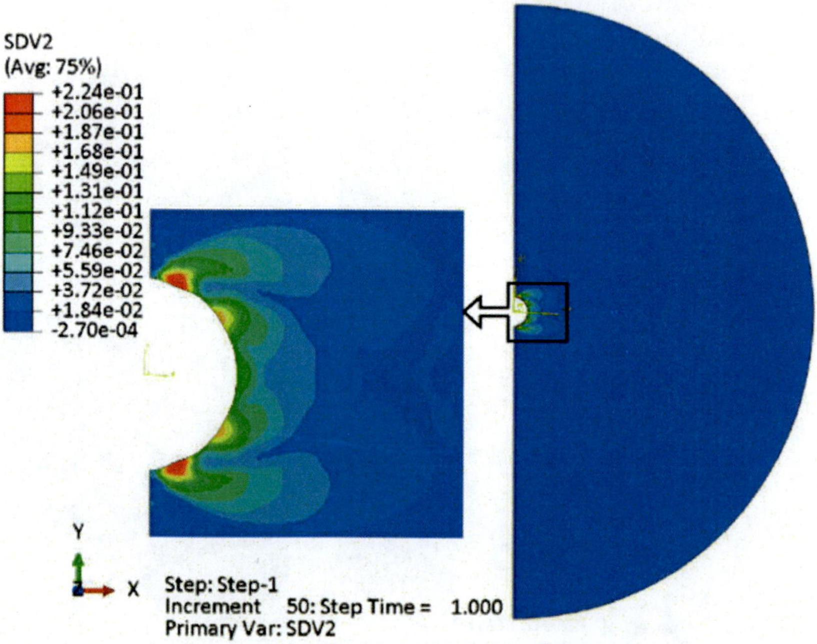

(c)

Fig. 10 (continued)

7 Conclusions

From the systematic review of literature, it is clear that although the use of Hoek–Brown criterion (Hoek et al. 2002) in FE analysis of rock mass is common in practice, very few FE studies are available in the literature for tunnels having large depth of overburden in poor rock mass which considers EPP formulation of actual Hoek–Brown criterion.

Therefore, a novel framework has been proposed for the FE analysis of tunnels in poor rock mass. The behaviour of the rock mass was modelled by nonlinear Hoek–Brown criterion in conjunction with non-associated plastic flow rule. The plasticity of rock mass was considered by adopting a plastic potential function which is similar to the HB yield criterion.

Subsequently, a finite element model of Maneri Bhali stage-I HEP was analysed using the proposed plasticity framework. A detailed investigation of tunnel convergence and equivalent plastic strains developed along the tunnel periphery is presented. Three types of support conditions, viz. unsupported tunnel, PCC-supported tunnel and PCC with steel set-supported tunnel, have been considered in this analysis. The convergence obtained through the proposed framework was found to be in close agreement with the field-measured values of convergence. Besides, ground response curves were developed for the tunnel under equivalent hydrostatic in situ stress condition. In conclusion, it can be stated that present FE-based elasto-plastic framework is

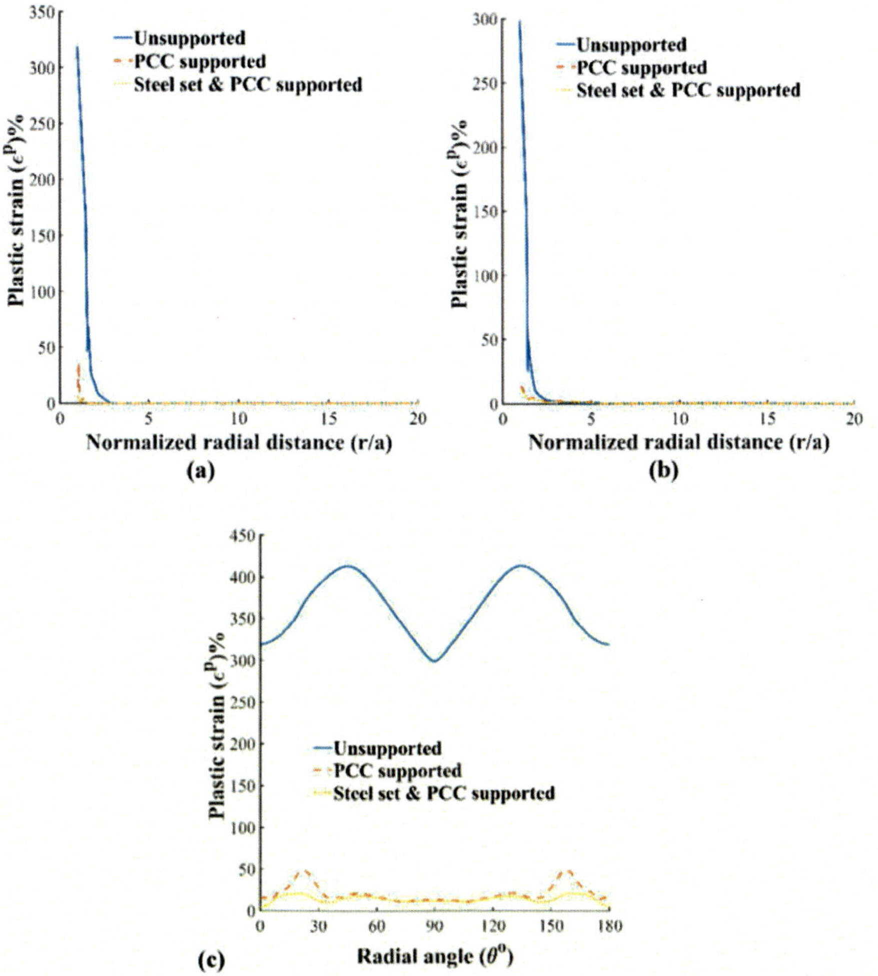

Fig. 11 Distribution of plastic strain (equivalent) at **a** crown, **b** sidewall and **c** periphery with normalized radial distance. Radius of tunnel cavity, $a = 2.9$ m

a convenient and competent tool for comprehensive investigation of the rock mass-tunnel support interaction and for design of tunnels and other underground structures in poor-quality rock masses displaying the squeezing ground condition.

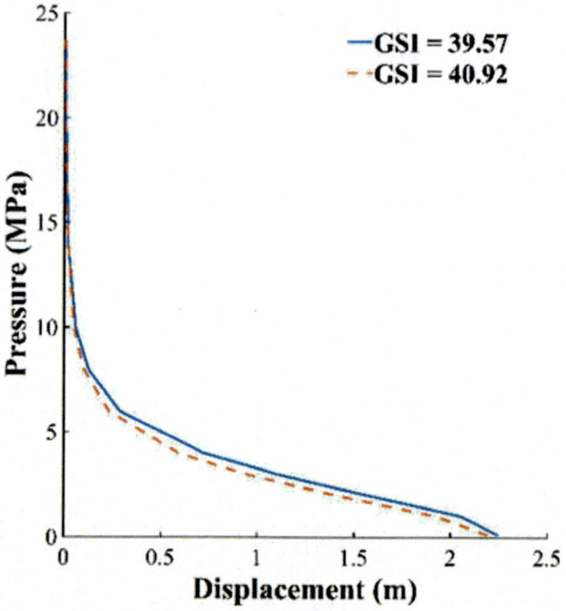

Fig. 12 Ground reaction curve of tunnel sec. at chainage 5350 m of Maneri stage-I HEP

References

Basarir H, Ozsan A, Karakus M (2005) Analysis of support requirements for a shallow diversion tunnel at Guledar dam site, Turkey . Eng Geol 81:131–145. https://doi.org/10.1016/j.enggeo.2005.07.010

Choi SO, Deb D (2005) Supplementation of generalized hoek-brown yield surface through the singularity adjustment in elastic-plastic analysis. Geosyst Eng 8:43–50. https://doi.org/10.1080/12269328.2005.10541235

Clausen J, Damkilde L (2008) An exact implementation of the Hoek-Brown criterion for elasto-plastic finite element calculations. Int J Rock Mech Min Sci 45:831–847. https://doi.org/10.1016/j.ijrmms.2007.10.004

Clausen JC (2007) Efficient non-linear finite element implementation of elasto-plasticity for geotechnical problems. Aalborg University, Esbjerg Institute of Technology, Niels Bohrs Vej 8, 6700 Esbjerg, Denmark

Crisfield MA (1996) Nonlinear finite element analysis of solids and structures, vol 2: advance topics. Wiley, West Sussex. https://doi.org/10.1017/CBO9781107415324.004

Dassault Systems Simulia corporation (2009) ABAQUS documentation and theory manual

Drucker DC, Prager W (1952) Soil mechanics and plastic analysis or limit design. Q Appl Math 10:157–165. https://doi.org/10.1090/qam/48291

Goel RK, Jethwa JL, Paithankar AG (1995) Tunnelling through the young Himalayas—a case history of the Maneri-Uttarkashi power tunnel. Eng Geol 39:31–44. https://doi.org/10.1016/0013-7952(94)00002-J

Gupta R, Shankar H (2012) Global energy observatory (WWW Document). https://globalenergyobservatory.org/

Gurocak Z, Solanki P, Zaman MM (2007) Empirical and numerical analyses of support requirements for a diversion tunnel at the Boztepe dam site, eastern Turkey. Eng Geol 91:194–208. https://doi.org/10.1016/j.enggeo.2007.01.010

Hoek E, Brown E (1997) Practical estimates of rock mass strength. Int J Rock Mech Min Sci 34:1165–1186. https://doi.org/10.1016/S1365-1609(97)80069-X

Hoek E, Carranza-torres C, Corkum B (2002) Hoek-brown failure criterion. Narms-Tac, pp 267–273

Hoek E, Marinos P (2000) Predicting tunnel squeezing problems in weak heterogeneous rock masses. Tunnels Tunn Int Part 1–2:1–20

Jain MS, Jaitle GS, Sondhi SN, Rajagopalan G (1976) Geotechnical note on the alternative alignments between Heena and Tiloth adits

Karaoulanis FE (2013) Implicit numerical integration of nonsmooth multisurface yield criteria in the principal stress space. Arch Comput Methods Eng 20:263–308. https://doi.org/10.1007/s11831-013-9087-3

Karaoulanis FE, Chatzigogos T (2010) Implicit numerical integration of the hoek-brown yield criterion in principal stress space. In: Fourth European conference on computational mechanics. Paris, France, pp 1–8

Kumar J, Mohapatra D (2017) Lower-bound finite elements limit analysis for Hoek-Brown materials using semidefinite programming. J Eng Mech 143:04017077. https://doi.org/10.1061/(ASCE)EM.1943-7889.0001296

Majumder D, Viladkar MN, Singh M (2017) A multiple-graph technique for preliminary assessment of ground conditions for tunneling. Int J Rock Mech Min Sci 100:278–286. https://doi.org/10.1016/j.ijrmms.2017.10.010

Merifield RS, Lyamin AV, Sloan SW (2006) Limit analysis solutions for the bearing capacity of rock masses using the generalised Hoek-Brown criterion. Int J Rock Mech Min Sci 43:920–937. https://doi.org/10.1016/j.ijrmms.2006.02.001

Owen DRJ, Hinton E (1980) Finite elements plasticity: theory and practice. Pineridge Press Limited, Swansea

Özsan A, Başarir H (2003) Support capacity estimation of a diversion tunnel in weak rock. Eng Geol. https://doi.org/10.1016/S0013-7952(02)00235-1

Pan XD, Hudson JA (1988) Plane strain analysis in modelling three-dimensional tunnel excavations. Int J Rock Mech Min Sci Geomech Abstr 25:331–337

Pan XD, Hudson JA, Cassie J (1989) Large deformation of weak rocks at depth—a numerical case study. Rock Gt. depth 613–620

Singh M, Singh B, Choudhari J (2007) Critical strain and squeezing of rock mass in tunnels. Tunn Undergr Sp Technol 22:343–350. https://doi.org/10.1016/j.tust.2006.06.005

Wan RG (1992) Implicit integration algorithm for Hoek-Brown elastic-plastic model. Comput Geotech 14:149–177. https://doi.org/10.1016/0266-352X(92)90031-N

Liquefaction Mitigation with Stone Columns for a Sewage Treatment Plant: A Case Study

K. Ganesh Deepak and Jinu Mary Jacob

Abstract Liquefaction of soil in the event of an earthquake is one among the major challenges in geotechnical engineering which may lead to catastrophic damages if proper mitigation measures are not provided. In this paper, a case study of a project site in Delhi (seismic zone IV) for a sewage treatment plant (capacity 318 MLD) has been considered. Based on the presence of poorly graded silty sand, low SPT *N* values and the presence of groundwater at shallow level, it has been worked out that the soil is susceptible to liquefaction. As a mitigation measure, ground improvement using vibro stone columns installed by dry bottom feed method has been adopted above which shallow foundations are proposed. CPTs have been performed on treated and untreated locations which have ascertained the improvement of soil, and the plate load tests performed have ensured the bearing capacity of the ground.

Keywords Liquefaction · Stone columns · SCPT

1 Introduction

Liquefaction is a phenomenon which occurs in the case of saturated soils, wherein the soil loses its strength and stiffness when subjected to an external stress like an earthquake. There have been notable failures of structures in history during earthquakes due to liquefaction. Hence in earthquake-prone areas, it is necessary to determine whether the soil is susceptible to liquefaction or not.

K. G. Deepak · J. M. Jacob (✉)
L&T Construction, Chennai, Tamil Nadu, India
e-mail: jinumary1992@gmail.com

K. G. Deepak
e-mail: kganeshdeepak@gmail.com

M. Latha Gali and P. Raghuveer Rao (eds.), *Geohazards*, Lecture Notes in Civil Engineering 86, https://doi.org/10.1007/978-981-15-6233-4_49

Among the mitigation techniques which can be adopted to overcome susceptibility to liquefaction, dynamic compaction, grouting, installation of stone columns, sand columns, adoption of driven piles, etc., are a few. In this case study, the use of stone columns to overcome liquefaction has been discussed. Stone columns have been selected as the method for ground improvement for this study considering the following advantages:

- Densification and increase of soil density while installation of stone columns
- Increase of lateral effective confining stress
- Act as drainage paths for dissipation of excess pore water pressures.

2 Literature Review

Reference was made to a study which has assessed the seismic hazard of Delhi region (Agarwal and Chawla 2006). The study also suggests the peak ground acceleration values that can be considered for Delhi region. Densification of loose to medium dense sands by the provision of granular piles is provided in Krishna and Madhav (2009). By the study, correlations have been obtained for improved standard penetration test (SPT) N value based on SPT value of untreated ground and area replacement ratio.

The evaluation of liquefaction potential of soils was presented in Priebe (1998) which also provides the use of vibro replacement to mitigate liquefaction by listing out the case histories where vibro replacement had been used to reduce the risk of liquefaction. The various methods of stone column installation and applications of stone columns in different types of soils had been presented in Raju et al. (2004).

The methods for evaluation of cyclic stress ratio and cyclic resistance ratio were presented by Youd et al. (2001). In their study, a range of values have been suggested for magnitude scaling factors giving upper and lower limits.

3 The Site

This paper deals with the ground improvement method chosen for a sewage treatment plant which is being constructed in Delhi. The site is in seismic zone IV. A detailed soil investigation was conducted to identify the soil type and its properties for design. Groundwater table was encountered at a depth of around 1.15–2.5 m below the existing ground level. The general soil profile has been indicated in Fig. 1, and average soil strata encountered have been briefed below:

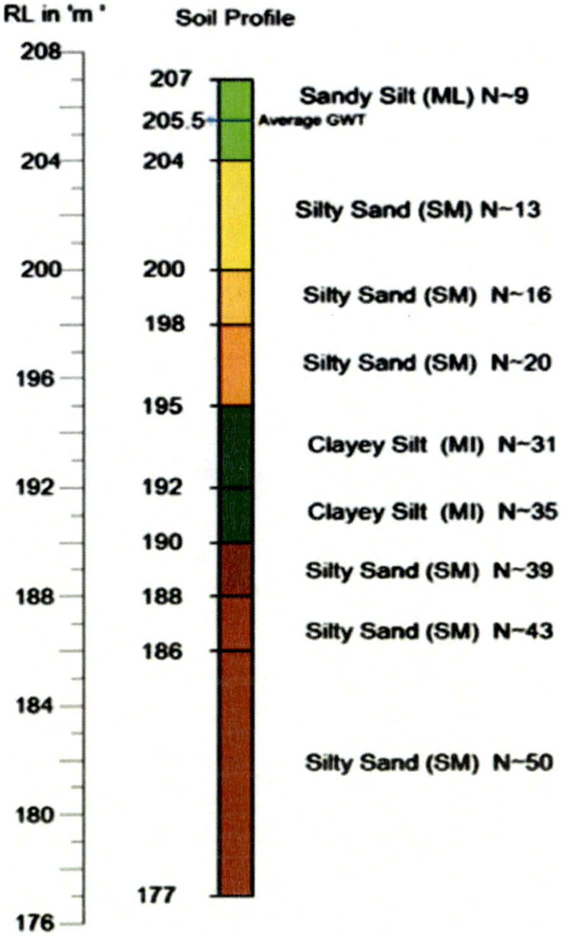

Fig. 1 General soil profile

Stratum-I: Fill material (sandy silt ML) of thickness around 3 m. Average (SPT) N value is 9.
Stratum-II: Medium dense silty sand (SM) with layer thickness of around 4 m. Average SPT N value is around 13.
Stratum-III: Medium dense silty sand (SM) with layer thickness of around 2 m. Average SPT N value is 16.
Stratum-IV: Medium dense silty sand (SM) with layer thickness of around 3 m. Average SPT N value is 20.

Stratum-V: Clayey silt of intermediate plasticity (MI) with layer thickness of 5 m. Average SPT *N* value is 31–35.
Stratum-VI: Dense silty sand (SM) with layer thickness of around 2 m. Average SPT *N* value is 39.
Stratum-VII: Dense silty sand (SM) with layer thickness of around 11 m. Average SPT *N* value is 43–50.

The soil parameters considered for design have been summarized in Table 1.

The susceptibility of in situ soil to liquefaction was checked based on the equations suggested by IRC 75-2015. The equations are used to evaluate cyclic stress ratio (CSR) and cyclic resistance ratio (CRR). Factor of safety against liquefaction is then calculated by dividing CRR by CSR. The following parameters were considered for the analysis:

- Zone factor—0.24
- Peak ground acceleration—0.18 g (Agarwal and Chawla 2006)
- Magnitude of earthquake—7.

Table 2 gives the factor of safety against liquefaction at various depths from existing ground level for the average soil strata.

Table 1 Average soil parameters for design

S. No.	Depth (m)	Description	SPT *N*	γ (kN/m^3)	c (kN/m^2)	Φ
1	0–3	Fill/sandy silt	9	17.6	45	0
2	3–7	Silty sand	13	19.2	0	30
3	7–9	Silty sand	16	19.2	0	30
4	9–12	Silty sand	20	19.3	0	31
5	12–17	Clayey silt	31–35	19.5	155	0
6	17–19	Silty sand	39	19.8	0	33
7	>19	Silty sand	43–50	19.8	0	33

Table 2 Calculation of factor of safety against liquefaction based on average soil parameters

Depth (m)	N_{untr}	Fines content (%)	CSR	CRR	FOS
1.5	9	78	0.27	0.285	>1
3.0	9	78	0.26	0.271	>1
4.5	13	4	0.25	0.226	0.90
6.0	13	4	0.24	0.226	0.92
7.5	16	3	0.24	0.258	>1
9.0	19	3	0.23	0.29	>1
10.5	20	3	0.22	0.305	>1
12	31	86	0.21	1.19	>1
13.5	33	86	0.20	1.19	>1

As per the above table, depth of liquefaction is 6 m below ground level when calculated based on average soil parameters. The depth of liquefaction was calculated for each of the boreholes covering the entire layout of the treatment plant (37 acres) and was found to be varying from 4.5 to 9 m below ground level. For mitigation of liquefaction, the option selected was vibro stone columns.

4 Design of Stone Columns

Stone columns were selected as the feasible option for this site considering the following points:

- Mitigates liquefaction potential of soil
- Improves bearing capacity and reduces settlement.

The installation technique adopted for vibro stone columns was dry bottom feed method. This technique was chosen since the soil type was sandy silt and silty sand in the plant area. In addition, since no water is involved in this process of installation, it leaves a dry land with no muck formation at site.

4.1 Specifications of Stone Columns and Expected Ground Improvement

The typical arrangement of stone columns provided for the location is shown in Fig. 2. Figure 3 shows a typical section of the stone columns. The following were the details finalized for stone columns:

- Pattern of stone columns—triangular
- Diameter of stone column—900 mm
- Centre-to-centre spacing—2.6 m
- Length from existing ground level—6 to 10.5 m (depth of stone columns has been arrived based on the depth of liquefaction)
- Area replacement ratio—11.04%.

A 500-mm-thick granular blanket was provided above the stone columns to act as a drainage layer to dissipate the excess pore water pressure generated at the event of an earthquake.

There are different methods to estimate improved SPT N values. For current discussion, as an example, the correlations suggested by Krishna and Madhav (2009) have been considered. The equation used has been given as Eq. 1. Using the estimated improved SPT *N* values, the possibility of liquefaction for the treated ground has been assessed.

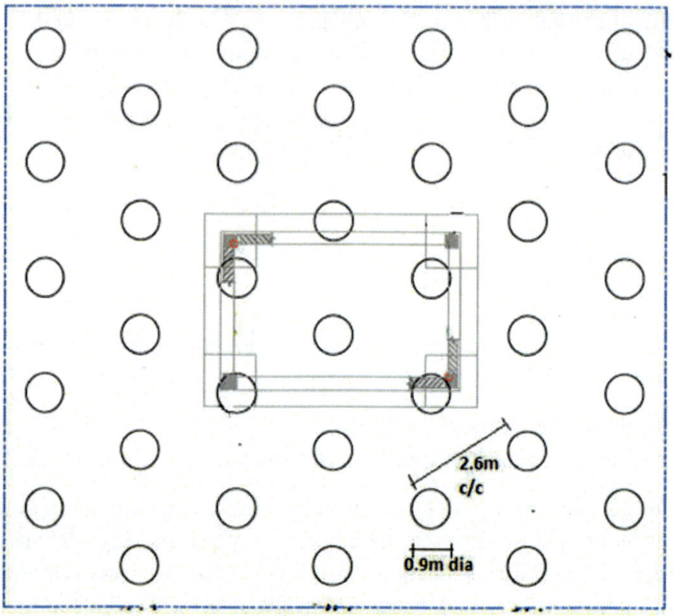

Fig. 2 Typical arrangement of stone columns

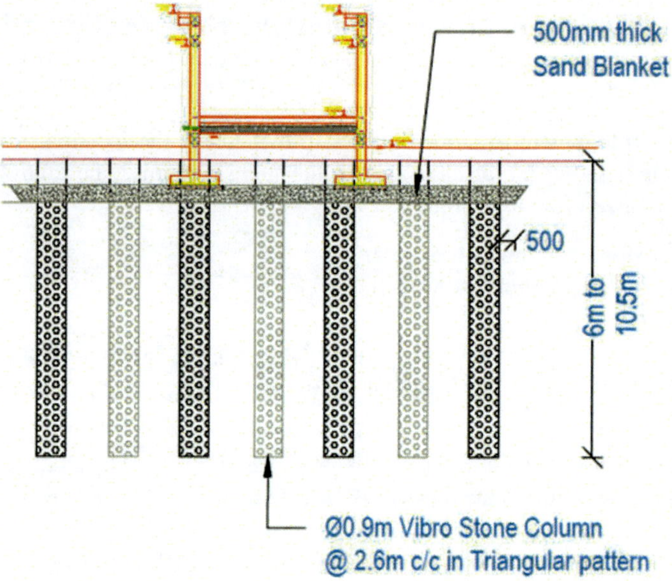

Fig. 3 Typical section of stone columns

$$N_{1\_tr} = N_{1\_untr} + (a_s N_{1\_untr}/(A + Ba_s)) \quad (1)$$

where N_1 refers to modified N value after applying correction for overburden pressure of 100 kPa. N_{1_untr} is the modified N value before treatment. N_{1_tr} is the modified N value after treatment. *as* is the area replacement ratio. A and B are parameters that depend on N_{1_untr}.

$$A = 1.23e^{0.13N1\_untr} \quad (2)$$

$$B = 0.03N_{1\_untr} \quad (3)$$

Net downward base pressures from most of the structures of the treatment plant are 80–160 kN/m^2. Net SBC of unimproved ground is in the range of 100–120 kN/m^2. The SBC after ground improvement has been calculated based on IS 15284 (Part 1): 2013 by which capacity based on bulging of stone column, support from intervening soil and surcharge effect are calculated and added together. The estimated net SBC has been found to be higher than 160 kN/m^2 satisfying both shear and settlement criteria.

Material specifications. Stones used for the construction of stone columns were hard, durable, chemically inert stones of size ranging from 40 to 10 mm.

The material used for granular blanket consisted of clean medium to coarse sand compacted in layers to relative density of 75%.

5 Liquefaction Check for the Treated Ground

Improved SPT N values of the treated ground were estimated as per Eq. 1 using the stone column design parameters. Based on the improved N values, liquefaction check was performed and has been shown in Table 3.

6 Field Tests for Confirming the Improvement

The stone columns installed must be tested for confirming whether the required SBC is attained or not. For confirming the densification, cone penetration tests (CPTs) were performed before and after ground treatment.

Improved bearing capacity due to the installation of stone columns has been ascertained by performing single-column and group column load tests as per IS 15284 (Part 1): 2013. Both initial and routine load tests were performed. Figure 4 shows the test set-up for single-column load test at site.

Table 3 Estimation of factor of safety against liquefaction after ground improvement

Depth (m)	N_{untr}	N_{1_untr}	N_{1_tr}	N_{tr}	CSR	CRR	FOS
1.5	9	15	26	15	0.27	1.19	>1
3.0	9	13	25	16	0.26	1.19	>1
4.5	13	17	27	20	0.25	0.417	>1
6.0	13	15	26	21	0.24	0.483	>1
7.5	16	19	27	24	0.24	1.19	>1
9.0	19	19	28	27	0.23	1.19	>1
10.5	20	27	27	28	0.22	1.19	>1
12	31	27	32	36	0.21	1.19	>1
13.5	33	32	32	38	0.20	1.19	>1

Fig. 4 Single-column load test set-up

7 Results

7.1 Cone Penetration Test Results

The results of CPTs performed before and after installation of stone columns for confirming the densification of soil have been shown in Figs. 5, 6 and 7.

From CPT value comparison for pre- and post-treatment, it can be noted that in the top approximate 2 m, lesser values of cone resistance are reported after ground treatment. This behaviour is due to the disturbance caused to the ground while using a vibroflot. However, this loosening of soil at the top is found not to affect the bearing

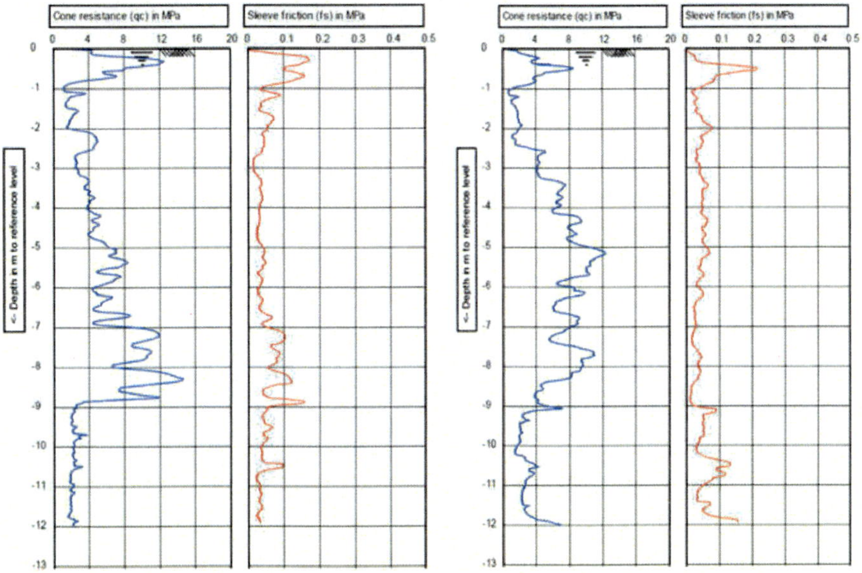

Fig. 5 Results of CPT (1) of untreated and treated ground

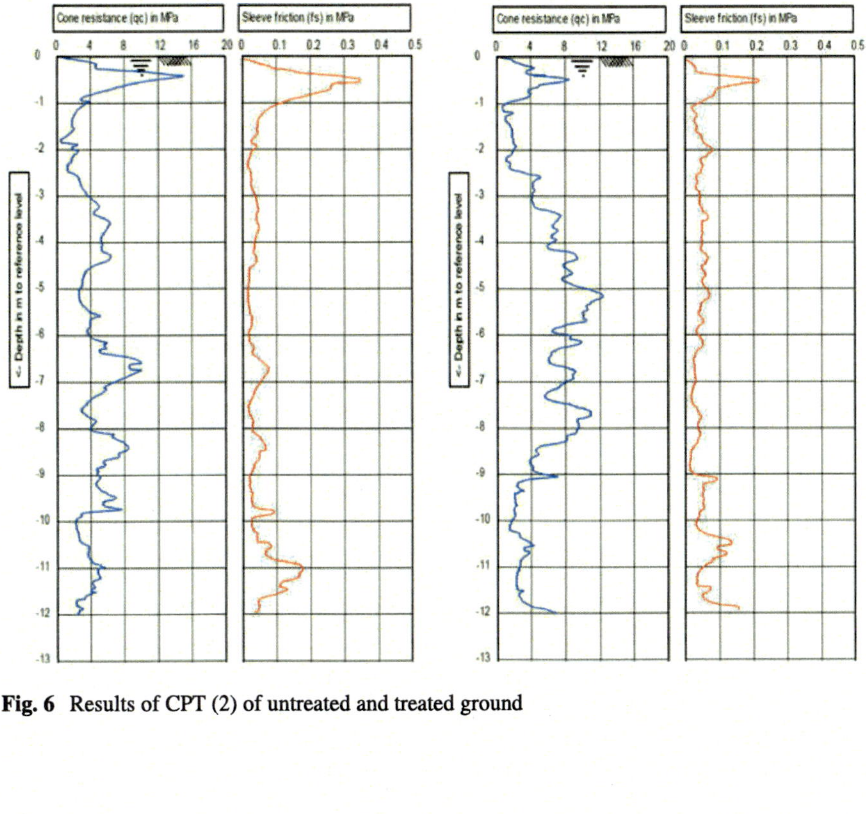

Fig. 6 Results of CPT (2) of untreated and treated ground

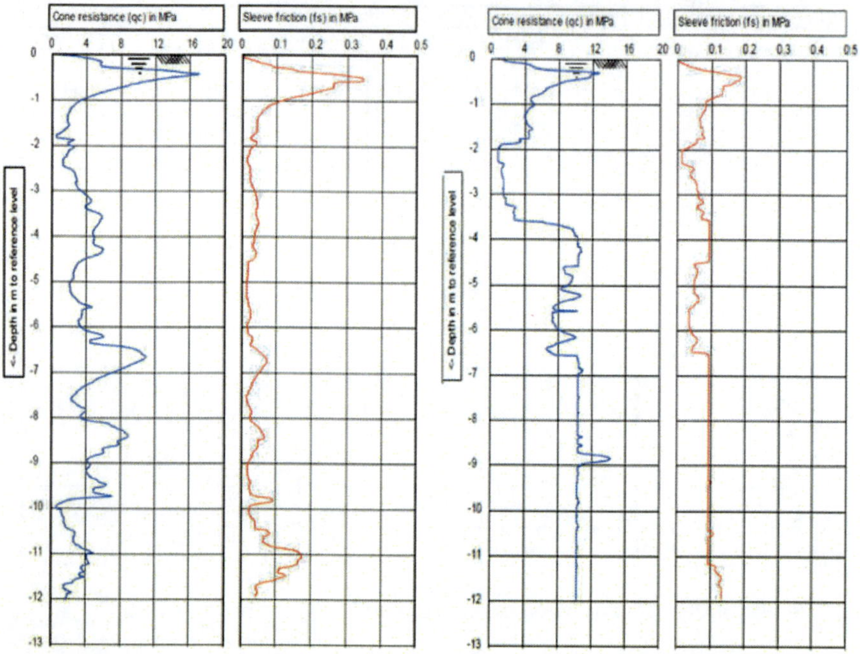

Fig. 7 Results of CPT (3) of untreated and treated ground

capacity of the ground since the overall capacity of the stone columns depends on bulging, contribution of intervening soil and surcharge effect.

From the pre- and post-treatment CPT values, SPT *N* values were correlated and the comparison has been shown in Figs. 8, 9, 10 and 11. From the SPT *N* value comparisons, as inferred from CPT results, the *N* values after treatment are lesser in the top approximate 2 m and thereafter there is increase in SPT *N* values confirming the densification of in situ soil. As observed in Table 1, fines content is higher in the top zone and hence even though a decrease in *N* values is observed, it does not induce liquefaction potential to the top zone. The sandy soil beneath has been densified, and this clearly overcomes the liquefaction potential of sandy soil.

7.2 Plate Load Tests

One initial single-column load test was performed near to the proposed ground improvement area to confirm whether the design is enough to cater the safe bearing capacity requirement.

Routine single-column and group column load tests were performed at the location of all major structures of the treatment plant like clarifiers, digester, etc. For initial load test, the test load is 1.5 times the design load, and for routine load test, the test

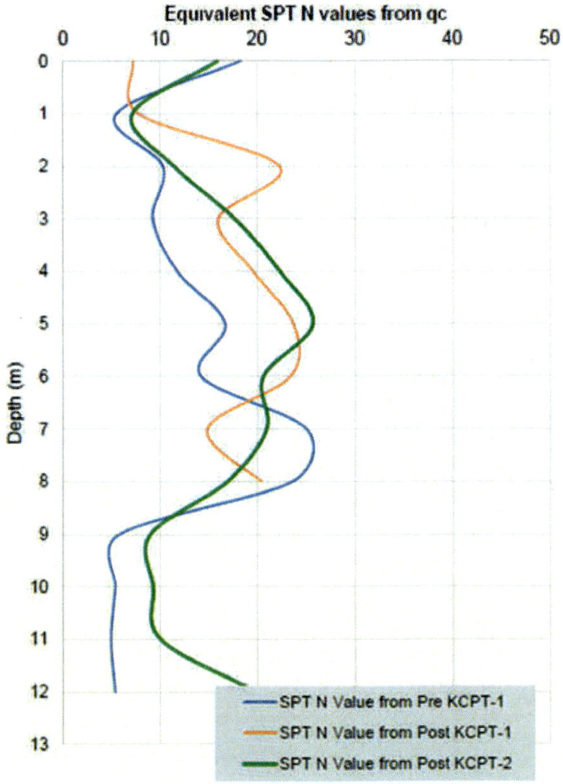

Fig. 8 Comparison of correlated SPT (location 1) of treated and untreated ground

load is 1.1 times the design load. The loading versus settlement graphs plotted for routine single-column load test-1 and routine group column load test-1 have been indicated in Figs. 12 and 13.

Table 4 shows settlements at design loads for all the load tests performed at the treatment plant location. At the design loads, settlements reported are found to be within the allowable limits of 12 and 30 mm for single-column and group column load tests, respectively, as per IS 15284 (Part 1): 2013.

8 Conclusions

Stone columns were adopted as the ground improvement measure to mitigate liquefaction in the SPT site located at Delhi. The densification of soil after stone column installation was confirmed from the CPTs performed before and after ground improvement. In addition to densification, the drainage of the stone columns will also mitigate the liquefaction problem.

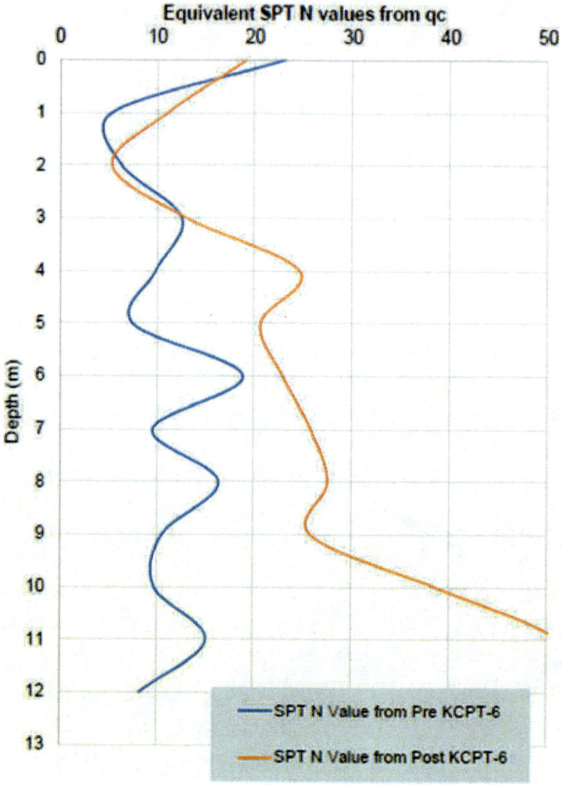

Fig. 9 Comparison of correlated SPT (location 2) of treated and untreated ground

Initial single-column as well as routine single-column and group column load tests were performed to ensure the attainment of the required bearing capacity. The settlements at the design loads were found to be within allowable limits. By this case study, it has been reaffirmed that stone columns can be used as an effective measure to overcome susceptibility to liquefaction and can improve the bearing capacity of soil.

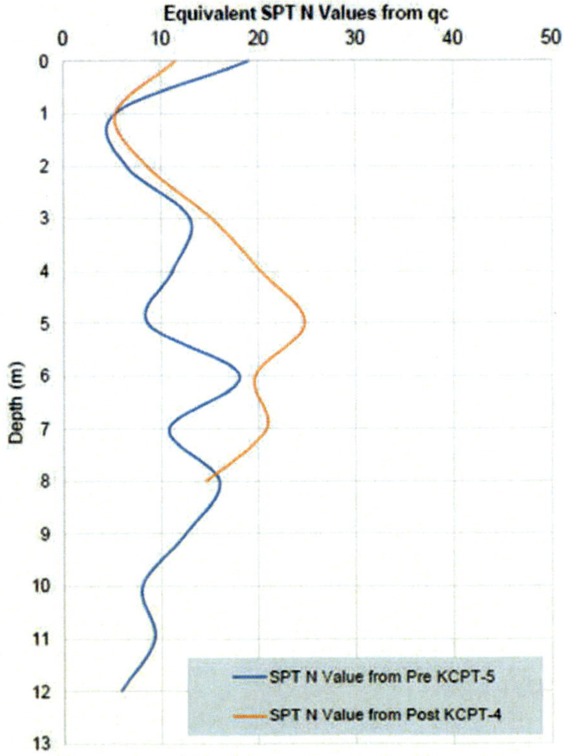

Fig. 10 Comparison of correlated SPT (location 3) of treated and untreated ground

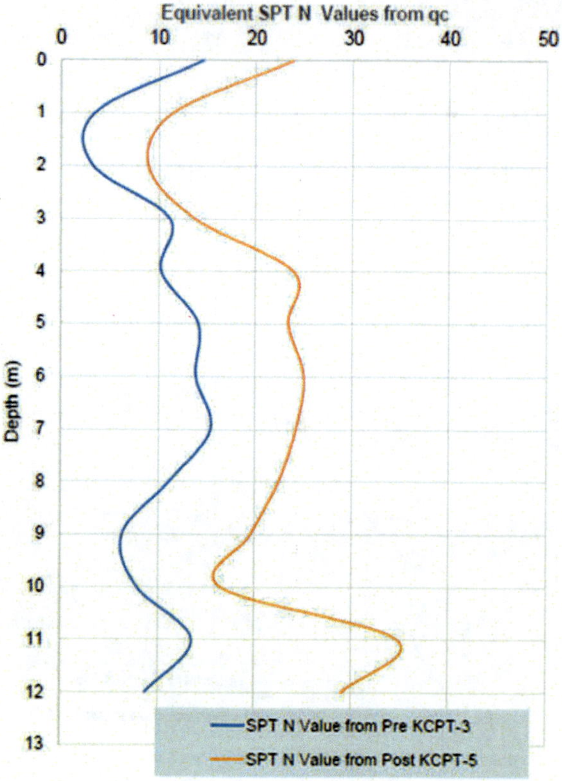

Fig. 11 Comparison of correlated SPT (location 4) of treated and untreated ground

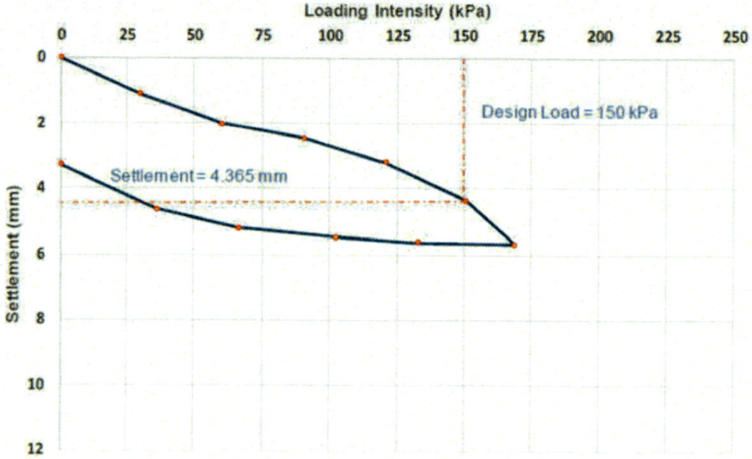

Fig. 12 Load versus settlement graph for routine single-column load test-1

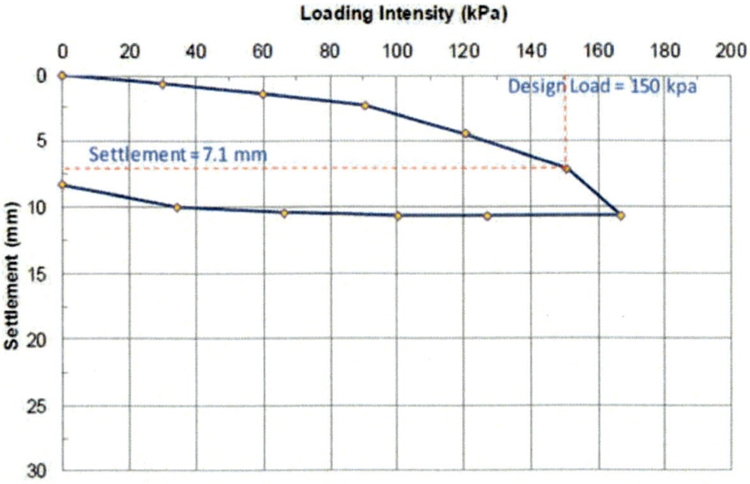

Fig. 13 Load versus settlement graph for routine group column load test-1

Table 4 Settlements at design loads from load tests

S. No.	Type of test	Design load (kN/m^2)	Settlement at design load (mm)
1	Initial single-column test	150	7.817
2	Routine single-column test-1	150	4.365
3	Routine single-column test-2	150	3.390
4	Routine single-column test-3	150	4.070
5	Routine single-column test-4	120	5.280
6	Routine single-column test-5	150	2.400
7	Routine single-column test-6	120	3.860
8	Routine single-column test-7	150	2.290
9	Routine single-column test-8	150	4.110
10	Routine group column test-1	150	7.100
11	Routine group column test-2	150	3.390
12	Routine group column test-3	160	3.890

References

Agarwal SK, Chawla J (2006) Seismic hazard assessment for Delhi region. Curr Sci 91(12):1717–1724

Idriss IM, Boulanger RW (2004) Semi-empirical procedures for evaluating liquefaction potential during earthquakes. In: 11th international conference on soil dynamics & earthquake engineering and the 3rd international conference on earthquake geotechnical engineering, Berkeley, CA, USA, pp 32–56

IRC:75-2015 (2015) Guidelines for the design of high embankments

IS 15284 (Part 1) (2003) Design and construction of ground improvement—guidelines, Part 1 stone columns
Krishna AM, Madhav MR (2009) Treatment of loose to medium dense sands by granular piles: improved SPT 'N_1' Values. Geotech Geol Eng 27:455–459
Priebe DIHJ (1998) Vibro replacement to prevent earthquake induced liquefaction. In: Proceedings of the geotechnique-colloquium, 12-57E. Darmstadt, Germany, pp 1–13
Raju VR, Krishna RH, Wegner R (2004) Ground improvement using vibro replacement in Asia 1994 to 2004—a 10 year review. In: 5th international conference on ground improvement techniques, 10-60E. Kuala Lumpur, Malaysia, pp 1–29
Youd TL, Idriss IM, Andrus RD, Arango I, Castro G, Christian JT, Dobry R, Finn WDL, Harder LF, Hynes ME, Ishihara K, Koester JP, Liao SSC, Marcuson WF, Martin GR, Mitchell JK, Moriwaki Y, Power MS, Robertson PK, Seed RB, Stokoe KH (2001) Liquefaction resistance of soils: summary report from the 1996 NCEER and 1998 NCEER/NSF workshops on evaluation of liquefaction resistance of soils. J Geotech Geoenviron Eng 127(10):817–833

Behavior of 3D-Reinforced Granular Trenches Under Cyclic Loading

M. Hema and N. Unnikrishnan

Abstract Reinforced soil has been successfully used in many geotechnical engineering applications where the soil as such is not strong enough to take up the loads. Many types of geosynthetics and fibers have been used as reinforcing elements. The study focuses on the use of multi-oriented reinforcements in improving the properties of the soil. Hexapods, horizontal–vertical orthogonal elements and 3D reinforcements are the three different types of 3D reinforcements used for soil stabilization. In the study, soil is reinforced with granular trenches that are reinforced with an optimum percentage of 3D inclusions. Triangular trenches having base width equal to twice the height (i.e., $W/H = 2$) are used for the study. Loads were applied through an automated and computer controlled hydraulic cyclic plate load testing equipment. Cyclic behavior of the reinforced soil is studied by conducting the tests at various amplitudes, i.e., 20, 40, 60 and 80% of ultimate static load. The frequency is maintained at 3 Hz. Similar study is carried out by varying the frequency, keeping the amplitude constant. It is seen that the reinforcements add an apparent cohesion to the cohesionless soil and also increase the angle of internal friction.

Keywords 3D reinforcement · Trenches · Footings · Settlement

1 Introduction

Soil reinforcement is the technique of improving the strength of the soil in order to enable it to support or carry more loads. Two common methods to improve the soil strength are mixing of a soil amendment agent like lime into weak clayey soil, then recompacting them to improve the bearing capacity of soil and producing a stronger

M. Hema (✉)
Department of Civil Engineering, Mar Baselios Institute of Technology and Science, Nellimattom, India
e-mail: hema03nov92@gmail.com

N. Unnikrishnan
Department of Civil Engineering, Government Engineering College, Thrissur, India

M. Latha Gali and P. Raghuveer Rao (eds.), *Geohazards*, Lecture Notes in Civil Engineering 86, https://doi.org/10.1007/978-981-15-6233-4_50

Table 1 Properties of river sand

Property	Value
Specific gravity	2.63
Natural moisture content (%)	0.44
Maximum dry density (kN/m^3)	17.43
Minimum dry density (kN/m^3)	14.62
Maximum void ratio	0.76
Minimum void ratio	0.5
Dry density (kN/m^3)	15.36
Φ (at $D_r = 35\%$)	36°

earth structure by installing alternate layers of plastic or composite inclusions like geosynthetics.

Many types of geosynthetics and fibers have been used as reinforcing elements. Replacement of the weak zone by a stronger soil is generally adopted for the purpose of ground improvement. However, it is not often feasible to replace the entire weak zone. Studies have been conducted on the use of granular trenches for soil improvement (Madhav and Vitkar 1978, Unnikrishnan et al. 2010). Here, the soil within the trench is reinforced with 3D reinforcements (Zhang et al. 2006, 2008; Harikumar et al. 2016).

2 Test Materials

2.1 *River Sand*

Locally available poorly graded river sand was used for the present study. The engineering and index properties of sand are given in Table 1.

2.2 *3D Reinforcements*

Reinforcing elements were manufactured from injection molding of Acrylonitrile Butadiene Styrene (ABS) plastic granules. ABS is derived from acrylonitrile, butadiene and styrene. Acrylonitrile is a synthetic monomer produced from propylene and ammonia; butadiene is a petroleum hydrocarbon obtained from the C4 fraction of steam cracking; styrene monomer is made by dehydrogenation of ethyl benzene, which is a hydrocarbon. ABS combines the strength and rigidity of acrylonitrile and styrene polymers with the toughness of polybutadiene rubber. ABS has superior properties in terms of hardness, gloss, toughness and electrical insulation. The reinforcements consisted of four legs or protrusions in a single plane (x-y), and two

Fig. 1 3D Reinforcements

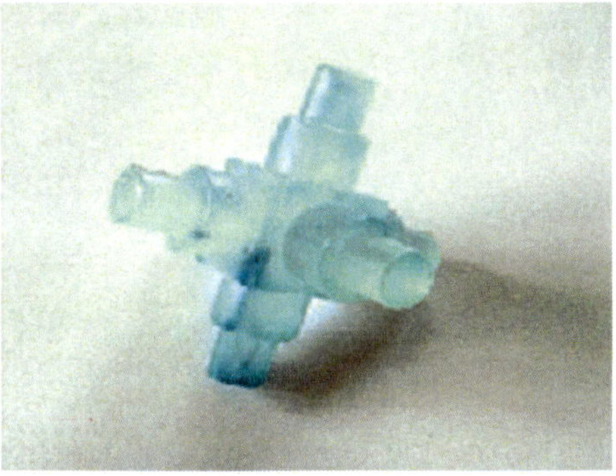

Table 2 Properties of 3D reinforcements

Material	ABS plastic
Specific gravity	1.04
Density	0.997 g/cc
Glass transition temperature	105 °C
Volume	2531.5 mm^3
Mean length	30 mm
Mean diameter	5 mm

protrusions in plane perpendicular to this plane (z), with an average length of 30 mm and a diameter of 5 mm were used for the study, as shown in Fig. 1. The engineering properties of the reinforcing elements are given in Table 2.

2.3 Geosynthetics

Two types of geosynthetics were used for the encapsulation of the trenches. Nova curtain (Grid 1) was used as the separator and Garware polymer geogrid (Grid 2) are used for the reinforcement of the trench. Figure 2 shows the geogrids used for the test. Grid 1 is a geotextile, and Grid 2 is a geogrid. Both the geosynthetics are laid within the trench over its entire length and width, according to the shape of the trench.

3 Methodology

The soil is reinforced with granular trenches that are reinforced with an optimum percentage of 3D inclusions ($V_r = 1.65\%$), where V_r is defined as the volume ratio and is given by,

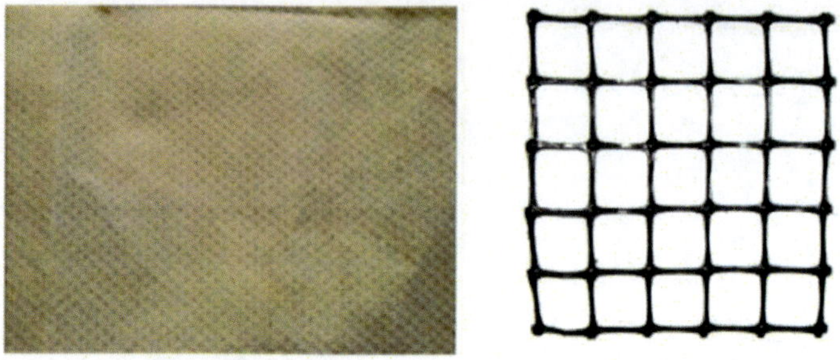

(a) Grid 1 (b) Grid 2

Fig. 2 Geosynthetics

$$V_r = \frac{V_g}{V_s} \tag{1}$$

where V_g is the volume of hexapods and V_s is the volume of the soil sample in the mold (Hema and Unnikrishnan 2017). Triangular trenches having base width equal to twice the height (i.e., $W/H = 2$) are used for the study (Hema and Unnikrishnan 2017). The dimensions of the trench are taken as width, $W = 30$ cm; height, $H =$ 15 cm. The soil within the trench was randomly mixed with the optimum percentage of reinforcements. Loads are applied through an automated and computer controlled hydraulic cyclic plate load testing equipment. Two types of geogrids are used for the encapsulation of the trench. Cyclic behavior of the reinforced soil is studied by conducting the tests at various amplitudes, i.e., 20, 40, 60 and 80% of ultimate static load. The frequency is maintained at 3 Hz. Similarly, tests are conducted at constant amplitude by varying the frequency, i.e., 1, 2 and 3 Hz. The relative density of the bulk soil was maintained at 25% and that of the trench infill was 85%.

4 Test Setup

The test setup used for the laboratory study consisted of a tank of dimensions 0.8 m × 0.8 m × 0.8 m. Loads were applied through an automated and computer controlled hydraulic cyclic plate load testing equipment on a footing 0.135 m wide and 0.78 m long. The test setup is shown in Fig. 3.

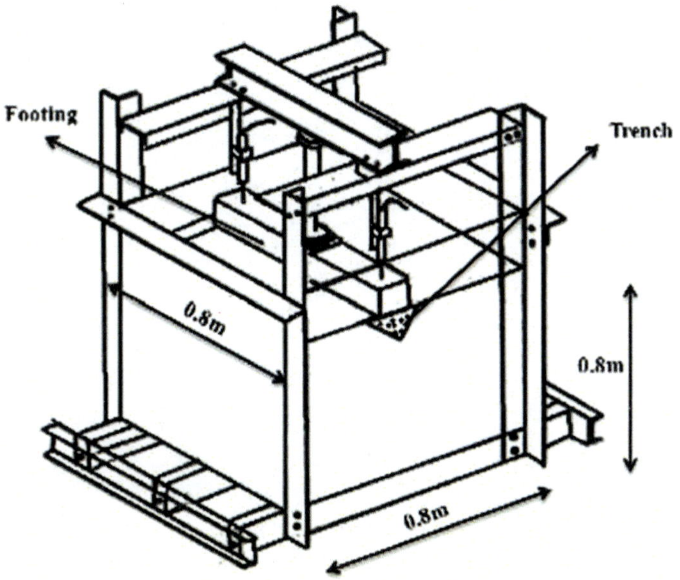

Fig. 3 Test Setup

5 Results and Discussions

5.1 Effect of Variation of Amplitude

Figure 4 shows the variation of monotonic load with settlement ratio (SR) for a strip footing supported on bulk soil having an unreinforced trench. The settlement ratio or SR is defined as the ratio of settlement to the width of the footing expressed as percentage. In the present investigation, width of the model footing was kept constant at 135 mm.

Fig. 4 Variation of SR with monotonic load for soil with unreinforced trench

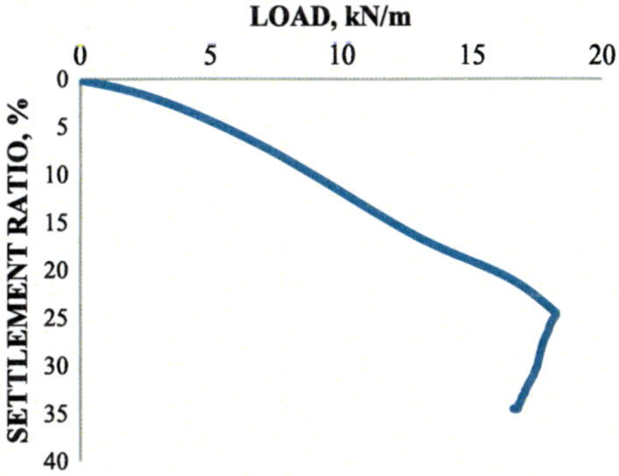

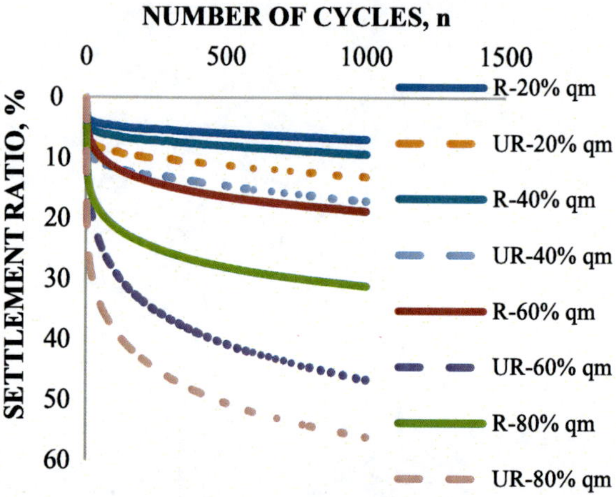

Fig. 5 Cyclic strength at different amplitudes at 3 Hz

This test on unreinforced test was regarded as a pioneer reference test, for comparisons with the upcoming repeated load tests on footings over granular trench reinforced with 3D inclusions. Load corresponding to 25% SR is considered to be the ultimate load. The ultimate load was thus obtained as 18.23 kN/m, which corresponds to a load of 1450.706 kg, applied through the cyclic hydraulic actuator. This ultimate monotonic load is designated as q_m, and all the amplitudes chosen for the cyclic tests are taken as a percentage of the q_m. In this phase of the study, repeated load tests were performed on the soil reinforced with reinforced triangular trenches. The behavior of footing resting on these reinforced trenches at different amplitudes was studied. Repeated load tests were conducted for load amplitudes of 20, 40, 60 and 80% q_m. While studying the effect on amplitude, the frequency was fixed as 3 Hz. The cyclic tests were followed by an equal number of static tests to study the post-cyclic behavior of the reinforced soil.

Figure 5 shows the variation of SR with number of cycles, *n*. It is seen that the initial settlement occurred within the first 50 cycles.

Here, *R* represents soil reinforced with reinforced granular trench, and UR represents soil with unreinforced trench. Considerable improvement was observed when reinforcement was randomly mixed with the soil within the granular trench. At an amplitude of 60% q_m, SR was found to decrease by almost 2.7 times in case of reinforced trenches when compared to the unreinforced ones, at the end of 1000 cycles.

Post-cyclic static load tests were conducted on these samples to determine the post-cyclic behavior of these specimens. The significance of this test is that it gives an idea about the residual strength of the soil after 1000 cycles of loading.

The results of these post-cyclic tests were plotted in the form of load deformation graphs. Figure 6 gives the post-cyclic behavior of the soil with reinforced trenches after 1000 cycles of loading.

In all the cases with reinforcement, a distinct phase of stiff resistance and brittle failure could be noticed. The ultimate post-cyclic load improved with increased load

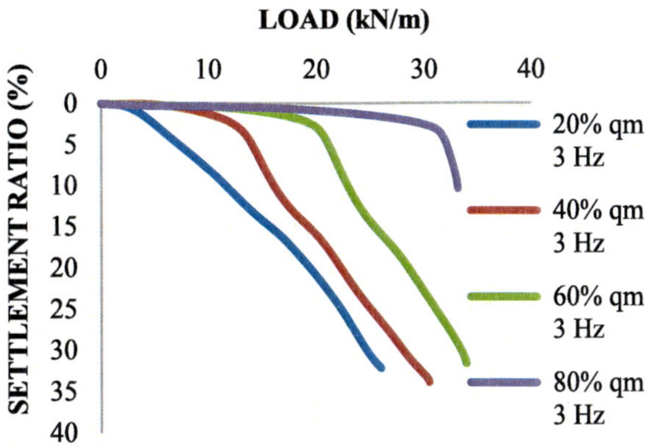

Fig. 6 Post-cyclic strength of soil reinforced with reinforced granular trenches at different amplitudes

amplitude. This may be due to the increase in relative density of the reinforced soil mass. This is also evident from the transformation to general shear failure as noticed from the curve corresponding to 80% q_m.

5.2 *Effect of Variation of Frequency*

In this phase of the study, experiments were conducted on soil at a constant amplitude of 60% q_m (Rajan 2010). Behavior at three different frequencies, viz. 1, 2 and 3 Hz, is studied. Figure 7 shows the behavior of soil with reinforced trenches under different frequencies. The frequency is found to have very little effect on the strength behavior under repeated loading. A trend of slightly improved performance at higher frequencies is noticed.

Figure 8 gives a comparison between the soil with unreinforced and unreinforced trenches. The settlement ratio is found to be much higher in case of unreinforced trenches, compared to the reinforced ones.

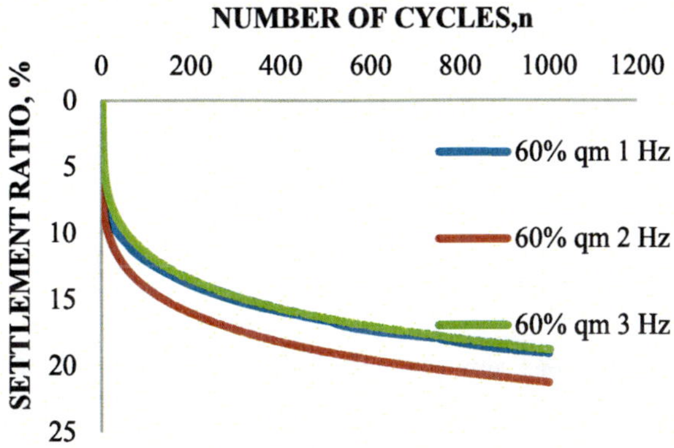

Fig. 7 Cyclic strength at varying frequencies for soil reinforced with reinforced trenches

Fig. 8 Variation of SR with frequency for soil with unreinforced and reinforced trenches

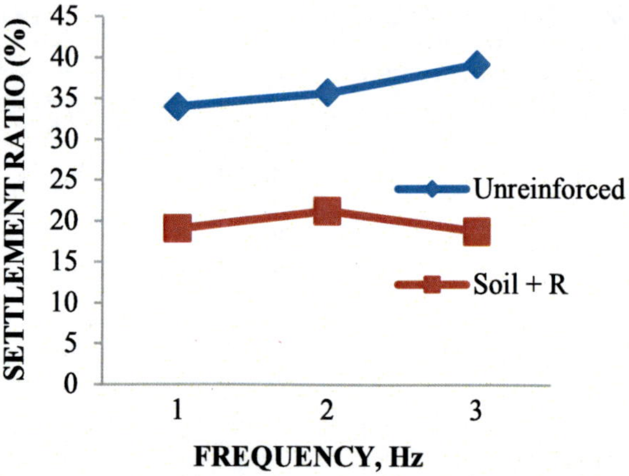

It can be noted from the curves that frequency of loading has only very little effect in the settlement behavior of footings placed over reinforced trench, unreinforced or unreinforced EGT. This inference is consistent with previous investigations that the loading frequency has very little effect on the strength behavior under cyclic loading (Sawwaf and Nazir 2010). A trend of higher efficiency at higher frequency is observed.

6 Conclusions

The effect of addition of three-dimensional inclusions on the cyclic behavior of reinforced trenches was studied. The results showed that while introducing reinforcement in the trench decreases the settlement of bulk soil by about 61%. The trend is found to continue for higher amplitudes of loading as well.

Load carrying capacity in post-cyclic monotonic tests increased with increased load amplitude. The nature of failure changed to general shear failure as the load amplitude increased. The post yield failure was brittle at 80% q_m. Hence, load amplitude up to 60% q_m is recommended to be applied in the interest of avoiding sudden failures.

From the study thus conducted, it was concluded that reinforcing the soil within the trench using 3D inclusions significantly improves the strength of the soil.

References

Harikumar M, Sankar N, Chandrakaran S (2016) Behaviour of model footing resting on sand bed reinforced with multi directional reinforcing elements. Geotext Geomembr 44(2016):568–578

Hema M, Unnikrishnan N (2017) Behaviour of footings resting on 3D reinforced trenches. In: Indian geotechnical conference 2017

Madhav MR, Vitkar PP (1978) Strip footing on weak clay stabilized with a granular trench or pile. Can Geotech J 15

Mathai B (2015) Footings on encapsulated granular trenches under repeated loading. Post Graduate thesis, College of Engineering, Trivandrum

Rajan S (2010) Behaviour of foundation on geosynthetic encapsulated granular trench. Post Graduate thesis, College of Engineering Trivandrum

Sawwaf M, Nazir AK (2010) Behavior of repeatedly loaded rectangular footings resting on reinforced sand. Alexandria Eng J 49:349–356

Unnikrishnan N, Johnson AS, Rajan S (2010) Response of strip footings supported on granular trench. In: Indian geotechnical conference 2010

Zhang MX, Javadi AA, Min X (2006) Triaxial tests on sand reinforced with 3D inclusions. Geotext Geomembr 24:201–209

Zhang MX, Zhou H, Javadi AA, Wang ZW (2008) Experimental and theoretical investigation of strength of soil reinforced with multi-layer horizontal-vertical orthogonal elements. Geotext Geomembr 26:1–13

A Review on the Role of Geosynthetics in Preventing the Excessive Settlement and Mud Pumping of Ballasted Railway Track

Ramesh Gedela and Rajagopal Karpurapu

Abstract Railways are the primary mode of public transport around the world and play a pivoting role in the day-to-day transportation needs of commuters. Hence, an improved railway track infrastructure demands high-quality control. Excessive settlement in rail tracks is a common failure phenomenon incurring an extra maintenance cost. Installation of railway tracks on soft grounds is a daunting task for any railway engineer. Inefficient drainage results in storage of impregnated water beneath the railway tracks. Consequently, under repeated wagon loading results in a phenomenon known as mud pumping. The present paper mainly reviews the role of geosynthetics to mitigate the mud pumping, thereby controlling the excessive settlement and degradation of the track geometry.

Keywords Mud pumping · Geosynthetics · Ballast breakage index · Deterioration · Interlocking

1 Introduction

Developing countries like India have a huge demand for a high-speed train and freight corridors due to a rapid increase in population and urbanisation. Most of these developing countries have conventional ballasted railway track framework containing sleepers and rails resting on the layer system due to its low construction cost (Selig and Waters 1994). In addition, the maintenance of this track is economical and cheap (Chrismer 1985; Esveld 2001). In order to construct high-speed railway tracks and freight corridors, there is a need to improve the existing track geometry and subgrade properties to prevent excessive settlements and lateral spreading of ballast. Geosynthetics are widely used in the railway tracks from the past one decade to increase the serviceability of the track due to its ease of installation and low cost.

R. Gedela (✉) · R. Karpurapu
Indian Institute of Technology Madras, Chennai, Tamil Nadu 600036, India
e-mail: rameshgedela1992@gmail.com

M. Latha Gali and P. Raghuveer Rao (eds.), *Geohazards*, Lecture Notes in Civil Engineering 86, https://doi.org/10.1007/978-981-15-6233-4_51

1.1 *Factors Influencing the Performance of Ballast*

The settlement and degradation of the ballast are influenced by the following factors: the strength of parent rock, the frequency of loading, particle size distribution (PSD), degree of compaction, confining pressure, subgrade soil properties, etc. Degradation of ballast is more in case of poor ballast material and leads to the excessive settlement of the ballast layer as well as drainage problems (Salim 2004; Indraratna et al. 2004). The loose packing of ballast due to improper degree of compaction leads to excessive settlement. Also, it accelerates the particle degradation (Knuston 1976). The vertical settlement of ballast is significantly affected by the confining pressure. The deformation of ballast decreases with increase in confining pressure. At low confining pressure less than 30 kPa, ballast exhibits dilation behaviour. As the confining pressure increases from 30 to 240 kPa, ballast exhibits compressive behaviour. With the increase in confining pressure, the ballast breakage index (BBI) initially decreases rapidly up to a certain limit, then remains constant for a small interval and thereafter increases. Based on the variation of BBI with respect to confining pressure, the optimum degradation zone is observed to be in the range of confining pressure 30–75 kPa (Indraratna et al. 2004).

Deformation and degradation of ballast are significantly affected by the amplitude of loading. Ballast deformation rapidly increases with the increase in the amplitude of deviatoric stress (Knuston 1976; Stewart 1982; Lackenby et al. 2007). Deformation of ballast increases logarithmically with an increase in the number of the load cycles (Raymonds and Williams 1978). The rate of increase in plastic strain in ballast increases with the load cycles up to 10,000, and thereafter, the increment was observed at a slow rate (Indraratna et al. 2010a). The frequency of loading on track increases with respect to increment in train speed, which in turn induces dynamic stresses in the ballast layer (Shenton 1975). Ballast densification is observed without any significant change in ballast breakage index in the frequency range of 20–30 Hz. Higher confining stresses are required to minimise the settlement and degradation of ballast under the high frequency of cyclic loading (Indraratna et al. 2010a).

2 Problems Associated with Ballasted Railway Track

Lateral spreading, excessive settlement of ballast and mud pumping are three significant problems in railway track due to poor subgrade soil, increase in train speeds and axle load, etc. The poor subgrade soil interacts with ballast under cyclic loading, which creates problems and affects the track performance. The performance of track majorly depends on the performance of the ballast layer. The performance of the ballast layer comprises deformation and degradation characteristics of ballast (Alias 1984).

Fig. 1 Lateral spreading of ballast in ballasted railway track (sourced from Hussaini et al. 2013)

2.1 *Lateral Spreading*

Lateral flow of ballast is due to less lateral confining stress (<30 kPa) (Baessler et al. 2003; Indraratna et al. 2010a & b). Lateral flow of ballast reduces the stability of the track, increases breakage of particle and the particle splitting and leads to the excessive vertical settlement of ballast layer (Chrismer 1985; Dash and Shivdas 2012). Figure 1 illustrates the lateral spreading of ballast in the railway track.

2.2 *Differential Settlement*

The track settlement and degradation of ballast are not uniform along the track length due to improper compaction and non-uniform subgrade soil. Differential track settlement occurs, which seriously affects the safety of the track and train speed (Chrismer 1985). The track deterioration and differential settlements of rails are shown in Fig. 2.

2.3 *Mud Pumping*

Mud pumping is one of the serious problems in railway track. The track constructed on poor subgrade soil experiences excessive settlement, and subgrade soil interacts with ballast. The fines are filled in voids present in the ballast layer and cause further abrasion, which leads to reduce the permeability of the ballast layer. These fines hold the moisture and form slurry under repeated loads. Shear strength of ballast reduces significantly due to mud formation in ballast which in turn causes excessive settlement and lateral spreading of ballast (Chrismer 1985). The effect of mud pumping in railway track is shown in Fig. 3. The mud pumping was observed for tracks on a

Fig. 2 Differential settlement of track (sourced from Suiker 2002)

Fig. 3 Mud pumping (sourced from Lieberenz et al. 2009)

typical silty soil subgrade. Due to the application of the cyclic load, highly permeable silty soil allows the pore water and fines from inside the subgrade to rise quickly. Also, the low plasticity of the silty soil makes it easier for rising pore water to dislodge the fine particles. The slurry water then softens the subgrade surface, thus resulting in the easy penetration of granular particles of the sub-ballast into the subgrade, which contributes to excessive settlement.

3 Role of Geosynthetics

Geosynthetics made from polymeric materials have been used widely in railways for the past ten years to improve the ballast performance. It has been found that additional confinement offered by the geosynthetic material used as reinforcement in ballast leads to improvement of the track performance. Lateral confinement occurs due to particle interlocking, residual compaction stresses and sleeper resistance. Indraratna et al. (2009) reported that using the geosynthetic material for ballast stabilisation is more suitable, easy and economical solution in comparison with the remaining ground improvement methods. The additional interlocking of ballast particle is due to the interaction between the ballast and geosynthetic, which in turn reduces the particle movement and increases the performance of ballast and track stability.

3.1 The Function of Geosynthetics in Railways

1. Geotextile prevents the rising of subgrade soil which in turn reduces the fouling of ballast.
2. Geogrid acts as a reinforcement to provide better interlocking and increases the confinement.
3. Geo-composite acts as both separator and reinforcement.
4. Geocell acts as reinforcement and provides three-dimensional confinement to the infill soil.

3.2 Reinforcement, Filtration and Separation Functions of Geosynthetics in a Ballasted Railway Track

3.2.1 Reinforcement

Generally, woven geotextile, geogrid and geocell serve as reinforcement in ballasted railway track. These materials are placed at the interface of ballast–sub-ballast and sub-ballast–subgrade layers to increase the bearing capacity. Geogrids are planar materials having a different aperture opening size and shape, which provides additional confinement to the ballast through interlocking effect. Geocells are three-dimensional honeycombed structure offering confinement three-dimensionally and improve the track performance and stability of the track. The main advantage of providing reinforcement such as geotextile, geogrid and geocells is that they reduce the magnitude of stresses from sleeper to subgrade by distributing the loads over a wider area. The mechanism of particle interlocking in geogrid aperture openings and three-dimensional confinement offered by geocells are shown in Figs. 4 and 5, respectively.

Fig. 4 Mechanism of particle interlocking within the geogrid aperture opening (sourced from Wrigley 1989)

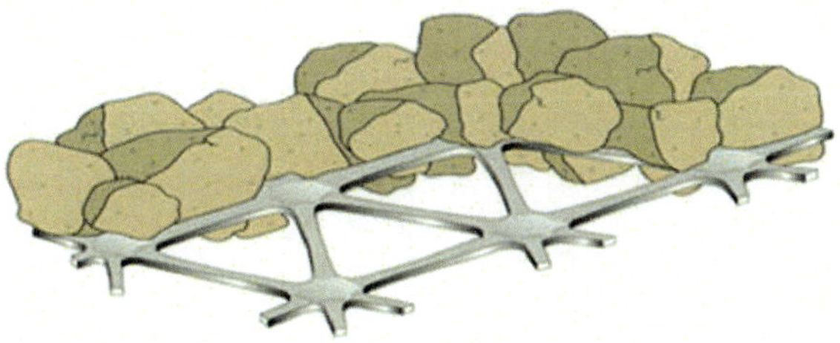

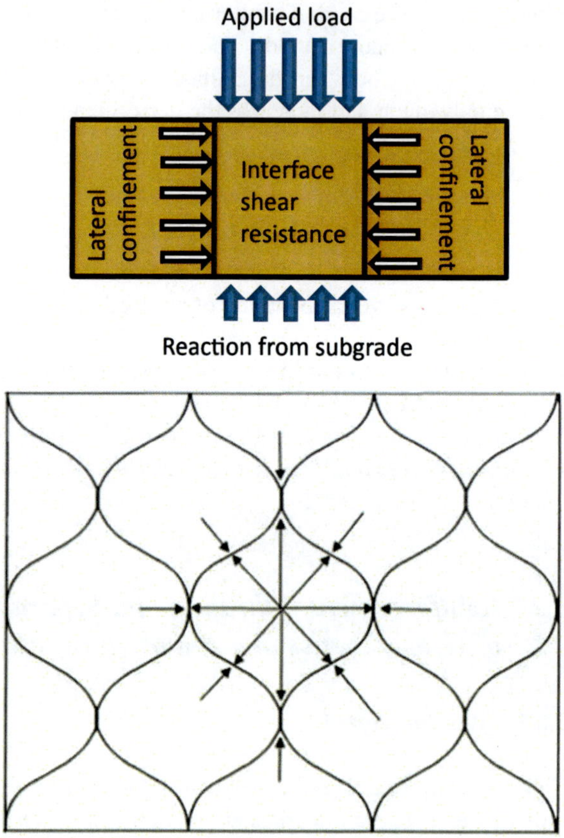

Fig. 5 Three-dimensional confinement and frictional resistance offered by geocell

3.2.2 Separation and Filtration

A non-woven geotextile used at the interface of subgrade–sub-ballast and sub-ballast–ballast layers prevents the rising fines from subgrade into sub-ballast and ballast layers. The aperture opening size (AOS) of geotextile must be less than that of the smallest particle size in subgrade and sub-ballast, which in turn satisfy the filtration and separation criteria. The geotextile should have enough permeability to dissipate the excess pore water pressure under repeated loading. The filtration and separation mechanism of the geotextile layer is illustrated in Fig. 6.

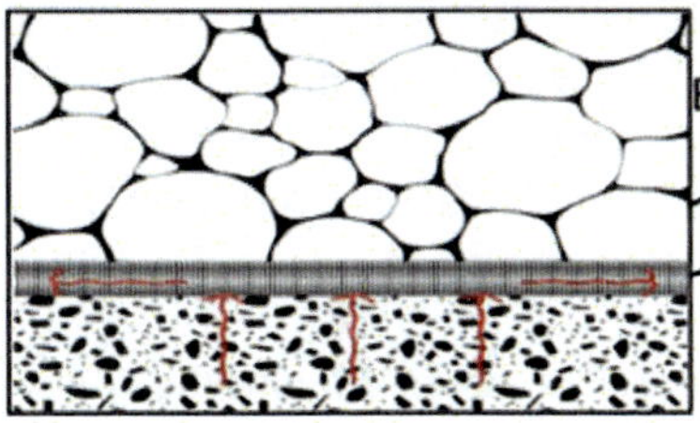

Fig. 6 Mechanism of filtration and separation

4 The Behaviour of Reinforced Ballast

4.1 Laboratory Studies

Several researchers have investigated the behaviour of geosynthetic-reinforced ballast under cyclic loading by using large-scale testing facilities in laboratory and field trials for the past two decades. Many factors, viz. aperture opening, type of geogrid (biaxial or uniaxial), depth of placement of for grid from the soffit of the sleeper, subgrade stiffness, etc., influence the track performance. The interaction between the geogrid and ballast particles plays a vital role in the interlocking mechanism. Tang et al. carried out a series of direct shear and pull-out test to investigate the interfacial shearing behaviour of the aggregates–geogrid surface. From the results of the direct shear test, the authors concluded that geogrid aperture size does not follow any trend with interface shear strength. However, from the results of the pull-out test, it has been reported that geogrid aperture size plays an essential role in its interaction with the aggregate. According to Indraratna et al. (2013), the optimum aperture opening (*A*) should be 1.2 times that of mean particle size (D_{50}) of ballast. The minimum and maximum aperture sizes of geogrid to obtain the beneficial effects of geogrid reinforcement should be $0.9D_{50}$ and $2.5D_{50}$.

The geogrid placed in ballast significantly reduces the settlement. However, the effect of reinforcement in reducing permanent deformation of ballast is more effective in the case of track constructed on poor subgrade soils having less CBR value less than 2. Inclusion of biaxial geogrid within ballast layer reduces the permanent deformation up to 50% after 100,000 load cycles. The number of load cycles required to cause permanent vertical deformation of 50 mm increased by a factor of 10 when geogrid was used as reinforcement compared to the unreinforced section (Bathurst and Raymond 1987; Matharu 1994). Shin et al. reported that the most beneficial effect of reinforcement to reducing track deformation is observed when geo-composite is used at the interface of subgrade soil and sub-ballast layer. The geogrid reduces the settlement of ballast, and the optimum depth for the inclusion of geogrid in the ballast is 125 mm from the sleeper soffit (Raymond et al. 1975).

Indraratna et al. (2006) had carried out a series of tests in the large-scale prismoidal triaxial chamber to investigate the deformation and degradation behaviour of railway ballast under cyclic loading by varying geosynthetic material. Three types of geosynthetic products (woven geotextile, geogrid and geo-composite) were used in their experimental study. The authors reported that the presence of geo-composite at the interface of sub-ballast and subgrade improved the deformation characteristics of the track. The influence of type of geosynthetic material on the performance of geosynthetic-reinforced ballast under cyclic loading is illustrated in Fig. 7. The track reinforcement increases the stiffness of track which in turn reduces the settlement significantly.

The track reinforcement increased with the stiffness of track by about 55–65% which in turn reduced the settlement to 99% compared to unreinforced track. Geo-composite increased stiffness by 9–12% and reduced settlement by 25% (Kennedy 2011). According to Brown et al. (2007), the stiffness and aperture size of geogrid are two key parameters that significantly affect the ballast settlements. Figures 8 and 9 illustrate the influence of geogrid stiffness and aperture opening on the settlement of ballast.

Chawla and Shahu (2016) have investigated the performance of modal track constructed on the compacted soil subgrades under static and cyclic loading. They have simulated a model track conditions after the rainfall and reported that the settlement of reinforced track under cyclic loading is not reached to a maximum limit of 25 mm even after 40,000 load cycles, but the unreinforced track is terminated at 10,500 cycles and reached to maximum permissible settlement. The stable behaviour of reinforced track is due to the better distribution of stresses inside the track. Similarly, Biabani et al. (2016) reported that the shear strength of sub-ballast improved substantially with increment in relative density and also at higher relative density, both unreinforced and reinforced sections show similar performance.

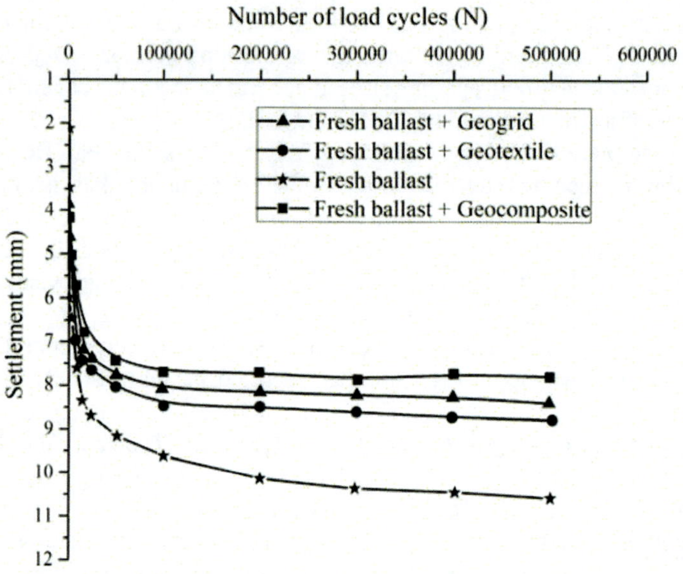

Fig. 7 Variation settlement with the number of load cycles (figure reproduced from Indraratna et al. 2006)

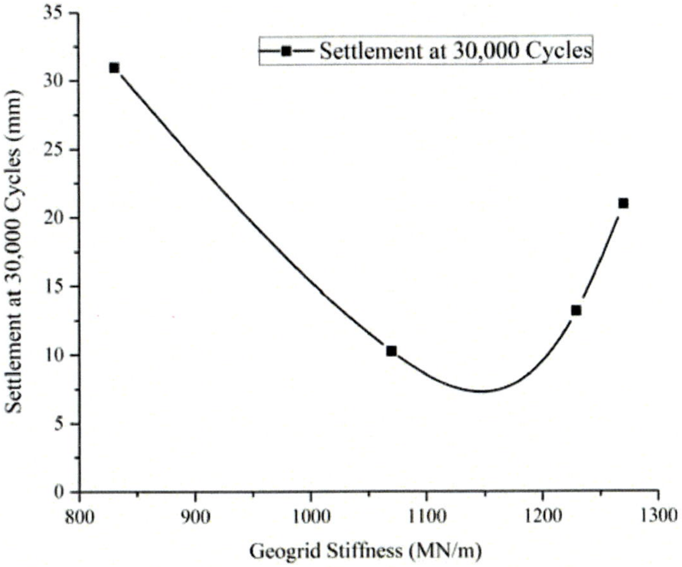

Fig. 8 Variation of settlement with varying geogrid stiffness (Figure reproduced from Brown et al. 2007)

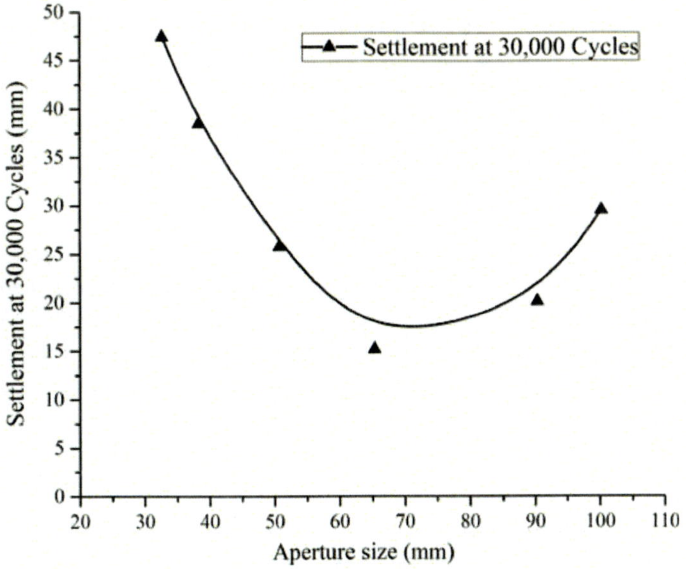

Fig. 9 Variation of settlement with varying aperture (Figure reproduced from Brown et al. 2007)

Ballast degradation occurred due to the number of passes which in turn led to considerable changes to ballast particle size, shape, as well as ballast packing. Due to degradation, fines were accumulated in the coarser particle, and spillage of coal from rail wagons also contributed to the ballast fouling (Qian et al. 2011). Indraratna et al. (2013) reported the stress–strain degradation response of railway ballast stabilised with geosynthetics. Geogrid provided at the interface of ballast and sub-ballast caused lesser damage compared to geotextile-reinforced track. The ballast bed reinforced with double layer of geosynthetics gives better interlocking and load dispersion over a large area and reduces the vertical, lateral and volumetric strains in ballast. Dual-layer reinforcements, i.e. geogrid at the ballast–sub-ballast interface and geo-composite at the sub-ballast–subgrade interface, are better at reducing settlement than single-layer reinforcements. According to Indraratna et al. (2013), the benefits of geogrid are more when it is placed at a depth of 130 mm above the sub-ballast layer causing less degradation of ballast under repeated loads.

4.2 Field Studies

Walls and Galbreath (1987) conducted a series of field trials and reported that the presence of geogrid at the interface of ballast and sub-ballast shows significant improvement on poor subgrade soils. It has been observed that the specific portion of track constructed on poor subgrade soil suffered from severe problems and it demands frequent maintenance operation about every 2–3 weeks. Due to this reason, the train speed is restricted to 8 km/h in affected portions. They have concluded that the inclusion of geogrid at the interface of ballast and sub-ballast increases the restricted train speed from 8 to 56 km/h.

Ashpiz et al. (2002) carried out field tests and concluded that the track performance is significantly improved in the case of geosynthetic-reinforced ballasted track compared to unreinforced track. Shape et al. (2006) carried out full-scale field studies in the UK. The track is constructed on soft subgrade soils, and it has a long history of problems and demanding frequent maintenance operations. From the outcomes of field trials, the authors reported that the rate of settlement of reinforced track is reduced to 0.4 mm/year as compared to unreinforced track 1.4 mm/year which in turn increases the time period between two successive maintenance operations.

5 Conclusion

The inclusion of geosynthetics during the construction of railway tracks shows a significant reduction of excessive settlement and mud formation in comparison with the unreinforced section. The performance of reinforced ballast depends on the type of reinforcement, aperture opening size of geogrid, depth from the soffit of the sleeper, the degree of ballast compaction and type of subgrade. In silty soil subgrades, mud pumping is avoided by providing the geotextile. Meantime, in the case of clay subgrades, mud pumping is not the primary issue; rather, a reinforcement is more relevant to provide better bearing capacity. Geo-composite gives combined benefits of both filtration and reinforcement. Meanwhile, geogrid at the ballast–sub-ballast interface and geo-composite at the sub-ballast–subgrade interface are better at reducing settlement than single-layer reinforcements. Providing geo-composite at ballast–sub-ballast interface satisfies both the reinforcement and filtration criteria.

References

Alias J (1984) La voie ferrée-Tome 1: Techniques de Construction et d'Entretien. Ecole Nationale des Ponts et Chaussées, Eyrolles, Paris, France, p 269 (in French)

Ashpiz ES, Deiderich R, Koslowski C (2002) The use of spunbonded geotextile in railway track renewal on the St. Petersburg—Moscow line. In: Proceedings of 7th international conference on geosynthetics, Nice, France, pp 14–19

Baessler M, Rucker W (2003) Track settlement due to cyclic loading with low minimum pressure and vibrations. System dynamics and long-term behaviour of railway vehicles, Track and subgrade, Springer, Berlin, pp 337–356

Bathurst RJ, Raymond GP (1987) Geogrid reinforcement of ballasted track. Transportation Research Record 1153, National Research Council, Transportation Research Board, Washington, DC, pp 8–14

Biabani MM, Indraratna B, Ngo NT (2016) Modelling of geocell-reinforced subballast subjected to cyclic loading. Geotext Geomembr 44(4):489–503

Brown SF, Kwan J, Thom NH (2007) Identifying the key parameters that influence geogrid reinforcement of railway ballast. Geotext Geomembr 25(6):326–335

Chawla S, Shahu JT (2016) Reinforcement and mud-pumping benefits of geosynthetics in railway tracks: numerical analysis Geotext Geomembr 44(3):344–357

Chrismer SM (1985) Considerations of factors affecting ballast performance. Bulletin 704 AREA—AAR Research and test department, Report No. WP-110

Dash SK, Shivdas AS (2012) Performance improvement of railway ballast using geocells. Indian Geotech J 42(3):186–193

Esveld C (2001) Modern railway track. MRT-Productions, Zaltbommel

Indraratna B, Khabbaz H, Salim W, Lackenby J, Christie D (2004) Ballast characteristics and the effects of geosynthetics on rail track deformation. In: International conference on geosynthetics and geoenvironmental engineering, ICGGE, Bombay, India, pp 3–12

Indraratna B, Khabbaz H, Salim W, Christie D (2006) Geotechnical properties of ballast and the role of geosynthetics in rail track stabilization. Ground Improv 10(3):91–101

Indraratna B, Hussaini SKK, Vinod JS (2013) The lateral displacement response of geogrid-reinforced ballast under cyclic loading. Geotext Geomembr 39:20–29

Indraratna B, Nimbalkar S, Christie D (2009) The performance of rail track incorporating the effects of ballast breakage, confining pressure and geosynthetic reinforcement. In: 8th international conference on the bearing capacity of roads, railways, and airfields. Taylor and Francis Group, London, pp 5–24

Indraratna B, Thakur PK, Vinod JS (2010a) Experimental and numerical study of railway ballast behavior under cyclic loading. Int J Geomech 10(4):136–144

Indraratna B, Nimbalkar S, Christie D, Rujikiakamjorn C, Vinod J (2010b) Field assessment of the performance of a ballasted rail track with and without geosynthetics. J Geotech Geoenviron Eng 136(7):907–917

Kennedy J (2011) A full-scale laboratory investigation into railway track substructure performance and ballast reinforcement. Ph.D. thesis, Heriot-Watt University, UK

Knuston RM (1976) Factors influencing the repeated load behaviour of railway ballast. Ph.D. thesis, University of Illinios, at Urbana, IL

Lackenby J, Indraratna B, McDowell G, Christie D (2007) Effect of confining pressure on ballast degradation and deformation under cyclic triaxial loading. Geotechnique 57(6):527–536

Lieberenz K, Kipper R (2009) Zum Einfluss des Unterbaus und des Tragsystems auf die Gleislage, EI Eisenbahningenieur

Matharu MS (1994) Geogrid cut ballast settlement rate on soft substructures. Railw Gazette Int 150(3):165–166

Qian Y, Tutumler E, Huang H (2011) A validated discrete element modeling approach for studying geogrid-aggregate reinforcement mechanisms. In: Han J, Alzamora DA (eds) Geo-Frontiers 2011: advances in geotechnical engineering, pp 4653–4662

Raymond GP, Williams DR (1978) Repeated load triaxial tests on a dolomite ballast. J Geotech Eng Div, ASCE, 104(GT7):1013–1029

Raymond GP, Gaskin PN, Svec O (1975) Selection and performance of railroad ballast. In: Proceedings of symposium railroad track mechanics and technology, Princeton University, NJ, 21–23 Apr. Pergamon Press, pp 369–387

Salim W (2004) Deformation and degradation aspects of ballast and constitutive modelling under cyclic loading. Ph.D. thesis, University of Wollongong, Australia

Selig ET, Waters JM (1994) Track geotechnology and substructure management. Thomas Telford Services Ltd., London, p 428

Shenton MJ (1975) Deformation of railway ballast under repeated loading conditions. In: Kerr (ed) Railroad track mechanics and technology, Proc. Of a symposium held at Princeton univ., pp 387–404

Stewart HE (1982) The prediction of track performance under dynam-ic traffic loading. PhD thesis, University of Massachusetts, Amherst, Mas-sachusetts, USA

Suiker ASJ (2002) The Mechanical Behaviour of Ballasted Railway Tracks. Delft University Press, The Netherlands

Walls JC, Galbreath LL (1987) Railroad ballast reinforcement using geogrids. In: Proceedings of geosynthetics, 87(1), New Orleans, pp 38–45

Strain Energy-Based Modeling of Soil Liquefaction Using Data-Driven Techniques

S. Raj Athira and S. Adarsh

Abstract This paper presents the application of Gaussian Process Regression (GPR) and M5 Model Tree as two alternative data-driven modeling practices for prediction of soil liquefaction. The initial effective mean confining pressure (σ'_{mean}), initial relative density after consolidation (D_r), percentage of fines content (FC), uniformity coefficient (C_u), Coefficient of curvature (C_c), mean grain size (D_{50}), etc. are used as model inputs to predict strain energy density (W) required for triggering the liquefaction. The performance evaluation criteria like mean absolute relative error (MARE), coefficient of correlation (R), root mean square error (RMSE) for the validation datasets are found to be 6.381, 0.849, 0.266, respectively. Use of multiple statistical criteria and graphical plots confirmed the superiority of PuK Kernel-based Gaussian Process Regression (GPR) model over five different empirical models, two linear genetic programming (LGP)-based expressions, artificial neural network (ANN) and M5 Model Tree-based predictions. Further, a parametric sensitivity analysis performed on input parameters showed that σ'_{mean} is the most influencing predictor to explain the variations of the capacity energy than other input parameters.

Keywords Liquefaction · Strain energy · Kernel · GPR · Data-driven techniques

1 Introduction

Soil liquefaction is one of the most complex phenomena studied by the geotechnical investigators. Liquefaction usually occurs when the pore water pressure increases to carry the overburden stress and Darve (1996) considered liquefaction as a specific feature of loose and saturated sandy soils. The available procedures for evaluation of liquefaction can be categorized broadly as stress-based procedures (Seed and Idriss 1971), strain-based procedures (Dobry et al. 1982) and energy-based procedures

S. R. Athira (✉)
Department of Civil Engineering, College of Engineering, Trivandrum, Kerala, India
e-mail: athirasraj1995@gmail.com

S. Adarsh
Department of Civil Engineering, TKM College of Engineering, Kollam, Kerala, India

M. Latha Gali and P. Raghuveer Rao (eds.), *Geohazards*, Lecture Notes in Civil Engineering 86, https://doi.org/10.1007/978-981-15-6233-4_52

(Figueroa et al. 1994; Green 2001; Liang 1995; Ostadan et al. 1996).The stress-based method, which uses the shear stress level and number of cycles as the main criteria for assessment, is the most popular approach, but it is mainly empirical and need laboratory and field observations. In the alternative strain-based method, the major hypothesis involved is that the pore pressure starts developing when shear strain crosses a threshold shear strain. It considers the mechanics of two interacting idealized sand grains and then generalized for natural soil deposits. But the uncertainty related to random loading is the common limitation in both of these methods. The basic principles of both the stress and strain methods are incorporated in the formulation of the energy-based method. In this method, the amount of total strain energy at the triggering of liquefaction is obtained from laboratory testing or field recorded data. The instantaneous energy and its summation overtime intervals are computed until the onset of liquefaction, which is used as the measures of the capacity of the soil sample against initial liquefaction occurrence in terms of the strain energy (capacity energy).Over the years, the researchers presented many analytical models for prediction of liquefaction triggering and based on experimental observations. These models include simple linear relations and the evolution of soft computing methods was a major breakthrough in this field. Artificial neural networks (ANN), perhaps the most popular data-driven technique, used for the evaluation of the liquefaction potential (Goh 1994, 2002) and many of them used stress- or strain-based database for the evaluation. Baziar and Jafarian (2007) developed an ANN model for evaluation of the liquefaction potential based on the energy concepts. They utilized an evolutionary approach based on genetic programming (GP) for estimation of capacity energy of liquefiable soils. Some of the researchers used different variants of GP for prediction of soil liquefaction including linear GP and multi-expression programming (MEP) for evaluation of soil liquefaction (Alavi and Gandomi 2011, 2012). This paper presents the application of Gaussian Process Regression (GPR) and M5 Model Tree as two alternative data-driven modeling practices for prediction of soil liquefaction.

In this present study, the usefulness of two recently developed data-driven techniques, namely M5 Model Tree and Gaussian Process Regression (GPR), is investigated to obtain generalized relationships between the energy per unit volume dissipated during liquefaction and the soil initial parameters. This level of imparted energy density, denoted as capacity energy of the soil, indicates whether liquefaction is triggered in the soil deposit. Further, the prediction performance of the derived correlations was compared with that of different models already available.

2 Materials and Methods

In this section, the basic principles of two algorithms—M5 Model Tree and GPR—used in this study are presented.

2.1 M5 Model Tree

M5 Model Tree (MT) is a popular machine learning technique proposed by Quinlan (1992) used for solving regression problems through classification and decision making. It follows a modular approach so that the entire domain is divided into sub-domains and separate multi-linear regression models are developed for each of them. Therefore, it formulates many piecewise linear models to approximate the nonlinear relationship between the input variables and output variable. In the first stage, a decision tree is created following a splitting criterion. The one which uses standard deviation reduction (SDR) as the splitting criteria is called as M5 learning algorithm (Witten and Frank 2005). In this method, the standard deviation of the class values that reach a node is treated as a measure of the error at that node and the expected reduction in this error as a result of testing each attribute at that node is calculated. The computation of SDR can be represented as follows:

$$\sigma_R = \sigma(N) - \sum \frac{|T_i|}{|N|} \sigma(T_i) \tag{1}$$

where σ_R = standard deviation reduction; N is the total number of training samples; T_i is the training samples of ith sub-domain; $\sigma(N)$ and $\sigma(T_i)$ are the standard deviations of complete training samples, and ith sub-domain samples, respectively. The computational process and decision making are depicted in Fig. 1. Figure 1 represents the modular division of domains to sub-domains followed by model selection (tree building) corresponding to a 2D input domain of parameters x_1 and x_2.

The splitting continues till the class values of all the instances that reach a node varies negligibly or only a few instances remain. Then the model improvement is done by two ways: by performing 'pruning' and 'smoothing' operations, which reduce the effect of 'over fitting' and sharp discontinuities between different sub-classes, which happen especially when the dataset is very small (Jothiprakash and Kote 2011). A more detailed description of the theory behind model trees and pictorial representations can be found elsewhere (Witten and Frank 2005). The use of M5 Model Tree does not involve the setting of any algorithm-specific user-defined parameters

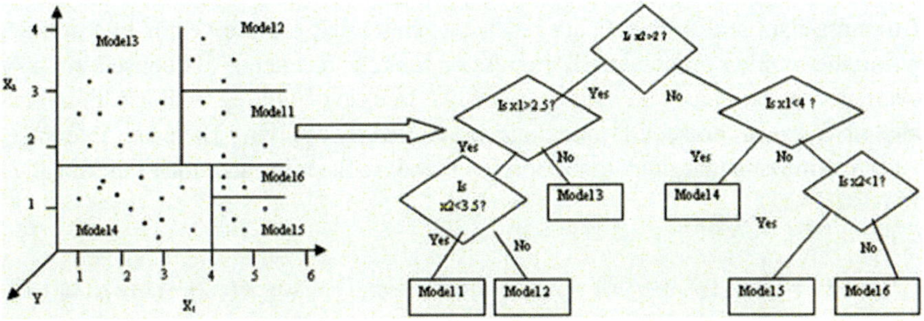

Fig. 1 Data division and example of Model Tree (Solomatine and Xue 2004)

(Singh et al. 2007) and another major advantage of M5 Model Tree is that it combines several simple linear relations and hence more transparent and acceptable by decision makers.

2.2 *Gaussian Process Regression*

GP regression provides probabilistic models work on the assumption that adjacent observations should convey information about each other (Williams and Rasmussen 1996; Rasmussen and Williams 2006). This is a natural generalization of the Gaussian distribution, whose mean and covariance are a vector and matrix, respectively. A GP is defined as a collection of random variables, any finite number of which has a joint multivariate Gaussian distribution. Let $\chi \times \gamma$ represent the domains of inputs and outputs, respectively, from which n pairs (x_i, y_i) are drawn independently and identically distributed. For regression, assume that $y \subseteq \Re$; then, a GP on χ is defined by a mean function μ: $\chi \rightarrow \Re$ and a covariance function k: $\chi \times \chi \rightarrow \Re$. The main assumption of GP regression is that y is given by $y = f(x) + \xi$, where $\xi \sim N(0, \sigma^2)$. The symbol ~ in statistics means sampling for. In GP regression, for every input x there is an associated random variable $f(x)$ drawn from the Gaussian process on χ specified by k. i.e., $Y = (y_1, \ldots, y_n) \sim N(0, K + \sigma^2 I)$, where $K_{ij} = k(x_i, x_j)$, and I is the identity matrix. As $Y/X \sim N(0, K + \sigma^2 I)$ is normal, so is the conditional distribution of test labels given the training and test data of $p(Y_*/Y, X, X_*)$. Then, one has $Y_*/Y, X, X_* \sim N(\mu, \Sigma)$, where

$$\mu = K(X_*, X)\left(K(X, X) + \sigma^2 I\right)^{-1} Y \tag{2}$$

and

$$\Sigma = K(X_*, X_*) - \sigma^2 I - K(X_*, X)\left(K(X, X) + \sigma^2 I\right)\right)^{-1} K(X, X_*) \tag{3}$$

If there are n training data and n^* test data, then $K(X, X_*)$ represents the $n \times n^*$ matrix of covariance evaluated at all pairs of training and test data sets and similarly for the other values $K(X, X)$, $K(X_*, X)$, $K(X_*, X_*)$ where X and Y is the vector of training data and training data labels y_i, whereas X^* is the vector of test data. During the training process of GP regression models, one needs to choose a suitable covariance function as well as its parameters. In case of GP regression with a fixed value of Gaussian noise, GP model can be trained by applying Bayesian inference, that is maximizing marginal likelihood. This leads to the minimization of the negative log-posterior:

$$p(\sigma^2, k) = \frac{1}{2} y^T \left(K + \sigma^2 I\right)^{-1} y + \frac{1}{2} \log\left|K + \sigma^2 I\right| - \log p(\sigma^2) - \log p(k) \tag{4}$$

To find the hyperparameters, the partial derivative of Eq. (4) can be obtained with respect to σ^2 and k, and minimization can be achieved by gradient descent algorithm. A specified covariance function is required to generate a positive semi-definite covariance matrix K, where $K_{ij} = k(x_i, x_j)$. The Radial Basis function (RBF), which has a Gaussian base and the polynomial are the two popular Kernels used in GPR. The width of the Gaussian Kernel and the polynomial order (exponent) are the algorithm-specific parameters in these popular Kernels (Pal and Deswal 2010). Üstün et al. (2006) proposed Pearson VII function universal Kernel (PuK) and suggested it to be an alternative to the linear, polynomial and RBF kernels. The mathematical form of which is

$$\left(1 \Big/ \left[\left[1 + \left(\frac{2\sqrt{|x_i - x_j|^2}\sqrt{2^{\left(\frac{1}{\omega}\right)} - 1}}{\sigma}\right)^2\right]\right]^{\omega}\right) \tag{5}$$

The Kernel specific parameter σ is the Pearson width, whereas ω is the tailing factor of the peak of the Kernel function.

3 Database

The energy required to trigger liquefaction of a soil deposit is reported to be independent of loading pattern (random or harmonic) and stress history (Liang et al. 1995). Therefore, unlike for the case of strain and stress-based approaches, in the energy-based approach there exists a linkage between field and laboratory behavior. Based on the literature on energy-based models proposed in the past, dissipated strain energy density (W), pore water pressure increment ratio (r_u), loading parameters such as cyclic stress ratio (CSR) or strain level, initial parameters of soils such as initial effective confining pressure (σ_0'), initial relative density (D_r) and some of the calibration parameters obtained from curve fitting of experimental data are recognized to be suitable input variables for prediction of liquefaction.

The evaluation is done by using the data (284) collected from previously published cyclic tests available in a published data (Baziar and Jafarian 2007). The database contains 217 cyclic triaxial, 61 cyclic torsional shear and 6 cyclic simple shear tests on Monterey, Yatesville, Reid Bedford, on clean and silty sands. The two criteria that indicate the liquefaction triggering are: (1) initial liquefaction ($r_u = 1$) and (2) double amplitude of strain of 55% ($\varepsilon_{DA} = 5\%$), whichever occurs first (Baziar and Jafarian 2007). The complete database was partitioned into training (70%) and testing (30%) sets. The data division is performed by following trial and error method to ensure that the statistical properties [such as minimum, maximum, mean, standard deviation and the coefficient of variation (CV)] between training and testing datasets are nearly same. It is reported that normalization of the variables helps in faster learning irrespective of the range and method chosen (Alavi et al. 2010; Swingler 1996).

Table 1 Statistics of training and testing datasets

Property	σ'_{mean}(kPa) (kPa)	D_r (%)	FC (%)	C_u	D_{50} (mm)	C_c	W (kJ/m^3)
Training							
Max	294.0	105.1	100.0	5.880	0.46	1.610	34.9
Min	41.10	−44.5	0.000	1.570	0.03	0.740	0.3
SD	27.33	33.17	26.67	0.996	0.13	0.185	5.40
Mean	98.77	47.21	19.44	2.347	0.24	0.938	3.48
CV	0.276	0.70	1.371	0.424	0.55	0.197	0.001
Testing							
Max	294.0	104.3	100.0	5.88	0.44	1.610	23.3
Min	41.10	−36.5	0.000	1.570	0.03	0.74	0.31
SD	28.85	34.75	25.57	1.080	0.12	0.203	4.14
Mean	97.44	47.70	22.76	2.434	0.21	0.949	3.05
CV	0.29	0.733	1.123	0.443	0.56	0.214	0.001

Therefore, in this study, the output and input variables were normalized between zero and 1 and is shown in Table 1.

4 Results and Discussions

First, the different empirical relations for the evaluation of liquefaction potential reported in the literature are used. Different energy-based models considered in the study are presented in Table 2.

In Table 2, n stands for normalized and hence $\sigma'_{\text{mean};n}$ $D_{r;n}$, FC_n, $C_{u;n}$ and $D_{50;n}$ respectively, denote the soil initial effective mean confining pressure, initial relative density after consolidation, percentage of fines content, coefficient of uniformity, and mean grain size in their normalized forms as given below:

$$\sigma'_{\text{mean};n} = \sigma'_{\text{mean}}/300 \tag{6}$$

$$D_{r;n} = (D_r + 40)/150 \tag{7}$$

$$\text{FC}_n = (\text{FC} + 40)/15 \tag{8}$$

$$C_{u;n} = (1/6) * C_u \tag{9}$$

$$D_{50;n} = 2 * D50 \tag{10}$$

Table 2 Empirical models for evaluating liquefaction potential

$\log(W) = 2.002 + 0.00477\sigma'_{\text{mean}} + 0.0116D_r$	Figueroa et al. (1994) (Model 1)
$\log(W) = 2.062 + 0.0039\sigma'_{\text{mean}} + 0.0124D_r$	Liang (1995) (Model 2)
$\log(W) = 2.484 + 0.00471\sigma'_{\text{mean}} + 0.00052D_r$	Liang (1995) (Model 3)
$\log(W) = 1.164 + 0.0124\sigma'_{\text{mean}} + 0.0209D_r$	Dief and Figueroa (2001) (Model 4)
$\log(W) = 2.4597 + 0.00448\sigma'_{\text{mean}} + 0.00115D_r$	Dief and Figueroa (2001) (Model 5)
$\log(W) = 2.1028 + 0.004566\sigma'_{\text{mean}} + 0.005685D_r + 0.001821\text{FC} - 0.02868C_u + 2.0214D_{50}$	Baziar and Jafarian (2007) (Model 6)
$\log(W)_{\text{LGP};1} = \frac{5}{4}(2\sigma'_{\text{mean};n}D_{r;n} + D_{r;n}D_{50;n} + D_{r;n}D^2_{50;n} * (\sigma'_{\text{mean};n} + D_{50;n} - (3\sigma'_{\text{mean};n} - 6\text{FC}_n + 4C_{u;n})^2 - 1) + 2)$	Alavi and Gandomi (2012) (Model 7)
$\log(W)_{\text{LGP};11} = \frac{5}{2} + 5\sigma'_{\text{mean};n}D_{r;n} - 5(\sigma'_{\text{mean};n}D_{r;n})^2$	Alavi and Gandomi (2012) (Model (8))

$$\log(W) = 0.2 * \log(W) \tag{11}$$

Different model performance evaluation measures like correlation coefficient (R), coefficient of efficiency (E), root mean square error (RMSE), mean bias error (MBE), mean absolute relative error (MARE) are used for comparing the efficacy of different models used in this study. The performance evaluation of predictions by different energy-based models stated in Table 2 is presented in Table 3.

For analyzing the result, log of the values of the strain energy density, obtained by applying the empirical equations, was used. Analyzing the coefficient of correlation, we can see the performance given by various empirical equations. The first five equations used (Figueroa et al. 1994; Liang et al. 1995; Dief and Figueroa 2001) were derived by performing a multiple linear regression (MLR) analysis and developed a relation taking into account only two of the initial soil parameters (σ'_{mean}, D_r). Baziar and Jafarian (2007) used an ANN model and developed a relation taking into account five of the initial soil parameters (σ'_{mean}, D_r, FC, C_u and D_{50}). LGP model by Alavi and Gandomi (2012) developed two relations taking into account five of

Table 3 Performance evaluation of different energy-based models

PEC	M1	M2	M3	M4	M5	M6	M7	M8
Training set								
R	0.54	0.55	0.22	0.54	0.31	0.8	0.84	0.6
E	−0.14	−0.14	−0.41	−1.07	−0.42	0.6	−4.86	0.3
RM	0.48	0.48	0.53	0.64	0.54	0.3	1.08	0.38
MB	−0.25	−0.24	−0.31	0.09	−0.32	−0.003	−1.03	−0.05
MA	1.02	1.02	11.6	22.8	20.4	12.1	32.5	8.6
Testing set								
R	0.59	0.58	0.40	0.61	0.47	0.8	0.82	0.6
E	0.04	0.01	−0.19	−1.03	−0.19	0.6	−5.04	0.4
RM	0.44	0.45	0.49	0.64	0.49	0.7	1.11	0.3
MB	−0.21	−0.19	−0.26	0.15	−0.28	0.002	−1.05	−0.01
MA	11.45	11.82	10.69	16.49	10.64	6.5	33.41	8.7

PEC performance evaluation criteria, *M* refers model, *RM* root mean square error; *MB* mean bias error; *MA* mean absolute relative error

the initial soil parameters (LGP(I)) (σ'_{mean}, D_r, FC, C_u and D_{50}) and two of the soil parameters((LGP (II)) (σ'_{mean}, D_r), but in their normalized form.

It was seen from the output analysis that, with the exception of the training data, the equation created by LGP(II) produced a better result than those generated by MEP. Also LGP(I) model developed a better result than that developed by Baziar and Jafarian (2007). The result demonstrated that the LGP and MEP-based formulas with five inputs significantly outperformed those using two inputs. From the above tables, it is clear that the several empirical or LGP model equations have different results and this will introduce considerable uncertainties in liquefaction assessment. ANN and LGP-based model involves complex step of control parameter setting and the complexity enhances as there are many such control parameters in this methods. So it is logical to investigate in this direction and solicit experiments using alternative predictive methods. In this context, the M5 Model Tree and a Kernel-based evaluator—the Gaussian Process Regression (GPR)—are chosen as two candidate methods. In these, first the M5 Model Tree algorithm is considered, in which no step of control parameter setting except the steps of pruning and smoothing.

The GPR modeling is experimented with different Kernals linear, PuK and the Gaussian. Several trials were done for each type of Kernel varying the Kernel specific parameters. The noise was kept fixed for each trial and the trial which gave maximum correlation coefficient for both training and testing data sets and minimum error was selected for each type of kernel. The results of performance evaluation of different data-driven models used in this study are summarized in Table 4. The performance evaluation of two best models M5 Model Tree and GPR Gaussian Kernel is presented in Fig. 2. From Table 4, it is clear that Model Tree and GPR have shown a fairly good predictions and a comparison with the statistics given in Table 3 confirms the

Table 4 Performance evaluation of data-driven models

Criteria	Model tree		GPR–PuK	
	Training	Testing	Training	Testing
R	0.789	0.745	0.887	0.849
E	0.632	0.619	0.687	0.651
RMSE	0.222	0.278	0.119	0.266
MARE	13.11	6.904	12.89	6.381
MBE	0.434	0.119	0.212	0.084

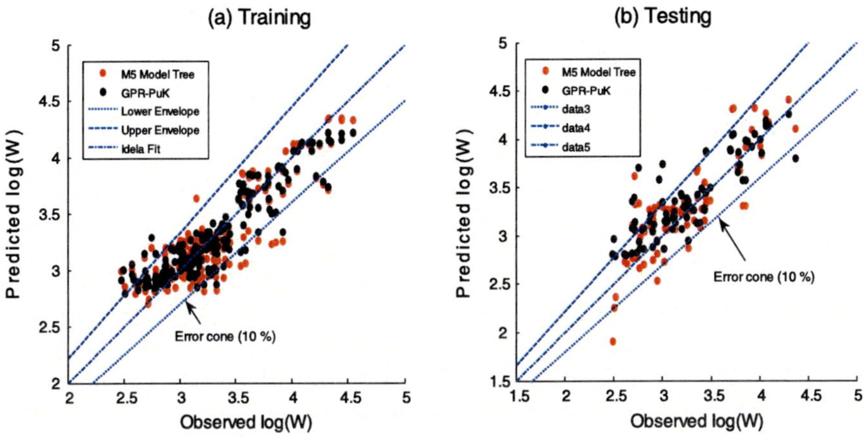

Fig. 2 Scatter plots of predictions of training and testing datasets

superiority of these alternative methods. The graph plotted with 10% error line shows that majority of the points lie between that error lines so it can be concluded that predictions made with these tools are reliable.

The results showed that the GPR-based method performs better in terms of fitting statistics and error measures with less bias, when compared with other empirical expressions. The performance is close or equally good with ANN model by Baziar and Jafarian (Figueroa et al. 1994). The Kernel-based estimate in GPR involves very less number of control parameters to optimize, when compared with ANN or LGP methods, which is an added advantage of the present approach.

In the next, the model sensitivity (S, in %) is done to find the influence of different input parameters. In this process, the sensitivity is computed upon the best model (GPR–PuK). It is found out by changing the variables one by one by 30% and determining the sensitivity as follows (Samui 2012);

$$S = \frac{1}{N_t} \sum_{j=1}^{j=N_t} \left(\frac{\%\,\text{Change in output}}{\%\,\text{Change in input}} \right)_j * 100, \tag{12}$$

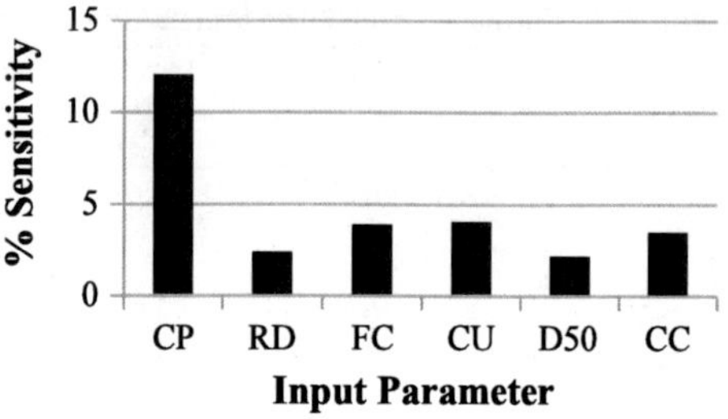

Fig. 3 Sensitivity of input parameters (*CP* confining pressure; *RD* relative density; *FC* fine content; *CU* uniformity coefficient; D_{50} 50% finer; *CC* coefficient of curvature)

where N_t is the number of testing datasets.

The results of sensitivity analysis are presented in Fig. 3. From the plot, it is clear that the confining pressure is the most influencing parameter to trigger soil liquefaction. The effect of fine content, uniformity coefficient and coefficient of curvature is similar.

5 Conclusion

This study proposed the application of M5 Model Tree and Gaussian Process Regression (GPR) for prediction of soil liquefaction. Six input parameters such as soil initial effective mean confining pressure, initial relative density after consolidation, percentage of fines content, coefficient of uniformity, mean grain size and coefficient of curvature are considered to predict the strain energy density required for triggering liquefaction. Rigorous performance evaluation based on multiple statistical performance evaluation showed that GPR-based method performs better in terms of fitting statistics and error measures with less bias, when compared with many empirical expressions and data-driven-based expressions reported in literature. The parametric sensitivity study showed that σ'mean is the most influencing predictor for liquefaction triggering.

References

Alavi AH, Gandomi AH (2011) A robust data mining approach for formulation of geotechnical engineering systems. Eng Comput 28(3):242–274

Alavi AH, Gandomi AH (2012) Energy-based numerical models for assessment of soil liquefaction. Geosci Front 3(4):541–555

Alavi AH, Gandomi AH, Sahab MG, Gandomi M (2010) Multi expression programming: a new approach to formulation of soil classification. Eng Comput 26(2):111–118

Baziar MH, Jafarian Y (2007) Assessment of liquefaction triggering using strain energy concept and ANN model: capacity energy. Soil Dyn Earthquake Eng 27:1056–1072

Darve F (1996) Liquefaction phenomenon of granular materials and constitutive instability. Eng Comput 13(7):5–28

Dief HM, Figueroa JL (2001) Liquefaction assessment by the energy method through centrifuge modeling. In: Zeng XW (ed) Proceedings of the NSF international workshop on earthquake simulation in geotechnical engineering. CWRU, Cleveland, OH
Dobry R, Ladd RS, Yokel FY, Chung RM, Powell D (1982) Prediction of pore water pressure build-up and liquefaction of sands during earthquakes by the cyclic strain method. National Bureau of Standards, US Department of Commerce, US Governmental Printing Office, Building Science Series, Washington, DC, p 138
Figueroa JL, Saada AS, Liang L, Dahisaria MN (1994) Evaluation of soil liquefaction by energy principles. J Geotech Geoenviron Eng 20(9):1554–1569
Goh ATC (1994) Seismic liquefaction potential assessed by neural networks. J Geotech Eng 120(9):1467–1480
Goh ATC (2002) Probabilistic neural network for evaluating seismic liquefaction potential. Can Geotech J 39(1):219–232
Green RA (2001) Energy-based evaluation and remediation of liquefiable soils. Ph.D. dissertation, Virginia Polytechnic Institute and State University, Blacksburg, VA
Jothiprakash V, Kote AS (2011) Effect of pruning and smoothing while using M5 Model tree technique for reservoir inflow prediction. J Hydrol Eng 16(7):563–574
Liang L (1995) Development of an energy method for evaluating the liquefaction potential of a soil deposit. Ph.D. dissertation, Department of Civil Engineering, Case Western Reserve University, Cleveland, OH
Liang L, Figueroa JL, Saada AS (1995) Liquefaction under random loading: a unit energy approach. J Geotech Geoenviron Eng 121:776–781
Ostadan F, Deng N, Arango I (1996) Energy-based method for liquefaction potential evaluation, Phase I. Feasibility study. U.S. Department of Commerce, Technology Administration, National Institute of Standards and Technology, Building and Fire Research Laboratory
Pal M, Deswal S (2010) Modelling pile capacity using Gaussian process regression. Comput Geotech 37:942–947
Quinlan JR (1992) Learning with continuous classes. In: Adams, Sterling (eds) Proceedings of 5th Australian joint conference on artificial intelligence. World Scientific, Singapore, pp 343–348
Rasmussen CE, Williams CKI (2006) Gaussian processes for machine learning. The MIT Press, Cambridge
Samui P (2012) Application of statistical learning algorithms to ultimate bearing capacity of shallow foundation on cohesionless soil. Int J Numer Anal Meth Geomech 36:100–111
Seed HB, Idriss IM (1971) Simplified procedure for evaluating soil liquefaction potential. J Soil Mech Found Div 97(SM8):1249–1274
Singh KK, Pal M, Singh VP (2007) Estimation of mean annual flood in Indian catchments using back propagation neural network and M5 model tree. Water Resour Manage 24:2007–2019
Solomatine DP, Xue Y (2004) M5 model trees and neural networks: application to flood forecasting in the Upper Reach of the Huai River in China. J Hydrol Eng 9:591–501
Swingler K (1996) Applying neural networks a practical guide. Academic Press, New York
Üstün B, Melssen WJ, Buydens LMC (2006) Facilitating the application of support vector regression by using a universal Pearson VII function based kernel. ChemometrIntell Lab Syst 81:29–40
Williams CKI, Rasmussen CE (1996) Gaussian process for regression. Adv Neural Process Syst 8:514–520
Witten IH, Frank E (2005) Data mining: practical machine learning tools and techniques. Morgan Kaufmann, San Francisco

Numerical Simulation of Tiered Reinforced Soil Retaining Wall Subjected to Dynamic Excitations

Seema Kumari and Arup Bhattacharjee

Abstract This paper presents the study carried out to analyze the response of multi-tiered reinforced earth walls with vertical (zero-tier) subjected to seismic/dynamic excitations. Plaxis 2D is a finite element program accomplishes the analysis of RE walls in different conditions. A numerical approach is selected to examine the safety of RE walls during dynamic excitation of 0.4 g Kobe earthquake (1995) and results of the response spectrum of finite element models are compared with the result of shake table test. A two-tiered RE wall of height nine meters designed as per FHWA (2010) is simulated using the different parameters of validated model. The lateral displacements of facing, maximum reinforcement load and acceleration amplification factor of wall without offset and two-tiered walls are compared.

Keywords Geosynthetic-reinforced wall · Seismic analysis · Tiered RE wall · FEM

1 Introduction

Reinforced earth (RE) retaining walls are combination of earth and reinforcements that are capable of bearing last tensile stresses; these reinforced enable the mass to resist the extension which earth alone could not bear. It is becoming popular nowadays in context of design methodology. Several benefits of RE retaining walls include sound performance, esthetics, cost and expediency of construction. The frictional resistance generates between soil and geogrids provides the stability to the soil to withstand at suitable height. The application of concrete is very less in reinforced earth retaining wall, as small facing blocks/panels are provided to retain the structure. Although many RE walls have been safely constructed and are performing

S. Kumari (✉)
Department of Civil Engineering, Faculty of Engineering and Technology, Assam Down Town University, Guwahati, Assam 781026, India
e-mail: civil1992seema1@gmail.com

A. Bhattacharjee
Department of Civil Engineering, Jorhat Engineering College, Jorhat 785007, India
e-mail: bhatta_arup@yahoo.com

M. Latha Gali and P. Raghuveer Rao (eds.), *Geohazards*, Lecture Notes in Civil Engineering 86, https://doi.org/10.1007/978-981-15-6233-4_53

excellent, there are many areas that need extensive studies with respect to better understand physical behavior of RE retaining walls under more devastating and harsh environments. There are many situations where RE walls are constructed in a tiered combination for a variety of problems such as stability, and construction limitations etc., which not only advances the structural performance and financial side of construction but also allows wall construction with multifaceted geometries. It is found from the analysis that the performances of multitiered walls are better than that of vertical walls.

The investigation on seismic behavior of multi-layered fortified soil divider is still in preliminary stage. Hardly any scientists introduced their work on multi-layered fortified divider (Liu et al. 2014; Mohamed et al. 2014). The rules recommended by NCMA (1997), AASHTO (1998), FHWA (2010) are not adequate for the execution of multi-layered reinforced earth wall in various condition. The best possible procedure for the sound plan of multi-tired RE wall is yet to accomplished.

The investigations led for watching the conduct RE wall exposed to dynamic stacking can be characterized into three classifications: test contemplates basically dependent on shaking table tests, analytical studies dependent on pseudo-static and pseudo-dynamic methodology and numerical approach.

2 Simulation of FEM Model of Reinforced Earth (RE) Wall

In paper study, the numerical model of RE wall by finite element program specifically intended for the analysis of geotechnical engineering problems. This finite element program is 2D plane strain with 15 node elements are used during analysis. The element stiffness matrix of a three gauss points or stress points is evaluated by numerical integration.

2.1 Validation Analysis

The arrangements got contrasted with the outcomes appeared in Fig. 1 show the physical scaled-up model of RE wall for shake table test. The actual RE wall constructed is 2.8 m high, 4 m long and 2 m wide and this wall is scaled down to build a physical model for laboratory testing having dimension 20 cm thick soil foundation and the facing squares are 20 cm high, 30 cm profound and 45 cm wide. Length of geogrids for miniature test model is taken as 205 cm which are set at a vertical interim of 60 cm. To forestall waves shimmering from the steel dividers during shuddering, 10 cm thick extended-polystyrene (EPS) sheets are set at the front and back parts of the bargains compartment. The production period of the wall is defined with staged construction phenomena. The properties of the various materials for the model are taken and shown in Table 1.

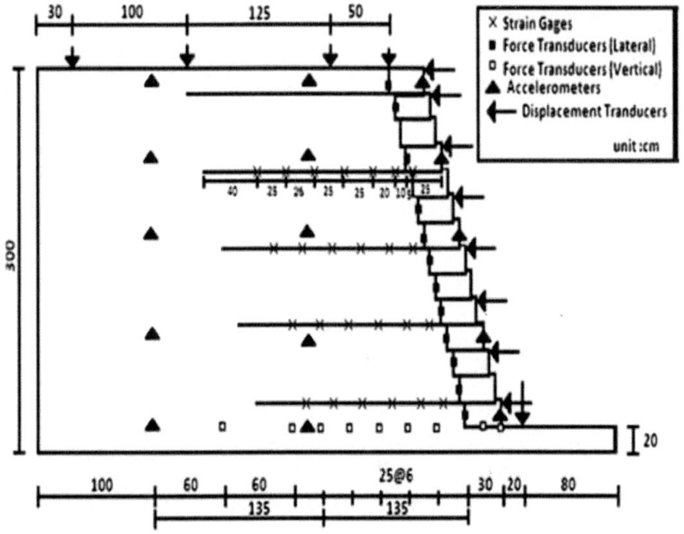

Fig. 1 The schematic of the numerical model

Table 1 Material properties used in numerical analysis

Properties	Backfill	EPS board	Facing wall	Geogrid
Material model	Mohr–Coulomb	Linear elastic	Linear elastic	Linear elastic
Elastic Modulus (kN/m^2)	90E3	2E6	2E6	
Mass-density (kN/m^3)	14.3	1	23	
Poisson's ratio (ν)	0.33	0.2	0.2	
Angle of friction (°)	38			
Dilatancy angle (ψ)	8			
EA (kN/m)				680

Fig. 2 Input seismic motion at PGA of 0.4 g

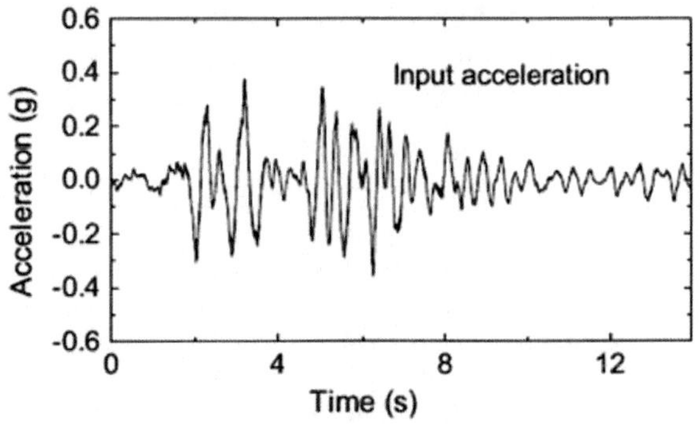

Modeling of backfill material—the establishment soil has indistinguishable properties from the inlay soil. Mohr–Coulomb model is utilized to speak to the sand utilized in the physical model. The information comprises of angle of shearing resistance and dilatancy angle along with modulus properties. Modeling of EPS board and concrete facing: To avoid box effect from the outerwall of physical model during shaking, 10 cm thick extended-polystyrenesheets and solid confronting squares are displayed as direct versatile material. The information comprises of modulus of flexibility, harms proportion and thickness. The input consists of modulus properties. When facing panels are used for facing of GRS walls, panels are modeled as plate element with its axial and bending stiffness.

Modeling of geogrid reinforcement: Reinforcements are models as geogrid components with its axial stiffness (EA). In Plaxis 2D program, properties of geogrid are characterized by parameters of tensile stiffness in the axial direction. Modeling of interface between dissimilar materials: The interface can be portrayed with an elasto-plastic mode to reenact the collaboration. The transfer of stress is decided by the strength of the interface, which equals to the strength of surrounding soil multiplied by the friction coefficient (R_{inter}) between soil and the interface element. So, R_{inter} can reflect the interaction degree. The strength reduction factor is taken as (7/10) and (5/10) for interface sandwiched between the backfill and facing blocks, geogrid and backfill soil and between facing blocks. The wall is subjected to a peak acceleration of 0.4 g shown in Fig. 2.

The maximum horizontal deformation of the wall is shown in Fig. 3. The wall deformations are seen to RE wall by FE analysis with facia of axial stiffness 700E+03 and flexural stiffness 7146 and poisons ratio 0.17. The maximum horizontal displacement measured in physical model and the finite element model is plotted and shown

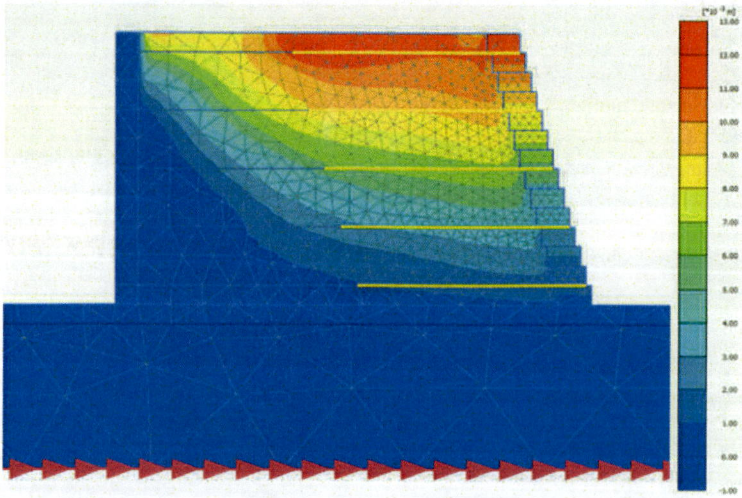

Fig. 3 Deformation characteristic of 2.8 m high wall model

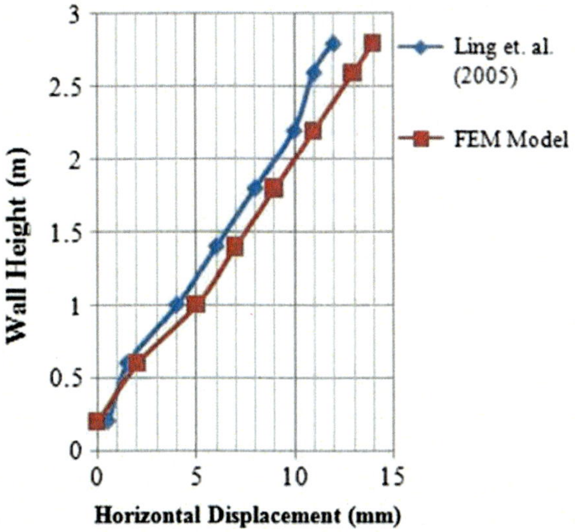

Fig. 4 Maximum horizontal displacements measured FE model

in Fig. 4. The maximum displacement at top of wall is 12 mm for physical test and 14 mm for FE model.

The tensile forces in the geogrid reinforcements both the FE model and physical model are shown in Fig. 5. The maximum tensile forces in the geogrids are 1.35 kN/m and 1.25 kN/m for numerical and physical model, respectively. It is clear from the figure that the tensile forces assessed by the FE models are in close treaty with the physical test. The likeness of the deformations and reinforcement tensile forces behaviors obtained from numerical and experimental results shows the FE model can predict the behavior of a RE retaining wall under seismicexcitation.

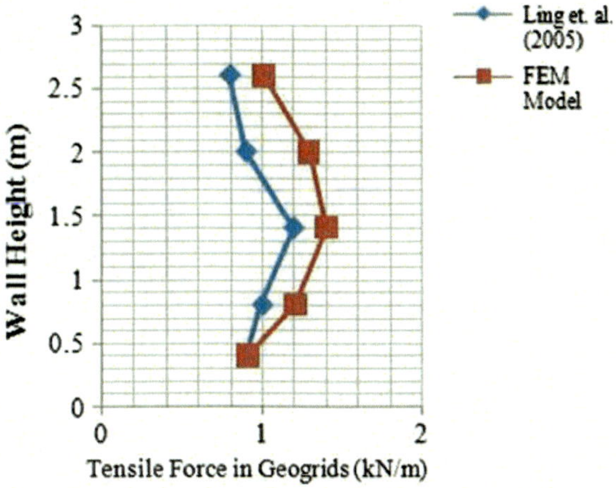

Fig. 5 Tensile forces in geogrids measured by Ling et al. (2005a, b) and FE model

3 Numerical Model of Two-Tiered Reinforced Soil Wall

A 9 m high wall is selected for numerical analysis of tiered GRS walls to study the different response of the walls. The validated modal parameters are used to model the 9 m high wall in two-tier configurations. Four different offset lengths of 1.2, 1.8, 2.4 and 3.0 m are considered for the analysis as per FHWA (2010). The reinforcement lengths are calculated as per FHWA (2010) given in Table 2.

For facing of these RE wall, facing panels are used and these are modeled as plate element in Plaxis 2D (Fig. 6). The thickness of the facing wall and size of foundation in each case are considered as 0.15 m and 30 m × 5.0 m, respectively. A total of 12 number of geogrid layers at a spacing of 0.7 m used in all models of tiered RE walls. The model parameters considered for vertical wall and two-tiered walls are same as that of validated model. The length of the model considered as 30 m beyond the toe of the wall in order to minimize the box effect. The models are all fixed at the base and have roller boundaries at the sides. For seismic analysis, the wall is excited with maximum horizontal acceleration of 0.4 g by applying prescribed displacement at the base of foundation. The geometry of vertical wall of zero offset and two-tiered wall of 1.2 m offset is shown in Figs. 6 and 7.

Table 2 Reinforcement lengths for different tiers as per FHWA 2010

No. of tier	Location of tier	Length of reinforcement
Vertical RE wall ($H = 9$ m)		$0.7H = 6.3$ m
RE wall with two-tier ($H = 9$ m, $H_1 = 4.5$ m)	Upper tier	$0.7H_1 = 3.15$ m
	Lower tier	$0.6H = 5.4$ m

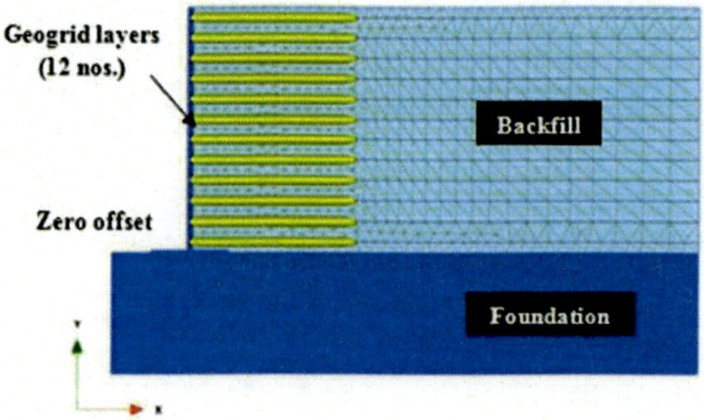

Fig. 6 FE model mesh of vertical wall (zero offset)

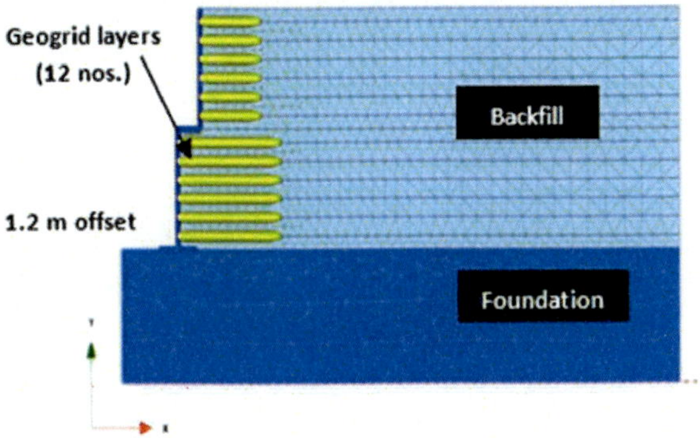

Fig. 7 FE model mesh of two-tiered wall (offset length = 1.2 m)

4 Result and Discussions

4.1 Lateral Facing Displacement in RE Wall by FE Analysis in Seismic Condition

The horizontal displacement of facing is studied in tiered walls and examined the seismic responses from each of the RE wall models. Figure 8 shows the contour of lateral displacements of the zero offset wall and two-tiered wall for 1.2 m offset length. The contours show that the maximum deformation is at the top of zero offset wall and moves toward the mid height of two-tiered wall of offset length 1.2 m.

Figure 11 shows the comparison of horizontal displacement of facing wall for different tier offset of two-tiered walls with zero offset vertical wall. The maximum wall displacements are found to be 18.62 mm, 17.13 mm, 16.6 mm, 15.8 mm and 16.04 mm for wall with zero offset, 1.2 m offset, 1.8 m offset, 2.4 m offset and 3 m offset lengths, respectively (Figs. 8, 9 and 10). The maximum deformation reduces with the increasing tier offset. Reinforcement length plays an important role in GRS

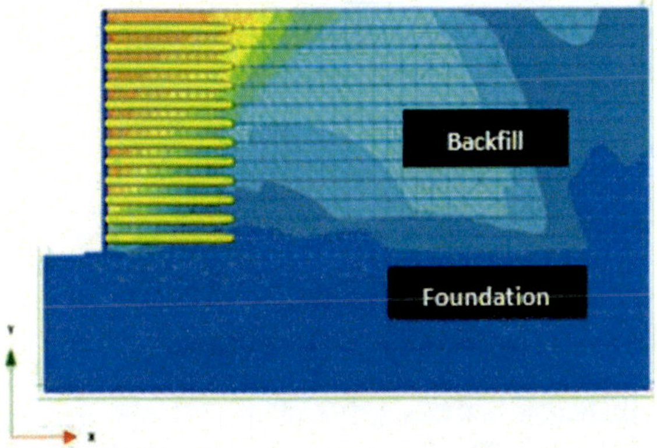

Fig. 8 Maximum horizontal displacements of vertical wall (zero offset)

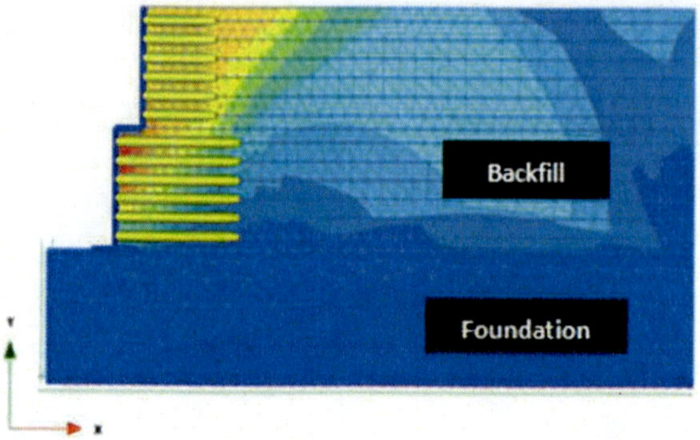

Fig. 9 Maximum horizontal displacements of two-tiered RE wall with 1.2 m offset

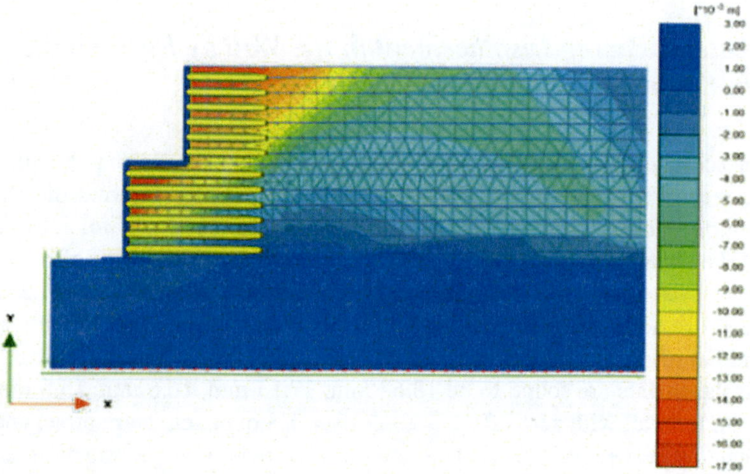

Fig. 10 Maximum horizontal displacements of two-tiered RE wall with 2.4 m offset

walls; due to uneven reinforcement length in lower and upper tier, the deformation pattern is not linear.

4.2 Reinforcement Load in Geogrid of RE Wall by FE Analysis

The RE wall with in tiered construction has great influence of offset of tier and reinforcement length with respect to height of wall. The reinforcement load is maximum at the bottom of RE wall. The ultimate reinforcement load is found to be 11.56 kN/m, 9.29 kN/m, 7.97 kN/m, 7.17 kN/m and 6.41 kN/m for zero offset, 1.2 m offset, 1.8 m

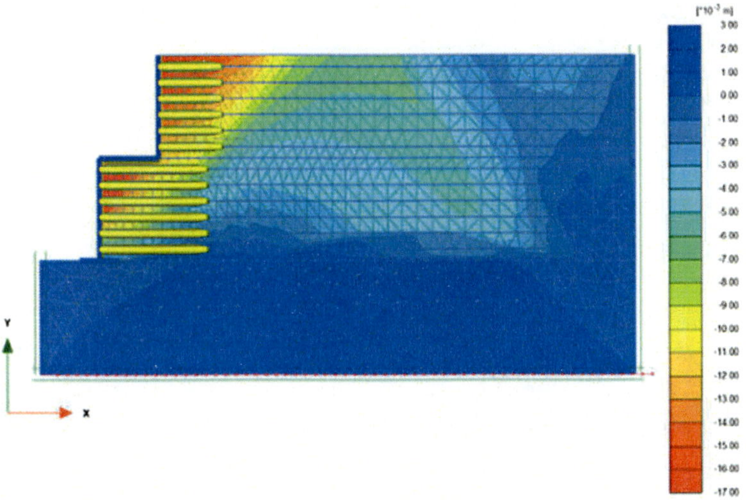

Fig. 11 Maximum horizontal displacements of two-tiered RE wall with 3 m offset

Fig. 12 Wall displacement for without offset wall and two-tiered RE walls for diverse tier offset

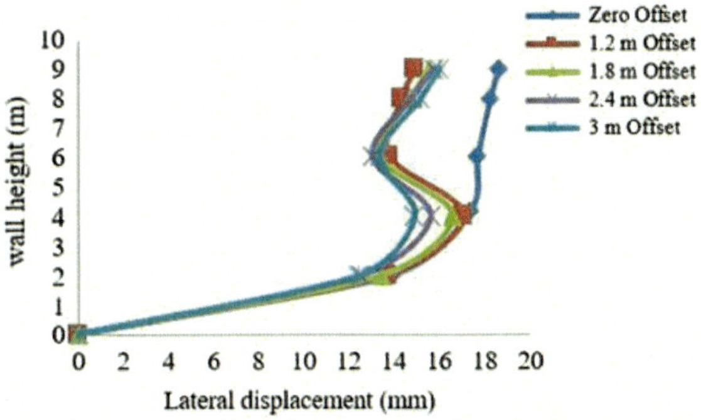

offset, 2.4 m offset and 3 m offset lengths, respectively. With the increase in number of offsets, the maximum reinforcement load in the wall decreases in comparison to zero offset wall. The maximum reinforcement loads at the top geogrids are 6.27 kN/m, 3.67 kN/m, 1.42 kN/m, 1.307 kN/m and 1.92 kN/m for zero offset, 1.2 m offset, 1.8 m offset, 2.4 m offset and 3 m offset lengths, respectively.

The reinforcement loads in uppermost geogrids are first decreased and then increased due to irregular reinforcement lengths in upper tiers. It can be seen that an increase of offset distance reduces the required reinforcement load from Fig. 13.

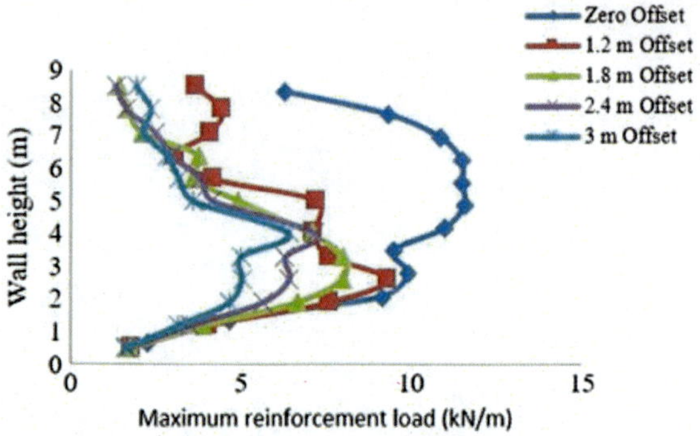

Fig. 13 Reinforcement loads: without offset RE wall and two-tiered RE walls for diverse offsets

4.3 *Acceleration Amplification Factor*

The height of wall is 9 m and acceleration amplification from the bottom of wall for two-tiered wall is shown in Fig. 14.

The amplification factors at the top of backfill are (0.77 for zero offset, 2.35 for 1.2 m offset, 1.925 for 1.8 m offset, 1.31 for 2.4 m offset) and 1.257 for 3 m offset length. The acceleration amplifications are high in two-tiered walls due to facing effect and smaller length of reinforcement in the upper tier compared to lower tiers for all the walls. The amplifications can be reduced by providing same reinforcement length or by increasing the stiffness of reinforcements in the upper tiers.

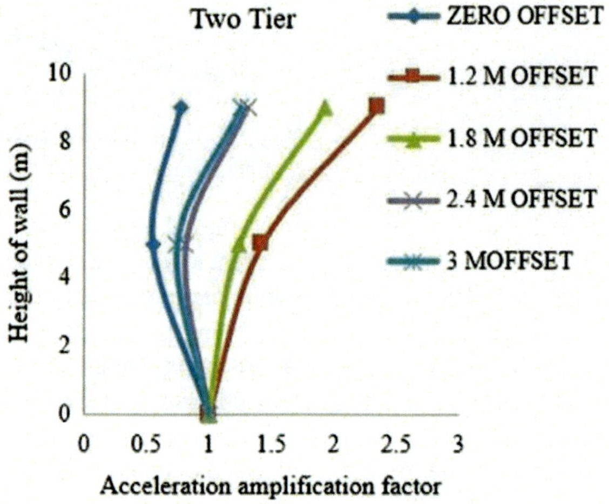

Fig. 14 The backfill of vertical wall and two-tiered walls for different offset with acceleration amplifications factor

5 Conclusion

A vertical geosynthetic RE wall is numerically developed by the finite element program and validated with the shake table test. A 9 m high GRS wall in two-tier configuration is numerically established in the finite element program using the validated model parameters of vertical wall for different offsets. The wall is subjected to (0.4 g) Kobe earthquake (1995) in the analysis. The walls are studied for different tier offset and the results are compared with the vertical wall.

The lateral movement of facing wall is maximum at the top of (zero offset) vertical wall and moves near the center of the wall for different offsets of two-tiered walls. The reinforcement length has importance in RE walls; due to irregular reinforcement length in bottom and top tier, the deformation pattern is nonlinear. With the increase in number of offsets, the maximum reinforcement load in the wall decreases in comparison to zero offset wall. The increase of offset distance reduces the required reinforcement load. The amplification is given for the ratio of maximum acceleration in the backfill, typically at the top of backfill, to the acceleration at the foundation level. The acceleration amplifications are high due to facing effect and smaller length of reinforcement in the uppermost tier compared to lower tiers for all the walls. The amplifications can be reduced by providing same reinforcement length or by increasing the stiffness of reinforcement in the upper tiers.

References

AASHTO (1996) Standard specifications for highway bridges. American Association of State Highway and Transportation Officials, Washington, D.C., USA

FHWA (2010) Mechanically stabilized earth walls and reinforced soil slopes design and construction guidelines, US Department of Federal Highway Administration (FHWA), Vol. I & II. Publication No. FHWA-NHI-10-024

Kumar J, Madhusudhan BN (2010) A note on the measurement of travel times using bender and extender elements. Soil Dyn Earthquake Eng 30:630–634. https://doi.org/10.1016/j.soildyn.2010.02.003

Ling HI, Liu H, Mohri Y (2005). Parametric studies on the behavior of reinforced soil retaining walls under. https://doi.org/10.1061/(ASCE)0733-9399(2005)131:10(1056)

Ling HI, Mohri Y, Leshchinsky D, Burke C, Matsushima K, Liu H (2005) Large-scale shaking table tests on modular-block reinforced soil retaining walls. J Geotech Geoenviron Eng 131(4):465–476. https://doi.org/10.1061/(ASCE)1090-0241(2005)131:4(465)

Liu H, Ling HI (2011) Seismic response of reinforced soil retaining walls and strain softening of backfill soil. Int J Geomech 12:579–593. https://doi.org/10.12989/gae.2017.12.4.579

Liu H, Yang G, Ling HI (2014) Seismic response of multi-tiered reinforced soil retaining walls. Soil Dyn Earthquake Eng . https://doi.org/10.1016/j.soildyn.2014.01.012

Mohamed SBA, Yang K-H, Hung W-Y (2014) Finite element analyses of two-tier geosynthetic-reinforced soil walls: Comparison involving centrifuge tests and limit equilibrium results. Comput Geotech 61(67–84):67–84. https://doi.org/10.1016/j.compgeo.2014.04.010

NCMA (1997) In: Collin J (ed) Design manual for segmental retaining walls, 2nd edn. National Concrete Masonry Association, Herndon, Virginia, USA. https://www.rcpblock.com/

Trandafir AC, Ertugrul OL (2011) Earthquake response of a gravity retaining wall with geofoam inclusion. In: Proceedings of the Geo-Frontiers, ASCE GSP.211

Yadav P (2018) Analytical and experimental analysis of retaining wall in static and seismic conditions: a review. 9(2):522–530. Article ID: IJCIET_09_02_051

Improvement of Seismic-Bearing Capacity of Foundation on Soft Clay by Granular Material

Paramita Bhattacharya and Puja Dutta

Abstract The undrained bearing capacity of the footing in soft clay is very low. In order to increase the bearing capacity, different techniques, such as usage of granular fills over the soft clay or inclusion of granular trench/column or stone column, are followed. A good amount of increase in bearing capacity can be achieved by choosing any of the aforesaid techniques. Although earthquake can cause significant amount of reduction of the bearing capacity of the foundation on layered soil or reinforced soil by granular trench, most of the studies available in the literature are based on the consideration of static loading. Therefore present research work have studied the seismic-bearing capacity of the foundation resting on soft clay and its improvement with (i) either the granular fill or (ii) granular trench by employing the lower bound limit analysis with finite elements and linear programing. The analysis has been performed for various values of $c_u/\gamma B$ values and medium dense and dense sands with soil friction angle equal to 40° and 45°, respectively, where B is the width of the foundation, c_u is the undrained cohesion of the clayey soil, and γ is the unit weight. The magnitude of horizontal earthquake acceleration ($\alpha_h g$) has been varied from 0 to 0.4 g. The normalized bearing capacity value p/c_u (i.e. Q_u/Bc_u) has been found to decrease with an increase in α_h for foundation resting on (i) only soft clay, (ii) granular fills lying over the soft clay and (ii) soft clay reinforced by the granular trench. It has been noted that Q_u/Bc_u increases with an increment in the depth of the overlying granular fills and becomes constant after an optimum depth of granular fills. The magnitude of the optimum depth of granular fills remains same for all values of α_h. The optimum depth of the granular fill depends on (i) the value of $c_u/\gamma B$ and (ii) the soil friction angle of the granular fill. Similar kinds of observations are made for the soft clay reinforced with granular trench under seismic loading. However the optimum depth of granular trench depends on the magnitude of the earthquake acceleration up to some extent. The failure patterns of the soil have also been studied for different values of α_h considering both the improvement techniques.

Keywords Bearing capacity · Granular fill · Granular trench

P. Bhattacharya (✉) · P. Dutta
Indian Institute of Technology Kharagpur, Kharagpur 721302, India
e-mail: paramita@civil.iitkgp.ac.in

M. Latha Gali and P. Raghuveer Rao (eds.), *Geohazards*, Lecture Notes in Civil Engineering 86, https://doi.org/10.1007/978-981-15-6233-4_54

1 Introduction

More land for building new infrastructure is the outcome of the rapid growth in urbanization and infrastructure development. Therefore, building infrastructure on good quality of soil is becoming not only a costly affair but also a difficult task to make it available. So construction on weak and soft soils is an inevitable choice for civil engineering industries. The bearing capacity of foundation on such soil is very low, so there is need to increase the strength of soil in such areas. The improvement of soil in such areas is an issue of great interest. There are different techniques available for improvement of soft soil. One of the easiest and less costly methods is to use the granular material over the soft soil, and the rest footing over this granular fill. Another technique which is very common in civil engineering industry is to use the granular column/trench to strengthen the week soil.

Several studies were carried out in past few decades to examine the increase in the bearing capacity of foundations with either (i) an insertion of granular trenches for example Madhav and Vitkar (1978), Schweiger and Pande (1986), Bouassida and Hadhri (1995), Bouasssida et al. (1995), Bouassida and Porbaha (2004), Stuedlein and Holtz (2012), Bhattacharya and Kumar (2017) or (ii) usage of granular fills over the soft soil such as Love et al. (1987), Das et al. (1994), Shiau et al. (2003), Basudhar et al. (2007), Deb et al. (2007). Madhav and Vitkar (1978) evaluated the upper bound of ultimate bearing capacity of strip footings stabilized with granular trenches placed over soft clay bed based on an assumed failure mechanism. Schweiger and Pande (1986) performed an elasto-plastic finite element analysis to estimate settlements and failure loads of raft foundations placed over soft clays reinforced with stone columns. Hamed (1986) conducted small-scale model tests to compute the bearing capacity of rectangular footings placed on clays reinforced with a trench filled with dense sand. Bhattacharya and Kumar (2017) provided ultimate bearing capacity of foundations on soft clay strengthened by granular trenches for various values of $c_u/\gamma B$ from 0.25 to 1. Love et al. (1987) conducted model tests and analytical studies on behavior of strip footing placed over soft clay subgrade overlain with geogrid-reinforced granular fill. Das et al. (1994) conducted model experimental tests for rough-based strip footing resting on geogrid-reinforced sand and saturated clay. Numerical lower bound limit analysis was performed by Shiau et al. (2003) to investigate the bearing capacity of strip footing on soft clay strengthened by overlying sand layer. Basudhar et al. (2007) conducted a series of model tests and numerical analysis on circular footing resting on geotextile reinforced sand bed over soft soil. Later Deb et al. (2007) performed numerical studies on strength improvement of soft clay by placing multi-layer geosynthetic reinforced granular soil. The major findings from the literature confers that the bearing capacity increases significantly with an increase in thickness of overlying sand layer. The usage of geosynthetic reinforcement imparts additional load carrying capacity and reduces the settlement of the foundation up to some extent.

However, no research works are found in the literature addressing the improvement of the bearing capacity of footings in the presence of seismic forces. Current research work have tried to address that issue for a range of soft clay by using (i) either the granular fill or (ii) granular trench with soil friction angle $\phi = 40°$ and 45° in the presence of seismic body forces with horizontal earthquake acceleration 0–0.4 g.

Lower bound limit analysis always furnishes a safe estimate of the collapse load for a material obeying associated flow rule. Researchers have used the numerical lower bound limit analysis successfully to determine optimal failure load including bearing capacity of footings (Lysmer 1970; Sloan 1988; Singh and Basudhar 1993; Shiau et al. 2003; Kumar and Khatri 2008; Kumar and Bhattacharya 2011, 2013). Therefore, lower bound finite elements limit analysis (LBFELA) with linear programming is used in the present research work. A comparative study between case-I and case-II has also been made towards the increase of the bearing capacity of the foundation.

2 Statement of the Problem

A strip footing having width B is placed over soft clayey deposit with $\phi = 0$. In one case (say Case-I), a granular fill of depth D_s is placed below the footing but above the soft clay stratum, and in second case (say Case-II), a vertical strip granular trench having width B with depth D_t is placed below a strip footing in such a way that the vertical axis of the footing and the trench coincide with each other. For clayey deposit, the undrained shear strength is c_u and for granular material, the internal friction angle is ϕ. The clayey soil and granular material are modelled to follow Tresca and Mohr–Coulomb failure criteria, respectively, with an associated flow rule. The goal of the present research is to determine the magnitude of ultimate load (Q_u) for different values of D_s/B for case-I and D_t/B for case-II, $c_u/\gamma B$ and ϕ in the presence of seismic body forces. The interface surface has been assumed to be rough between the footing and underlying soil mass. Both granular media and clayey deposit are assumed to have same unit weight.

3 Discretization of Domain with Stresses on Boundaries

A rectangular domain JKLM, as shown in Fig. 1a, b for case-I and case-II, respectively, represents the problem domain. The horizontal extent from the edge of the footing towards the right and left boundaries are represented as L_r and L_l, respectively, varies from 10 to 40B for different values of ϕ of the granular materials. The vertical reach of the domain (L_d) is varied as 10B and 25B depending upon the values of D_t/B or D_s/B,α_h and ϕ. The extent of horizontal and vertical boundaries are selected in a way such that there is no change in collapse with further increase in the size of the domain and the yielded elements are contained within the domain boundaries.

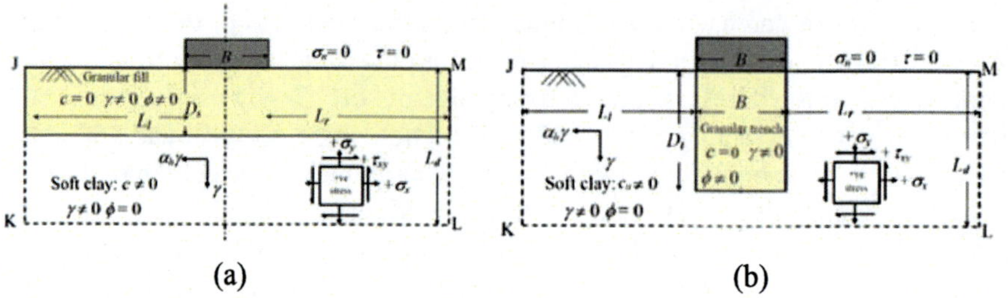

Fig. 1 Schematic diagram for strip footing on **a** granular fill over soft clay and **b** soft clay reinforced by granular trench

The prescribed stress boundary conditions are presented in Fig. 1a, b. Both normal (σ_n) and shear (τ) stresses along the ground surface (JM) are kept equal to 0. To impose a rough interface condition between the footing and soil, a shear slip have been permitted at the interface by enforcing the following stress boundary condition:

$$\left|\tau_{xy}\right| \leq \left(c\cot - \sigma_y\right)\tan\delta \tag{1}$$

where δ is the friction angle between footing and soil on which it is resting.

The tensile stresses are taken positive in the analysis, due to which normal stress σ_y is associated with negative sign; τ_{xy} is the shear stress along the footing–soil interface for strip footing. The soil domain has been discretized by 3-noded triangular elements. The size of the elements decreases continuously when approaching towards the footing edge. Discretized soil domains for a strip footing placed over a fill/trench of granular material and $D = L_d$ are shown in Fig. 2 where E, N, N_i and D_c imply total number of elements, nodes, nodes along the footing–soil interface and stress discontinuities, respectively.

4 Analysis

The present analysis is performed by LBFELA as proposed by Sloan (1988) for solving plane strain problem with linear programming. The basic unknown variables are the nodal stresses σ_x, σ_y and τ_{xy}.

$$\frac{\partial \sigma_x}{\partial x} + \frac{\partial \tau_{xy}}{\partial y} = -\alpha_h \gamma \tag{2}$$

$$\frac{\partial \tau_{xy}}{\partial x} + \frac{\partial \sigma_y}{\partial y} = \gamma \tag{3}$$

where γ is the unit weight of soil mass and α_h is the coefficient of horizontal earthquake acceleration.

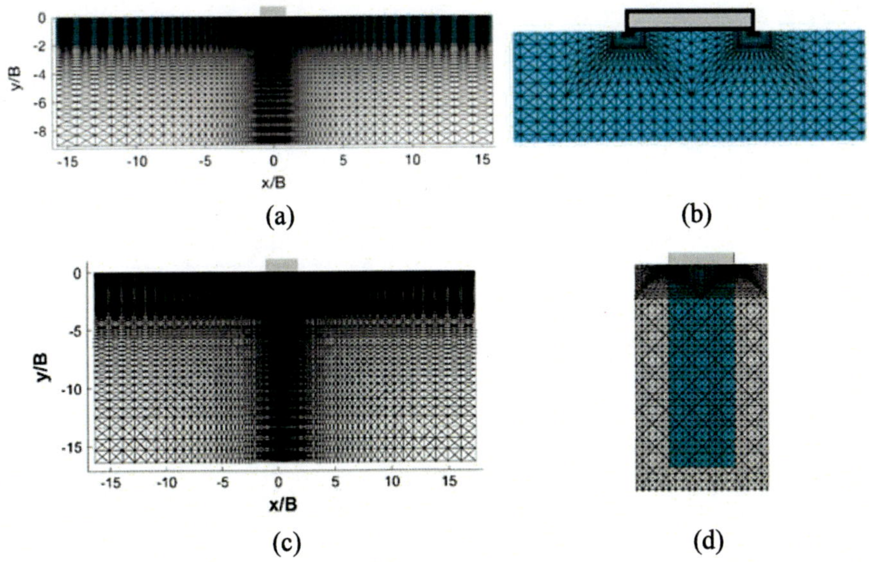

Fig. 2 Discretized soil domain for **a, b** soft clay with overlying granular fill of $D_s/B = 2$ and its zoomed view nearby the footing and **c, d** soft clay strengthened with granular trench with $D_t/B =$ 3and its zoomed view nearby the footing

Statically admissible discontinuities are allowed across all the common edges shared by two adjacent elements by ensuring continuity of shear and normal stresses at the common edges. Along the interface of granular fill and clay layer in case–I and trench/column material and surrounding clay in case-II, the continuity of normal and shear stresses is always required to be ensured. The yield criteria is satisfied at every node in the soil domain. The Mohr–Coulomb yield criteria under plane strain condition can be represented as:

$$F = \left(\sigma_x - \sigma_y\right)^2 + 4\tau_{xy}^2 - \left[2c\cos - \left(\sigma_x + \sigma_y\right)\sin\right]^2 \leq 0 \quad (4)$$

where c and ϕ refer to the cohesion and soil internal friction angle, respectively.The value of ϕ becomes equal to 0 for soil obeying Tresca failure criterion in Eq. 4.

The nonlinear Mohr–Coulomb yield function has been linearized in the present study by following an approach suggested by Bottero et al. (1980). In X–Y plane where $X = \sigma_x - \sigma_y$ and $Y = 2\tau_{xy}$, the yield function as given in Eq. 4 becomes a circle. The linearization is accomplished by inscribing a regular polygon of n sides into the parent yield circle. The imposition of yield criteria as shown in Eq. 4 generates nonlinear inequality condition, which now has been converted into a set of n numbers of linear equality conditions at each node (j) as shown below:

$$A_k\sigma_{xj} + B_k\sigma_{yj} + c_k\tau_{xyj} \leq D_k \quad k = 1,\ 2,\ 3, \ldots, n \quad (5)$$

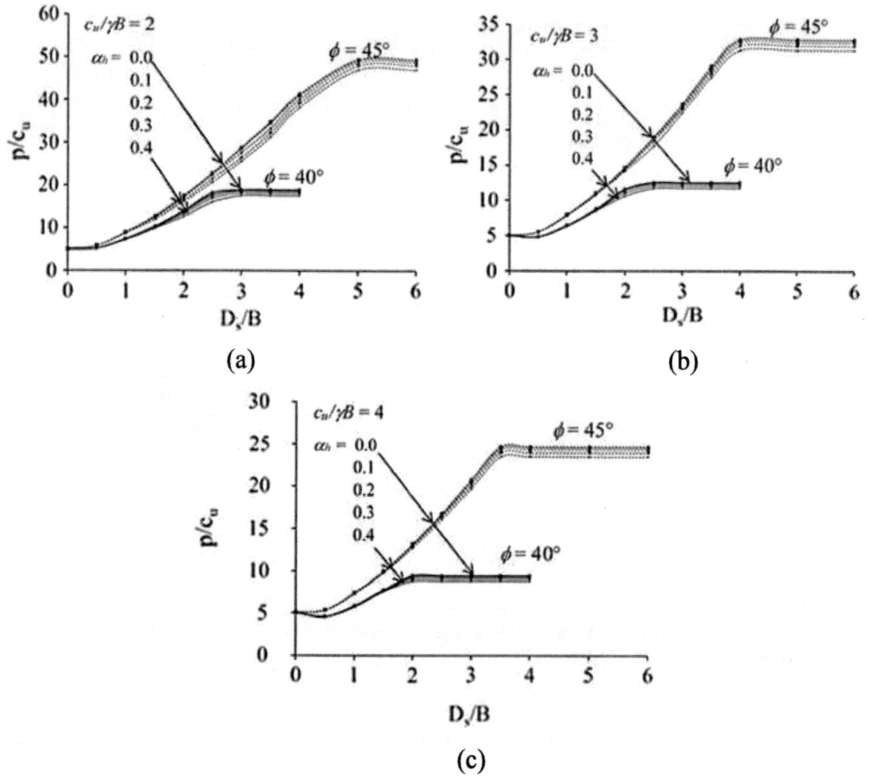

Fig. 3 Variation of p/c_u with D_s/B for **a** $c_u/\gamma B = 2$, **b** $c_u/\gamma B = 3$ and **c** $c_u/\gamma B = 4$

where $A_k = \cos\left(\frac{2\pi k}{n}\right) + \sin\phi\cos\left(\frac{\pi}{n}\right)$; $B_k = \sin\phi\cos\left(\frac{\pi}{n}\right) - \cos\left(\frac{2\pi k}{n}\right)$; $C_k = 2\sin\left(\frac{2\pi k}{n}\right)$; $D_k = 2c\cos\cos\left(\frac{\pi}{n}\right)$.

The above inequality constraint is satisfied at every node in the soil domain.The measure of the ultimate load (Q_u) is evaluated by integrating the normal stresses (σ_n) along the interface between the surrounding soil mass and strip footing in any statically admissible stress field by using the following expression:

$$Q_u = \int_{\text{Footing-soil interface}} (-\sigma_n \mathrm{d}A) \tag{6}$$

where σ_n is the normal stress acting on an element of area dA on the surface of the footing.

The objective is to maximize the collapse load (Q_u) subjected to a set of equality and inequality constraints. Equality constraints are comprised of element equilibrium equations, discontinuity equilibrium conditions and stress boundary conditions. Inequality constraints are formed due to incorporation of stress boundary condition at footing–soil interface and imposition of yield criteria at every node. The linear optimization problem is represented by the following canonical form:

Maximize:

$$-\{g\}^{\mathrm{T}}\{\sigma\} \tag{7}$$

subjected to:

$$\text{(i) } \{A_{\mathrm{eq}}\}\{\sigma\} = \{b_{\mathrm{eq}}\} \tag{8}$$

$$\text{(ii) } \{A_{\mathrm{ineq}}\}\{\sigma\} \leq \{b_{\mathrm{ineq}}\} \tag{9}$$

The linear optimization has been carried out by using LINPROG function available in MATLAB 2013.

5 Definition of Bearing Capacity

The bearing capacity (p) of the footing resting on the soil surface is described as the ratio of the ultimate load carrying capacity per unit length to the width of the footing as shown below:

$$p = \frac{Q_u}{B \times 1} \tag{10}$$

However, the results are presented in terms of p/c_u where c_u is the undrained cohesion value of the original clay soil.

6 Results and Comparison

In the present numerical works, the results are plotted in terms of variation of p/c_u with (i) normalized thickness of the overlying granular fill (i.e. D_s/B of sand layer) in case-I and (ii) normalized depth of the granular trench (i.e. D_t/B) in case-II. The analysis has been performed for (i) $c_u/\gamma B$ equal to 2, 3 and 4, (ii) α_h equal to 0 to 0.4 and (iii) two different values of ϕ say 40° and 45°. The results are discussed below:

6.1 Variation of p/c_u with D_s/B for Soft Clay with Overlying Granular Fill

The variation of p/c_u with D_s/B has been plotted in Fig. 3a–c. It has been observed from the analysis that the value of normalized bearing pressure, i.e. p/c_u, increases with an increase in D_s/B for all combination of (i) $c_u/\gamma B$, (ii) α_h and (iii) ϕ.

The increase in p/c_u continues till it reaches to a maximum or constant value. After reaching the maximum value, the value of p/c_u decreases slightly, and thereafter, the value of p/c_u becomes constant. The value of D_s/B at which the normalized bearing pressure attains maximum or constant value is called the optimum normalized depth of granular fill, i.e. $(D_s/B)_{\text{opt}}$. The value of $(D_s/B)_{\text{opt}}$ depends on the ϕ value of the granular fill. It does not depend on the magnitude of α_h and $c_u/\gamma B$. However the magnitude of the normalized bearing pressure p/c_u has been found to decrease with an increase in α_h for all combination of (i) $c_u/\gamma B$ and (ii) ϕ. The maximum 9.09% and 11.10% decrease in p/c_u due to the presence of seismic body forces has been found in two layered soil system comprising of soft clay of $c_u/\gamma B = 2$ with overlying granular fill for $\phi = 40°$ and 45°, respectively. The maximum value of p/c_u also depends on the soil friction angle of granular fills.

6.2 Variation of p/c_u with D_t/B for Soft Clay Strengthened by Granular Trench

The variation of p/c_u with D_t/B has been plotted in Fig. 4a–c. For the clay layer strengthened with granular trench, p/c_u increases with an increase in D_t/B for all combinations of (i) $c_u/\gamma B$, (ii) α_h and (iii) ϕ. The increase in p/c_u takes place until it reaches an optimum normalized depth $(D_t/B)_{\text{opt}}$ and thereafter becomes constant. The value of $(D_t/B)_{opt}$ depends on α_h and ϕ value of granular trench. Although there is a slight increase in the value of $(D_t/B)_{\text{opt}}$ with an increase in α_h, the magnitude of p/c_u has been found to increase with a decrease in α_h for all combination of (i) $c_u/\gamma B$ and (ii) ϕ.

The maximum 17% and 26.2% reduction in p/c_u due to the presence of seismic body forces has been found in soft clay of $c_u/\gamma B = 2$ strengthened with granular trench of optimum depth for $\phi = 40°$ and 45°, respectively. The maximum value of p/c_u also depends on the soil friction angle of granular fills.

6.3 Proximity of the Stress State to Failure

The failure plots have been generated to determine the proximity of the stress state to shear failure in an optimized statically admissible stress field by a ratio called a/d where $a = (\sigma_x - \sigma_y)^2 + (2\tau_{xy})^2$ and $d = (2 \cos\phi - (\sigma_x + \sigma_y) \sin\phi)^2$. The failure

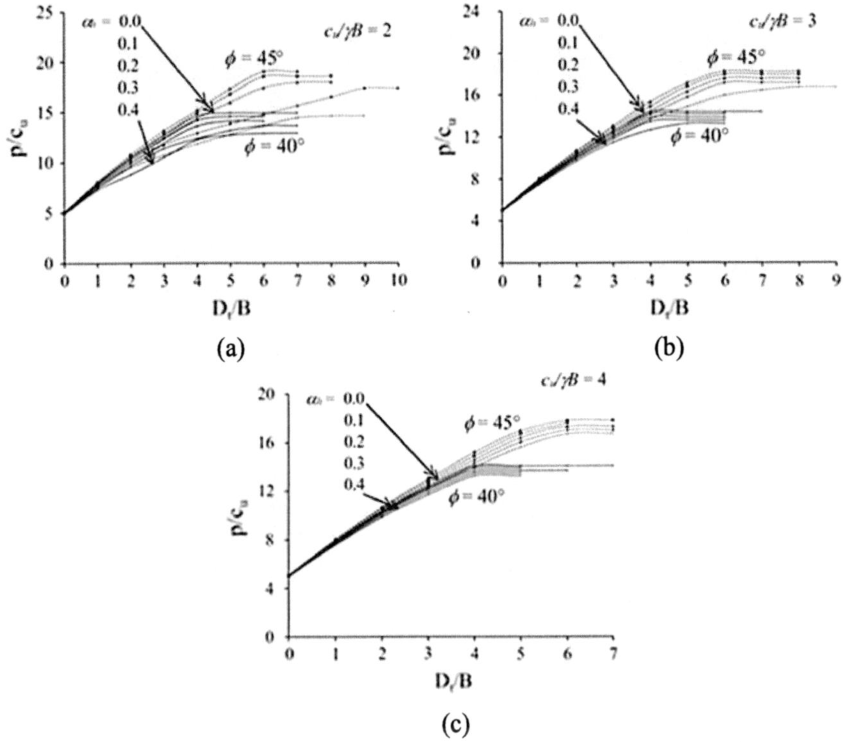

Fig. 4 Variation of p/c_u with D_t/B for **a** $c_u/\gamma B = 2$, **b** $c_u/\gamma B = 3$ and **c** $c_u/\gamma B = 4$

patterns are shown in Fig. 5a–f for $c_u/\gamma B = 3$. The soil friction angle of granular material in this case is 40°.

The dark red color zone represents the zone with $a/d = 1$ represents full plastic failure. The size of the failure zone increases in both (i) case-I where the overlying granular fill is used over soft clay and (ii) case-II where the granular trench is used to strengthen the soft clay for all values of α_h. The failure zone is found to be symmetric about the centre line of the footing in the absence of seismic body forces, i.e. $\alpha_h = 0$. However, the size of the failure zone has been found to decrease with a decrease in α_h in both the cases along the direction of the horizontal seismic force. The size of the failure zone has been found to be large in case-I in comparison to the case-II in the presence of seismic body forces of same magnitude.

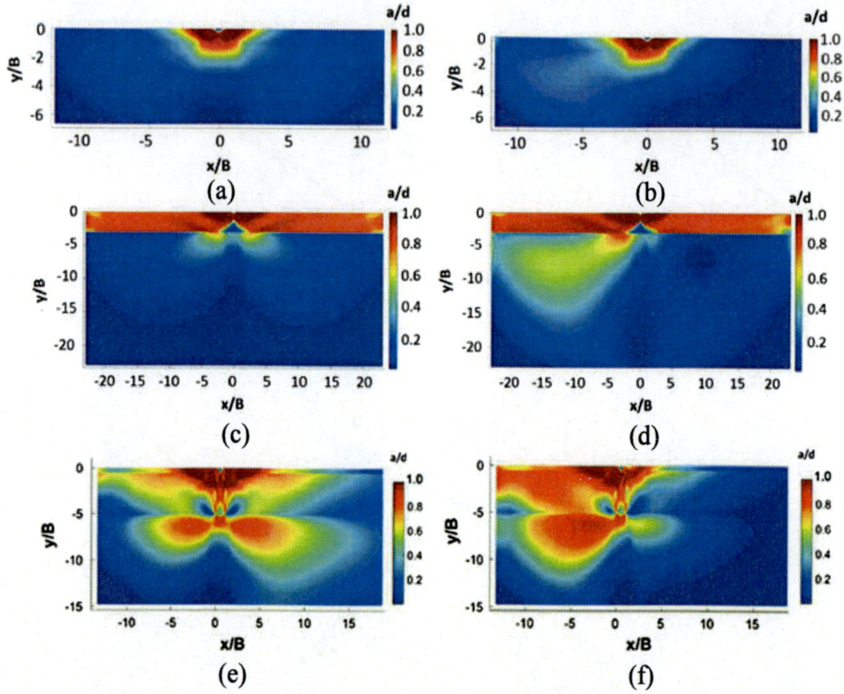

Fig. 5 Proximity of the stress state to failure for **a** only clay stratum with $c_u/\gamma B = 3$ and $\alpha_h = 0$, **b** only clay stratum with $c_u/\gamma B = 3$ and $\alpha_h = 0.2$, **c** clay stratum with overlying granular fill with $D_s/B = 3$, $c_u/\gamma B = 3$, $\phi = 40°$ and $\alpha_h = 0$, **d** clay stratum with overlying granular fill with $D_s/B = 3$, $c_u/\gamma B = 3$, $\phi = 40°$ and $\alpha_h = 0.2$, **e** clay stratum strengthened by granular trench with $D_t/B = 5$, $c_u/\gamma B = 3$, $\phi = 40°$ and $\alpha_h = 0$ and **f** clay stratum strengthened by granular trench with $D_t/B = 5$, $c_u/\gamma B = 3$, $\phi = 40°$ and $\alpha_h = 0.2$

6.4 Comparison Between Case-I with Granular Fill Over Soft Clay and Case-II with Granular Trench In Soft Clay

A comparison between case-I and case-II has been made in terms of maximum value of p/c_u and $(D_s/B)_{opt}$ or $(D_t/B)_{opt}$ for $\alpha_h = 0$ and $\alpha_h = 0.4$ for all combinations of (i) $c_u/\gamma B$ and (ii) ϕ. The comparison is presented in Table 1. It has been noted that in the absence or the presence of seismic force, the usage of granular fill is better choice in comparison to the usage of granular trench in soft clay with $c_u/\gamma B = 2$ for all ϕ values. However, in case of $c_u/\gamma B > 2$, the usage of granular fill is appeared to be better over the usage of granular trench for $\phi = 45°$. In this context, it is worthy to mention that the usage of denser sand can cause significant change in p/c_u value for case-I where granular fill is used over the soft clay in comparison to the case-II where soft clay is strengthened by inclusion of granular trench. However if medium dense sand is used as granular fill over the clay stratum with $c_u/\gamma B > 2$, it cannot be proved more effective in comparison to the inclusion of granular trench with medium dense

Table 1 Comparison of p/c_u between case-I (granular fill over soft clay) and case-II (granular trench in soft clay)

α_h	$c_u/\gamma B$	Case-I: granular fill over soft clay stratum		Case-II: soft clay stratum with granular trench	
		$\phi = 40^o$	$\phi = 45^o$	$\phi = 40^o$	$\phi = 45^o$
0.0	2	18.91 [3][a]	49.42 [5]	14.92 [5]	19.05 [6]
	3	12.61 [2.5]	32.95 [4]	14.37 [4]	18.24 [6]
	4	9.45 [2]	24.71 [3.5]	14.10 [4]	17.83 [6]
0.4	2	17.53 [3]	46.99 [5]	12.88 [6]	14.61 [8]
	3	11.69 [2.5]	31.33 [4]	13.26 [5]	16.72 [8–9]
	4	8.77 [2]	23.5 [3.5]	13.27 [4–5]	16.72 [7]

[a] Values within bracket represents $(D_s/B)_{\text{opt}}$ or $(D_t/B)_{\text{opt}}$

sand ($\phi = 40°$). However, the optimum depth for granular fill, i.e. $(D_s/B)_{\text{opt}}$ required for particular $c_u/\gamma B$ is less than the required optimum depth of granular trench, i.e. $(D_t/B)_{\text{opt}}$.

6.5 *Comparison of Present Work*

The present work has been compared with the results provided by Shiau et al. (2003) in absence of seismic forces. The comparison has been made in terms $p/\gamma B$ for various values of D_s/B and ϕ and presented in Tables 2 and 3. It is found that the value of $p/\gamma B$ provided by the present code matched well with the work of Shiau et al. (2003) with $\alpha_h = 0$. Both works have been performed with lower bound limit analysis with finite elements.

Table 2 Comparison of $p/\gamma B$ for $D_s/B = 1$ and $\phi = 40°$

$c_u/\gamma B$	Shiau et al. (2003) $D_s/B = 1$ ($\phi = 40°$)	Present work $D_s/B = 1$ ($\phi = 40°$)
2	14.68	14.96
3	19.29	19.4
4	23.003	23.12

Table 3 Comparison of $p/\gamma B$ for $D_s/B = 2$ and $\phi = 45°$

$c_u/\gamma B$	Shiau et al. (2003) $D_s/B = 2$ ($\phi = 45°$)	Present work $D_s/B = 2$ ($\phi = 45°$)
2	34.95	35.04
3	44.24	44.18
4	52.69	52.63

The results of the work of granular trench used to strengthen the soft clay is also validated with the results of Bhattacharya and Kumar (2017) for $c_u/\gamma B = 1$ which is not shown here as that case is not analyzed here under seismic forces. However, all the meshes used for this analysis have been validated only for the case of bearing capacity of strip footing on soft clay, and the normalized bearing capacity factor p/c_u has been found to be equal to 5.09 which is same as reported by Sloan (1988).

7 Conclusion

The present research work provides an estimate of increase in normalized bearing capacity of foundation on soft undrained clay with an increment in the depth of the granular fill over the soft clay or the granular trench used to reinforce the original soft clay under static and seismic forced. The analysis has been carried out considering different values of $c_u/\gamma B$ of original clay soil, friction angle of granular materials used for strengthening the capacity and four different magnitudes of seismic forces. It has been observed that the seismic forces cause significant decrease in the bearing capacity. The magnitude of p/c_u has been found to be more while using granular fill over soft soil. However, the present works provides a few chart on variation of p/c_u with D_s/B or D_t/B for different values of (i) $c_u/\gamma B$, (ii) α_h and (iii) ϕ which may be useful to choose appropriate techniques to increase bearing capacity of foundation over soft soil in seismically active and non-active zones.

References

Basudhar PK, Saha S, Deb K (2007) Circular footings resting on geotextile-reinforced sand bed. Geotext Geomembr 25:377–384

Bhattacharya P, Kumar J (2017) Bearing capacity of foundations on soft clays with granular column and trench. Soils Found 57(3):488–495

Bottero A, Negre R, Pastor J, Turgeman S (1980) Finite element method and limit analysis theory for soil mechanics problem. Comput Methods Appl Mech Eng 22(1):131–149

Bouassida M, Hadhri T (1995) Extreme loads of soils reinforced by columns: the case of an isolated column. Soils Found 35(1):21–35

Bouassida M, de Buhan P, Dormieux L (1995) Bearing capacity of a foundation resting on a soil reinforced by a group of column. Geotechnique 45(1):25–34

Bouassida M, Porbaha A (2004) Ultimate bearing capacity of soft clays reinforced by a group of columns-application to a deep mixing technique. Soils Found 44(3):91–101

Deb K, Chandra S, Basudhar PK (2007) Response of multi-layer geosynthetic-reinforced bed resting on soft soil with stone columns. Comput Geotech 35(3):323–330

Das BM, Shin EC, Omar MT (1994) The bearing capacity of surface strip foundation on geogrid reinforced sand and clay—a comparative study. Geotech Geol Eng 12:1–14

Kumar J, Bhattacharya P (2011) Reducing the computational effort for performing linear optimization in the lower-bound finite elements limit analysis. Int J Geomech 11(5):406–412

Kumar J, Bhattacharya P (2013) Bearing capacity of two interfering strip footings from lower bound finite elements limit analysis. Int J Numer Anal Meth Geomech 37(5):441–452

Kumar J, Khatri VN (2008) Effect of footing roughness on lower bound N_γ values. Int J Geomech 8(3):176–187

Hamed JTAH (1986) Bearing capacity of strip foundation on a granular trench in soft clay. Master of Science Thesis Dissertation, The University of Texas at El Paso, USA

Lysmer J (1970) Limit analysis of plane problems in soil mechanics. J Soil Mech Found Div 96(SM4):1311–2133

Love JP, Burd HJ, Milligan GWE, Houlsby GT (1987) Analytical and model studies of reinforcement of a layer of granular fill on a soft clay subgrade. Can Geotech J 24:611–622

Madhav MM, Vitkar PP (1978) Strip footing on weak clay stabilized with a granular trench or pile. Can Geotech J 15:605–609

Schweiger HF, Pande GN (1986) Numerical analysis on stone column supported foundations. Comput Geotech 2:347–372

Singh DN, Basudhar PK (1993) Optimal lower bound bearing capacity of strip footing. Soils Found 30(4):18–25

Shiau JS, Lyamin AV, Sloan SW (2003) Bearing capacity of a sand layer on clay by finite element limit analysis. Can Geotech J 40:900–915

Sloan SW (1988) Lower bound limit analysis using finite elements and linear programming. Int J Numer Anal Meth Geomech 12(1):61–77

Stuedlein AW, Holtz RD (2012) Analysis of footing load tests on aggregate pier reinforced clay. J Geotech Geoenviron Eng 138(9):1091–1103